wave generation and shaping

mcgraw-hill electrical and electronic engineering series

Frederick Emmons Terman, Consulting Editor
W. W. Harman and J. G. Truxal, Associate Consulting Editors

Angelakos and Everhart · *Microwave Communications*
Angelo · *Electronic Circuits*
Angelo · *Electronics: BJTs, FETs, and Microcircuits*
Aseltine · *Transform Method in Linear System Analysis*
Atwater · *Introduction to Microwave Theory*
Beranek · *Acoustics*
Bracewell · *The Fourier Transform and Its Application*
Brenner and Javid · *Analysis of Electric Circuits*
Brown · *Analysis of Linear Time-invariant Systems*
Bruns and Saunders · *Analysis of Feedback Control Systems*
Carlson · *Communication Systems: An Introduction to Signals and Noise in Electrical Communication*
Chen · *The Analysis of Linear Systems*
Chen · *Linear Network Design and Synthesis*
Chirlian · *Analysis and Design of Electronic Circuits*
Chirlian · *Basic Network Theory*
Chirlian and Zemanian · *Electronics*
Clement and Johnson · *Electrical Engineering Science*
Cunningham · *Introduction to Nonlinear Analysis*
D'Azzo and Houpis · *Feedback Control System Analysis and Synthesis*
Elgerd · *Control Systems Theory*
Eveleigh · *Adaptive Control and Optimization Techniques*
Feinstein · *Foundations of Information Theory*
Fitzgerald, Higginbotham, and Grabel · *Basic Electrical Engineering*
Fitzgerald and Kingsley · *Electric Machinery*
Frank · *Electrical Measurement Analysis*
Friedland, Wing, and Ash · *Principles of Linear Networks*
Gehmlich and Hammond · *Electromechanical Systems*
Ghausi · *Principles and Design of Linear Active Circuits*
Ghose · *Microwave Circuit Theory and Analysis*
Glasford · *Fundamentals of Television Engineering*
Greiner · *Semiconductor Devices and Applications*
Hammond · *Electrical Engineering*
Hancock · *An Introduction to the Principles of Communication Theory*
Harman · *Fundamentals of Electronic Motion*
Harman · *Principles of the Statistical Theory of Communication*
Harman and Lytle · *Electrical and Mechanical Networks*
Hayashi · *Nonlinear Oscillations in Physical Systems*

HAYT · *Engineering Electromagnetics*
HAYT AND KEMMERLY · *Engineering Circuit Analysis*
HILL · *Electronics in Engineering*
JAVID AND BROWN · *Field Analysis and Electromagnetics*
JOHNSON · *Transmission Lines and Networks*
KOENIG AND BLACKWELL · *Electromechanical System Theory*
KOENIG, TOKAD, AND KESAVAN · *Analysis of Discrete Physical Systems*
KRAUS · *Antennas*
KRAUS · *Electromagnetics*
KUH AND PEDERSON · *Principles of Circuit Synthesis*
KUO · *Linear Networks and Systems*
LEDLEY · *Digital Computer and Control Engineering*
LEPAGE · *Complex Variables and the Laplace Transform for Engineering*
LEPAGE AND SEELY · *General Network Analysis*
LEVI AND PANZER · *Electromechanical Power Conversion*
LEY, LUTZ, AND REHBERG · *Linear Circuit Analysis*
LINVILL AND GIBBONS · *Transistors and Active Circuits*
LITTAUER · *Pulse Electronics*
LYNCH AND TRUXAL · *Introductory System Analysis*
LYNCH AND TRUXAL · *Principles of Electronic Instrumentation*
LYNCH AND TRUXAL · *Signals and Systems in Electrical Engineering*
MCCLUSKEY · *Introduction to the Theory of Switching Circuits*
MANNING · *Electrical Circuits*
MEISEL · *Principles of Electromechanical-energy Conversion*
MILLMAN AND HALKIAS · *Electronic Devices and Circuits*
MILLMAN AND TAUB · *Pulse, Digital, and Switching Waveforms*
MINORSKY · *Theory of Nonlinear Control Systems*
MISHKIN AND BRAUN · *Adaptive Control Systems*
MOORE · *Traveling-wave Engineering*
NANAVATI · *An Introduction to Semiconductor Electronics*
PETTIT · *Electronic Switching, Timing, and Pulse Circuits*
PETTIT AND MCWHORTER · *Electronic Amplifier Circuits*
PFEIFFER · *Concepts of Probability Theory*
REZA · *An Introduction to Information Theory*
REZA AND SEELY · *Modern Network Analysis*
RUSTON AND BORDOGNA · *Electric Networks: Functions, Filters, Analysis*
RYDER · *Engineering Electronics*
SCHILLING AND BELOVE · *Electronic Circuits: Discrete and Integrated*
SCHWARTZ · *Information Transmission, Modulation, and Noise*
SCHWARZ AND FRIEDLAND · *Linear Systems*
SEELY · *Electromechanical Energy Conversion*
SEELY · *Electron-tube Circuits*
SEIFERT AND STEEG · *Control Systems Engineering*
SHOOMAN · *Probabilistic Reliability: An Engineering Approach*
SISKIND · *Direct-current Machinery*
SKILLING · *Electric Transmission Lines*
STEVENSON · *Elements of Power System Analysis*
STEWART · *Fundamentals of Signal Theory*
STRAUSS · *Wave Generation and Shaping*

Su · *Active Network Synthesis*
Terman · *Electronic and Radio Engineering*
Terman and Pettit · *Electronic Measurements*
Thaler · *Analysis and Design of Nonlinear Feedback Control Systems*
Thaler · *Elements of Servomechanism Theory*
Thaler and Brown · *Analysis and Design of Feedback Control Systems*
Thaler and Pastel · *Analysis and Design of Nonlinear Feedback Control Systems*
Tou · *Digital and Sampled-data Control Systems*
Tou · *Modern Control Theory*
Truxal · *Automatic Feedback Control System Synthesis*
Tuttle · *Electric Networks: Analysis and Synthesis*
Valdes · *The Physical Theory of Transistors*
Van Bladel · *Electromagnetic Fields*
Weeks · *Antenna Engineering*
Weinberg · *Network Analysis and Synthesis*

brooklyn polytechnic institute series

ANGELO · *Electronic Circuits*

ANGELO · *Electronics: BJTs, FETs, and Microcircuits*

LEVI AND PANZER · *Electromechanical Power Conversion*

LYNCH AND TRUXAL · *Signals and Systems in Electrical Engineering*

Combining *Introductory System Analysis*

Principles of Electronic Instrumentation

MISHKIN AND BRAUN · *Adaptive Control Systems*

SCHILLING AND BELOVE · *Electronic Circuits: Discrete and Integrated*

SCHWARTZ · *Information Transmission, Modulation, and Noise*

SHOOMAN · *Probabilistic Reliability: An Engineering Approach*

STRAUSS · *Wave Generation and Shaping*

wave generation and shaping

SECOND EDITION

Leonard Strauss

Professor of Electrical Engineering
Polytechnic Institute of Brooklyn

McGraw-Hill Book Company

New York St. Louis San Francisco London Sydney Toronto Mexico Panama

This book was set in Monotype by The Maple
Press Company, and printed on permanent
paper and bound by The Maple Press Company.
The designer was Paula Tuerk; the drawings were
done by B. Handelman Associates, Inc. The editors
were Michael Elia and Molly Scully. Les Kaplan
supervised the production.

wave generation and shaping

Library of Congress Catalog Card Number 74-90024
ISBN 07-062161-6
4567890MAMM7654321

to Rosemary, Karen, and Neil

preface

In the decade since the publication of the first edition of this text, the coming of age of solid state technology has revolutionized active circuit design. Integrated circuits, which are diverse, plentiful, and economical, have replaced many of the standard discrete component circuits. The circuit designer has become more systems oriented, devoting a significant portion of his activity to the interconnection of digital and analog integrated circuit modules where whole circuit functions are treated as a single element. Much of standard instrumentation, equipment used for information processing, on-line computers, and generators of signals of specified geometries incorporate large numbers of integrated circuits.

Many circuits, especially those containing energy-storage elements, cannot be readily integrated. Prototypes and special-purpose circuits must still be fabricated out of discrete components. But even discrete component design has changed over the past decade. Because technological advances have reduced the cost of bipolar and field-effect transistors to the point where they are among the less expensive components, economics dictates the use of more and more transistors and fewer and fewer energy-storage elements. By combining discrete components with integrated modules, a new dimension in design can be obtained.

Since this book is written primarily as a text, its objective is to present a logical unified approach to the analysis of those circuits in which the nonlinearity of the active device is the significant factor. A developmental treatment is followed throughout as we focus on the essential features of practical wave generating, shaping, and logic circuits. Discrete bipolar transistors, FETs, digital, and analog integrated circuit building blocks are used interchangeably to support the contention that the basic mode of operation is independent of the type of active device employed.

To this end the text is arbitrarily divided into five parts: Models and Logic, Timing, Switching, Memory, and Oscillations. While it is, of course, impossible not to step across these bounds in discussing specific circuits, the examples in each section have been selected so that the basic ideas arise naturally from the discussion. In most cases, the analysis is sufficiently detailed so that the techniques may be applied to other existent, and not as yet existent, circuits.

The organization, which is described in some detail below, is a result of experimentation with class notes over several years. Because this work is primarily a text, designed for a senior or graduate course of one or two semesters, it is assumed that the reader is familiar with the transient analysis of linear networks and with simple bipolar transistor and FET models and circuits. However, since the viewpoint presented may be somewhat different from that previously encountered, some review material is incorporated in the body of the text at the point where it is first needed.

Part 1. Models and Logic. Since we are interested in the transient response of piecewise-linear models to various excitations, the first chapter presents approximation methods which can be employed for the rapid solution of linear circuits.

Chapter 2 introduces a grossly nonlinear element, the ideal diode. The model of an actual diode is then derived from the physics of the semiconductor junction and compared with the measured diode characteristic. Once the model is approximated by one of several different ordered series of line segments, linear algebraic equations may be written to define the behavior in each region. Most of the problems met in more complex circuits make their initial appearance when the diode is combined with an energy-storage element. Combinations of diodes and resistors can be used to represent, in piecewise manner, almost any arbitrary characteristic. For solution, the complex circuit is reduced to a sequence of linear networks and the complex problem reduces to finding the boundaries between the various regions (breakpoint analysis). Finally, an analysis of the dynamic behavior of the diode illustrates the internal time delays inherent in switching a semiconductor junction.

Because of the demands for faster and faster switching in the past decade, time scales have been improved by several orders of magnitude, from microseconds down to nanoseconds. Delays inherent in the operation of the active device have become the limiting factor. Since more sophisticated models are needed to explain the phenomena of interest, the engineer must keep returning to the physics of the solid state devices to develop these models.

Transistor models are derived from the semiconductor physics in Chapter 3. Both large signal limitations and small-signal models are treated as special cases of the two diode Ebers-Moll model. The dynamic response of a saturating inverter is used to define the rise, fall, and storage delay times typical of all switching circuits.

Chapter 4 combines diodes and transistors into basic logic circuits. Starting from simple diode AND and OR gates, the circuitry is made progressively more complex as the RTL, DTL, TTL, and ECL logics are developed. Integrated circuit geometries favoring specific logic classes illustrate the similarity between the diode AND gate and the TTL logic. Problems posed by the interconnection of logic arrays lead to the special output stages of the integrated circuit logics.

The junction field-effect transistor and the MOSFET are described in Chapter 5. It is indeed fortuitous that a voltage-controlled solid-state device does in fact exist. If it did not, the necessity of illustrating voltage controlled circuit behavior would either require the postulating of an ideal device or the inclusion of the essentially obsolete vacuum tube. The derived MOSFET model proves to be highly nonlinear. But combining one nonlinear active device with another connected as a load results in an extremely linear inverter stage where the gain is a function solely of the geometries of the MOSFETs. The geometry of MOSFET logic circuitry follows naturally. In solving for the dynamic response of the inverter pair, which we find is not of the exponential form, we introduce a new look at the solution of the transient response problem.

Part 2. Timing. The next two chapters deal with the linear sweep. Since the essential portion of any timing circuit is the exponential charging of an energy-storage element, the problem of analysis resolves itself into finding the time-constant, the initial, the steady-state, and the final sweep voltages.

In the simple voltage sweep of Chapter 6, an SCR is used as the switching element. This choice makes it possible to treat synchronization in a very straightforward manner.

Linearization of the sweep always involves some form of feedback. The aim is to approximate constant-current charging of the sweep capacitor. Calculation of the initial jumps, overshoots, and recovery exponentials from the several piecewise-linear circuit models lays the groundwork for the later analysis of the multivibrator and the blocking oscillator. Furthermore, the separate consideration of the recovery time and the switching problem leads to the construction of a sweep system.

Chapter 7 applies the same techniques of analysis to the current sweep.

Part 3. Switching. When a closed-loop system contains active elements, and when the loop gain is positive and greater than unity, we call it a switching circuit. If any timing networks are included in the transmission path, the circuit exhibits quasi-stable behavior. The same effect can be obtained by shunting a negative-resistance device with an energy-storage element.

Chapters 8 and 9 discuss a closed-loop regenerative switching circuit, the multivibrator. This circuit can be synthesized either with discrete components or by cross connecting two integrated logic circuits. We are mainly concerned with where and how to begin the analysis. A guess serves as the convenient starting point, and the circuit calculations are checked to see if they yield consistent results. If a contradiction arises, it simply indicates that the first guess was wrong. In the multivibrator, both transistors or FETs are active only during the switching interval. At all other times, one is cutoff or the other is saturated. Hence the exponential timing, which sustains the limiting, usually occurs within the cutoff zone, and recovery usually depends on the saturated transistor. After the appropriate models are drawn, the calculations involve finding the initial and switching points of the timing network.

Chapter 10 extends the concept of regenerative switching to the voltage comparator. External positive feedback converts any difference amplifier into a regenerative comparator exhibiting hysteresis.

Many devices exist whose driving-point volt-ampere characteristic exhibits a negative-resistance region. The most useful treatment of switching circuits containing these elements is from the viewpoint expressed in Chapter 11. Switching and timing are first treated with the aid of a postulated ideal device. Even though the waveshapes and trajectories obtained are approximately correct, the use of an ideal negative resistance leads directly to a stability criterion that may be inconsistent with the physical device. Since the reasons that the contradictory results arose are pointed out later, this chapter also illustrates the dangers inherent in over-idealizing a model.

The ideas presented in Chapter 11 may well be considered as some of the unifying concepts. These are presented rather late in the text so that the student may better appreciate the limitations as well as the advantages of this viewpoint. The negative-resistance treatment is useful for understanding the operation of all sweeps and switching circuits even where it is difficult to isolate the two terminals across which the negative resistance is developed. It is, however, a convenient design method only for those circuits which employ such special devices as the unijunction transistor, the *pnpn* transistor, and the tunnel diode.

Chapter 12 applies the methods developed in the earlier sections of the text to the blocking oscillator. Since extremely narrow pulses are generated, the rise and fall times become critical. Timing may depend on such diverse circuit factors as current buildup in a transistor and transformer core saturation. Care must be taken to select the significant one.

Part 4. Memory. Bistable multivibrators, formed by the interconnection of basic logic elements, serve as elementary memory cells. Chapter 13 shows how the *SR*, *JK*, and *D* type bistables are synthesized. Chains of these bistables can be interconnected to form shift registers, counters, and frequency dividers.

The structure of active memory arrays is left to Chapter 14. Here the subsystems of steering logic, storage, and read-write circuitry are examined in connection with specific memory configurations. Typical examples of short-time dynamic, long-time static, random access, and read only memories are examined in detail. This is where the MOSFET integrated circuit comes into its own, since from 1,000 to 4,000 active devices can be formed in an iterative structure on a single chip measuring less than $\frac{1}{10}$ in. on a side.

A hysteresis loop indicates memory and it therefore follows that magnetic and dielectric materials are ideally suited for information storage. Chapter 15 examines briefly the terminal characteristics of a core constructed out of ideal square-loop material. Furthermore, a clearly defined time is needed to saturate the core, and although not representable in a piecewise-linear manner, the driving-point impedance exhibits two distinct regions. This enables us to use either a core or a ferroelectric device as the timing element in a switching circuit.

Part 5. Oscillations. Although most of the circuits previously discussed depended for their timing on a single energy-storage element, at least two are necessary in order for the system to have a pair of complex conjugate poles located in the right half plane.

The sinusoidal oscillator is first treated as an almost linear feedback system (Chapter 16) which allows the separation of the frequency-, amplitude-, and gain-determining circuit elements. The role of each essential part of the oscillator is individually examined, with emphasis on minimizing the distortion and maximizing the stability. Since the signal produced is almost sinusoidal, the amplitude is readily found by plotting a system-describing function.

Chapter 17 returns to the negative-resistance viewpoint and describes the solution of a specific nonlinear differential equation. Here topological constructions, which supplement analytic methods, yield the waveshape and amplitude of oscillations. Hence the text ends by introducing a new topic and not by saying the last word on an old one.

While preparing this revision, I received a great deal of assistance from those of my colleagues who taught sections of courses for which this text was written. In stimulating technical discussions, I received the benefit of their ideas and thinking. They also corrected errors in the original class notes and they even contributed some of the problems. The typing and drafting assistance made available by the Polytechnic Institute of

Brooklyn is greatly appreciated. Finally I wish to thank all faculty members and students of the Electrical Engineering Department for their encouragement during this project. The friendly and cooperative atmosphere made the writing of this book almost a pleasure.

Leonard Strauss

contents

Preface xi

Part 1 Models and Logic

1. Linear Wave Shaping 3

1-1. Initial conditions 4
1-2. Solution of periodically excited circuits 11
1-3. Differentiation 13
1-4. Integration 15
1-5. Summation 17
1-6. Approximate solutions of the single-pole case 18
1-7. Double-energy-storage-element systems 20
1-8. Compensated attenuators 23

2. Diode Wave-shaping Techniques 33

2-1. Passive-nonlinear-circuit representation 34
2-2. Semiconductor diode characteristics 40
2-3. Physical diode models 50
2-4. Voltage clipping circuits 54

2-5. Single diode and an associated energy-storage element—some general remarks 58
2-6. Diode voltage clamps 61
2-7. Current clippers 68
2-8. Current clamps 69
2-9. Arbitrary transfer and volt-ampere characteristics 71
2-10. Dead zone and hysteresis 79
2-11. Depletion-region capacitance 83
2-12. Diode dynamic response 85
2-13. Summary 87

3. Transistor Models and Simple Circuits 99

3-1. Transistor physics 100
3-2. The Ebers-Moll model 103
3-3. Large-signal models—quiescent and boundary conditions 108
3-4. Small-signal models 114
3-5. Circuit applications—small-signal calculations 119
3-6. Composite devices 122
3-7. Transistor dynamic behavior 124
3-8. Integrated-circuit devices 131

4. Digital Logic and Amplitude Gating 149

4-1. Digital logic statements 150
4-2. Diode logic gates 154
4-3. Diode transistor logic (DTL) 159
4-4. Direct-coupled transistor logic (DCTL) 166
4-5. Resistor-transistor logic (RTL) 168
4-6. Transistor-transistor logic (TTL or T^2L) 171
4-7. Output circuits 178
4-8. Nonsaturated logic (ECL) 183
4-9. Wired logics 187
4-10. Amplitude gating (Choppers) 189

5. Field-effect Transistors 205

5-1. Junction field-effect transistors 205
5-2. MOS field-effect transistors 208
5-3. Small-signal models 216
5-4. Biasing and small-signal applications 220

5-5. MOSFET used as a load resistor 224
5-6. Inverter transient response—part I, piecewise-linear approximation 230
5-7. Inverter transient response—part II, exact response 234
5-8. The FET as a logic element 242
5-9. Analog switching applications—transmission gates 249
5-10. Composite bipolar-unipolar and other specialized FET devices 253

Part 2 Timing

6. Linear Voltage Sweeps 267

6-1. Basic voltage sweeps 268
6-2. A simple periodic-voltage-sweep generator 270
6-3. Sweep linearity 273
6-4. The transistor as a current source 275
6-5. Current-feedback linearization—Miller sweep 280
6-6. Voltage-feedback linearization—Miller sweep 286
6-7. Bootstrap sweep 296
6-8. Switching 302
6-9. Synchronization 303
6-10. Regions of synchronization 308

7. Linear Current Sweeps 325

7-1. Basic current sweeps 325
7-2. Switched inductive loads 331
7-3. Periodic switched sweeps 334
7-4. Current-sweep linearization 338
7-5. A transistor bootstrap sweep 341
7-6. Constant-current-source current sweep 345

Part 3 Switching

8. Collector-base and Drain-gate Coupled Multivibrators 353

8-1. Basic multivibrator considerations 353
8-2. Bistable multivibrators 358
8-3. Triggering of bistables 365

8-4. Multivibrator rise time 367
8-5. Monostable multivibrators—bipolar transistor 369
8-6. Monostable multivibrators—FET active element 380
8-7. Astable multivibrators—sequential toggle mode 385
8-8. Integrated-circuit monostable and astable multivibrators 392
8-9. Incomplete circuit recovery 394
8-10. Astable multivibrator—single energy-storage element 397
8-11. Multivibrator synchronization 400
8-12. Summary 401

9. Emitter-coupled Multivibrators 408

9-1. Transistor emitter-coupled multivibrator—monostable operation 408
9-2. Modes of operation of the emitter-coupled multivibrator 415
9-3. Monostable pulse variation 424
9-4. Emitter-coupled astable multivibrator 426

10. Voltage Comparators 434

10-1. Difference-amplifier comparator 434
10-2. High-gain differential amplifier 439
10-3. Regenerative comparators 445
10-4. The differential amplifier as an astable multivibrator 447
10-5. Schmitt-trigger circuits 449
10-6. Modes of operation—Schmitt Trigger 456
10-7. Some comparator applications 458

11. Negative-resistance Switching Circuits 464

11-1. Basic circuit considerations 464
11-2. Basic switching circuits 466
11-3. Calculation of waveshapes 473
11-4. Composite terminal characteristics 478
11-5. Voltage-controlled NR switching circuits 481
11-6. NR characteristics—collector-base-coupled monostable multivibrator 483
11-7. Negative-impedance converters 488
11-8. Devices possessing NR characteristics 497
11-9. Frequency dependence of NR characteristics 507
11-10. Improvement in switching time through the use of a nonlinear load 513

12. The Blocking Oscillator 523

12-1. Some introductory remarks 524
12-2. Switching times 525
12-3. Inductively timed pulse duration 529
12-4. Recovery region 531
12-5. Transformer core properties and saturation 533
12-6. Capacitively timed blocking oscillator 535
12-7. Astable operation 538

Part 4 Memory

13. The Bistable as a Logic Element 547

13-1. *S-R* multivibrator switching: NOR inputs 548
13-2. *S-R* bistable switching: NAND inputs 551
13-3. *S-R* flip-flop: toggle-mode operation 553
13-4. Other toggling techniques 556
13-5. *J-K* flip-flops 563
13-6. Type-*D* edge-triggered flip-flop 569
13-7. Bistable applications 572

14. Active-element Memories 585

14-1. Characteristics of a memory system 585
14-2. Short-time MOSFET memory 589
14-3. A bipolar-bistable matrix 598
14-4. Coding and decoding 602
14-5. MOSFET read-only memory 606

15. Magnetic and Dielectric Devices as Passive Memories and Switching Elements 613

15-1. Ferromagnetic properties 613
15-2. Terminal response of cores 617
15-3. Magnetic counters 626
15-4. Core-transistor counters and registers 632
15-5. Magnetic-core memory arrays 634
15-6. Core-transistor multivibrator 640
15-7. Properties of ferroelectric materials 645

15-8. Ferroelectric terminal characteristics 648
15-9. The ferroelectric counter 650
15-10. Coincident memory arrays 653

Part 5 Oscillations

16. Almost Sinusoidal Oscillations—The Linear Approximation **663**

16-1. Basic feedback oscillators 663
16-2. Characteristics of some *RC* and *LR* frequency-determining networks 668
16-3. Current-controlled oscillations 676
16-4. Tuned-circuit oscillators 677
16-5. Frequency stabilization 688
16-6. Amplitude of oscillation—part I, active-element limiting 694
16-7. Amplitude of oscillation—part II, automatic gain control 706
16-8. Amplitude stability 713

17. Negative-resistance Oscillators **721**

17-1. Basic circuit considerations 721
17-2. A FET tuned oscillator as a negative-resistance circuit 726
17-3. First-order solution for frequency and amplitude 727
17-4. Frequency of oscillation to a second approximation 733
17-5. Introduction to topological methods 739
17-6. The Liénard diagram 746
17-7. Summary 753

Index **759**

wave generation and shaping

PART 1
models and logic

1. linear wave shaping

Often a communication or instrumentation engineer finds he is faced with the problem of producing a designated complex waveform. Usually his simplest approach is to divide the problem into two parts: first, the generation of simple waveshapes and, second, some type of operation on them to achieve the desired final result. If the required waveform is of a sufficiently complex nature, several substages of shaping may be needed, with combination taking place in the final stage. Generation always involves the use of active elements, and although they also perform a function, shaping is predominantly effected by the circuit's passive components. One might even go so far as to say that a major portion of the process of wave generation itself consists in passive-element shaping.

In this chapter we shall examine the behavior of linear passive circuits with a view to solving for their time response to various input waveshapes. Exact analysis is, of course, possible, but it is often so tedious that the competent engineer looks instead for reasonable approximations. Our aim, therefore, is to set up conditions whereby solutions may be obtained rapidly, almost by inspection, avoiding involved algebraic manipulation. This, in fact, becomes possible only under very specific circumstances, which must be clearly delineated. Particular emphasis will be placed on the treatment of circuits containing only a single energy-storage element, not only because these appear quite frequently, but also because it is here that the approximation techniques are most fruitful. Sometimes even multiple-energy-storage circuits may be reduced to a number of separate single-energy circuits, which are then individually solved. The solutions are finally combined, yielding the approximate total response of the original complex circuit.

In our effort to develop facility in utilizing simplifying approximations, i.e., how to make them and when to make them, we shall start by considering some very basic circuits. Since we already know methods of finding their exact response to simple inputs, we can concentrate attention on the methodology of the solution.

1-1 Initial Conditions

We shall first consider the transient response of a passive linear network, i.e., the response to some abrupt change in either the external excitation or the internal circuit. We know that the solution will be directly related to the poles of the system transfer function, and for poles at s_1, s_2, and s_3, the response to a step will always have the form

$$r(t) = A_0 + A_1 e^{s_1 t} + A_2 e^{s_2 t} + A_3 e^{s_3 t}$$

One exponential term is due to each independent energy-storage element included in the network.

The exact response to any excitation is readily found by the classical method of writing and solving the system's integrodifferential equations or by modern transformation analysis. If, however, we restrict our interest to relatively simple systems, i.e., those containing one, or sometimes two, widely separated poles, then these may often be found by direct inspection of the circuit. They will correspond to the negative reciprocal of the time constants.

The techniques of rapid analysis draw heavily on the physical limitations imposed on the problem by the energy-storage-circuit elements. The application of these constraints can be seen during the solution of a particular circuit.

As an example, we shall consider the time response of the circuit of Fig. 1-1. The external conditions are the following: at $t = 0$, switch 1 is closed; once the voltage at the output falls to the predetermined final value $V_f = 10$ V (at time t_f), switch 2 is closed.

Rather than attempt to solve the entire problem in a single step, it is advisable to divide it into two separate parts and solve each in proper time

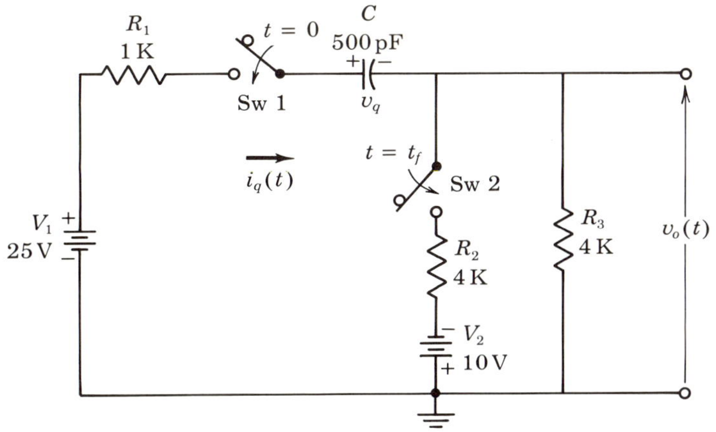

Fig. 1-1 Single-energy-storage-element circuit.

sequence. The complete solution will then be the aggregate of the individual answers. The two parts are:

1. (*a*) Response from $t = 0^-$ to $t = 0^+$, indicating circuit changes as switch 1 is closed
   (*b*) Response from $t = 0^+$ to the time t_f at which $v_o(t_f) = V_f$
2. (*a*) Response from $t = t_f^-$ to $t = t_f^+$, giving the circuit and the voltage changes as switch 2 is closed
   (*b*) Final response, from $t = t_f$ to $t = \infty$

Note that changes in the circuit state determine the problem division, with each part holding for a specific circuit configuration. In proceeding with the solution, each part will start from its own zero time as if it were a completely independent problem. Numerical subscripts corresponding to the particular part are used to indicate the time interval under examination. Within each interval, the form of the solution is well known. It is given by

$$v_o(t) = A + Be^{-t/\tau} \tag{1-1}$$

The constant A in Eq. (1-1) represents the steady-state response and may be found from a consideration of the response at $t = \infty$. From Eq. (1-1) we see that the constant A corresponds to the steady-state voltage

$$v_o(\infty) = A = V_{ss}$$

B is related to the behavior at $t = 0^+$ and must be found from the conditions holding just before and just after switching. Solving Eq. (1-1) at $t = 0^+$, we obtain

$$B = -[V_{ss} - v_o(0^+)] = -(V_{ss} - V_i)$$

where the initial voltage $V_i = v_o(0^+)$. By substituting A and B, Eq. (1-1) can be rewritten as

$$v_o(t) = V_{ss} - (V_{ss} - V_i)e^{-t/\tau} \tag{1-2}$$

The time constant τ is simply the product of C and the total effective resistance seen by the capacitor. With switch 1 closed,

$$\tau_1 = (R_1 + R_2)C \qquad \text{and} \qquad s_1 = -\frac{1}{\tau_1} \tag{1-3}$$

Substituting numbers, $\tau_1 = 5 \times 10^3 \times 500 \times 10^{-12} = 2.5\ \mu s$.

Since a prerequisite to the complete solution is the value of the coefficients A and B, a slight digression as to the means of their evaluation is in order before proceeding to the general solution of this circuit. But these coefficients depend upon the initial conditions; therefore the first question to ask is: What determines these conditions? Obviously, one part of the answer is the known external constraints imposed on the circuit, i.e., the excitation function and the various circuit changes. The second half of the answer is the internal constraints determined by the energy-storage elements present within the circuit, in this case a single capacitor.

The voltage drop across any capacitor is related to circuit current flow as expressed in the differential equation

$$V_q(t) = \frac{1}{C}\int i\,dt = \frac{q(t)}{C} \tag{1-4}$$

This equation says that, in a physical system, the accumulation of charge takes finite time unless an infinite current flows. Consequently, the terminal voltage cannot change instantaneously. In an ideal system, such instantaneous changes may be forced by placing a short circuit or an ideal battery across the capacitor terminals or by injecting an infinite pulse of current. These do not exist in practice, but sometimes the charge or discharge time is so very fast compared with the total time range of interest that the small time interval may be ignored, and we can say that the capacitor voltage has jumped to its final value. From the above discussion, the first circuit constraint can now be stated: *The voltage across a capacitor must be continuous during a circuit or excitation change.*

$$v_q(0^+) = v_q(0^-) \tag{1-5}$$

Equation (1-5) does not impose any constraints on the initial current flow. In an RC circuit, current not only can but must change instantaneously to satisfy Kirchhoff's voltage laws. Carrying this argument a step further to include the constraint imposed by Eq. (1-5) leads directly to the corollary that *all abrupt changes in circuit voltages will be distributed proportionately across the various noncapacitive portions of the circuit.*

Part 1—Response from $t = 0$ to $t = t_f$ at which $v_o = V_f$

Closing switch 1 in Fig. 1-1 applies an abrupt voltage to the RC network. The input voltage jumps from 0 to V_1, causing an initial charging-current flow of

$$i_{q1}(0^+) = \frac{V_1 - v_q(0^+)}{R_1 + R_2}$$

Whenever the capacitor is initially uncharged, $v_q(0^+) = 0$, and the output voltage jumps from 0 to

$$v_{o1}(0^+) = \frac{R_2}{R_1 + R_2} \Delta V_1(0) \tag{1-6}$$

$\Delta V_1(0)$ is the $0 \rightarrow V_1$ positive input transition. For the circuit of Fig. 1-1,

$$V_{o1}(0^+) = V_{i1} = \frac{4\text{ K}}{1\text{ K} + 4\text{ K}} 25 = 20\text{ V}$$

The remaining 5 V appears across R_1.

One can derive a second condition from Eq. (1-4) by noting that the voltage across C continues to change so long as any current flows. The final steady-state capacitor voltage is reached when $i_q = 0$. In a simple network it will always be the dc Thévenin equivalent appearing across the capacitor.

With the switch in position 1, the steady-state conditions become

$$v_{o1}(\infty) = 0$$
$$v_{q1}(\infty) = V_1 = 25\text{ V}$$

Substituting the initial and steady-state conditions into the response equation, we obtain

$$v_{o1}(t) = V_1 e^{-t/\tau_1} = 20e^{-t/\tau_1} \tag{1-7a}$$
$$v_{q1}(t) = 25 - 25e^{-t/\tau_1} \tag{1-7b}$$
$$i_q(t) = 5e^{-t/\tau_1} \quad \text{mA} \tag{1-7c}$$

where $\tau_1 = 2.5\ \mu\text{s}$. The first region terminates once $v_{o1} = 10$ V. From Eq. (1-7*a*),

$$v_{o1}(t_f) = V_{f1} = 10 = 20e^{-t_f/\tau_1}$$

Upon solving for t_f, we find that the first interval lasts for

$$t_f = \tau_1 \ln \frac{V_{i1}}{V_{f1}} = 2.5 \ln \frac{20}{10} = 1.73\ \mu\text{s} \tag{1-8}$$

A more general expression for the time required for an exponential response to reach a predetermined final value can be found by solving Eq. (1-2) for the time t_f at which $v_o(t_f) = V_f$:

$$t_f = \tau_1 \ln \frac{V_{ss} - V_i}{V_{ss} - V_f} \tag{1-9}$$

Equation (1-9) reduces to Eq. (1-8) when $V_{ss} = 0$.

Now that we know the time t_f, the final values of the charging current and capacitor voltage can be found directly by substituting t_f into Eqs. (1-7*b*) and (1-7*c*). The same results can be obtained directly from the circuit (Fig. 1-1) by recognizing that once we know the output voltage (10 V), we also know the charging current. This current determines the total *IR* drop. Any remaining voltage must appear across the capacitor.

$$i_{q1}(t_f) = \frac{V_{f1}}{R_2} = \frac{10}{4\text{ K}} = 2.5\text{ mA}$$

$$v_{q1}(t_f) = V_1 - (R_1 + R_2)i_{q1}(t_f) = 12.5\text{ V}$$

The waveshape of the first portion of the total response is clearly seen in the sketch of the total response in Fig. 1-3.

Part 2—Response from $t = t_f$ to $t = \infty$ or, by redefining $t = 0$, from $t = 0_2$ to $t = \infty$

Figure 1-2 shows the circuit holding in this interval. The network on the right-hand side of C has been replaced by its Thévenin equivalent circuit. The voltage across the capacitor must remain invariant across the circuit change. Consequently, the first restraint on the response is given by

$$v_{q2}(0^+) = v_{q2}(0^-) = v_{q1}(t_f) = 12.5\text{ V}$$

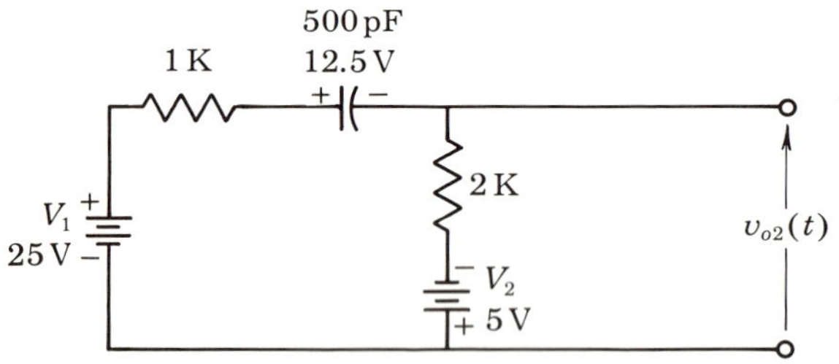

Fig. 1-2 Model of Fig. 1-1 when both switch 1 and switch 2 are closed. The output branch is replaced by its Thévenin equivalent.

Using the fixed capacitor voltage in the circuit of Fig. 1-2, we find that the initial capacitor charging current is given by

$$i_{q2}(0^+) = \frac{V_1 - v_{q1}(0^+) + V_2}{R_1 + R_2 \parallel R_3} = 5.83 \text{ mA}$$

and by using $i_{q2}(0^+)$ we obtain the new initial output voltage

$$v_{o2}(0^+) = i_{q2}(0^+)R_2 \parallel R_3 - 5 = 6.66 \text{ V}$$

The parallel lines indicate that the adjacent elements are in parallel; for example, $R_1 \parallel R_2 = R_1R_2/(R_1 + R_2)$. This notation will be used throughout in an attempt to keep the circuit-element relationships apparent.

The steady-state conditions for this time interval are found from the circuit of Fig. 1-2 by setting $i_{q2}(\infty) = 0$:

$$v_{o2}(\infty) = V_{ss2} = -5 \text{ V}$$

$$v_{q2}(\infty) = 30 \text{ V}$$

Substituting into the basic equation (1-2), we obtain

$$v_{o2}(t) = -5 + 11.66e^{-t/\tau_2}$$

where $\tau_2 = C(R_1 + R_2 \parallel R_3) = 1.5\ \mu\text{s}$. The charging current decays to 0, and the capacitor voltage builds up to 30 V at the same rate. We assume that the exponential response is complete in four time constants.

Figure 1-3 is the plot of the total solution, with the response of each portion beginning immediately after that of the previous part.

In the solution for part 2, the circuit responded as if no change were scheduled. It could not predict the future, and therefore, in solving the problem, we must be careful lest our foreknowledge lead to fallacious reasoning, i.e., the incorrect substitution of V_{f1} instead of V_{ss1} for the steady-state value. Upon entering any region, the response is toward the final value calculated as if the circuit were invariant.

When the circuit contains an ideal inductance, additional constraints appear:

$$v_L = L\frac{di}{dt} \tag{1-10}$$

From Eq. (1-10) it may be seen that an infinite voltage is required to change the current instantaneously. As this is an impossible situation in any physical circuit, our first conclusion is that *the current through the coil must be continuous across circuit or excitation changes.* A second constraint, also derivable from the above equation, is that steady state is finally reached when there is no further time rate of change of current, i.e., when $v_L = 0$. These conditions are just the duals of the ones found for the capacitor. The overlapping constraints of the two energy-storage elements

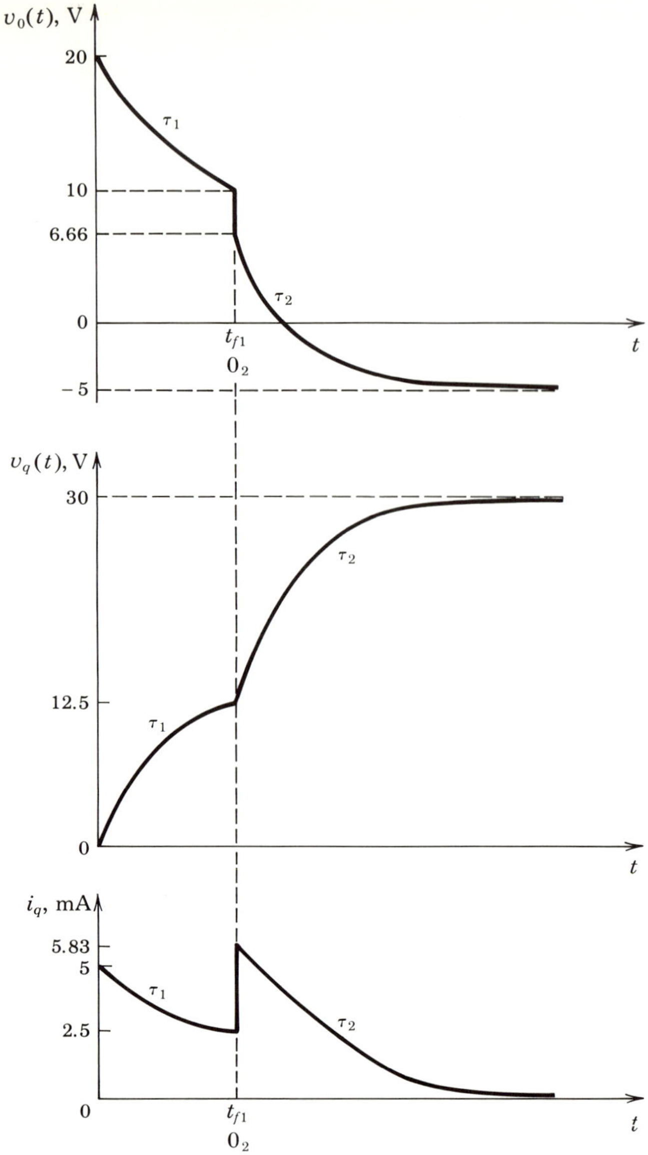

Fig. 1-3 Waveshapes showing the response of the circuit of Fig. 1-1.

forbid any instantaneous changes of current, and of capacitor or resistor voltage, in a series *RLC* circuit. Because of the physical nature of the various circuit elements, they will all have associated stray capacity and lead inductance. Therefore all circuits contain more than one type of energy-storage element. These parasitic elements act to slow down the rapid changes found when we consider a simple ideal *RL* or *RC* circuit. But if the time range of interest is sufficiently long, the second-order effects may be neglected.

1-2 Solution of Periodically Excited Circuits

After the sudden application of a periodic input to a passive network, the output itself will eventually become periodic in nature. The transient dies out in four time constants, leaving the steady-state response. The time required for this, in terms of the number of cycles of the input waveshape, depends upon the relative values of the time constant and the period of the excitation function. If the time constant is much smaller than the input period, the periodic output appears within one or at most a few cycles and it can be found by following the behavior cycle by cycle.

On the other hand, when the time constant is on the order of or greater than the period of the input, periodicity will not be reached until an extremely large number of cycles have passed. During the buildup, the capacitor charges to the average level of the voltage appearing across its terminals and the direct current through a coil increases from zero to the average circuit value. To attempt to start at zero time and follow the response cycle by cycle until final periodicity is a ridiculous approach to the solution of this class of problems—it will only lead to frustration.

A more judicious treatment would be the application of the results of the previous section, with the addition of the known periodicity of the solution. This enables us to write recurrent initial conditions; i.e., the starting point of any one cycle is the same as that of the next cycle. As an example, we shall now examine the *RL* circuit of Fig. 1-4. The rectangular input shown

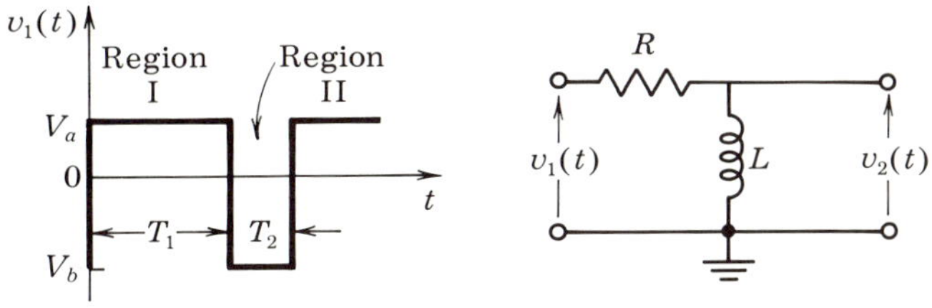

Fig. 1-4 Circuit for the study of recurrent boundary conditions.

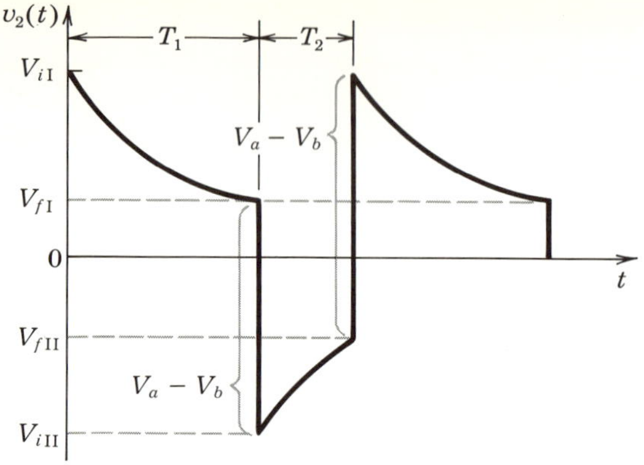

Fig. 1-5 Output voltage across L in the circuit of Fig. 1-4.

has been applied for a sufficiently long time so that the output has already reached the final periodic form shown in Fig. 1-5.

The solution during either portion of the input signal is

$$v_2(t) = A + Be^{-t/\tau} \qquad \tau = \frac{L}{R}$$

Except for the time constant, this equation is of the same form as the one found for the RC circuit [Eq. (1-2)]. Since the inductance prevents abrupt changes in the current, the resistor voltage will be invariant across the discontinuities of the input and any abrupt jump in the driving voltage will be reflected across the coil. We can show this by considering the voltage drops around the loop:

$$v_1(t) = v_R + v_L = iR + v_L \tag{1-11}$$

Across the boundary of the two regions the input voltage jumps, and by defining this change as $\Delta v_1(t)$, we see that

$$\Delta v_1(t) = \Delta iR + \Delta v_L$$

But since the inductance constrains Δi to be zero, the complete jump of the input voltage must immediately appear across the inductance.

Starting the solution at the beginning of region I, which holds for $0_1 < t < T_1$, the first set of initial conditions is

$$V_{ssI} = 0 \qquad V_{iI} = \text{unknown}$$

Therefore the solution in this region is

$$v_{2I}(t) = V_{iI}e^{-t/\tau} \tag{1-12}$$

and the final value, occurring at $t = T_1$, may be expressed as

$$V_{f\text{I}} = v_{2\text{I}}(T_1) = V_{i\text{I}}e^{-T_1/\tau} \tag{1-13}$$

Upon entering region II, the input voltage drops from V_a to V_b, with the same change immediately appearing across the inductance. The final output of region I was $V_{f\text{I}}$. The initial value of region II must be

$$V_{i\text{II}} = V_{f\text{I}} - (V_a - V_b)$$

In region II, holding for $0_2 < t < T_2$, $V_{ss\text{II}} = 0$. The equation for the time response can be written as

$$v_{2\text{II}}(t) = [V_{f\text{I}} - (V_a - V_b)]e^{-t/\tau} \tag{1-14}$$

At $t = T_2$,

$$V_{f\text{II}} = v_{2\text{II}}(T_2) = [V_{f\text{I}} - (V_a - V_b)]e^{-T_2/\tau} \tag{1-15}$$

But the initial voltage of region I and all other voltages which have been written in terms of it are still unknown. We must now apply the known conditions of periodicity in order to find the final answer. At the end of region II, the input voltage jumps from V_b to V_a and the output voltage changes by the same amount. Periodicity tells us that the new value upon reentering region I is the original assumed $V_{i\text{I}}$.

$$V_{i\text{I}} = V_{f\text{II}} + (V_a - V_b) \tag{1-16}$$

Substituting Eq. (1-16) into Eq. (1-13) results in

$$V_{f\text{I}} = [V_{f\text{II}} + (V_a - V_b)]e^{-T_1/\tau} \tag{1-17}$$

Equations (1-15) and (1-17) each involve only the two unknowns $V_{f\text{I}}$ and $V_{f\text{II}}$; all other terms are constants of the input waveshape or the circuit. The simultaneous solution of the above equations completes the analysis.

If the output of interest were taken across the resistor instead of the inductance, the technique of solution would still be the same, the principal difference being in the initial conditions over each portion of the cycle. Current is continuous, and therefore resistor voltage must also be continuous in the *RL* circuit. *RC* series circuits are treated in a similar manner, applying, however, their special boundary conditions; i.e., all voltage jumps appear across the circuit resistance.

Even when the inputs are nonrectangular, the shape and equation of their response over any period or portion thereof usually can be obtained by many convenient methods. The known periodicity supplies the additional information necessary for a complete solution.

1-3 Differentiation

A series *RC* circuit (Fig. 1-6) will, contingent on the satisfaction of certain conditions, have an output approximately proportional to the derivative

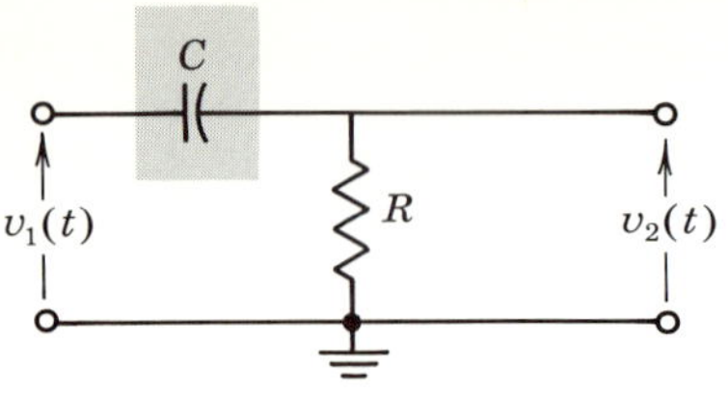

Fig. 1-6 Differentiator.

of the input. These conditions may best be determined by examination of the circuit differential equation

$$v_2(t) = Ri = v_1(t) - \frac{1}{C} \int i(t)\, dt \tag{1-18}$$

Since the voltage across a resistor is directly proportional to the current flow through it, for the output to be proportional to the derivative, the current must also have this relationship. Satisfaction of this necessary condition is made possible only by maintaining the output voltage small with respect to the input. Then, from Eq. (1-18),

$$v_1(t) \approx \frac{1}{C} \int i\, dt$$

Therefore

$$i \approx C \frac{dv_1(t)}{dt}$$

and

$$v_2(t) \approx RC \frac{dv_1(t)}{dt} \tag{1-19}$$

The major portion of the circuit voltage drop is developed across C only when the time constant is small compared with the time range of interest. At discontinuities, the total change of input voltage appears at the output and the circuit only roughly approximates a differentiator. With a sufficiently small time constant, the exact output waveshape is reasonably close, except at the discontinuities, to the one found by assuming a perfect differentiator. The exact derivative at a discontinuity of the input waveshape is infinity. Since instantaneous jumps do not exist in nature, the full change in the input voltage, which appears across the output at the input "discontinuities," is so very much larger than the normal small output signal that the circuit may well be said to approximate the derivative even here. Thus, after checking the time constant against the input waveshape, the output can be drawn by inspection and the voltage values calculated later. Figure 1-7 illustrates this process for an RC circuit having a time constant of 400 μs.

Because the capacitor will not pass direct current, the periodic response to periodic inputs must have zero average value. The capacitor will

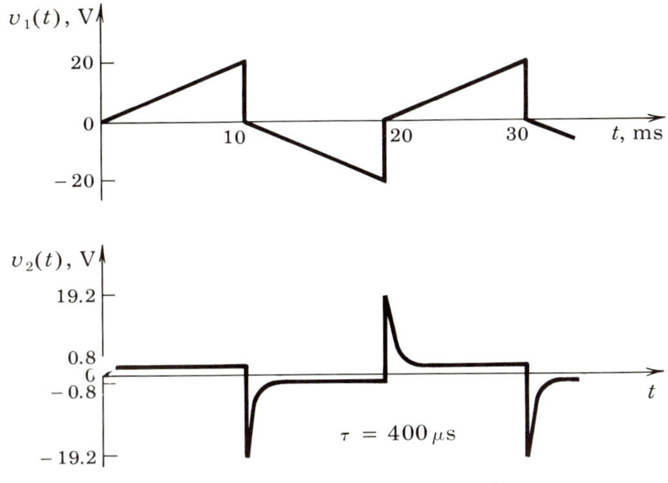

Fig. 1-7 Differentiator input and output waveshapes.

store a charge proportional to the average level of the input signal, and its voltage will fluctuate almost between the input extremes.

An *RL* series circuit having a small time constant also differentiates. Here the output must be taken across the inductance because the basis of differentiation in this circuit is the proportionality of the inductive voltage to the time rate of the circuit current change. By maintaining almost the full input voltage drop across the series resistor, we ensure the current's dependence only on this resistance and guarantee good differentiation.

1-4 Integration

Provided that the time constant is very long, the capacitor voltage in a series *RC* circuit is proportional to the integral of the input voltage (Fig. 1-8). Again the required condition may be found from an examination of the circuit differential equation:

$$v_2(t) = \frac{1}{C}\int i\,dt = v_1(t) - Ri \qquad (1\text{-}20)$$

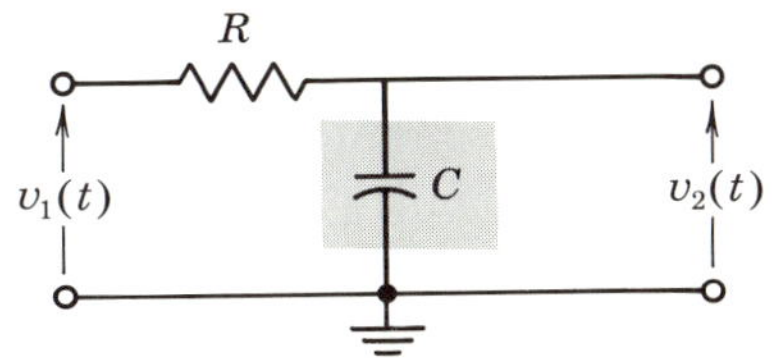

Fig. 1-8 Integrator.

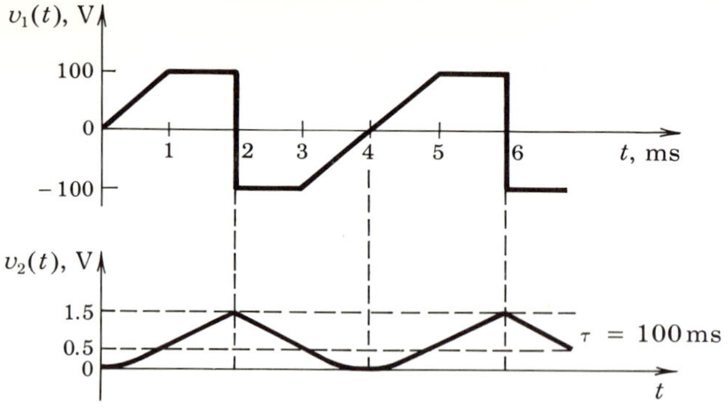

Fig. 1-9 Input signal and integrated output. Input applied at $t = 0$.

When the output voltage is small compared with the input, the approximate relationship

$$v_1(t) \approx Ri$$

results, and therefore

$$v_2(t) \approx \frac{1}{RC} \int v_1 \, dt \tag{1-21}$$

If the output voltage rises above a relatively small value, the current, $i = (v_1 - v_2)/R$, becomes dependent on v_2 and the circuit will no longer integrate satisfactorily. An excessively rapid rise in the output voltage within any period is directly dependent on the rate of charging of C. A slow charging rate remains possible only through keeping the time constant long compared with the time over which the integral is desired. C eventually charges to the dc level of the input with the circuit time constant, and the output voltage shifts accordingly. For a signal impressed at $t = 0$, Fig. 1-9 shows both the input and its integrated output. Eventually the output will shift down until it has a zero average value.

The output voltage, at any time, is proportional to the area under the curve of the input signal from zero to that time. Consequently its shape may be sketched directly from the curve and the coordinates found from Eq. (1-21). In order to calculate the voltage change at the output over any time interval, the input volt-time area is simply divided by the time constant.

A large-time-constant RL circuit also integrates with the small output voltage now appearing across R. Physical inductances always have associated winding resistance and, for large coils, appreciable stray capacity. The impossibility of obtaining an ideal element usually precludes the use of RL circuits for many wave-shaping functions. On the other hand, almost ideal capacitors are readily available. Using the new plastic dielectrics, leakage resistance, the major parasitic element, is of the order of $10^{12}\ \Omega$, a value so large that it may be ignored in most circuit calculations.

1-5 Summation

After independent wave-shaping operations have been performed on various signals, combination is accomplished by summing the individual results. Of the several available methods, the simplest, shown for two inputs, uses the resistor network of Fig. 1-10. From the circuit we see that the two input currents are

$$i_1 = \frac{v_1 - v_o}{R_1} \qquad i_2 = \frac{v_2 - v_o}{R_2}$$

Provided that the output voltage is kept small compared with each of the inputs, the current through each input resistor is approximately independent of the output voltage. But this is possible only if $R \ll R_1$ and $R \ll R_2$. Almost the complete voltage drop appears across the input resistors, and

$$i_1 \approx \frac{v_1}{R_1} \qquad i_2 \approx \frac{v_2}{R_2} \tag{1-22}$$

The output becomes

$$v_o \approx R(i_1 + i_2) = \frac{Rv_1}{R_1} + \frac{Rv_2}{R_2} \tag{1-23}$$

Even when R is not small compared with the other resistors, the output voltage can still be found by the linear superposition of the individual terms. The circuit operates by converting the input voltage into a proportional current and then summing all the individual independent currents. Equation (1-23) shows the output's proportionality to the sum of the individual inputs, each of which is multiplied by a scale factor. Determination of this factor is through a choice of the input resistors R_1 and R_2. Summation may be extended to multiple inputs by simply adding additional input branches.

The condition for summation, i.e., the output voltage small compared with each input, is exactly the same as that required for integration and

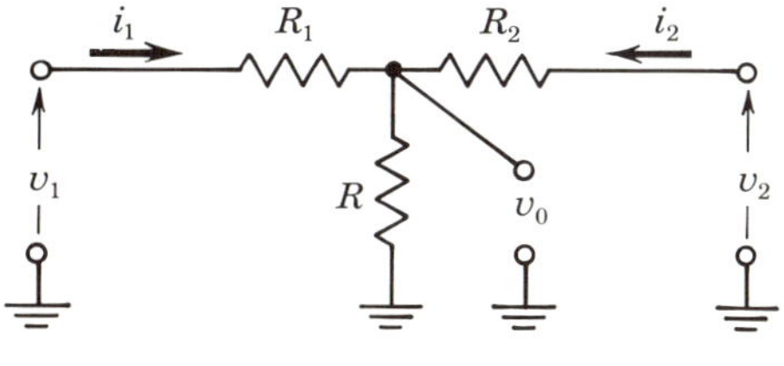

Fig. 1-10 Summation circuit.

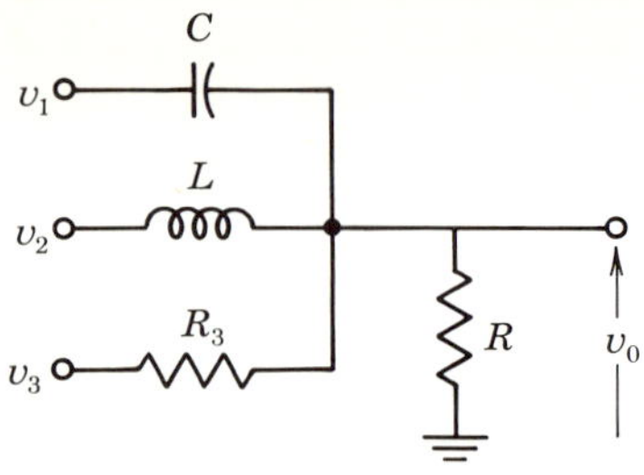

Fig. 1-11 Combined operations (differentiation, integration, summation).

differentiation, and all three operations may conveniently be combined in a single circuit as illustrated in Fig. 1-11. The output expressed by Eq. (1-24) is found by calculating and sketching each term individually and summing the results; this is often carried out graphically.

$$v_o(t) = CR\frac{dv_1}{dt} + \frac{1}{L/R}\int v_2\,dt + \frac{R}{R_3}v_3 \tag{1-24}$$

CR must be small and L/R very large compared with the time interval of the various signals. The inputs, if desired, may be supplied from the same source.

1-6 Approximate Solutions of the Single-pole Case

The results of the previous sections are quite valuable in that they open up two classes of single-energy-storage-element circuits to rapid solution. The first group consists of those with short time constants, the differentiators; the second, those with long time constants, the integrators. These two cases, representing only about 10 percent of the possible range of RC or RL circuits, include, however, a much larger portion of the important engineering problems.

In handling a problem of this nature, our first step is to compare the time constant with the input signal and on the basis of this comparison to classify the circuit response. If it is a differentiator, the voltage across the resistor or inductance may be sketched by inspection, and the drop across the other circuit element is simply the difference between the applied input and the now known branch voltage. In an RC integrator, the voltage across C is the easily found quantity, with the resistor voltage remaining the single unknown. This is illustrated in Fig. 1-12, where the signal is assumed to be applied at zero time and where we shall apply the techniques of integration in finding the voltage across both R and C. Because C cannot pass direct current and v_R has a dc level, the output waveshapes sketched

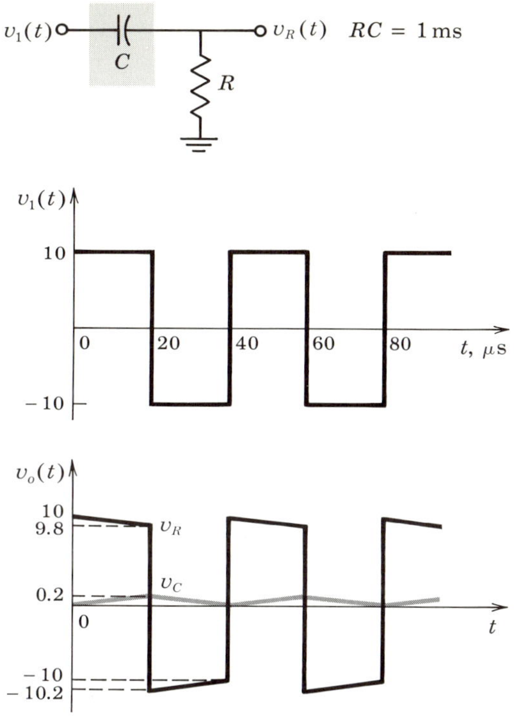

Fig. 1-12 Integrator-type circuit and approximate output.

(Fig. 1-12) require a minor correction. The voltages must be shifted slightly until both have zero average value. Figure 1-13 shows the final periodic solution.

It is often convenient to be able to express the response of a long-time-constant circuit to square or rectangular inputs as an equation. The

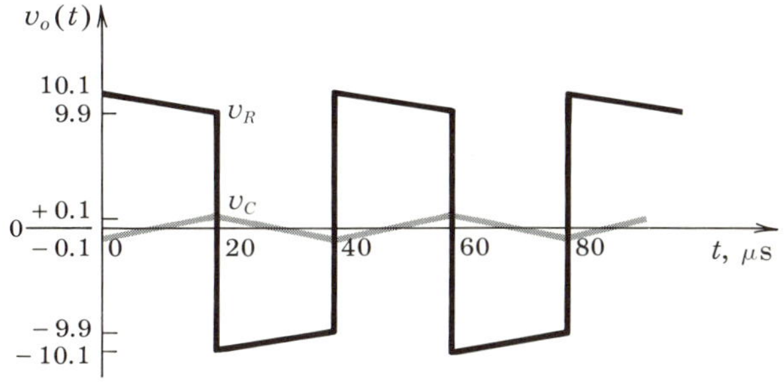

Fig. 1-13 Final periodic output of the circuit of Fig. 1-12.

exact solution of any single-energy-storage-element circuit to a step excitation is, of course,

$$v_o(t) = A + Be^{-t/\tau} \tag{1-25}$$

Substitution of the series expansion [Eq. (1-26)] for the exponential simplifies the final result. Since the time constant is long, provided that we restrict the ratio of the maximum interval to the time constant, t_m/τ, to less than 0.1, an error of less than 5 percent is introduced by neglecting the higher-order terms of the expansion.

$$e^{-t/\tau} = 1 - \frac{t}{\tau} + \frac{1}{2!}\left(\frac{t}{\tau}\right)^2 - \frac{1}{3!}\left(\frac{t}{\tau}\right)^3 + \cdots \tag{1-26}$$

Therefore, for $t/\tau < 0.1$, Eq. (1-25) may be approximated by

$$v_o(t) \approx A + B\left(1 - \frac{t}{\tau}\right) \tag{1-27}$$

which is recognizable as the equation of a straight line having a slope $-B/\tau$.

If the exact expression of the response [Eq. (1-25)] were differentiated and evaluated at $t = 0$, the slope found would be the same as that obtained from the first term in the series expansion. The interpretation which follows is that if the circuit response is restricted in time to the beginning of the exponential decay, this small portion of the curve may be represented by a linear decay.

1-7 Double-energy-storage-element Systems

Many circuits containing two energy-storage elements have poles separated widely enough so that the transient response can be approximated by treating them as two isolated single-energy circuits. Of course, there must be continuity across the boundary between the two individual response curves and they must satisfy the original system. The pole which is located far from the origin (small time constant) will determine the circuit's behavior with respect to any fast changes in the input excitation. It is related to the amplifier's high-frequency response and is predominant in controlling the initial rise. The pole which is located close to the origin is related to the low-frequency response of the system and will contribute a slow exponential to the output. It will determine the holding power, i.e., the decay rate.

In effecting the separation of the system response into two time regions, we shall have to depend on the physical characteristics of the energy-storage elements for clues as to the permissible approximations. This technique is best illustrated by an example, and to this end we shall analyze the simple *RC*-coupled amplifier whose incremental model is shown in Fig. 1-14.

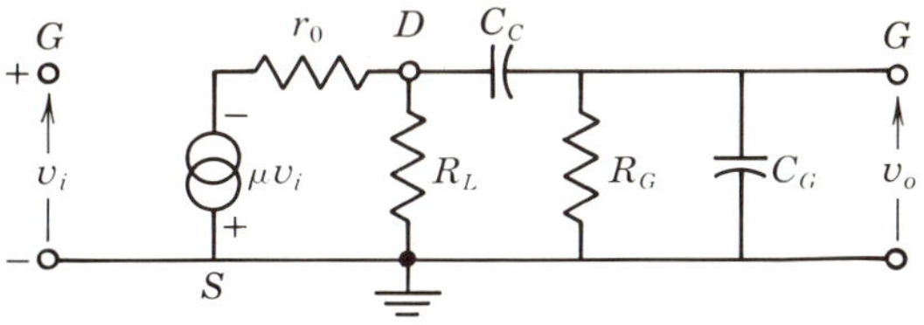

Fig. 1-14 Incremental model of an *RC*-coupled FET amplifier.

Consider the application of a voltage step of height $-V_1$ at the gate of the FET. The output is constrained in its time rate of rise primarily by C_G. We see from the circuit (Fig. 1-14) that the full charging current of C_G together with any current through R_G must also flow through C_C. If, as is normally the case, R_G is very large, the total charge flow just after the excitation is applied is essentially determined by the uncharged gate input capacity C_G. The same charge is also accumulated in C_C. Since $C_C \gg C_G$, the voltage across C_C will change but slightly while C_G charges fully. Thus the coupling capacitor may be assumed to be a short circuit during this entire interval and the equivalent circuit is reduced to one containing a single-energy-storage element (Fig. 1-15).

The final steady-state output and the circuit time constant are found by taking the Thévenin equivalent across C_G. They are

$$V_{ss1} = \frac{\mu(R_L \parallel R_G)}{r_o + R_L \parallel R_G} V_1 \qquad \tau_1 = C_G(R_G \parallel R_L \parallel r_o) \tag{1-28}$$

We are now in a position to write the equation defining the initial portion of the output response:

$$v_{o1} = V_{ss1}(1 - e^{-t/\tau_1}) \tag{1-29}$$

In four time constants, the output rises to 98 percent of the steady-state value V_{ss1} and the initial rise may be assumed complete.

During this whole interval, C_C is charging, even though it is doing so very slowly. The relatively large current required (because of the large value

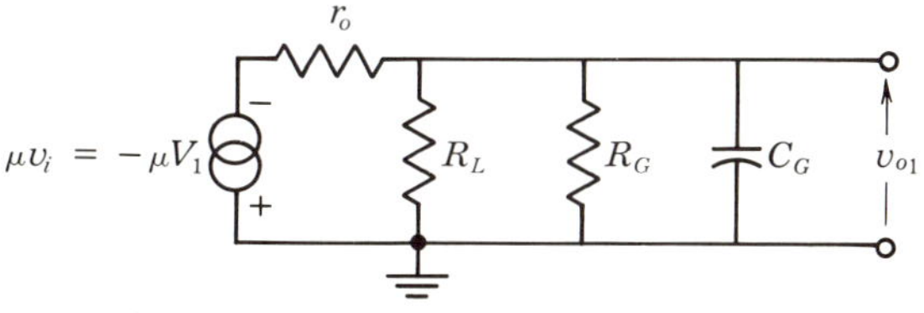

Fig. 1-15 Rise equivalent circuit.

of C_C) will now control the output voltage and swamp any contribution from the discharge of C_G. We are justified in ignoring C_G and in removing it from the circuit. If the initial value of the output across R_G upon the sudden excitation of the system is now calculated, we also find it to be V_{ss1}. As C_C charges, the output decays toward zero, with the new time constant

$$\tau_2 = C_C(R_G + R_L \parallel r_o) \tag{1-30}$$

Thus the equation defining the final portion of the response may be written by inspection:

$$v_{o2} = V_{ss1}e^{-t/\tau_2} \tag{1-31}$$

The only question remaining unanswered is: At what point will the decay take over from the initial rise? If we compare Eqs. (1-28) and (1-30), we see that both the capacity and the resistance terms of τ_2 are much larger than τ_1. Since the decay-time constant is so very much longer, the error introduced by starting the decay anywhere in the vicinity of zero will be negligible. We might just as well choose the most convenient point, the end of the initial rise, that is, $4\tau_1$. The total output is sketched in Fig. 1-16.

The exact solution, the broken line, shows the greatest deviation at the peak, asymptotically approaching the approximate solution at both long and short times. By the very nature of the assumption made, i.e., complete charge before any discharge, the exact answer will always lie below the approximate one. As the separation of time constants increases, agreement between the two answers improves. For $\tau_2 > 100\tau_1$, as is normal in this type of amplifier, agreement is extremely good.

When an amplifier is used in pulse applications, the time response, rather than the related high-frequency 3-dB point, is used to characterize its quality. Even if the amplifier transmits all signals faithfully, it may still delay them by a fixed time interval; this delay will also be considered

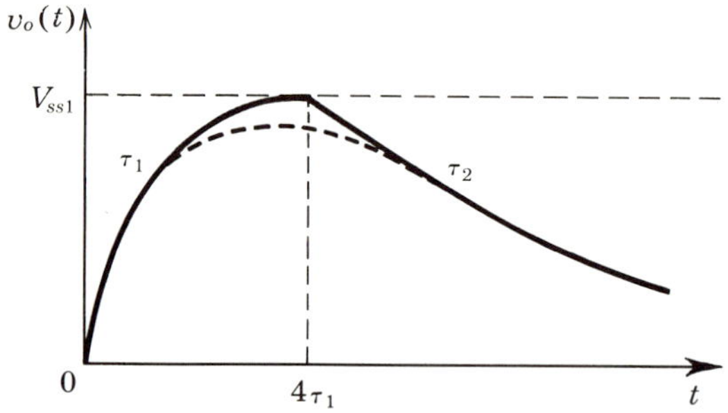

Fig. 1-16 *RC*-coupled amplifier output—step excitation input.

as a defining quantity. Generally, the shape of the output will be affected, and in order to separate the rise time from the inherent delay, we choose the 10 percent value as our lower reference point. Moreover, the applied pulse makes its presence known long before the output reaches steady state; the 90 percent point is therefore arbitrarily chosen as the upper reference value. An amplifier's rise time is defined as the time required for the output to rise from 10 to 90 percent of the final steady-state voltage, after the excitation by a unit step. Substituting these limits into and solving Eq. (1-29) gives the rise time of a single stage as

$$t_r = 2.2\tau_1 \tag{1-32}$$

The smaller the time constant, the faster the initial rise rate. The rise time ranges from several microseconds in an audio amplifier down to a few nanoseconds in systems designed primarily for pulse amplification.

When the input is a square wave, the amplifier output must eventually become periodic. The stray capacity present slows the rise of the leading edge, and the coupling capacity introduces some tilt on the flat top. The rise time of each half period will be the same as that given by Eq. (1-32) for the unit step. However, the amount of tilt depends on the duration of the half cycle and must be defined in terms of the input period. If the half period $T/2$ is small compared with the decay-time constant τ_2, only a very small decay will occur over each half cycle, and this may be approximated by a straight line. From Eq. (1-27), the linear representation of this region, we find that the initial value of the top is B (A represents the steady-state value, which in this case is zero). The final value, at the end of each half cycle, becomes $B(1 - T/2\tau_2)$. Tilt is defined as the relative slope of the top of the square wave and is

$$S = \frac{V_i - V_f}{V_i} = \frac{T}{2\tau_2} \tag{1-33}$$

S is often expressed as a percentage. Measurement of the tilt for a given square-wave input serves as a convenient method of evaluating the holding response of an amplifier, just as a measurement of the rise time characterizes its high-frequency response.

1-8 Compensated Attenuators

Attenuation of signals through pure resistive networks generally proves unsatisfactory because of the stray capacity loading of the output (C_2 in Fig. 1-17). Instead of the rapid response to a step expected from a resistive attenuator, the output rises with a time constant

$$\tau = C_2[(R_1 + R_s) \parallel R_2]$$

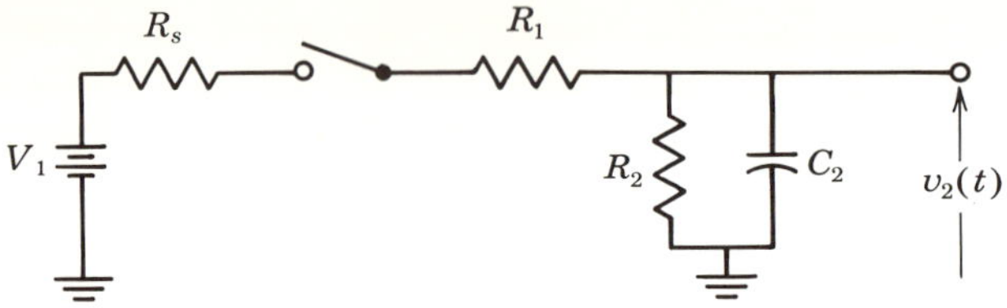

Fig. 1-17 Uncompensated attenuator.

toward the final value

$$V_{ss} = \frac{R_2}{R_1 + R_s + R_2} V_1$$

In order to prevent the source impedance from affecting the signal division, usually $R_1 \gg R_s$; otherwise, a change from one network driving source to another would also result in a new attenuation ratio.

The attenuator response can be greatly improved by shunting R_1 with a small capacitor C_1. Since it furnishes a low-impedance path to the initial rise, the output will reach steady state much sooner. At the instant of closing the switch, the two uncharged capacitors are effectively short circuits. Since most of the current will flow through them, rather than through the parallel resistors, we shall remove R_1 and R_2 from the circuit while investigating the initial rise (Fig. 1-18). The two capacitors form a capacitive divider, and the applied voltage divides across them inversely proportional to their capacities. The output rises toward a steady-state voltage of

$$V_{ss1} = \frac{C_1}{C_1 + C_2} V_1 \tag{1-34}$$

with the time constant

$$\tau_1 = R_s \frac{C_1 C_2}{C_1 + C_2} \tag{1-35}$$

We implicitly assume that the steady-state voltage of Eq. (1-34) is reached before the resistive divider need be considered. We can now draw the model of Fig. 1-19 to represent the final portion of the response. Since

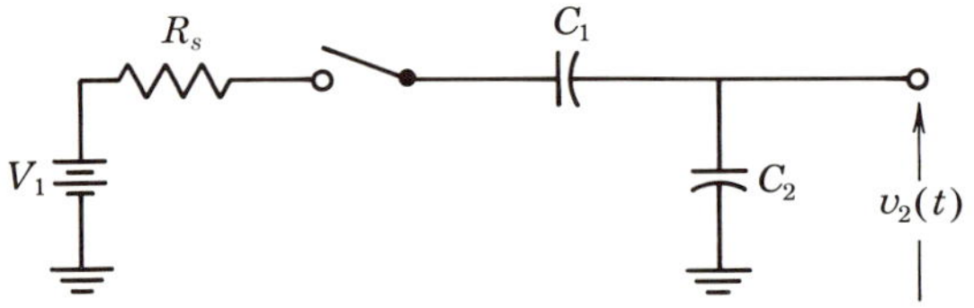

Fig. 1-18 Circuit of a compensated attenuator used for the calculation of the initial rise.

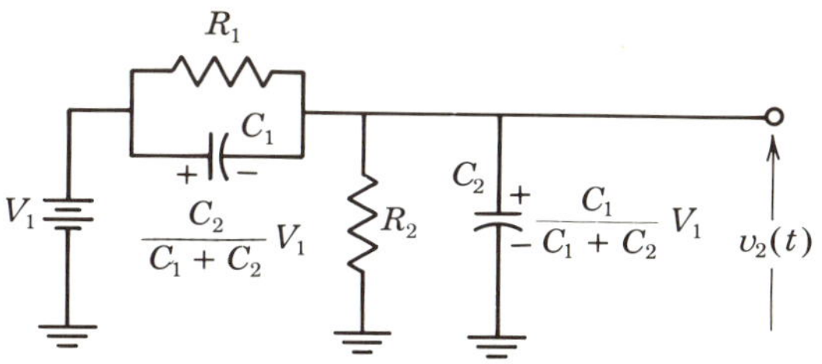

Fig. 1-19 Decay model—compensated attenuator.

$R_1 + R_2 \gg R_s$, R_s can be omitted from this model. In this circuit the initial charge on each capacitor is indicated. The output now rises or decays from its capacitor voltage-divider value toward the new steady-state voltage V_{ss2}:

$$V_{ss2} = \frac{R_2}{R_1 + R_2} V_1 \tag{1-36}$$

The new time constant becomes

$$\tau_2 = (R_1 \parallel R_2)(C_1 + C_2) \tag{1-37}$$

Comparison of Eqs. (1-35) and (1-37) indicates that τ_2 is much longer than τ_1. C_1 and C_2 in parallel are obviously larger than the same two capacitors in series. R_s had been assumed small, so that the parallel combination of R_1 and R_2 will probably be the bigger term.

If the two steady-state values [Eqs. (1-34) and (1-36)] are equal, there will be only one transient, the initial rise. By equating these two voltages and solving, we arrive at the conditions for the optimum attenuator response,

$$R_1C_1 = R_2C_2 \tag{1-38}$$

Compensating the attenuator greatly improves the rise time, as shown in Fig. 1-20. When $R_1C_1 < R_2C_2$, the attenuator is overcompensated, introducing an overshoot. If we are interested in attenuation of triggers (sharp pulses), the overcompensated case is often desirable since it gives the fastest rise and the largest initial amplitude. Much to our regret, the RC-coupled amplifier cannot be compensated because r_o does not in fact exist as an entity across which a capacitor can be connected. Any external capacity added only parallels the existing strays. This simply increases the rise time, with a corresponding degeneration in the overall response.

The initial rise time of the compensated attenuator is usually so short that it may be ignored. This leads to the assumption that the output instantaneously jumps to the point from which it starts decaying toward the final steady-state value.

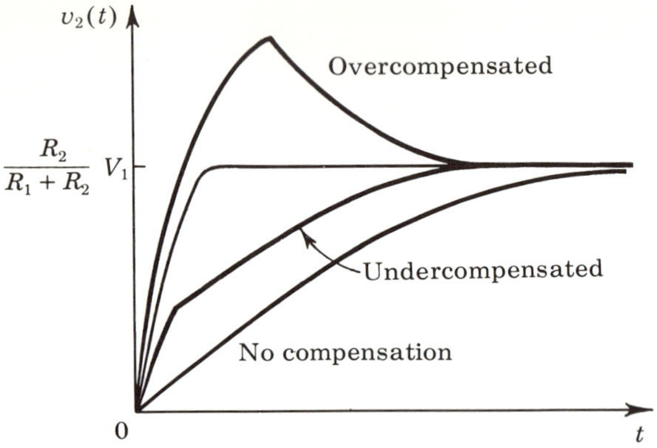

Fig. 1-20 Compensated-attenuator output (slightly exaggerated for purposes of illustration).

Treating the same problem from the pole-and-zero viewpoint and neglecting the initial rise ($R_s = 0$), we see that the placing of C_1 across R_1 introduces a zero along the real axis. Proper adjustment moves it into coincidence with the pole determined by R_2C_2, annihilating that pole. Since the circuit no longer contains any poles, the response cannot be of exponential form but must be the constant value associated with a purely resistive attenuator.

Problems

1-1 The switch in the circuit of Fig. 1-21 is closed at $t = 0$ and reopened 1 μs later.

(*a*) Sketch the voltage waveshape appearing across the capacitor, giving the values of all time constants and voltages.

(*b*) Sketch the waveshape of the current flowing through the switch. Evaluate the initial and final values of the current either directly from the circuit or by using the information found in part *a*. Do not solve the exponential-response equation for these current values.

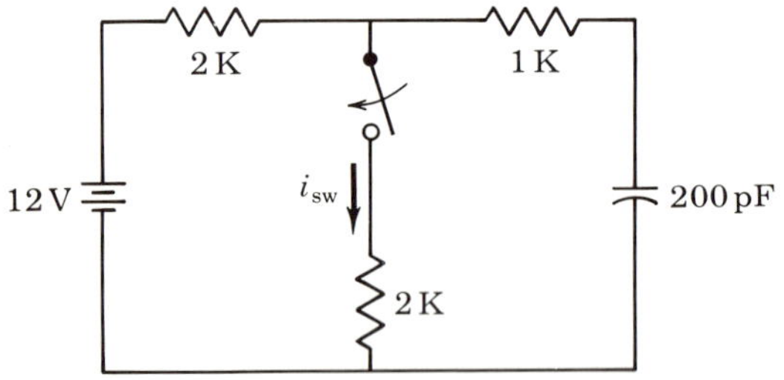

Fig. 1-21

1-2 In Fig. 1-22, switch 1 is closed at $t = 0$ and switch 2, 100 μs later. Evaluate and sketch on the same axis the current response of both inductances. Make whatever reasonable approximations are necessary to simplify the calculations, including changing the ideal elements to nonideal elements which become ideal in the limit. Repeat the calculations when switch 2 is closed 1 ms after switch 1.

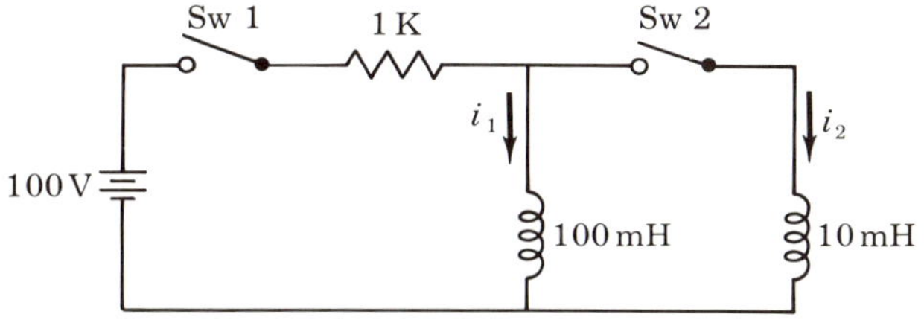

Fig. 1-22

1-3 Sketch the waveshapes of both output voltages, with respect to ground, and the charging current in the circuit of Fig. 1-23. Label your sketches with all significant voltage and current values.

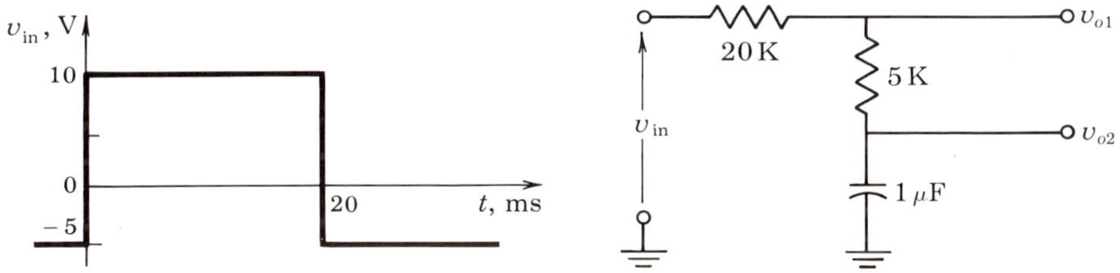

Fig. 1-23

1-4 A single 10-μs-wide pulse is applied to the circuit of Fig. 1-24 at $t = 0$. The switch shown is voltage-controlled; it automatically closes whenever the output voltage is positive and opens whenever the output voltage is negative. Draw the voltage waveshape appearing across the capacitor. Give all significant times, time

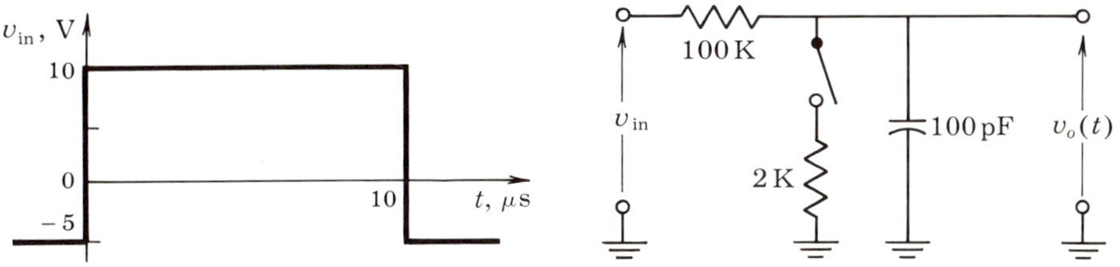

Fig. 1-24

constants, and voltage values. On the same sketch, indicate the duration of the input pulse and the interval that the switch is closed.

1-5 Repeat Prob. 1-4 when the switch is closed for $v_o(t) < 0$ and opened for $v_o(t) \geq 0$.

1-6 Switch 1 in the circuit of Fig. 1-25 is thrown from position 1 to position 2 and back again periodically. Switch 2 is voltage-controlled; it is open whenever $v_o(t) < 0$ and closed whenever $v_o(t) \geq 0$. Sketch the steady-state output waveshape, labeling all significant voltage values, times, and time constants when the position of switch 1 is changed every 20 μs.

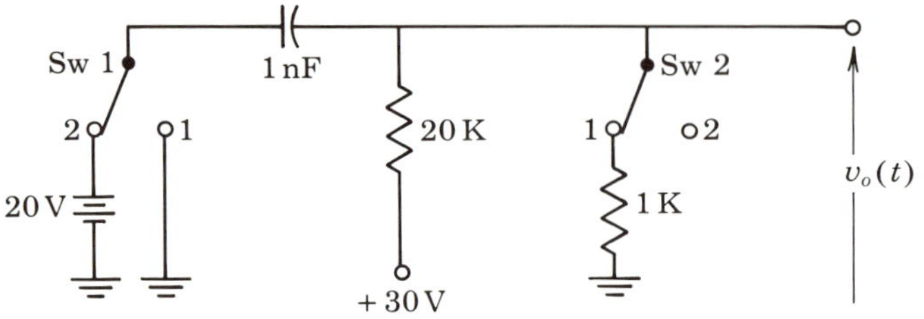

Fig. 1-25

1-7 A periodic rectangular voltage waveshape, such as the one shown in Fig. 1-4, is the driving signal applied to a series RC circuit having a time constant of 10 ms. The 10-V positive peak lasts for 30 ms, and the -5-V negative peak for only 5 ms.

(*a*) Sketch the steady-state voltage appearing across the capacitor, giving the maximum and minimum values.

(*b*) What voltage would be read on a dc voltmeter connected across the capacitor? Across the resistor? Explain your answers.

1-8 A 10-V peak-to-peak 20-μs-period square wave is applied to a series RC circuit ($\tau = 50\ \mu$s). The output is taken across the resistor.

(*a*) Sketch and give values for the steady-state output.

(*b*) If at some $t = 0$ we add a $+20$-V step to the periodic square-wave input, how long does it take for the instantaneous output voltage to reach zero? Do not follow the response cycle by cycle. Prove that the zero crossing point must occur at a negative transition of the input square wave.

1-9 The excitation of a parallel RL circuit is a square wave of current having a peak-to-peak amplitude of 50 mA and a period of 200 μs. The inductance is 1 H, and the resistance is 5 K.

(*a*) Sketch the steady-state node voltage and evaluate the maximum and minimum values.

(*b*) Find the power dissipated in the resistance.

1-10 The periodic current flowing in the series RLC circuit of Fig. 1-26 is given. What waveshape and values of excitation voltage must be applied to cause this current flow? Is your solution unique? If your answer is "yes," explain. If your answer is "no," what must be specified to make the solution unique?

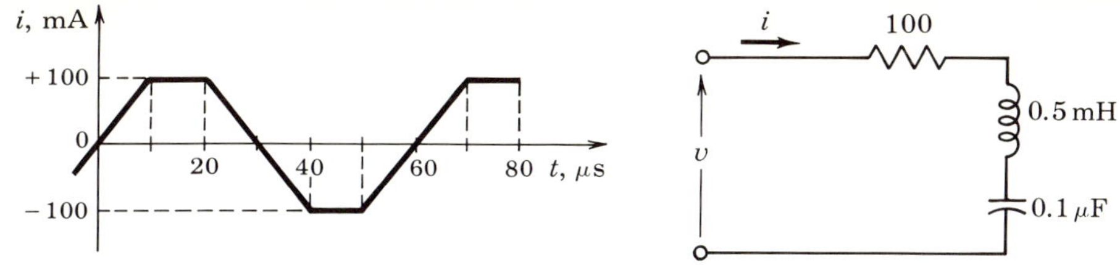

Fig. 1-26

1-11 A periodic triangular voltage of 200 V peak to peak appears across the parallel RLC circuit shown in Fig. 1-27.

(*a*) Calculate the inductance that will make the peak inductive current equal to 20 mA. (The dc level of the input current is zero.)

(*b*) Sketch the current waveshape and give the values at all break points.

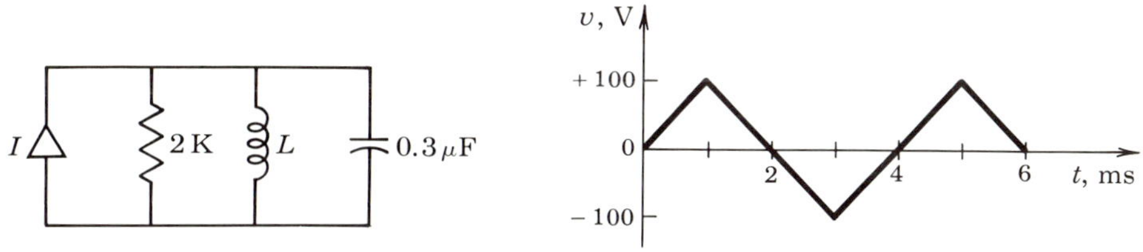

Fig. 1-27

1-12 The nonperiodic signal shown in Fig. 1-28 is applied to a series RC circuit. Calculate the approximate response under the following conditions:

(*a*) The voltage across C when the time constant is 10 ms

(*b*) The voltage across R when the time constant is 10 ms

(*c*) The voltage across R when the time constant is reduced to 10 μs

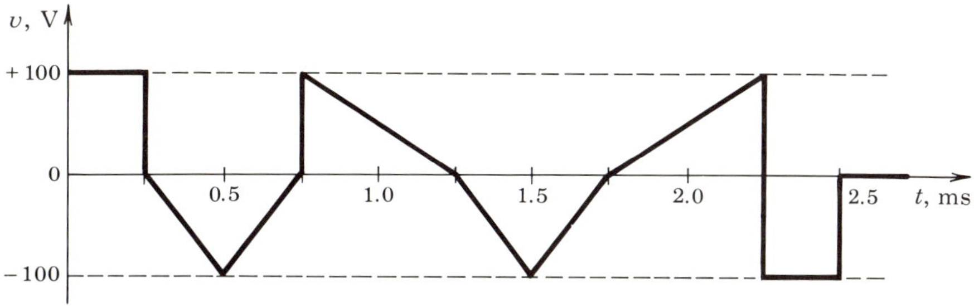

Fig. 1-28

1-13 (*a*) The triangular voltage of Fig. 1-27 is applied to an integrator circuit having a time constant of 25 ms. Sketch the steady-state voltage appearing across the capacitor.

(*b*) If a signal consisting of only the positive portion of the triangular wave of Fig. 1-27 were applied to this integrator, how long would it take for the output to rise to 9 V?

1-14 (*a*) Under what conditions will one of the branch currents of a parallel *RC* circuit be proportional to the integral of the driving current signal? Which branch?

(*b*) Show that a parallel *RC* circuit can also be used for current differentiation.

1-15 Repeat Prob. 1-14 with respect to a parallel *RL* circuit. Explain what further restrictions, if any, must be imposed as a result of using a nonideal inductance (winding resistance).

1-16 (*a*) A sawtooth signal of the type shown in Fig. 1-7 but with each segment lasting for 100 μs is applied as the input to an integrator having a time constant of 10 ms. The output is coupled through an ideal isolation amplifier which multiplies the voltage by a factor of 50. The amplifier output is then passed through a differentiator having a time constant of 1 μs. Sketch the ultimate output.

(*b*) The circuits and input of part *a* are again used except that the applied signal is first differentiated, then integrated, and finally amplified. Sketch the output appearing in this case and explain any discrepancy between this answer and the one for part *a*.

1-17 Two signals are combined in the summation circuit of Fig. 1-29. Draw the output, giving voltage values under the following conditions:

(*a*) The circuit is as shown in the figure.

(*b*) We replace the 1-K resistor by a 0.05-μF capacitor.

(*c*) The circuit is further modified by also replacing the 50-K resistor with a 0.005-μF capacitor.

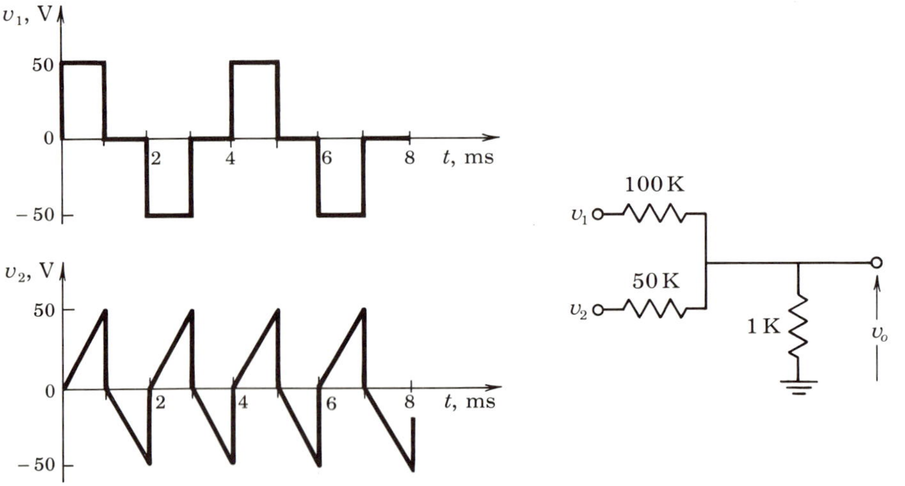

Fig. 1-29

1-18 We are interested in generating a reasonable approximation to the signal shown in Fig. 1-30. Under the conditions listed, find the circuit required and the values of components needed. Make your smallest resistor 1 K.

(*a*) The only signal available is v_2 of Fig. 1-29.

(*b*) The only signal available is a 100-V peak-to-peak square wave of the proper period.

(*c*) Repeat parts *a* and *b* but use a series combination of circuit elements for the output branch instead of a summation-type circuit.

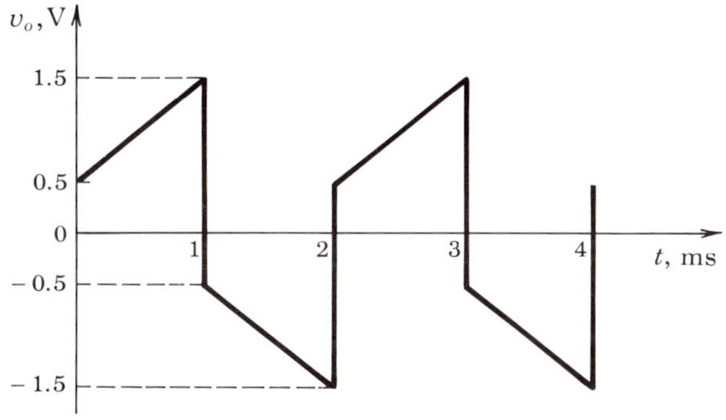

Fig. 1-30

1-19 A single-stage field-effect-transistor amplifier having the equivalent circuit shown in Fig. 1-14 is excited by a voltage step at the gate. The FET parameters and the circuit components are $r_0 = 20$ K, $\mu = 25$, $R_L = 20$ K, $R_G = 100$ K, $C_G = 200$ pF, and $C_C = 0.01\ \mu$F.

(*a*) Find the output response, giving all time constants and steady-state values, by the approximation method discussed in the text (time yourself).

(*b*) Solve for the exact response, using either classical or transformation methods, and compare both the time required and the accuracy of solution with those found in part *a*.

1-20 (*a*) Derive Eqs. (1-32) and (1-33).

(*b*) Calculate the rise time of the amplifier of Prob. 1-19.

(*c*) What is the lowest-frequency square wave that would be transmitted without introducing more than 5 percent tilt? Sketch the output if the peak-to-peak voltage is 10 V. Give the actual voltage values.

(*d*) In order for a pulse to register its presence at some later point in the system, at least 80 percent of it must be transmitted by the amplifier of Prob. 1-19. What is the narrowest pulse that can be used, and how will it appear after passing through the amplifier?

1-21 In the circuit of Fig. 1-31, a 1-mA step of current is applied at $t = 0$. Find i_2. Sketch your answer and give all time constants and steady-state values. Justify the approximations made in the course of your solution.

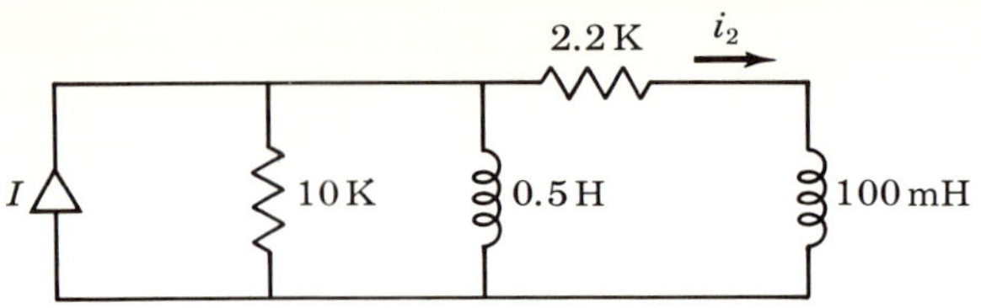

Fig. 1-31

1-22 The two input signals of Fig. 1-29 are applied to the circuit of Fig. 1-32.

(*a*) Sketch the approximate steady-state output, giving all voltage values when v_1 is connected to terminal X and v_2 to terminal Y.

(*b*) Repeat part *a* if a 20-pF capacitor is connected across the 5-M resistor.

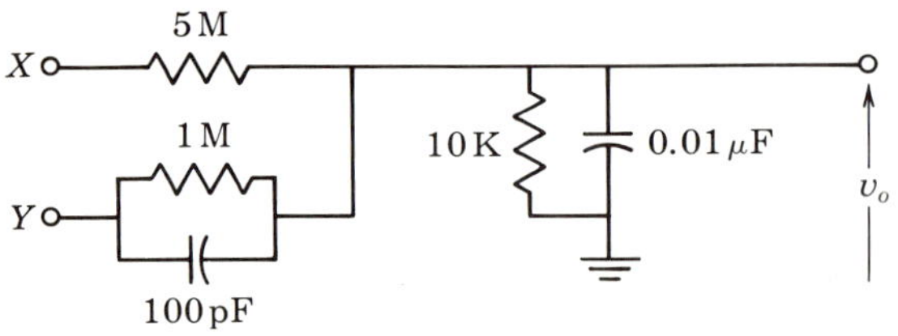

Fig. 1-32

2. diode wave-shaping techniques

Wave-shaping functions performed by purely linear circuits—differentiation, integration, summation, and not much else—are relatively limited. If at some point within any operation we could, at will, change the value of even one circuit component, there would result a tremendous increase in the possible shapes of the output. When considering periodic inputs, the component changes should be voltage- or current-controlled, e.g., a bivalued resistor—a high resistance at voltages below a threshold value and low resistance thereafter, or vice versa. Otherwise, the circuit change would not occur at the same voltage point of each cycle and a periodic output would not result from the periodic input. One method of obtaining this response is through the use of two resistors and a switch, which is thrown from one to the other and back again as the circuit voltage rises or drops past the threshold value. Of course, for fast complex waveshapes, mechanically operated switches could not possibly operate rapidly enough, and how could we throw the switch continuously?

Conveniently for us, there exist a number of voltage-controlled nonlinear devices with approximately the properties desired. One of the most common of these devices is the diode, and in this chapter we propose to examine its application to various circuits, making use of its inherent nonlinearity.

2-1 Passive-nonlinear-circuit Representation

The response of any network to a forcing function is determined by the individual characteristics of the various components comprising that network. In the analysis of so-called linear circuits, ideal elements (resistors, capacitors, inductors, and voltage and current sources) are convenient fictions introduced to simplify engineering calculations. The actual physical elements not only are slightly nonlinear but also include various parasitics; e.g., a capacitor contains lead and winding inductance and its capacity varies somewhat with the charge storage. Analogous characteristics may be detected in all other "linear" components. These minor second-order effects will cause at most a slight correction in the calculated ideal-network response. Their inclusion in the problem will, however, obscure the phenomena under examination beneath the complexity of the additional mathematics. Thus, in the interests of simplicity, we are completely justified in treating only ideal linear elements.

Following a similar argument, we can postulate the existence of an ideal nonlinearity, which we shall call the ideal diode, to account for the response of a large class of grossly nonlinear systems. This element has as much validity as our other circuit components and will be utilized in networks in much the same manner. It has the bivalued characteristics expressed both graphically and in equation form in Table 2-1. The ideal diode is an absolute short circuit for any positive current flow and an open circuit when the polarity across its terminals is negative (back-biased). Its volt-ampere plot consists of two connected segments with the break between the two regions occurring at $v_d = 0$ (or $i_d = 0$). Thus the ideal diode is identical with the voltage-controlled switch previously discussed, the switching value being 0 V.

Table 2-1 illustrates the schematic representation of the branch elements with which we shall be concerned, together with their defining equations and volt-ampere plots.

The combined circuit response of these elements is found directly from Kirchhoff's voltage and current laws. Any number of elements in series must have the same current flow, and thus the total terminal voltage will be the sum of the individual drops measured at this current. A parallel arrangement has an input current that is the sum of the current flow in each branch evaluated at the common terminal voltage. The diode introduces constraints dictated by its nonlinearity.

Let us now consider the two examples shown in Fig. 2-1. In Fig. 2-1*a*, the series diode limits the current flow to positive values, the battery shifts the break point to the right, and the resistor introduces a finite slope in the conduction region. Summing the individual equations given in Table 2-1 leads directly to the defining equations for the compound branch:

$$v_a = v_d + v_R + v_s$$

$$v_{a1} = V + i_a R \qquad \text{for } i_a \geq 0 \text{ or } v_a \geq V \qquad (2\text{-}1a)$$

$$v_{a2} = v_d + V \qquad \text{for } i_a = 0 \text{ or } v_a \leq V \qquad (2\text{-}1b)$$

Table 2-1

ELEMENT	SYMBOL	DEFINING EQUATIONS	VOLT-AMPERE CHARACTERISTIC
Ideal diode	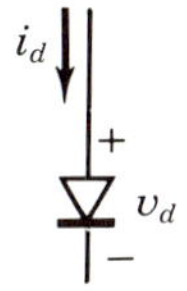	$v_d = 0$ or $i_d \geq 0$ and $v_d \leq 0$ or $i_d = 0$	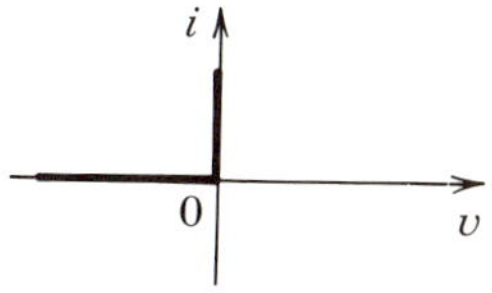
Resistor	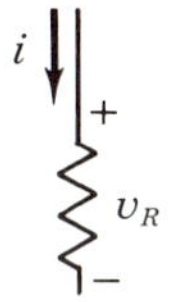	$v_R = iR$ or $i = Gv_R$	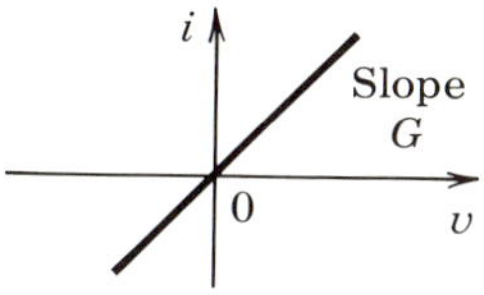
Voltage source	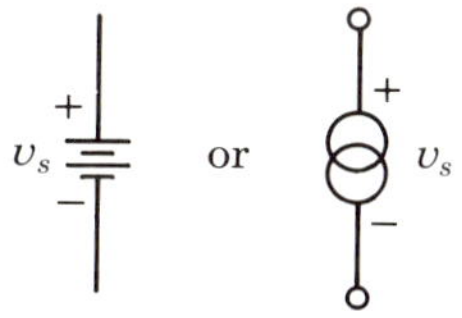	$v_s = \text{const} = V$ or $v_s = v(t)$	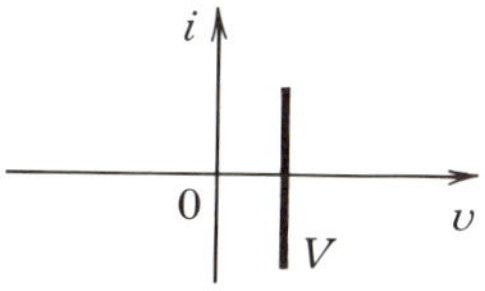
Current source	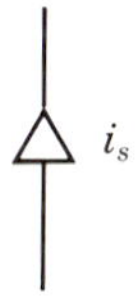	$i_s = \text{const} = I$ or $i_s = i(t)$	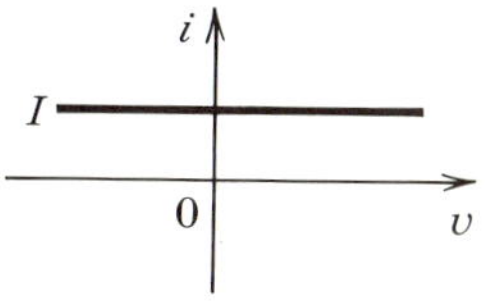

Two equations are necessary because of the bivalued diode properties: one holds for the conduction region (2-1a), and the other for the back-biased zone (2-1b). They agree at the boundary point, which may be expressed in terms of the applied terminal voltage by setting $i_a = 0$ in Eq. (2-1a) and $v_d = 0$ in Eq. (2-1b). Figure 2-1a illustrates the graphical summation of the individual curves, i.e., the superposition of the element voltage drops. We might note that the voltage drops add normally in the positive-current region but that the back-biased diode ($v_d < 0$) absorbs the total loop voltage.

Reversal of the diode would establish the conduction zone for negative rather than for positive currents, and reversal of the battery would shift the break point to the left instead of to the right. The diode connection determines the permissible current-flow direction, and the battery voltage only sets the break-point location.

In Fig. 2-1b, the currents must be summed over the voltage range per-

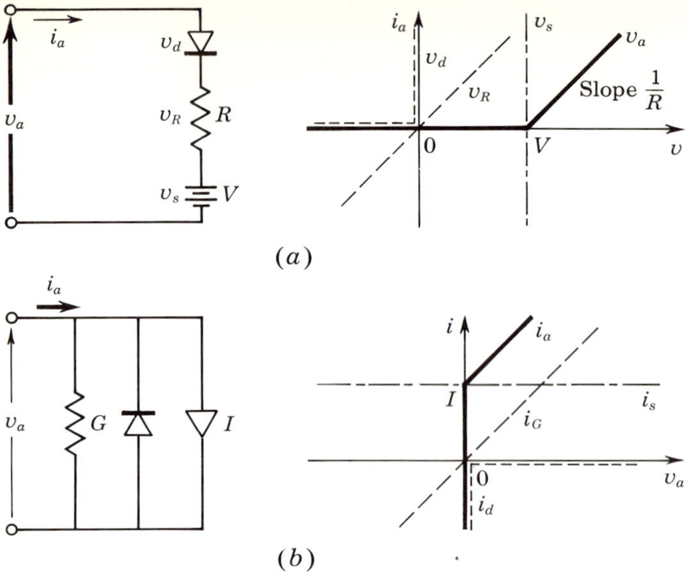

Fig. 2-1 Biased ideal-diode circuits and their volt-ampere characteristics. (*a*) Series combination of elements; (*b*) parallel combination of elements.

mitted by the diode (for negative voltages it is a short circuit forcing $v_a = 0$).

$$i_a = i_G + i_d + i_s$$

$$i_{a1} = Gv_a + I \qquad v_a \geq 0 \text{ or } i_a \geq I \tag{2-2a}$$

$$i_{a2} = i_d + I \qquad v_a = 0 \text{ or } i_a \leq I \tag{2-2b}$$

The shunt current source shifts the current coordinate of the break point to a positive value, as shown in the graphical characteristic of Fig. 2-1*b*. Again, the diode restricts the region where graphical addition is necessary, in this case by introducing a short circuit upon conduction.

Even if these circuits did not contain a nonlinear element, the intercepts of the series-resistance and parallel-conductance curves would still differ from zero by the value of the voltage and current biases, respectively. The addition of the diode only limits the portion of the total shifted curve which may be used.

In general, we can say that:

1. If there is no current source included across the diode, then one coordinate of the branch break point will always be the bias voltage contained in the diode branch.
2. If there is no voltage source included in a parallel arrangement of elements or branches, then one coordinate of the break point will always be the value of the current bias across the diode.

Combining parallel current and series voltage biasing with the single diode permits us to locate the break point anywhere in the volt-ampere plane. Moreover, by including both series and parallel padding resistors, the slopes in the two zones may be individually controlled (Fig. 2-2).

Suppose that we are interested in evaluating the input volt-ampere characteristic of any single diode circuit. How would we do so? We could, of course, solve the problem graphically, by summing the response of each element. For more complex circuits, such as the one shown in Fig. 2-2, this process may become somewhat tedious.

For an algebraic solution, the simple diode circuit may be replaced by the two implicit resistive models, one of which holds when the diode conducts and the other when it is back-biased. By superposition from the model of Fig. 2-3*a*, where the current source is short-circuited by the conducting diode,

$$i_a = \frac{v_a}{R_1 \parallel R_2} + \frac{V_1}{R_2} \qquad i_d > 0 \tag{2-3}$$

From the second model (Fig. 2-3*b*), which holds for $v_d < 0$, we obtain the other defining equation,

$$i_a = \frac{v_a}{R_1} + I \tag{2-4}$$

The boundary between the regions is the intercept of the two straight lines whose equations are given by (2-3) and (2-4). This is shown graphically in Fig. 2-3*c*. Equating and solving for the terminal break value v_{ab} yields

$$v_{ab} = IR_2 - V_1$$

The current break coordinate i_{ab} can now be found by substituting v_{ab} into either defining equation.

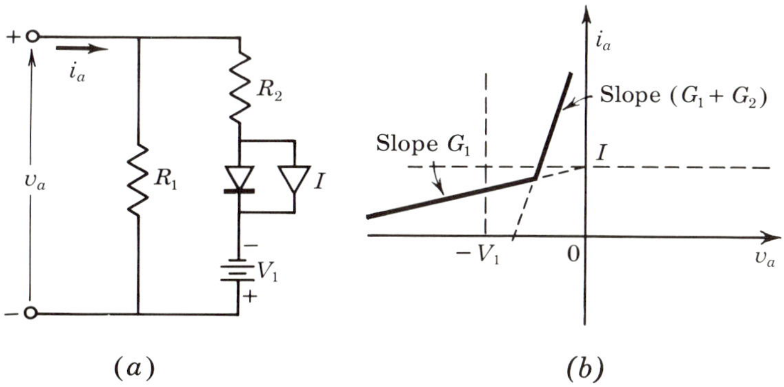

Fig. 2-2 (*a*) Combined biasing of piecewise-linear circuit; (*b*) volt-ampere characteristic.

We might observe that the break does not occur at the intersection of $i_a = I$ and $v_a = -V_1$. This is a direct consequence of the additional current flow through R_1 (due to V_1) and the additional voltage drop across R_2 (due to I). The reader might remove each resistor in sequence and observe the effect on the break-point location.

Each of the two models may be reduced to the Thévenin or Norton equivalent circuit, shown in Fig. 2-3, by algebraic manipulation. The open-circuit voltage or short-circuit current source is the value of the appropriate intercept, and the slope of the line segment is the resistance or conductance.

Since it is known that the volt-ampere curve consists of two linear contiguous segments, it follows that three pieces of information are sufficient to define these regions. Instead of writing the complete equation, we can simply solve for the coordinates of the break point and the two adjacent slopes. While doing so, attention must remain focused on the diode because its state controls the circuit behavior. Example 2-1 illustrates this technique for the circuit of Fig. 2-2.

Example 2-1

If we select the conducting state in Fig. 2-2 as our arbitrary starting point, the diode will short the current source and reduce the network to one containing two resistors and a voltage source (Fig. 2-3*a*). This occurs where the diode current is zero and where the current through R_2 equals I. Thus the break voltage is the total branch drop at this current flow,

$$v_{ab} = IR_2 - V_1$$

For $R_1 = 10\text{ K}$, $R_2 = 1\text{ K}$, $V_1 = 25\text{ V}$, and $I = 5\text{ mA}$,

$$v_{ab} = 1\text{ K} \times 5\text{ mA} - 25\text{ V} = -20\text{ V}$$

The other coordinate becomes the sum of the current through R_1, because of the applied break voltage v_{ab}, and the bias current I. From Eq. (2-4),

$$i_{ab} = 5\text{ mA} + \frac{-20}{10\text{ K}} = 3\text{ mA}$$

Within each region, the slope is found by calculating the incremental conductance, first when the diode is conducting ($v_a > -20$) and next when it is back-biased ($v_a < -20$). These values are already indicated in Fig. 2-3*c*. In the forward region it is 1.1 mmhos, and in the back-biased zone it is only 0.1 mmho.

We conclude that once the nonlinear circuit is represented by a model containing a diode, we can write two sets of linear equations to describe

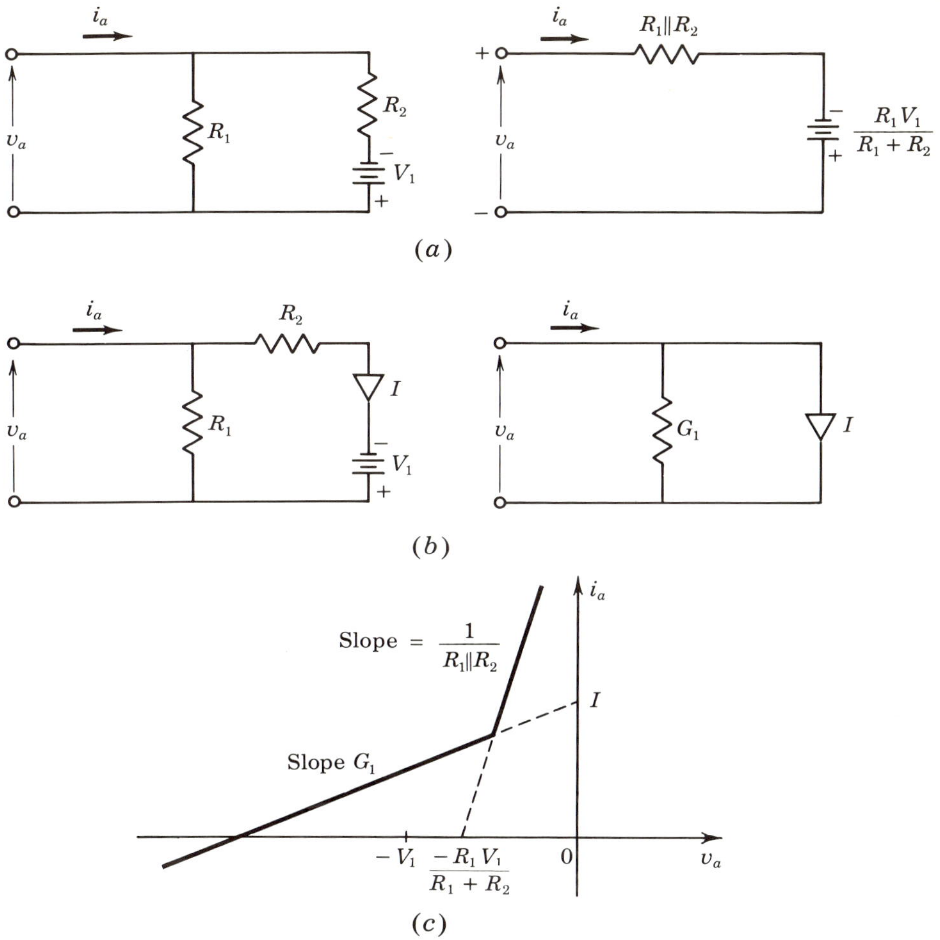

Fig. 2-3 Models illustrating the piecewise solution of the circuit of Fig. 2-2. (*a*) Reduced circuit when diode conducts and the Thévenin equivalent circuit; (*b*) reduced circuit when diode is back-biased and the Norton equivalent circuit; (*c*) composite volt-ampere characteristic.

the system response, one equation holding while the diode conducts and another when it is back-biased. These equations must coincide at the boundary point ($v_d = 0$); the continuous diode characteristic introduces no voltage or current discontinuity, and neither should the model or the equation. By using two regions to represent the nonlinear device, all theorems and techniques of the solution of linear circuits become applicable in each region. The only complexity introduced is that upon completion of a solution we must check to see that we have not left one linear region and entered another. If the circuit has done so, then the original equation, written for one region, will not hold and the answer will be incorrect. The problem must subsequently be reformulated and re-solved.

2-2 Semiconductor Diode Characteristics

In this section we shall briefly review the semiconductor-device physics which gives rise to the volt-ampere characteristic of an actual diode shown in Fig. 2-8. Section 2-3 will then discuss various circuit models which can be used to approximate the actual response.

Silicon and germanium are group-IV elements which crystallize in a diamond structure. Each atom is tied to its four nearest neighbors with strong covalent bonds. The spacing of 2.5 Å between neighboring atoms corresponds to a density of about 5×10^{22} atoms per cm^3. Thermal energy causes the disruption of some of the bonds, and every bond broken makes available one electron-hole pair as charge carriers. In the intrinsic semiconductor, the density of thermally produced holes (p_i) will equal the density of electrons (n_i). The total density of carriers may be expressed by the product

$$n_i p_i = A_o T^3 e^{-V_G/V_T} \qquad V_T = \frac{kT}{q} \tag{2-5}$$

where T = lattice temperature, °K
A_o = constant of the material
k = Boltzmann's constant
q = electronic charge
V_G = energy gap between the valence and conduction bands

At room temperature, $T \approx 300$°K and the thermal potential $V_T \approx 26$ mV. Table 2-2 gives the solution of Eq. (2-5) as well as other important semiconductor constants. We note from the solution of Eq. (2-5) that since only about 1 in every 10^{11} bonds in intrinsic silicon is broken at room temperature, relatively few free carriers are available to carry current.

The intrinsic semiconductor is doped by replacing about 1 in 10^6 atoms with a column-III element. Certain dopants, such as boron and aluminum,

Table 2-2 Constants of intrinsic semiconductors at room temperature (300°K)

CONSTANT	Si	Ge	UNITS
A_o	1.5×10^{33}	3.1×10^{32}	Carriers² per cm⁶-°K³
V_G	1.106	0.67	Volts
$n_i = p_i$	1.2×10^{11}	2.4×10^{13}	Carriers per cm^3
μ_n	1,350	3,900	cm^2/V-s
μ_p	480	1,900	cm^2/V-s

have activation energies close to the edge of silicon's valence band (Fig. 2-4). It is very easy for an electron in the valence band to be trapped into a tight covalent bond with an acceptor atom. Each acceptor atom suppresses one of the valence electrons, leaving an extra hole as a charge carrier. For this reason we call column-III doped silicon a *p*-type semiconductor. Charge neutrality is maintained by the negative charge tied to the lattice in forming the covalent bonds.

Doping with column-V elements (phosphorus or arsenic) creates an *n*-type semiconductor. In forming the required covalent bonds, each donor atom gives up an electron to the conduction band (Fig. 2-4). An equal positive charge remains behind in the nucleus of the donor atom.

In a sufficiently heavily doped semiconductor, the density of majority carriers will be determined solely by the dopant. Consider *p*-type silicon created by doping with boron, 1 part in 10^7. The density of acceptor atoms will be $N_A = 5 \times 10^{15}$ atoms per cm^3, which is appreciably higher than the intrinsic $p_i = n_i = 1.2 \times 10^{11}$ electrons per cm^3. The acceptor activation energy is very small. Therefore electrons which do not have sufficient thermal energy to jump to the conduction band can still form bonds, leaving holes behind in the valence band as charge carriers (Fig. 2-4). At room temperature, the majority hole density will be approximately equal to the

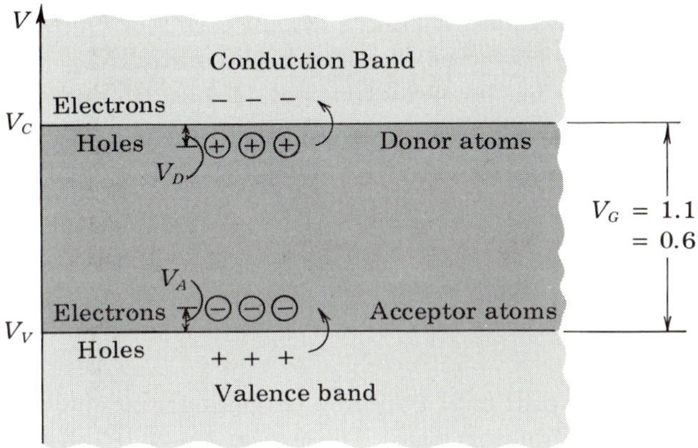

Fig. 2-4 Doping of a semiconductor with donors or acceptors. Circled charges are bound to the lattice by the dopants.

Activation energy, in electron volts

TYPE-III ACCEPTORS	S	Ge	TYPE-V DONORS	S	Ge
Boron (B)	0.045	0.010	Phosphorus (P)	0.044	0.012
Aluminum (Al)	0.057	0.010	Arsenic (As)	0.049	0.013

total acceptor-atom density. Thus

$$p \approx N_A = 5 \times 10^{15} \text{ holes per cm}^3$$

and from the solution of Eq. (2-5) the minority electron density becomes

$$n = \frac{n_i p_i}{N_A} = \frac{1.4 \times 10^{22}}{5 \times 10^{15}} = 2.8 \times 10^6 \text{ electrons per cm}^3$$

Since the product $n_i p_i$ is strongly temperature-dependent, at a sufficiently high temperature the electron density will again equal the hole density and the material will again behave like an intrinsic semiconductor.

Drift current flow

The free charge carriers in the bulk semiconductor can be accelerated by an applied electric field E_x, resulting in a current density

$$J_x = q(n\mu_n + p\mu_p)E_x = \sigma E_x \tag{2-6}$$

where μ_p and μ_n are the mobility of the charge carriers. Holes and electrons are accelerated in opposite directions by the applied field, but since they carry opposite charges, the current contributions add. Equation (2-6) is simply an expression for Ohm's law in terms of charge density and charge mobility. The coefficient σ is the bulk conductivity of the semiconductor. Intrinsic silicon has a conductivity of 17.6×10^{-6} mho/cm at room temperature, compared with 60×10^4 mhos/cm for copper.

Mobility is strongly temperature-dependent. For silicon, over the range of 150 to 400°K, the empirically found variation is

$$\mu_n = 2.1 \times 10^9 T^{-2.5} \qquad \mu_p = 2.3 \times 10^9 T^{-2.7} \tag{2-7}$$

Carrier diffusion

Suppose that additional minority carriers are injected at one point into an n-type semiconductor bar. These carriers will diffuse from the point of injection in the same manner as an ideal gas. The one-dimensional diffusion equation in terms of the excess minority-carrier density Δp is expressed by

$$D_p \frac{\partial^2 \Delta p}{\partial x^2} = \frac{\partial \Delta p}{\partial t} + \frac{\Delta p}{\tau_p} \tag{2-8}$$

The left-hand side of Eq. (2-8) is the spatial diffusion of the excess holes. The value of D_p, the diffusion constant, depends on the type of carrier and on the particular semiconductor. Since the injected holes carry charge,

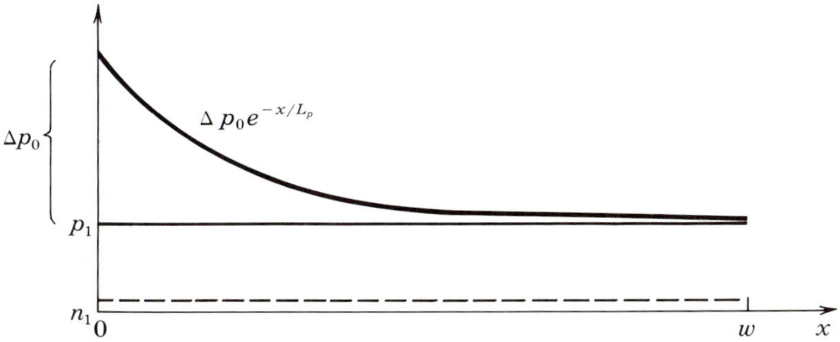

Fig. 2-5 Diffusion of excess holes along a semiconductor bar. p_1 and n_1 are normal carrier levels.

the time rate of injection ($\partial \Delta p/\partial t$) will affect the local field and hence the spatial buildup of excess carriers. In addition, the diffusing holes recombine with drift electrons. τ_p is called the carrier lifetime and represents the mean time before a hole disappears through recombination with an electron. The total recombination is proportional to the density of excess holes. Impurities, surface defects, and deliberately injected recombination centers (such as gold) all act to reduce the carrier lifetime.

If we assume steady-state injection, then $\partial \Delta p/\partial t = 0$ and Eq. (2-8) reduces to a second-order differential equation having the well-known exponential solution for the spatial variation of excess charge,

$$\Delta p = \Delta p_o \, e^{-x/L_p} \tag{2-9}$$

where $L_p = \sqrt{D_p \tau_p}$ is the diffusion length. The solution of Eq. (2-9) was found by considering the two boundary conditions $\Delta p = \Delta p_o$ at $x = 0$ and $\Delta p \to 0$ as $x \to \infty$. Thus, in four diffusion lengths L_p, the excess holes have been swept out and the hole distribution is back to normal. The excess hole distribution is shown in Fig. 2-5.

Now let us consider the charge. If we multiply Eq. (2-8) by $q\,dx$ and integrate over the bar, we obtain

$$qD_p \frac{\partial \Delta p}{\partial x}\bigg|_0^w = \frac{\partial}{\partial t}\int_0^w q\,\Delta p\,dx + \frac{1}{\tau_p}\int_0^w q\,\Delta p\,dx \tag{2-10}$$

where $\int_0^w q\,\Delta p\,dx = Q_x$ is the total excess charge per unit cross-section area stored in the bar. Since the spatial distribution of excess charge [Eq. (2-9)] goes to zero in four diffusion lengths, then for a long bar where $w \gg 4L_p$, $\partial \Delta p/\partial x \big|_w = 0$ and the solution of Eq. (2-10) is given by

$$J_D = -qD_p \frac{\partial \Delta p}{\partial x}\bigg|_0 = \frac{\partial Q_x}{\partial t} + \frac{Q_x}{\tau_p} \tag{2-11}$$

Equation (2-11) states that the injection of excess charge corresponds to a one-dimensional diffusion current. This current can be determined solely from the slope of the excess-charge distribution at any point. Diffusion current has two components: (1) the current due to the buildup of the excess-charge distribution and (2) the recombination current, which is proportional to the total excess charge. Under steady-state conditions, where $\partial \Delta p/\partial t = 0$, only the recombination current remains. The entering diffusion current is exactly balanced by drift electrons entering the excess-hole region. It is convenient to evaluate the diffusion current at $x = 0$ because at this spatial point excess charge is entering the bar through diffusion and the total current is diffusive. To satisfy current continuity, electron drift current must increase as the hole-diffusion term goes to zero in four diffusion lengths.

An abrupt *pn* junction

Let us now consider the abrupt p^+n junction shown in Fig. 2-6. The plus indicates that the p material is more heavily doped and is of higher conductivity than the n material. Because the carrier concentrations are unequal on the two sides of the junction, electrons will diffuse from the n to the p^+ region while holes will diffuse in the opposite direction, from the p^+ to the n region.

The diffusion of electrons will leave behind the bound positive charge associated with the donor atoms, while the hole diffusion leaves behind bound acceptor atoms (Fig. 2-6*c*). At the two edges of the metallurgical junction, the bound charge will reach maximums of qN_D and $-qN_A$. By integrating once across the total space-charge region, we can obtain the local field strength. A second integration yields the potential distribution.

The built-in potential barrier (Fig. 2-6*d*) is in the direction to cause a drift current of holes from the n region into the p^+ region and of electrons from the p^+ to the n region. Since there is no external energy source, the total electron and hole currents must separately sum to zero across the junction:

$$J_p = \overbrace{p\mu_p E_x}^{\text{Drift}} - \overbrace{D_p \frac{\partial p}{\partial x}}^{\text{Diffusion}} = 0 \tag{2-12a}$$

$$J_n = n\mu_n E_x + D_n \frac{\partial n}{\partial x} = 0 \tag{2-12b}$$

Integrating Eq. (2-12*a*) across the space-charge region, we obtain

$$V_j = -\int_{x_1}^{x_2} E_x\, dx = \frac{-D_p}{\mu_p} \int_{p_1}^{p_2} \frac{dp}{p} \tag{2-13}$$

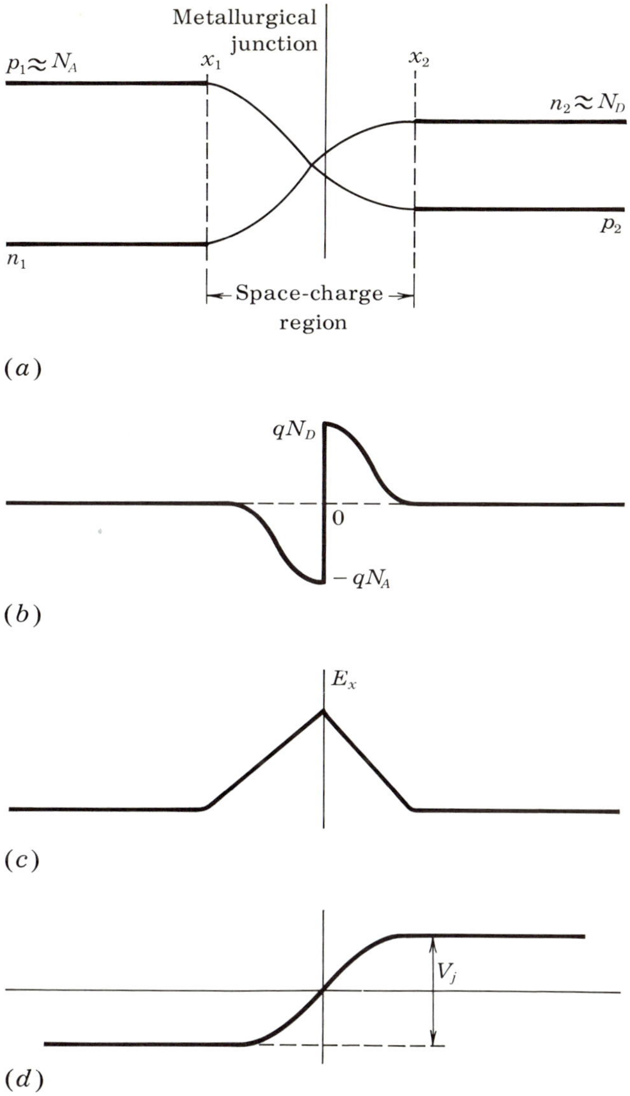

Fig. 2-6 An abrupt *pn* junction. (*a*) Semiconductor carrier distribution; (*b*) space charge; (*c*) electric field; (*d*) potential distribution.

The solution of Eq. (2-13) is the built-in voltage which back biases the junction:

$$V_j = \frac{D_p}{\mu_p} \ln \frac{p_1}{p_2} = V_T \ln \frac{p_1}{p_2} \tag{2-14}$$

where by the Einstein equation we have the identity

$$V_T = \frac{kT}{q} = \frac{D_p}{\mu_p} = \frac{D_n}{\mu_n}$$

We can also express the solution given in Eq. (2-14) as

$$p_2 = p_1 e^{-V_j/V_T} \tag{2-15}$$

Biased junction

We shall now forward bias the junction by making the p material positive with respect to the n material. The heavily doped n and p regions are highly conductive. Consequently, we may assume that the full applied voltage drop V appears right across the junction in series with the built-in junction voltage. The external voltage causes a drift current, which is carried by the majority carriers, by holes in the p material and by electrons in the n material. These majority carriers will build up at the edge of the space-charge region, where they become excess carriers. For example, the excess holes at the edge of the p material, Δp_{10}, will cause the minority hole concentration at the edge of the n material to become [from Eq. (2-15)]

$$p_2 + \Delta p_{20} = (p_1 + \Delta p_{10}) e^{-(V_j - V)/V_T} \tag{2-16}$$

Excess minority holes (Δp_{20}) will be swept out in four diffusion lengths. Holes are the majority carriers in the p material, and their density is more than 10^6 greater than the density of the minority holes in the n material. For low-level hole injection into the p material, that is, $\Delta p_{10} \ll p_1$, we have

$$p_1 + \Delta p_{10} \approx p_1$$

Using this approximation and substituting Eq. (2-15) into Eq. (2-16), we obtain the excess minority-carrier density in the n material in terms of the normal hole density:

$$\Delta p_{20} = p_2 (e^{V/V_T} - 1) \tag{2-17}$$

In a similar manner, the excess minority electron density at the edge of the p material will be found to be

$$\Delta n_{10} = n_1 (e^{V/V_T} - 1) \tag{2-18}$$

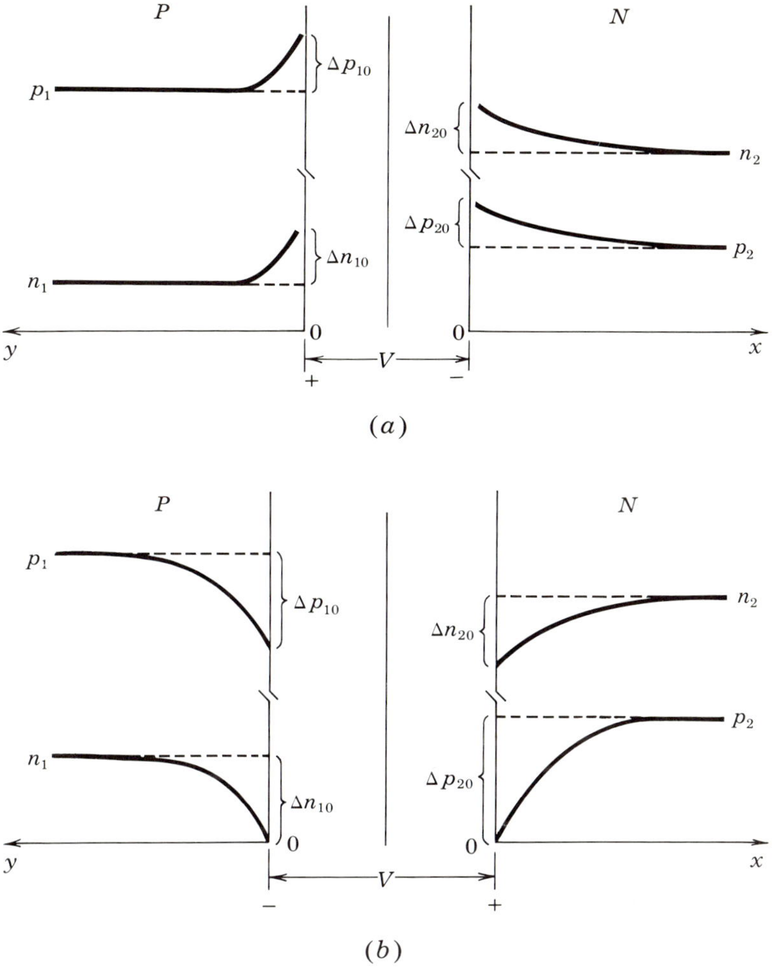

Fig. 2-7 Charge distribution in biased junction. (*a*) Forward-biased; (*b*) back-biased. Majority and minority carrier densities are not drawn to the same scale.

After diffusion, the minority carriers are swept out by the drift field. The carrier distributions are shown in Fig. 2-7*a*.

Since the two types of carriers are of opposite charge and diffuse in opposite directions, the current components add. Applying Eq. (2-11) to the charge distribution shown in Fig. 2-7*a* and taking into account the defined x and y directions, we obtain the total diffusive current density:

$$J_{T_D} = -qD_n \frac{\partial}{\partial y} [n_1(e^{V/V_T} - 1)e^{-y/L_n}] - qD_p \frac{\partial}{\partial x} [p_2(e^{V/V_T} - 1)e^{-x/L_p}] \quad (2\text{-}19)$$

Evaluating Eq. (2-19) at the two edges of the depletion region, where the current is solely diffusive, results in the junction current

$$J_T = q\left(\frac{D_n}{L_n} n_1 + \frac{D_p}{L_p} p_2\right)(e^{V/V_T} - 1) \tag{2-20}$$

If we back bias the junction with more than $5V_T$, or -125 mV, the exponential term becomes negligible compared with -1. Thus the junction reverse current becomes constant and equal to a reverse saturation current I_{rs}:

$$I_{rs} = Aq\left(\frac{D_n}{L_n} n_1 + \frac{D_p}{L_p} p_2\right) \tag{2-21}$$

A is the cross-sectional area of the junction. Since we are considering a p^+n junction where $p_2 \gg n_1$, Eq. (2-21) reduces to

$$I_{rs} \approx Aq\frac{D_p}{L_p} p_2$$

From Eqs. (2-17) and (2-18), with $V < -125$ mV,

$$\Delta p_{20} = -p_2 \qquad \text{and} \qquad \Delta n_{10} = -n_1$$

We see in Fig. 2-7*b* that the back bias causes the minority-carrier concentration to drop to zero at the edges of the transition region.

Substituting the reverse saturation current into Eq. (2-20) finally results in the general diode volt-ampere equation

$$I_D = I_{rs}(e^{V/V_T} - 1) \tag{2-22}$$

This is plotted for a typical silicon diode in Fig. 2-8.

We might note that intrinsic semiconductors with wide energy gaps, such as silicon, have fewer charge carriers than semiconductors with narrower gaps, such as germanium. When doped, silicon will have about 10^{-4} times fewer majority carriers, and hence the reverse current will be about four orders of magnitude smaller than in germanium.

In order to obtain the same forward-current density in a silicon diode as in a germanium diode, we would have to have a much larger forward bias. Silicon diodes operate with a normal forward bias of 650 to 750 mV, compared with 150 to 200 mV across a germanium diode. In both cases, the exponential term is large compared with -1, and we can approximate the forward diode current by the exponential response

$$I_D \approx I_{rs}e^{V/V_T} \tag{2-23}$$

Diode breakdown

When we back bias a heavily doped junction, the high density of majority charge carriers can result in high peak field strengths even with a small

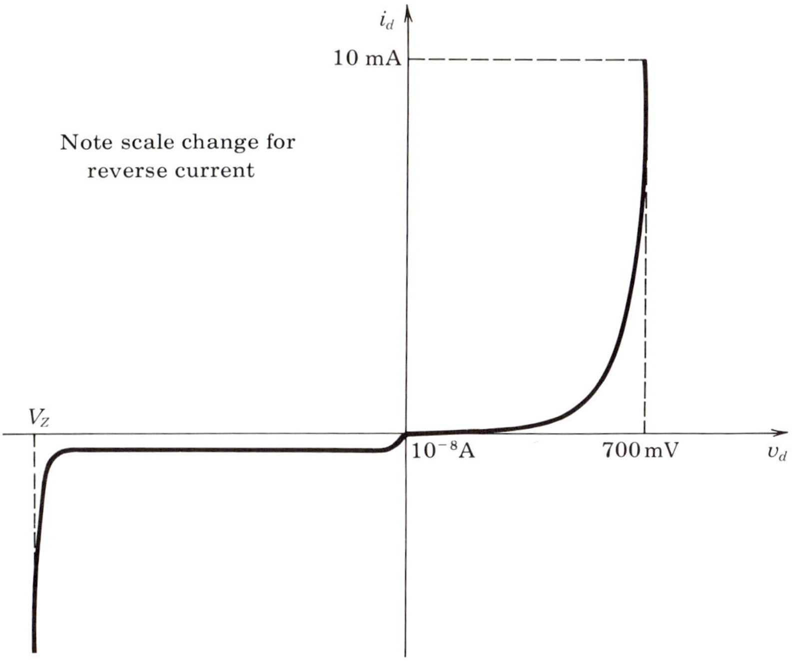

Fig. 2-8 Typical silicon-diode volt-ampere characteristic.

applied voltage. Charge carriers entering the transition region are accelerated by the high field and acquire sufficient energy to create new electron-hole pairs when they collide with the lattice atoms. The newly created carriers will themselves be accelerated and in colliding with the lattice may cause the generation of further electron-hole pairs. Since the reverse current depends on the number of charge carriers present, the avalanche effect results in a rapid increase of current at the given applied breakdown voltage (see Fig. 2-8). S. L. Miller found an empirical expression for the current multiplication:

$$M = \frac{1}{1 - (V/V_B)^\gamma} \tag{2-24}$$

where V_B is the avalanche breakdown voltage. γ is 2 for an n^+p silicon junction and is 4 for a p^+n junction. The breakdown voltage depends on the doping level, ranging down from almost 300 V for an impurity concentration of 10^{15} carriers per cm^3 on the lightly doped side of a silicon junction to about 6 V when the concentration increases to 10^{19} carriers per cm^3.

At lower reverse voltages, in certain junctions, a phenomenon can exist where the electrons tunnel from the valence band through the junction into the conduction band. This is called *zener breakdown.* The current increases as in the avalanche mechanism, limited only by the external resistance. Both V_B and V_Z are generally lumped together and referred to as the zener-breakdown voltage (Fig. 2-8).

2-3 Physical Diode Models

There are two techniques of establishing a model for a physical device. The first is empirical. After experimentally determining the volt-ampere characteristics of an actual device, the plotted characteristics are approximated by a series of straight lines. These lines should be chosen to straddle the expected range of operation. In Fig. 2-9 one possible piecewise-linear approximation is superimposed on the measured diode characteristics. When the diode conducts, there is only one obvious possibility. This line is equivalent to a resistor R_f biased with a series battery V_D. R_f is called the forward resistance and is usually between 5 and 60 Ω. V_D is about 700 mV for a silicon diode and 150 to 250 mV for a germanium diode.

For the back-biased region, we can also draw a straight line and approximate the reverse operation by the very large resistor R_r. If we draw this line to intersect with R_f at $i_d = 0$, $v_d = V_D$, we have a very conveniently delineated break point. The reverse region is also represented by a biased resistor. Thus the model formulated in Fig. 2-9 makes use of the ideal diode as the switching element; it shunts the very large R_r with the very much smaller R_f upon conduction at $i_d \geq 0$.

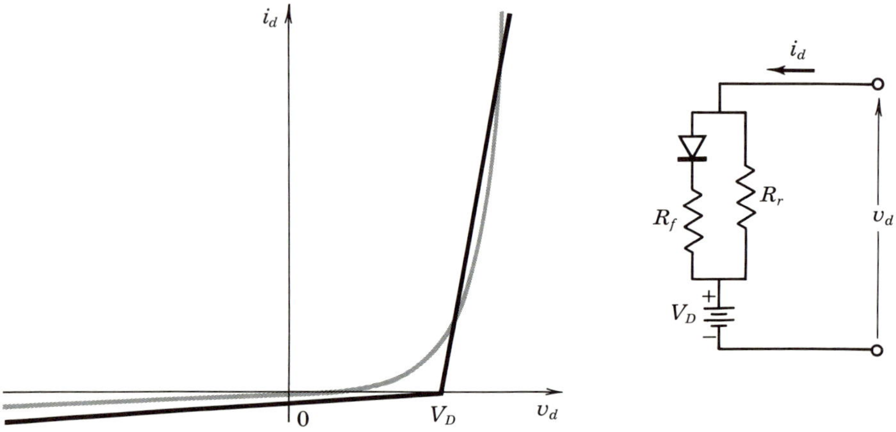

Fig. 2-9 Piecewise-linear diode characteristics superimposed on actual diode characteristic and an equivalent model.

The second technique used in formulating the models shown in Table 2-3 starts from the derived equations defining the diode operation. Of course, any equations must also be verified by checking against measured characteristics. We start by considering the orders of magnitude and nature of the changes of magnitude of the voltages and currents given by the forward-biased equation (2-25*a*) and the back-biased equation (2-25*b*) (also see Fig. 2-8):

$$I_D = I_{rs}(e^{V/V_T} - 1) \approx \begin{cases} I_{rs}e^{V/V_T} & V > 5V_T \quad \text{(2-25a)} \\ -I_{rs} & V < -5V_T \quad \text{(2-25b)} \end{cases}$$

The exponential nature of the forward volt-ampere characteristic means that equal-ratio increases in I_D are accompanied by equal-magnitude increases of the diode's terminal voltage. For example, if a diode has a voltage drop of $V_1 = 660$ mV at $I_{D1} = 100\ \mu$A, then for a current

$$I_{D2} = 10I_{D1} = 1 \text{ mA}$$

the voltage drop will be $V_1 + \Delta V$. By taking the ratio of I_{D2}/I_{D1} from Eq. (2-25*a*),

$$\frac{I_{D2}}{I_{D1}} = 10 = e^{\Delta V/V_T} \tag{2-26}$$

At room temperature, $V_T \approx 26$ mV. Solving Eq. (2-26), we find that

$$\Delta V = V_T \ln \frac{I_{D2}}{I_{D1}} = 26 \ln 10 = 60 \text{ mV}$$

Thus at room temperature, for every decade increase in current, the terminal voltage increases by only 60 mV. ΔV is small compared with the normal signal or circuit-supply voltages. For silicon diodes, the 60-mV change is an order of magnitude less than the normal forward drop of 650 to 750 mV.

Further examination of Eq. (2-25*a*) indicates that the forward drop of 720 mV at 1 mA in a silicon diode corresponds to

$$I_D = I_{rs}e^{720/26} = 10^{12}I_{rs} \tag{2-27}$$

I_{rs} is vanishingly small when compared with I_D. In a germanium diode forward-biased to the normal level of 180 mV,

$$I_D = I_{rs}e^{180/26} = 10^3I_{rs}$$

I_{rs} is much more significant and usually cannot be neglected.

Now that we know orders of magnitude, we can linearize the actual exponential characteristics by various piecewise-linear approximations. Table 2-3 shows several. The one selected will depend upon the level of

Table 2-3 Various possible piecewise-linear models for a diode

Physical diode

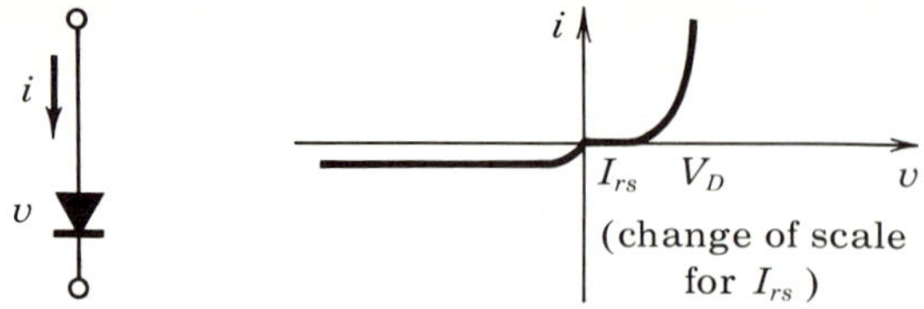

First-order approximation

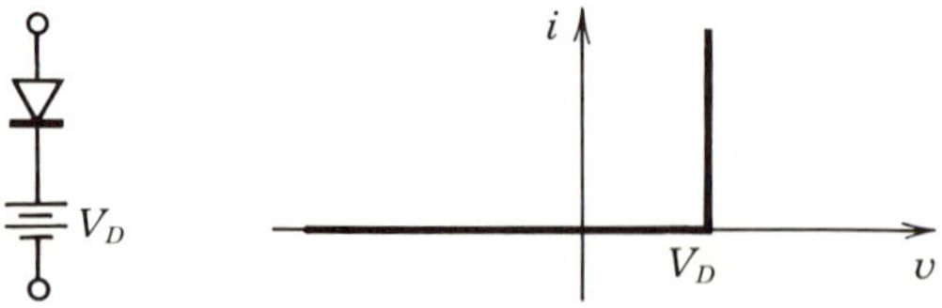

Second-order approximation

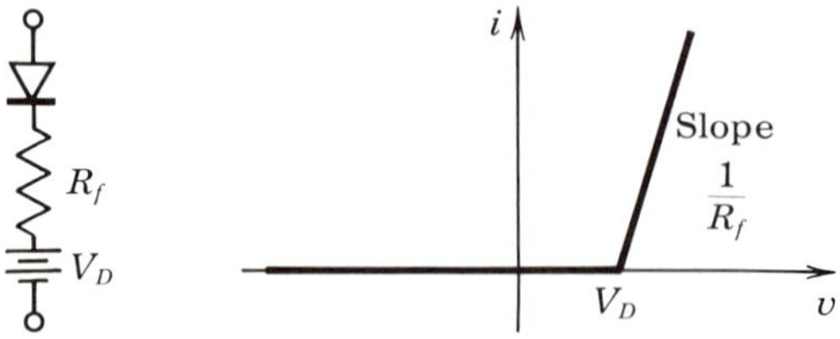

Third-order approximation

(a)

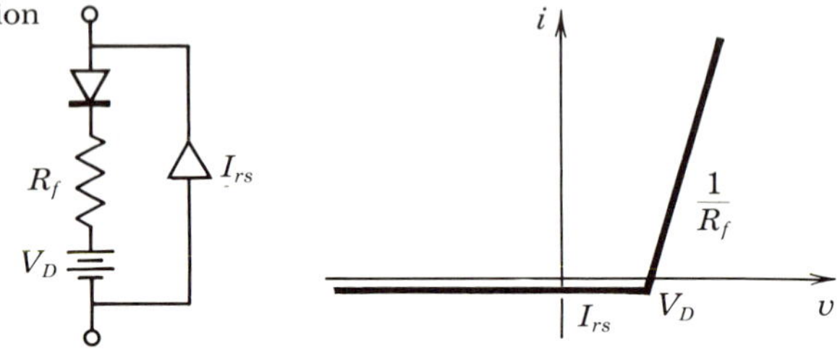

(b)

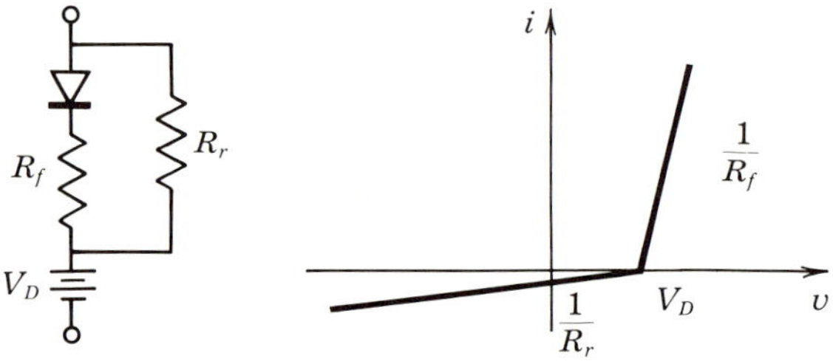

Breakdown voltage added to first-order model (zener diode)

Symbol 

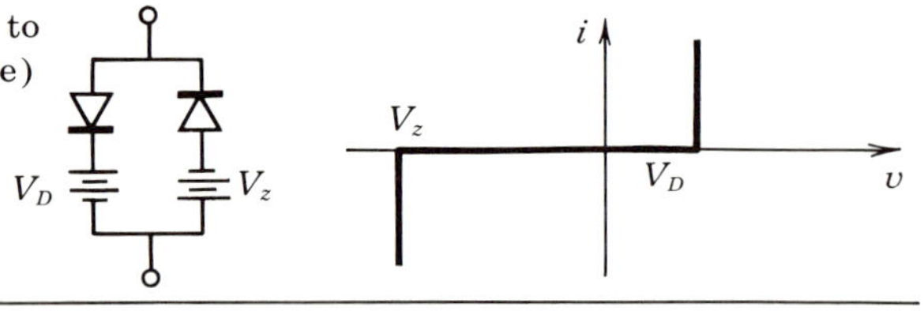

approximation which can be tolerated in the analysis of a particular problem. The ideal diode of Sec. 2-1 can be considered to be the zeroth-order approximation. The first-order model neglects the small reverse current flow and represents the forward drop by the constant voltage V_D. This approximation is usually used in the solution of dc circuits and for the first look at more complex switching circuits.

The second-order approximation takes cognizance of the change in the forward drop with changes in current by including a series resistance R_f. But what value resistor shall we use to approximate the actual exponential curve? One reasonable choice is to define an equivalent conduction resistance based on the 60-mV voltage change accompanying the 10:1 current change:

$$R_f \approx \frac{60 \text{ mV}}{I_D} = \frac{60}{I_D \text{ in mA}} \qquad \text{ohms} \tag{2-28}$$

In Eq. (2-28), I_D is the normal diode current. The value of R_f defined above will give reasonable answers for a 20:1 current change: from 0.1 to $2I_D$.

An incremental diode resistance can also be defined at the operating current I_D. By differentiating Eq. (2-25*a*), we obtain

$$r_f = \frac{dV_D}{dI_D} = \frac{V_T}{I_D} = \frac{26}{I_D \text{ in mA}} \qquad \text{ohms} \tag{2-29}$$

By comparing Eqs. (2-28) and (2-29), we see that the incremental resistance will always be smaller than the piecewise-linear forward resistance.

In making the third-order approximation, we add the constant-current generator I_{rs} in parallel with the forward-diode model. In certain applications it is more convenient to approximate this small reverse current flow by a large resistance (model 3*b*). R_r is defined by

$$R_r = \frac{V_{Dr} + V_D}{I_{rs}}$$

where V_{Dr}, the reverse diode voltage, is usually an order of magnitude or more greater than the forward drop V_D. Since it is convenient to approximate the complete reverse operation by a constant resistance, we shall arbitrarily choose a reverse voltage 10 times the forward drop. This leads to

$$R_r \approx \frac{11V_D}{I_{rs}} \qquad \text{ohms}$$

Any particular application may dictate a different choice for R_r. Since $R_r \gg R_f$, we do not have to modify the value of the diode conduction-region resistance.

Any of the elements shown in the separate models of Table 2-3 may be combined to suit a particular circuit. In schematics, a filled-in diode symbol will now be interpreted as a physical diode which can be replaced, for circuit analysis, by one of the models shown. Open symbols represent ideal diodes.

The reverse breakdown voltage can also be included in our model by incorporating a second biased diode path. Zener diodes designed to operate in the breakdown region are indicated by the modified diode symbol accompanied by the actual breakdown voltage.

2-4 Voltage Clipping Circuits

One of the most widely used nonlinear circuits utilizes the bivalued properties of the diode to control signal excursions. In the ideal shunt clipper of Fig. 2-10*a*, when the diode is back-biased, the shunt branch is opened and the input voltage is transmitted through R_1 to the output, unchanged and unaffected by the network.

On the other hand, once the rising input forces the diode into conduction, it may be replaced by a short circuit. The subsequent output is V_B, provided, however, that it is exceeded by the input. Thus the two zones of operation are, first, a transmission region where the output equals the input ($v_{o1} = v_{\text{in}}$) and, next, a clipped region where the output is determined solely by the bias ($v_{o2} = V_B$). These zones are plotted as the ideal transfer characteristic in Fig. 2-10*d*.

Figure 2-10*b* and *c* show how the clipping circuit would be modified to account for the nonideal nature of an actual diode. In Fig. 2-10*b* the diode is replaced by diode model 3*b*, and in Fig. 2-10*c* we use model 3*a* of Table 2-3. The first effect is that the diode conduction drop adds to the external bias and shifts the break point up and to the right. V_B is transformed to $V'_B = V_B + V_D$. The second effect is due to the introduction of resistance in the output branch. The forward diode resistance R_f forms a voltage divider with the series dropping resistor R_1. A small fraction of the input will be transmitted in the clipping region. The third effect is that the reverse resistance R_r (or the reverse current I_{rs}) shifts the transfer characteristic. Because of this shift, there will be a small output offset voltage when $v_{\text{in}} = 0$. In each of the operating regions, only one of the diode resistors need be considered, and the actual output may be found by the linear superposition of the contributions from both voltage sources. When $v_{\text{in}} < V'_B$ and the diode is back-biased,

$$v_{o1} = \frac{R_r}{R_1 + R_r} v_{\text{in}} + \frac{R_1}{R_1 + R_r} V'_B \qquad (2\text{-}30)$$

When conducting ($v_{in} > V'_B$), the diode may be replaced by its forward resistance and the output voltage may be written as

$$v_{o2} = \frac{R_1}{R_1 + R_f} V'_B + \frac{R_f}{R_1 + R_f} v_{in} \tag{2-31}$$

Both equations, each defining the operation in an individual region, are straight lines, as shown in Fig. 2-10*d*. At $v_{in} = V'_B$, the break voltage,

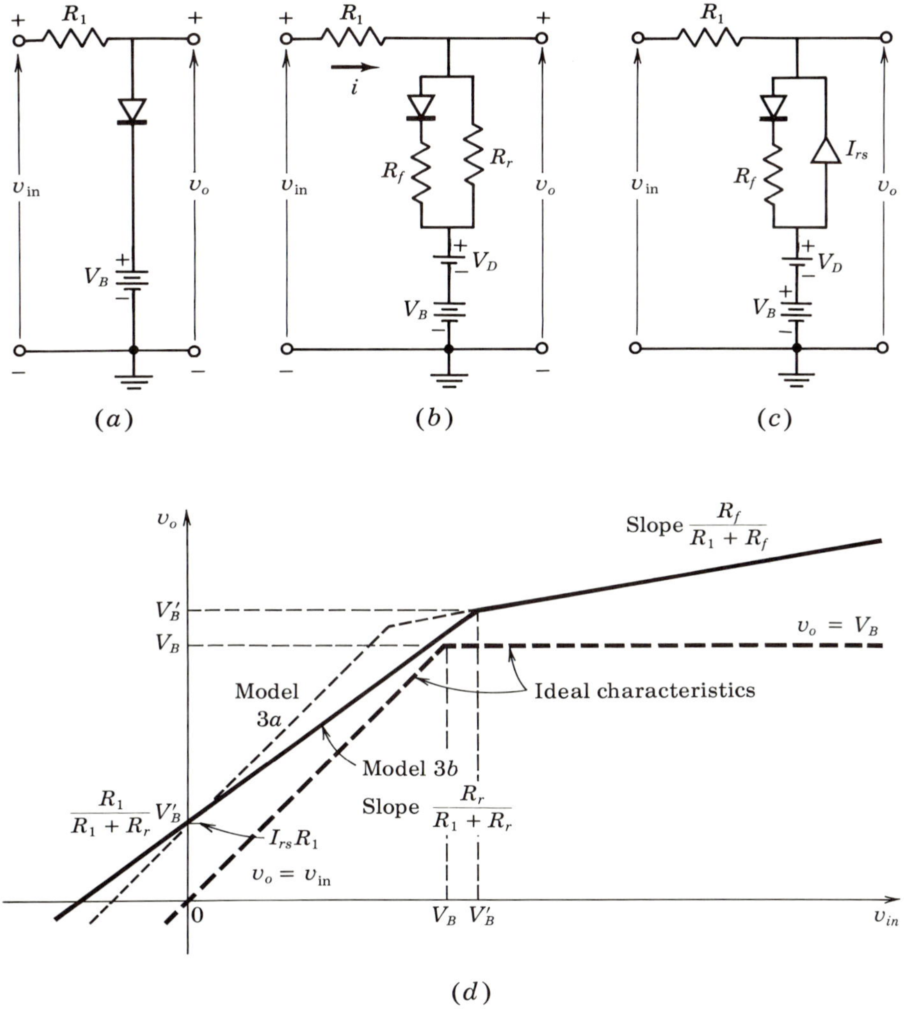

Fig. 2-10 Shunt diode voltage-clipping circuit. (*a*) Ideal model; (*b*) physical model; (*c*) alternate model; (*d*) clipper transfer characteristics (slightly exaggerated).

both Eqs. (2-30) and (2-31) give the output as V'_B, proving the continuity of the transfer function.

A plot of Eqs. (2-30) and (2-31), the transfer characteristic of the clipper, appears in Fig. 2-10*d*, with the ideal characteristic also presented for comparison. Examination of Eq. (2-30) indicates that within the transmission region the finite diode reverse resistance results in an undesirable finite output at zero input. This contribution comes from V'_B and is the second term of the equation. In the clipping region, the second term of Eq. (2-31) represents the output contribution from the now undesired input. We are faced with the necessity of optimizing this circuit, helped in this by our knowledge of the behavior of the ideal circuit.

Freedom of choice is limited to selecting an optimum value for R_1. Since the complete second terms of both equations represent the proportional error, they should, for optimum response, be made small. The conditions on R_1 are contradictory: in Eq. (2-30) the optimum value becomes $R_1 = 0$, and in Eq. (2-31), $R_1 = \infty$. To resolve the contradiction, a compromise choice for R_1 will be designated the optimum value. We can write the two contradictory conditions as

$$R_1 \ll R_r \qquad R_1 \gg R_f \tag{2-32}$$

If we assume that as a compromise we let the sign $\gg$ mean the same thing in both inequalities, then by multiplying together both right and both left sides, we obtain $R_1{}^2 = R_rR_f$. The compromise choice for R_1 turns out to be the geometric mean of R_r and R_f:

$$R_1 = \sqrt{R_fR_r} \tag{2-33}$$

Suppose that we use the alternative diode model, where the back-biased diode is treated as a current source (Fig. 2-10*c*). The forward-biased-diode equation will be identical with Eq. (2-31). When the diode is back-biased, the defining equation becomes

$$v_o = v_{\text{in}} + I_{rs}R_1 \tag{2-34}$$

The second term of Eq. (2-34) is a constant offset voltage. Now, for optimum response, we must satisfy the two contradictory inequalities

$$I_{rs}R_1 \ll V'_B \qquad R_1 \gg R_f$$

and, as before, by multiplying through,

$$R_1{}^2 = R_f \frac{V'_B}{I_{rs}} \tag{2-35}$$

If we interpret V'_B/I_{rs} as the equivalent to the reverse resistance $R_{r'}$ the condition of Eq. (2-35) reduces to that given in Eq. (2-33).

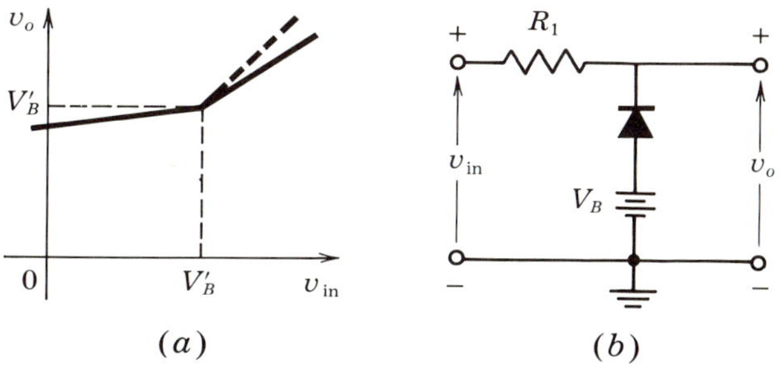

Fig. 2-11 Negative clipper transfer characteristic and circuit. In this circuit, the diode and battery drops subtract; therefore $V'_B = V_B - V_D$.

We conclude that for clipping circuits, a diode with the smallest reverse current represents the best choice. In any individual circuit, the series resistance used may not be the one found above; particular constraints may dictate a value closer to zero or to infinity.

When the diode in the clipper of Fig. 2-10 is reversed, the clipping region is below the break point; the diode now conducts through its forward resistance R_f until $v_{in} \geq V'_B$. After conduction ceases, a transmission path through the reverse resistance still exists, and the final result is the transfer characteristic given in Fig. 2-11. Here the signal is clipped below the bias voltage and transmitted above. Reversing the polarity of the bias voltage only shifts the break point of the transfer characteristic to the third quadrant. We conclude *that the direction of the diode connection determines whether clipping will be below or above the bias value and that the bias voltage determines the actual clipping point.*

Series clipping, a possible circuit variation, gives results similar to those found for the previously discussed shunt clipping. Insertion of the diode in series with the transmission path (Fig. 2-12) allows transmission while it conducts, in this case for $v_{in} \leq V'_B$. When back-biased, the diode introduces a high series impedance as the means of limiting the signal output. In this respect, its behavior is exactly opposite to that of the shunt clipper, which stops transmission upon conduction. The shape of the transfer

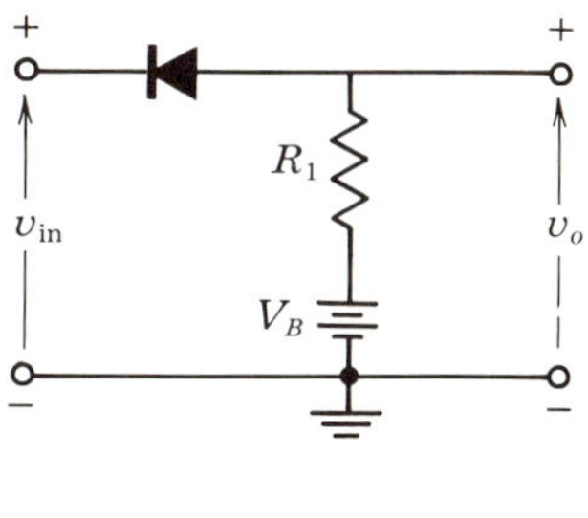

Fig. 2-12 Series clipper circuit.

characteristic of the circuit of Fig. 2-12 is nearly identical to that of the shunt clipper of Fig. 2-10 (differing in the slopes), and the optimum value of R_1 will also be the same as that discussed above.

The use of the zener region of the diode in addition to the forward-conduction region sets two clipping values instead of one. One point is located at V_D and the other is at $-V_z$, as indicated in the volt-ampere characteristic (Table 2-3). Biasing this diode shifts both break points in the same direction while still maintaining their original separation. It follows that one difficulty facing us is the lack of individual control over the slopes in the two conduction regions. Any series resistance added contributes equally to the zener region and the diode forward-current region. Generally, for versatility of adjustment, two individual diodes are preferred to one zener diode. However, the constant voltage drop in the zener region does avoid the necessity for inserting a separate bias source.

2-5 Single Diode and an Associated Energy-storage Element—Some General Remarks

In the solution of circuits containing both ideal diodes and energy-storage elements, notice must always be taken of the immediate past history as well as of the present excitation. The storage element should be considered as a memory which contains information stored in the past and which permits history to play a role in determining the present and future states of the circuit.

Our mode of attack follows the principle of time-zone superposition developed in Chap. 1. First, the initial state of the circuit is determined; next, the network is solved for the time at which the diode will change state.

Thereafter we simply treat the new circuit with the new time constant and initial conditions. The energy-storage element adds constraints to the continuity of current or voltage across the circuit and/or excitation changes and in this manner contributes to the new initial values.

Example 2-2

We shall now solve the problem posed in Fig. 2-13, where the output branch is shunted by 100-pF capacity. At $t = 0$, this capacitor is charged to the -5-V dc level of the input. The diode is back-biased and may be removed from the circuit. When the 15-V 1-μs-wide pulse is applied, the

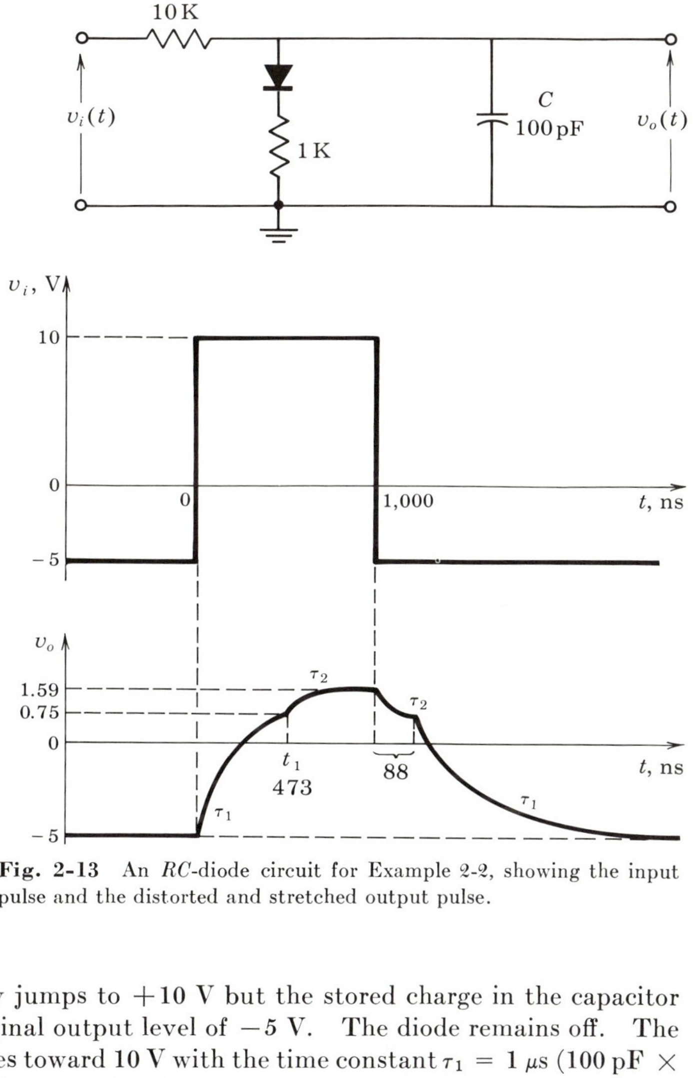

Fig. 2-13 An *RC*-diode circuit for Example 2-2, showing the input pulse and the distorted and stretched output pulse.

input immediately jumps to +10 V but the stored charge in the capacitor maintains the original output level of −5 V. The diode remains off. The output now charges toward 10 V with the time constant $\tau_1 = 1\ \mu s$ (100 pF × 10 K). The defining equation becomes

$$v_{o1}(t) = 10 - 15e^{-t/\tau_1}$$

Eventually the output reaches 750 mV, and the silicon diode conducts. The time required is found from the above equation:

$$t_1 = \tau_1 \ln \frac{10 - (-5)}{10 - 0.75} = 473 \text{ ns}$$

After the diode conducts, we must solve the new resulting circuit. The diode branch introduces a 1-K resistor, in series with the small R_f, and the 750-mV diode drop. This establishes a Thévenin equivalent voltage seen by the capacitor of

$$V_{\mathrm{Th}_1} = \frac{10 - 0.75}{11\ \mathrm{K}}\, 1\ \mathrm{K} + 0.75 = 1.59\ \mathrm{V}$$

Charging continues toward the new steady state with the much faster time constant of $\tau_2 = C(10\ \mathrm{K} \,\|\, 1\ \mathrm{K}) = 91$ ns. Charging is virtually complete in $4\tau_2$.

When the pulse terminates after 1 μs, the charge stored in the capacitor keeps the diode turned on. Conduction continues until the output drops below 750 mV. The capacitor charges toward the new Thévenin equivalent voltage found with -5 V at the input,

$$V_{\mathrm{Th}_2} = \frac{-5 - 0.75}{11\ \mathrm{K}}\, 1\ \mathrm{K} + 0.75 = 230\ \mathrm{mV}$$

Since the diode is on, the first portion of the decay is governed by the fast time constant τ_2. We can write

$$\begin{aligned} v_{o2} &= V_{\mathrm{Th}_2} - (V_{\mathrm{Th}_2} - V_{\mathrm{Th}_1})e^{-t/\tau_2} \\ &= 0.23 + 1.36e^{-t/\tau_2} \end{aligned}$$

This interval lasts until the output decays to 750 mV and the diode turns off:

$$t_2 = \tau_2 \ln \frac{0.23 - 1.59}{0.23 - 0.750} = 88\ \mathrm{ns}$$

After the diode turns off, final recovery is toward the original input level of -5 V with the initial time constant τ_1.

An inductance in series with the diode would cause an effect similar to that discussed in Example 2-2 by sustaining the diode current flow. If, instead of shunting the diode by a capacitor, we place a capacitor in series, it would couple all input-voltage transitions and the diode could immediately switch from a conducting to a nonconducting state. *We conclude that the location as well as the type of energy-storage element employed directly influences the nonlinear behavior of a single-diode system.* In any analysis, great care should be taken to see that the controlling factor, the diode state, is always kept in view.

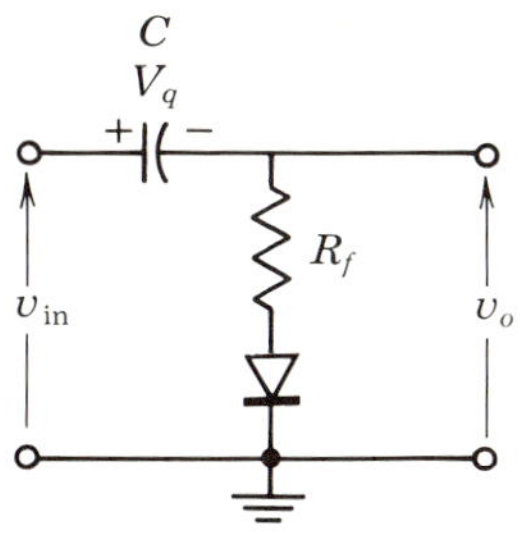

Fig. 2-14 Diode voltage clamp.

Some practical, widely used circuits employing various energy-storage elements will be discussed in detail in the sections below (Secs. 2-6, 2-8, and 2-10), and they will further illustrate the techniques of solution.

2-6 Diode Voltage Clamps

The function of the diode clamp of Fig. 2-14 is to shift the input voltage in such a manner that the resultant maximum value of the output will be maintained at a fixed voltage, without at the same time distorting the waveshape. The output voltage is then said to be clamped negatively. Sketches of typical input and output voltages appear in Fig. 2-15.

In the initial discussion we shall idealize the clamp at least to the extent of neglecting the small diode voltage drop and assuming that $I_{rs} = 0$. Upon application of the input signal, C charges to the peak positive value of the input with the time constant CR_f. Provided that this time constant is sufficiently short, charging will be completed within the first cycle; otherwise it may take several cycles. Upon completion of the charging, the output voltage becomes

$$v_o = v_{in} - v_q = v_{in} - V_1 \tag{2-36}$$

The conclusion drawn from Eq. (2-36) is that the output will always be less than or equal to zero since the voltage across the capacitor, V_1, is the maximum positive value of v_{in}. Diode conduction ceases once C is fully charged, with the diode subsequently remaining back-biased. The output never exceeds zero, and it is said to be clamped there. For a rectangular-wave input, the clamping action is illustrated in Fig. 2-15. The diode serves first to introduce a low-resistance path for charging and then a very-high-resistance one, preventing the discharge of C. Since the output is simply the sum of the capacitor and the input voltages, the signal-level shift in the clamp is provided through the charge accumulated in the capacitor during the initial buildup. The behavior is the same as if a battery whose voltage is equal to the peak positive value of the input replaced the capacitor in the series transmission path.

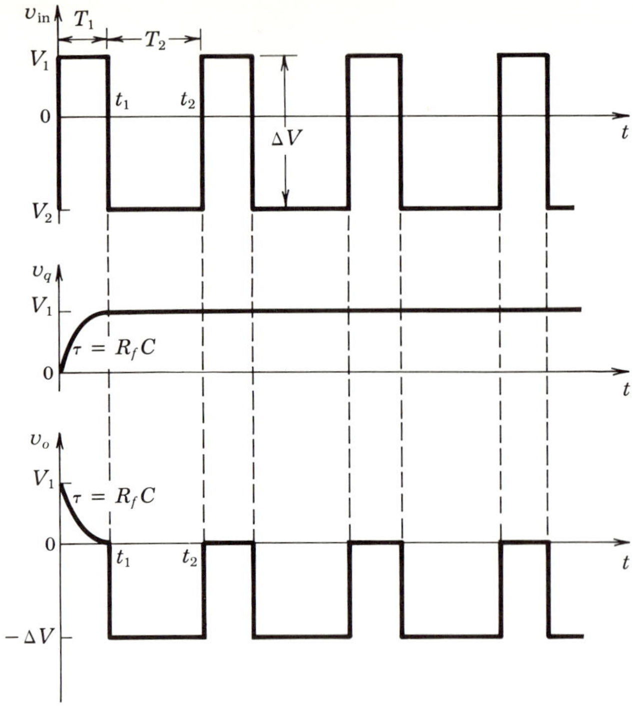

Fig. 2-15 Ideal-diode clamping waveshapes.

Suppose that the amplitude of the input signal suddenly changes. For proper negative clamping, the capacitor must recharge to the new positive peak voltage. Any increase of the input signal will cause the clamping diode to conduct on the positive peak, reestablishing the required charging path. The capacitor will charge to the new peak voltage with the fast time constant

$$\tau_1 = R_fC$$

On the other hand, if the input-signal amplitude decreases, the previously stored charge in the capacitor keeps the diode back-biased. The reverse current in an actual diode, I_{rs}, will linearly discharge the capacitor. The constant-current discharge will be a linear voltage ramp starting at the previously stored voltage V_1:

$$v_q(t) = V_1 - \frac{I_{rs}t}{C}$$

Since I_{rs} is extremely small, the time required for the capacitor voltage to decay to the new storage level may be excessively long.

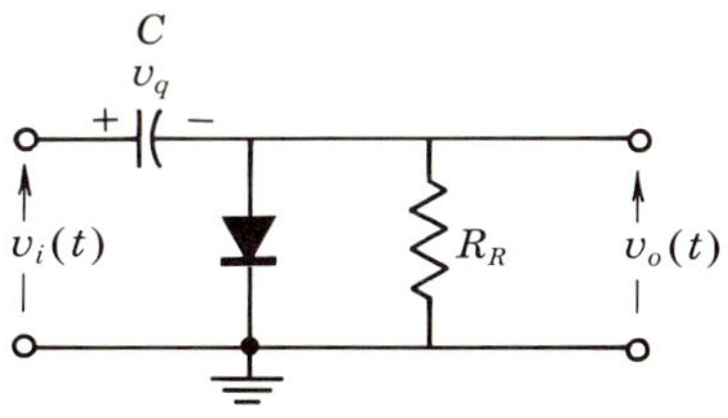

Fig. 2-16 Practical voltage clamp.

Shunting the diode with a relatively large external resistor R_R, as shown in Fig. 2-16, provides a method of controlling the clamp's time response to signal-amplitude variations. Whenever the diode is back-biased, the output will decay toward zero at a rate determined by R_R instead of at the slower rate set by the reverse current flow. The current through R_R predominates, and I_{rs} can be neglected in all calculations. We conclude that an ideal diode will not function properly in a clamp and that the ideal diode must be changed to a nonideal element through the addition of an external leakage path.

Figure 2-16 shows the practical form of the clamp. We see that the output waveshape of Fig. 2-17 differs in several important respects from the ideal-clamp output of Fig. 2-15. Assuming a constant-conduction diode drop of V_D, on the positive input peak the capacitor will charge, with the fast time constant $\tau_1 = R_fC$, to

$$v_q = V_1 - V_D$$

Substituting into Eq. (2-36), the output will now be given by

$$v_o = (v_{\text{in}} - V_1) + V_D \tag{2-37}$$

As in the clipper, the diode drop causes an additional offset at the output; the clamping point shifts from zero to V_D.

V_D is not really a constant. The exponential nature of the diode characteristics means that V_D drops toward zero as the capacitor charges and the diode current drops. But the reduced diode current also means a greatly increased forward resistance [Eq. (2-28)]. The final voltage depends on the

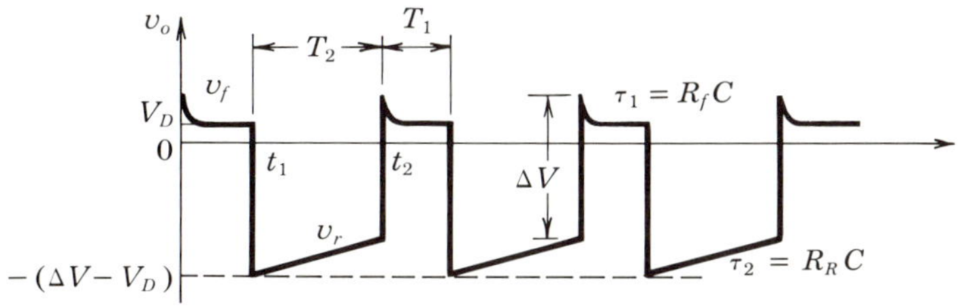

Fig. 2-17 Clamped-output waveshape for rectangular input signal.

duration of the conduction interval and will be above zero but somewhat less than the assumed constant V_D.

During the interval that the diode is back-biased, C discharges slightly through the shunt resistor R_R with the long time constant CR_R. The energy dissipated in R_R is supplied from the charge previously stored, and of course it must be replaced once the diode again conducts. The discharge introduces tilt over the negative portion of the cycle. Since the full change of input voltage appears across the diode (because of continuity of charge in C), the output becomes slightly positive when the input voltage rises by ΔV at the end of each cycle. Therefore the output appears as shown in Fig. 2-17. Note that the signal has become somewhat distorted because of the cyclical charging and discharging.

If the recharge-time constant is excessively long, the circuit will not recover to zero within the recharge period. A solution for this case would follow the discussion of Sec. 1-2, with, however, different time constants used over each portion of the response.

Some general conclusions about the circuit behavior may be drawn by realizing that for a periodic solution, the net change in the charge stored in C over any cycle must be zero. The charge flowing into the capacitor is

$$Q_f = \int_0^{T_1} i_f \, dt = \int_0^{T_1} \frac{v_f}{R_f} \, dt \tag{2-38}$$

where v_f is the time-varying voltage across the output when the diode conducts. It is the voltage above the V_D level. The charge flowing out of the capacitor is

$$Q_r = \int_0^{T_2} i_r \, dt = \int_0^{T_2} \frac{v_r}{R_R} \, dt \tag{2-39}$$

where v_r is the voltage across the output when the diode does not conduct; it is the voltage below the V_D level. Since there can be no net change in charge over the cycle,

$$Q_f = Q_r$$

Equating (2-38) and (2-39) and solving for R_f/R_R yields

$$\frac{R_f}{R_R} = \frac{\int_0^{T_1} v_f \, dt}{\int_0^{T_2} v_r \, dt} = \frac{\text{net forward-biased area}}{\text{net back-biased area}} \tag{2-40}$$

Equation (2-40) leads to the conclusion that we must make a trade-off between minimum clamp-recovery time (small R_R) and minimum clamp-waveshape distortion (large R_R). Observe that in deriving this equation, the only assumption made is that the input signal is periodic. Even though the waveshape used to illustrate the clamp's behavior is rectangular, its shape appears neither explicitly nor implicitly in the charge equations (2-38) and (2-39). The final result, that the ratio of areas equals the ratio of diode resistances, must be independent of the applied waveshape.

Reversing the diode clamps the input signal positively at $-V_D$, instead of negatively, with the capacitor recharge showing itself as a slight negative excursion. As in the clipping circuit, biasing this clamp sets the output voltage at any desired level, diode direction determining positive or negative clamping. Figure 2-18 shows a triangular input clamped positively at -10 V, with the distortion exaggerated for purposes of illustration. V_B' is adjusted to allow for the shift due to the diode drop.

One necessary condition for properly biased operation is the presence of the external shunt resistance, which establishes a path for the initial charging of C to the bias voltage. Consider the behavior of the circuit of Fig. 2-18 when an ideal diode ($R_r = \infty$) is used. If the negative peak of the input voltage never reaches -10 V, the diode always remains back-biased and, since C is uncharged, the input appears at the output, unchanged and unclamped. Signals with negative peaks greater than -10 V forward bias the diodes on the peaks, and this provides the charging path. However, by shunting the diode with the resistance R_R, C always has a charging path and the clamping action becomes independent of the signal amplitude.

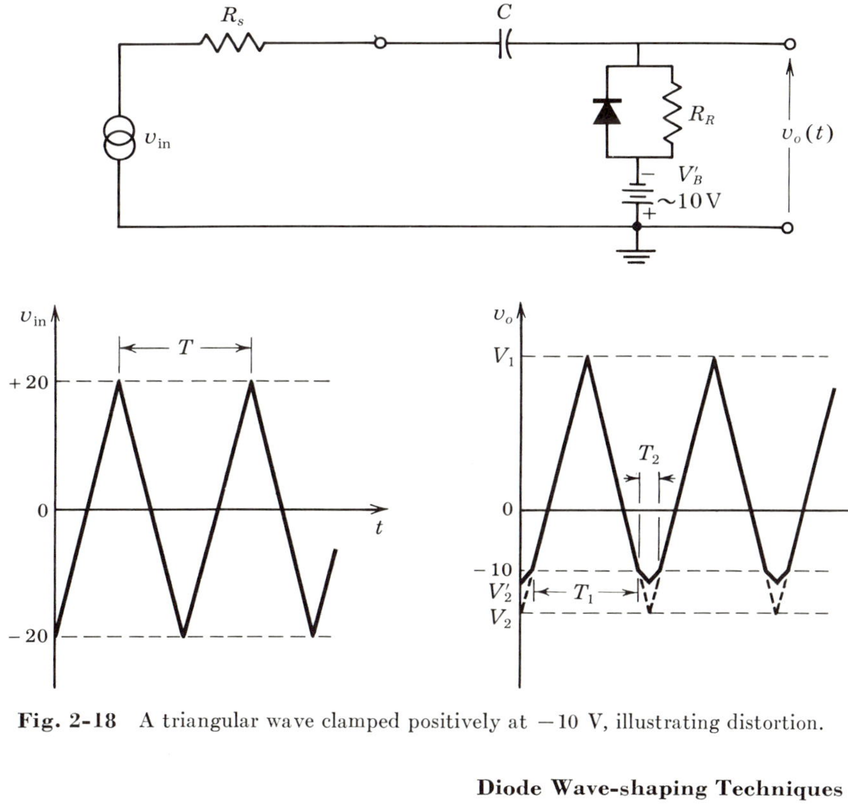

Fig. 2-18 A triangular wave clamped positively at -10 V, illustrating distortion.

Referring back to Eq. (2-40), we see that for optimum clamping with any given diode and R_R, the input signal should be clamped so as to minimize the net area on either side of the clamping bias voltage. When the area where the diode is back-biased is large, then the distortion area during conduction will also be large. When the area is small, then the net distortion will also be small.

Looking into a clamp, the signal source sees a different complex impedance during diode conduction than when reverse diode current flows. If the source has any internal impedance, the loading by the forward resistance of the diode introduces additional waveshape distortion. The total recharge-voltage excursion divides proportionately across the source resistance R_s and the diode forward resistance. Since the output is taken only across R_f, the voltage in this region will be smaller than expected, as seen in Fig. 2-18. During recharge, the diode forward voltage drop at the output, measured from the bias value, becomes

$$v_{fd} = \frac{R_f}{R_f + R_s} V_f$$

V_f is the peak voltage drop across the total series resistance while the diode conducts.

In addition, when using Eq. (2-40), the source resistance must be added to the diode forward resistance and may even become the dominant term. We can see that this large increase in resistance decreases the ratio of areas, with correspondingly poorer clamping. When the diode is back-biased, the output is also attenuated slightly because of R_s. But if R_s were large enough to have an appreciable effect, it would so distort the conduction signal that the clamp would be unusable.

For clamped triangular and sinusoidal signals (Figs. 2-18 and 2-19), we are often interested in knowing the diode conduction period. Because the source impedance of the input generator causes division only of the recharge-voltage drop, a logical first step in the calculation is to lump all forward resistors and ignore the voltage division. Furthermore, in a close-to-ideal

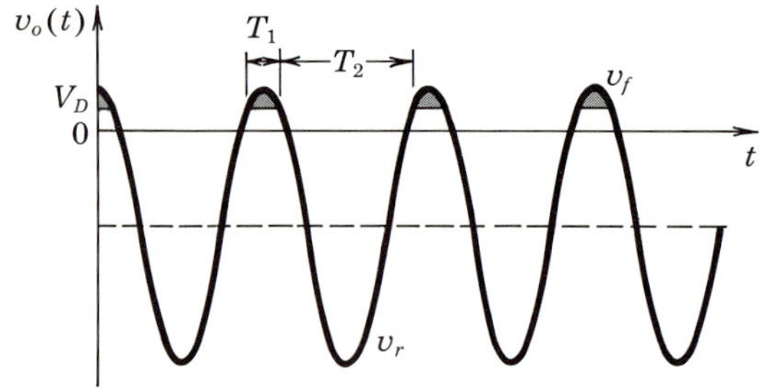

Fig. 2-19 A negatively clamped sine wave.

circuit, the waveshape distortion also may be neglected; the clamp is treated as a device which shifts the output waveshape only to a point where it satisfies Eq. (2-40). The two areas can now be calculated in terms of the unknown conduction angle, and their ratio can be equated to the known ratio of forward to reverse resistance. For the triangular wave of Fig. 2-18, neglecting distortion, the area above the clamping voltage is

$$A_1 = \tfrac{1}{2}T_1(V_1 + 10) \tag{2-41}$$

The area below, measured from the bias voltage, becomes

$$A_2 = \tfrac{1}{2}T_2|V_2 + 10| \tag{2-42}$$

But from the input wave,

$$V_1 - V_2 = 40 \qquad T_1 + T_2 = T \tag{2-43}$$

In addition, from the geometry, by the ratio of similar triangles,

$$\frac{T/2}{40} = \frac{T_1/2}{V_1 + 10} \qquad \frac{T/2}{40} = \frac{T_2/2}{|V_2 + 10|} \tag{2-44}$$

Substituting first Eq. (2-44) and then Eq. (2-43) into Eqs. (2-41) and (2-42) and equating to the ratio of resistance,

$$\frac{R_f + R_s}{R_R} = \frac{T_2^2}{T_1^2} = \frac{T_2^2}{(T - T_2)^2} \tag{2-45}$$

T_2 is the only unknown, and this equation is readily solved. We reject the negative root because it has no physical meaning. Next, V_2 may be found by substituting back into Eq. (2-44). Using the resistive voltage divider of R_s and R_f and recognizing that V_2 is developed across both resistors finally give V_2'.

The transfer characteristic of a clamp is given in Fig. 2-20. Note that it is a straight line of unity slope, with the locus of the end point the bias voltage line. The y intercept depends on the input-signal amplitude.

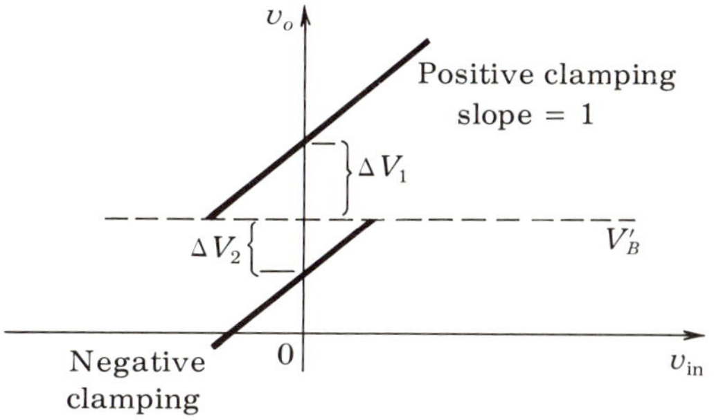

Fig. 2-20 Clamp transfer characteristics.

This is obvious if we consider that for positive clamping the output must always (neglecting the capacitor recharge) be equal to or greater than V'_B. ΔV_1 is the peak negative voltage applied to a positive clamp, and ΔV_2 is the peak positive voltage applied to a negative clamp. These transfer characteristics have no value or meaning except for periodic inputs. When presented on an oscilloscope, they do, however, show whether the clamp is operating properly.

2-7 Current Clippers

We find the concept of duality extremely helpful in developing the circuits used for current clamping and clipping. By simply taking the dual of a voltage clipper, we arrive at the current clipper (Fig. 2-21*a*). Just as loop equations were ideally suited for the analysis of voltage circuits, the node equations are ideal for treating the current-activated devices. To complete the dual relationships, where previously the break point of the ideal-diode piecewise-linear model was taken at $v_d = 0$, now it might well be regarded as occurring when $i_d = 0$. Figure 2-21*b* shows the transfer characteristic of the current clipper. Except for the change in axis (current instead of voltage), it is identical with that given in Fig. 2-11 for the voltage clipper. Introduction of a bias current I_B sets the break point of the transfer characteristic wherever desired.

Analysis of this circuit starts with the writing of the node equation

$$i'_d = i_1 - (I_B - I_{rs}) - i_R \qquad (2\text{-}46)$$

When $i'_d \geq 0$, the diode conducts. From Eq. (2-46) we see that the conduction region corresponds to $i_{\text{in}} > I'_B$ and the nonconduction region to $i_{\text{in}} < I'_B$, where $I'_B = I_B - I_{rs}$. The break point turns out to be, as expected, the value of the bias current. Since the regions of conduction

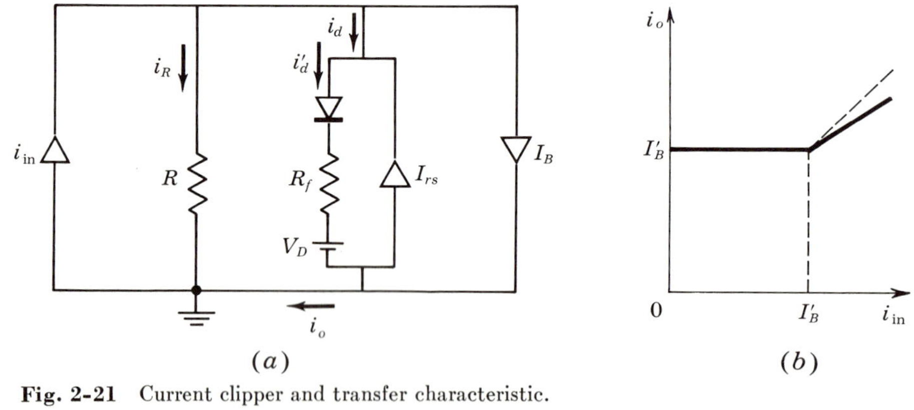

Fig. 2-21 Current clipper and transfer characteristic.

and nonconduction are known, the two equations defining operation are, by superposition,

$$i_{o1} = I_B - I_{rs} = I'_B \tag{2-47}$$

$$i_{o2} = \frac{R}{R + R_f} i_{\text{in}} + \frac{R_f}{R + R_f} (I_B - I_{rs}) \tag{2-48}$$

Equation (2-47) holds when the diode is nonconducting, and Eq. (2-48) when it conducts. The first term on the right-hand side represents the desired transmission, and the second term, the error component of load current, i.e., the transmission of undesired current to the output. If we used the diode model containing R_r, we could obtain a set of equations analogous to the set used in describing the operation of a voltage clipper [Eqs. (2-30) and (2-31)]. Therefore, identical reasoning leads to the same optimum value of the shunt resistor R as that previously found for R_1 in Sec. 2-4:

$$R = \sqrt{R_r R_f}$$

Reversing the diode changes the clipping characteristics, now allowing conduction for $i_{\text{in}} \leq I'_B$ and clipping when $i_{\text{in}} \geq I'_B$.

2-8 Current Clamps

In the treatment of voltage clamps (Sec. 2-6), we saw that the prime constituents of a clamp are an energy-storage element supplying the additional energy (previously stored) necessary in shifting the signal level and a diode providing short storage (charge) and long-decay-time constants. When we must clamp current waveshapes, we logically turn to the inductance as our storage element. This circuit, the dual of the voltage clamp of Fig. 2-14, appears in Fig. 2-22. Since its basic behavior is independent of waveshape, the excitation chosen might as well be the simplest, e.g., a current square wave. For the initial discussion, assume its application at $t = 0$, as shown in Fig. 2-23.

By writing the node equation, the short-circuit load current i_o may be expressed as

$$i_o \approx i_d = i_{\text{in}} - i_L \tag{2-49}$$

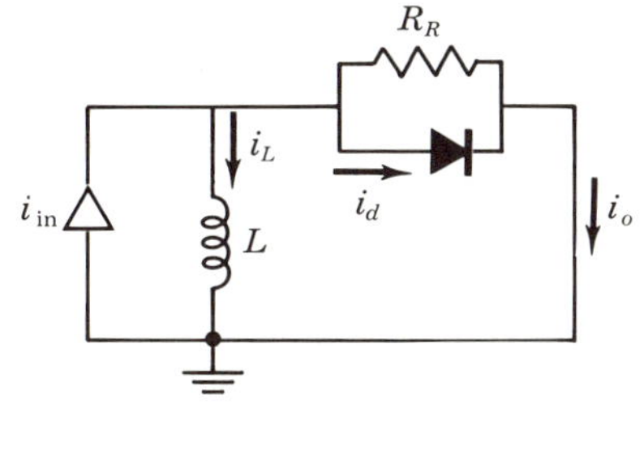

Fig. 2-22 Current clamp.

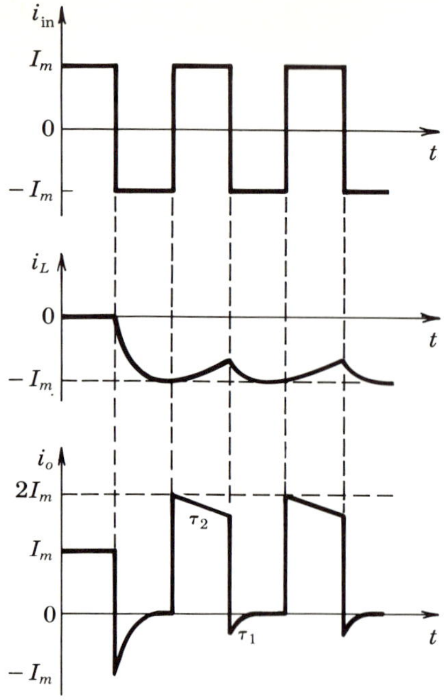

Fig. 2-23 Current-clamp waveshapes.

During the first half cycle of the input square wave, the diode conducts and, assuming $R_f = 0$, the full input current flows in the output lead. However, in the second half cycle, $i_{in} < 0$ and forward conduction ceases. The input current must flow somewhere. Since there was zero initial current in L, the only place left for the current to flow is through the diode's shunt resistance. Inductive current starts building up toward $-I_m$, with a corresponding decay in the reverse diode current. Buildup is rapid because of the small time constant

$$\tau_1 = \frac{L}{R_R}$$

For a half period sufficiently long compared with τ_1, that is, $T/2 \geq 4\tau_1$, charging will be completed within the half cycle; otherwise it will be completed within several cycles. R_R also limits the peak reverse voltage developed across the diode to $R_R I_m$. This must be below the diode breakdown voltage; otherwise, we would also have to account for breakdown-region operation.

The input again becomes positive, and with no circuit dissipation, the inductive current remains at $-I_m$. From Eq. (2-49), the output subsequently becomes

$$i_o = i_{in} - (-I_m)$$

But since i_{in} takes on only one of two values, $\pm I_m$, i_o may be either zero or positive but never negative. The output current is composed of two components. On the positive peak, the contribution of I_m from the signal source and the additional contribution of I_m previously stored in the inductance add. On the negative peaks, these components subtract. We have thus clamped the output current positively at zero; it varies between zero and $2I_m$.

The inductance acts as the circuit's memory, remembering always the peak value of the input excitation. Once this information has been stored, the output will be shifted to satisfy the circuit conditions established by the diode.

Physically, $R_f > 0$, resulting in an inductive-current decay during diode conduction; this decay corresponds to the energy dissipated in the forward resistance. Since the decay-time constant τ_2 is very long, the load current droops only slightly:

$$\tau_2 = \frac{L}{R_f}$$

We must periodically restore the dissipated energy in the inductor's magnetic field. On alternate half cycles (negative input), recharge takes place, manifesting itself as a small negative excursion of output current. The final periodic response is sketched in Fig. 2-23. These results are exactly analogous to those obtained in Sec. 2-6 for the voltage clamps. However, whereas recharge in the voltage clamp took place during diode conduction, here it occurs when the diode is back-biased. The same condition of a large ratio of shunt resistance to diode forward resistance is necessary for proper current clamping, and the signal will shift until the ratio of the output current areas equals the ratio of resistance.

Bias currents may be introduced in parallel with the diode, shifting the clamp point wherever desired. As with the voltage clamp, reversing the diode reverses the direction of clamping. The nonideal inductance has associated coil resistance, which adds to the other circuit resistance and decreases both τ_1 and τ_2, having the most pronounced effect in shortening the long discharge-time constant.

2-9 Arbitrary Transfer and Volt-Ampere Characteristics

Many devices and systems exhibit grossly nonlinear responses which, for analysis, we should like to approximate by an equivalent network of biased diodes and resistors. These nonlinearities may inadvertently arise from the physical properties of the device and will thus represent an undesirable characteristic. Alternatively, they may be deliberately created and inserted in the signal transmission path in order to perform various operations. In

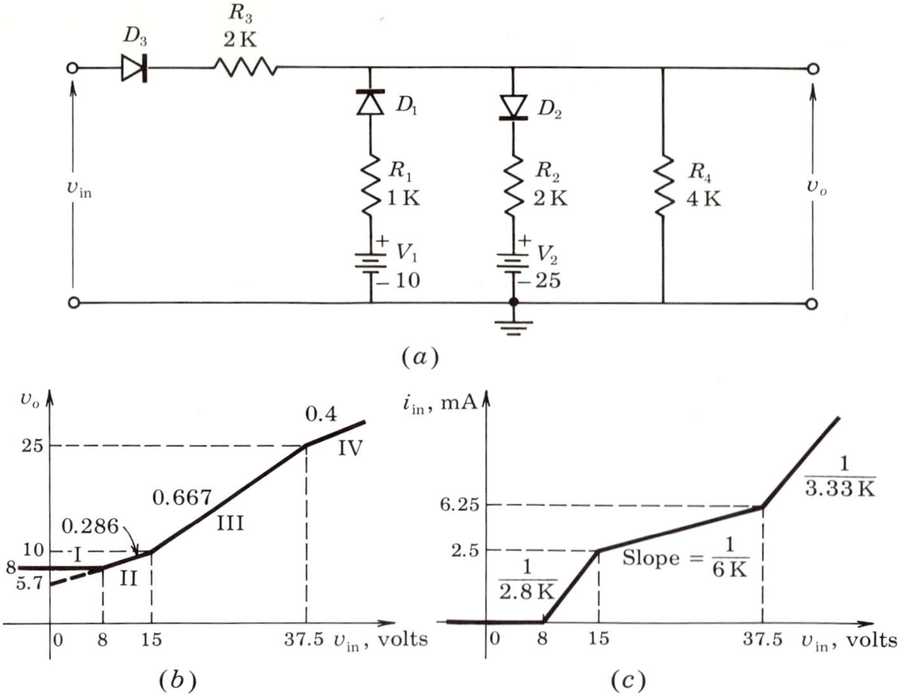

Fig. 2-24 An example of a three-diode network together with the transfer and input volt-ampere characteristic. (*a*) Three-diode circuit; (*b*) transfer characteristic; (*c*) input volt-ampere characteristic.

both cases, generally more than two linear segments are necessary for a satisfactory piecewise-linear representation. This simply means that additional diodes must be incorporated into the network—each break point corresponds to the change of state of an ideal diode—from conduction to cutoff, or vice versa. One simple example of arbitrary transfer and volt-ampere curves and the equivalent diode model appears in Fig. 2-24.

If the network contains at most two or three diodes in a relatively simple configuration, then in solving for the break coordinates of any one diode, we can always assume the condition of the others. In Fig. 2-24 we may assume D_1 and D_2 conducting and solve for the break values of D_3. Any inconsistency arising when checking the answer (e.g., a negative current flow through a diode that was assumed conducting) means that the problem must be re-solved from the new starting point. Three diodes afford eight possibilities as to the circuit condition, but these would be halved by the particular diode and biasing arrangement. With more than three diodes, in other than a trivial circuit, the trial-and-error method becomes unreasonably tedious and we must seek a more organized method of solution.

Example 2-3

In the case where we can readily identify the diode states, as in the circuit of Fig. 2-24*a*, we reduce the complex circuit to a number of simpler models. Each region is defined by the resistive network obtained when all back-biased diodes are replaced by open circuits and the conducting ones by short circuits. If these regions are selected by assuming the diode states, then each model must be checked for consistency. In the circuit considered, the region limits are easily specified in terms of the output voltage: D_1 conducts when $v_o < 10$, D_2 conducts when $v_o > 25$, and D_3 is forward-biased when $v_o < v_{\text{in}}$. In order to simplify the calculations, we shall assume ideal diodes. The diode drops can be accounted for by modifying the series bias voltages.

Region I: D_2 and D_3 back-biased (open circuits)
D_1 conducts (short circuit)

$$v_{o1} = \frac{R_4}{R_1 + R_4} V_1 = \frac{4\text{ K}}{1\text{ K} + 4\text{ K}} 10 = 8\text{ V}$$

This output remains constant at this value until D_3 conducts once $v_{\text{in}} > 8$ V.

Region II: D_1 and D_3 conducting $(8 \le v_{o2} \le 10)$
D_2 back-biased

$$v_{o2} = \frac{R_1 \parallel R_4}{R_1 \parallel R_4 + R_3} v_{\text{in}} + \frac{R_3 \parallel R_4}{R_3 \parallel R_4 + R_1} V_1 = 0.286 v_{\text{in}} + 5.7$$

Region III: D_3 conducts $(10 \le v_{o3} \le 25)$
D_1 and D_2 nonconducting

$$v_{o3} = \frac{R_4}{R_3 + R_4} v_{\text{in}} = 0.667 v_{\text{in}}$$

Region IV: D_3 and D_2 conducting $(25 \le v_{o4})$
D_1 back-biased

$$v_{o4} = \frac{R_2 \parallel R_4}{R_2 \parallel R_4 + R_3} v_{\text{in}} + \frac{R_3 \parallel R_4}{R_3 \parallel R_4 + R_2} V_2 = 0.4 v_{\text{in}} + 10$$

The limits of the output are inserted into the individual equations to find the limits expressed in terms of input voltage. These are indicated in Fig. 2-24*b*. Note that in each region the transfer characteristic is a straight line with a slope found from the incremental model (all battery voltages set to zero).

A further application of the type of circuit shown in Fig. 2-24*a* comes from recognizing that its input volt-ampere characteristic (Fig. 2-24*c*) changes as diodes start and stop conducting. The inverse slope of this characteristic is simply the resistance seen looking into the input terminals. Therefore diode circuits are useful in developing almost any required nonlinear resistance characteristics. Piecewise approximations result in a discontinuous rather than continuous incremental input resistance, but by taking enough segments we can approximate any continuous resistance as closely as necessary.

A general approach

Although we are developing more explicit methods of analysis and the implicit techniques of synthesis of multiple-diode circuits, we shall be forced to restrict the discussion to a single reasonably general network. This is essential if the treatment is to be kept within bounds: the model used in describing an arbitrary nonlinear characteristic is not necessarily unique, and therefore many feasible solutions exist. It is neither possible nor practicable to examine all of them.

Each branch of the general network of Fig. 2-25*a* consists of a parallel combination of biased diodes and resistors, as shown in Fig. 2-25*b*. The overall response may be determined by noting that:

1. The number of break points in the transfer or input volt-ampere characteristic equals the number of diodes contained in the two branches.
2. The piecewise-linear response is completely defined by finding the coordinates of the break points and the slopes of the two unbounded segments.

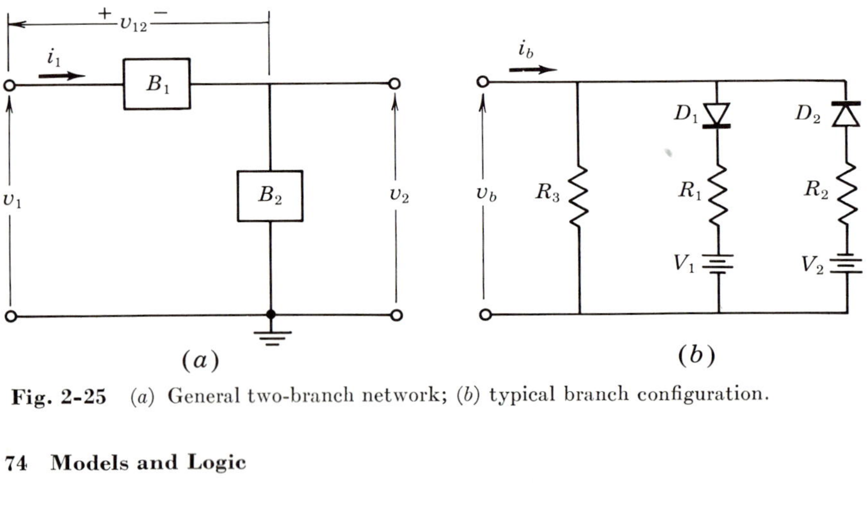

Fig. 2-25 (*a*) General two-branch network; (*b*) typical branch configuration.

3. The maximum possible transfer slope is unity since the network is completely dissipative. (We restrict R to positive values.)

In calculating these break points, each branch may be considered separately and then their individual responses combined. Under the rules developed in Sec. 2-3, the voltage break values of any branch are simply the bias voltages appearing in series with the internal diodes. Once these voltages are known, the corresponding currents are readily found by direct evaluation. It is important to determine which diodes are operative in any region since each one that starts to conduct introduces additional shunt resistance and each one that stops conducting reduces the number of elements in the branch.

As an aid to our inconsistent memory, the behavior of each branch and the state of each diode should be tabulated as a function of the controlling terminal voltage. This not only enables us to construct the response characteristics in the most organized manner but in itself reflects the many resistive models inherent in the diode network.

The input volt-ampere characteristic is the totality of the branch responses. Under the series arrangement shown, the current must be continuous and the voltage drops simply add. The break-point current coordinates comprise the set of those found separately for the two branches. The corresponding voltage coordinates must be found by direct evaluation and summation of the drops at these known currents. For one branch the drop is identically the break value, and for the other it may be determined from the graph or from the equivalent resistive network.

Finally, the overall transfer characteristic is obtained by cross-plotting the output branch voltages corresponding to the input break values. In each region the slope is identical with that found from the purely resistive network holding. Voltage sources only shift the location of the segment; they cannot alter its slope. To reiterate, the complete network response is determined by the location of the break points and the incremental slopes of the line segments terminating on them. Only this information is necessary for a complete description of the piecewise-linear network.

Example 2-4

The circuit of Fig. 2-26 utilizes series- and parallel-resistance padding of a zener diode for its series branch and two individually adjustable diodes for its shunt branch. As in Example 2-3, we shall assume ideal diodes. The volt-ampere response of each branch may be tabulated with respect to decreasing terminal voltage, as indicated in Table 2-4*a* and *b*. The characteristics of each branch are presented graphically in Fig. 2-27*a* and *b*.

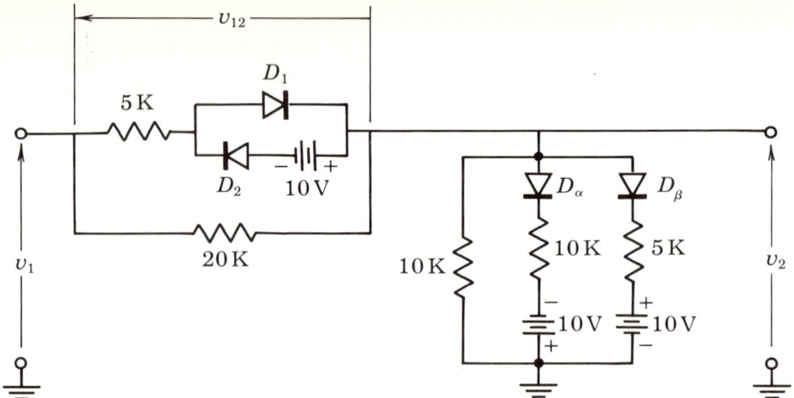

Fig. 2-26 Circuit to illustrate the method of analysis of multiple-diode networks.

Table 2-4*a* Series-branch response

DIODE STATES	v_{12}	i, mA	PIECEWISE RESISTANCE* R_{12}, K
D_1 on, D_2 off	$v_{12} > 0$	. . . . .	4
D_1 turns off†	0	0	
D_1 off, D_2 off	$-10 < v_{12} < 0$	. . . . .	20
D_2 turns on†	-10	-0.5	
D_1 off, D_2 on	$-10 > v_{12}$	. . . . .	4

Table 2-4*b* Shunt-branch response

DIODE STATES	v_2	i, mA	PIECEWISE RESISTANCE* R_2, K
D_α on, D_β on	$v_2 > 10$	. . .	2.5
D_β turns off†	10	3	
D_α on, D_β off	$-10 < v_2 < 10$	. . .	5
D_α turns off†	-10	-1	
D_α off, D_β off	$v_2 < -10$	. . .	10

* The incremental resistance of each region is simply the parallel combination of all resistors inserted by the diodes conducting in that region.

† The individual branch break points, which are found first, serve as the critical points of the table. The voltage values are determined by inspection of the branch, and the current by direct evaluation.

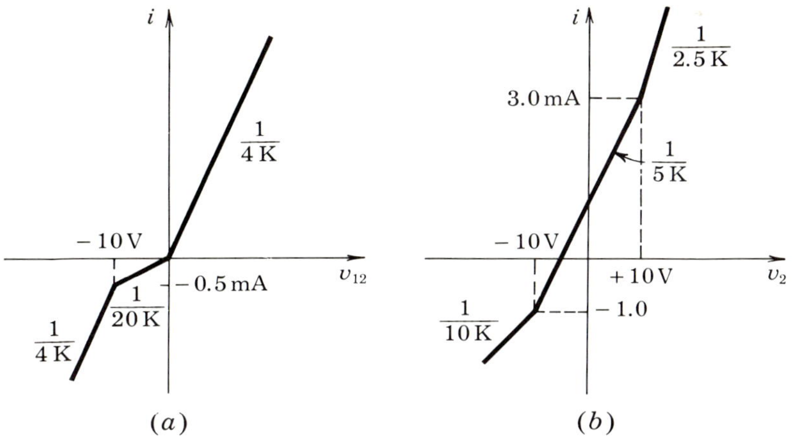

Fig. 2-27 Individual-branch characteristic for the circuit of Fig. 2-26. (*a*) Series branch; (*b*) shunt branch.

Now that the response of each branch is known, we are in a position to find the total input volt-ampere characteristic. This can be done by direct summation of the two graphs of Fig. 2-27*a* and *b* or by considering the data presented in Table 2-4*a* and *b*. We can construct a new table by interleaving the known break currents in descending order (Table 2-4*c*).

Table 2-4*c* Total circuit response

REGION	(1) CONDUCTING DIODES AND DIODE CHANGING STATE	(2) i, mA	(3) v_{12}*	(4) v_2*	(5) $v_1 = v_{12} + v_2$	(6) PIECE-WISE INPUT RESIST-ANCE, K	(7) TRANS-FER SLOPE
I	D_1, D_α, D_β on	. . .	. . . . . .	. . . . .	$v_1 > 22$	6.5	2.5/6.5
	D_β turns off†	3	(12)	10	22		
II	D_1, D_α on	. . .	. . . . . .	. . . . .	$-5 < v_1 < 22$	9	5/9
	D_1 turns off†	0	0	(−5)	−5		
III	D_α on	. . .	. . . . . .	. . . . .	$-17.5 < v_1 < -5$	25	5/25
	D_2 turns on†	0.5	−10	(−7.5)	−17.5		
IV	D_α, D_2 on	. . .	. . . . . .	. . . . .	$-22 < v_1 < -17.5$	9	5/9
	D_α turns off†	−1	(−12)	−10	−22		
V	D_2 on	. . .	. . . . . .	. . . . .	$v_1 < -22$	14	10/14

* The derived values are enclosed in parentheses.
† Circuit break points.

These represent the totality of break points and separate the linear input regions. Furthermore, the incremental resistances previously found are in series and may be directly added. The reader should compare the appropriate columns of Table 2-4*c* with Table 2-4*a* and *b*.

The only data still unknown are the values of the input-voltage coordinates. For these, we must find the voltage drops across the individual branches at the known current values. The derived voltages are given in parentheses in columns 3 and 4 of Table 2-4*c*. They may be found by linear interpolation from the tables of the individual branch responses, from the branch graphics, or directly from the circuit.

The voltage break value may also be determined by writing and solving the equation of a straight line holding within each region. In the region defined by the bottom line of Table 2-4*a*, the segment is a straight line of slope ¼ K passing through the point (−0.5 mA, −10 V). Thus

$$v_{12}(i) = -10 \text{ V} + (i + 0.5)4 \text{ K}$$

At the known break current of −1 mA, this yields

$$v_{12}(-1) = 12 \text{ V}$$

Similar equations may be written for the other unknown terms. Figure 2-28*a* is the plot of the complete input volt-ampere characteristics.

The transfer curve becomes immediately apparent if we recognize that the incremental transfer slope of each region is

$$\frac{\Delta v_2}{\Delta v_1} = \frac{R_2}{R_{12} + R_2}$$

where R_2 and R_{12} are the piecewise-linear resistances in the appropriate regions. Table 2-4*c* also includes these various slopes. Figure 2-28*b* shows the cross-plotting of the input and output voltage coordinates of the break points leading to the network transfer characteristic. The regions are those given in Table 2-4.

We can draw the following general conclusions from the above example:

1. Each time a series diode starts conducting, it reduces the net series resistance and increases the slope of the transfer characteristic. Each time a series diode stops conducting, it reduces this slope.
2. The conduction of a shunt diode reduces the branch resistance and decreases the transfer-characteristic slope. As a shunt diode stops conducting, both the shunt resistance and the incremental transmission increase in value.

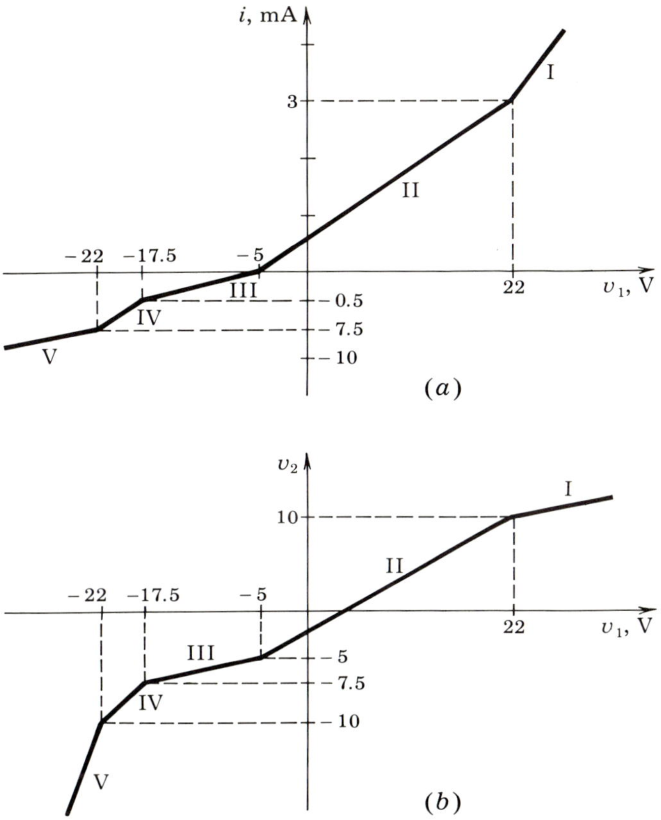

Fig. 2-28 (*a*) Input volt-ampere response; (*b*) transfer characteristic of the circuit of Fig. 2-26.

For reference purposes, some commonly used diode-resistor combinations and their volt-ampere responses are tabulated in Fig. 2-29. The first two diagrams illustrate a diode biased to start conducting and one which stops conducting as the voltage rises above the threshold value. Figure 2-29*c* represents the response of a two-diode branch.

2-10 Dead Zone and Hysteresis

Two interesting examples of diode wave shaping which are used in the analog solution of electrical, chemical, and mechanical problems are the generation of the transfer characteristics of dead zone and hysteresis. We shall first examine dead zone, which appears whenever a threshold value must be exceeded before transmission is possible. The steering system of

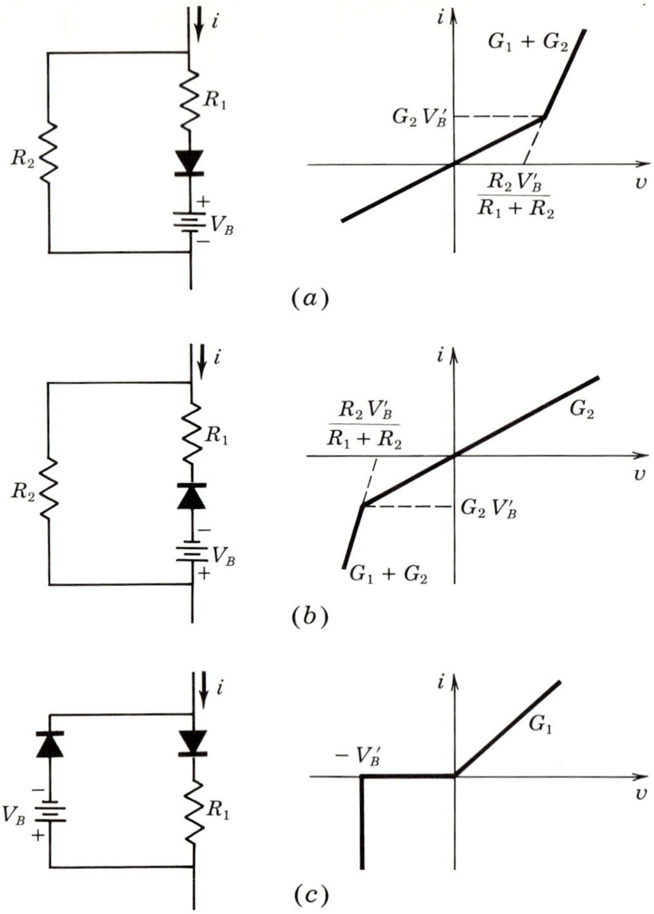

Fig. 2-29 Various diode-resistor combinations and their volt-ampere responses.

an automobile has a built-in dead zone for safety; small disturbances of the wheel are not transmitted into a turning movement. Some threshold value of steering-wheel rotation must be exceeded before the automobile responds.

The freezing-and-melting process of pure materials presents a second example of dead zone. Here the addition of heat produces a linear temperature rise at temperatures below the critical point. But at the melting and boiling points, a finite amount of heat must be added to a constant temperature before there will be a change of state (solid to liquid, liquid to gas). In the new state, temperature again increases linearly with heat flow.

The simplest diode network that can be used to simulate a fixed-width dead zone employs, as the series branch, two zener diodes connected back to back (Fig. 2-30*a*). Their constant-voltage drop in the zener region avoids the addition of any series-bias source (a difficult problem since this voltage must be isolated from ground).

We shall consider the response of the nonlinear series branch first. On positive voltages, conduction eventually becomes possible through the forward diode of D_1 (D_{1f}) and the zener region of D_2 (D_{2z}). However, this will not occur until the branch voltage exceeds $V_{z2} + V_D$. The coordinates of the break point are

$$v_1 = V'_{z2} = V_{z2} + V_D \qquad i = 0 \tag{2-50}$$

At zero current, there will be no drop in R_2 and the input break coordinates will be identical with those given in Eq. (2-50).

By symmetry, we can see that the other break point occurs at

$$v_1 = -V'_{z1} = -(V_{z1} + V_D) \qquad i = 0$$

Consequently, within the limits

$$-(V_{z1} + V_D) < v_1 < V_{z2} + V_D$$

only the extremely limited transmission through the high inverse resistance of a physical diode is possible. Outside these limits, the incremental transfer slope is determined by the voltage divider composed of R_1 and R_2:

$$S = \frac{R_2}{R_1 + R_2} \tag{2-51}$$

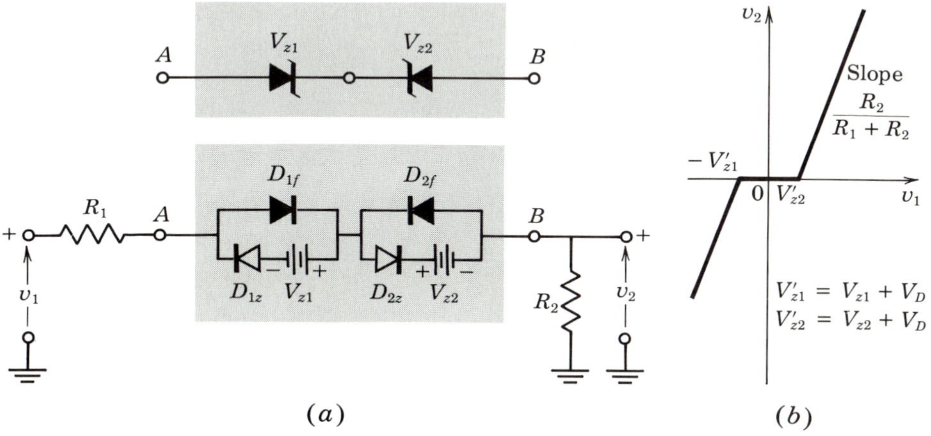

Fig. 2-30 Dead-zone circuit and transfer characteristic.

In the transfer characteristic of Fig. 2-30*b*, we note that the width of the dead zone is a function of the individual zener-diode break voltages and that the transmission slope is controlled by the series padding resistor. The dead zone may be shifted to any output voltage by suitably biasing the output resistor.

Hysteresis arises in many physical properties of materials, such as the magnetization curve of iron and the stress-strain relationship. If the forcing function is increased from zero to some final value and then removed, the forced-system parameter will remain fixed at a value other than zero. But this is simply the object's memory, which shows up as the magnetic flux still present after the removal of the magnetizing current or as the residual stress in a steel rod after the stretching force is released. Of course, limits are imposed by the physical nature of the materials. Only so much flux can be supported, regardless of the magnitude of the magnetization current; this value is called the *saturation flux*. In a stressed bar, excessive force will cause fracture.

To return a system (one which may be saturated but not fractured) to its original state requires the application of a forcing function in the opposite direction, usually of somewhat smaller magnitude than the original signal. If the new input is too large, it will simply leave the system in an excited state of the opposite magnetic polarity or with the strain acting in the opposite direction (Fig. 2-31*b*).

A smaller-amplitude excitation will generate smaller-area loops, with regard to both the displacement from the origin and the maximum output amplitude. These are extremely difficult to simulate, and therefore we shall concentrate our attention on the circuit of Fig. 2-31*a*, which gives the response shown in Fig. 2-31*b*. The two diodes D_3 and D_4 limit the maximum output excursion to $\pm V'_m$ and establish the saturation values so necessary in many hysteresis systems. (Fracture may be simulated by fuses.)

Operation is similar to that of the dead-zone circuit. On a rising output, the forward diode D_{1f} and the zener diode D_{2z} eventually conduct, providing

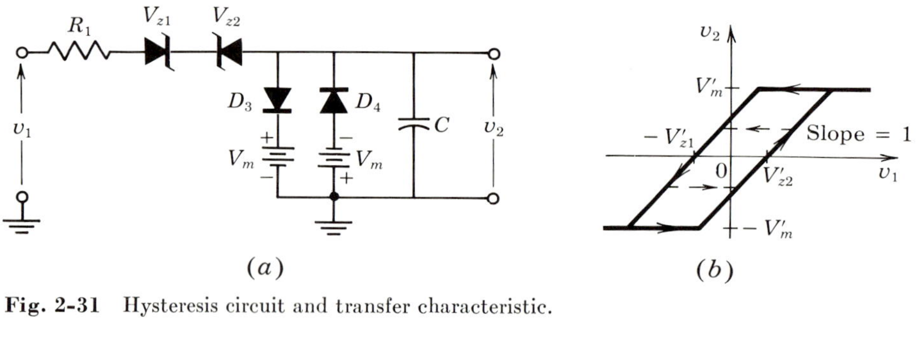

Fig. 2-31 Hysteresis circuit and transfer characteristic.

an output across C. Capacitor voltage then follows the input to a maximum value determined by the clipping diode D_3. As the input falls, the charge on the capacitor (the memory of previous maximum excitation) back biases the formerly conducting diodes, and since no discharge path exists, the output remains at the maximum value previously reached. Eventually the falling input brings D_{1z} and D_{2f} into conduction, establishing a discharge path, and again the output follows the input. The transfer characteristic now traces the left-hand portion of the curve. If the input continues falling and then rises, the cycle repeats, with the roles of the series diodes interchanged.

There is no necessity to drive the output into saturation. For generation of hysteresis or backlash curves, smaller inputs generate the minor loops indicated by the dashed lines of Fig. 2-31*b*.

The slowest input signal for proper response is determined by the leakage of the capacitor charge through the diode inverse resistance, and the fastest signal by the charging-time constant

$$\tau_c = CR_1$$

We choose C for reasonable response to the expected time range of the input signals.

2-11 Depletion-region Capacitance

When the diode is back-biased, there is, as we saw, a depletion of the mobile carriers from the junction vicinity. The transition zone contains the charge bound to the lattice during the doping process. On the p side of the metallurgical junction, electrons are tied to the acceptor atoms in forming the covalent bonds. The n side of the junction has a net positive charge located at the nucleus of the donor atoms. The distribution of the bound charge in the depletion region is shown in Fig. 2-6.

For purposes of calculation, we can approximate the actual bound-charge distribution by the uniform distribution of Fig. 2-32. The total charge in each area must be equal:

$$Q_B = qAl_nN_D = qAl_pN_A \tag{2-52}$$

The location of the edges of the depletion region (l_p and l_n) can be found by integrating the charge distribution twice and setting the resultant potential function equal to the known total junction voltage, $V_j - V$.

$$l_p = \left[\frac{2\epsilon}{q}\frac{(V_j - V)N_D}{N_A(N_A + N_D)}\right]^{1/2} \qquad l_n = \left[\frac{2\epsilon}{q}\frac{(V_j - V)N_A}{N_D(N_A + N_D)}\right]^{1/2} \tag{2-53}$$

We see from Eq. (2-53) that the location of the edges of the depletion region and hence the charge stored vary as $\sqrt{V_j - V}$. A plot of the bound-charge variation with voltage appears in Fig. 2-32.

We define the incremental depletion-region capacitance by

$$C_d = -\frac{dQ_B}{d(V_j - V)}$$

Substituting Eq. (2-53) into Eq. (2-52) and differentiating, we obtain

$$C_d = \frac{A}{2}\left(\frac{2q\epsilon}{V_j - V}\frac{N_A N_D}{N_A + N_D}\right)^{1/2} = \frac{K}{\sqrt{V_j - V}} \tag{2-54}$$

This capacitance is identical to that of a parallel-plate capacitor having a plate separation of $l_n + l_p$.

When the diode junction is graded instead of abrupt, the depletion-region capacitance will vary inversely with the cube root of the voltage instead of inversely with the square root [Eq. (2-54)]. Through controlling the diffusion of the dopants, diodes are fabricated with depletion capacitances

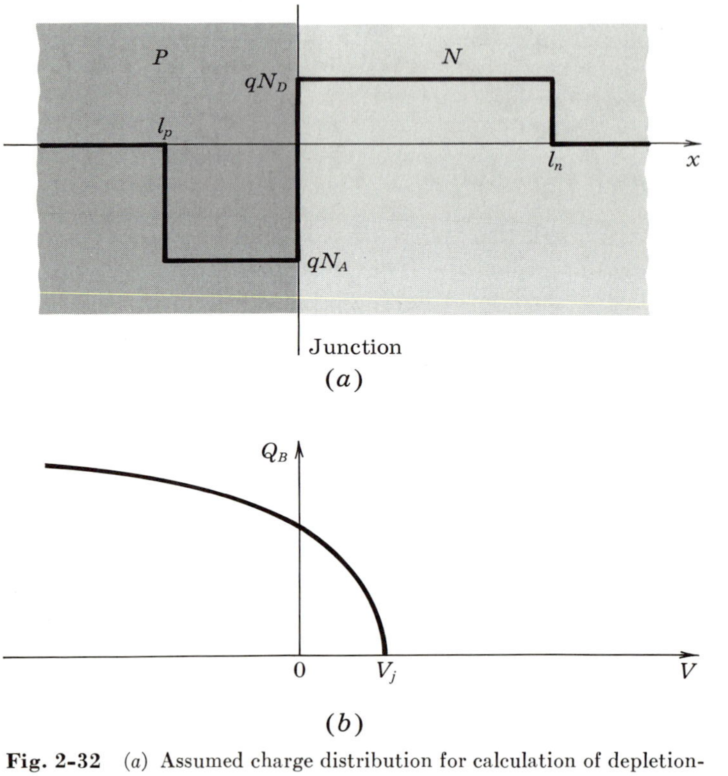

Fig. 2-32 (*a*) Assumed charge distribution for calculation of depletion-region capacitance; (*b*) bound charge as a function of applied voltage.

varying as K/V^γ, where $\frac{1}{5} < \gamma < 7$. These special devices are not very good diodes, but they are very good voltage-controlled capacitances. They are generally known as *varicaps*.

2-12 Diode Dynamic Response

In the forward-biased diode, an increase in terminal current is accompanied by the buildup of excess charge on the two sides of the junction and an increased diode voltage drop [Eq. (2-17) and Figs. 2-7 and 2-33]. This charge storage corresponds to a forward diode capacitance which appears in parallel with the forward incremental resistance. The diode's pulse response will be similar to that given in Example 2-2. Because of the effective shunt capacity, we expect turn-on and turn-off delays. Since it can easily be shown that the incremental forward resistance varies inversely and the forward capacitance varies directly with diode current (Prob. 2-28), the forward time constant itself is invariant with the operating point. In solving for the dynamic response, we therefore consider the time constant rather than a fixed R and C equivalent model:

In the circuit of Fig. 2-33*a*, when the switch is thrown to position 1, the positive current flow immediately forward biases the diode. Current is limited by the external resistor R. If $V_D \ll V_F$, then $I_F \approx V_F/R$. We have previously shown, in Eq. (2-16), that for a p^+n junction the current can be completely defined in terms of the minority hole distribution at the junction edge of the n material:

$$I_F = -qAD_p \frac{\partial \Delta p_2}{\partial x}\bigg|_{x=0} = \frac{\partial \Delta Q_x}{\partial t} + \frac{\Delta Q_x}{\tau_p} \tag{2-55}$$

where ΔQ_x is the total excess minority charge in the n material. Solving Eq. (2-55), we find that the excess charge builds up exponentially with the recombination-rate time constant τ_p:

$$\Delta Q_x(t) = I_F \tau_p (1 - e^{-t/\tau_p}) \tag{2-56}$$

From Eq. (2-55), we see that the constant input current implies that the slope of the hole distribution at the edge of the transition region remains constant during the complete charge buildup. During the buildup interval, the excess charge must also diffuse spatially. If as a first approximation we assume a triangular charge distribution having a constant-slope hypotenuse, then the x dimension will be proportional to $\Delta p_{20}(t)$ and

$$\Delta Q_x(t) \approx K[\Delta p_{20}(t)]^2 \tag{2-57}$$

We already know the relationship between Δp_{20} and V_D from Eq. (2-17). In addition, the final steady-state voltage across the diode is given by

$$e^{V_{DSS}/V_T} = \frac{I_F}{I_{rs}} \tag{2-58}$$

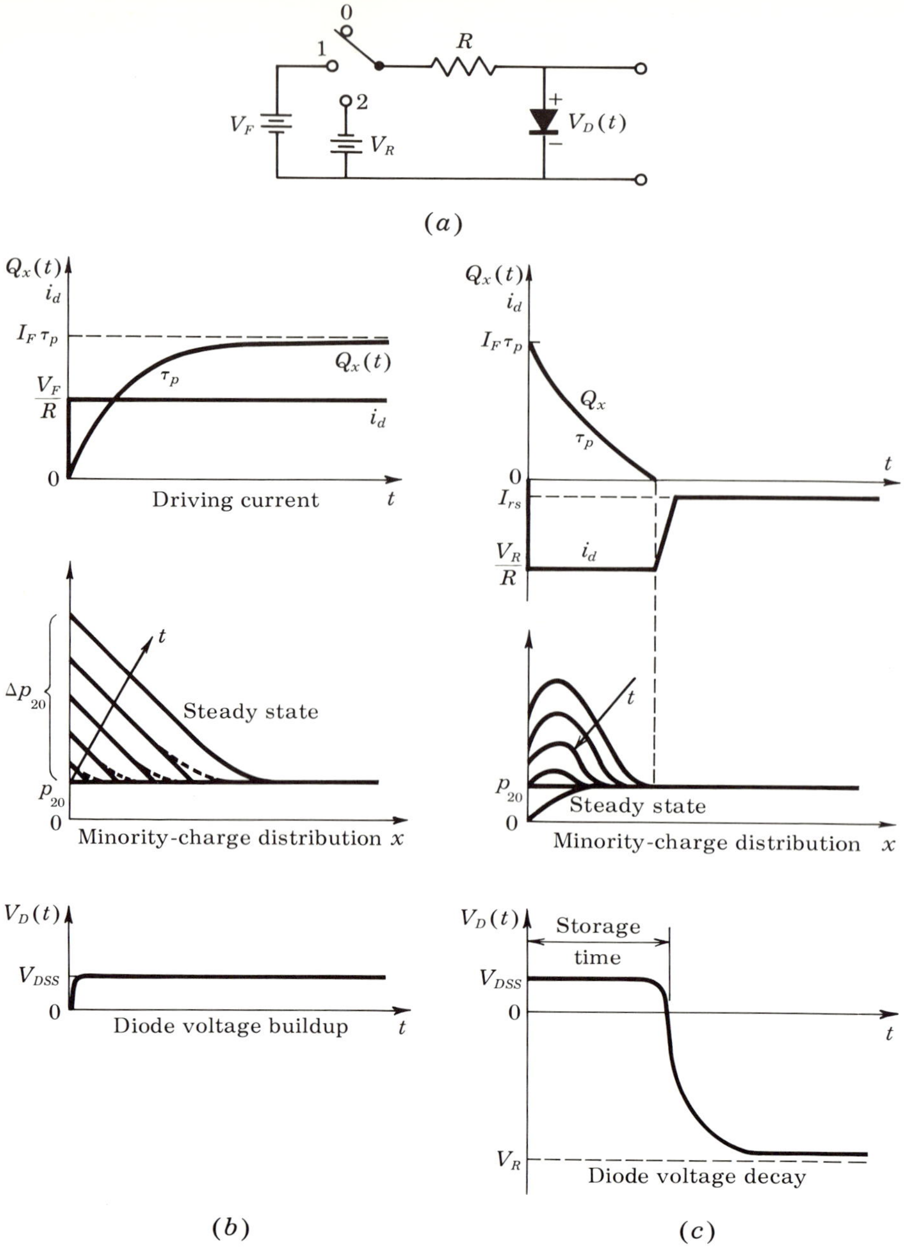

Fig. 2-33 (*a*) Circuit for calculation of turn-on and turn-off times. (*b*) Current, minority-charge spatial distribution, and voltage buildup during turn-on interval—switch in position 1. (*c*) Current, minority-charge spatial distribution, and diode voltage during turn-off interval—switch in position 2.

Combining the time variation of $\Delta p_{20}(t)$ from Eq. (2-57), the final steady-state condition from Eq. (2-58), and Eq. (2-17), we finally obtain

$$e^{V_D(t)/V_T} = \frac{I_F}{I_{rs}}\sqrt{1 - e^{-t/\tau_p}} \tag{2-59}$$

Equation (2-59) indicates an extremely rapid diode-voltage buildup. This is to be expected from the exponential relationship between excess charge and voltage.

To turn off the diode, the switch is thrown to position 2. The external current reverses. But the diode does not turn off immediately; the previously stored charge, $I_F\tau_p$, keeps the diode turned on. This is similar to the capacitive delay illustrated in Example 2-2. Solving Eq. (2-55) by substituting a current of $-I_R$ for I_F, and subject to the new initial conditions, we obtain the solution

$$\Delta Q_x(t) = -I_R\tau_p + \tau_p(I_F + I_R)e^{-t/\tau_p}$$

Once $\Delta Q_x(t_s) = 0$, the stored charge has been swept out and the diode voltage changes sign. Thus the storage delay time is

$$t_s = \tau_p \ln \frac{I_F + I_R}{I_R}$$

If we use a square-wave excitation such that $I_F = -I_R$, then $t_s = 0.693\tau_p$. t_s can be readily measured for the diode, and this is a method for finding the recombination-time constant τ_p. The voltage variation during the turn-off interval can be found as discussed above. We would find that the diode voltage remains almost constant over the full storage-time interval and then very rapidly drops to zero (Fig. 2-33).

After the diode reverses state and turns off, the series R and the voltage-dependent depletion-region capacitance determine the time rate of voltage buildup to the final steady-state value. For simplicity, we usually estimate the final portion of the response by the single recovery-time constant $\tau_r = C_dR$.

2-13 Summary

Many nonlinear systems whose externally measured characteristics are available may be treated from the viewpoint of their piecewise-linear representation. The nonlinear system, difficult to treat in all its complexity, is reduced to a sequence of linear problems, each with its own response and limits of operation. Simple mathematical tools already at our disposal from studies of linear systems are sufficient to reach acceptable engineering answers in those cases amenable to this treatment. We can always handle circuits involving only dissipative elements, no matter how nonlinear their characteristics may be. When energy-storage elements are also contained in the nonlinear system, the problem becomes much more difficult.

If in each region the boundary conditions are independent (able to be calculated only from the model and the boundaries of the region), we can solve the system in a straightforward manner. If the constraints involve past history, as in the hysteresis circuit, then when only a single energy-storage element is present, we are able to complete the analysis. But if multiple modes of energy storage exist, the circuit will not yield so readily to simple methods. Except in very special cases, such a problem is extremely difficult and beyond the scope of this text.

The first step in any analysis is the construction of the appropriate models together with the clear delineation of their boundaries. Generally, boundaries are determined by the energy-storage elements and points of conduction. The second step is to consider the conditions of continuity across the region's boundaries. These conditions arise primarily from the original characteristics and the type of energy-storage elements involved, with, however, the energy-storage elements often determining the diode conditions. Then, finally, we are ready to proceed to the solution.

The same techniques are used in solving many chemical, thermal, and mechanical systems, where, for example, an ideal mechanical "diode" might be postulated rather than an electrical one. A velocity-operated clutch which engages or disengages at preset shaft speeds is one mechanical example. Its slip on engagement is analogous with the diode forward resistance, while its imperfect disengagement represents reverse resistance. In fluid flow, on-off pressure valves behave similarly, with their pressure loss and leakage modifying the ideal valve in much the same way that resistance modifies the ideal diode. A metal plate polished on one side and black on the other allows radiant heat to flow more readily in one direction than in the other; it may be used as our thermal diode under the applicable conditions.

In many cases where it is difficult to formulate proper models, making the grossest approximations leads to a circuit which can be solved. The answer, admittedly incorrect, often gives some insight. This enables us to refine the analysis later, eventually coming to an acceptable engineering solution.

Problems

2-1 (*a*) By direct graphical construction, evaluate the input volt-ampere characteristic of the ideal-diode circuit of Fig. 2-34. Specify the individual slopes and the coordinates of the break point. Indicate the segment corresponding to diode conduction.

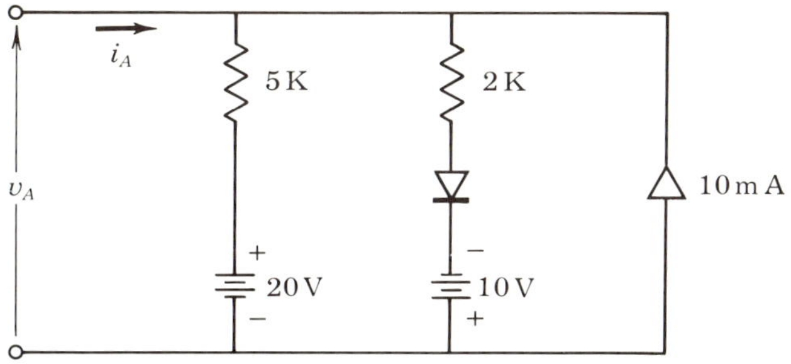

Fig. 2-34

(*b*) How will the characteristic of part *a* be modified if the diode is reversed? Do not solve this part by a second construction; just look carefully at the steps taken in your solution of part *a*.

(*c*) Convert the circuit holding for each state of the diode into both Thévenin and Norton equivalent circuits. Find the single break point from each set of defining equations.

2-2 Consider an abrupt-junction silicon diode (Fig. 2-6) where, at 300°K, the *n*-type material has a conductivity of 1 mho/cm and the *p*-type material has a conductivity of 100 mhos/cm. The junction has an area of 1×10^{-3} cm^2 and is located 0.1 cm from each of the end ohmic contacts. Take $D_p = 12$ cm^2/s and $D_n = 30$ cm^2/s. $\tau_p = \tau_n = 100$ ns.

(*a*) Plot the carrier densities and bulk conductivity of both the *n* and *p* materials as a function of temperature over the range of −100 to 200°C.

(*b*) Find the built-in junction back-bias voltage at the two temperature extremes.

(*c*) Plot the volt-ampere characteristic of the abrupt junction at −100, 0, and 100°C. (Maximum V_D at 0°C = ±750 mV.)

(*d*) Modify the plots of part *c*, with a different color, to account for the bulk conductivity of the semiconductor between the junction and the ohmic contacts. How would this change be accounted for in a diode model? Would the diode performance be improved if the ohmic contacts were located closer to the junction, say at 50 μm?

(*e*) Specify the approximate temperature coefficients for the *n* and *p* bulk conductivity and for the diode forward junction drop in the range of normal operating temperatures from 0 to 100°C.

2-3 Using the results of Prob. 2-2*c*, specify a series of models (Table 2-3) for the diode (at room temperature, 23°C). Give the values of I_{rs}, R_f, and R_r where necessary.

2-4 (*a*) Specify the parameters to be used in the models of Table 2-3 for a silicon diode where we measure a forward drop of 700 mV at a current of 1 mA.

(*b*) This diode is now used in a circuit where the current through it is 20 mA. What model parameter must be changed? To what new value?

(*c*) Repeat parts *a* and *b* for a germanium diode where we measure a forward drop of 120 mV at a current of 1 mA.

2-5 In this problem we shall consider the level of approximations made when using each of the models of Table 2-3. Tabulate the two output voltages and the loop current in the circuit of Fig. 2-35, using each model. Take $v_{in} = +1, +10, +100, -1, -10, -100$ V. The diode used has a measured voltage drop of 720 mV at a forward current of 1 mA. Give the figure of merit for each model by expressing the ratio of the difference between the exact solution and the model solution to the input voltage v_{in}. Under what conditions must the greatest care be taken in selecting a diode model?

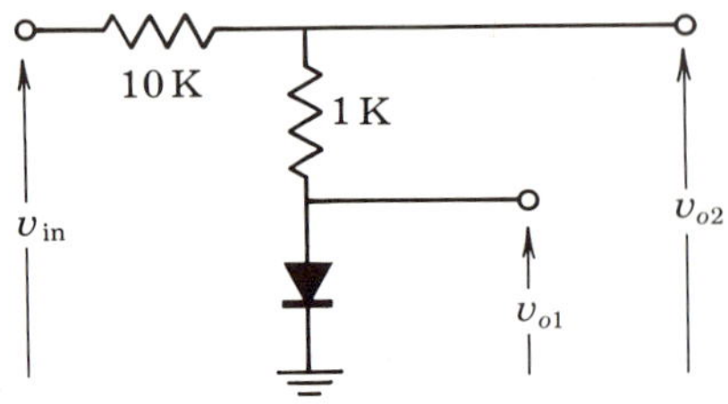

Fig. 2-35

2-6 Two diodes are available for use in a clipping circuit. The first one, which is germanium, has a drop of 100 mV at a forward current of 1 mA. The second diode is silicon and has a forward drop of 700 mV at 1 mA. Draw the circuit of a negative clipper having a clipping level of -5 V. Give the value of the series resistor which might be used with each diode. Which one is best suited for use as a clipper? Why? Does the other diode afford any special advantages?

2-7 We wish to calculate the diode currents and drops in the circuit of Fig. 2-36. D_1 has a measured drop of 760 mV at 5 mA, while D_2 has a drop of only 700 mV at 1 mA. Make all reasonable approximations in solving this problem.

(*a*) Find the current in each diode.

(*b*) Repeat part *a* for reversed diodes.

(*c*) Repeat part *a* when separate 50-Ω resistors are inserted in series with each diode.

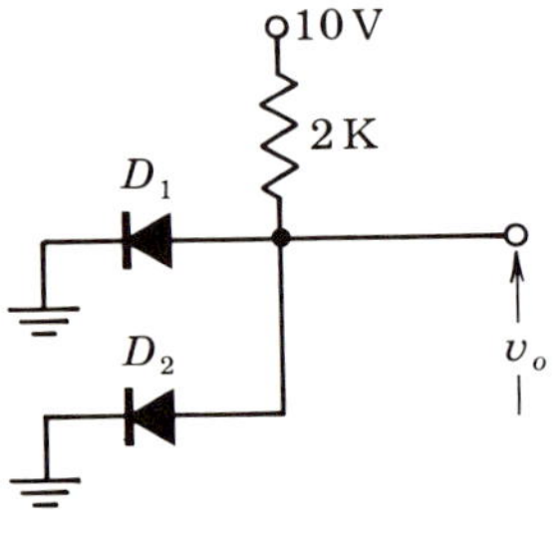

Fig. 2-36

2-8 A symmetrical, triangular, 40-V peak-to-peak voltage waveshape is the input to the series clipper of Fig. 2-12. The bias battery is 5 V, and the diode has a drop of 700 mV at 1 mA. Sketch the transfer characteristic and draw the input and output waveshapes to scale when $R_1 = 25$, 5, and 1 K. In all cases, calculate the output for zero input signal. Repeat when the diode is shunted by a 100-K resistor.

2-9 A diode having a forward drop of 720 mV at 5 mA and a zener breakdown of 15 V is to be used in a shunt clipper. This circuit must transmit signals above 5 V and clip below this point. The series resistor is restricted to $R_1 \leq 200$ K.

(*a*) Define an optimum value of clipping resistance and draw the circuit.

(*b*) Sketch the transfer characteristic, indicating the values of all slopes and the coordinates of all break points.

(*c*) Draw the output waveshape when the input is a 100-V peak-to-peak sinusoidal signal.

2-10 (*a*) Sketch the voltage across and the current through the diode of Fig. 2-37 to a reasonable scale. Use the second-order model of Table 2-3 for the diode. (The actual diode characteristic is defined by $V_D = 700$ mV at $I_D = 1$ mA.)

(*b*) During the interval that the diode conducts, what are the actual limits of the diode drop?

(*c*) Should the approximate model used in part *a* be modified on the basis of the answer of part *b*? Will any modification significantly affect the results of part *a*?

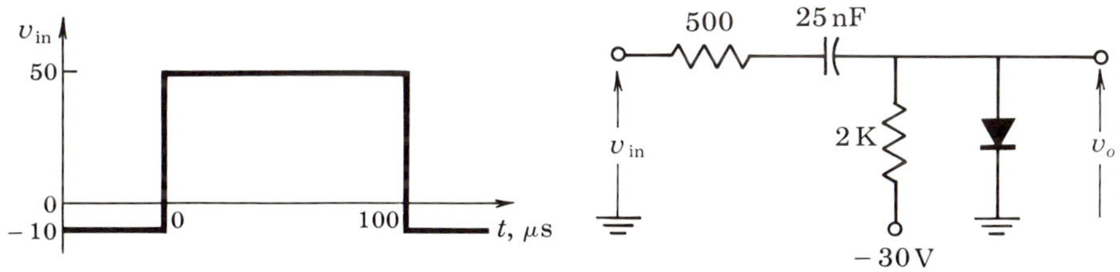

Fig. 2-37

2-11 The single pulse shown excites the circuit of Fig. 2-38 at $t = 0$. Evaluate the complete output response from $t = 0$ until the return to steady state. Specify

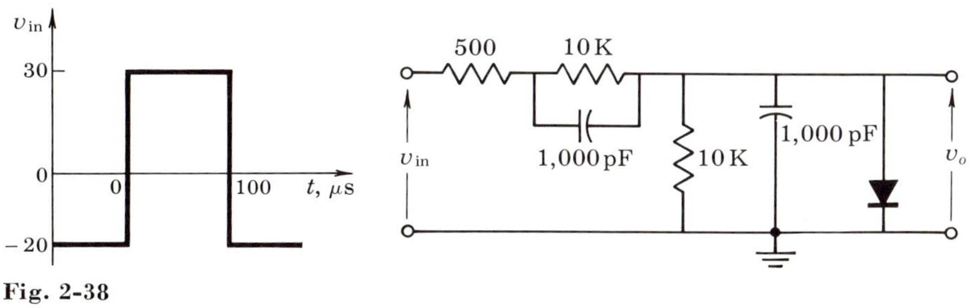

Fig. 2-38

all voltage and time-constant values. You may make all reasonable approximations, but you must make a statement justifying your approximation. Compare this circuit and solution with the solution of Example 2-1. (The diode has a drop of 720 mV at I_D = 1 mA.)

2-12 (*a*) Show that the silicon-diode counting circuit of Fig. 2-39 will have an output proportional to the number of applied input pulses (v_1) provided that they are of equal amplitude and duration. Also discuss the possibility of erasing this stored information upon the injection of a sufficiently large negative pulse at v_2.

(*b*) If the amplitude of each pulse is 100 V and its duration is only 2 μs, what size capacitor must we use so that the output will rise by approximately 1 V per pulse for up to 10 pulses? Sketch the output waveshape.

(*c*) Under the conditions of part *b*, how many pulses have been applied if the output reads slightly less than 15 V?

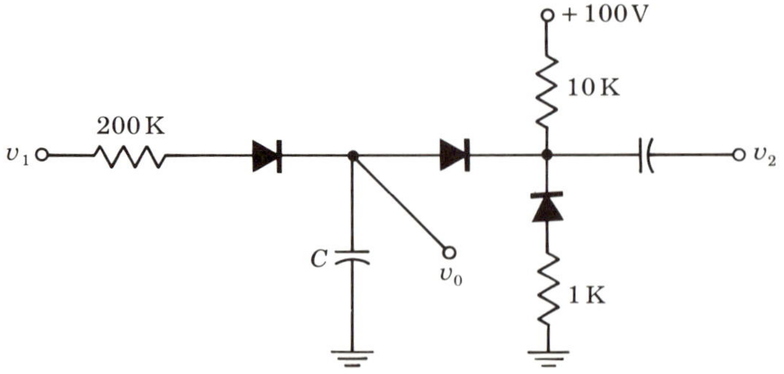

Fig. 2-39

2-13 A 30-V peak-to-peak 100-μs-period square wave is clamped negatively at +5 V. In order to establish a discharge path, the silicon diode is shunted by 100 K. The diode used may be modeled by a forward resistance of 60 Ω in series with a break voltage of 660 mV. We want to use the smallest possible clamp capacitor so that the clamp will have a fast response to changes in the input-signal amplitude, but we are restricted in that the output must never exceed the clamping level by more than 2 percent of the peak-to-peak input signal.

(*a*) What is the smallest-clamp capacitor we may use?

(*b*) How much energy must be restored in the capacitor over each cycle?

(*c*) If the peak-to-peak signal is suddenly increased to 50 V, how many periods of the input signal will the circuit take to recover its clamping action?

(*d*) Repeat part *c* when the signal is suddenly reduced to 10 V peak to peak. (*Hint:* Consider the superposition of the square-wave input and the initial voltage across *C*. Find the time at which the diode again conducts. Also see Prob. 1-8.)

2-14 The current waveshape of Fig. 2-40 is to be clamped negatively at −5 mA with a maximum of 2 percent overshoot. The diode available has a drop of 700 mV at 10 mA. The 50-mH coil approximates an ideal inductance in that its resistance is only 10 Ω.

(*a*) Draw the circuit used, showing how the bias current is injected. Give resistor values used for optimum current clamping and for satisfactory recovery to changes in the input-signal level.

(*b*) Sketch the output waveshape, giving values for all important currents and time constants.

(*c*) What effect does the coil resistance R_L of a nonideal inductance have on the current-clamping action?

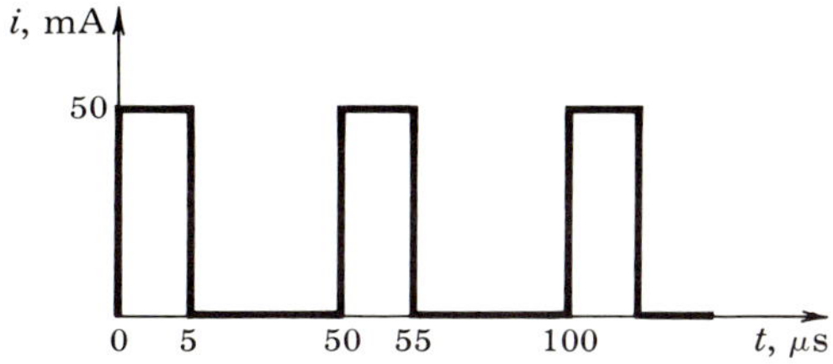

Fig. 2-40

2-15 Prove that the output waveshape of a current clamp will shift until the ratio of areas is equal to the ratio of diode resistances.

2-16 A 20-V peak-to-peak 1-ms square wave is the input to the clamp of Fig. 2-16. The diode has a forward resistance of 20 Ω and is shunted by 100 K; $C = 0.1\ \mu F$. The signal source impedance is 100 Ω.

(*a*) Sketch the output waveshape for the first two cycles if the square wave is initially applied when it is zero, going positive.

(*b*) Repeat part *a* if a 0.01-μF capacitor is shunting the diode. Make all reasonable approximations.

2-17 Given a dc voltmeter with a 10-M input impedance, our problem is to construct an ac voltmeter. The dc unit reads correctly in conjunction with a 1-M probe. Two different ac probes are under consideration (Fig. 2-41*a* and *b*). In each case, they will be connected directly to the voltmeter, i.e., not through the dc probe.

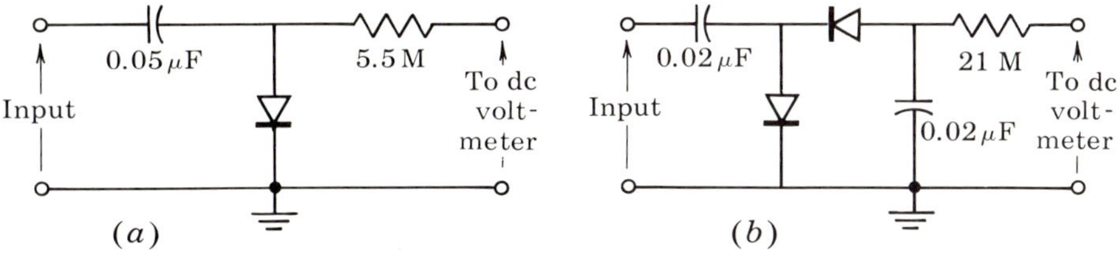

Fig. 2-41

Calculate the dc-voltmeter reading for each probe for the various inputs listed below. In each case, relate the reading to some parameter of the input signal (rms, peak, or average).

(*a*) 10-V-peak 1,000-Hz sine wave.
(*b*) 10-V-peak 1,000-Hz square wave.
(*c*) 20-V peak-to-peak 1,000-Hz triangular wave having a +10-V average value.
(*d*) Repeat part *c* for a signal having a −10-V average value.
(*e*) 10-V 10-μs-wide positive pulses with a period of 100 μs.
(*f*) Repeat part *e* for negative pulses.

2-18 Plot the transfer and input volt-ampere characteristics of the circuit given in Fig. 2-42. Sketch the output voltage if $v_{in} = 100 \sin \omega t$, giving all important values (assume ideal diodes).

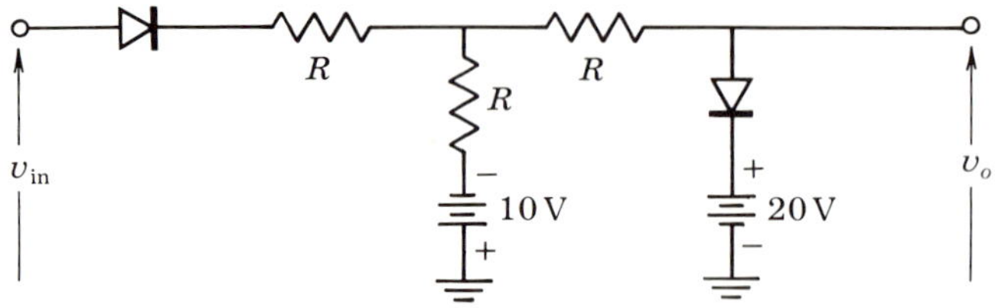

Fig. 2-42

2-19 Draw the input volt-ampere characteristic and the transfer characteristic of the circuit of Fig. 2-43 and sketch the output when a 50-V peak-to-peak triangular wave is impressed at the input.

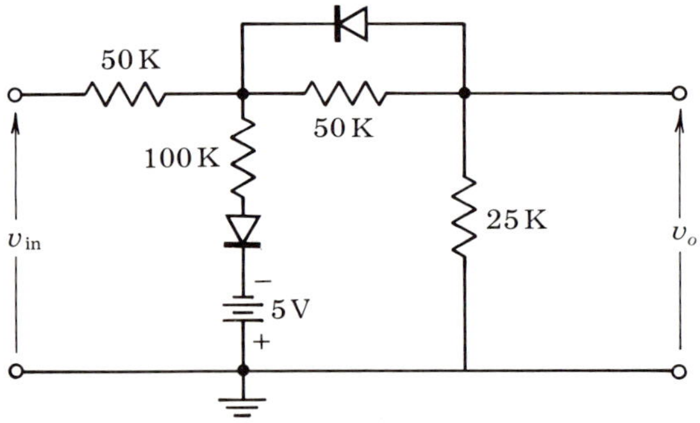

Fig. 2-43

2-20 (*a*) Set up the equations representing each region of operation given in Table 2-4 (Sec. 2-9) and verify the values of all break points.
(*b*) Show the purely resistive model (Thévenin or Norton equivalent) implicit within each region and specify the various component values. By direct solution of the intersection of the models found, evaluate the break points bounding these regions.

2-21 Design a circuit which will have the transfer characteristic given in Fig. 2-44. Specify all batteries and resistors, taking the smallest resistor as 10 K.

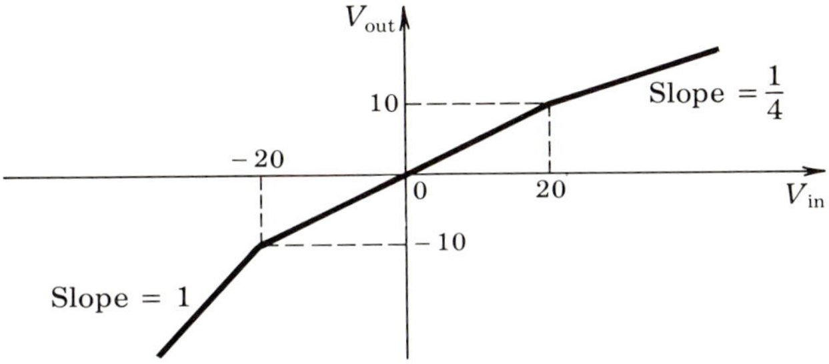

Fig. 2-44

2-22 In the circuit of Fig. 2-45, the input voltage is a symmetrical triangular wave having a period of 16 s, a peak-to-peak amplitude of 80 V, and an average value of zero.

(*a*) Plot the output voltage as a function of time and label all break points with their time and voltage values.

(*b*) The output signal is to be a rough approximation of a sine wave. What is the peak amplitude of the sine wave we are approximating, and to what angles do the break points correspond?

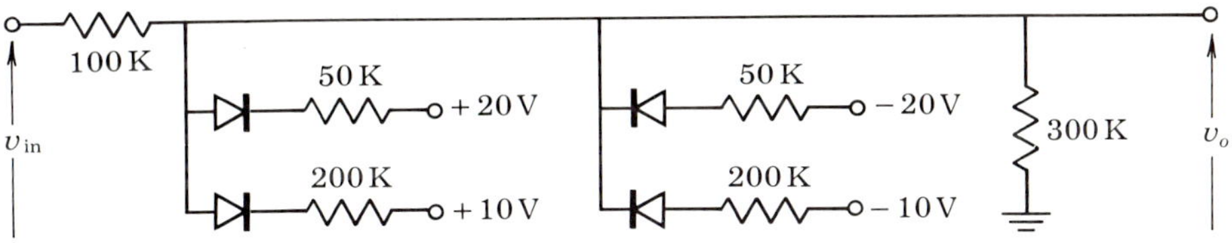

Fig. 2-45

2-23 One method of multiplying two voltages together is to convert each voltage signal into a current proportional to its logarithm and then to add the two currents in an arrangement similar to the summer of Chap. 1. If the output voltage is kept small compared with the input, then we might employ a circuit whose volt-ampere characteristics are logarithmic to reconvert $\log A + \log B$ to the product AB. This basic circuit is shown in block-diagram form in Fig. 2-46.

(*a*) As the first step, design a diode circuit which will approximate the logarithmic volt-ampere response for positive input signals. Except for a scale factor, the input current should exactly agree with $\log v_{in}$ at the following integer voltages: 10, 20, 40, 70, and 100 V. Use 100 K as your largest resistance value and specify all resistors and bias sources.

(*b*) If we use tables to reconvert the sum of the logs found from the two circuits of part *a*, what error is introduced when multiplying the following numbers: 20×65, 20×50, 12×30?

(*c*) If we use a circuit similar to the one developed in part *a* (except that the largest resistor is now 1 K) for reconversion of the sum of the logs to the product of the input voltages, then

$$v_o = mv_1v_2$$

Draw the circuit and give resistance and battery values for $m = 0.01$. To what values would the errors of the products of part *b* increase as a result of this additional circuit?

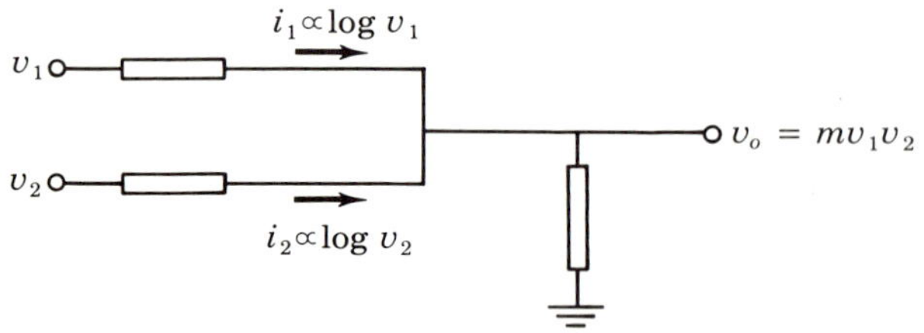

Fig. 2-46

2-24 (*a*) Discuss the effects on the transfer characteristic of the circuit of Fig. 2-30 when we return the bottom of R_2 to a variable bias instead of directly to ground.

(*b*) A 100-V peak-to-peak sinusoidal signal is applied at the input of this circuit. Sketch the input and output to scale and determine the transmission angle if the following circuit components are used:

$$V_D = 700 \text{ mV} \qquad |V_{z1}| = |V_{z2}| = 20 \text{ V} \qquad R_1 = 10 \text{ K} \qquad R_2 = 100 \text{ K}$$

2-25 Prove that the silicon-diode circuit of Fig. 2-47 also simulates a dead zone. Calculate the transfer characteristic and specify all break points and slopes. What

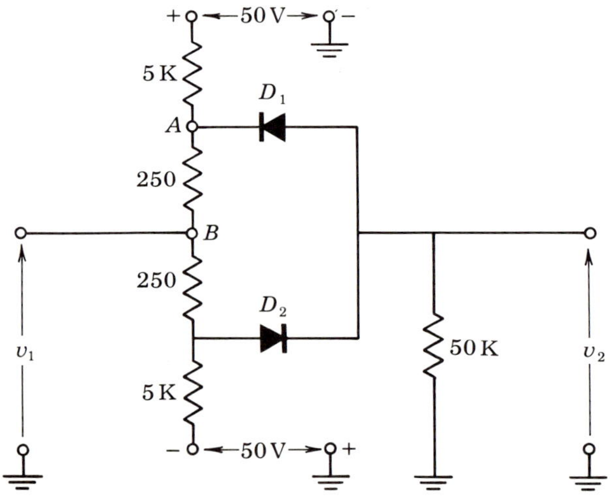

Fig. 2-47

would happen to the transfer characteristic if diode D_1 were transferred from point A to point B?

2-26 The circuit of Fig. 2-31 may also be employed as a memory, as may any device exhibiting hysteresis. If the saturation region crosses the vertical axis, then once the input drives the output into saturation, the output will remain there even after the removal of the input pulse. Only by injecting a pulse of the opposite polarity can the circuit be reset to the other saturation value.

(*a*) Design a circuit which will saturate at ± 10 V and which will cross the $v_1 = 0$ line at this value. The only two available voltages are ± 45 V. Specify all resistors and draw the transfer characteristic.

(*b*) Explain the disadvantages of setting the crossover point on the sloping side of the transfer characteristic instead of in the center or near the center of the saturation region.

(*c*) Sketch the output waveshape if the input pulses are those shown in Fig. 2-48.

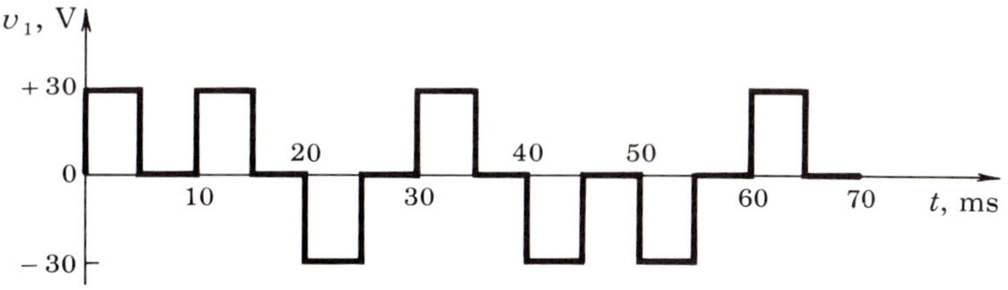

Fig. 2-48

2-27 The temperature-vs.-heat-content curve of many pure substances exhibits one or more dead zones. These occur where heat of fusion or heat of vaporization must be added to cause a change of state: solid to liquid or liquid to gas. For example, the temperature of ice increases linearly with heat content from −20 to 0°C and then remains at 0°C until sufficient heat has been added to melt all the ice. The temperature again increases with the addition of heat, but at a new rate, to the boiling point.

(*a*) We are interested in developing a circuit to represent this action, where the input voltage would correspond to the heat content (1 V/2 cal) and with the output voltage corresponding to the temperature (5°C/V). Choose water at 10°C as the origin of the axis and calculate the slopes, intercepts, and voltages needed to represent a temperature range of −20 to 90°C. Design the diode circuit which will give the required transfer characteristic. Try to use the minimum number of components.

(*b*) Repeat part *a* for a temperature range of −20 to 110°C. Choose a new calorie-temperature conversion ratio.

2-28 (*a*) Starting from the diode equations, evaluate the incremental r_f and C_f in terms of the operating current and the diode constants. (Assume a high-conductivity n material and a low-conductivity p material.)

$$r_f = \frac{dV_D}{dI_D} \qquad C_f = \frac{d\Delta Q_x}{dV_D}$$

(*b*) Show that the product $r_f C_f$ is independent of the diode current. Express this product in terms of the semiconductor constants.

2-29 In the circuit of Fig. 2-49, the abrupt-junction silicon diode has a measured depletion-region capacitance of 20 pF at 5 V reverse bias. This capacitance varies as given by Eq. (2-54). The diode forward drop is 700 mV at 1 mA, and $\tau_p = 5$ ns. Make a piecewise-linear model for the diode representing the depletion-region capacitance by a constant equal to the average capacitance. Solve for the circuit response to the applied pulse. Indicate all delay times. What physical conditions give rise to these intervals?

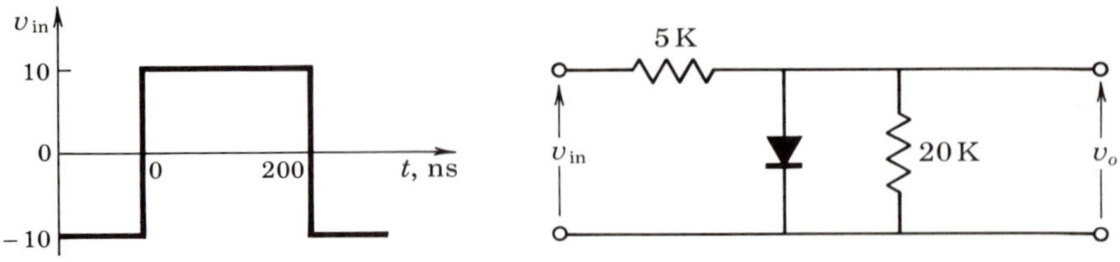

Fig. 2-49

2-30 (*a*) Repeat Prob. 2-29 when the 5-K series resistor is shunted by a capacitor equal to four times the average value of the depletion-region capacitance.

(*b*) Repeat Prob. 2-29 when the 5-K series resistor is shunted by a capacitance C_1 such that the product $(5\text{ K})C_1 = \tau_p = 5$ ns.

Bibliography

Adler, R. B., et al.: "Introduction to Semiconductor Physics," SEEC vol. 1, John Wiley & Sons, Inc., New York, 1964.

Conwell, E. M.: Properties of Silicon and Germanium: II, *Proc. IRE,* vol. 46, pp. 1281–1300, June, 1958.

Gray, P. E., et al.: "Physical Electronics and Circuit Models of Transistors," SEEC vol. 2, John Wiley & Sons, Inc., New York, 1964.

Miller, S. L.: Avalanche Breakdown in Germanium, *Phys. Rev.,* vol. 99, pp. 1234–1241, August, 1955.

3.
transistor models and simple circuits

An active element such as a transistor, with its controlled signal-transmission path, its internal amplification, and the related conversion of energy from the power supply to the signal output, adds a third dimension to our capabilities for wave generation and shaping. Where previously we had only passive linear and nonlinear elements at our disposal, in this chapter we shall discuss some simple uses for a controlled energy source which can be incorporated within, instead of remaining external to, the wave-forming network.

The transistor is a multiregion device, bracketing the active region with an open-circuited cutoff region and a short-circuited saturation region. We shall first develop models which define the total device behavior and then draw a reduced model for each region of operation. For the simplest analysis of subsequent circuits, our model should be a piecewise-linear one, allowing us to write a separate linear defining equation for each region of operation. In formulating our models, we shall find that this is not, in fact, completely possible. The active-region model will differ from an ideal linear device. However, when examining the macroscopic operation, linearization, which does lead to simplified results, will be employed. At the same time we shall indicate the limitations of any approximations made.

3-1 Transistor Physics

We can explain the operation of the bipolar junction transistor and can derive adequate circuit models by considering the very simple one-dimensional idealized representation of Fig. 3-1*a*. A very thin *n* base region is sandwiched between two much more heavily doped *p* regions, forming two back-to-back *pn* junctions.

Figure 3-1, the schematic representation of the *pnp* transistor, is also used to define the circuit polarities which we shall employ in the remainder of the text. The ready availability of complementary units, i.e., both *npn* and *pnp* transistors, creates some confusion when a current direction is arbitrarily assigned. In an effort to avoid any ambiguity, we shall choose the emitter arrow as the reference and assume that all current flows in the direction in which it points. If the emitter current flows in (as shown in Fig. 3-1*a* for the *pnp* unit), then the base and collector currents flow out of the device. Alternatively, in the *npn* transistor the emitter arrow is pointed in the opposite direction, out of the transistor, and all currents are reversed. Thus the transistor may be considered a simple node where

$$i_e = i_b + i_c$$

Transistor voltage polarities are defined in the direction to support the current flow.

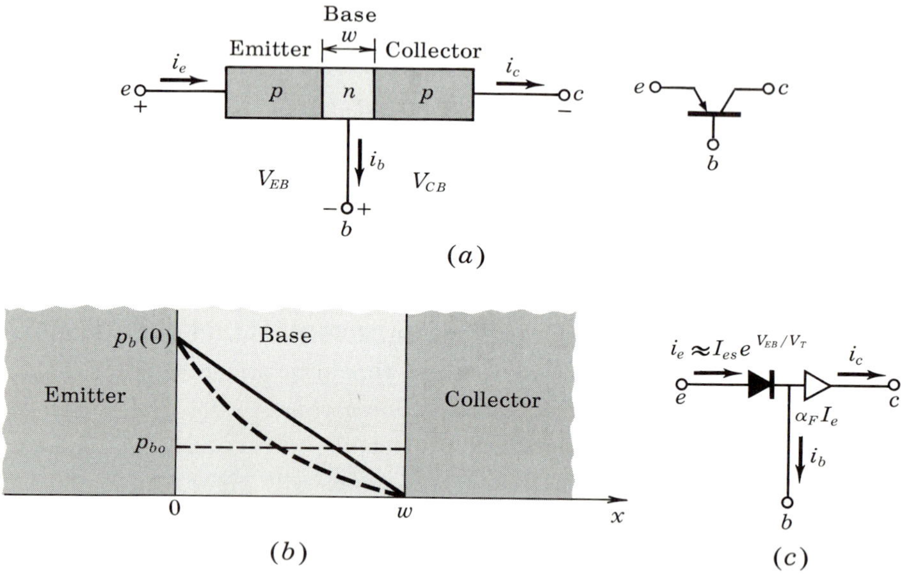

Fig. 3-1 (*a*) One-dimensional representation of *pnp* transistor and circuit schematic; (*b*) minority hole distribution in the base; (*c*) transistor active-region model.

The emitter-base junction is normally forward-biased. It is identical to the conducting diode of Sec. 2-2. A positive emitter-base voltage causes the diffusion of holes from the emitter into the base region. As in the diode, the emitter current increases exponentially with the junction voltage V_{EB}. However, unlike the diode, the minority carriers that diffuse across the junction are not all neutralized by recombination with the base electron current. In fact, the base is made extremely thin ($w \ll L_p$ and $w \ll L_n$) to minimize recombination. The minority carriers diffuse right through the base to the second junction. Because the collector-base junction is back-biased, the electric field is in the direction to collect all the holes that diffuse through the base. This junction acts as a back-biased diode.

At the forward-biased emitter edge of the base, the hole density is given by Eq. (2-17). At the back-biased collector edge, we have a depletion region, and the hole density goes to zero. So long as the two junction-voltage magnitudes are greater than $5V_T$, the boundary conditions of the minority hole distribution within the base become

$$p_b(0) = p_{bo} + p_{bo}e^{V_{EB}/V_T} \qquad p_b(w) = 0 \tag{3-1}$$

where p_{bo} is the normal rest-state base-hole density. p_{bo} is very small compared with the excess density $\Delta p_b(0)$ diffused from the emitter.

Initially, by assuming that there is no recombination, we find that the hole distribution will be linear, decreasing from $p_b(0)$ at the emitter edge to 0 at the collector edge (Fig. 3-1*b*). But we know from Eq. (2-11) that the diffusion current is proportional to the slope of the minority-carrier distribution. Thus the hole component of the emitter current is

$$I_{ep} = qD_pA\,\frac{p_b(0)}{w} \approx \frac{qD_pA}{w}\,p_{bo}e^{V_{EB}/V_T} \tag{3-2}$$

Because of the assumed constant slope of the excess-hole distribution, the hole-diffusion current remains constant throughout the complete base region up to the collector edge of the base, across the depletion zone, and into the collector.

In a similar manner, we can find the electron current diffusing from the base to the emitter:

$$I_{en} \approx q\,\frac{D_nA}{L_n}\,n_{eo}e^{V_{EB}/V_T} \tag{3-3}$$

n_{eo} is the normal rest-state emitter-electron density. But since we have a p^+n junction, $n_{eo} \ll p_{bo}$, and as a consequence, $I_{en} \ll I_{ep}$.

We shall now calculate the recombination current within the base to see if we have to modify the assumed hole distribution. The component of electron flow into the base which accounts for the recombination is proportional to the total excess-hole density within the base. From the assumed linear distribution and with $p_b(0) \gg p_{bo}$, we have

$$\Delta p_b(x) = p_b(0)\left(1 - \frac{x}{w}\right) \tag{3-4}$$

The base recombination current is

$$I_{bR} = qA \int_0^w \frac{\Delta p_b}{\tau_b}\,dx = \frac{qA}{\tau_b}\frac{p_b(0)w}{2} \tag{3-5}$$

where τ_b is the recombination-time constant in the base. By using Eq. (3-1), we can express Eq. (3-5) as

$$I_{bR} = \frac{qAp_{bo}w}{2\tau_b}e^{V_{EB}/V_T} \tag{3-6}$$

If the base recombination current is more than 1 or 2 percent of the total hole current I_{ep}, we would have to modify the assumed linear hole distribution and recalculate. But a high recombination current means that fewer carriers arrive at the collector and that we have a poor transistor.

The collector current due to the holes injected at the emitter junction and eventually collected at the collector is

$$I_c = I_{ep} - I_{bR} = TI_{ep} \tag{3-7}$$

where T, the *transport factor*, represents the efficiency of the hole transfer from the emitter to the collector. Solving Eq. (3-7) for T and substituting for I_{ep} [Eq. (3-2)] and I_{bR} [Eq. (3-6)], we obtain

$$T = 1 - \frac{w^2}{2\tau_b D_p} \tag{3-8}$$

where $L_p{}^2 = D_p\tau_b$. Equation (3-8) points up the importance of base width in transistor operation. Only when $w \ll L_p$ will the transport factor be even close to unity.

The total emitter current consists of the minority electron current diffusing from the base, given in Eq. (3-3), as well as the majority hole current which diffuses to the base. But only the holes eventually reach the collector. Thus the emitter efficiency is

$$\eta_e = \frac{I_{ep}}{I_e} = \frac{I_{ep}}{I_{ep} + I_{en}} \tag{3-9}$$

In a p^+n junction, $I_{ep} \gg I_{en}$ and the efficiency approaches unity.

We can now express the collector current in terms of the total emitter current. From Eqs. (3-7) and (3-9),

$$I_c = \eta_e T I_e = \alpha_F I_e \tag{3-10}$$

The factor $\alpha_F = \eta_e T$ is the forward-current-amplification factor of the transistor.

Equation (3-10) should also include the Miller multiplication factor of Eq. (2-24). The carriers reaching the collector are accelerated by the reverse junction voltage. Collisions with the lattice can cause carrier multiplication and breakdown. But since the Miller multiplication factor is unity everywhere except near the reverse breakdown voltage of the collector junction, M is usually not explicitly included in Eq. (3-10). But it is implicitly understood to be the operative term for voltages near V_B.

The collector current must also include the reverse current of the back-biased collector-base junction. We call this current I_{co}.

As an alternative to Eq. (3-10), the collector current can be expressed in terms of the base current. Recognizing that

$$I_e = I_c + I_b$$

and substituting $I_e = I_c/\alpha_F$, we obtain

$$I_c = \frac{\alpha_F}{1 - \alpha_F} I_b = \beta I_b \tag{3-11}$$

β is the base-current-amplification factor. For $\alpha_F = 0.99$, $\beta = 99$. We see from Eq. (3-11) that small base currents control large collector currents.

The idealized transistor of Fig. 3-1 is completely symmetrical: only the biasing determines which p region is the emitter and which is the collector. If the biasing is reversed, then the roles of emitter and collector are interchanged. An actual transistor is not symmetrical; base geometry and the doping levels are different at the emitter and the collector edges. The device will still operate backward but with a reverse current amplification α_R, which will be about one-half to one-tenth of α_F. β_R is usually less than unity.

Figure 3-2 shows the collector volt-ampere characteristics of a typical silicon transistor. The nonsymmetrical reverse operation is clearly illustrated.

The transported emitter current reaches the back-biased collector junction, where it is swept out in the same manner as any junction reverse current. We thus might interpret the collector volt-ampere characteristics of Fig. 3-2 as a parallel set of shifted back-bias diode curves where the base (or emitter) current is the shifting parameter.

3-2 The Ebers-Moll Model

We have seen that the transistor is essentially two back-to-back junctions with the base layer thin enough to permit the diffusion of carriers from the

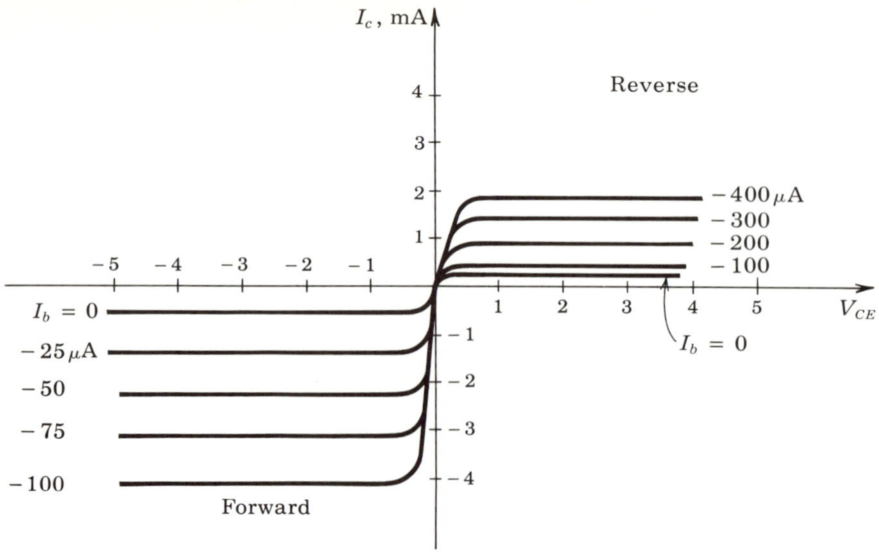

Fig. 3-2 Typical collector volt-ampere characteristics for a *pnp* silicon transistor.

forward-biased junction through the base region to the back-biased junction, where the minority carriers are collected. Ebers and Moll proposed the two-diode model of Fig. 3-3 to explain the transistor's behavior. Each diode has an associated controlled source representing the current due to the transported charge carriers. If we compare this model with the active-region model of Fig. 3-1, we see that the Ebers-Moll model essentially consists of two transistors with common terminals; one operates in the forward direction, and the other in the reverse direction.

The terminal currents of the model of Fig. 3-3*a* may be written in terms of the diode currents and the controlled sources:

$$I_e = I_F - \alpha_R I_R \tag{3-12a}$$

$$I_c = \alpha_F I_F - I_R \tag{3-12b}$$

$$I_b = I_e - I_c \tag{3-12c}$$

With the sign convention chosen to agree with the normal transistor-current flow, the forward transistor currents are all positive terms and the reverse transistor currents all have negative signs. In Eqs. (3-12*a*) and (3-12*b*), I_F and I_R are the emitter-base and collector-base diode currents:

$$I_F = I_{es}(e^{V_{EB}/V_T} - 1) \tag{3-13a}$$

$$I_R = I_{cs}(e^{V_{CB}/V_T} - 1) \tag{3-13b}$$

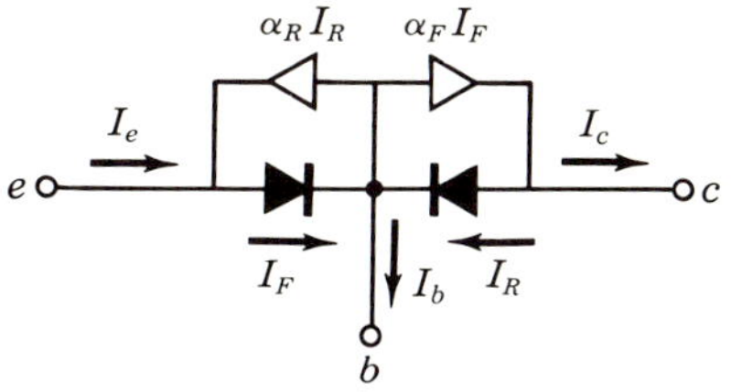

Fig. 3-3 Ebers-Moll two-diode model of a transistor.

By expressing α_F, α_R, and the reverse junction saturation currents in terms of the base geometry and the physical constants of the semiconductor, we can show that

$$\alpha_R I_{cs} = \alpha_F I_{es} \tag{3-14}$$

Even though symmetrical transistors exist, in general, $\alpha_R \neq \alpha_F$. The reverse transistor will have a much smaller current gain.

We occasionally find it more convenient to represent the model in terms of the terminal currents I_c and I_e instead of the diode currents I_F and I_R. Solving Eq. (3-12*a*) for I_F and substituting into Eq. (3-12*b*),

$$I_c = \alpha_F I_e - (1 - \alpha_F \alpha_R) I_R \tag{3-15}$$

and in a similar manner,

$$I_e = \alpha_R I_c + (1 - \alpha_F \alpha_R) I_F \tag{3-16}$$

We define a reverse collector current I_{co} measured with $I_e = 0$ and with the collector junction back-biased. Under these conditions $I_R = -I_{cs}$, and from Eq. (3-15),

$$I_{co} = (1 - \alpha_F \alpha_R) I_{cs} \tag{3-17}$$

By defining a reverse emitter current I_{eo} in an analogous manner, we obtain from Eq. (3-16)

$$I_{eo} = -(1 - \alpha_F \alpha_R) I_{es} \tag{3-18}$$

Equations (3-15) and (3-16), together with Eqs. (3-17) and (3-18), can be rewritten as

$$I_c = \alpha_F I_e - I_{co}(e^{V_{CB}/V_T} - 1) \tag{3-19a}$$

$$I_e = \alpha_R I_c - I_{eo}(e^{V_{EB}/V_T} - 1) \tag{3-19b}$$

The two-diode model of Fig. 3-3 indicates four possible transistor operating conditions; each diode can be either forward-biased or back-biased. Figure 3-4 shows the four possibilities together with the reduced models and the minority base-charge distribution. Just as we did in Chap. 2, in analyzing transistor circuits we can use various levels of approximation in treating the diodes of the transistor model.

The first possible device condition is where the external circuit voltages cut off the transistor (Fig. 3-4*a*). Both junctions are back-biased and exhibit charge-depletion regions. All the holes in the base have been swept out. We see that only the very small reverse leakage currents I_{co} and I_{eo} flow. Because of the exponential nature of the diode volt-ampere characteristic, it is difficult to designate an absolute cutoff voltage. It usually

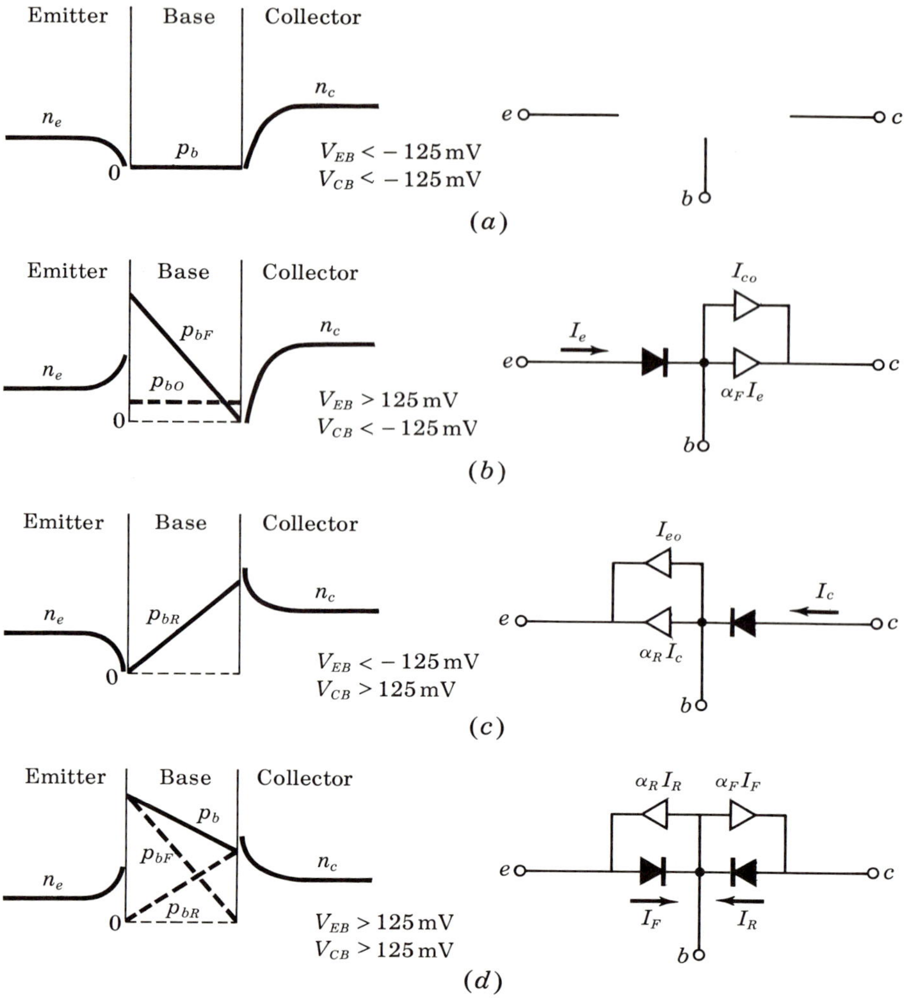

Fig. 3-4 Possible transistor conditions, base-charge distributions, and approximate models. (*a*) Cutoff operation—both junctions back-biased; (*b*) normal operation—emitter junction forward-biased, collector junction back-biased; (*c*) reverse transistor operation—emitter junction back-biased, collector junction forward-biased; (*d*) saturation operation—both junctions forward-biased.

suffices to say that if the junction voltage is somewhat below the assumed constant diode drop, that particular diode is cut off. In a silicon transistor, a junction voltage 120 mV below the normal level of 700 mV reduces the current flow to only 1/100 of the normal level. Certainly, in this case, the transistor may be approximated by the cutoff model. If we may assume an ideal diode characteristic, then cutoff corresponds to any junction voltage below the sharp knee of the curve.

The second condition, Fig. 3-4*b*, is the normal operating condition of the transistor. Here the emitter junction is forward-biased and the collector junction is back-biased. Since only the small reverse I_{co} flows through the collector-junction diode, we can omit the controlled source $\alpha_R I_c$ shunting the emitter diode, as well as the collector diode, and simply include I_{co} in the output. From Eq. (3-19*a*),

$$I_c = \alpha_F I_e + I_{co} \tag{3-20}$$

For many purposes, the model is further simplified by neglecting the extremely small I_{co}.

For the third case, Fig. 3-4*c*, we back bias the emitter junction and forward bias the collector junction. This situation is essentially the same as in the preceding case except that the transistor operates backward. The equation for I_e would be similar to Eq. (3-20), but the signs would differ. From Eq. (3-19*b*),

$$I_e = \alpha_R I_c - I_{eo}$$

With our sign convention, for a forward-biased collector I_c is a negative current and hence I_e is also negative. The total reverse transistor current flows out of the emitter.

Figure 3-4*d* shows the final device condition, where both junctions are forward-biased. We call this *saturation*. Both diodes are conducting and will have approximately equal drops. We might note that since the two diodes are connected back to back, the collector-to-emitter drop will be close to zero.

The hole density in the base is readily found from the diode equations. At the emitter edge, the excess-hole density is given by

$$\Delta p_b(0) \approx p_{bo} e^{V_{EB}/V_T} \tag{3-21}$$

At the collector edge,

$$\Delta p_b(w) \approx p_{bo} e^{V_{CB}/V_T} \tag{3-22}$$

The base-charge distribution under saturation conditions (solid line in Fig. 3-4*d*) can be constructed by the superposition of the two transistors of Fig. 3-4*b* and *c* (dashed lines). As a consequence of saturation, we observe that a relatively large amount of charge is stored in the base. This charge must be swept out before the saturated transistor can be turned off. In Sec. 3-7 we shall discuss in much greater detail the dynamic charge behavior during the turn-on and turn-off intervals.

3-3 Large-signal Models—Quiescent and Boundary Conditions

Macroscopic behavior of a transistor circuit extends from beyond cutoff well into the saturation region. The limits of each region as well as the quiescent operating point are found with the aid of the models of Fig. 3-4. Just as with the diode circuits, different levels of approximation can be used. In order to demonstrate different valid solutions, we shall analyze several circuits.

The first circuit shown in Fig. 3-5 is a generalized amplifier stage which will illustrate the problems of transistor biasing. Quiescent conditions are always most conveniently expressed in terms of the emitter current I_e. We can write the base-emitter loop equation directly from the circuit:

$$V_{\mathrm{Th}} = (1 - \alpha)I_e R_B - I_{\mathrm{co}} R_B + V_{BE} + I_e R_E \tag{3-23}$$

V_{Th} and R_B are the Thévenin equivalent terms seen looking back from the base. Solving Eq. (3-23) for I_e,

$$I_e = \frac{V_{\mathrm{Th}} + I_{\mathrm{co}} R_B - V_{BE}}{R_E + (1 - \alpha)R_B} \tag{3-24}$$

As the first-order approximation, we assume that V_{BE} is a constant. In small silicon transistors, I_{co} is completely negligible; the product $I_{\mathrm{co}}R_B$ will be on the order of magnitude of (or even much less than) a few micro-

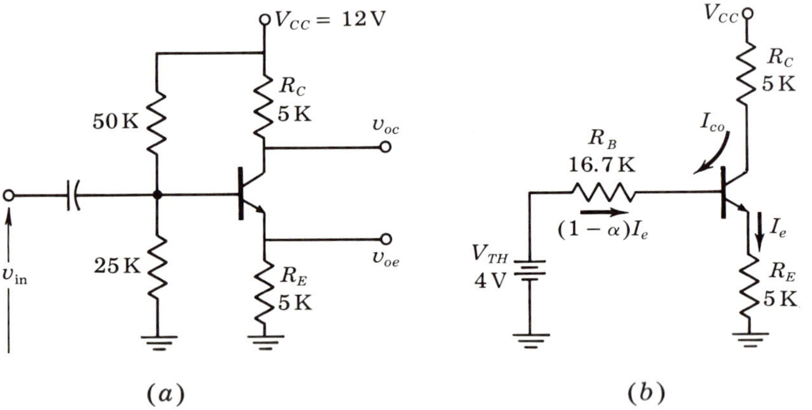

Fig. 3-5 (*a*) Generalized transistor amplifier; (*b*) Thévenin equivalent base circuit used for bias calculations. Transistor used has $\alpha_F = 0.98$ and $V_{BE} = 700$ mV at $I_e = 1$ mA.

volts. Substituting values into Eq. (3-24), the quiescent current becomes

$$I_e = \frac{4 - 0.7}{5\text{ K} + (1 - 0.98)16.7\text{ K}} = 0.62\text{ mA}$$

Thus $V_E = I_e R_E = 3.1$ V and $V_c = V_{CC} - \alpha I_e R_c = 8.95$ V. The drop across the transistor becomes $V_{CE} = 5.85$ V. We see that the transistor is properly biased within the active region.

We can draw the following conclusions from Eq. (3-24): First, because R_B is reduced by the factor $(1 - \alpha)$ in reflecting it into the emitter loop, we normally have $R_E \gg R_B(1 - \alpha)$. The emitter current is almost completely independent of α. Small changes in α will have practically no effect on the quiescent conditions but will have large effects on the base current. On the other hand, if the series base resistance R_B is significant, then the transistor α influences the quiescent point. We conclude that for most stable circuit operation, the transistor should always be biased as if the base were effectively grounded.

Second, we assumed a fixed V_{BE}. The value used corresponded to $I_e = 1$ mA. But I_e turned out to be only 0.62 mA. Solving the exponential diode characteristics, the actual V_{BE} is given by

$$V_{BE} = V_T \ln \frac{I_e}{I_{es}}$$

This equation can be used to find I_{es} for any transistor where we know the V_{BE} drop at a given I_e. The change in V_{BE} with I_e can be expressed by

$$\Delta V_{BE} = V_{BE_2} - V_{BE_1} = V_T \left(\ln \frac{I_{e_2}}{I_{es}} - \ln \frac{I_{e_1}}{I_{es}} \right)$$

$$= V_T \ln \frac{I_{e_2}}{I_{e_1}} = -26 \ln \frac{1}{0.62} = -12.4\text{ mV} \qquad (3\text{-}25)$$

Our assumption that V_{BE} remained a constant 700 mV was in error by only 2 percent. Certainly this is an acceptable approximation. It is well within the limits of the expected nominal V_{BE} resulting from variations in the diffusion process.

Equation (3-24), defining the quiescent current, includes two temperature-dependent terms. V_{BE} decreases with increasing temperature at the rate of about 2 mV/°C. I_{co} increases exponentially with increasing temperature, doubling approximately every 10°C. In general, we can represent the temperature dependence of I_{co} by

$$I_{co}(T_1) = I_{co}(T_0)e^{\Delta T/K}$$

where $I_{co}(T_0)$ is the reverse current measured at temperature T_0. K is the constant of the rate of increase, and the temperature change is $\Delta T = T_1 - T_0$. In silicon transistors, I_{co} doubles approximately every 10°C, requiring us to set $K = 14$°C.

The changes in both V_{BE} and I_{co} cause I_c to increase with increasing temperature:

$$I_c = \alpha I_e + I_{co} = \alpha \frac{V_{Th} + I_{co}R_B - V_{BE}}{R_E + (1-\alpha)R_B} + I_{co}$$

We can define a stability factor S relating changes in I_c to changes in I_{co}:

$$S = \frac{\partial I_c}{\partial I_{co}} = \frac{R_E + R_B}{R_E + (1-\alpha)R_B} \tag{3-26}$$

Since $S > 1$, any change in I_{co} will have a proportionally larger effect on the quiescent operating point. In order to minimize S, R_B should be made small compared with R_E. Consequently, for acceptable temperature stability, we also want the circuit to be biased as if it were in the grounded-base configuration.

As our second example, we shall construct the transfer function of the simple inverter stage of Fig. 3-6*a*. Initially, we shall assume a constant V_{BE}, and after drawing the linearized characteristic, we shall then see how the actual exponential response modifies the idealized curve. In the active or saturation region, the base-emitter diode conducts and the base current is given by

$$I_b = \frac{v_{in} - V_{BE}}{R_B} > 0 \tag{3-27}$$

The inequality of Eq. (3-27) defines the upper limit of the cutoff region and the lower limit of the active region. Due to our assumed piecewise-linear approximation, the transistor is cut off whenever $v_{in} < 700$ mV. For

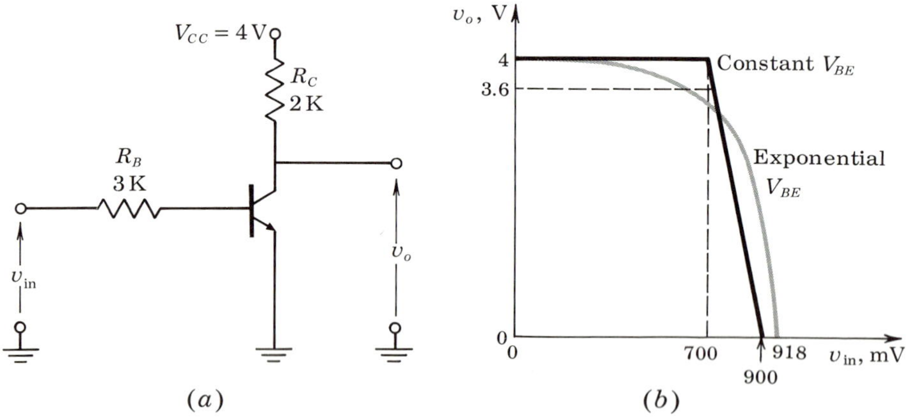

Fig. 3-6 (*a*) Simple inverter stage; (*b*) transfer functions using both linear approximation and the exponential diode characteristic ($\alpha = 0.97$ and $V_{BE} = 700$ mV at 1 mA).

$v_{\text{in}} > 700$ mV, the base current increases linearly with v_{in}. The output voltage decreases from the cutoff voltage V_{CC} as

$$v_o = V_{CC} - \beta I_b R_C \tag{3-28}$$

Substituting Eq. (3-27) and the circuit values into Eq. (3-28), we obtain the linearized transfer equation

$$v_o = 4 - 20(v_{\text{in}} - 0.7)$$

Figure 3-6*b* shows the plot of v_o vs. v_{in}. The slope of these characteristics is the linearized gain.

In this ideal case, we assume saturation to occur when both junctions are forward-biased by 700 mV. Since the junctions are back to back, the junction drops cancel and saturation corresponds to $V_{CE} \approx 0$. Setting $v_o = 0$ in Eq. (3-28), for saturation,

$$I_b \geq \frac{V_{CC}}{\beta R_C} \tag{3-29}$$

and, from Eq. (3-27), we find that an input voltage of 900 mV drives the inverter to the edge of saturation.

For a more accurate plot of the transfer function, account must be taken of the exponential nature of the emitter-base junction characteristics. An exact solution requires the solution of Eq. (3-27) after the substitution of V_{BE} given by the diode equation. The result is

$$I_b = (1 - \alpha)I_e = \frac{v_{\text{in}} - V_T \ln (I_e/I_{es})}{R_B}$$

But this is a transcendental equation, which is not easy to solve explicitly. Of course, for any given v_{in}, we can readily find the emitter current.

If we work backward, we can, for any given v_o, define the required I_e and solve for the corresponding V_{BE}. v_{in} is then given by summing V_{BE} and the $I_b R_B$ drop. For example, at the edge of saturation, $V_{CE,\text{sat}} \approx 0$ and

$$I_{c,\text{sat}} = \beta I_{b,\text{sat}} = \frac{V_{CC}}{R_C} = 2 \text{ mA}$$

Since we know that $V_{BE} = 700$ mV at $I_e = 1$ mA, for

$$I_e = \frac{I_{c,\text{sat}}}{\alpha} = 2.06 \text{ mA}$$

we find from Eq. (3-25) that $\Delta V_{BE} = 18$ mV and $V_{BE,\text{sat}} = 718$ mV. The corresponding driving-point voltage in the base circuit is given by

$$v_{\text{in,sat}} = I_{b,\text{sat}} R_B + V_{BE,\text{sat}} = 200 + 718 = 918 \text{ mV}$$

This is quite close to the value found from the linearized approximation.

For a second point on the transfer characteristic, we arbitrarily choose a value near cutoff which can be used to define the width of the active region. $I_c = 0.1I_{c,\text{sat}}$ is a convenient value, which sets $v_o = 0.9V_{CC}$ close to the cutoff voltage V_{CC}. Since the ratio of currents is 10:1, V_{BE} drops by 60 mV from the saturation voltage of 718 mV. I_b, proportional to I_e, will be $0.1I_{b,\text{sat}}$. Thus at the 10 percent point,

$$v_{\text{in},0} = 0.1I_{b,\text{sat}}R_B + (V_{BE,\text{sat}} - 60 \text{ mV}) = 678 \text{ mV}$$

From the linear approximation, we find that $v_{\text{in},0} = 720$ mV (almost 10 percent too high) corresponds to the same collector current of $0.1I_{c,\text{sat}}$.

The actual curve superimposed on the linearized transfer function in Fig. 3-6*b* clearly indicates the limitations of our approximations. We can see that the active-region transfer characteristic reflects the exponential nature of the junction characteristics. Its bounds are somewhat wider than found from an assumed constant V_{BE}. The slope of the actual characteristics and hence the incremental gain increase with increasing collector current; the gain is a minimum near cutoff and a maximum at the edge of saturation.

A very large R_B will cause the I_bR_B drop in the base circuit to predominate. The change in voltage across R_B as the collector current is driven from $0.1I_{c,\text{sat}}$ to $I_{c,\text{sat}}$ greatly overshadows the exponential change of 60 mV of V_{BE}. Hence the transfer function will be linearized at a sacrifice of gain and by broadening the active region. A large R_B effectively converts the input driving voltage to a current source, and we know that I_c is a linear function of I_b; V_{BE} just changes to accommodate the changing drive current. Current amplification remains equal to β.

Saturation-region operation

As a final application of the Ebers-Moll model, we shall investigate the voltage drops across a saturated transistor. Consider the circuit of Fig. 3-7*a*, where the biasing is adjusted to solidly saturate the transistor. The external base and collector resistors R_B and R_C limit the current flow. Initially assuming that $V_{BE} = 700$ mV and, because of the two bucking junction drops, $V_{CE} \approx 0$, we have

$$I_c = \frac{V_{CC} - V_{CE}}{R_C} = \frac{10}{2 \text{ K}} = 5 \text{ mA}$$

$$I_b = \frac{V_{CC} - V_{BE}}{R_B} = \frac{10 - 0.7}{2 \text{ K}} = 4.65 \text{ mA}$$

and $I_e = I_c + I_b = 9.65$ mA. The Ebers-Moll model expresses I_c and I_e as [Eqs. (3-12*a*) and (3-12*b*)]

$$I_e = I_F - \alpha_R I_R$$

$$I_c = \alpha_F I_F - I_R$$

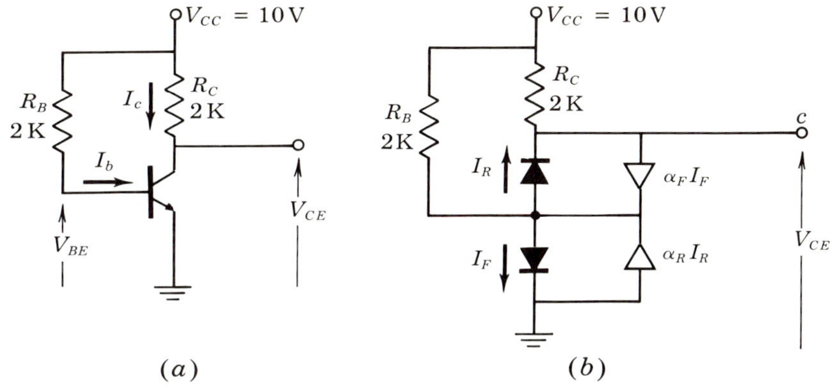

Fig. 3-7 *npn* transistor circuit and model used for calculation of saturation ($\alpha_F = 0.98$, $\alpha_R = 0.1$, and $I_{es} = 2 \times 10^{-15}$ A).

Solving this set of equations, I_F and I_R can be expressed in terms of the externally limited I_e and I_c by

$$I_F = \frac{I_e - \alpha_R I_c}{1 - \alpha_F \alpha_R} \qquad (3\text{-}30a)$$

$$I_R = \frac{\alpha_F I_e - I_c}{1 - \alpha_F \alpha_R} \qquad (3\text{-}30b)$$

By substituting I_e and I_c and with $\alpha_F = 0.98$ and $\alpha_R = 0.1$, Eqs. (3-30*a*) and (3-30*b*) yield the equivalent diode currents

$$I_F = 10.1 \text{ mA}$$

$$I_R = 4.95 \text{ mA}$$

From the diode equations we obtain the drops

$$V_{BE} = V_T \ln \frac{I_F}{I_{es}} \qquad (3\text{-}31a)$$

$$V_{BC} = V_T \ln \frac{I_R}{I_{cs}} = V_T \ln \frac{\alpha_R I_R}{\alpha_F I_{es}} \qquad (3\text{-}31b)$$

To obtain the second form of Eq. (3-31*b*), we made use of the identity $\alpha_F I_{es} = \alpha_R I_{cs}$. With $I_{es} = 2 \times 10^{-15}$ A, Eqs. (3-31*a*) and (3-31*b*) yield

$$V_{BE} = 760 \text{ mV}$$

$$V_{BC} = 683 \text{ mV}$$

The saturated collector-to-emitter voltage drop is only 77 mV. Both V_{BE} and V_{CE} are close enough to our assumed saturation voltages to accept the initially calculated I_e and I_c as the correct solutions. If either voltage turned out to differ widely from the assumed $V_{BE} = 700$ mV or $V_{CE} = 0$, we would have to recalculate I_b and I_c using the first trial solution for the diode drops. By calculating the second-order values of I_b and I_c, we find the new junction drops. Again we must check our assumed starting points. The solution for V_{BE} and V_{CE} will converge very rapidly.

3-4 Small-signal Models

Transistor small-signal active-region models can be derived directly from the large-signal Ebers-Moll model of Fig. 3-3 by evaluating the incremental response. In the case of the common-base connection, the emitter input is simply a forward-conducting diode. We can immediately write

$$I_e \approx I_{es}e^{V_{EB}/V_T}$$

As we did with the diode, by differentiating I_e with respect to V_{EB} and taking the reciprocal we obtain the incremental junction resistance

$$r_e = \frac{dV_{EB}}{dI_e} = \frac{V_T}{I_e} = \frac{26}{I_e \text{ in mA}} \qquad \text{ohms} \qquad (3\text{-}32)$$

The forward-current-transfer factor α_F depends primarily on the geometry. Except at very low currents, we can assume that it remains a constant.

Figure 3-8*a* shows the simplest model for active-region operation of an *npn* transistor. But this model is not quite correct. Our second approximation, Fig. 3-8*b*, includes two additional resistors. The base padding resistance r_b represents the ohmic portion of the path from the base connection on the surface of the silicon chip to the active region of the base between the emitter and collector. Because the base region is quite narrow, r_b will be about 200 Ω in a typical low-current transistor. r_b should also have been included in the large-signal model. Usually, however, the external bias resistance is so very much larger than r_b that it will swamp the internal base resistance.

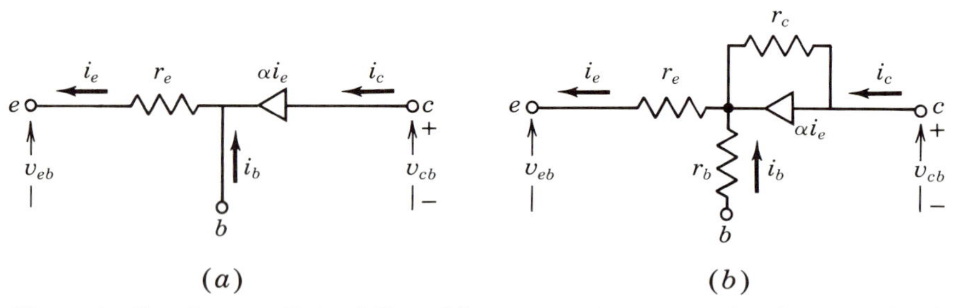

Fig. 3-8 Transistor small-signal T models—common-base connection (*npn* transistor).

In Sec. 2-11 we saw that the width of the back-biased collector-base depletion region increases with an increasing back-bias voltage. Any broadening of the depletion region reduces the effective base width, and as a consequence, α increases. We can account for the increased collector current produced by the base-width modulation by shunting the forward-controlled source with the resistance r_c. In a typical silicon transistor, r_c will be very large. Since several megohms can be expected in a small transistor, r_c is often neglected. In germanium transistors, r_c will be an order of magnitude smaller and its effect may have to be accounted for during circuit calculations.

It is often convenient to define the small-signal-model response in terms of the independent and dependent terminal currents and voltages. The model's h parameters become the coefficients of the independent driving-point signals. For the grounded-base configuration, one possible set of equations is

$$v_{eb} = -h_{ib}i_e + h_{rb}v_{cb} \tag{3-33a}$$

$$i_c = h_{fb}i_e + h_{ob}v_{cb} \tag{3-33b}$$

The second subscript of the h parameters indicates the reference terminal—in this case, the base. Figure 3-9 shows the model defined by Eqs. (3-33a) and (3-33b).

Because the T model of Fig. 3-8 and the h-parameter model of Fig. 3-9 are two different ways of representing the behavior of the same device, they must yield equivalent solutions under identical terminal conditions. We can therefore express the h parameters in terms of the T-model elements. For a short-circuited output, that is, for $v_{cb} = 0$, from Eqs. (3-33a) and (3-33b) we identify the short-circuit output constants as

$$h_{ib} \equiv \left.\frac{v_{eb}}{-i_e}\right|_{v_{cb}=0} = \text{input impedance}$$

$$h_{fb} \equiv \left.\frac{i_c}{i_e}\right|_{v_{cb}=0} = \text{forward-current amplification}$$

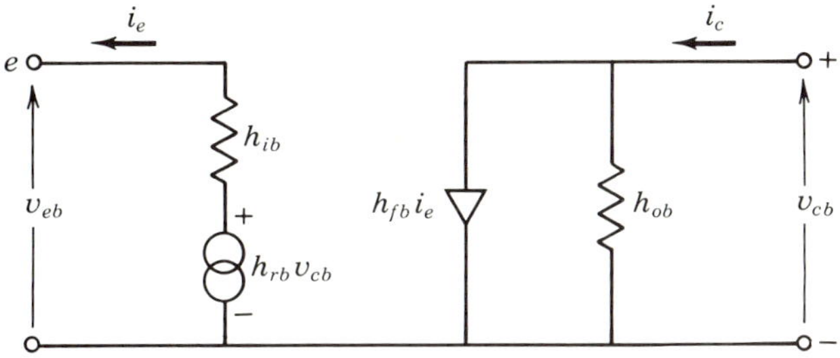

Fig. 3-9 Transistor h-parameter model—grounded-base connection.

These terms appear in the reduced model of Fig. 3-10*a*. From Fig. 3-8*b*, under the same condition of a short-circuited output and by recognizing that $r_c \gg r_b$, we obtain the reduced T model of Fig. 3-10*a*. For this reduced model, v_{eb} is given by

$$v_{eb} = -r_e i_e - r_b(1 - \alpha)i_e \tag{3-34}$$

By associating the $(1 - \alpha)$ term with r_b instead of with i_e, we can identify the effective input impedance as

$$Z_{\text{in}} = \frac{v_{eb}}{-i_e} = r_e + (1 - \alpha)r_b = h_{ib} \tag{3-35}$$

We note that the controlled current source αi_e, which shunts r_b, by reducing the current flow in the base resistance effectively reduces the branch resistance by the factor $1 - \alpha$.

From the same model, we also have $i_c = \alpha_f i_e$ or

$$\alpha = \alpha_f = h_{fb} \tag{3-36}$$

The second set of identities is found by setting $i_e = 0$, which corresponds to an open-circuited emitter. From Eqs. (3-33*a*) and (3-33*b*), we obtain the open-circuit input constants

$$h_{rb} \equiv \left.\frac{v_{eb}}{v_{cb}}\right|_{i_e=0} = \text{reverse-voltage transfer ratio}$$

$$h_{ob} \equiv \left.\frac{i_c}{v_{cb}}\right|_{i_e=0} = \text{output admittance}$$

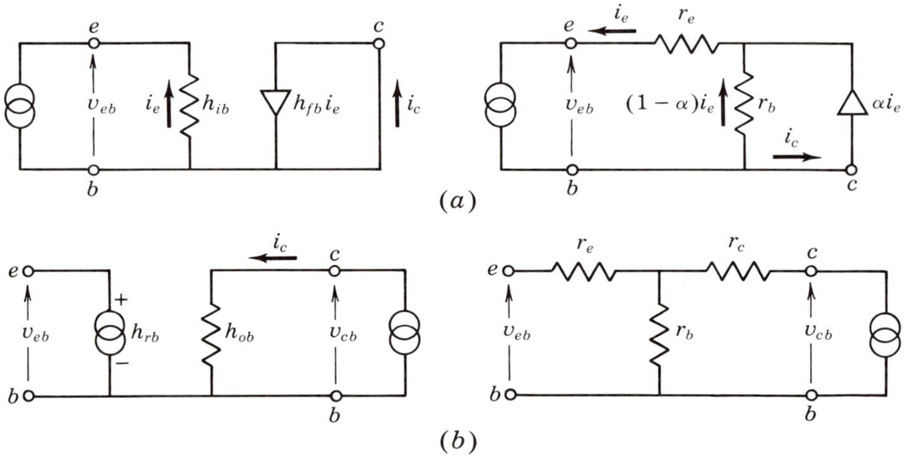

Fig. 3-10 Equivalent *h*-parameter and T models. (*a*) Short-circuited output ($v_{cb} = 0$); (*b*) open-circuited input ($i_e = 0$).

With an open-circuited input, both the h-parameter model (Fig. 3-9) and the T model (Fig. 3-8b) reduce to Fig. 3-10b. We obtain the following identities:

$$h_{ob} \equiv \frac{1}{r_c + r_b} \approx \frac{1}{r_c} \tag{3-37a}$$

$$h_{rb} \equiv \frac{r_b}{r_b + r_c} \approx \frac{r_b}{r_c} \tag{3-37b}$$

Common-emitter models

Many transistor circuits exist where the input current is injected into the base and where the emitter replaces the base as the reference terminal. Consequently, it is of interest to construct alternative models in which the base current is the controlling term.

This transformation follows directly from the model of Fig. 3-8b. We might first note that

$$i_b = i_e - i_c = (1 - \alpha)i_e \tag{3-38}$$

Solving Eq. (3-38) for i_e and multiplying the results by α yield

$$\alpha i_e = \frac{\alpha}{1 - \alpha} i_b = \beta i_b \tag{3-39}$$

where $\beta \approx h_{fe}$ is identified as the base-to-collector current-amplification factor. Since α is close to unity, β normally ranges between 30 and 100.

To complete the transformation from the emitter-controlled model of Fig. 3-8b to the base-controlled models of Fig. 3-11, we must ensure that all the remaining terminal characteristics of the two models are identical. In both, the emitter-to-base voltage drop must be the same for a given emitter current. Because of the resultant model simplification, the short-circuit condition, $v_{ce} = 0$, is a very convenient starting point. By substituting the value of i_e from Eq. (3-38) into Eq. (3-34), we obtain the expression for v_{be} in terms of i_b:

$$v_{be} = r_b i_b + \frac{r_e}{1 - \alpha} i_b = [r_b + r_e(\beta + 1)]i_b = h_{ie} i_b \tag{3-40}$$

In this case, we see that the controlled source increases the current flow through r_e and thus multiplies the effective resistance.

The output conductance is computed by setting $i_b = 0$ in our models. Equating the collector currents from Fig. 3-11a and b, we find that

$$i_c = h_{oe} v_{ce} = \frac{v_{ce}}{r_d} \tag{3-41}$$

But with $i_b = 0$ in the model of Fig. 3-8b, $i_e = i_c$ and

$$v_{ce} = r_e i_c + (1 - \alpha) r_c i_c \approx (1 - \alpha) r_c i_c$$

Thus by comparing the coefficients of i_c, we can write the identity

$$r_d = (1 - \alpha)r_c \tag{3-42}$$

r_d is only a one-twentieth to a one-hundredth part of r_c.

The one remaining h parameter, h_{re}, is found by setting $i_b = 0$. From Fig. 3-11*a*,

$$h_{re} \equiv \left.\frac{v_{be}}{v_{ce}}\right|_{i_b=0} = \frac{r_e}{r_d + r_e} \approx \frac{r_e}{r_d} \tag{3-43}$$

Just as with h_{rb}, h_{re} is small enough to be neglected in the usual circuit calculations.

An alternative form of the h-parameter model relates the collector current to the base-to-emitter voltage. We call this a hybrid π model (Fig. 3-12). The controlled source βi_b is replaced by the voltage-controlled current source $g_m v_{be}$. Since $v_{be} = h_{ie} i_b$, substituting for v_{be} in the voltage-controlled source and equating coefficients,

$$g_m = \frac{\beta}{h_{ie}} \approx \frac{1}{r_e} \tag{3-44}$$

The approximation holds for a high-β transistor where $\beta + 1 \approx \beta$.

The model of Fig. 3-12 also includes the device capacitances which limit the high-frequency response of a transistor circuit. C_{bc} is the depletion-region capacitance of the back-biased base-collector junction (see Sec. 2-11).

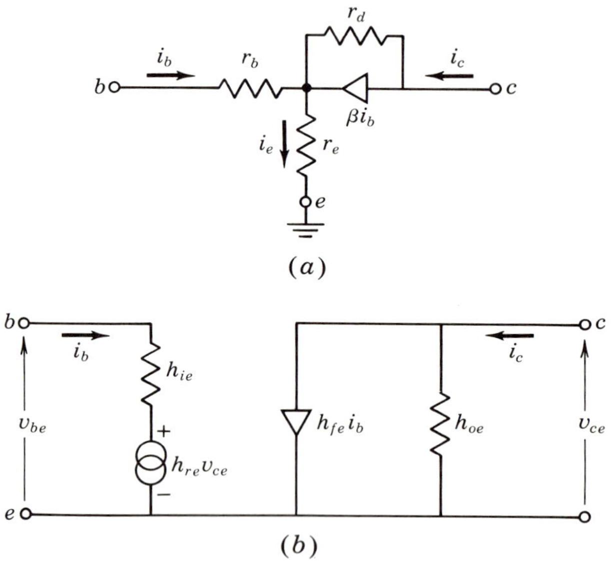

Fig. 3-11 (*a*) Common-emitter T model for *npn* transistor; (*b*) equivalent *h*-parameter model for common-emitter configuration.

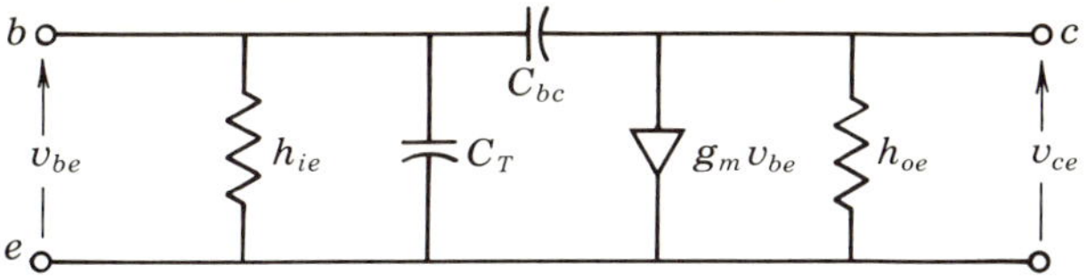

Fig. 3-12 Hybrid π model—common-emitter configuration. This approximate model, which shows the transistor capacitors, is used for high-frequency calculations.

C_T is the capacitance associated with the storage of minority charge carriers in the base region. It can be found from the change of the base-charge storage with voltage. Starting with the charge distribution given in Fig. 3-4*b* and solving for C_T, we obtain

$$C_T = \frac{\partial Q_B}{\partial V_{BE}} = \frac{w^2 \tau_b}{2V_T L_p^2} I_e \tag{3-45}$$

Because C_T varies directly with the operating point while r_e varies inversely with I_e [Eq. (3-32)], the $r_e C_T$ product remains independent of I_e.

3-5 Circuit Applications—Small-signal Calculations

As an example of how the small-signal models are applied in the analysis of specific circuits, we shall analyze the behavior of the circuit of Fig. 3-13*a*. The quiescent operating point of this generalized amplifier was found in the beginning of Sec. 3-3. We shall be concerned only with the midband response, where the coupling capacitors can be considered short circuits and the shunt capacitors are open circuits. Figure 3-13*b* shows the complete T model.

In Fig. 3-13*c*, we separate the input from the collector output circuits. The controlled current source is first converted into two equal-series-current generators. The center of these generators is connected to the ground node. Since we have equal currents flowing into and out of the ground node, the transformation shown leaves the circuit operation unchanged.

Figure 3-13*c* and *d* illustrate the further reduction of the input circuit. The βi_c controlled source is a function of the common-node (b') input current i_b. The two current components sum, and $(\beta + 1)i_b$ flows through the resistors $r_e + R_E$, producing a branch voltage drop of

$$v_b' = (r_e + R_E)(\beta + 1)i_b$$

If, as before, the $(\beta + 1)$ factor is interpreted as being part of the resistor term, then the same drop will be produced by the input current i_b flowing through the equivalent resistance $(r_e + R_E)(\beta + 1)$. When looking into the base, the controlled source effectively multiplies any resistance in the

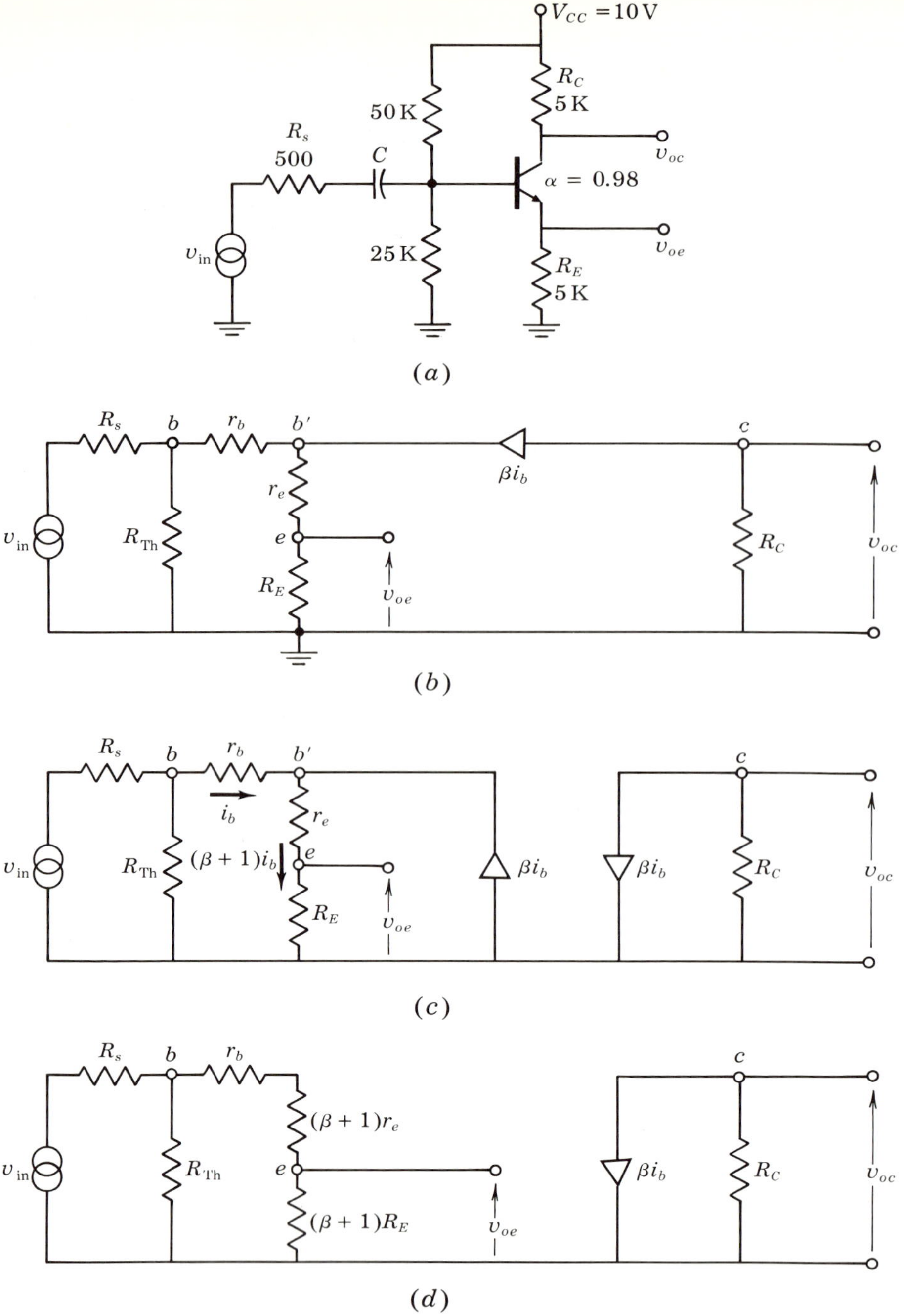

Fig. 3-13 (*a*) Generalized transistor amplifier; (*b*) equivalent small-signal T model; (*c*) illustrating the separation into separate input and output circuits; (*d*) reduction of the base circuit to an equivalent resistor network.

emitter lead by $\beta + 1$. The high input impedance is

$$Z_{\text{in}_b} = r_b + (r_e + R_E)(\beta + 1) = h_{ie} + (\beta + 1)R_E \tag{3-46}$$

At the operating point of $I_e = 0.62$ mA, Eq. (3-32) gives $r_e = 42\ \Omega$. With $r_b = 100\ \Omega$ and $\beta = 49$, we find that $h_{ie} = 2{,}200\ \Omega$. The total input impedance looking into the base is 252 K. This is shunted by the bias network Thévenin resistance of 16.7 K, yielding the actual input impedance of 15.6 K.

From the reduced circuit of Fig. 3-13d, we can find i_b and v_{oe}. At the base

$$v_b = \frac{Z_{\text{in}} \parallel R_{\text{Th}}}{R_s + (Z_{\text{in}} \parallel R_{\text{Th}})} v_{\text{in}}$$

with $R_s = 500\ \Omega$, $v_b = 0.97v_{\text{in}}$. Knowing v_b, i_b is given by

$$i_b = \frac{v_b}{h_{ie} + (\beta + 1)R_E} \tag{3-47}$$

In Fig. 3-13d, we see that the output voltage v_{oe} is found from the voltage divider composed of h_{ie} and $(\beta + 1)R_E$:

$$v_{oe} = \frac{(\beta + 1)R_E}{(\beta + 1)R_E + h_{ie}} v_b \tag{3-48}$$

In this circuit, since $v_b = 0.97v_{\text{in}}$, $v_{oe} = 0.964v_{\text{in}}$. For a high-$\beta$ transistor, where $R_E(\beta + 1) \gg h_{ie}$, the positive voltage gain to the emitter will be close to unity. As we see from this circuit, the primary reason for the less-than-unity voltage gain is the drop across the input voltage divider (from v_{in} to v_b) due to the signal-source impedance R_s. If $R_s = 0$, then $v_b = v_{\text{in}}$ and $v_{oe} = 0.992v_{\text{in}}$.

The voltage gain from the base to the collector is found by substituting i_b from Eq. (3-47) into the controlled source in the collector circuit (Fig. 3-13d):

$$A_{V_c} = \frac{v_{oc}}{v_b} = -\frac{\beta R_C}{h_{ie} + R_E(\beta + 1)} \approx -\frac{R_C}{R_E} \tag{3-49}$$

The approximation holds for a high-β transistor. Equation (3-49) indicates that in order to obtain a high voltage gain, R_E should be made very small. In a capacitively coupled amplifier, R_E can be effectively short-circuited by bypassing it with a large capacitor. By doing this, we also reduce the input base impedance from that given in Eq. (3-46) to h_{ie} and increase the voltage-divider attenuation in the base circuit.

Output impedance

The load at the collector sees the controlled source βi_b shunted by the base-width-modulation resistance r_d. So long as the transistor remains in its active region, the controlled source is independent of anything happening at the collector. Consequently, the output impedance will be r_d.

A different situation exists at the emitter. The controlled source in the reduced model of Fig. 3-13*c* is not an independent generator; it depends on the emitter current $[i_b = (1 - \alpha)i_e]$. Figure 3-14*a* points up this dependency by including the controlled source as αi_e. The current through $R_s \parallel R_{\mathrm{Th}}$ is only $(1 - \alpha)$ times as large as the emitter terminal current. As before, if we associate the $(1 - \alpha)$ term with the resistors, the controlled source can be removed without changing the circuit response. Figure 3-14*b* shows the reduced circuit. By inspection, we see that the output impedance looking into the emitter is

$$Z_{oe} = r_e + [(R_s \parallel R_{\mathrm{Th}}) + r_b](1 - \alpha) \tag{3-50}$$

Even though Z_{oe} is a function of the source impedance, the multiplication by $(1 - \alpha)$ means that the output impedance will be quite small. In the circuit of Fig. 3-13*a*, Z_{oe} is only 52 Ω.

3-6 Composite Devices

In an effort to approach an extremely high β, and at the same time reduce the dependency of the circuit behavior on the individual transistor, Darlington proposed using a compound arrangement in place of a single transistor. Figure 3-15 illustrates the suggested configuration; for our convenience

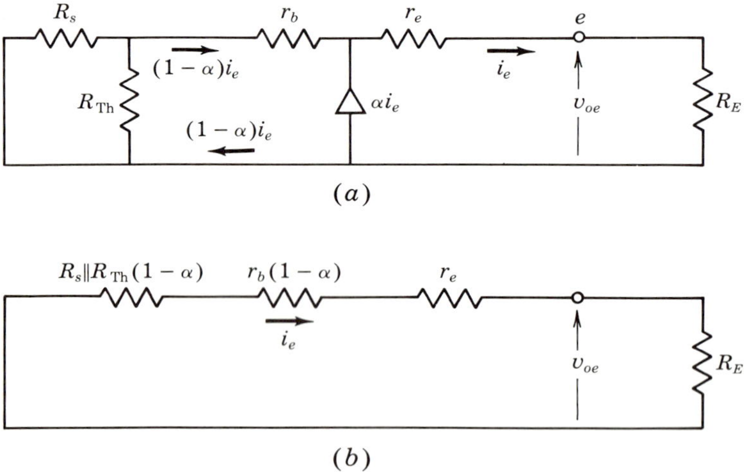

Fig. 3-14 Illustrating the reduction of the circuit by removal of the controlled source when calculating the output impedance seen at the emitter.

in the later discussion, the approximate current flow into and out of each transistor is indicated. The arrows indicate the actual direction of the current in the *pnp* transistors shown.

The small base current of the primary transistor Q_1, $i_{e1}(1 - \alpha_1)$, is amplified by the correction transistor Q_2 and added to the output of Q_1. Thus the composite collector current is composed of two terms:

$$i_c = i_{c1} + i_{c2} = a_1 i_{e1} + (1 - \alpha_1)\alpha_2 i_{e1}$$

The first term is the normal transmission through any transistor, and the second term represents a small additional correction current flow from Q_2 that raises the overall output to a value much closer to the input driving current. So long as the individual transistor α's are less than unity, the total output current will always be less than i_{e1}. The two transistors, taken together, can be said to act as a single compound device having

$$\begin{aligned} \alpha_c &= \alpha_1 + (1 - \alpha_1)\alpha_2 \\ &= 1 - (1 - \alpha_1)(1 - \alpha_2) \end{aligned} \tag{3-51}$$

For example, when $\alpha_1 = 0.98$ and $\alpha_2 = 0.97$, the composite current gain is 0.9994.

Stability of the composite element with respect to the individual α's is also much better than that of a single transistor. If only small variations are considered, then

$$\frac{d\alpha_c}{d\alpha_1} = 1 - \alpha_2 \qquad \frac{d\alpha_c}{d\alpha_2} = 1 - \alpha_1 \tag{3-52}$$

Using the figures given above, a 1.0 percent change in α_1 causes only a 0.03 percent change in the overall current gain α_c. Equation (3-52) is not valid with respect to large variations of α_1 or α_2, and in this case, the effect on α_c would have to be found by evaluating Eq. (3-51) over the expected values of α_1 and/or α_2.

The composite transistor is not an unmixed blessing; further examination of the circuit of Fig. 3-15 exposes serious drawbacks, which severely limit

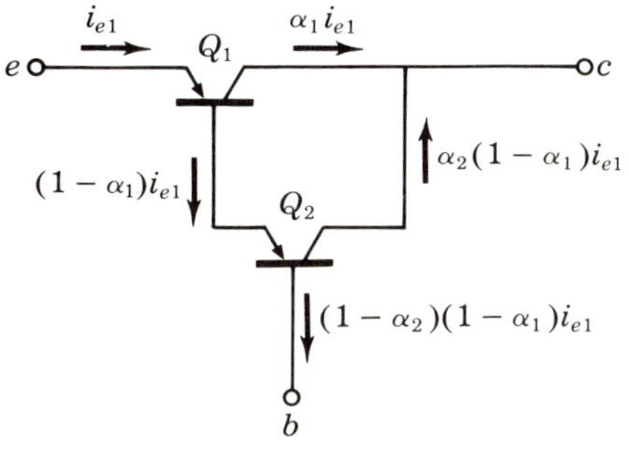

Fig. 3-15 Composite transistor circuit.

the possible applications. For example, the temperature-dependent reverse collector current I_{co} of Q_2 still flows unchanged at the base of the composite unit. Since β_c is very large, only a very small base current is needed to control the complete collector-current flow. But I_{co} is of the same order of magnitude, making the problem of temperature stabilization extremely difficult.

Furthermore, it can be shown that the input impedance at the composite base may be approximated by

$$h_{ib} \approx [r_{e1}(1 + \beta_1) + r_{b1}](1 + \beta_2) \tag{3-53}$$

β_1 and β_2 are very large, and therefore the base input will no longer approximate the ideal short circuit that is desirable in a current-controlled device. But for just this reason, this circuit configuration is more convenient for use as a voltage amplifier; the loading of the external source decreases with the increase in β_c. If ever a higher impedance were needed, additional padding resistance would be inserted in series with the emitter. It would be multiplied by the product $(1 + \beta_1)(1 + \beta_2)$ as it is reflected into the base circuit.

3-7 Transistor Dynamic Behavior

We now need a model that will describe the dynamic behavior of the transistor. Since we know that the total current flow can be expressed in terms of the minority-carrier distribution in the base, our model must describe the time rate of change of charge. In a *pnp* transistor, the slope of the excess-hole distribution at the two edges of the base determines the steady-state emitter and collector currents while the base current is proportional to the total stored excess minority carriers. Figure 3-16*a* shows the most general distribution, where the total excess base charge Q_B is resolved into the two components Q_c and Q_s. Q_c is the normal active region charge up to the edge of saturation, and Q_s is the excess charge stored in the base after the transistor saturates.

I_b consists of two terms; the first is due to the time rate of charge buildup within the base, and the second is due to recombination:

$$I_b = \frac{\partial Q_B}{\partial t} + \frac{Q_B}{\tau_B} \tag{3-54}$$

The nature of Eq. (3-54) indicates that the solution of base-charge storage is of exponential form, having a time constant τ_B and subject to initial and steady-state conditions established by the external circuit.

Before solving for the dynamic response, let us first see what information we can obtain from the steady-state solution of Eq. (3-54). In the special

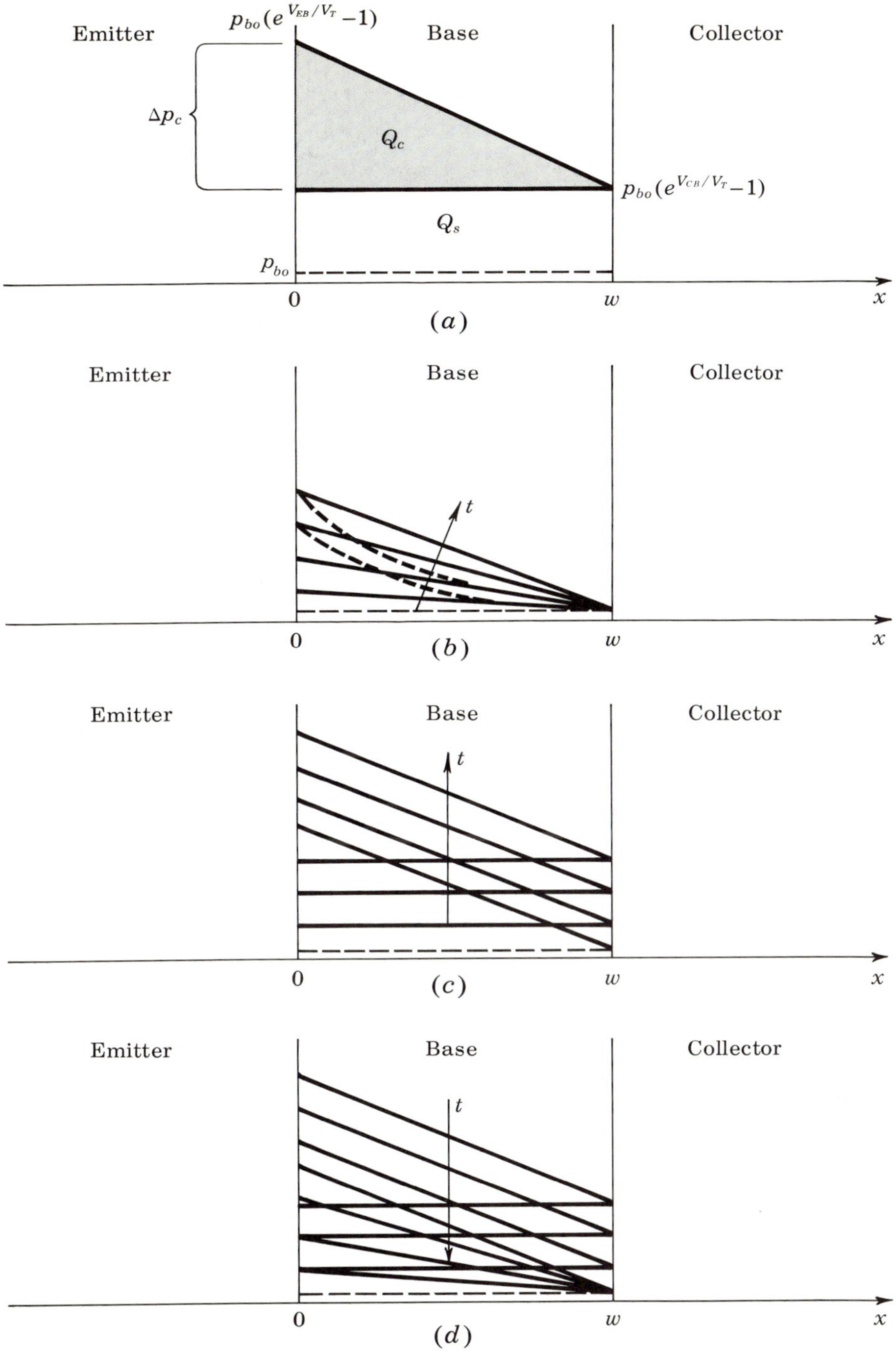

Fig. 3-16 Base-minority-charge distribution. (*a*) Steady-state charge distribution when both junctions are forward-biased; (*b*) charge buildup in the active region; (*c*) charge buildup during saturation; (*d*) approximate charge decay during turn-off interval.

case of no saturation (collector back-biased), no excess charge is stored, that is, $Q_s = 0$, and $Q_B = Q_c$. Equation (3-54) is reduced to

$$I_{b_{ss}} = \frac{Q_c}{\tau_B} \tag{3-55}$$

The steady-state collector current is proportional to the slope of the hole distribution. From Eq. (2-11), we have

$$I_c = -qD_pA \left.\frac{\partial \Delta p}{\partial x}\right|_{x=w} = qD_pA \frac{\Delta p_c}{w} \tag{3-56}$$

But because of the triangular geometry, the slope and hence I_c are also proportional to the charge Q_c contained in the shaded triangular area of Fig. 3-16*a*. We can write

$$I_c = \frac{Q_c}{\tau_c} = \frac{Aq\,\Delta p_c\,w}{2\tau_c} \tag{3-57}$$

where τ_c is the constant of proportionality. Equating (3-56) to (3-57) and solving for τ_c, we obtain the collector-current time constant

$$\tau_c = \frac{w^2}{2D_p}$$

So long as the transistor remains active, $I_c = \beta I_{b_{ss}}$. It follows from Eqs. (3-55) and (3-57) that

$$\frac{I_c}{I_{b_{ss}}} = \beta = \frac{\tau_B}{\tau_c}$$

Because the collector time constant is fast compared with the base recombination rate, within the active region the collector current immediately follows all changes in the base current. The minority-charge storage depends on the slow time constant τ_B. We can therefore represent the total dynamic response by a single time-constant system.

On the basis of the above argument, for slowly varying excitations we are able to treat the base-charge distribution from a steady-state viewpoint. The solid lines of Fig. 3-16*b* illustrate the increasing storage of base charge. Spatial diffusion of the injected carriers must be taken into account whenever the input changes are fast compared with the transit time of carriers through the thin base. The dashed curves indicate the modified charge distribution. For our purposes, we shall use the triangular area to indicate the charge storage up to the edge of saturation.

After the transistor saturates, we can represent the continuing charge storage by the addition of rectangular areas containing the excess base charge (Q_s) shown in Fig. 3-16*c*. Since I_c remains constant, limited by the external circuit, the triangular area Q_c does not change. The excess charge Q_s may even become orders of magnitude greater than Q_c.

Because of the nonuniform base diffusion, the recombination rate is slower at the collector edge of the base. While in saturation, appreciable charge is stored near the collector. To account for the increased contribution of the slower recombining charge contained in Q_s, we should replace τ_B with τ_{Bs} in Eq. (3-54). τ_{Bs} is usually from 1.5 to $5\tau_B$. For many purposes, the analysis becomes simpler if we approximate the full response by a single equivalent recombination-time constant.

Figure 3-16*d* represents the withdrawal of the stored base charge as we turn off a saturated transistor. First, the excess charge Q_s is removed and the transistor reenters its active region. Then the normal base charge is swept out and the transistor turns off.

Let us now calculate the complete dynamic response of the inverter of Fig. 3-17 as the input signal is switched between ± 10 V. We shall solve for the switching times assuming a single base-time constant $\tau_B = 5$ ns.

At $t = 0^-$, the transistor is off and consequently the stored base charge is zero. The base-emitter junction is depleted, and this depletion-region capacitance C_{BE} is charged to $+10$ V. (C_{BE} varies inversely with voltage, but for simplicity we are assuming a single average value of 1.5 pF.) After the input voltage is switched at $t = 0^+$, C_{BE} starts charging from $+10$ toward -10 V, with the turn-on-delay time constant of 7.5 ns ($\tau_D = R_B C_{BE}$). The base voltage is given by

$$v_b(t) = -10 + 20e^{-t/\tau_D} \tag{3-58}$$

The transistor turn-on point occurs at $v_b(t_D) = -700$ mV. Substituting the turn-on voltage into Eq. (3-58) and solving for t_D, we obtain the turn-on delay interval

$$t_D = \tau_D \ln \frac{-10 - 10}{-10 + 0.7} = 5.7 \text{ ns}$$

After the 5.7-ns delay, we say that the transistor has turned on. Since the base-emitter junction is now forward-biased, the depletion-region

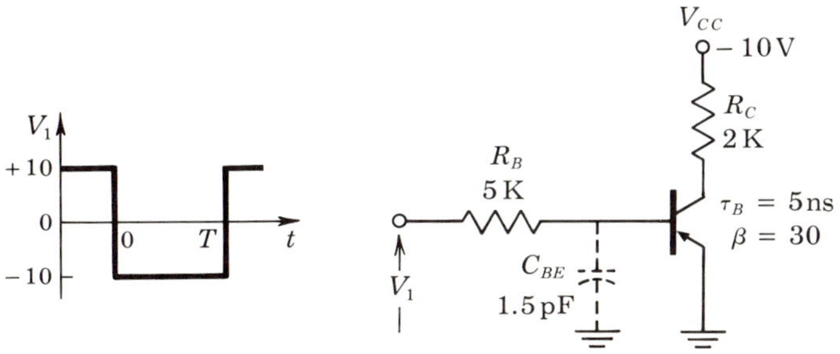

Fig. 3-17 Saturating inverter.

capacitance no longer exists. The base current becomes

$$I_{b1} = \frac{10 - 0.7}{5\ \text{K}} = 1.86\ \text{mA}$$

Equation (3-54) now defines the buildup of the base charge supplied by I_{b1}. Only one question remains unanswered: What is the effective steady-state charge? From Eq. (3-54) we have $Q_{B_{ss}} = I_{b1}\tau_B$. Solving Eq. (3-54) for $Q_B(t)$, the initial portion of the base-charge buildup, consistent with the known steady-state condition, is given by

$$Q_B(t) = I_{b1}\tau_B(1 - e^{-t/\tau_B}) \tag{3-59}$$

Equation (3-59), plotted in Fig. 3-18, shows how the initial portion of base current primarily stores charge in the base. We shall assume that the storage increases by the triangular areas depicted in Fig. 3-16*b*. Collector current follows the increase in slope up to the edge of saturation, which corresponds to $Q_B(t_{\text{ON}}) = Q_{c,\text{sat}}$. Using $V_{CE,\text{sat}} = 0$, we obtain the current boundary conditions from the circuit of Fig. 3-17:

$$I_{c,\text{sat}} = \frac{V_{CC}}{R_C} = 5\ \text{mA} \qquad I_{b,\text{sat}} = \frac{I_{c,\text{sat}}}{\beta} = 166\ \mu\text{A}$$

Substituting the saturation condition into Eq. (3-59) and solving for t_{ON}, we find that with $\tau_B = 5$ ns, it takes

$$t_{\text{ON}} = \tau_B \ln \frac{Q_{B_{ss}}}{Q_{B_{ss}} - Q_{c,\text{sat}}} = \tau_B \ln \frac{I_{b1}}{I_{b1} - I_{b,\text{sat}}} = 0.45\ \text{ns} \tag{3-60}$$

for the transistor to switch through the active region into saturation. The harder the transistor is driven into saturation, the shorter the turn-on time. Equation (3-60) must not be interpreted as indicating an exponential current change; the currents are simply proportional to the charge-storage levels ($Q = I\tau$).

Within the saturation region, base storage continues until the total excess charge $Q_s = Q_{Bss} - Q_{c,\text{sat}}$ is stored. Storage here corresponds to the increasing rectangular areas shown in Fig. 3-16*c*. In this region, the slope of the triangular area remains constant, indicating a constant collector current.

Turning off the transistor requires the withdrawal or recombination of all the previously stored base charge. This process can be speeded by establishing a negative base current. Connecting the conducting base to +10 V causes a large negative base turn-off current to flow:

$$I_{b2} = -\frac{10 + 0.7}{5\ \text{K}} = -2.14\ \text{mA}$$

The negative I_{b2} helps to sweep out the stored charge.

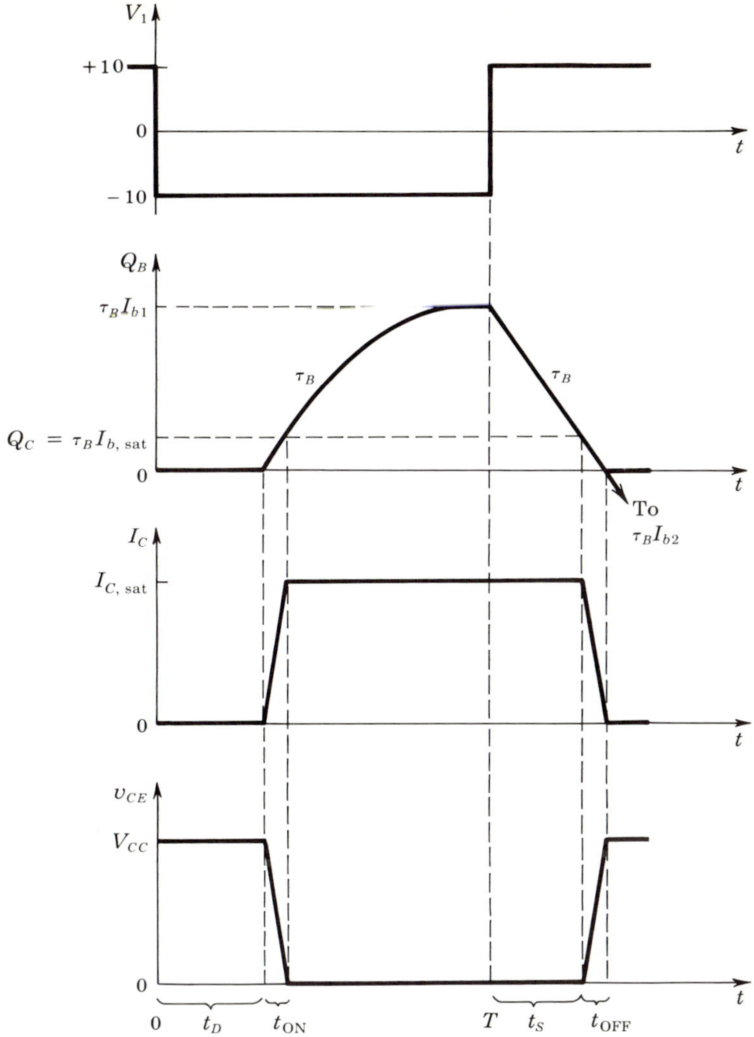

Fig. 3-18 Switching conditions for the saturated inverter of Fig. 3-17 showing the waveshapes of base-charge storage, collector current, and collector voltage (output capacity assumed ≈ 0).

The stored base charge decays with the time constant τ_B from $Q_{B_{ss}}$ toward the new steady-state charge level $I_{b2}\tau_B$. The new solution of Eq. (3-54) holding over the turn-off interval is

$$Q_{B2}(t) = I_{b2}\tau_B - (I_{b2}\tau_B + Q_{B_{ss}})e^{-t/\tau_B} \tag{3-61}$$

Once the total excess base charge has been swept out and $Q_{B2}(t_S) = Q_{c,\text{sat}}$, the transistor again enters the active region. Solving Eq. (3-61), we find that the saturation-delay time can be expressed in terms of the currents corresponding to the various stored-charge levels:

$$t_S = \tau_B \ln \frac{I_{b2} - I_{b1}}{I_{b2} - I_{b,\text{sat}}} = 2.77 \text{ ns} \tag{3-62}$$

Equation (3-62) shows that the saturation-delay time can be minimized by using a large negative current to speed the recombination process. If, instead of $I_{b2} = -2.14$ mA, we set $I_{b2} = 0$ by grounding the base, the saturation-delay time becomes 12.7 ns, almost a fivefold increase.

After the transistor becomes active, stored base charge continues to decay toward $I_{b2}\tau_B$. Only after all charge has been swept out can we consider the transistor off. The defining equation will be of the same form as Eq. (3-61), though starting from $Q_B = Q_{c,\text{sat}} = I_{b,\text{sat}}\tau_B$:

$$t_{\text{OFF}} = \tau_B \ln \frac{I_{b2}\tau_B - Q_{c,\text{sat}}}{I_{B2}\tau_B} = \tau_B \ln \frac{I_{b2} - I_{b,\text{sat}}}{I_{B2}} = 0.35 \text{ ns}$$

The dynamic behavior can also be described by dividing the base charge into the forward and reverse distributions shown for the Ebers-Moll model in Fig. 3-4. By doing this, we split the base into two nodes instead of the one used above. Since the storage and recombination at each node are different, we would have to find several time constants instead of one and the solution would become correspondingly more difficult.

The foregoing analysis leads us to think of the switching response in terms of charge transfer. Fixed quantities of charge must be stored or removed during the turn-on, saturation-delay, and turn-off intervals. During the transfer, the base is treated as a single time-constant node.

To speed the switching process, compensated-attenuator techniques discussed in Sec. 1-8 can be applied in the base driving circuit. In Fig. 3-19, the base is represented by the time constant τ_B. Even though fixed

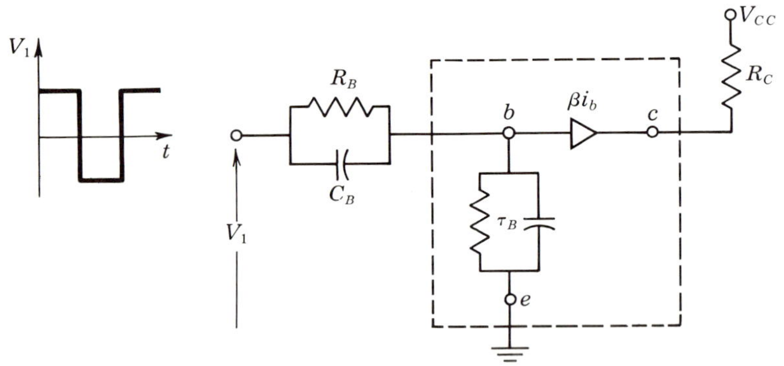

Fig. 3-19 Compensated base resistor for faster rise and fall times.

values cannot be assigned to these elements, we do specify their product. Since the internal node resistor, which is h_{ie}, decreases with increasing base current and the capacity C_T increases with I_b, the time constant does not change. Figure 3-12, which illustrates the high-frequency model, introduced the storage capacitor C_T.

If the base drive resistor R_B is now shunted with a capacitor C_B, then the series network and the internal base model are converted into a compensated attenuator. Choosing the external network such that $R_BC_B = \tau_B$ adjusts the network for fastest charge transfer, and the transistor will be switched on and off in the shortest possible time. We cannot, of course, exactly compensate simultaneously for the turn-on delay caused by the charging of the back-biased emitter-base junction capacitor and for the saturation turn-off delay due to the base time constant τ_B. By observing the collector-voltage response, we adjust the R_BC_B product for the best compromise response. $R_BC_B = \tau_B$ serves as a convenient starting point for the experimental adjustment.

3-8 Integrated-circuit Devices

Monolithic integrated circuits are fabricated by the controlled diffusion of impurities into a single chip of silicon. Not only transistors and diodes but also resistors and small capacitors are formed within the chip. Individual-circuit-element values are determined by the geometry. Isolation between elements is provided by means of the intervening back-biased junctions.

The fabrication of any integrated circuit starts with a 1- to 2-in.-diameter wafer of p-type silicon sliced about 8 to 10 mils thick (1 mil $\approx$ 25 μm). After multiple identical circuits are fabricated, the wafer is sliced into separate dice, each containing a single circuit. Final die sizes vary from 25×25 mils for very simple circuits up to about 100×100 mils for much more complicated arrays.

The wafer of p-type silicon, which serves as the substrate for the monolithic circuit, is usually doped to a resistivity of 10 Ω-cm. To form a transistor on top of the substrate, regions of n-, p-, and n-doped silicon must be sequentially diffused into the substrate. These layers are formed either completely by multiple diffusions or first by growing an epitaxial n layer and then following with p and n diffusions.

Epitaxy is the growth of a thin doped layer of single-crystal silicon upon a suitable substrate. Since the p substrate is already a single crystal, silicon atoms from the high-temperature environment readily grow onto the basic lattice. When dopants are introduced during the growth process, they will become uniformly distributed. Bulk resistivities of the heavily

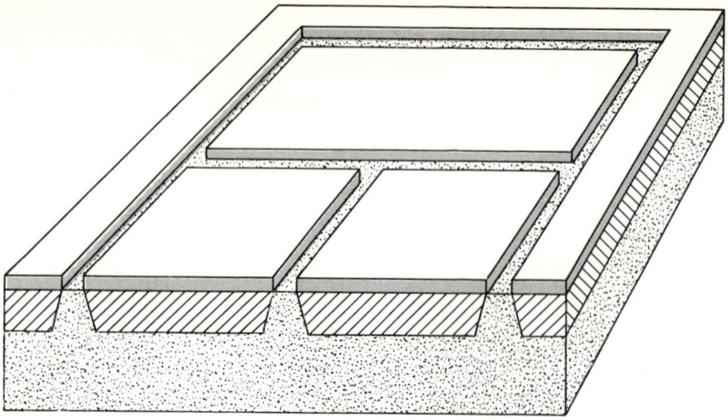

(a)

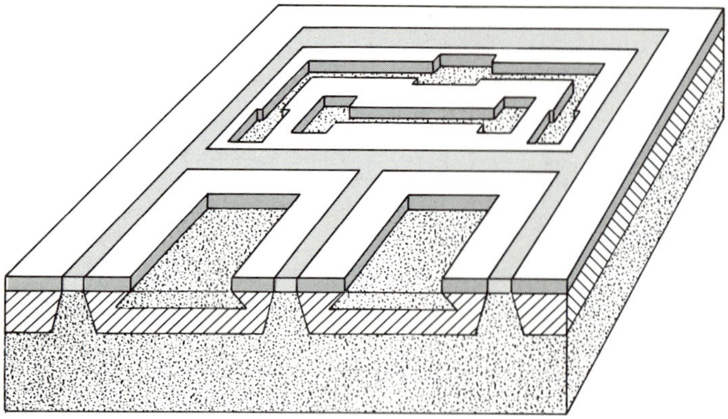

(b)

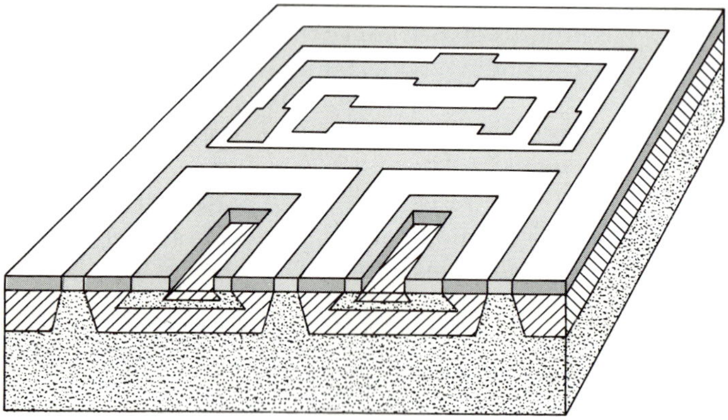

(c)

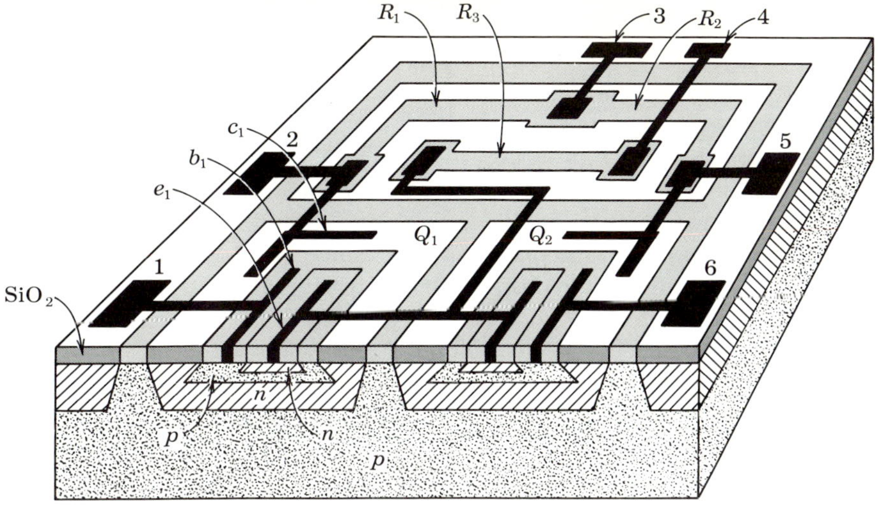

(*d*)

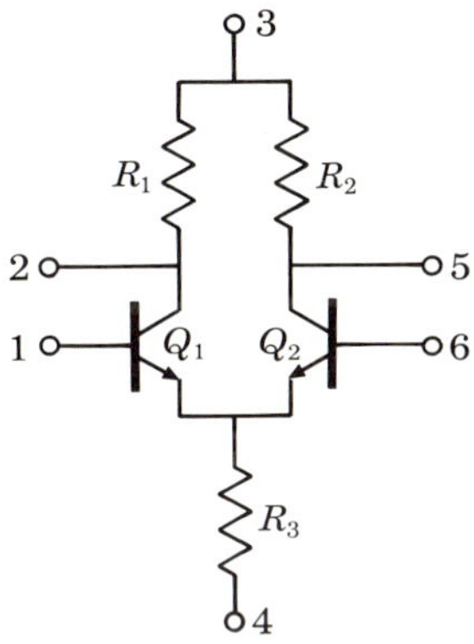

(*e*)

Fig. 3-20 Major steps in the production of an integrated circuit. (*a*) Diffusion of *p* regions through epitaxial layer to form isolation islands; (*b*) SiO_2 masking, followed by *p*-base-region and resistor diffusion; (*c*) SiO_2 masking, followed by *n*-emitter-region diffusion; (*d*) final metallization; (*e*) circuit of differential amplifier pair formed in step *d*.

doped 10- to 25-μm *n*-type epitaxial layer are in the range of 0.1 to 0.5 Ω-cm. (0.5 Ω-cm corresponds to a majority-carrier concentration of 1.2×10^{16} electrons per cm^3.)

The next step in the production of an integrated circuit is to form a layer of SiO_2, roughly 5,000 Å thick, on the surface of the epitaxial layer. Photographic techniques are used to form a mask on the top of the oxide. By use of an etchant, the oxide is removed in selected areas down to the surface of the n layer (see Fig. 3-20a). p-type impurities (boron) are diffused completely through the n layer into the p substrate, forming an island of n material.

It appears that the npn structure looks just like a transistor, with the p isolation region serving as the base. Under certain bias conditions, the α_F or α_R direct transmission from n island to n island would defeat the purpose of isolation. In practice, however, since the 1-mil p stripe is much wider than the diffusion length, the transport factor is vanishingly small.

Each island of n material is thus separated from its neighbors by the two back-to-back np junctions. No matter at what potentials the n regions are, at least one junction will be back-biased, affording excellent isolation between adjacent islands. If the p substrate is returned to the most negative circuit potential, both junctions will always be back-biased; isolation will be greatly improved. All islands will also be isolated from the substrate.

We next start forming the transistors within several of the separate islands and forming the resistors within others. The epitaxial n layer serves as the collector for the transistors. After again forming an SiO_2 layer, photoetching exposes the areas into which we shall diffuse the base impurities (Fig. 3-20b). By carefully controlling the duration and the rate of the diffusion of a p-type impurity, usually boron, we can form a p-type region extending about 2.7 μm into the n layer. Because the diffusion is from the surface into the wafer, the impurity concentration will decrease with depth, as shown in the profile of Fig. 3-21. At the surface we have a net majority-p-carrier concentration of 3×10^{18} holes per cm^3, and this drops to the uniform majority-n-carrier concentration of 1.2×10^6 electrons per cm^3 at the interface with the epitaxial n collector region.

Since the impurities are not uniformly distributed in the base region, it makes no sense to attempt to specify a bulk resistivity. Instead, we define the sheet resistivity of the total p diffusion in terms of ohms per square. This specifies the resistance measured between two opposite sides of any square area the full depth of the region. This measurement is independent of the dimensions of the square. The particular p-diffusion profile of Fig. 3-21 yields a typical sheet resistivity of 200 $\Omega/\square$.

Once more, a SiO_2 layer is formed across the wafer surface. Masking and photoetching expose the area within the previous base diffusion in which we shall form the emitter (Fig. 3-20c). A heavy diffusion of n-type impurities (phosphorus) into the p base changes a 1-mil-wide stripe, 2.1 μm deep, into the n emitter region. The profile of this diffusion is also shown in Fig. 3-21. 2 $\Omega/\square$ is the normal sheet resistivity of the emitter.

In Fig. 3-21, we see that the emitter and base diffusion profiles intersect at a depth of 2 μm. The p region between the n emitter and the n collector is only 0.7 μm thick, and this thin layer is the active base region. With

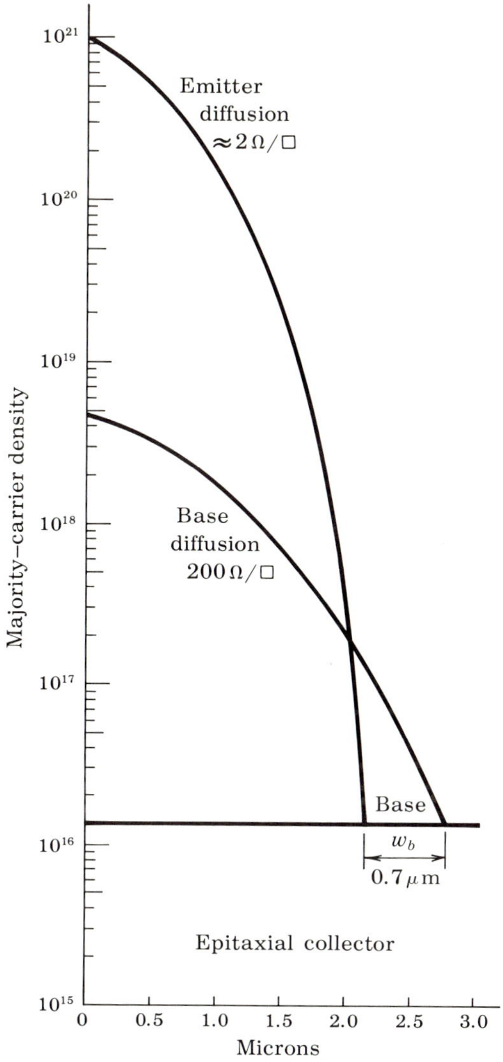

Fig. 3-21 Diffusion profiles in integrated-circuit transistor of Fig. 3-20.

this geometry we can expect β's of 30 to 60. Other expected parameters are given in Table 3-1.

The area of the transistor is selected on the basis of the emitter current flow. A transistor having a single-stripe emitter contact designed for 1-mA emitter current will require a chip area approximately 8×9 mils. This includes a 1-mil-perimeter isolation diffusion. A 9,000-mil^2 chip can

Table 3-1 Parameters of a typical 1-mA integrated-circuit transistor—collector epitaxial layer doped to 0.1 Ω-cm

BV_{CBO}	Collector-base reverse breakdown voltage	20 V
BV_{EBO}	Emitter-base reverse breakdown voltage	6.5 V
BV_{CEO}	Collector-emitter reverse breakdown voltage	12 V
h_{fe}	Forward-current-amplification factor (β) (measured at 1 mA)	50 A/A
r_{sc}	Collector saturation resistance	15 Ω
r_b	Base padding resistance	200 Ω
V_{BE}	Base-emitter forward drop (measured at $I_e = 1$ mA)	750 mV
$V_{CE,\text{sat}}$	Collector-emitter saturation drop (measured at $I_e = 1$ mA)	215 mV
C_{cs}	Parasitic collector-to-substrate isolation capacitance (1-V back bias)	3 pF
C_{cb}	Collector-base depletion-region capacitance (1-V back bias)	2 pF
C_{be}	Base-emitter depletion-region capacitance (1-V back bias)	1.5 pF
C_T	Base storage capacitance (at $I_e = 1$ mA forward bias)	1 pF
f_T	Common-base cutoff frequency (measured at $I_e = 1$ mA)	600 MHz

easily accommodate 20 or more transistors along with associated resistors and wiring. For higher currents, the active area of the transistor can easily be increased.

Resistors

In order to obtain the highest possible values of resistance in the smallest possible area, we must use high-resistivity materials. The highest available sheet resistivity is the base diffusion, which has a bulk resistivity of 200 Ω/□, and this diffusion is used to form the resistors. Since the resistance is a function of the geometry, as shown in Fig. 3-22, we can form any required resistance by selecting the appropriate width and then making the resistor as long as necessary.

Neglecting contact areas, the resistance is given by

$$R = \rho_s \frac{L}{W} \tag{3-63}$$

where ρ_s is the sheet resistivity in ohms per square. From Eq. (3-63), we see that the thinner the base diffusion line, the shorter the line length needed

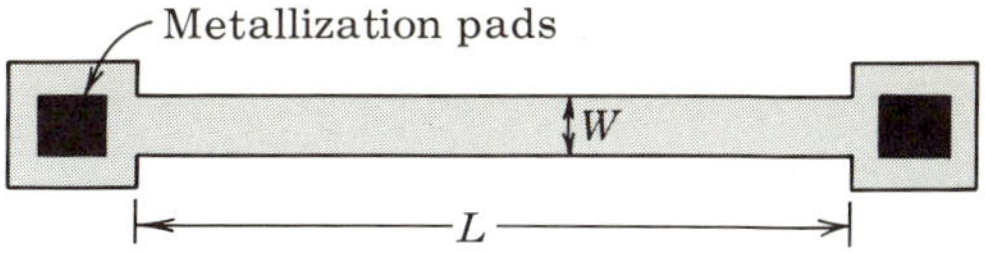

Fig. 3-22 Geometry of an integrated-circuit resistor.

for any given resistance. The line thickness is limited by the photographic process. Stripes are usually ½, 1, or 2 mils wide. Because the diffusion extends under the edges of the mask, the thinner the line, the poorer the resistance tolerance. Even though the sheet resistance can be held to 5 percent, the absolute accuracy of any resistor decreases from 10 percent with a 2-mil stripe to 20 percent for a 1-mil stripe and 30 percent for a ½-mil stripe. We trade off chip surface area for improved tolerance. But since all resistors are formed at the same time, by the same base diffusion, all tolerances will vary in the same way. Resistor ratios will be much closer than individual resistor values. It follows that design by ratios, instead of by absolute values, will improve expected integrated-circuit performance.

The maximum practical resistor values are determined by the surface area that can be assigned to resistors. For example, with a ½-mil-stripe geometry, a 5-K resistor requires a length of 12.5 mils and a 6-mil^2 surface area, not including contact areas or isolation. A 5-K resistor of a 1-mil stripe would occupy 25 mil^2. This is a fourfold increase in area and is only one-third as much as the area devoted to a single transistor. Low values of resistors always use wider stripes. Practical values of resistors formed by base diffusion range from 200 Ω up to 10 K, with 2 K a reasonable design center value. If very low resistance values are required, they can be formed from the 2 Ω/□ emitter-diffusion sheet resistance.

Several resistors can be diffused within each isolation area. Since the *n* epitaxial layer forms a *pnp* structure between adjacent resistors, the resistors are isolated by at least one back-biased junction. Spacing, of course, must be wide enough to prevent operation as a parasitic transistor. By returning the *n* resistor island to the most positive potential of the circuit, both resistor-to-resistor and resistor-to-substrate isolation will be improved.

Metallization

The final step in producing the circuit on the chip is the interconnection of all the discrete components diffused into the chip. As before, the surface is coated with a layer of SiO_2. Holes are etched to reach into the transistor and resistor contact areas. A thin layer of aluminum is evaporated across the whole wafer, forming ohmic contacts at each of the etched holes. A second masking and selective etching remove the unwanted metallization, leaving aluminum lands connecting the contact areas of the circuit elements (Fig. 3-20*d*). The SiO_2 under the aluminum lands insulates the wafer from the metallization.

Aluminum is, itself, a column-III element which can be used as a dopant in forming *p*-type silicon. Since we must have ohmic contacts, we cannot permit the slight diffusion accompanying the high-temperature metallization step to change the surface layers of the *n* emitter or collector contacts to *p* regions. Usually, before the metallization step, additional column-V impurities are diffused into the emitter and collector contact areas, leaving them highly doped n^+ regions. Any diffusion of aluminum will not change the nature of the contact region; it will only reduce the n^+ carrier density. We thus ensure ohmic contacts.

The interconnecting lands can cross over circuit components since they are insulated by the SiO_2 layer. One land cannot, however, cross another land. The component layout problem becomes paramount; areas must be minimized, connections must not intersect, long lead paths add series resistance, and final isolation must be preserved.

Figure 3-23 shows the circuit and microphotograph of an integrated-circuit amplifier. The transistors, resistors, and circuit interconnections are clearly visible in the photograph, and the circuit can be traced out. The particular elements are labeled with the circuit numbers in Fig. 3-23*a* for identification purposes. This circuit is a differential-input operational amplifier similar to the type discussed in Chap. 10.

Capacitors

A capacitor can be included in the circuit either through use of a back-biased junction or by forming a physical capacitor between a low-resistance emitter-diffusion area and an overlay metallization area. The SiO_2 layer is the dielectric. Since in both cases the maximum capacity is only about 0.3 pF/mil^2, we must sacrifice large amounts of surface area to obtain even a small capacitor. Using 25 percent of the full chip area of 60×70 mils will provide only 300 pF of capacity. Ten transistors occupy the same 1,000-mil^2 area. It is easier and cheaper to include extra transistors in a circuit to avoid the use of even a small capacitor. A similar problem arises with the use of inductance.

Diodes

Whenever we make a transistor, we have available two separate junctions, either of which may be used as a diode. The question is: Which one

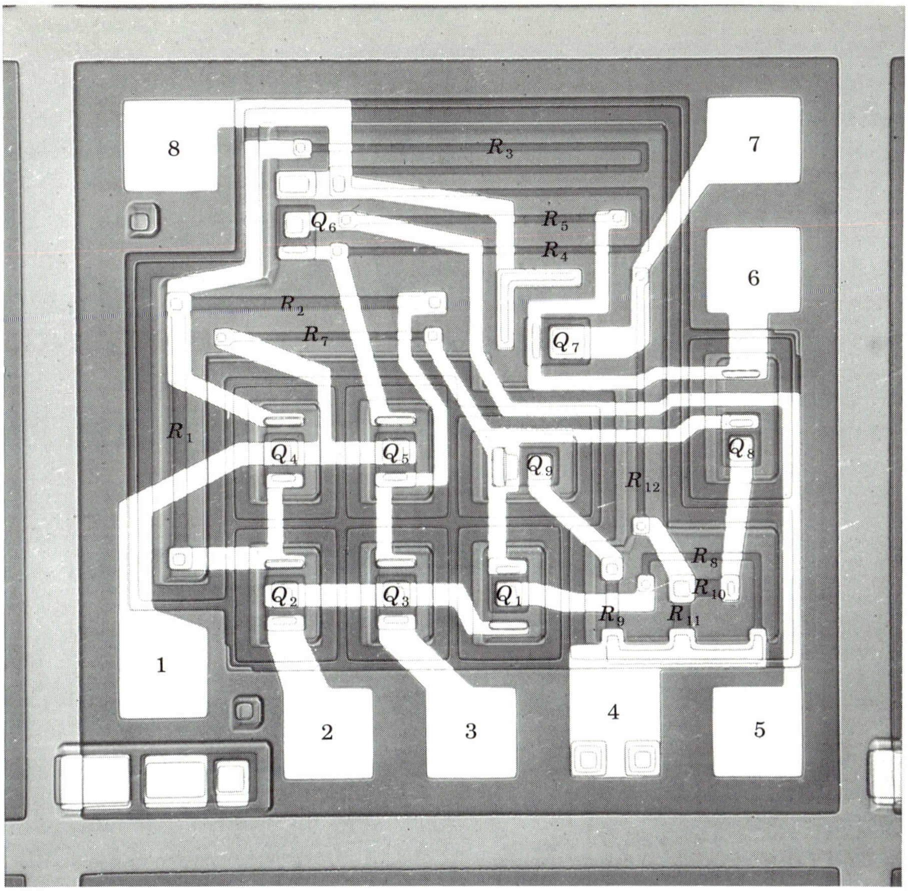

(*a*)

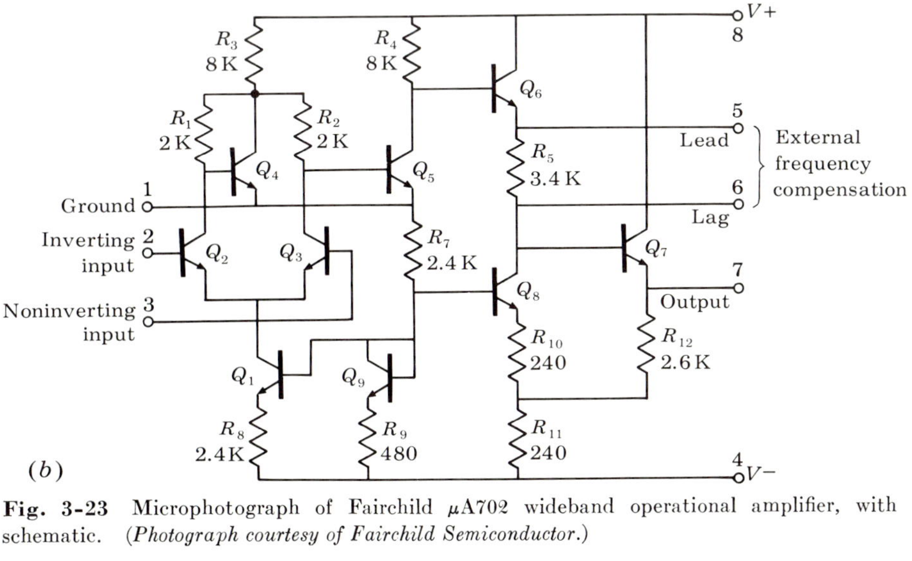

(*b*)

Fig. 3-23 Microphotograph of Fairchild μA702 wideband operational amplifier, with schematic. (*Photograph courtesy of Fairchild Semiconductor.*)

should be chosen? It is well known that for fastest switching, the charge stored in the base must be minimized. If we use the base-emitter junction as our diode and during metallization short the collector to the base, then it can be shown that the minimum charge will be stored in the base. Of course, this requires use of the whole transistor connected as a diode.

If speed is not the determining factor, surface area can be saved by using the collector-base junction. In this case, there is no need for the emitter diffusion; hence the base area can be reduced in size.

Because of the graded doping of the base, there are fewer majority charge carriers at the collector edge of the base than at the emitter edge. Since the junction breakdown voltage varies inversely with the doping level, the collector breakdown will be much larger than that at the emitter junction: $BV_{CB} \approx 20$ V compared with $BV_{EB} \approx 6$ V. We see that we are forced to use the collector junction whenever we expect large reverse diode voltages.

Complementary integrated devices

Several different techniques are employed in fabricating *pnp* transistors on the same chip as the *npn* transistors. The simplest method is to make use of the parasitic transistor formed by the *p* base diffusion, the *n* collector layer, and the *p* substrate. The substrate is the collector, and the *n* layer acts as the base of the complementary transistor.

The *pnp* transistor is poorly designed. First, because the *n* layer is relatively thick, the transport factor will be small. Second, the emitter efficiency is poor. Instead of the high-conductivity emitter, low-conductivity base required for high efficiency [Eq. (3-9)], the *p* emitter is the low-conductivity base diffusion of the *npn* transistor and the *n* base is the high-conductivity *n* epitaxial layer. Finally, we see that the *pnp* transistors are not completely independent; all the collectors are connected together within the substrate. The β of the *pnp* device will be only 3 or 4, compared with 40 or 50 for the *npn* transistor.

An alternative technique for forming a *pnp* transistor is to diffuse an additional *p* layer into the *n* emitter diffusion. This final layer serves as the emitter of the *pnp* transistor, while the high-conductivity *n* layer serves as the base. Again, the emitter efficiency will be poor. In this case, however, since we have a five-layer structure, all the *pnp* devices are independent.

We have seen that it is impossible to produce both good *npn* and good *pnp* transistors on the same chip. Normally, only *npn* transistors are used. In the few circuits employing both types, we must settle for low-β *pnp* devices.

Problems

3-1 The p and n silicon semiconductors of Prob. 2-2 are now made into a symmetrical pnp transistor having a base width of 0.5 μm. Calculate α_F, I_{co}, and V_{EB} for a forward current of $I_e = 2$ mA. Take $D_p = 12$ cm²/s, $D_n = 30$ cm²/s, and $\tau_p = \tau_n = 100$ ns ($L_p = \sqrt{D_p\tau_p}$).

3-2 Show from the Ebers-Moll model that a transistor's collector characteristics may be interpreted as a set of shifted back-biased diode curves, where I_e is the shifting parameter.

3-3 By direct substitution from the diode equations, prove that $\alpha_F I_{es} = \alpha_R I_{cs}$ [Eq. (3-14)].

3-4 Find the quiescent operating point (I_e, V_{ce}) of the circuits shown in Fig. 3-24. Make all reasonable approximations. In all cases, $\alpha_F = 0.98$, $\alpha_R = 0.75$, $r_b = 100\ \Omega$, $V_{BE} = 700$ mV at $I_e = 1$ mA, and $V_{CE,\text{sat}} = 150$ mV.

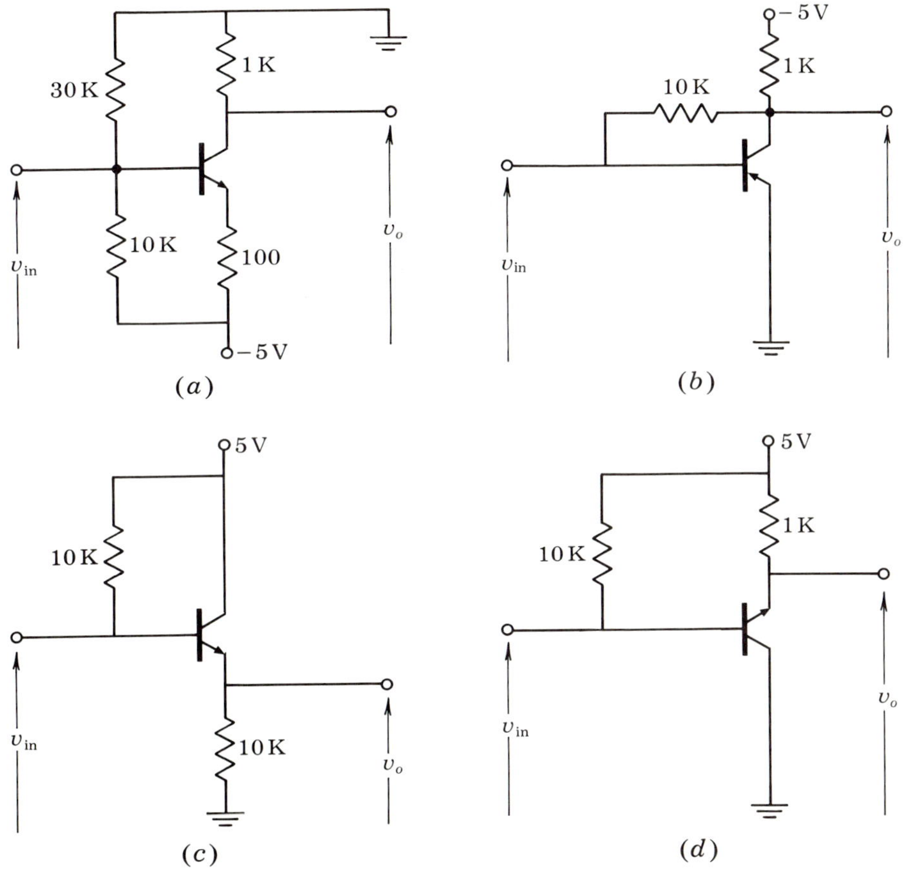

Fig. 3-24

3-5 In the circuit of Fig. 3-25, the transistor has $\beta_F = 50$ and $\beta_R = 2$. We take $V_{BE} = 700$ mV. V_B can be either positive or negative as needed.

(*a*) Over what range of V_B will this transistor be saturated?

(*b*) Over what range of V_B will this transistor be cut off?

(*c*) Can this transistor ever operate in the reverse node?

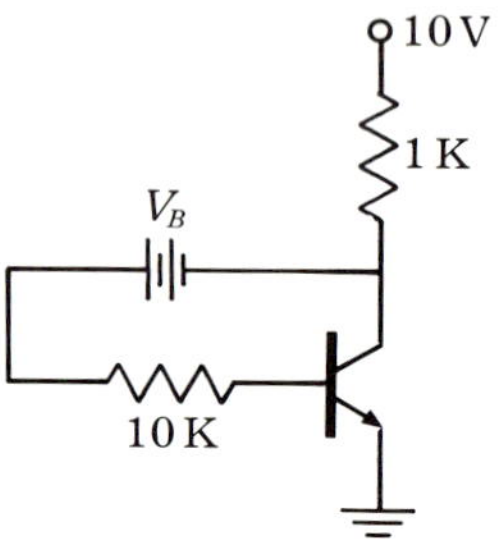

Fig. 3-25

3-6 Calculate the temperature-stability factor of each of the circuits of Fig. 3-24.

3-7 In the circuit of Fig. 3-24*a*, the transistor used is germanium ($V_{BE} = 120$ mV) having $I_{co} = 1\ \mu$A at 23°C. Find the shift in operating point as the temperature increases from 23 to 100°C.

3-8 Figure 3-26 is the circuit of a constant-current-drain stage widely used in integrated circuits. For example, this stage can replace R_3 in Fig. 3-20*e*. Assume $V_{BE} = V_D = 750$ mV and $\beta = 50$.

(*a*) Calculate the constant-current flow through this transistor.

(*b*) Prove that any changes in the junction drop due to changes in temperature will have a negligible effect on this current. (The transistor and diode drops change equally with temperature; both drop by 2.5 mV/°C.)

(*c*) What is the stability factor of this circuit?

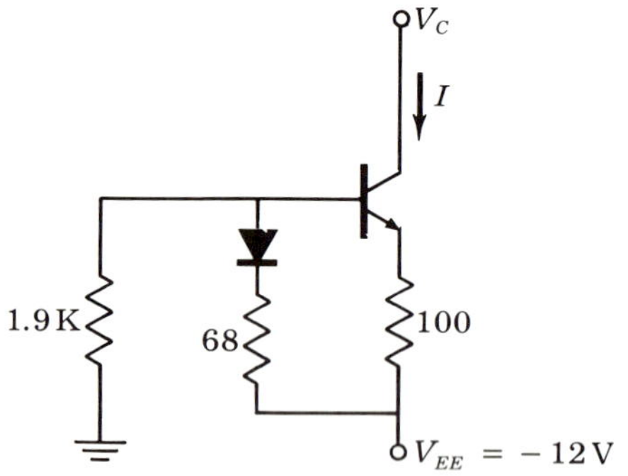

Fig. 3-26

3-9 (*a*) Calculate the quiescent operating conditions of the circuit of Fig. 3-27 ($i_{in} = 0$).

(*b*) Draw the small-signal model holding in the active region. From this model calculate the midband transfer resistance (that is, v_o/i_{in}).

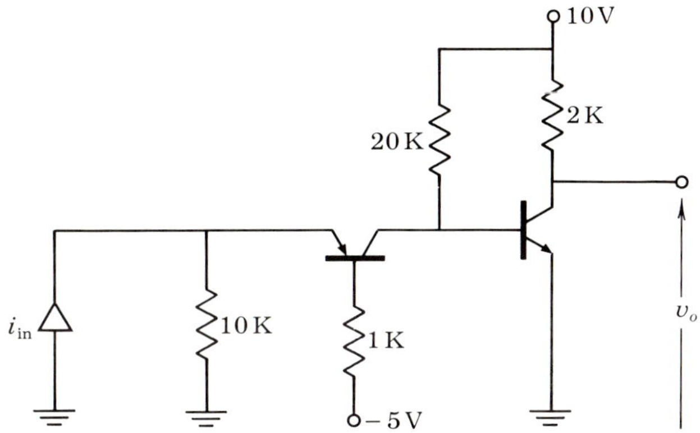

Fig. 3-27

3-10 The transistor of Fig. 3-28 is saturated ($\beta_F = 100$, $\beta_R = 10$, $V_{BE} = 720$ mV at $I_e = 1$ mA).

(*a*) Specify the parameters needed for the Ebers-Moll model.

(*b*) Find the terminal voltages of the transistor.

(*c*) Specify the incremental model seen looking into the collector of the saturated transistor.

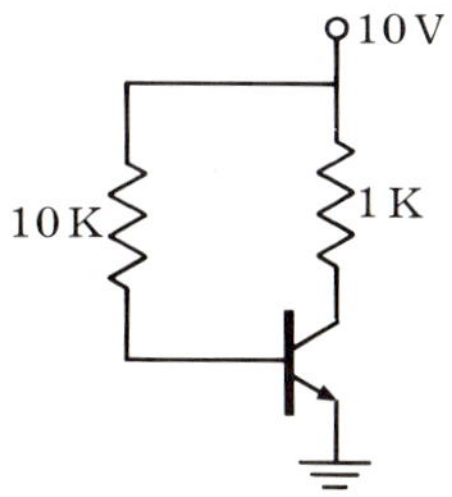

Fig. 3-28

3-11 The transistor of Fig. 3-28 is inserted into its socket backward, i.e., with collector and emitter interchanged. Repeat Prob. 3-10 for this case.

3-12 (*a*) Plot both the linearized and the exact transfer characteristic of the circuit of Fig. 3-29. ($\beta = 50$, $r_b = 100$, $V_{BE} = 700$ mV at $I_e = 1$ mA.)

(*b*) Define the small-signal gain at $I_e = 0.5$ mA.

(*c*) Define the small-signal gain at $I_e = 3$ mA.

(*d*) From your plot of part *a*, define what constitutes a small signal for this circuit. Choose your limits so that the gain at the peaks will not vary by more than ±5 percent from the center gain.

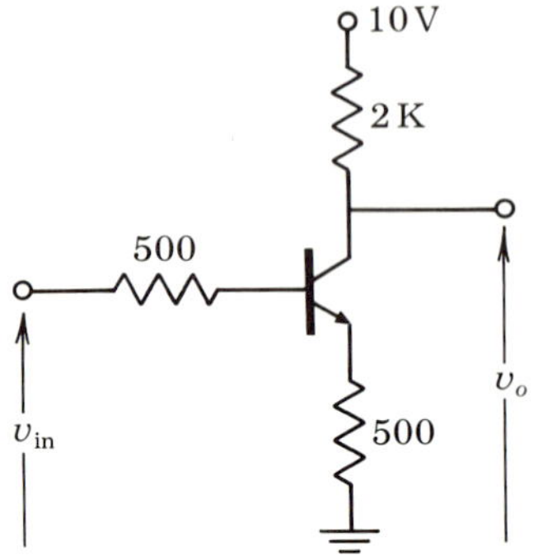

Fig. 3-29

3-13 The 500-Ω R_E in Fig. 3-29 is bypassed with a very large capacitor. Recalculate the small-signal gain at $I_e = 3$ mA. For this case, redefine what is meant by "small signal."

3-14 In integrated circuits it is impossible to synthesize the large capacitor used to bypass R_E in Fig. 3-29 (Prob. 3-13). Instead, we can shunt R_E with a second transistor, as shown in Fig. 3-20*e*. The base of Q_2 is returned to $V_{CC} = 10$ V through a 5-K resistor ($R_2 = 0$).

(*a*) Find the quiescent conditions for this case.

(*b*) Repeat Prob. 3-13 for this circuit and compare the results with those of Prob. 3-13.

3-15 Define the small-signal gain in the emitter-follower circuit of Fig. 3-24*c*. What amplitude of signal is a "small signal"? (See Prob. 3-12*d*.)

3-16 Draw the small-signal model for the circuit of Fig. 3-24*b* (Prob. 3-4). Give values for your model parameters. Calculate the small-signal gain, the input impedance, and the output impedance.

3-17 (*a*) Draw the small-signal model of the differential amplifier of Fig. 3-20*e*.

(*b*) Calculate the gain from terminal 1 to output 5 and from terminal 6 to output 5 (the unused input is grounded). What is the gain from the push-pull input (1–6) to the push-pull output (2–5)?

(*c*) What is the common-mode gain? This is the gain with the same signal applied to both inputs. ($R_1 = R_2 = 2.5$ K, $R_3 = 2$ K, $V_{CC} = 6$ V, $V_{EE} = -3$ V, and transistor parameters are as given in Table 3-1.)

3-18 (*a*) Define all the parameters of the hybrid π model of Fig. 3-12 in terms of the physical constants of the transistor and the operating point I_e.

(*b*) Starting from the linear base minority-charge distribution, derive Eq. (3-45). By what must C_T be replaced in a grounded-base model?

(*c*) Express the product $C_T h_{ie}$ in terms of the transistor's physical constants.

3-19 (*a*) Express the grounded-emitter h parameters of the composite transistor of Fig. 3-15 in terms of the individual-transistor h parameters.

(*b*) Repeat part *a* for a grounded-base configuration.

3-20 The transistor in the emitter follower of Fig. 3-24*c* is replaced by a composite transistor where Q_1 has a $\beta = 50$ and Q_2 has a $\beta = 100$. Show the resultant circuit, including whatever additional biasing is needed to set the proper operating point. Calculate the new gain and input impedance.

3-21 (*a*) Sketch and label completely the base and collector waveshapes if the switch in Fig. 3-30 is closed at $t = 0$ and opened at $t = 25$ ms. ($\beta = 30$, $V_{BE} = 700$ mV at $I_e = 1$ mA, $V_{CE,\text{sat}} = 100$ mV.) Specify the cutoff and saturation intervals.

(*b*) How will the results of part *a* be modified if the base-emitter–junction reverse breakdown voltage is 8 V?

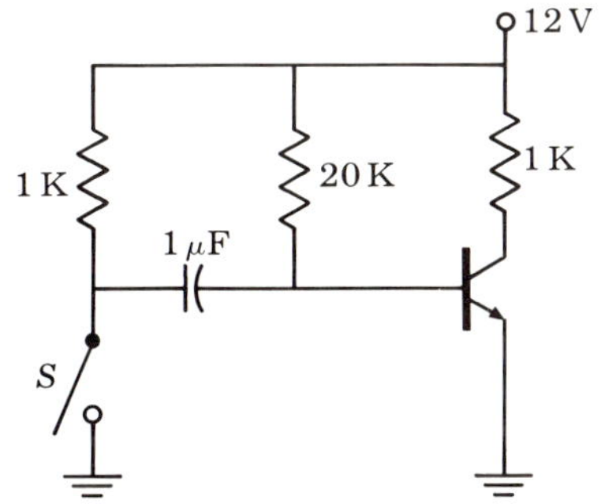

Fig. 3-30

3-22 The circuit of Fig. 3-31 is excited by a single 1-μs pulse. $\beta = 50$, $V_{BE} = 700$ mV.

(*a*) Sketch and label completely the waveshape at the collector when the collector is loaded by a 500-pF capacitor.

(*b*) Repeat part *a* for the emitter output when the 500-pF capacitor is transferred to the emitter.

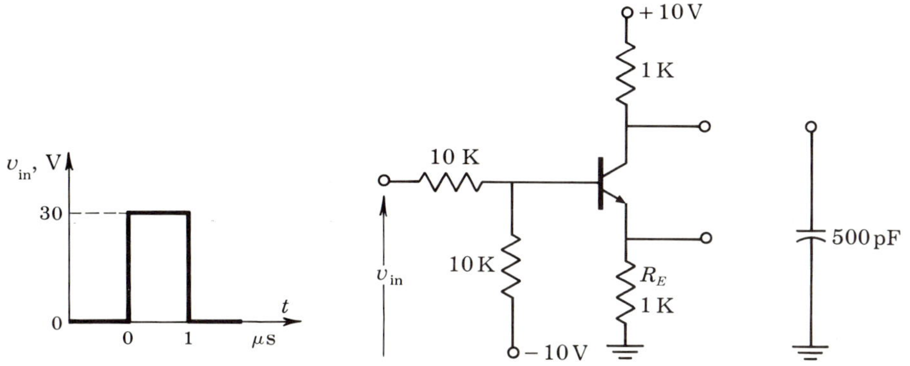

Fig. 3-31

3-23 Sketch and label the collector voltage in the circuit of Fig. 3-32 if the switch is opened at $t = 0$ and closed just after steady state is reached. Make all reasonable approximations. ($\beta = 50$, $r_b = 50\ \Omega$, $V_{BE} = 700$ mV.)

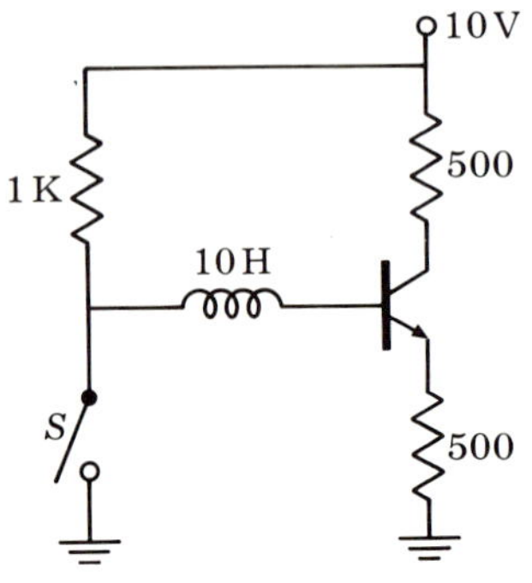

Fig. 3-32

3-24 As the emitter current of a transistor goes toward zero, the current gain also decreases. Figure 3-33 shows an approximate view of this decrease for a silicon *npn* transistor ($r_b = 200\ \Omega$). This transistor is used in the simple inverter stage shown.

(*a*) Plot the voltage gain as a function of V_{CC}. Show where the transistor saturates.

(*b*) Plot the input impedance as a function of V_{CC}.

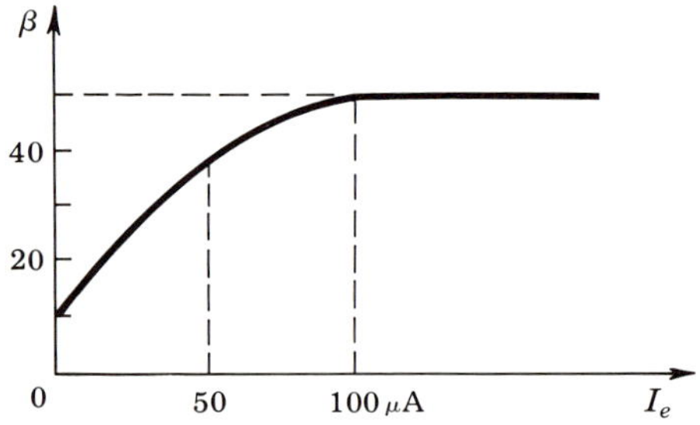

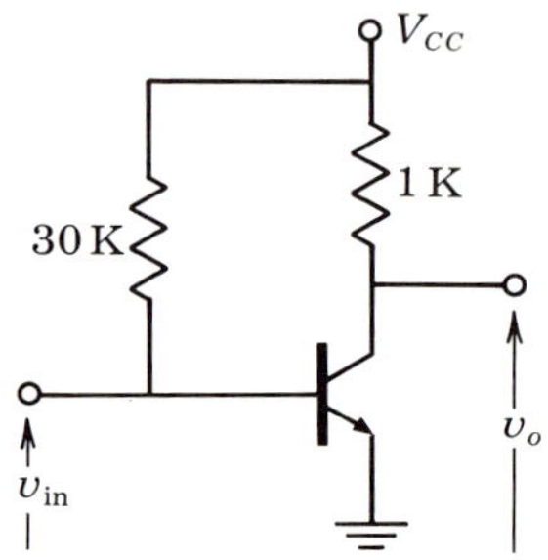

Fig. 3-33

3-25 The inverter of Fig. 3-34 is excited by a single 1-μs 6-V pulse.

(*a*) Sketch and label both v_o and v_{be}, taking into account the turn-on, turn-off, and storage-delay times. ($\beta = 50$, $V_{BE} = 700$ mV at $I_e = 1$ mA, $\tau_B = 10$ ns, $C_D \approx 10$ pF at $V_{BE} = -1$ V.)

(*b*) Repeat part *a* when the 20-K resistor is returned to ground instead of to -10 V.

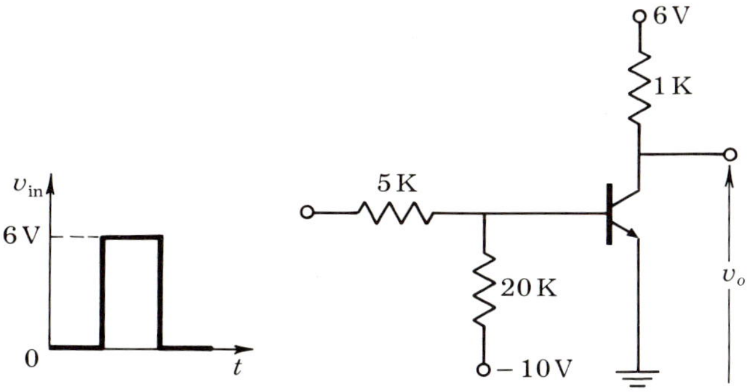

Fig. 3-34

3-26 Explain the effects of C_{bc} (5 pF) on the transient response of the inverter of Fig. 3-34 (Prob. 3-25).

3-27 Show a layout to scale, on a 50- by 50-mil chip, of the circuit of Fig. 3-35. Each transistor requires a 7 × 8 mil^2 area, not including isolation area. Use 1- or 2-mil stripes for the resistors (200 Ω/□). Show metallization and connection pads.

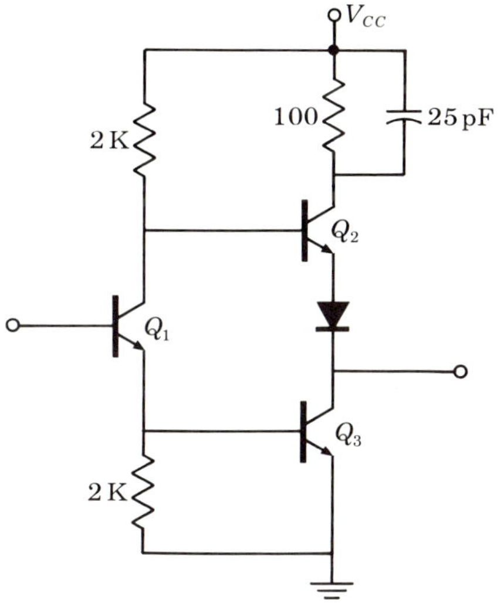

Fig. 3-35

Bibliography

Angelo, E. J., Jr.: "Electronics: BJTs, FETs, and Microcircuits," McGraw-Hill Book Company, New York, 1969.

Camenzind, H. R.: "Circuit Design for Integrated Electronics," Addison-Wesley Publishing Company, Inc., Reading, Mass., 1968.

Clarke, K. K., and M. V. Joyce: "Transistor Circuit Analysis," Addison-Wesley Publishing Company, Inc., Reading, Mass., 1961.

Ebers, J. J., and J. L. Moll: Large-signal Behavior of Junction Transistors, *Proc. IRE*, vol. 42, pp. 1761–1772, December, 1954.

Gray, P. E., et al.: "Physical Electronics and Circuit Models of Transistors," SEEC vol. 2, John Wiley & Sons, Inc., New York, 1964.

Jonscher, A. K.: "Principles of Semiconductor Device Operation," John Wiley & Sons, Inc., New York, 1960.

Moll, J. L.: Large-signal Transient Response of Junction Transistors, *Proc. IRE*, vol. 42, pp. 1773–1784, December, 1954.

Motorola Semiconductor Products Division (R. M. Warner, Jr., ed.): "Integrated Circuits," McGraw-Hill Book Company, New York, 1965.

Research Triangle Institute: Integrated Silicon Device Technology, XI, *Bipolar Transistors* (*AD*632556), Clearinghouse for Federal Scientific and Technical Information, Washington.

Searle, C. L., et al.: "Elementary Circuit Properties of Transistors," SEEC vol. 3, John Wiley & Sons, Inc., New York, 1964.

4. digital logic and amplitude gating

This chapter will discuss circuits designed for operation with multiple inputs. Where previously the complete circuit response was characterized by the transmission and volt-ampere curves, this simple representation now becomes quite inadequate. The multiplicity of inputs interact, and consequently the state of any particular output, at any particular time, cannot be defined in terms of one input but rather depends on the relative effects of the several sources. It is completely possible for a signal applied at one set of terminals to control the transmission path between another terminal and the output. Since it is impossible to represent these circuits in completely general terms, we shall at least try to answer the following three questions in the course of the analysis of specific circuits:

1. Does a transmission path exist from any one terminal to the output?
2. What conditions must be satisfied elsewhere in the network to establish (or to interrupt) this path?
3. How does the circuit behave in the two regions separated by the transmission threshold?

Gates are circuits which make use of bistate-device properties for switching purposes. They are defined, not in the narrow sense of turning something on and off, even though this is included, but as the means of establishing information-transmission paths upon the application of a proper stimulus, and they find wide application in digital computers, control and measuring instruments, and stimulation systems.

In computers, the gate output amplitude is usually of little or no importance provided that it at least exceeds a preset threshold value. We consider the output's presence as representing a "yes," or 1, answer, with its absence as a "no," or 0, answer. Almost any desired information for computing or control can be conveyed and operated upon by passing a coded time sequence of pulses through various combinations of gates, each pulse carrying one item of information.

In special-purpose instruments, an external control signal, or proper sequence of signals, opens and closes the gate, thus allowing or preventing transmission of the information signal. Here the amplitude response becomes important because this information may later be used for control purposes.

Gate inputs are often short pulses. Consequently, to avoid introducing reading errors, the gate should open rapidly to transmit the individual pulse and close rapidly upon its removal to prevent false information from appearing at the output. Any delay in opening and closing creates an ambiguity in performance; therefore our investigation of the particular circuits must be concerned with both their amplitude and their time response.

4-1 Digital Logic Statements

Digital logic networks process binary information. The circuits respond to two separated voltage levels which represent, respectively, a logical 0 and a logical 1. For example, in a positive-logic system, we might arbitrarily define the logical-1 level as a signal equal to or greater than 2 V and the logical-0 level as a signal less than 0.3 V. As shown in Fig. 4-1, this wide separation makes it easy to distinguish between the two input signals. A different positive logic might be based on a logical-1 level at or above -0.75 V and a logical-0 level below -1.55 V. For a negative-logic system, we define the higher level as the 0 state and the lower level as the 1 state. We see that it is the separation between the voltage values assigned to the logic states and not the particular levels that is the significant feature of a logic system.

The logic inputs are applied to digital gates, which respond by producing a logical-0 or logical-1 output when the particular input logic requirements are satisfied.

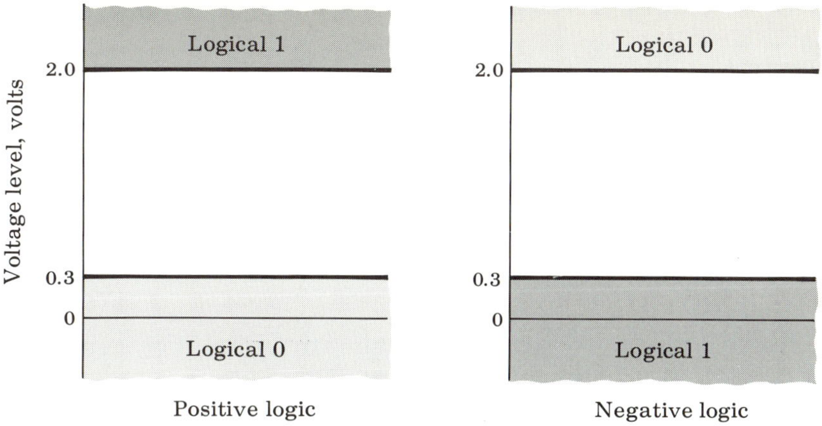

Fig. 4-1 Logic levels, illustrating separation between assigned logical-0 and logical-1 levels.

There are two basic forms of digital gates. The first, the OR gate shown schematically in Fig. 4-2*a*, satisfies the statement that a logical-1 output exists when either the A *or* the B input, *or* both, is equal to a logical 1. This logic statement is expressed by the equation

$$Y = A + B \tag{4-1}$$

In Eq. (4-1), the plus may be interpreted as a sum in a restricted sense; it means that only when A *or* B is 1 will the right-hand side be other than zero, i.e., be a logical 1.

A tabulation of all possible combinations of inputs and outputs of a logic block is called a *truth table*. The truth table given in Fig. 4-2*a* simply expresses Eq. (4-1) in tabular form.

The second basic gate (Fig. 4-2*b*) is called an AND gate. Here we obtain a logical-1 output only when A and B are both simultaneously present and at, or above, the logical-1 level. Thus the AND-gate logic statement becomes

$$Y = A \cdot B \qquad \text{or} \qquad Y = AB \tag{4-2}$$

In Eq. (4-2), the multiplication dot may be interpreted as meaning *and*. If either A or B is a logical 0, the product remains 0; only when A *and* B are both logical 1's will the product be a logical 1.

If the output of a basic gate is fed into a transistor inverter, then the high-level logical-1 signal will cause the inverter to saturate. The output will be driven toward ground, into the logical-0 state. Thus an inverter may be represented as shown in Fig. 4-2*c*. It simply inverts the input

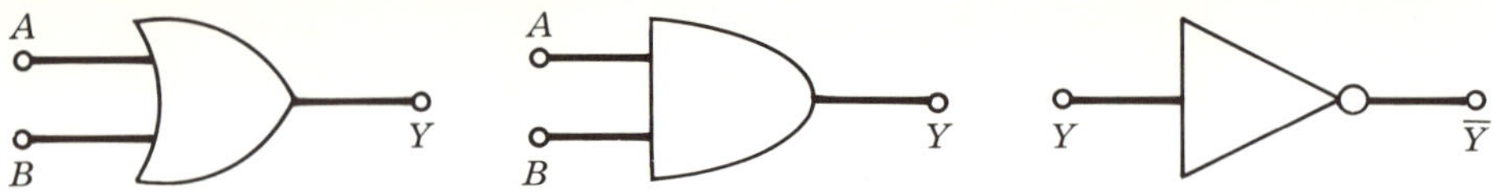

$Y = A + B$ $\qquad$ $Y = A \cdot B$

Input A	Input B	Output Y
0	0	0
1	0	1
0	1	1
1	1	1

(*a*)

Input A	Input B	Output Y
0	0	0
1	0	0
0	1	0
1	1	1

(*b*)

Input Y	Output $\overline{Y}$
0	1
1	0

(*c*)

Fig. 4-2 Basic logic elements and truth tables. (*a*) OR gate; (*b*) AND gate; (*c*) inverter.

logic statement. This is expressed by placing a bar over the statement. An inverter is often drawn as a plain triangle, omitting the circle.

By combining an OR gate and an inverter, we obtain the NOR gate of Fig. 4-3*a*. Here the small circle indicates the physical location of the inverter. Figure 4-3*b* shows a NAND gate, which is simply an AND gate combined with an inverter output stage.

If we examine the truth tables, we note that the following identities hold:

$$AB \equiv \overline{\bar{A} + \bar{B}} \tag{4-3}$$

$$\overline{A + B} \equiv \bar{A}\bar{B} \tag{4-4}$$

$$\bar{\bar{A}} \equiv A \tag{4-5}$$

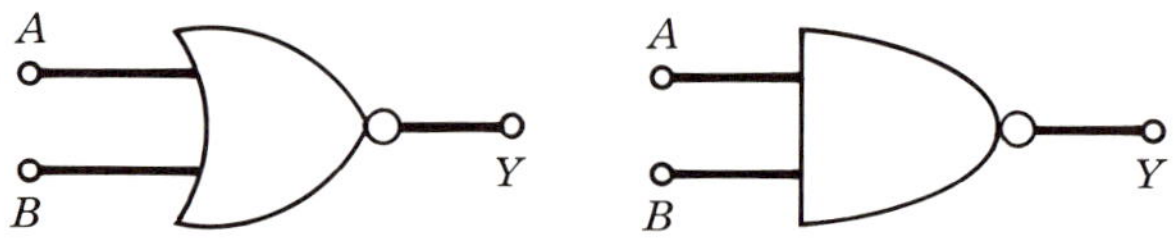

$Y = \overline{A + B}$ $\qquad$ $Y = \overline{AB}$

A	B	Y
0	0	1
1	0	0
0	1	0
1	1	0

(*a*)

A	B	Y
0	0	1
1	0	1
0	1	1
1	1	0

(*b*)

Fig. 4-3 (*a*) NOR and (*b*) NAND gates and truth tables.

All logic statements can be synthesized from either NAND or NOR gates. All four types of gates may be used in combination to reduce the number of required circuit elements.

An example of the synthesis of a logic statement is shown in Fig. 4-4, where a half adder is constructed out of NOR gates. A half adder sums two binary numbers and yields the sum and, if present, a carry forward to the next higher binary position. We first express all possible combinations of inputs and the desired sum and carry outputs in the form of a truth table (Fig. 4-4). The known outputs are next expanded in terms of a series of input-excitation OR statements or in a series of AND statements. From the truth table we see that the direct AND expansions of the rows containing logical 1's in the sum and carry columns yield the sum and carry equations

$$S = A\bar{B} + \bar{A}B \tag{4-6a}$$

$$C = AB \tag{4-6b}$$

We can also expand the truth table through the negation (or inversion) of the logical-0 outputs. For the sum and any carry, expanding on the remaining rows,

$$S = \overline{\bar{A}\bar{B} + AB} \tag{4-7a}$$

$$C = \overline{\bar{A}\bar{B} + A\bar{B} + \bar{A}B} \tag{4-7b}$$

Equations (4-7*a*) and (4-7*b*) are directly equivalent to the statements of Eqs. (4-6*a*) and (4-6*b*). It is obvious that the carry statement of Eq.

A	B	S	C
0	0	0	0
1	0	1	0
0	1	1	0
1	1	0	1

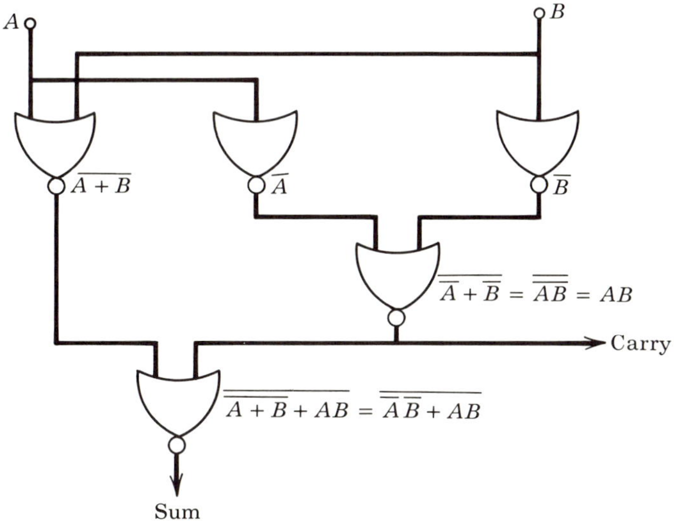

Fig. 4-4 Half adder.

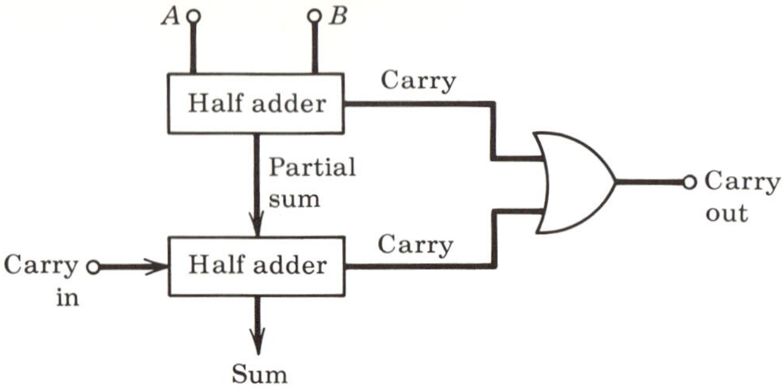

Fig. 4-5 Full adder.

(4-6*b*) can be synthesized with fewer logic elements than the statement of Eq. (4-7*b*).

Equation (4-7*a*) contains the carry output of Eq. (4-6*b*) as an element of the sum statement. We shall therefore synthesize the sum expressed in Eq. (4-7*a*) and extract the carry as a partial output. Figure 4-4 shows one possible synthesis, which uses only NOR gates. Two single-input gates invert the A and B inputs. The combination of $\bar{A}$ and $\bar{B}$ followed by a second inversion results in the AB carry output. A single NOR gate produces the statement $\overline{A + B}$, which is equivalent to $\bar{A}\bar{B}$ [Eq. (4-4)]. From Eq. (4-7*a*), we see that NORing $\bar{A}\bar{B}$ with AB yields the desired sum output. Ingenuity can be profitable in reducing the number of required logic elements.

Provision for a carry-in from the next lower binary position can be made by using two half adders, as shown in Fig. 4-5. A separate full adder would be used for each binary column except in the least significant position, where a half adder suffices.

4-2 Diode Logic Gates

OR gates

The circuit of a typical three-diode OR gate designed to operate with positive logic is shown in Fig. 4-6. The time and amplitude relationships between the various inputs and the output are illustrated in the same figure.

When all inputs are at the logical-0 level ($V_0 \leq 300$ mV), the diodes are effectively nonconducting and the output remains zero. Once any single input is raised to the logical-1 level (say, $V_A = 2$ V), the associated diode D_A conducts, establishing a transmission path between the input and the output. Because of the diode drop, the output will be roughly 700 mV less than the input ($V_Y = 1.3$ V). All other diodes remain back-biased. Only the single source V_A supplies power; it sees the load $R_L = 4$ K.

Suppose that two inputs are simultaneously present and greater than the minimum logical-1 level. Only the diode connected to the highest voltage conducts. To prove this, let us apply the inputs $V_A = 0$ V, $V_B = 2.5$ V, and $V_C = 2.0$ V and let us guess that initially only diode D_C conducts. The resultant output V_Y is 700 mV lower than V_C, or 1.3 V. With 1.3 V at the output, the other excited diode D_B is forward-biased. Thus there appears to be a transmission path for each of the inputs. However, this argument is fallacious: Once D_B conducts, the output reaches one diode drop below V_B, or 1.8 V. The forward drop across diode D_C will be only 200 mV compared with the normal conduction drop of 700 mV. From the diode equation we find that the current through D_C will be more than 10^6 below the normal range of forward current, and we can safely say that diode D_C is cut off. Diode D_A is actually back-biased by 1.8 V. We maintain the single transmission path only through D_B. *At every instant, the output of any diode OR gate will automatically become equal to one diode drop below the largest excitation signal.*

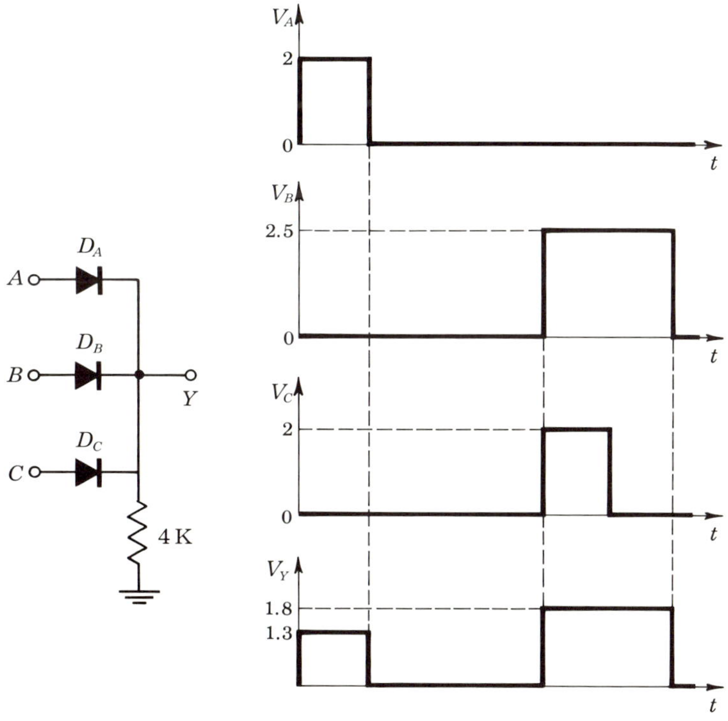

Fig. 4-6 Diode OR gate for positive logic.

For the OR gate to be closed under logical-0 excitations, all diodes must remain back-biased. It follows that the edge of diode conduction sets the limit of the maximum possible logical-0 level. When the diode conducts on a logical-1 input, the series drop in the transmission path must still leave the output identifiable as a logical 1. The input signal must be at least 700 mV above the minimum acceptable output logical-1 level.

The OR gate may be considered analogous to a group of normally open parallel relays or switches connected from various signal sources to the common output. Closing any one provides an output regardless of the state of any of the others. A mechanical system of several power sources coupled through clutches to the same drive shaft behaves similarly. The engagement of any single clutch furnishes rotational power, and if ratchet or slip clutches are used, only the one connected to the fastest-rotating power source transmits power. All others slip (are back-biased).

AND gates

An AND gate determines the coincidence of inputs, presenting a logical-1 output when, and only when, all inputs are simultaneously excited. Its operation may be compared to a parallel combination of normally closed switches shorting the output to ground; only after opening all switches will the output be other than zero. In fluidics, the analogous circuit is a series of check valves which must all be opened before any fluid can flow from the source to the sink.

A basic three-diode AND operating on positive logic appears in Fig. 4-7*a*. In contrast to the OR gate, where the diodes were normally off, the power supply biases the AND-gate diodes on. Thus if any single input is at the logical-0 level, the associated diode conducts. We can consider the 700-mV output at point x to be a shifted logical-0 level. Below we shall discuss a technique for downshifting back to a true logical 0.

In order to raise the output to a logical 1, all inputs must be high and all diodes should be off. The output will then be equal to the supply voltage V_{CC}, which is set well above the logical-1 level. Our AND statement becomes: *Only when A AND B AND C are simultaneously at the logical-1 level will the output become a logical 1.*

In the AND gate of Fig. 4-7*b*, an additional diode D_y is inserted in the transmission path between point x and the output. This diode is always conducting. As shown in the models of Fig. 4-8, its forward drop is used to compensate for the drop across the input-gate diode and thus downshift the voltage at point x by 700 mV. The output amplitude will now be a true logical 0 for logical-0 inputs.

If all inputs are at the logical-0 level, which diodes are conducting? Suppose for the moment that the three logical-0 inputs are all different:

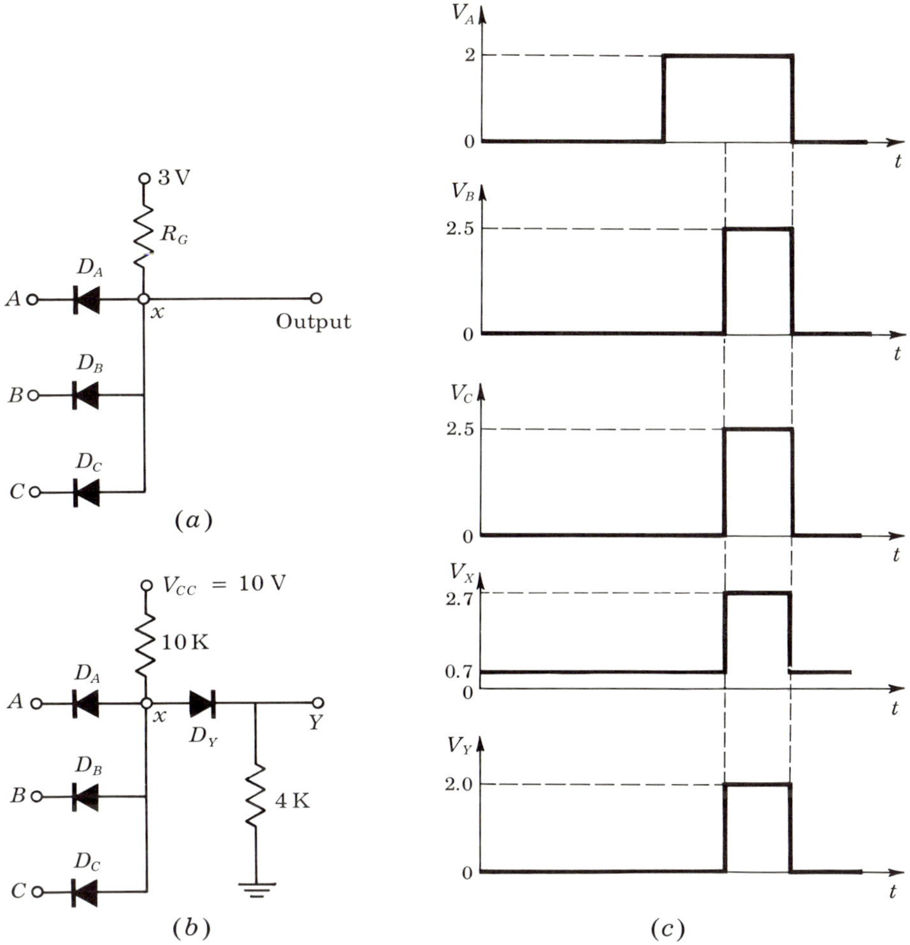

Fig. 4-7 Diode AND gate for positive logic. (*a*) Basic AND gate; (*b*) modified gate illustrating downshifting; (*c*) gate response.

$V_A = 0.3$ V, $V_B = 0.2$ V, and $V_C = 0$ V. Let us initially assume that diode D_A conducts. The output at point x will be one diode drop above V_A, or 1.0 V. This obviously forward biases the other two diodes. But the conduction of diode D_C pulls the output down to 700 mV. The voltage across D_B falls to 500 mV, and across D_A the forward drop is only 400 mV. In both cases, these diodes are effectively cut off. Our original assumption was incorrect. But our argument led to the consistent statement: *In an AND gate, only the single diode connected to the lowest potential conducts; all others are back-biased.*

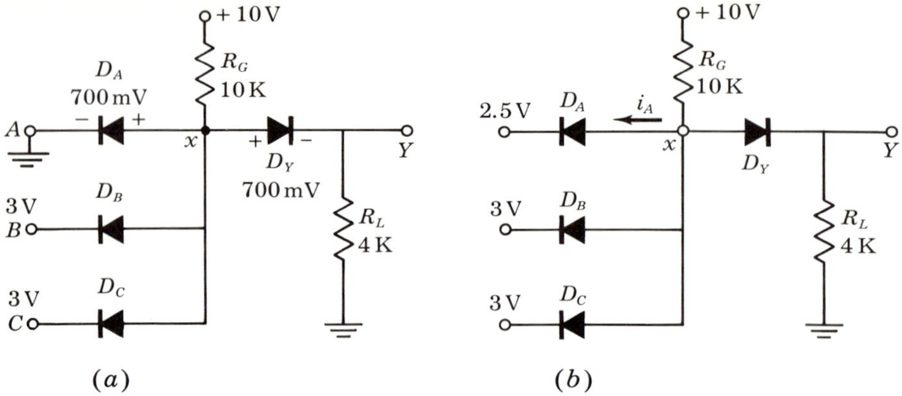

Fig. 4-8 Models for diode AND gate. (a) Logical-0 output; (b) logical-1 output.

Let us now calculate the input-current requirements under both logical-0 and logical-1 excitations. The worst condition is the logical-0 input, where the diode is conducting heavily. From Fig. 4-8a, with a single input at a logical-0 level of 0 V, point x will be at 700 mV. Very little current flows through the 4-K load resistance. Therefore the gate-diode current under logical-0 excitation is given by

$$I_{A0} = \frac{V_{CC} - V_D}{R_G} = \frac{10 - 0.7}{10\text{ K}} = 930\ \mu\text{A} \tag{4-8}$$

The signal source must be able to act as a sink for the diode current. It must have a current-sink capability equal to NI_{A0}, where N is the number of gate diodes driven from the same source. Suppose that the source has appreciable internal impedance R_s. Then the total current NI_{A0} flowing through R_s may result in an input voltage above the maximum acceptable logical-0 level. If, for example, five gates were driven by a source having only 100-Ω internal impedance, the drop across the 100 Ω would be 465 mV, which is well above our acceptable 0 level of 300 mV.

Under logical-1 excitations, some or even all the diodes can even be driven into cutoff, and if they are, only a small reverse junction current flows. In the case shown in Fig. 4-8b, only the diode D_A connected to the lowest input conducts, while D_B and D_C are essentially open circuits. The output will be equal to V_A, or 2.5 V. Thus there is 625 μA flowing in R_L. The total current from the supply is given by

$$I_{s1} = \frac{V_{CC} - V_x}{R_G} = \frac{10 - 3.2}{10\text{ K}} = 680\ \mu\text{A}$$

Therefore the current through the single conducting input diode is only

$$I_{A1} = I_{s1} - I_L = 55\ \mu\text{A}$$

As we increase the logical-1 voltage levels, the AND gate eventually becomes self-limiting. Suppose we assume that the input levels are high enough to back-bias all gate diodes. Then the full supply current flows through the 4-K load, resulting in an output voltage of

$$V_Y = 4\text{ K } I_s = \frac{10 - 0.7}{14\text{ K}} 4\text{ K} = 2.66\text{ V}$$

Because of the cancellation by the bucking series diodes, when the logical-1 inputs rise above 2.66 V, the input diodes are all back-biased and the output remains at 2.66 V independent of the input levels. The AND gate has become self-limiting. To take advantage of this self-limiting, we shall now increase our minimum positive-logic logical-1 input level to 3 V.

If negative logic is used, where the high state is defined as logical 0 and the low state as logical 1, the AND gate of Fig. 4-7 functions as an OR gate. Any single input dropping from 3 V (logical 0) to 0 V (logical 1) will cause the associated diode to conduct. Any single conducting input diode pulls the output down to the logical-0 level.

In a similar manner, we can show that the OR gate of Fig. 4-6, with negative-logic inputs, will operate as an AND gate. All inputs must be raised from 0 V to the negative-logic level of 3 V before the output rises to the logical-0 level of 3 V.

When synthesizing complex logic statements, such as the half adder of Fig. 4-4 or the full adder of Fig. 4-5, the output of one gate must feed the input of one or more other gates. The output impedance of the simple OR and AND diode gates of Figs. 4-6 and 4-7 is very high, and consequently it is difficult to design them to maintain the minimum logical-1 level and not exceed the maximum logical-0 level when draining or supplying current. Furthermore, since the diode gate is completely passive, there is no internal provision for the regeneration of the signal. A logical-1 signal subjected to successive attenuation by the loading of even a few stages would soon be below the acceptable minimum level.

4-3 Diode Transistor Logic (DTL)

Several of the disadvantages of the simple diode gate may be overcome by directly coupling its output into a saturating transistor inverter similar to the one analyzed in Sec. 3-3. The saturation and cutoff states of the inverter reestablish the required logic levels. Figure 4-9 shows the resulting positive-logic three-diode NAND gate (NOR gate for negative logic).

We can identify the three input diodes of Fig. 4-9 as the basic diode AND gate. For a logical-0 level of 0 V, at least one diode conducts, bringing point x to 700 mV. But in order for the transistor to conduct, point x must be, at least, at three junction drops. Thus a logical-0 input certainly keeps the transistor off. The inverted output is at V_{CC}, which is well above our minimum logical-1 level.

When all inputs are at a logical 1, the input AND-gate conditions are satisfied and we can assume that all gate diodes are back-biased. Bias current now flows through the 4-K resistor and the decoupling diodes into the base of the transistor, driving it into saturation. The output drops to the low logical-0 level of $V_{CE,\mathrm{sat}}$.

The two break points of the linearized transfer characteristic of Fig. 4-9 establish the absolute limits of the logical-0 and logical-1 driving signals. At the lower knee (0), the transistor is off and just going on. Assuming a turn-on $V_{BE_0} = 600$ mV, the corresponding input signal is found by summing the two additive coupling diode voltages and subtracting the input gate-diode drop:

$$\underline{V_{\mathrm{in}_0}} = \underline{V_D} + \underline{V_{BE_0}} \approx 1.3\ \mathrm{V} \tag{4-9}$$

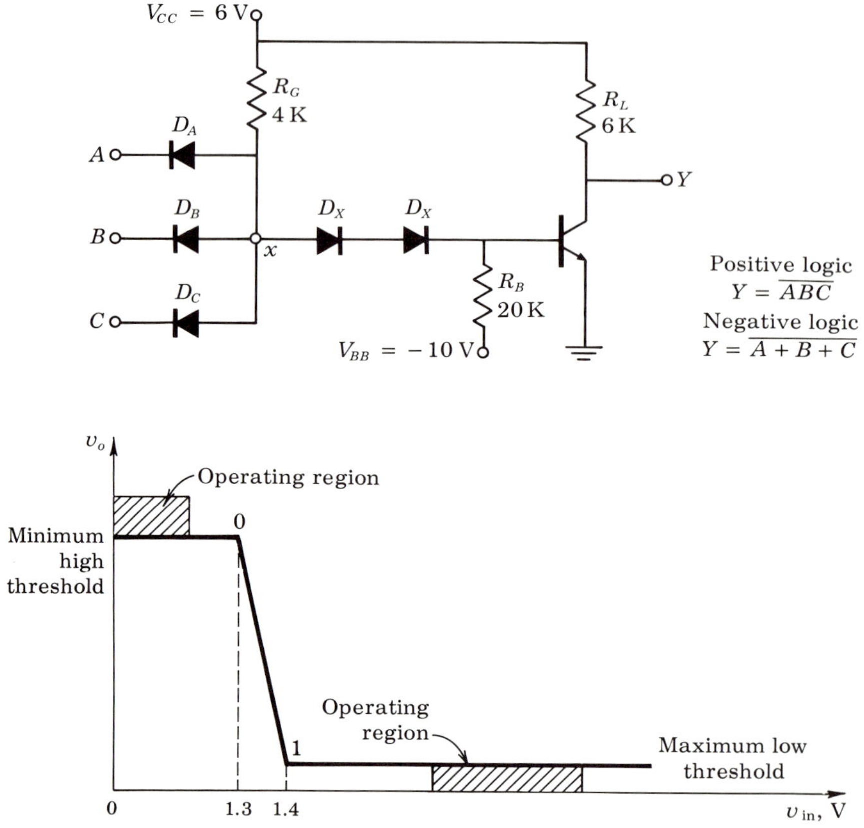

Fig. 4-9 A DTL NAND/NOR gate and transfer characteristic.

An underbar indicates that we are concerned with the minimum possible break voltage.

Transistor saturation is accompanied by an increased emitter-current flow and hence an increased V_{BE}. Taking $V_{BE,\text{sat}} = 700$ mV, the upper knee (1) occurs at

$$\overline{V_{\text{in}_1}} = \overline{V_D} + \overline{V_{BE,\text{sat}}} \approx 1.4 \text{ V} \tag{4-10}$$

In Eq. (4-10), the overbar indicates that the worst-case condition is the maximum expected value of the logical-1 break point.

Actual logical-0 and logical-1 signals are kept well away from the break points. The hatched regions of the transfer characteristic indicate the ranges of input and output signals expected in the DTL gate.

Once the transistor is driven into saturation, the output drops from V_{CC} to the small saturation voltage $V_{CE,\text{sat}}$. The base current of the transistor is found from the circuit of Fig. 4-9:

$$I_b = \frac{6 - 2.1}{4 \text{ K}} - \frac{10 + 0.7}{20 \text{ K}} = 440\ \mu\text{A}$$

Assuming a minimum $\beta = 30$, the collector current in an active transistor corresponding to this base drive is

$$I_c = 30 \times 440\ \mu\text{A} = 13.2 \text{ mA} \tag{4-11}$$

For collector currents less than 13.2 mA, the transistor will be saturated. Since we want the output to remain at the low logical-0 level, we must ensure that nothing causes the transistor to come out of saturation. Under no-load conditions, the actual collector current of the inverter is only

$$i_c = \frac{V_{CC} - V_{CE,\text{sat}}}{R_L} = \frac{6 - 0.15}{6 \text{ K}} \approx 1 \text{ mA} \tag{4-12}$$

where we assume $V_{CE,\text{sat}} = 150$ mV. Certainly the transistor is saturated. The inverter output stage can even act as a sink for 12.2 mA from the inputs of other gates before the total current flow into the collector brings the transistor out of saturation.

Fan-out capabilities

Figure 4-10 shows the output of the NAND gate of Fig. 4-9 connected to the inputs of N similar gates. For a logical-0 gate output, we have for each conducting input a diode current of

$$I_{D0} = \frac{V_{CC} - V_D - V_{CE,\text{sat}}}{R_G} = \frac{6 - 0.7 - 0.15}{4 \text{ K}} = 1.27 \text{ mA} \tag{4-13}$$

From Eqs. (4-11) and (4-12), we see that the external current which can flow into the output stage without causing the transistor to come out of

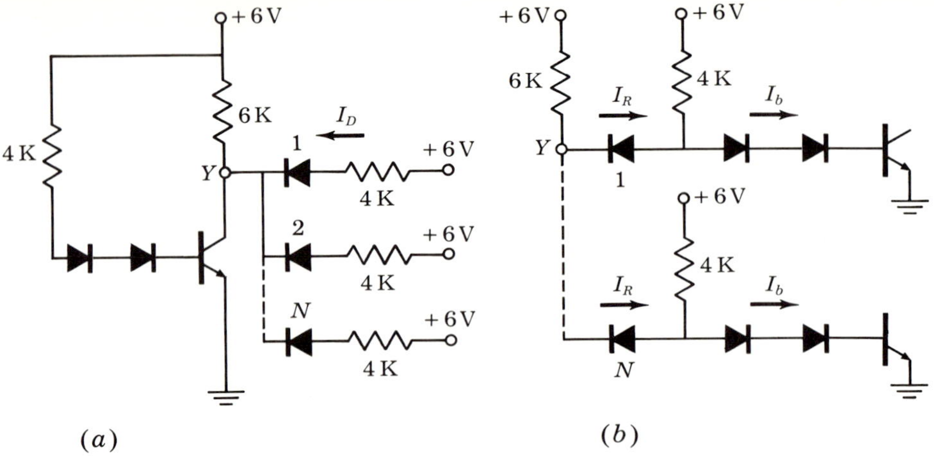

Fig. 4-10 DTL NAND gate fan-out models. (*a*) Logical-0 output; (*b*) logical-1 output.

saturation is 12.2 mA. The maximum number of gate inputs which can be coupled to the output of a NAND gate is

$$N = \frac{I_{c,\text{external}}}{I_{D0}} = \frac{12.2 \text{ mA}}{1.27 \text{ mA}} = 9 \qquad (4\text{-}14)$$

The acceptable solution of Eq. (4-14) is the smaller whole integer.

When the NAND output is in the logical-1 state (Fig. 4-10*b*), the input diodes of the following driven gates are back-biased. Only a small reverse current flows into each. The total output current in the 1 state is only

$$I_{D1} = I_R N$$

where N is the fan-out capability determined under the high-current logical-0 condition. I_{D1} is extremely small, and consequently the drop across R_L will be negligible; the loaded logical-1 output remains at V_{CC}. A resistive load at the output would, however, adversely affect the logical-1 voltage level.

DTL noise immunity

In order to prevent a gate from falsely switching from a 0 to 1, or vice versa, due to random noise pulses, we must ensure that the applied input levels are well inside the thresholds. These calculations should be performed under worst-case conditions.

Figure 4-10 shows the DTL NAND gate driven by a similar gate and loaded with the fan-out N. The maximum logical-0 output occurs under maximum fan-out. From the logic-0 model of Fig. 4-11, we can write

$$\overline{V_{\text{out}_0}} = (I_{R_L} + I_0\overline{N})r_{sc} + V_{CE,\text{sat}} \tag{4-15}$$

For the numbers used before and with $r_{sc} = 25\ \Omega$,

$$\overline{V_{\text{out}_0}} = (1 + 1.27 \times 9)25 + 150 = 460 \text{ mV}$$

The logical-0 noise immunity is the difference between the minimum value of the lower knee of the transfer characteristics, defined in Eq. (4-9), and the maximum logical-0 output given in Eq. (4-15). Under logical-0 conditions,

$$NI_0 = \underline{V_{\text{in}_0}} - \overline{V_{\text{out}_0}} = 1.3 - 0.46 = 0.84 \text{ V} \tag{4-16}$$

Equation (4-16) gives the minimum-size noise pulse which will cause the worst gate to just start changing state. It represents a safety factor in the logic design.

To calculate the noise immunity at the logical-1 level, we must subtract the worst-case (maximum) input required to drive the transistor just to the edge of saturation from the minimum expected logical-1 output obtained from a loaded cutoff transistor. The output logical 1 is found from the model of Fig. 4-10*b*:

$$\underline{V_{\text{out}_1}} = V_{CC} - R_L I_R \overline{N} \approx V_{CC}$$

and, from Eq. (4-10), defining the upper knee of the transfer curve,

$$NI_1 = \underline{V_{\text{out}_1}} - \overline{V_{\text{in}_1}} = 6 - 1.4 = 4.6 \text{ V} \tag{4-17}$$

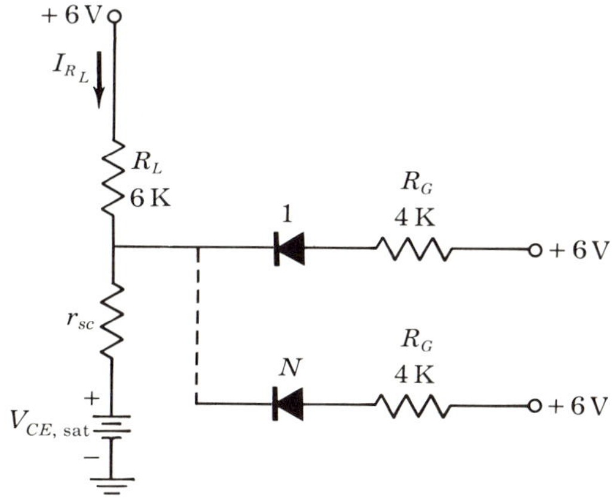

Fig. 4-11 Model of logical-0 output of a NAND gate loaded by *N* inputs.

A third diode in the decoupling path would shift the transfer characteristic 700 mV to the right, increasing NI_0 to 1.625 V and lowering NI_1 to 3.9 V.

Often a desired NI is one of the limiting design conditions. This may require a fan-out restriction much more severe than that given by Eq. (4-13). For example, if we specify that the worst-case noise immunity must be about 1 V, then by setting $NI_0 \approx 1$ V in Eq. (4-16), we find that we must ensure the restriction

$$\overline{V_{\text{out}_0}} \leq 0.3 \text{ V}$$

Solving Eq. (4-15) yields $N = 4$, compared with the $N = 9$ found previously from saturation limitations [Eq. (4-13)].

Switching times

Whenever any one input switches back to a logical 0, the associated diode conducts and pulls down the base of the output transistor. Since originally the transistor was strongly saturated, the appreciable charge stored in the base must be pulled out before the transistor first becomes active and then eventually cuts off. The 20-K resistor returned to a negative supply is used to speed up the recombination of this charge. (See Sec. 3-7 for a fuller description of the turn-off process.) Sometimes, in order to ensure that the transistor turns off first, slow-recovery diodes are used for coupling. The recombination taking place in the diode makes use of charge stored in the transistor's base. The 20-K resistor and negative power supply may then be replaced by a single resistor to ground.

The base-emitter depletion-layer capacitance must be charged in order to turn on a cutoff transistor. Stray circuit capacity further slows the turn-on time; and when a gate contains several transistors between the input and the output, the individual device delays add.

Since we normally apply the output of one gate as the input to an identical gate, we shall assume that the input and output waveshapes are similar. Figure 4-12 shows the input and the delayed output. The propagation-delay time is defined as

$$t_{pd} = \frac{t_{d1} + t_{d2}}{2} \tag{4-18}$$

where the leading and trailing delay times are measured at the 50 percent level of the pulse. t_{pd} represents the average time shift in the pulse position.

We also define the signal rise time and fall time, respectively, as the time required for the output to rise from 10 to 90 percent of the final value and to fall from the 90 percent to the 10 percent point. Standard DTL logics have rise and fall times of about 50 ns and propagation delays of approximately 30 ns.

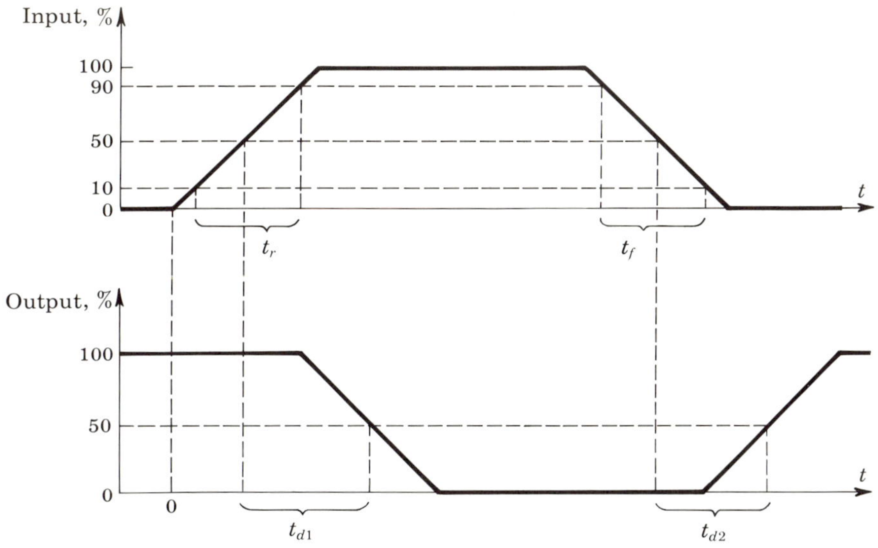

Fig. 4-12 Propagation delays.

An integrated-circuit DTL employs the configuration shown in Fig. 4-9. The input diodes are usually emitter-base junctions, with the collector short-circuited to the base by metallization. This configuration results in minimum charge storage and fastest diode turn-off times. Figure 4-18 shows the geometry of the input circuit diode AND gate.

Complementary-symmetry logic

If the collector epitaxial layer is thin enough, or if the transistors are made by triple diffusions, then we may use the base-collector-substrate layers as a *pnp* transistor. The extra gain of this stage results in a slight reduction in the input current requirements under logical-0 input. Figure 4-13 illustrates a NAND-gate input stage.

It is difficult to obtain both good *pnp* and good *npn* transistors in the same chip; *pnp* transistors in Fig. 4-13 have a β of only about 3.

The replacement of the diode by the input transistor reduces the input current under logical-0 excitation to

$$I_{D_0} \approx \frac{V_{CC} - V_{BE}}{R(\beta + 1)} \tag{4-19}$$

where β is only 3 or 4. When we compare Eq. (4-19) with Eq. (4-13), we see that the loading of the output by similar gate inputs is, consequently, also reduced by $1/(\beta + 1)$. The fan-out capability is improved. A second advantage of this circuit configuration is that the input transistors

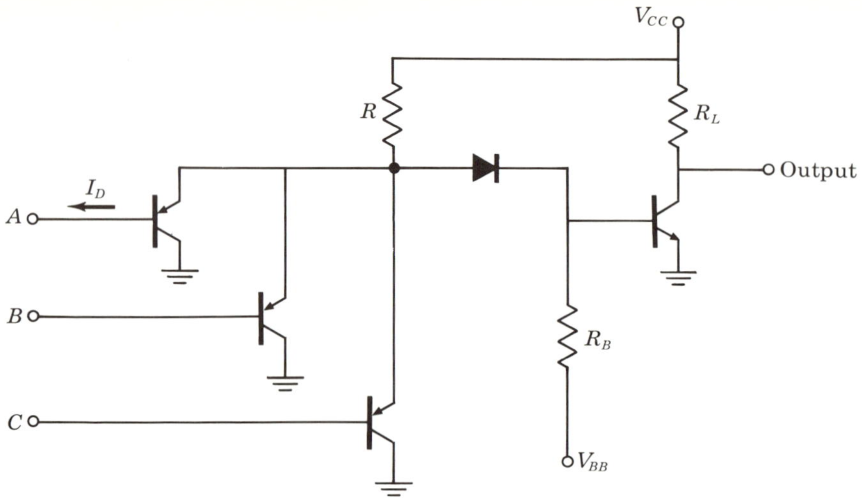

Fig. 4-13 Modified DTL logic (MDTL).

are not saturated under a logical-0 input. Excess charge is simply shunted into the substrate. Thus the switching time is improved and the propagation-delay time is slightly reduced. Rise and fall times of 25 to 50 ns and propagation-delay times of 30 ns are typical.

4-4 Direct-coupled Transistor Logic (DCTL)

It is extremely easy to make a good transistor in a monolithic integrated circuit. In fact, the diodes used in the input AND gate are usually transistors with the base and collector shorted during the metallization process. Resistors occupy almost as much area on the chip as do the transistors. Therefore let us consider the feasibility of direct-coupled transistor logic (DCTL). The two possible circuit arrangements appear in Fig. 4-14.

In the NOR gate of Fig. 4-14*a*, whenever one transistor is excited with a logical 1, it turns on and pulls point Y from V_{CC} to near ground. In the series NAND gate of Fig. 4-14*b*, only when all transistors are forward-biased will the conduction path be established, yielding a logical-0 output.

The direct-coupled transistor NAND gate has some obvious disadvantages:

1. The logical-0 output is very high. It is equal to the saturation drops of the three transistors in series. Furthermore, the saturation resistances of these three transistors sum, resulting in poor noise immunity and low fan-out capabilities.

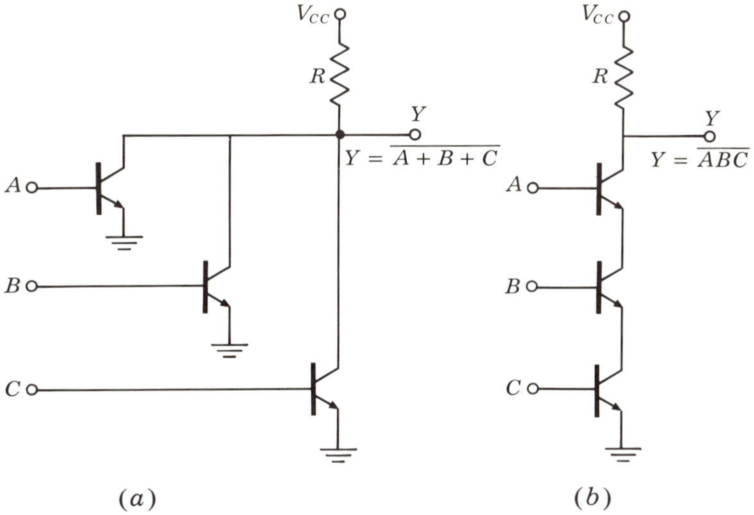

Fig. 4-14 Direct-coupled transistor logic (DCTL). (*a*) Parallel-transistor NOR gate; (*b*) series-transistor NAND gate.

2. If two or more gates are driven from the same source, the input transistors closest to ground will prevent the highest-input-level transistor from turning on. For example, consider an A input and a C input in parallel. The C input sets a single V_{BE} drop, and hence we would not have sufficient signal amplitude to properly excite the A input.

The direct-coupled NOR gate also has certain inherent disadvantages. Suppose that two gate inputs are being driven from the same logical-1 output, as shown in Fig. 4-15. The currents may be expressed by

$$I_1 = I_{rs_1} e^{V_{BE}/V_T} \qquad I_2 = I_{rs_2} e^{V_{BE}/V_T} \tag{4-20}$$

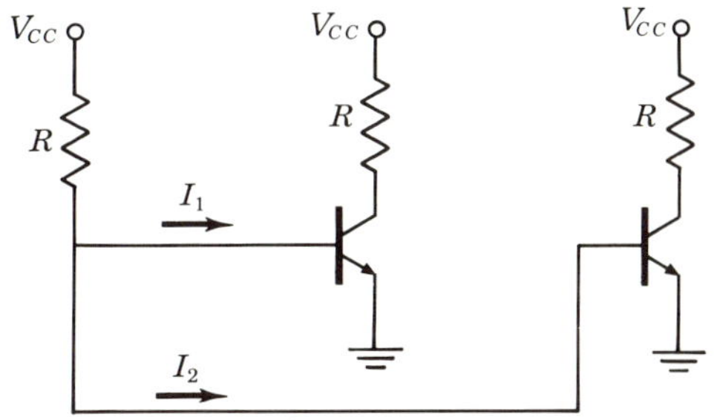

Fig. 4-15 DCTL inputs fed from the same logic output.

Thus, since V_{BE} must be the same for each transistor,

$$\frac{I_1}{I_2} = \frac{I_{rs_1}}{I_{rs_2}} \tag{4-21}$$

If the two chips are made at different times or are even located at different points on the same silicon wafer, there may be an appreciable difference in the doping and, consequently, in the reverse saturation current. This problem is further compounded when we parallel the inputs of N separate gates made at different times and even by different manufacturers. The high-current transistors starve the low-current devices: they prevent the low-current transistors from saturating. Within any one chip, however, the transistors are uniform enough to employ direct-coupled logic; both NOR and NAND are used. DTL or RTL input stages interface with the external circuits.

4-5 Resistor-Transistor Logic (RTL)

RTL logic is very simple, very cheap, and ideally suited for slow-speed industrial and commercial applications. As shown in Fig. 4-16*a*, the basic NOR gate incorporates a simple modification of the DCTL gate. Starvation of an input transistor is prevented by inserting a small resistor in series with the base. This resistor equalizes the base-current flow when many inputs are connected to the same source.

Logical-0 signals keep the gate transistors off, with the output going to the logical-1 level of V_{CC}. Any single logical-1 input causes the transistor to saturate, pulling the output down to the logical-0 level of $V_{CE,\text{sat}}$.

Limiting conditions for proper logic operation are found from the models of Fig. 4-16*b* and *c*. Worst-case loading occurs under a logical-1 output when the turn-on base current for N subsequent gates flows through the collector resistor R_2. From the model of Fig. 4-16*c*, the loaded logical-1 output voltage is

$$V_1 = \frac{(V_{CC} - V_{BE})(R_1/N)}{R_1/N + R_2} + V_{BE} > R_1 I_{b,\text{sat}} + V_{BE} \tag{4-22}$$

Equation (4-22) also indicates the inequality necessary to ensure saturation of the driven gates. The difference between the actual logical-1 driving signal and the edge of saturation, defined by the right-hand side of the inequality, is the logical-1 noise immunity

$$NI_1 = \underline{V_1} - \overline{(V_{BE} + I_{b,\text{sat}} R_1)} \tag{4-23}$$

Figure 4-16*b* shows the circuit when the output is at a logical 0. The only external current flow into the saturated transistor is the leakage current

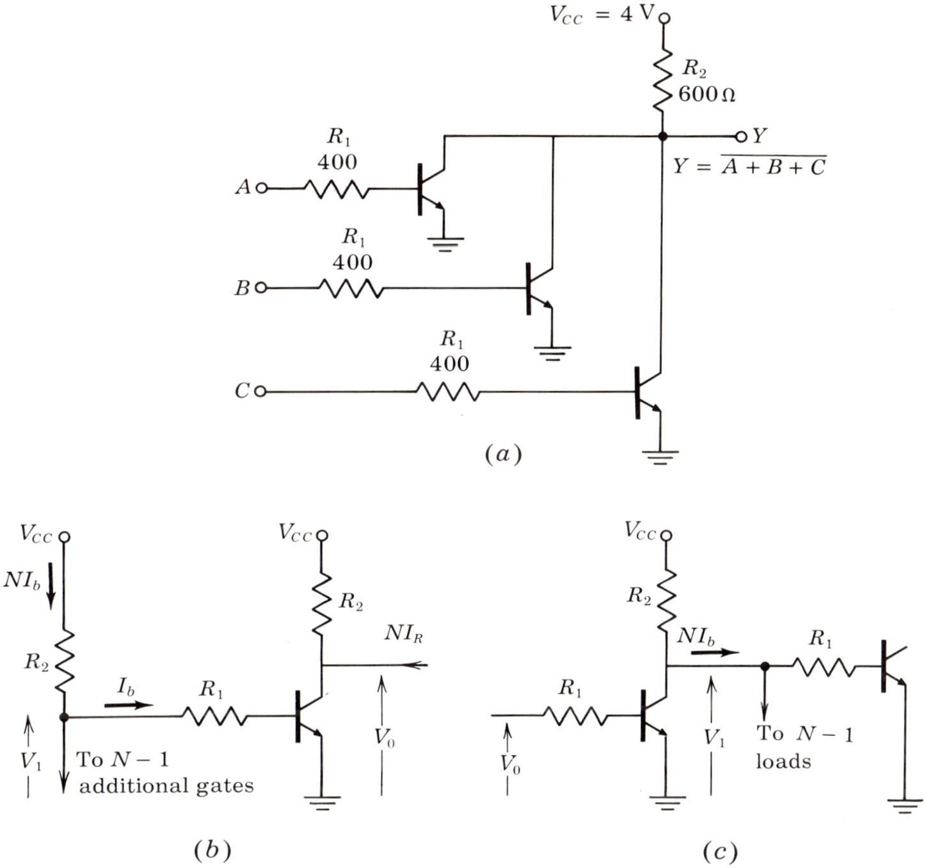

Fig. 4-16 (*a*) Basic RTL NOR gate; (*b*) loaded logical-1 input, logical-0 output; (*c*) logical-0 input, loaded logical-1 output.

of the back-biased junctions. I_0N, as well as the internal I_c, flows through r_{sc}, raising the logical-0 output voltage slightly above $V_{CE,\text{sat}}$:

$$V_0 = V_{CE,\text{sat}} + (I_0N + I_c)r_{sc} \tag{4-24}$$

We can, of course, neglect the very small I_0N current term in Eq. (4-24). As before, the noise immunity under logical-0 conditions can be identified as

$$NI_0 = \underline{V_{BE}} - \overline{(V_{CE,\text{sat}} + I_cr_{sc})} \tag{4-25}$$

We can improve NI_0 only through redesign of the RTL circuit. An increase in R_2 will reduce the I_cr_{sc} drop. V_{BE} can be effectively doubled by introducing a series diode drop.

Only V_1, and hence NI_1, is affected by external loading. For a specific noise immunity, the fan-out becomes

$$N = \frac{R_1(V_{CC} - V_{BE} - R_1 I_{b,\text{sat}} - NI_1)}{R_2(R_1 I_{b,\text{sat}} + NI_1)} \tag{4-26}$$

To solve Eq. (4-26), we first have to know $I_{b,\text{sat}}$. Starting from the saturated model of Fig. 4-16*b*,

$$I_c = \frac{V_{CC} - V_{CE,\text{sat}}}{R_2 + r_{sc}}$$

and $I_{b,\text{sat}} = I_c/\beta$.

For example, suppose that we require 300-mV noise immunity and choose the NOR gate of Fig. 4-16*a* as our basic logic. From the saturated model of a typical transistor whose parameters are $\beta_{\min} = 30$, $V_{BE,\text{sat}} = 750$ mV, $V_{CE,\text{sat}} = 100$ mV, and $r_{sc} = 25\ \Omega$, we find

$$I_c = \frac{4 - 0.10}{625} = 6.24 \text{ mA}$$

Substituting into Eq. (4-24), the logical-0 output becomes $V_0 = 256$ mV. Assuming that V_{BE} at the edge of conduction is 650 mV, $NI_0 \approx 400$ mV, which certainly satisfies our constraint.

The fan-out limitations are now found by solving Eq. (4-26). With $I_{b,\text{sat}} = 208\ \mu$A, we obtain $N = 5$. The actual logical-1 voltage with a fan-out of 5 is only 1.132 V (for $N = 1$, $V_1 = 2.05$ V).

Base current is also affected by fan-out. With $N = 5$, $I_b = 955\ \mu$A, increasing to 3.25 mA for $N = 1$. In both cases, I_b is well above the boundary value. The transistor is solidly saturated, with all that implies with respect to long storage-delay times. Expected propagation delays in an RTL logic range up to 50 ns.

RCTL logic

Figure 4-17 shows a modified RTL gate. In order to shorten the storage-delay interval, the series R_1 is shunted by a small capacitor. This change converts the driving-point network of the inverter into a compensated attenuator. In Sec. 3-7, it was shown that for optimum switching response, CR_1 should be adjusted to match the base time constant. Propagation delays of 10 ns are typical when one RCTL input is driven from the low-impedance output of a similar gate.

The necessity of including capacitors is a major disadvantage of the RCTL circuit. Each of the 25-pF capacitors occupies about 80 mil^2 of surface area, thus limiting the total number of components which can be made on a given-size die.

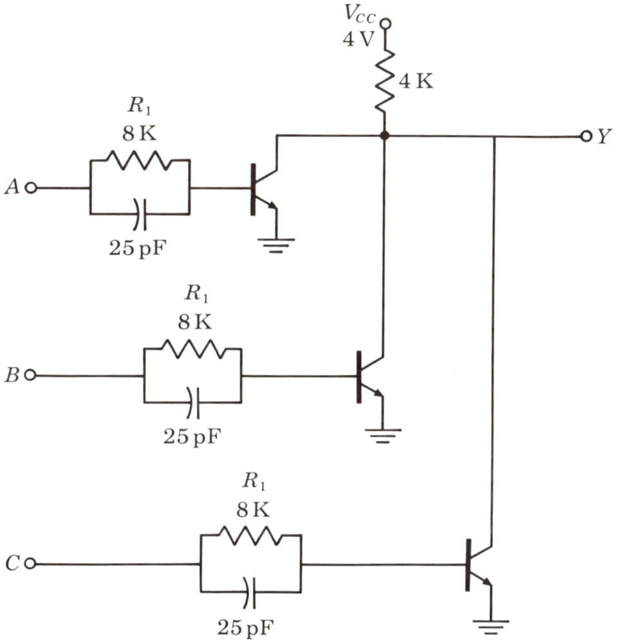

Fig. 4-17 RCTL positive-logic NOR gate.

4-6 Transistor-Transistor Logic (TTL or T^2L)

Switching speed of an integrated logic can be greatly improved by exercising ingenuity in the layout of the components. We cannot examine the properties of all possible geometries, but because they illustrate how a fast class of logic can be developed from another relatively slow type without really changing the circuits, three particular geometries will be discussed below.

Figure 4-18*a* shows the geometry of the diode input network of a two-input DTL logic (see also Fig. 4-9). The three diodes are formed by diffusing three separate base regions into a single *n* collector island. Three transistors connected together at the collector result from the final emitter diffusion. Each separate emitter-base junction is used as a diode. Two junctions become the basic two-diode AND gate. The third junction is used as the coupling diode which connects the AND logic to the base of the output-inverter stage. From the circuit we see that all three diodes of Fig. 4-18*b* have a common *p*-region connection. Therefore, in order to minimize the base-charge storage, each base can be shorted to the collector during the final metallization process.

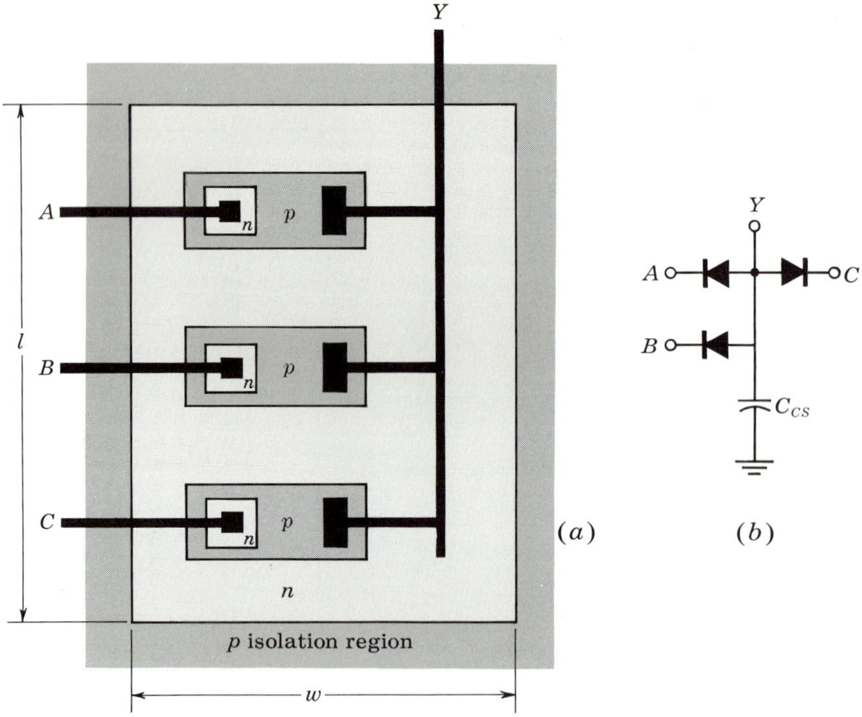

Fig. 4-18 Integrated-circuit diodes—separate base regions. (*a*) Geometric layout; (*b*) three-diode circuit.

Both the edge and the bottom of the collector island form a back-biased junction with the substrate. Assuming an epitaxial collector, the stray capacity of the isolation diode is given by

$$C_{CS} = \frac{2(w + l)dK}{\sqrt{V}} + \frac{wlK}{\sqrt{V}} \tag{4-27}$$

The first term of Eq. (4-27) is the capacity due to the perimeter area of the isolation region, where d is the thickness of the n epitaxial layer. The second term is the capacity at the bottom of the collector island. V is the back-bias voltage from substrate to collector, and K is a constant dependent on the material and the doping [see Eq. (2-54)].

In triple-diffused integrated circuits, the majority charge concentration varies with depth. We must, consequently, use the graded junction approximation for the perimeter contribution. The first term of Eq. (4-27) will vary as $V^{-1/3}$ instead of as $V^{-1/2}$. The bottom contribution remains the same.

During any change in the circuit state, C_{CS} must be either charged or discharged. Since C_{CS} loads the input-signal source, anything that will reduce C_{CS} will reduce proportionately the switching time.

Figure 4-19*a* shows the first improvement over the geometry of Fig. 4-18. We take advantage of the fact that in an AND gate, the *p* regions of all three diodes are connected together. The diodes are now formed within a common-base diffusion region. We eliminate the isolation areas between the separate base regions and thus greatly reduce the surface area. The new geometry results in a much smaller C_{CS}.

When the input diodes are simultaneously at a logical 1 or at a logical 0, they behave in exactly the same manner as do independent diodes. With both inputs at the logical-1 state, the diodes are back-biased; only the small reverse current I_R flows through each junction.

With both inputs at a logical 0, the diodes conduct heavily. Current is limited by the pull-up resistor R_B. The total supply current will be

$$I_0 = -\frac{V_{CC} - 0.7}{R_B} \tag{4-28}$$

I_0 divides between all conducting diodes. Equation (4-28) indicates, by the minus sign, that the current is flowing out of the gate. The external logical-0 driver must be able to sink the full diode current.

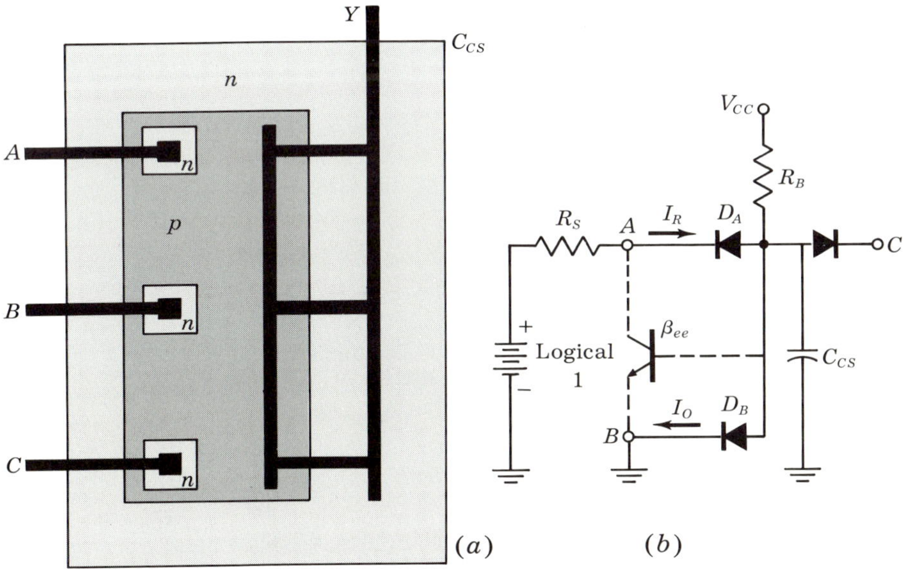

Fig. 4-19 (*a*) Integrated-circuit diodes using the same base region; (*b*) equivalent model (diode *A* at a logical 1 and diode *B* at a logical 0).

Suppose that one input is at the high logical-1 level and the adjacent one is at the low logical-0 level. The two diodes form the forward-biased parasitic *pnp* transistor shown schematically in Fig. 4-19*b*. The high-potential emitter-diffusion region acts as the collector, and the low-potential *p* region is the emitter. Because of the transistor action, the input current drain on the logical-1 source will be

$$I_{10} = \beta_{ee} \frac{V_{CC} - 0.7}{R_B} \tag{4-29}$$

β_{ee} is the β of the emitter-base-emitter transistor. By spacing the emitter regions further apart and by diffusing gold into the base (to decrease the recombination-time constant), β_{ee} can be made as small as 0.1 to 0.01. Thus I_{10} will be on the order of a few microamperes. Even though I_{10} is very small, it is still orders of magnitude greater than the diode reverse current I_R.

In the third modification, Fig. 4-20, we disconnect the metallization short circuit between the base region and the collector. The base-collector junction now replaces the separate coupling diode. Because one emitter region is eliminated, the reduction in area further reduces C_{CS}. Much more significant is the conversion of the simple diode array into a multiemitter transistor AND logic. The input transistor action affects switching speed and loading. In spite of the change, the resultant gate still operates in the ON-OFF mode at the appropriate logical-0 and logical-1 levels. It is often very convenient to think of the TTL input stage generically as a diode AND gate.

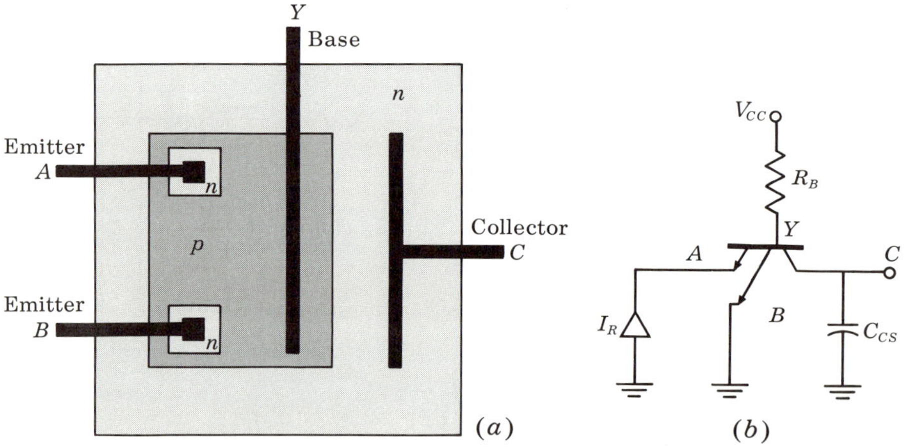

Fig. 4-20 (*a*) Development of TTL integrated circuit; (*b*) equivalent model (*A* at logical 1, *B* at logical 0). Note basic similarity to the diode AND gate of Fig. 4-18.

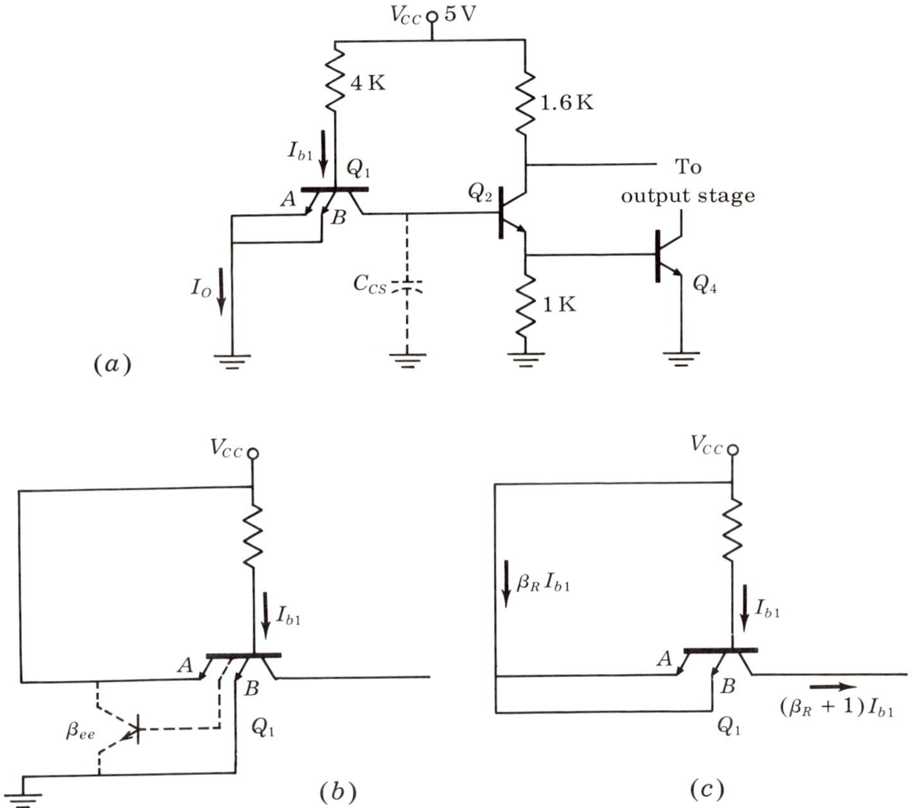

Fig. 4-21 TTL AND input stage (see also Fig. 4-28 for complete circuit). (*a*) Both inputs shown at logical 0; (*b*) *A* at logical 1, *B* at logical 0, showing parasitic emitter-to-emitter transistor; (*c*) both *A* and *B* at logical 1, showing reverse transistor operation.

The three cases shown in Fig. 4-21 represent all possible input conditions of a TTL gate. Both inputs can simultaneously be at a logical 1 or be at a logical 0, or one input can be at a logical 1 while the other is at a logical 0. For simplicity, we shall set the logical 1 = V_{CC} and the logical 0 = 0 V. Figure 4-21 also includes a portion of the output-inverter stage of Figs. 4-27 and 4-28. This circuit completes the path for the collector current of Q_1. Without it, we would be unable to define completely the switching levels and currents.

Case 1—Figure 4-21*a*, $A = B = 0$

Both input emitters are conducting heavily, pulling Q_1's base down to one junction drop. But in order to turn Q_2 on, the base of Q_1 must be at least at three junction drops (base-collector drop of Q_1 and V_{BE} of Q_2 and Q_4). Since no current flows out of the collector of Q_1, Q_2 is off. The total current through the two input emitters into the logical-0 source is the I_0 defined in Eq. (4-28).

Case 2—Figure 4-21*b*, $A = 1$, $B = 0$

When we apply a logical 1 to only a single input of an AND logic, we do not affect the state of the gate. Q_2 will still be off. The two TTL emitters and base form the same forward-biased parasitic transistor as that shown in Fig. 4-19*b*. Thus the A-terminal input current will be the I_{10} defined by Eq. (4-29). The total current out through the conducting emitter (B) will be

$$I_B = I_{10} + |I_0|$$

Case 3—Figure 4-21c, $A = B = 1$

Since in this case the input AND condition is satisfied, the base-current flow of Q_1 switches from the previously grounded emitter to the collector of Q_1 and into the base of Q_2. It drives Q_2 into saturation. But with a logical 1 at both emitters, the input transistor is correctly biased for operation in the reverse mode. The new base current of Q_1 is

$$I_{b1} \approx \frac{V_{CC} - 2.1}{R_B} = 725\ \mu\text{A}$$

The 2.1 V is the sum of the three forward-biased junction drops: the base-collector junction of Q_1 in series with the V_{BE} of Q_2 and of Q_3.

The total logical-1 driving current into the two emitters is the collector current of the reverse transistor:

$$I_{RT} = \beta_R I_{b1}$$

I_{RT} divides between the two emitters. β_R is usually on the order of 1 down to 0.1 or less. Our $I_{RT} \approx 72.5\ \mu\text{A}$ is relatively large compared with the two other reverse currents I_R and I_{10}. I_{RT} is the predominant term, and it must be the current used in the calculation of logical-1 fan-out limitations and noise immunity NI_1.

Turn-off time

The input transistor Q_1 must have a large enough base area to accommodate the multiple emitter regions comprising the AND gate. As a consequence of the larger area, more charge will be stored in Q_1 than in the following transistor Q_2. Under normal switching conditions, we can expect Q_2 to turn off before Q_1.

Suppose that one input of Q_1 is brought from the logical-1 to the logical-0 level. The charged-node capacity maintains the collector node at 1.4 V. From the model of Fig. 4-22, we see that Q_1 changes from reverse to forward transistor operation. The base current increases to $I_{b1} \approx 1$ mA. But for normal operation, the collector current must be $I_{c1} = \beta_F I_{b1} \approx 30$ mA. I_{c1} can be supplied only from the charge previously stored in the base of Q_2 and in the stray capacity of C_{CS}. It is obvious that the high collector current at the active Q_1 helps to pull out the stored charge, thus speeding the turn-off process.

If we approximate the complete storage process by combining the base-charge storage and C_{CS} into a single capacitor C_S, then the turn-off time can be found from the model of Fig. 4-22. The collector voltage of Q_1 runs down linearly as the constant current I_{c1} discharges C_S:

$$v_c(t) = 1.4 - \frac{\beta_F I_{b1} t}{C_S} \qquad (4\text{-}30)$$

Solving Eq. (4-30) for the time required for the removal of all stored charge i.e., for the time for v_c to drop to zero, we obtain

$$t_f = C_S \frac{1.4}{\beta_F I_{b1}}$$

Letting $C_S = 50$ pF and $\beta_F = 30$, the turn-off time of Q_2 turns out to be 2.3 ns.

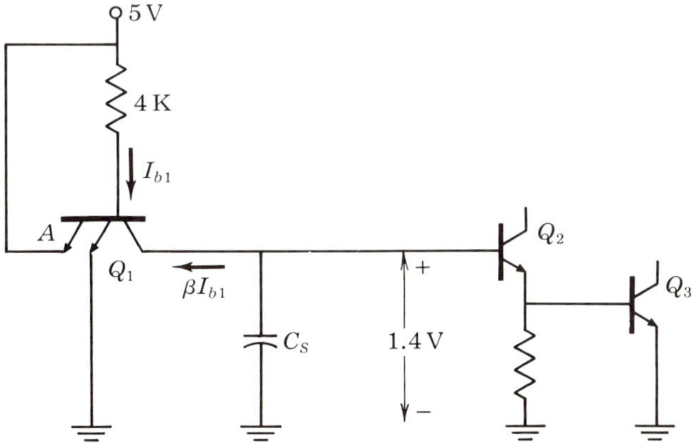

Fig. 4-22 Turn-off model for TTL logic of Fig. 4-21.

The TTL is the fastest of all saturating logics. Although propagation-delay times less than 1 ns have been achieved in laboratory prototypes, 3 to 10 ns are more realistic figures for a typical device.

4-7 Output Circuits

We have already shown that the steady-state fan-out and noise immunity of a saturated logic depend on the ability of the output, when in the logical-0 state, to act as a sink for the input currents of the driven gates. When the output is in the logical-1 state, it becomes a source supplying reverse current to the OFF inputs.

Under dynamic operation, the turn-off time of a logic is dependent on the removal of stored base charge and the turn-on time is controlled by the charging of the input-node capacitance. Each output is loaded by the capacitive inputs of other gates. For fast switching, an inverter output stage must be capable of supplying enough current to rapidly charge the capacitive load and must also be capable of acting as a sink to rapidly discharge the capacitive load.

Inverter output stage

The simplest possible output stage is the single-transistor inverter used in the DTL NAND logic of Fig. 4-9 and in the RTL NOR logic of Fig. 4-16. Figure 4-23 shows the applicable models when the inverter drives the load capacitor C_L.

So long as the input is at the logical-0 level, the inverter is cut off and the output is at V_{CC}. At $t = 0$, we apply a logical 1 (V_{CC}) sufficient to saturate the inverter. We are faced with two time-sequential problems: First, the depletion-region capacitance must be charged before the transistor turns on, and only then will the transistor discharge the load capacitor. This problem of internal transistor turn-on delay was previously solved in Sec. 3-7. In our solution for the complete response, we shall simply indicate a turn-on delay t_0.

The second problem is the discharge of C_L by the ON transistor. But which model should we use for transistor, the active or the saturated model? C_L, charged to V_{CC}, is connected directly across the transistor. During the discharge interval, a charged capacitor can supply any current required by the circuit connected across it. It follows that since the collector current is not externally limited, after the turn-on delay the transistor winds up in the active region. Figure 4-23*b* shows the equivalent model. From this model we see that the Thévenin equivalent voltage toward which the capacitor discharges is

$$V_{\text{Th}} = V_{CC} - \beta I_b R_L \tag{4-31}$$

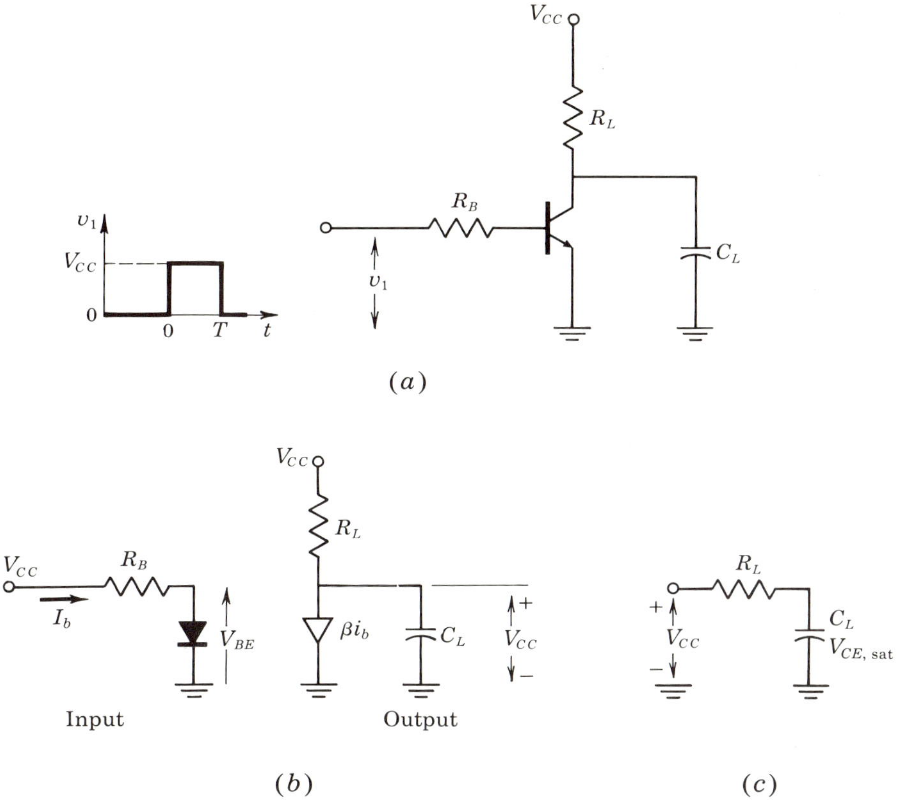

Fig. 4-23 Switching of a simple inverter output stage. (*a*) Inverter output with capacitive load; (*b*) model when input is at a logical 1 and the transistor is active; (*c*) final recovery model—input at a logical-0 transistor cutoff.

where $I_b = (V_{CC} - V_{BE})/R_B$. The time constant for the decay from V_{CC} is $R_L C_L$.

When the collector-voltage rundown reaches $V_{CE,\text{sat}}$, the transistor saturates and the discharge terminates. This takes

$$t_f = C_L R_L \ln \frac{V_{\text{Th}} - V_{CC}}{V_{\text{Th}} - V_{CE,\text{sat}}} \tag{4-32}$$

Saturated inverters are normally designed to be driven far into saturation. V_{Th} turns out to be a very large negative number, and the collector rundown from $V_{CE,\text{sat}}$ is only a small portion of the total exponential.

We can approximate the actual response by assuming that the βI_b-controlled source acts as a constant-current drain on the charged C_L. In making this approximation, we neglect the steady-state collector-current

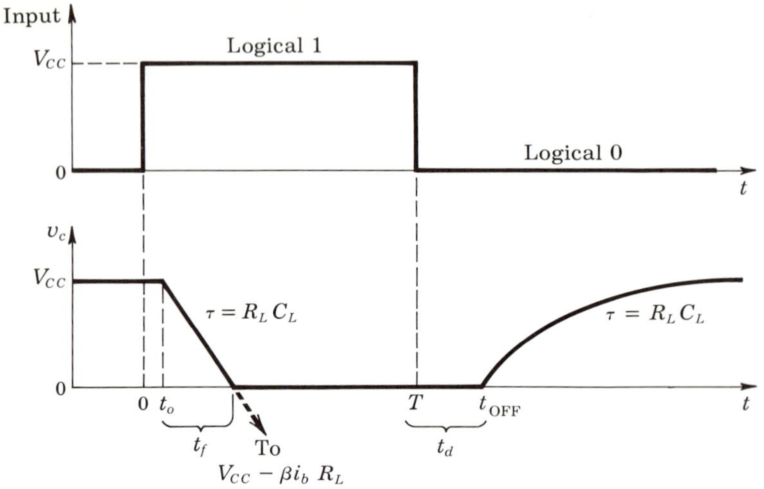

Fig. 4-24 Output waveshape of the inverter of Fig. 4-23.

component flowing through the load resistor R_L. But to ensure hard saturation of the inverter, we must satisfy the steady-state inequality

$$I_{c_{ss}} = I_{R_L} \ll \beta I_{b1}$$

Under this circumstance, the discharge of C_L is exactly the same process as that which speeds the turn-off of the TTL gate. The linear discharge equation is

$$v_c(t) \approx V_{CC} - \frac{\beta I_b t}{C_L}$$

As shown in Fig. 4-24, the inverter discharges the load capacitor extremely rapidly.

At time T, the input pulse drops back to the logical-0 level. The transistor remains saturated for the internal-storage-delay time t_d, defined in Sec. 3-7. Once the transistor is off, the capacitor recovers from $V_{CE,\text{sat}}$ back toward V_{CC} with the same time constant R_LC_L (Fig. 4-23*c*). Complete recovery is assumed in $5R_LC_L$.

Even though the same time constant governs both the discharge and the recharge exponentials, the discharge interval is orders of magnitude shorter. We conclude that the inverter is best suited for use as a current sink.

Emitter-follower output

Figure 4-25 shows the second basic output stage, an emitter follower. With an applied logical-1 input, the transistor turns on. Looking into the

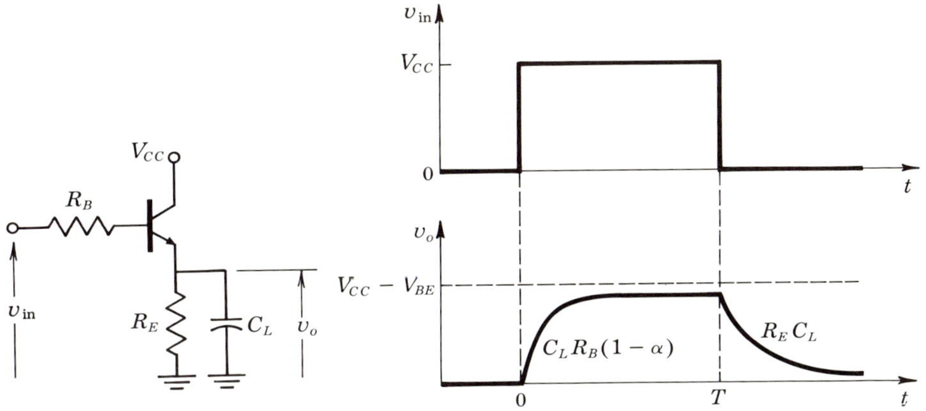

Fig. 4-25 Emitter-follower output circuit driving a capacitive load, and output waveshape.

emitter, the capacitor sees the very low impedance $R_B(1 - \alpha)$ [Eq. (3-50)]. C_L will charge to approximately $V_{CC} - V_{BE}$ with the fast time constant $C_L R_B(1 - \alpha)$. The exact final steady-state condition can readily be found from the model.

At $t = T$, the input drops back to a logical 0. But the charged load capacitor maintains the output voltage fixed across the switching interval. Consequently, the transistor becomes back-biased. Because the transistor never saturated, we do not observe any turn-off delay. C_L now discharges back to zero with the slow time constant $R_E C_L$.

By comparing the response shown in Figs. 4-24 and 4-25, we observe that the emitter follower and inverter complement each other. While the inverter discharges the capacitor rapidly and charges it slowly, the emitter follower charges C_L rapidly but discharges it slowly.

Totem-pole output

In order to obtain the speed of the inverter used as a sink as well as the speed of the emitter follower used as a source, we can combine the two circuits in the totem-pole output of Fig. 4-26. Two different drivers are shown. In the circuit of Fig. 4-26*a*, a logical-1 output from the input gate saturates Q_2 and turns on Q_4. Thus Q_3 is cut off, and the totem pole is operating as a simple inverter. The circuit may be represented by the model of Fig. 4-27*a*. Assuming an equal division of base current, we have

$$I_{b2} = I_{b4} = \frac{1}{2}\frac{V_1 - V_{BE}}{R} \tag{4-33}$$

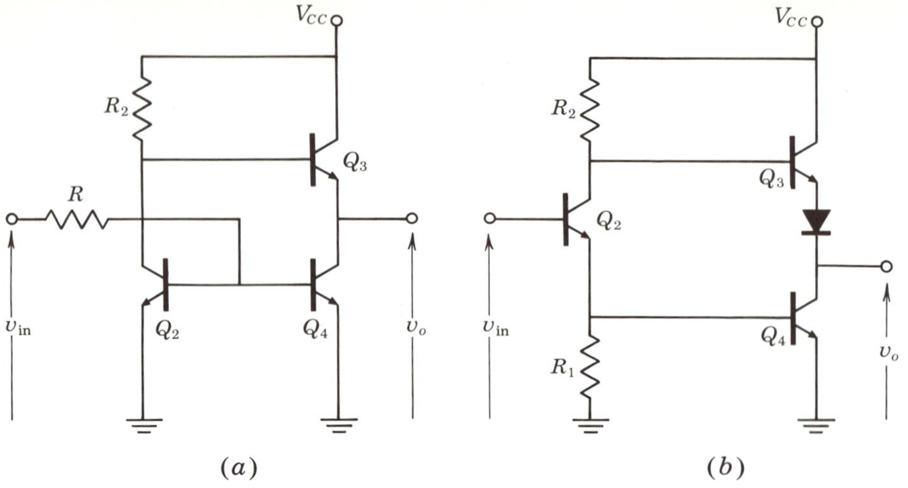

Fig. 4-26 Totem-pole output stages. (*a*) Direct drive; (*b*) phase-splitter drive.

V_1 is the logical-1 drive signal. So long as the external sinking current I_0 satisfies the inequality

$$I_0 < \beta I_{b4}$$

transistor Q_4 remains saturated and the output remains at a logical 0.

A logical-0 input cuts off Q_2 and Q_4. R_2 biases Q_3 on as an emitter follower, and the output rises to a logical 1. Q_3 supplies the necessary current to the external load. The model holding is shown in Fig. 4-27*b*. For any given load current I_1, the base current will be $I_1(1 - \alpha)$. By considering the drops in the base path, we find the logical-1 output voltage to be given by

$$V_1 = V_{CC} - V_{BE} - R_2 I_1(1 - \alpha) \tag{4-34}$$

The minimum allowable logical-1 output imposes the constraints on the maximum permissible load current I_1.

In the inverter stage of Fig. 4-26*b*, a phase splitter, Q_2, is used to drive both transistors of the totem pole simultaneously. When Q_2 is turned on (saturated), its emitter current drives Q_4 into saturation. The extra diode drop in series with the emitter of Q_3 guarantees that Q_3 will be off. Our equivalent model will be identical with that shown for the other circuit in Fig. 4-27*a*. However, because the ON biasing of Q_4 is from V_{CC} through the internal R_2, the maximum sinking current is now independent of the input drive signal.

On the contrary, a logical-0 input turns Q_2 off. Q_3 is biased on as an emitter follower through R_2. The model is as shown in Fig. 4-27*b*, but with the extra diode included in series with the output emitter.

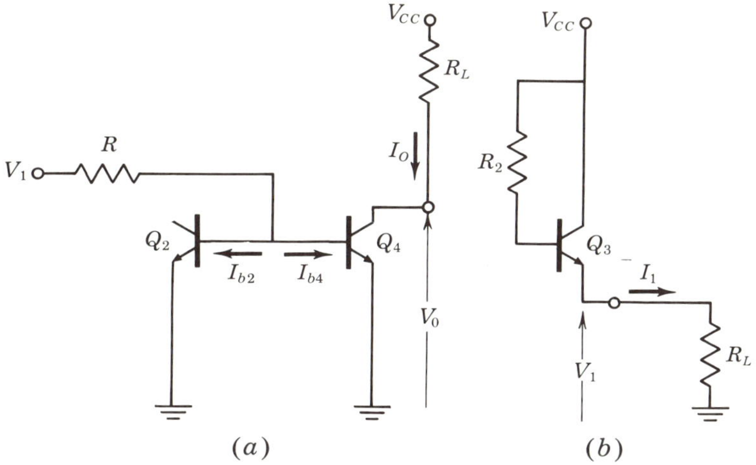

Fig. 4-27 Models of totem-pole output stage of Fig. 4-26*a* including external load R_L. (*a*) Output used as a sink when at a logical 0; (*b*) output used as a source when at a logical 1.

A common variation of the circuit of Fig. 4-26*b* replaces the diode in the emitter circuit with a slow-recovery diode in series with the base.

Figure 4-28*a* and *b* combine an input gate with the totem-pole inverter for a complete logic circuit. The first example is a standard high-speed TTL four-input NAND logic. The second example is a three-input DTL NAND gate designed for high fan-out capabilities. It incorporates an extra emitter follower (Q_2) to increase the base drive current of Q_3, thus making available a greater logical-1 output current.

4-8 Nonsaturated Logic (ECL)

The switching time of logic circuits can be greatly improved if one of the two logic states is established in the active region instead of the saturated region of the transistor. The other logic state remains in the cutoff region. By this means we eliminate the problems created by the excess charge stored in the base of the saturated logic element. In laboratory circuits, propagation delays of less than a nanosecond can be obtained. As with the TTL, delays of 3 to 5 ns are more realistic for production circuits. This class of logic employs current-mode switching. It is called current-controlled logic (CCL) or emitter-coupled logic (ECL).

A typical circuit which makes available both OR and NOR outputs appears in Fig. 4-29. The input circuit, consisting of the two gate tran-

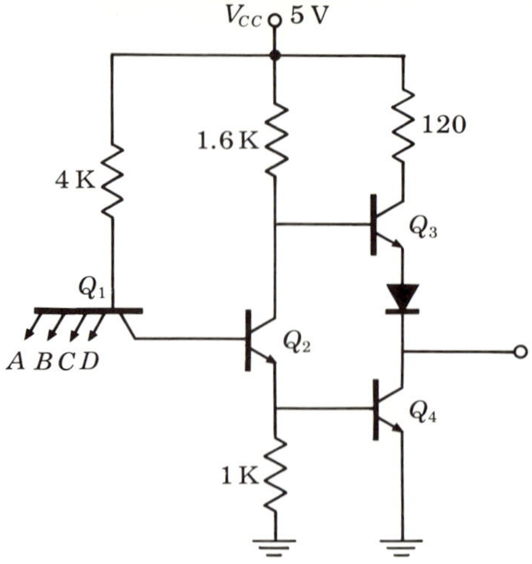

(a)

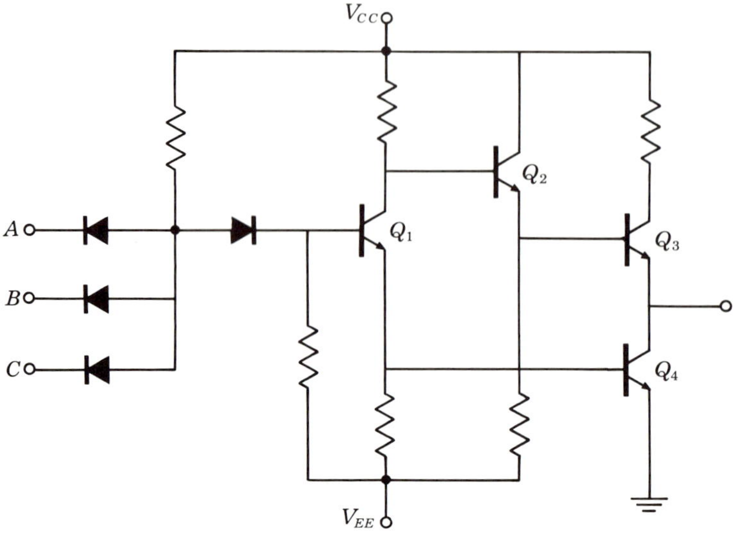

(b)

Fig. 4-28 Complete logic circuits using a totem-pole output stage. (*a*) Four-input TTL NAND logic; (*b*) three-input DTL NAND gate designed for high fan-out capabilities.

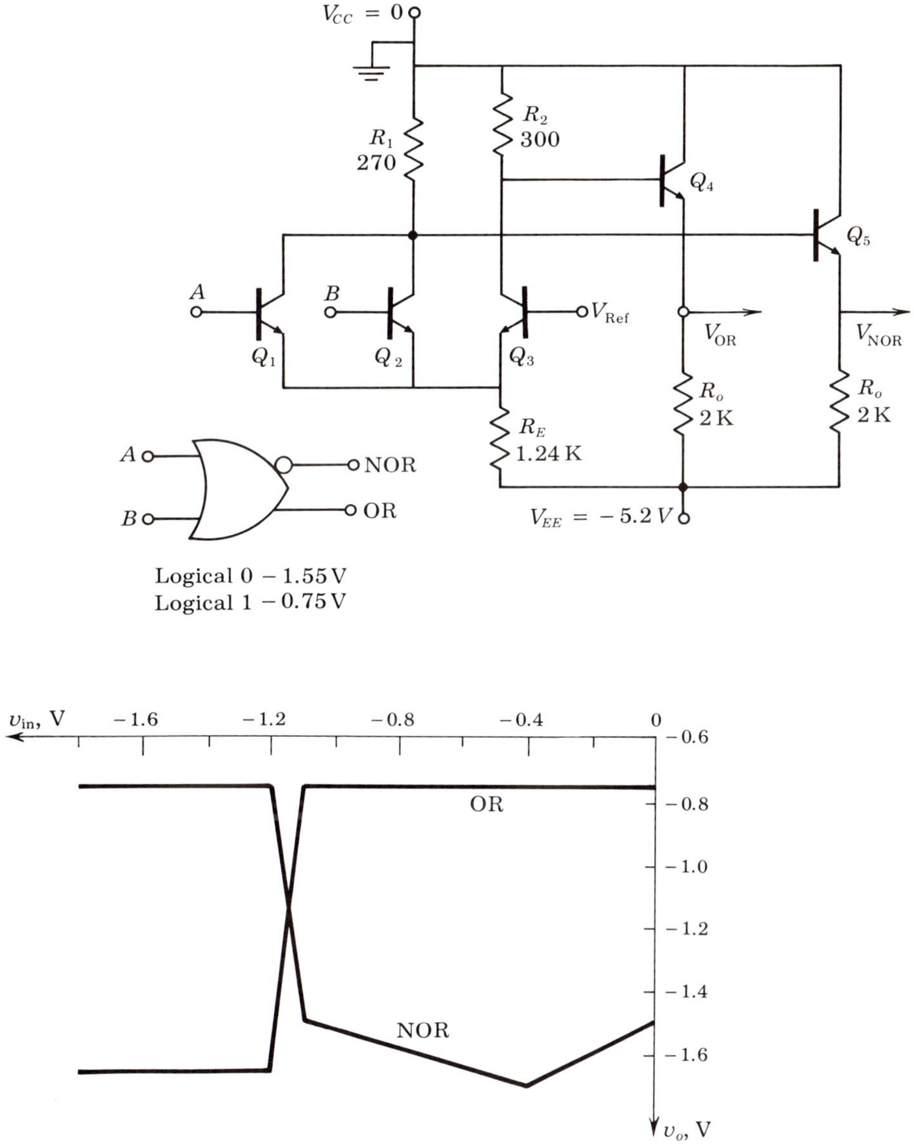

Fig. 4-29 Emitter-coupled logic (ECL) and transfer function.

sistors Q_1 and Q_2 and the reference transistor Q_3, is a differential amplifier. The reference voltage (V_{Ref}) applied at the base of Q_3 is chosen to be in the middle of the transistor's active region. In this case, $V_{\text{Ref}} = -1.15$ V while the logical levels are chosen at -0.75 and -1.55 V. Q_1 and Q_2 form a simple NOR gate, each acting as an inverter, with the NOR function performed at the common-collector and common-emitter connections. We shall assume a fixed V_{BE} drop of 750 mV for the conducting transistors. If the gate inputs are more negative than the reference voltage (say, at the logical $0 = -1.55$ V), the input transistors Q_1 and Q_2 are cut off. All of the differential amplifier's current flows through Q_3. This current is given by

$$I_3 = \frac{-V_{EE} - V_{BE} + V_{\text{Ref}}}{R_E} = \frac{5.2 - 0.75 - 1.15}{1.24 \text{ K}} = 2.66 \text{ mA} \qquad (4\text{-}35)$$

The collector of Q_3, at $-R_2 I_3 = -800$ mV, is directly coupled to the OR emitter follower. Q_4 introduces an additional 750-mV drop, resulting in an OR output at -1.55 V, which is the logical-0 level. A NOR output appears at the emitter of Q_5. Since only the small base current of Q_5 flows through R_1, the output will be one V_{BE} drop below zero, at the logical-1 level of -750 mV.

When we apply a logical-1 signal (-750 mV) at the base of Q_1 (or Q_2), Q_1 turns on and the common-emitter voltage rises to -1.5 V. This back biases the reference transistor. The current path has switched from Q_3 to Q_1:

$$I_1 = \frac{-V_{EE} - V_{BE} + V_1}{R_E} = \frac{5.2 - 0.75 - 0.75}{1.24 \text{ K}} = 2.98 \text{ mA} \qquad (4\text{-}36)$$

The collector voltage of Q_1 will be at $-R_1 I_1 = -805$ mV, and the NOR output will be one V_{BE} drop below this, or at -1.55 V. The OR output is at the logical-1 level of -750 mV.

Neither of the logic levels depends on β. For both NOR and OR outputs, the logical-1 output is simply minus one V_{BE} drop.

The logical-0 output is a function of the input voltage (logical-1 input or reference voltage) and the ratio of the appropriate collector resistor (R_1 or R_2) to R_E. Changes in the base-diffusion sheet resistivity leave the resistor ratios unchanged.

The noise immunity is approximately the difference between the reference signal and the logical-1 and logical-0 inputs as obtained from the output of a similar stage. For both 0 and 1, the NI is 400 mV.

Fan-out capability is high. The emitter follower has an output impedance of only $300\ \Omega/(\beta + 1)$. For a logical-1 output, it must only supply the small base current required by nonsaturated inputs. For logical-0 output, it must only act as a drain for the small reverse base current. In an ECL, switching speed and noise immunity, rather than current-handling capabilities, determine the maximum fan-out.

4-9 Wired Logics

Any gate output having a simple resistor load (but not a totem-pole stage) can be connected in parallel with a similar output to form a wired logic statement. When inverters using a collector pull-up resistor have their outputs paralleled (Fig. 4-30*a*), they form a NOR gate of the type shown in Figs. 4-14 and 4-16. Our logic statement becomes

$$Z = \overline{X + Y} \tag{4-37}$$

where X and Y are the logic statements at the input of the inverter stages. From Eq. (4-4), we obtain the equivalence of Eq. (4-37):

$$Z = \bar{X}\bar{Y} \tag{4-38}$$

We identify $\bar{X}$ and $\bar{Y}$ in Eq. (4-38) as the individual logic statements appearing at each separate inverter output. The paralleling of outputs results in a *wired* AND of the individual outputs. Or we can refer to the resultant output logic as the *wired NOR of the inverter inputs.*

Emitter-follower outputs, having emitter pulldown resistors (Fig. 4-30*b*), can also be connected together. The output in this case is the wired OR function

$$Z = X + Y \tag{4-39}$$

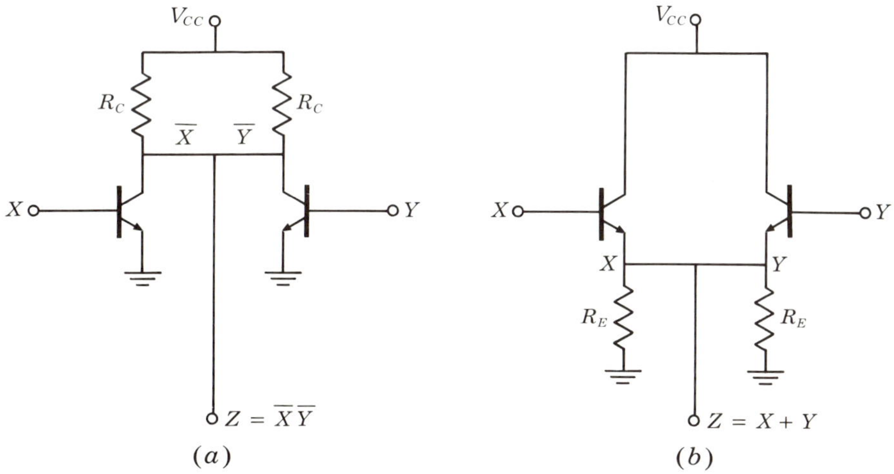

Fig. 4-30 Wired logic functions. (*a*) Common-collector *wired AND;* (*b*) common-emitter *wired OR.*

where, as before, X and Y are inputs to the wired stages. If the emitter follower is the final stage of an inverter, then X and Y will be the inverse of the input ($X = \bar{A}$ and $Y = \bar{B}$) and Eq. (4-39) may be written as

$$Z = \bar{A} + \bar{B} = \overline{AB} \tag{4-40}$$

Equation (4-40) states that the wired NAND function (referred to the input) is created by the paralleling of emitter-follower outputs.

Certain logic stages do not lend themselves to wired functions. For example, the totem-pole output is neither a true inverter nor a true emitter follower; it has some of the attributes of both. Paralleling two outputs results in an ambiguous state when one input is at a logical 0 and the other is at a logical 1; the output voltage will then lie somewhere between the 0 and 1 levels.

For some inexplicable reason, manufacturers' data sheets refer to all wired output functions as wired OR connections. Most of them are, in fact, the wired ANDs of the common-collector connection. Great care must be taken to examine the operation of the actual circuit. We must not depend solely on the logic symbols when interconnecting outputs. Blind faith in the data sheet will lead to the synthesis of wrong logic statements.

Wired logic statements are often included within the basic logic of the integrated-circuit die. In Fig. 4-31, the outputs of two TTL input AND gates Q_{1A} and Q_{1B} drive separate stages Q_{2A} and Q_{2B}. By paralleling Q_{2A} and Q_{2B}, we create a NOR function which controls the state of the

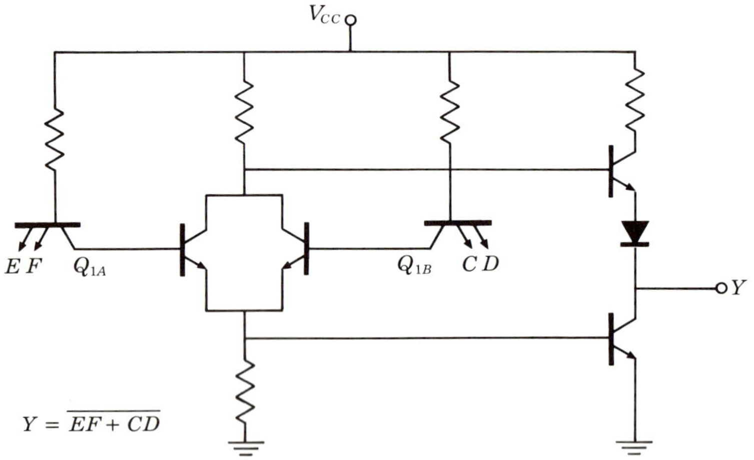

Fig. 4-31 Wired NOR logic within a TTL gate.

totem-pole output. Either Q_{2A} or Q_{2B}, driven on by its input AND statement, will set the output in a logical 0. The logic of the gate of Fig. 4-31 is the NOR combination of the AND inputs:

$$Y = \overline{EF + CD}$$

An obvious advantage in creating the wired function within the gate is that only one totem-pole output stage is needed.

4-10 Amplitude Gating (Choppers)

Often analog information must be periodically sampled. The transmission paths are gated on and off by time-sequential digital signals. Our requirements for the transmission gate are more severe than for the bilevel logic gates. The amplitude of the input signal contains significant information which must not be distorted during the gating process. Polarity is also important, and consequently the gate must be able to control equally well the transmission of analog signals of either polarity. Finally, none of the control signal should be coupled into the output.

One convenient method of satisfying these conditions is through the use of a balanced system driven by equal-amplitude pulses of opposite polarity. The error voltage produced by each signal will be equal and opposite, canceling out. Since we know that in linear circuits almost complete voltage cancellation takes place in a balanced bridge, we might well consider the application of a diode bridge for amplitude gating.

Figure 4-32 shows a diode bridge driven by a balanced gating voltage V_G. In this particular circuit, the drive is obtained from a transformer. Direct push-pull outputs from a balanced amplifier or multivibrator would work equally well. The sole function of the control or gating signal is the closing or opening of the gate, i.e., bringing the diodes into and out of conduction. For the polarity indicated in Fig. 4-32*b*, all diodes are conducting, and on a reversed gating polarity, all diodes become back-biased (Fig. 4-32*c*). However, this statement implicitly assumes that the contribution due to the analog signal v_{in}, which will be examined below, is not large enough to turn off a conducting diode or to turn on a nonconducting one.

In Fig. 4-32*b*, we see that once the gating signal has turned on all diodes, the effect on the transmission path can be found by looking at the equivalent model existing across the output from point b to point d. Each branch consists of two conducting diodes connected back to back. If the drops

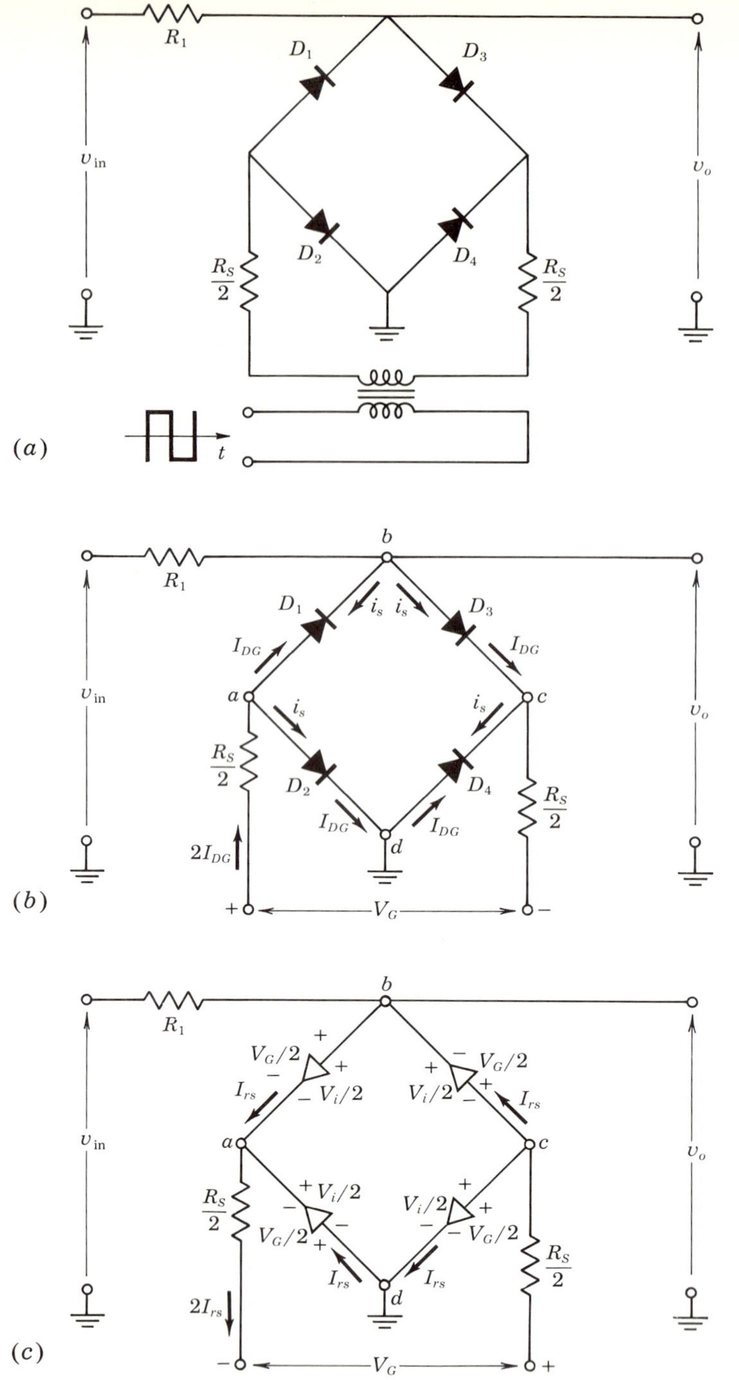

Fig. 4-32 Diode bridge for amplitude gating (or chopping). (*a*) Bridge circuit; (*b*) conditions holding in conducting bridge (currents shown are found independently from V_G and v_{in}); (*c*) conditions holding in back-biased bridge (voltages and polarities shown are found independently from V_G and v_{in}).

are equal they will cancel; any unbalance results in a small offset voltage. Our first balance condition becomes

$$V_{D_1} = V_{D_2} \qquad V_{D_3} = V_{D_4} \tag{4-41}$$

Since the diode drops are dependent on the current, to satisfy Eqs. (4-41) we must have an equal turn-on current through each diode. But this means that the bridge must also be balanced with respect to the gating signal. Thus we also must satisfy

$$V_{D_1} = V_{D_3} \qquad V_{D_2} = V_{D_4} \tag{4-42}$$

Equations (4-41) and (4-42) taken together imply that a matched set of diodes should be employed in the bridge. Each diode has a turn-on current contributed by V_G:

$$I_{DG} = \frac{1}{2}\frac{V_G - 2V_D}{R_S} \tag{4-43}$$

The input voltage v_{in} causes an additional current component to flow in each diode. For a positive input, the maximum signal current is

$$I_s = \frac{1}{2}\frac{V_{i,\max}}{R_1} \tag{4-44}$$

Referring back to Fig. 4-32*b*, we note that the two current components subtract in diodes D_1 and D_4 but add in D_2 and D_3. The former diode pair is in danger of being forced out of conduction on the positive input peaks of the analog signal, and the latter pair on the negative peaks. If all diodes are not identical, any initial unbalance will be compounded. To prevent the undesirable change of diode states, the current contribution supplied by the gating signal [Eq. (4-43)] must always be greater than the current supplied by the signal source [Eq. (4-44)]. By writing the required inequality, we arrive at one design relationship between the circuit parameters:

$$\frac{V_G}{R_S} > \frac{V_{i,\max}}{R_1} \tag{4-45}$$

For a second relationship, we naturally turn to the second bridge state: the open gate (Fig. 4-32*c*). All diodes are back-biased by the reversed-polarity control signal. The individual diode currents are I_{rs}. Thus the back-bias voltage across each diode due to V_G is

$$V_{DG} = -\frac{V_G - 2I_{rs}R_S}{2} \approx -\frac{V_G}{2} \tag{4-46}$$

Assuming identical diodes, each having a leakage resistance very large compared with R_1, the signal source sees a balanced high-impedance bridge. The maximum voltage across each back-biased diode due to v_{in} is

$$V_{DS} \approx \tfrac{1}{2}V_{i,\max}$$

V_{DS} is in phase with V_{DG} across one diagonal pair of diodes but is of opposite phase across the other pair. Conditions established may drive one diagonal pair of diodes into conduction on positive peaks and the other pair on the negative peaks of v_{in}. Of course, this would again seriously unbalance the bridge. To prevent this, we must restrict the input to satisfy a second inequality:

$$V_{i,\max} < V_G \tag{4-47}$$

From the viewpoint of the analog signal, the turned-on bridge looks just like a single conducting diode (two parallel branches each having two series diodes). The diode drops, however, cancel. In calculating the incremental response, the complete conducting gate can be replaced by a single diode's forward incremental resistance r_f.

When the control signal turns off the diode bridge, we can replace each back-biased diode by the equivalent incremental reverse resistance r_r. The balanced bridge presents the equivalent of a single resistor r_r from point b to ground.

Thus the gating signal causes sequential shunting of the output, first by r_f and then by r_r. The circuit related to v_i appears similar to the shunt clipper of Chap. 2, where the clipping diode shunted the output by its forward resistance to prevent undesired transmission and by its reverse resistance to allow desired signal transmission. This leads us to interpret bridge operation as controlled clipping, with its information-transmission path a function of the control-signal period (and voltage) rather than input-signal amplitude. Since the two circuit states shown in Fig. 4-33 are the same as those existing in the clipper, the problem posed in solving

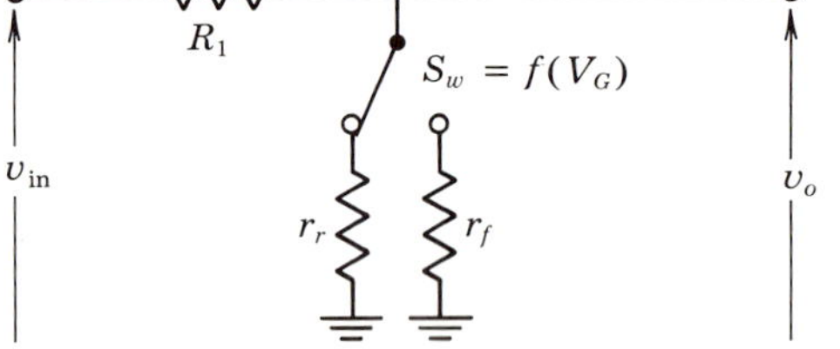

Fig. 4-33 Equivalent bridge circuit—the switch position is a function of the control signal.

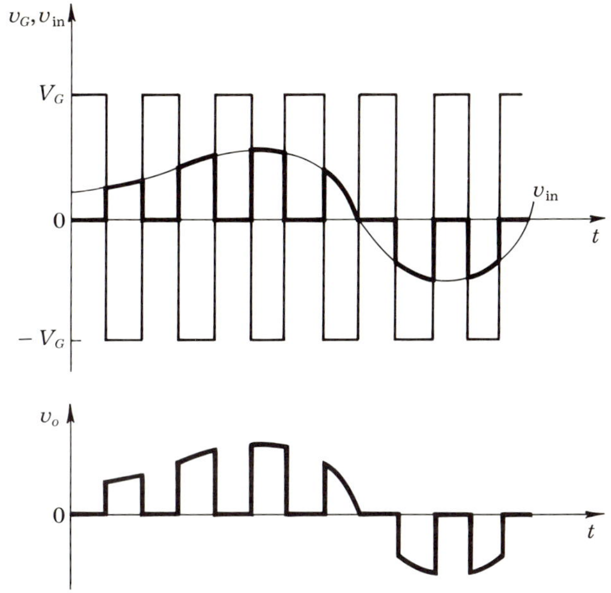

Fig. 4-34 Bridge input-output signal relationships.

for the optimum value of R_1 must lead to the identical value found in Eq. (2-33):

$$R_1 = \sqrt{r_r r_f}$$

Waveshapes of the input (square-wave control signal superimposed) and the resultant gate output, illustrating the gate's operation, are shown in Fig. 4-34. Note that the output appears only during the negative half period of the gating signal, when the bridge is nonconducting. The gating signal chops the input into a train of equally spaced varying-amplitude pulses. For this reason, the bridge gate is often referred to as a chopper or bridge modulator.

Many alternative circuit configurations are practicable, each solving a particular amplitude-gating problem. The bridge can be interposed in series with the transmission path rather than in shunt. It can be used in a push-pull system or in the single-ended one previously discussed, driven by pulses or even sinusoidal signals. The bridge might be used for bilateral as well as unilateral transmission since its resistance is independent of signal direction. One feasible circuit, illustrating many versatile attributes of the diode bridge, is shown in Fig. 4-35. The bilateral transmission path (v_1 to v_2, and vice versa) is instituted by forward biasing the diodes and is interrupted by back biasing them.

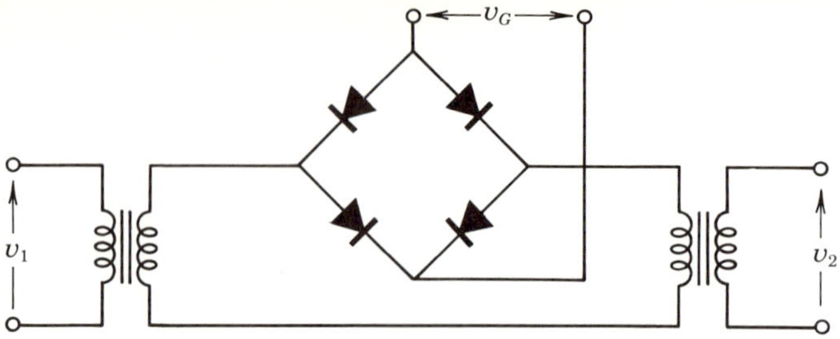

Fig. 4-35 Bilateral transmission bridge.

Transistor choppers

A transistor inserted either in series with the transmission path, Fig. 4-36*a*, or in shunt with the output, Fig. 4-36*b*, can also be used to sample or chop the analog signal. These gates function in much the same manner as the diode bridge. Control signals at the 0 level keep the gates turned off. The high resistance inserted in the series path interrupts the transmission. On the other hand, the high resistance in the shunt path permits the transmission.

Once the control signal forward biases the transistors and drives them into saturation, the state of the gate reverses. The series gate now permits transmission, while the shunt gate effectively shorts the output to ground. Proper gating for both polarity signals is predicated on the reverse transistor operation. For example, with a positive v_{in}, the transistor in the series gate of Fig. 4-36*a* is conducting in the forward direction but the transistor in the shunt gate of Fig. 4-36*b* is biased in the reverse direction. Special transistors having a relatively large β_R are available for chopper applications.

Both the series and the shunt gates of Fig. 4-36*a* and *b* introduce the dc offset voltage of the saturated transistor into the output. They are therefore not suitable for gating small signals.

Replacement of each separate transistor in Fig. 4-36 by a series *pnp-npn* pair (Fig. 4-36*c*) will reduce the dc offset to the difference between the saturation drops of the separate devices. The pair must be triggered on with opposite-polarity gating signals. Furthermore, when one transistor is conducting in the forward direction, the other is conducting as a reverse transistor. The saturation drops are slightly different in the two directions, and hence even for a matched pair there will not be complete dc cancellation.

Transient response

The response of the diode bridge to the leading and trailing edges of the control signal determines the switching time and, consequently, the output waveshape. Moreover, because of the interaction within the gate, the input signal will likewise play a role in determining the response of the bridge to the gating signal. An exact analysis of the total problem becomes unreasonably tedious, but the approximate response, sufficient to delineate the limitations of the bridge, is not difficult to calculate. We divide the problem into two parts: first, we find the time required to turn

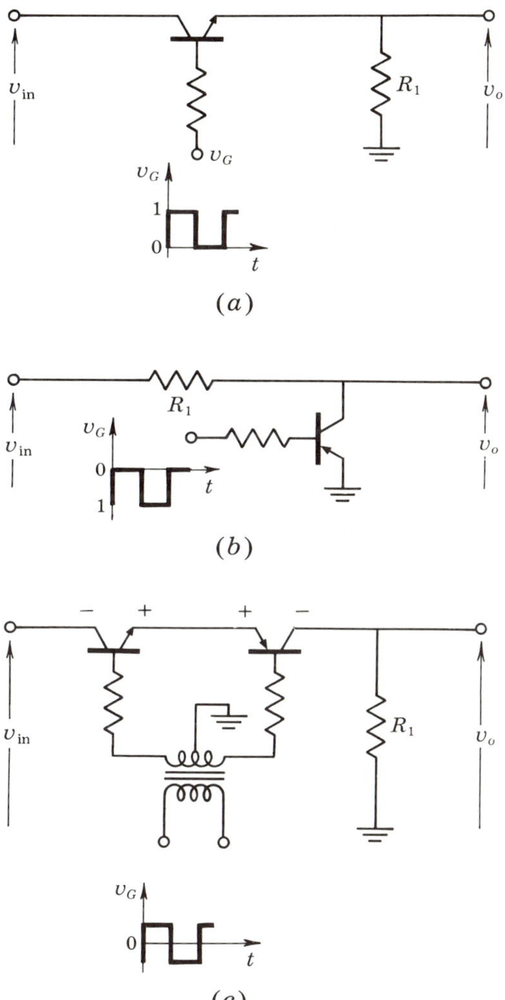

Fig. 4-36 Transistor choppers. (*a*) Series-transistor chopper; (*b*) shunt chopper; (*c*) complementary-transistor chopper.

the bridge on and off with the signal set at zero; second, we calculate the response of the bridge output, with respect to the input, while the bridge alternately conducts and opens.

Any load across the output terminals (*bd*) is across the balance points of the bridge with respect to the control signal and will not influence the switching process. The gating signal will divide equally across the two series diodes.

If we start with diodes off, being driven on, the back-biased depletion-region capacitance (C_D) of each diode will be charged to $-V_G/2$. Once the gate signal switches state, the individual-diode voltage charges toward $V_G/2$ with a time constant

$$\tau_1 = C_D R_S \tag{4-48}$$

Thus the turn-on delay is found from the exponential charging equation

$$v_D(t) = \frac{V_G}{2} - V_G e^{-t/\tau_1}$$

Solving for the turn-on time t_{ON} by setting $v_D = V_D$, we find that the charging delay lasts for approximately

$$t_{ON} = \tau_1 \ln \frac{2V_G}{V_G - 2V_D} \tag{4-49}$$

Figure 4-37 illustrates the turn-on delay defined by Eq. (4-49).

Upon the reversal of V_G, the stored charge keeps the diodes conducting. The charge is swept out as described in Secs. 2-7 and 3-7. After the recombination-time delay t_R, the diode turns off and C_D recharges to $-V_G/2$ with the time constant τ_1.

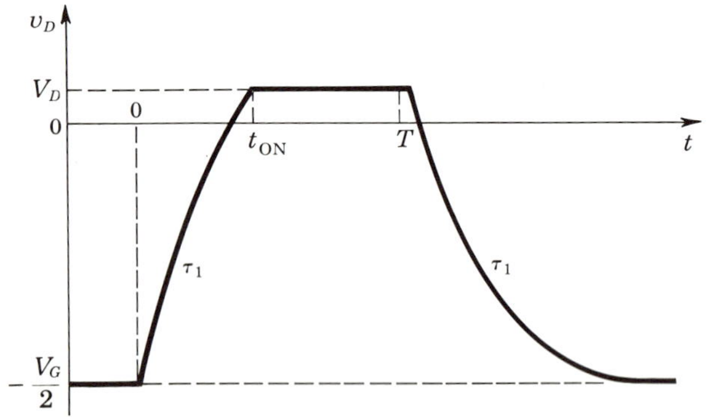

Fig. 4-37 Single-diode voltage waveshape produced by V_G.

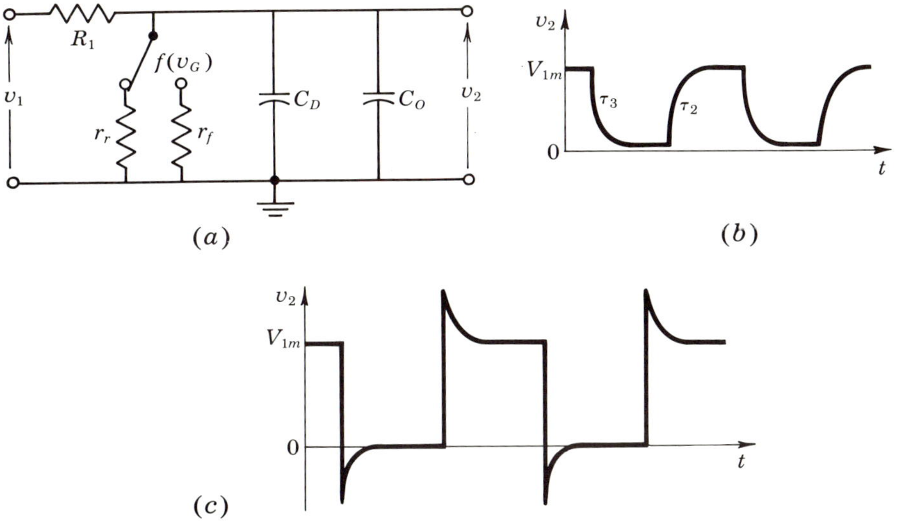

Fig. 4-38 (*a*) Diode-bridge input-output equivalent model; (*b*) ideal output waveshape (constant input voltage); (*c*) actual output response.

When the gating period is sufficiently long, these switching times become insignificant. They do, however, set a definite lower limit to the usable square-wave period. In addition, any capacitive unbalance causes unequal diode turn-on and turn-off times, unbalancing the bridge and producing sharp exponential pulses at the output. Sometimes additional trimmer capacitors are placed across each diode and adjusted for proper capacitive balance. These, of course, increase C_D and, consequently, the time required to switch the diodes from ON to OFF, and vice versa.

The solution of the second part of the response problem, i.e., the calculation of the output waveshape with respect to the input itself, becomes almost trivial once we draw the proper model. Since the bridge behaves as a controlled clipper, it might well be represented as two resistors and a switch controlled by V_G, alternately connecting one and then the other (Fig. 4-38). The circuit has two time constants, a long one for the rise (τ_2) and a short one for the decay (τ_3):

$$\tau_2 \approx R_1(C_o + C_D) \qquad \tau_3 \approx r_f(C_o + C_D) \tag{4-50}$$

We can expect τ_2 to be much longer than the bridge turn-off time constant τ_1 [Eq. (4-49)] as a consequence of the extra stray capacity at the output, and it will determine the fastest allowable gating signal.

If we now attempt to find the actual bridge response by combining the response of each signal taken individually, we find ourselves facing a rather difficult problem. Our calculation of turn-on and turn-off times ignored the presence of the input signal, which will, during the turn-off interval, contribute additional current to one diagonal pair and reduce that flowing in the other pair. A rising voltage appears across the diodes because of the gating signal (Fig. 4-37); therefore Eq. (4-45) will not remain satisfied until the gate is completely off. The two diagonal diodes D_2 and D_3 conduct first, unbalancing the bridge. A similar effect appears during the turn-on time; one pair of diagonal diodes cuts off before the other pair, again causing bridge unbalance.

Because of this unbalancing, exponential overshoots contributed by the gating signal are evidenced at the output (Fig. 4-38). By reconsidering the previous simplified discussion, the method of minimizing then becomes apparent. When the time constants τ_1 and τ_2 are small, the bridge recovers rapidly, remaining unbalanced for the smallest possible time and producing only narrow overshoots. Excessive output capacity may also slow the circuit response, preventing the rapid rise shown. However, these overshoots do improve the time response of the bridge, and if they are sufficiently narrow, they may not disturb us at all; they can certainly be filtered out.

Problems

4-1 (*a*) Design a half adder, using only NAND gates.

(*b*) Design a half adder, exercising your ingenuity to minimize the number of separate gates. You may use NAND, NOR, AND, and OR gates.

4-2 The outputs of a single full subtractor logic stage are defined by

$$\text{Difference} = [(A\bar{B} + \bar{A}B)\bar{C}] + [(AB + \bar{A}\bar{B})C]$$
$$\text{Borrow} = (AB + \bar{A}\bar{B})C + \bar{A}B$$

where A is the minuend, B the subtrahend, and C the borrow input.

(*a*) Starting from the truth table, prove that the above logic equations are true.

(*b*) Synthesize the above logic, using mainly NAND gates.

(*c*) Synthesize the above logic, using mainly NOR gates.

4-3 (*a*) Prove that, with negative logic, the diode OR gate of Fig. 4-6 functions as an AND gate.

(*b*) Prove that, with negative logic, the AND gate of Fig. 4-7 functions as an OR gate.

4-4 The three inputs of Fig. 4-39 are applied to the silicon-diode OR gate of Fig. 4-6 through relatively high signal-source impedances (1 K).

(*a*) Plot the current in each diode as a function of time.

(*b*) Plot the output voltage as a function of time.

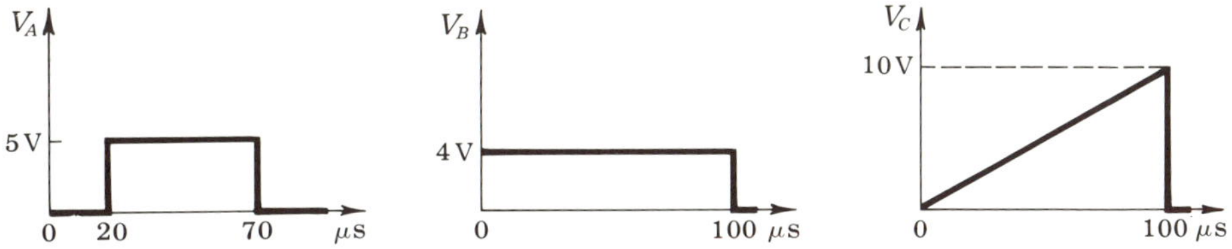

Fig. 4-39

4-5 Repeat Prob. 4-4 for the diode AND gate of Fig. 4-7*b*.

4-6 (*a*) If $V_{CC} > V_A > V_B > V_C > 0$ in the AND gate of Fig. 4-7*a* and if the signal-source impedance is small, show that the gate's excitation will result in a small jump in the output, after which it will rise exponentially to its final value. Calculate the value of this jump in terms of the circuit parameters. (*Hint:* Treat this gate as a compensated attenuator at $t = 0^+$.)

(*b*) Will such a jump appear in the OR gate? If it will, how large will it be? Explain your answer.

4-7 Figure 4-40 is a silicon-diode-controlled gate where the transmission characteristic is a function of the control signal v_c. For this problem, v_c is a 20-V 1-ms positive pulse which raises the normally negative control voltage from −20 to 0 V.

(*a*) A train of 10-V 20-μs-wide positive pulses spaced 100 μs apart (from the end of one pulse to the beginning of the next) is the input to v_1. Sketch the output waveshape from slightly before the control pulse is applied until slightly after.

(*b*) Repeat part *a* if we simultaneously apply a 25-V positive-pulse train of the same duration and period at v_2.

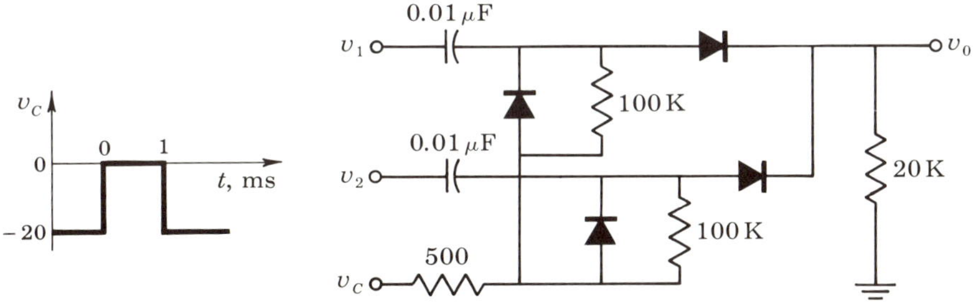

Fig. 4-40

4-8 (*a*) Plot the transfer characteristic of the DTL logic of Fig. 4-9 for a single input. Return all unused inputs to V_{CC}. Assume $\beta = 50$, $V_D = V_{BE} = 750$ mV at I_D or $I_c = 1$ mA, $V_{CE,\text{sat}} = 150$ mV, and $r_{sc} = 50\ \Omega$.

(*b*) Find the maximum fan-out for 1-V noise immunity (worst-case conditions). Repeat for 250-mV noise immunity.

(*c*) Suppose that a particular transistor has $\beta = 100$. How will this affect the gate fan-out capabilities and the noise immunity?

4-9 Repeat Prob. 4-8 for the modified DTL gate of Fig. 4-13. Assume that the input transistors have $\beta = 4$. All other conditions and circuit values are identical with those of the DTL of Fig. 4-9.

4-10 Consider the RTL NOR gate of Fig. 4-16, where $R_1 = 1.5$ K, $R_2 = 3$ K, $V_{CC} = 6$ V, $\beta = 30$, $V_{CE,\text{sat}} = 150$ mV, and $r_{sc} = 50\ \Omega$. The V_{BE} drop may be approximated by a constant 700 mV.

(*a*) Plot the transfer characteristic.
(*b*) Plot the fan-out vs. worst-case noise immunity. Make a decision as to a reasonable choice for N and NI. Justify your choice.
(*c*) What is the worst-case power dissipation for this circuit?

4-11 The signals of Fig. 4-39 are used as the inputs to the RTL gate of Fig. 4-16 and Prob. 4-10. Plot the output voltage as a function of time. Make all reasonable approximations.

4-12 Figure 4-41 represents an adaptation of the basic RTL gate of Fig. 4-16. In this circuit, $\beta_F = 40$, $\beta_R = 1$, $V_{BE} = 700$ mV, $V_{CE,\text{sat}} = 150$ mV, and $r_{sc} = 25\ \Omega$.

(*a*) Define the circuit positive-logic statement.
(*b*) Give the required logic levels for 0.5-V noise immunity.
(*c*) Calculate the input loading presented by this logic under worst-case conditions.
(*d*) Define the maximum fan-out when feeding similar gates (worst-case conditions).

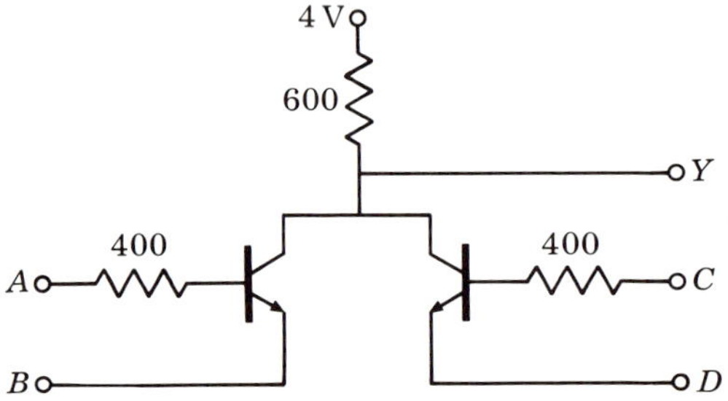

Fig. 4-41

4-13 (*a*) Draw the transfer function of the TTL gate of Fig. 4-28*a*. All unused inputs are returned to V_{CC} (5 V). (For all transistors, $\beta_F = 50$, $\beta_R = 1$, $\beta_{ee} = 0.1$, $V_{BE} = 750$ mV, $V_{CE,\text{sat}} = 100$ mV, and $r_{sc} = 25\ \Omega$.)

(*b*) Calculate the loading by the input under all possible input conditions.
(*c*) Find the worst-case fan-out. What is the limitation?
(*d*) Calculate the NI_0 and NI_1 under a fan-out of 10.

4-14 The signals of Fig. 4-39 are used as the inputs to the TTL gate of Fig. 4-28. Plot the output voltage as a function of time. Make all reasonable approximations to simplify your calculations (e.g., assume a constant V_{BE} drop of 750 mV).

4-15 Figure 4-42 shows an advanced version of the TTL gate of Fig. 4-28. In this circuit, Q_5 serves mainly to increase the drive current to the totem-pole output stage for improved fan-out under logical-1 output conditions. Q_6 improves the noise immunity by squaring the first knee of the transfer characteristic. The input diodes eliminate the high-frequency ringing of this high-speed switching circuit.

(*a*) Draw the transfer characteristic of this circuit and compare your results with the characteristic found in Prob. 4-13. $\beta = 50$, $V_{BE} = 700$ mV.

(*b*) What are the conditions on the design of the Q_6 stage to ensure the optimum shape of the transfer characteristic?

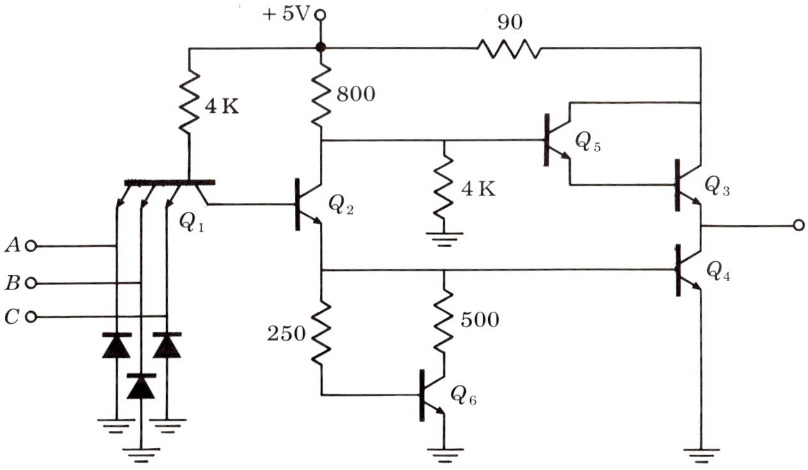

Fig. 4-42

4-16 An RTL inverter stage (Fig. 4-16) is loaded by N parallel inputs, where each input is approximated by 50 pF shunted by 500 Ω. The input to the inverter is a single 10-μs 4-V pulse. We shall assume a 10-ns internal turn-on delay and a 10-ns saturation-charge-storage delay. Plot the propagation-delay time as a function of the fan-out N.

4-17 Calculate the transient response of the TTL circuit of Fig. 4-28*a* when the output is loaded by 500 pF shunted by 2 K. A 10-μs 4-V logical-1 pulse is applied at input A. Inputs B, C, and D are returned to V_{CC} (5 V). Assume a 3-ns internal turn-on delay and a 5-ns saturation-charge-storage delay.

4-18 A complementary-symmetry transistor output stage is shown in Fig. 4-43. ($V_{CE,\text{sat}} = 150$ mV, $V_{BE} = 750$ mV, $r_{sc} = 50$ Ω, $\beta = 50$.)

(*a*) Sketch and label completely the transfer function from A to the output. (B and C are connected to ground.)

(*b*) What is the logic of this circuit (positive logic)?

(*c*) What is the maximum fan-out in driving similar gates ($NI = 0.5$ V)?

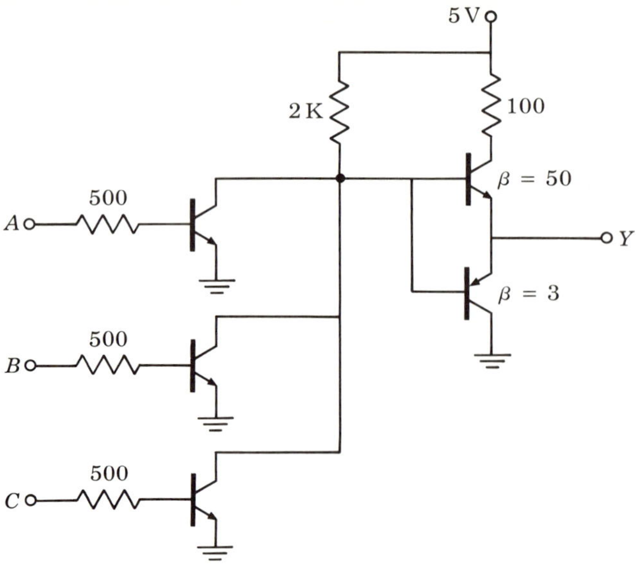

Fig. 4-43

4-19 Repeat Prob. 4-17 for the circuit of Fig. 4-43.

4-20 (*a*) Plot the transfer characteristic of the ECL of Fig. 4-29 from one input to both the OR and the NOR outputs. The unused input is left open. ($\beta = 40$, $V_{BE} = 750$ mV.)

(*b*) Find the fan-out capabilities for $NI = 0.5$ V.

(*c*) Over what range can we vary V_{Ref} without any transistor saturating? Consider logical-1 and logical-0 inputs separately.

4-21 Repeat Prob. 4-17 for the ECL logic of Fig. 4-29.

4-22 Design an interface stage which can translate the RTL levels of $0 < 0.25$ V and $1 > 2.5$ V ($V_{CC} = 4.5$ V) to the logical-0 and logical-1 levels of the ECL. This circuit must operate with the available RTL and ECL supply voltages.

4-23 Design an interface stage which will translate the ECL levels up to the levels required by the RTL or TTL stages (see Prob. 4-22).

4-24 Show how to modify the internal circuitry of the TTL of Fig. 4-28 to satisfy the positive-logic statement

$$Y = AB + CD$$

4-25 Figure 4-44 illustrates a simple diode amplitude gate. The germanium diodes used have a measured drop of 150 mV at a current of 1 mA. The gating signal v_g is a 30-V peak-to-peak square wave. Sketch the input and output to scale, giving significant values, for the following conditions:

(*a*) The input is a constant 1 V.
(*b*) The input is a 10-V-peak sine wave having a period one-tenth as long as the square wave.
(*c*) The input sinusoid of part *b* is increased to a 30-V peak value.

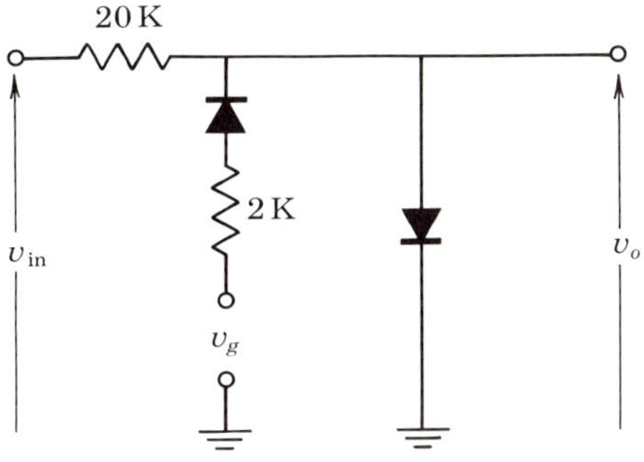

Fig. 4-44

4-26 Evaluate the percentage of the gating signal appearing at the output of a diode bridge if one diode, say D_1, differs by 10 percent from the other diodes. By this we mean that I_{rs} of D_1 is $0.9I_{rs}$ of the other diodes. In this bridge, $R_1 = 50$ K, $R_S = 2$ K, and the gating signal is a 20-V peak-to-peak square wave.

4-27 The transistor used in the shunt chopper of Fig. 4-45 has $\beta_F = 50$ and $\beta_R = 5$. Its state is changed by a 10-V peak-to-peak 100-kHz square-wave base excitation. When saturated, the collector-to-emitter circuit may be approximated by a 25-Ω resistor in series with the saturation drop of 50 mV.

(*a*) Sketch the output to scale when the input is a 5-V-peak 10-kHz sine wave.
(*b*) What is the maximum-possible-amplitude input signal which can be properly chopped? What is the limiting factor?
(*c*) Sketch the output for an input which is twice the limiting voltage found in part *b*. Explain your sketch.

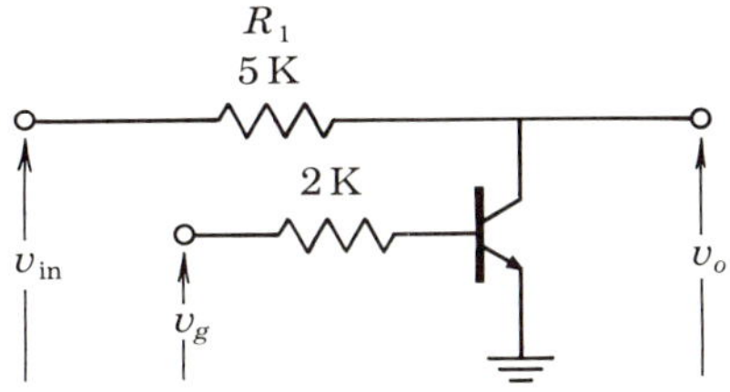

Fig. 4-45

4-28 We wish to investigate the transient response of the transistor chopper of Fig. 4-45 when the output is loaded with 500 pF shunted by 10 K. The input is at +10 V dc, while the base drive is the square wave of Prob. 4-27.

(*a*) Sketch the output waveshape, giving the values of all significant times and time constants.

(*b*) Will the response be improved if we shunt R_1 with a 1,000-pF capacitor? Explain your answer.

Bibliography

Harris, J. N.: "Digital Transistor Circuits," SEEC vol. 6, John Wiley & Sons, Inc., New York, 1966.

Motorola Semiconductor Products Division: Data Sheets and Application Notes on Various Classes of Logic, Phoenix, Ariz.

——— (D. K. Lynn, C. S. Meyer, and D. J. Hamilton, eds.): "Analysis and Design of Integrated Circuits," McGraw-Hill Book Company, New York, 1967.

Narud, J. A., and C. S. Meyer: Characterization of Integrated Logic Circuits, *Proc. IEEE*, vol. 52, pp. 1551–1564, December, 1964.

Texas Instruments Inc.: "TI Series 54/74 Integrated Circuits," Dallas, Tex., 1966.

5. field-effect transistors

The field-effect transistor differs from the regular or bipolar transistor in that its current flow is due solely to the drift of majority carriers. Since only one type of charge carrier is used, the field-effect transistor (FET) is also called a *unipolar* transistor, in contrast to the *bipolar* transistor of Chap. 3, where both holes and electrons contribute to the current flow. Current flow in an FET is voltage-controlled rather than current-controlled. In many respects its terminal response is the solid-state analog of the vacuum tube.

The FET is also easier to fabricate and occupies a very much smaller area on the chip. In integrated circuits, a single FET requires 1 mil^2 compared with 40 mil^2 for the bipolar device. Furthermore, no provision need be made for isolation regions.

5-1 Junction Field-effect Transistors

In order to explain the operation of the FET, we shall use a simplified geometry. Consider an n-type silicon bar biased as shown in Fig. 5-1. The end of the bar at which the charge enters is called the *source*, and the end at which the charge leaves is called the *drain*. Source and drain are interchangeable. They depend only on the polarity of the applied voltage.

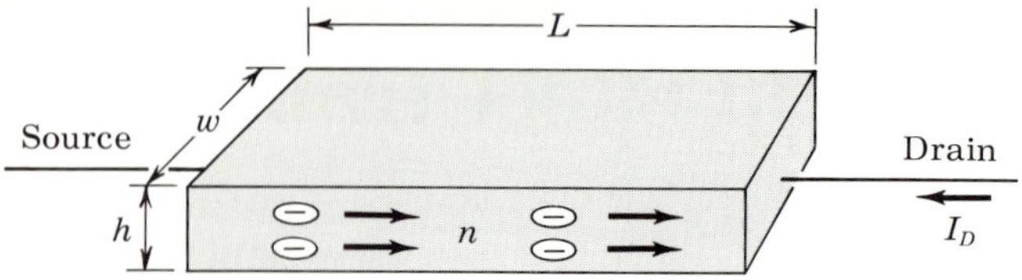

Fig. 5-1 An *n*-type silicon bar, illustrating drift-current flow.

The end-to-end conductivity of the rectangular bar is found from Eq. (2-6) to be

$$G_o = \frac{qn\mu_n wh}{L} \tag{5-1}$$

An examination of Eq. (5-1) indicates that the conductivity of the bar can be modified by controlling or modulating the bar geometry.

We establish the mechanism of voltage control over the effective geometry by forming junctions within the n bar. As shown in Fig. 5-2, p-type regions are diffused into the bar between the drain and the source, and these regions are called the transistor's gate. We now have pn junctions formed with the bulk material. Back biasing these junctions establishes a depletion region within the n material. The width of the depletion region increases with the increasing back bias [Eq. (2-53)]. It follows that in the vicinity of the junction there is a reduction of the effective cross-sectional area of the channel. This constriction increases the drain-to-source resistance (Fig. 5-2*b*). Eventually the depletion region extends through the whole channel and pinches off the current flow. At pinchoff,

$$V_{GS} = V_p \qquad I_D = 0 \tag{5-2}$$

The resistance of the bar increases from a low resistance value at $V_{GS} = 0$, where the channel occupies the whole cross-sectional area, to almost infinity at pinchoff. Figure 5-4*a* shows the resistance variation of the bar as a function of gate-to-source voltage.

Since the gate-to-channel junction remains back-biased from $V_{GS} = 0$ to pinchoff and beyond, the input impedance at the gate is extremely high. This is in contrast to the bipolar transistor, where, in the active region, current is injected into the low impedance of a forward-biased junction.

Forward biasing the gate-to-channel junction causes the injection of additional carriers into the channel. Because we increase n in the numerator of Eq. (5-1), the channel resistance is reduced and the conductivity is enhanced. In a junction FET, the disadvantage of operating in the enhancement mode is that the gate presents the low impedance of a forward-biased diode to the external circuit.

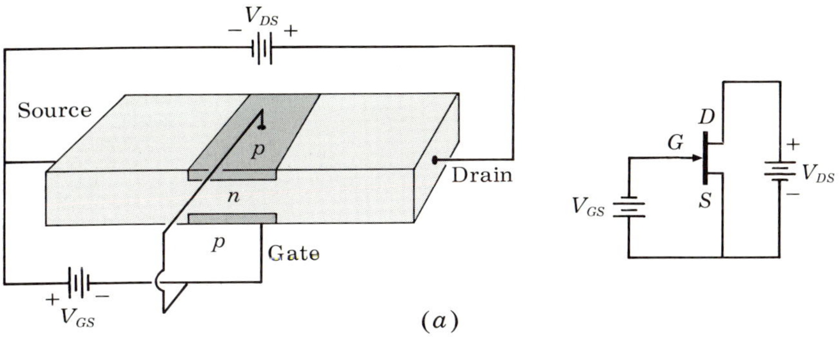

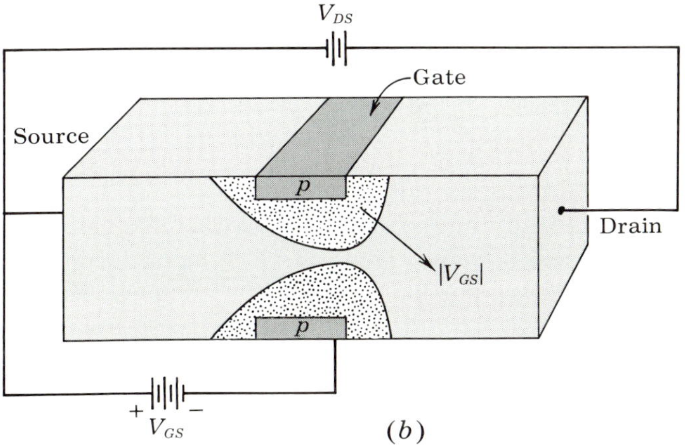

Fig. 5-2 (a) Junction FET and schematic model; (b) illustrating the buildup of the depletion region and consequent narrowing of the channel with increasing $|V_{GS}|$.

Now let us consider what happens as the drain-to-source voltage is increased. For a given $V_{GS} > V_p$, a channel exists. V_{DS} establishes a longitudinal field in the bar, and the majority carriers (electrons in the n bar) flow as drift current.

The amplitude of the current flow is proportional to the field strength. Figure 5-3a gives an idealized picture of how the potential distribution in the bar varies with increasing V_{DS}. In the vicinity of the drain, the channel is most positive. As V_{DS} increases, at some point the local potential difference between the gate and channel reaches the pinchoff value. Because of the geometry, the pinchoff region is located at the drain end of the gate. Pinchoff is defined by

$$V_p = V_{GS} - V_{DSp} \tag{5-3}$$

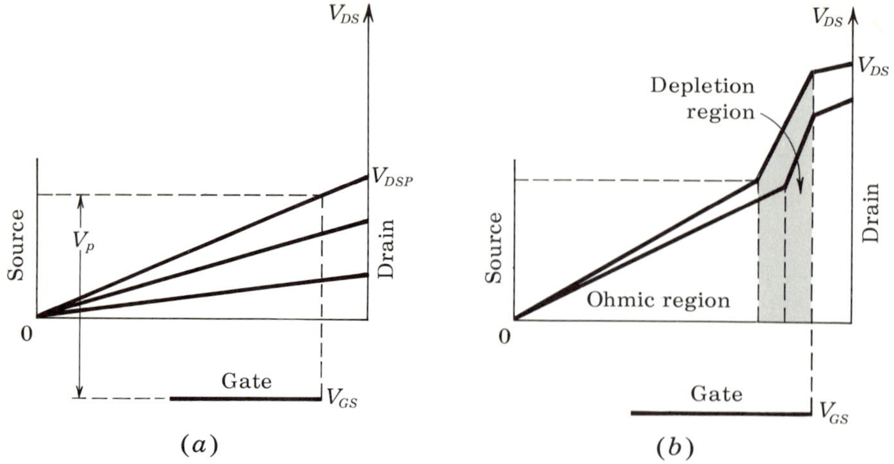

Fig. 5-3 (*a*) Idealized voltage distribution before pinchoff; (*b*) idealized voltage distribution after pinchoff.

V_p is shown schematically in Fig. 5-3*a*. Once the channel is pinched off, any further increase in V_{DS} simply causes a widening of the depletion region. The drop across the ohmic portion of the channel remains constant at

$$V_\Omega = V_{GS} - V_p$$

and consequently the drift current remains almost constant. From Fig. 5-3*b* we can see that the field in the depletion region can become very large. All electrons accelerated by the drift field in the ohmic portion of the channel are rapidly swept across the depletion region to the drain. The mechanism of current flow through the pinchoff region is similar to the collector-current flow through the back-biased collector-base junction.

In the pinchoff region, $V_{DS} > V_{GS} - V_p$, the drain current may be expressed as

$$I_D = I_{DSS}\left(\frac{V_{GS}}{V_p} - 1\right)^2 \tag{5-4}$$

where I_{DSS} is the drain current at $V_{GS} = 0$. The FET's volt-ampere and transfer characteristics appear in Fig. 5-4.

5-2 MOS Field-effect Transistors

The metal-oxide semiconductor field-effect transistor (MOSFET) operates in the enhancement rather than the depletion mode. Figure 5-5*a* illustrates

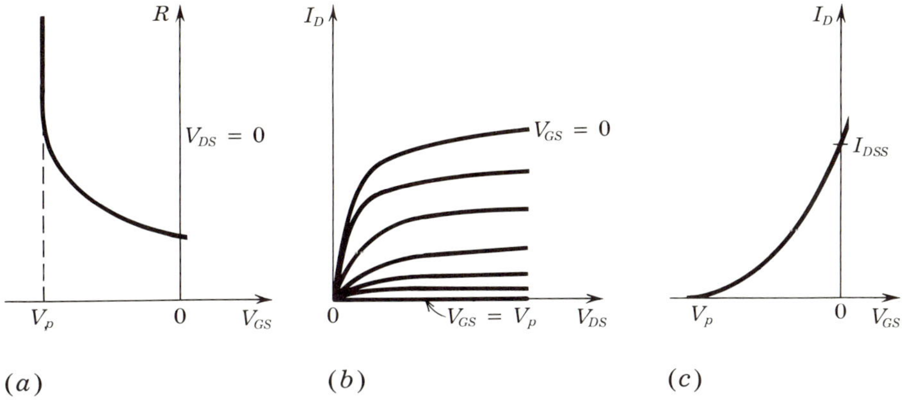

Fig. 5-4 (a) Resistance of channel in ohmic region; (b) junction-FET volt-ampere characteristics; (c) transfer characteristics.

the construction of the MOSFET. Two regions of *p*-type silicon are obtained by the diffusion of impurities into the *n*-type substrate. One of these regions is used as the source and the other as the drain. No matter which way V_{DS} is applied, at least one of the two *pn* junctions is back-biased. It follows that the normal drain current will only be the small current of a back-biased junction.

A 1,000-Å-thick layer of SiO_2 is formed on top of the *n* substrate. Finally, a metal gate region is laid down by the evaporation of aluminum on top of the oxide. The gate-to-substrate region is simply a capacitor where the SiO_2 layer serves as the dielectric.

If we apply a negative voltage between the gate and substrate, we shall induce a positive charge in the substrate, right under the gate electrode. The majority electrons in the *n* material move out of this region, leaving behind the donor atoms, each having a single positive charge. Since this charge is tightly bound in the lattice structure, it cannot be used to carry current.

As V_{GS} is made more negative, the gate-to-semiconductor capacitor requires a positive charge in the semiconductor greater than the charge due to the bound donor atoms. Minority holes from the bulk *n*-type substrate move into the region under the gate, where, because of the depletion of electrons, they become the majority carriers. When enough of these mobile charge carriers move into the gate region, they establish a conducting channel from the source to the drain. The gate potential at which sufficient mobile holes are present to form a complete channel is called the *threshold potential* V_T.

The thickness of the induced channel depends on the applied gate voltage. Thus the MOSFET operates in an enhancement mode, where the increasing

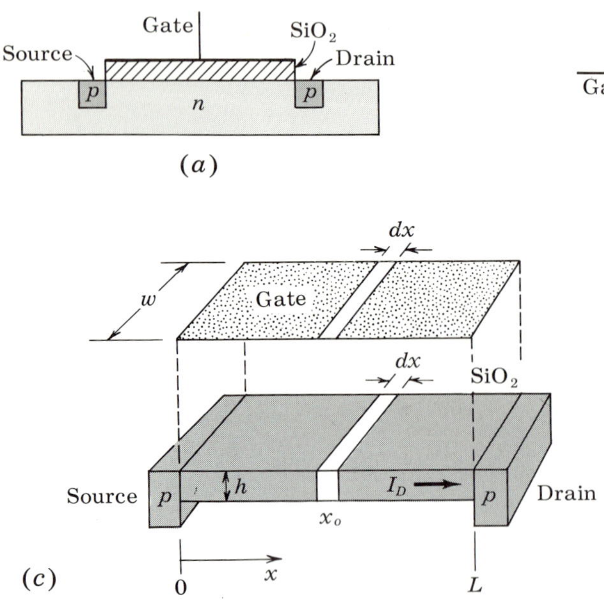

Fig. 5-5 (*a*) Induced *p*-channel MOSFET; (*b*) circuit schematic (for induced *n* channel, the source arrow is reversed); (*c*) induced channel geometry.

gate-to-drain bias increases the channel thickness and thus reduces the source-to-drain resistance. The source arrow in the schematic of Fig. 5-5 indicates the direction in which the current actually flows. An induced-*n*-channel MOSFET would have the arrow pointing out of the source.

Figure 5-6 illustrates the change in the induced-charge distribution along the channel as the drain-to-source voltage increases from zero. For $V_{GS} = -10$ V (where $V_T = -5$ V) and $V_{DS} = 0$, we have induced a uniform channel. As V_{DS} increases, the voltage drop along the bar causes the hole distribution to change. When $V_{DS} = -5$ V, the local voltage at the drain end is reduced to V_T and the induced channel pinches off. For higher values of V_{DS}, the device operates in the pinchoff mode with essentially constant drain current. The transfer and volt-ampere characteristics are shown in Fig. 5-8.

Let us now treat the current flow in the induced channel quantitatively. The simple model which we shall use for the channel has the geometry shown in Fig. 5-5*c*. Continuity of current flow along the channel requires that at any point $x = x_0$,

$$I(x_0) = I_D$$

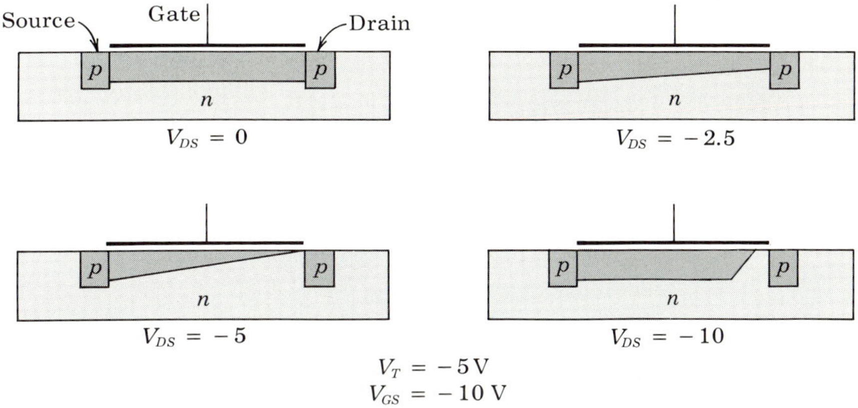

Fig. 5-6 Idealized channel distribution in MOSFET, showing change in distribution as a function of V_{DS}. At $V_{DS} = -5$, the channel begins to pinch off near the drain.

But the drain current can also be expressed in terms of the total charge motion in the induced channel:

$$I_D = \frac{dQ}{dt} = \frac{qpwh\,dx}{dt} \tag{5-5}$$

where p is the hole density and dx is the distance the charge moves in time dt. Starting with the carrier velocity, which is defined as

$$v_x = \frac{dx}{dt} = \mu_p E_x \tag{5-6}$$

and substituting into Eq. (5-6) for the electric field E_x the negative of the spatial derivative of the potential,

$$E_x = -\frac{dV_{xS}}{dx} \tag{5-7}$$

the channel current may be written as

$$I_D = whqp\mu_p E_x = -whqp\mu_p \frac{dV_{xS}}{dx} \tag{5-8}$$

Gauss' law states that the charge contained in the channel volume $(wh\,dx)$ is equal in magnitude but opposite in sign to the surface charge (σ_G) on the area $(w\,dx)$ of the gate electrode above this portion of the channel. The volume charge has two components:

1. Charge due to the current flow [Eq. (5-5)].
2. (*a*) Bulk charge of the semiconductor, which is associated with the impurity atoms in the lattice. For n-type silicon, this charge density is positive and equal to qN_D.
   (*b*) The net surface charge density of the semiconductor ($q\sigma_s$). This charge depends on the imperfections of the crystal.

Thus the statement of Gauss' law may be written as

$$-[(pq)wh\,dx + q(\sigma_s w\,dx + N_D wh\,dx)] = \sigma_G w\,dx \tag{5-9}$$

By using the identity $Q = CV$, the surface charge of the gate electrode can be expressed by

$$\sigma_G w\,dx = C_o w(V_{GS} - V_{xS})\,dx \tag{5-10}$$

where V_{GS} is the gate-to-source potential, V_{xS} is the channel potential at point x, and C_o is the capacitance per unit area. With an SiO_2 layer 1,000 Å thick, C_o turns out to be approximately 0.2 pF/mil^2. Substituting Eq. (5-10) into Eq. (5-9), we obtain

$$-[(pq)wh\,dx + q(\sigma_s w\,dx + N_D wh\,dx)] = C_o w(V_{GS} - V_{xS})\,dx \tag{5-11}$$

Now let us consider the first boundary condition. If there were no internal charge, the channel would become established at $V_{GS} = 0$. It does not become conducting until $V_{GS} = V_T$. We shall apply this boundary condition in the simple case where $V_{DS} = 0$. The channel voltage everywhere is zero, and consequently $I_D = 0$. Thus setting $V_{GS} = V_T$, $V_{xS} = 0$, and $pq = 0$, Eq. (5-11) is reduced to

$$q(\sigma_s w\,dx + N_D wh\,dx) = -C_o w V_T\,dx \tag{5-12}$$

Substituting Eq. (5-12) back into Eq. (5-11) and solving for the channel charge density pq, we obtain

$$pq = -\frac{C_o}{h}(V_{GS} - V_{xS}) - \frac{C_o V_T}{h} \tag{5-13}$$

We next substitute pq back into Eq. (5-8), with the result that

$$I_D = w\mu_p C_o[(V_{GS} - V_{xS}) - V_T]\frac{dV_{xS}}{dx} \tag{5-14}$$

Our second boundary condition is the statement of continuity of current flow: I_D must be independent of x. We first separate variables and integrate Eq. (5-14). As x goes from 0 to L, V_{xS} increases from 0 to V_{DS}.

Applying the boundary condition at $x = L$ results in the final expression for I_D:

$$I_D = \frac{C_o\mu_p w}{L}\left[(V_{GS} - V_T)V_{DS} - \frac{V_{DS}^2}{2}\right] \tag{5-15}$$

Equation (5-15) holds only up to the point of inflection, which occurs at the boundary of the ohmic region:

$$V_{DS\Omega} = V_{GS} - V_T$$

This is called the saturation voltage. At $V_{DS\Omega}$, the drain-to-gate potential has been reduced to the threshold voltage and the channel begins to pinch off near the drain. Above saturation, the voltage along the active portion of the channel is essentially constant at the saturation value. For $V_{DS} \geq V_{GS} - V_T$, I_D is independent of V_{DS} and is given by

$$I_D = \frac{K}{2}(V_{GS} - V_T)^2 \qquad \text{(saturated case)} \tag{5-16}$$

where

$$K = C_o\mu_p\frac{w}{L} \tag{5-17}$$

Using $\mu_p = 200\ \text{cm}^2/\text{V-s}$ and $C_o = 2\ \text{pF/mil}^2$, the product is

$$C_o\mu_p = 6.4 \times 10^{-6}\ \text{A/V}^2$$

The ratio of channel width to length scales this factor. In a typical n-channel device, μ_n will be about 300 cm²/V-s.

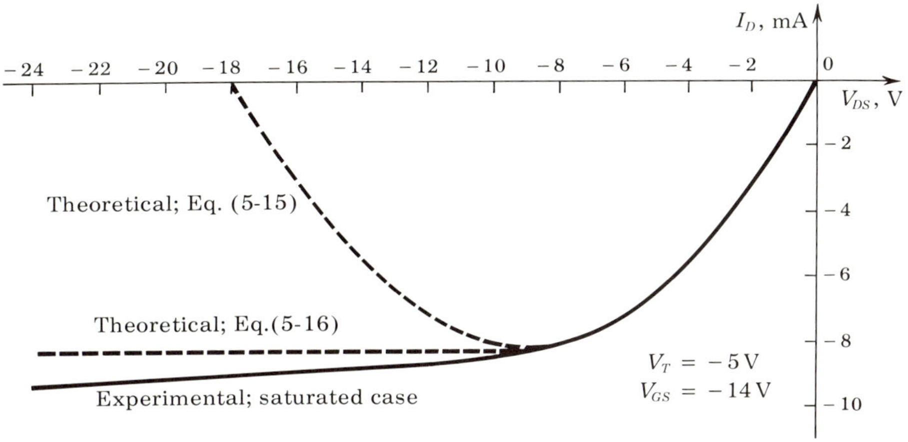

Fig. 5-7 A single theoretical and experimental volt-ampere characteristic curve for a MOSFET.

Equation (5-16) is often written in the form

$$I_D = \frac{KV_T^2}{2}\left(\frac{V_{GS}}{V_T} - 1\right)^2$$

This equation should be compared with Eq. (5-4) expressing I_D for the junction FET. The constant in both equations is usually determined experimentally.

Any increase in the drain voltage simply increases the voltage drop across the pinchoff region. Because the length of the ohmic portion of the channel is reduced as the depletion region broadens, dV_{xS}/dx of Eq. (5-14) increases. The coefficient K in Eq. (5-16) can no longer be assumed to be a constant. Since L decreases with increasing V_{DS}, K becomes a function of V_{DS}. We observe a small increase in drain current.

Equations (5-15) and (5-16) are plotted in Fig. 5-7 for one value of V_{GS}. The family of curves for a typical MOSFET appear in Fig. 5-8. They exhibit good agreement with the derived curve of Fig. 5-7. Note that the experimental curve shows the effect of the depletion-region broadening. As V_{DS} is increased, eventually the potential across the depletion region will reach the junction breakdown voltage. The curves will exhibit the sharp upward break characteristic of the avalanche breakdown. Because of the relatively low doping in the substrate, breakdown usually occurs well beyond the operating region.

A MOSFET can also be designed to operate in both the depletion and the enhancement modes. We start with a thin p channel diffused between

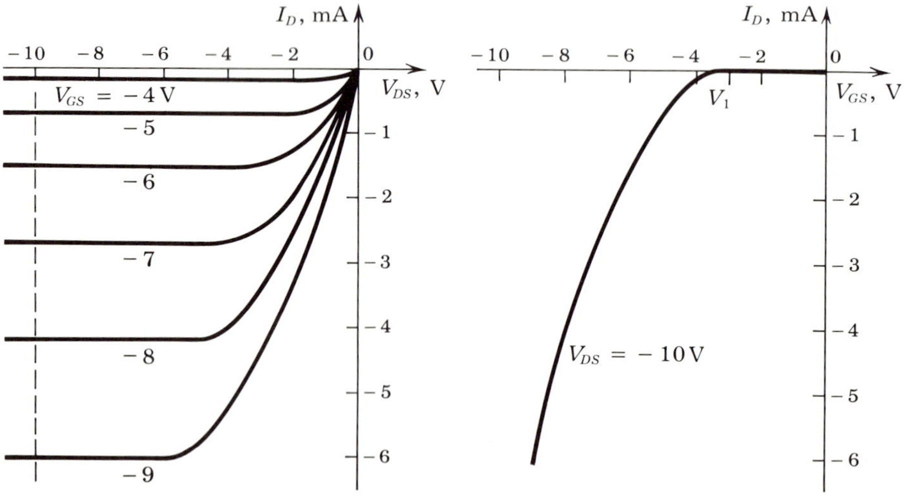

Fig. 5-8 MOSFET volt-ampere and transfer characteristics.

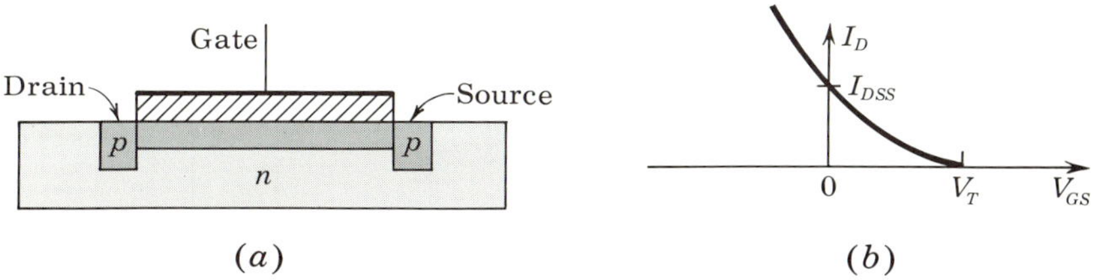

Fig. 5-9 (a) Depletion-enhancement MOSFET; (b) transfer characteristic.

the source and drain contacts, as shown in Fig. 5-9. A positive gate voltage depletes the existing p charge carriers and decreases the effective channel thickness. A negative gate voltage induces additional carriers and reduces the channel resistance. The shape of the transfer characteristic is the same as that of the enhancement-mode FET, only shifted to the right. All equations remain the same; we simply use a new value for V_T.

Role of the substrate

The substrate and the induced channel form a junction FET of the depletion type. If the substrate is always connected to the source, then, for the junction FET, V_{GSj} is always zero and the substrate will not affect the enhancement-mode operation of the MOSFET. But if many MOSFETs are formed on the same chip and connected to different portions of the circuit, we can no longer satisfy the $V_{GSj} = 0$ condition for all devices. Those MOSFETs whose source is above the substrate potential will have the parasitic junction FET biased on in a direction to deplete the channel. A higher threshold voltage will therefore be necessary to overcome these effects. Assuming that the substrate is at zero, we find that the increase in V_T can be approximated by

$$\Delta V_T \approx K \sqrt{|V_{S\text{-sub}}| + \psi}$$

where $V_{S\text{-sub}}$ is the source-to-substrate potential and ψ is the electrostatic potential on the silicon surface (about 0.8 V). K is a constant dependent on the substrate doping and on the thickness of the oxide; it ranges from 0.1 to about 1.5 in standard devices. Two measurements of threshold voltage at different source-to-substrate levels are sufficient to calculate K and ψ for the device.

MOSFET-gate protection

Excessive voltage on the MOSFET gate can punch through the thin SiO_2 dielectric and short-circuit the gate to the substrate. ($V_{GS} = 50$ V corre-

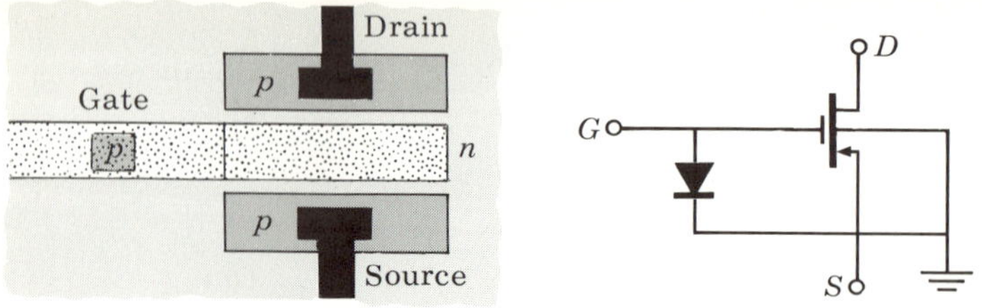

Fig. 5-10 Zener-diode–protected MOSFET.

sponds to a field strength of 5×10^6 V/m across the 1,000-Å layer.) Even an accumulation of static charge on the low-capacity gate can result in an excessive field. In order to protect the MOSFET, a zener diode can be shunted from gate to substrate (Fig. 5-10). Integrated circuits require protection only at the input stage.

The protective diode is fabricated by diffusing an additional p region into the n substrate (or n region into a p substrate). During metallization, the gate is connected to the diode. The junction breakdown is the zener voltage. Since this diode is back-biased by the normal gate voltages, it will not affect the MOSFET's characteristics.

5-3 Small-signal Models

The small-signal parameters of the FET can be obtained by differentiating the equations defining I_D. In the ohmic region, where $|V_{GS} - V_T| \geq |V_{DS}|$, we use Eq. (5-15) to obtain the gate-to-drain transconductance,

$$g_{m\Omega} = \left.\frac{\partial I_D}{\partial V_{GS}}\right|_{V_{DS}=\text{const}} = KV_{DS} \tag{5-18}$$

the drain-to-source small-signal output impedance,

$$\frac{1}{r_{o\Omega}} = g_{o\Omega} = \left.\frac{\partial I_D}{\partial V_{DS}}\right|_{V_{GS}=\text{const}} = K[(V_{GS} - V_T) - V_{DS}] \tag{5-19}$$

and the voltage-amplification factor,

$$\mu = \left.\frac{\partial V_{DS}}{\partial V_{GS}}\right|_{I_D=\text{const}} = g_m r_o \tag{5-20}$$

The derived parameters agree closely with the experimentally measured terms.

In the saturation region, where $|V_{GS} - V_T| \leq |V_{DS}|$, the small-signal transconductance is obtained by differentiating Eq. (5-16):

$$g_m = K(V_{GS} - V_T) = \sqrt{2KI_D} \tag{5-21}$$

The constant K depends solely on the geometry of the channel [Eq. (5-17)]. Currently available MOS devices have a transconductance at $I_D = 1$ mA of

$$g_m \approx 40 \sqrt{\frac{w}{L}} \qquad \mu\text{mhos} \tag{5-22}$$

Thus, by simply adjusting the ratio of channel width to length, we can control the g_m of the device.

We know that the gate forms a capacitance with the channel. The size of this capacitance depends on the gate (and channel) surface area (Lw). For good high-frequency response, it is necessary to minimize the capacitance while still maintaining a high g_m. The inherent limitation is the thickness of the line which can be deposited for the gate electrode. Current technology sets the lower limit of the channel length at 0.2 mil. Low-frequency devices normally have much longer channel lengths, on the order of 1 mil. Once we settle on L, the g_m is then determined by controlling the width of the channel.

Since in the saturation region, to a first approximation, current is independent of V_{DS},

$$r_o = \infty \tag{5-23}$$

and

$$\mu = \infty \tag{5-24}$$

Equation (5-21) agrees with the experimental results, but Eqs. (5-23) and (5-24) are incorrect. Our derivation assumed that the channel length remained constant. We know, however, that after pinchoff, the depletion region widens and the effective length of the channel is reduced. The change in channel length results in a slight increase of current, with increasing V_{DS}. Thus we actually have a large but finite value of r_o. This resistance varies approximately as

$$r_o \approx \frac{K_2}{I_D} \tag{5-25}$$

K_2 is an experimentally determined constant. Equation (5-25) is a relatively gross approximation but is nevertheless suitable for general circuit analysis. The amplification factor is still defined by $\mu = g_m r_o$.

High-frequency devices have wide, short channels [Eq. (5-22)]. Thus a small change in the depletion layer will have a relatively large effect on the effective channel length. This means that high-frequency MOS devices will have low values of r_o while audio-frequency devices have much longer channel lengths and hence larger values of r_o. They more nearly approximate the ideal device.

By comparing Eqs. (5-19) and (5-21), we observe that for any V_{GS}, the drain-to-source resistance in the ohmic region, near the origin ($V_{DS} \approx 0$), is equal to the reciprocal of g_m in the saturation region:

$$R_{\mathrm{ON}} \approx \frac{1}{g_m} \tag{5-26}$$

The junction FET and the MOSFET have the same square-law drain-current dependency on V_{DS} and V_{GS}. For the junction FET, in the saturation region

$$I_D = I_{DSS}\left(\frac{V_{GS}}{V_p} - 1\right)^2 \tag{5-27}$$

Thus

$$g_m = \frac{2I_{DSS}}{V_p}\sqrt{\frac{I_D}{I_{DSS}}} \tag{5-28}$$

We see that g_m varies as $\sqrt{I_D}$. The variation of r_o and μ of the junction FET is similar to that in the MOS device.

By comparing Eqs. (5-16) and (5-27), we note that the constant K used in the derivation of the MOS small-signal parameters can be replaced by

$$K_J = \frac{2I_{DSS}}{V_p^2} \tag{5-29}$$

When we do this, and when we also replace V_T by the pinchoff voltage V_p, all equations derived for the MOS become applicable to the junction FET.

In drawing the small-signal model for the junction FET, we start from the fact that the gate forms a diode junction with the channel. When operating in the saturation region, the depletion layer isolates the drain from the source. This isolation effectively separates the junction diode into two diodes, one connected between the gate and drain and the other connected between the gate and source. Both diodes are back-biased.

Figure 5-11*a* shows the small-signal model for a junction FET. In this model, R_{gd} and C_{GD} represent the reverse resistance and junction capacitance of the gate-to-drain diode. R_{gs} and C_{GS} are the resistance and junction capacitance of the gate-to-source diode. Since the back bias at the drain is usually larger than the back bias at the source, $C_{GD} < C_{GS}$. The controlled source, $g_m v_{gs}$, accounts for the forward amplification, and the dynamic output resistance is r_o.

Our simplified model of Fig. 5-11*b* illustrates the capacitive nature of the gate-to-channel coupling of a MOSFET. C_{GS} and C_{GD} are the capacitances from the gate electrode to the source and to the drain regions, respectively. These capacitances are smaller than the reverse diode capacitance in the junction FET. Because the gate is insulated from the body of the semiconductor, R_{gd} and R_{gs} are true leakage resistances. They are about 10^6 higher than the reverse diode resistance in the junction FET.

The depletion layer appears in the induced channel between the source and the drain. We incorporate its effects in the model by including the depletion-layer capacitance C_{DS} and depletion-layer broadening resistance

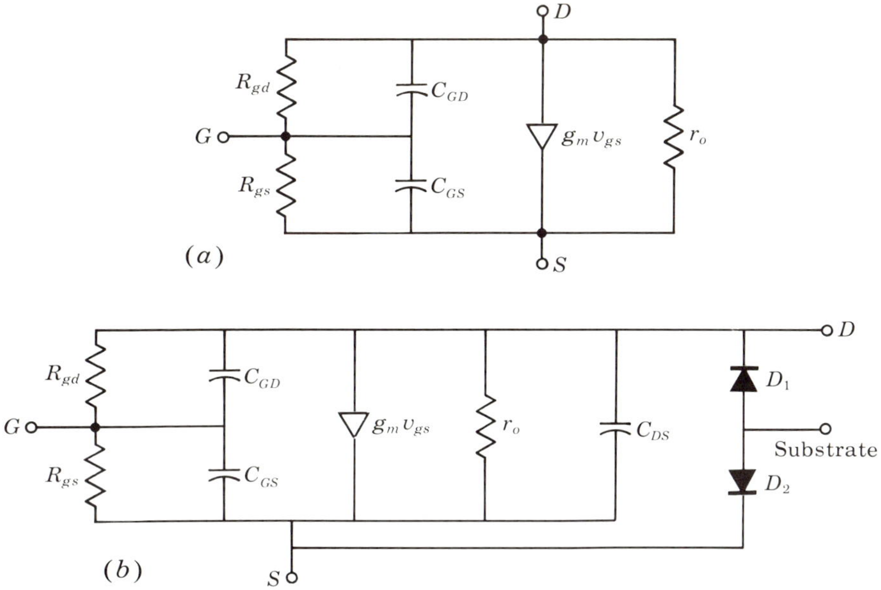

Fig. 5-11 (*a*) Junction-FET small-signal model; (*b*) MOSFET small-signal model.

Typical parameters

	JUNCTION	MOSFET
R_{gd}	= 350 M	10^{14} Ω
R_{gs}	= 350 M	10^{14} Ω
C_{GD}	= 5 pF	1 pF
C_{GS}	= 20 pF	1 pF
C_{DS}	= ———	2 pF
r_o	= 250 K	20 K
g_m	= 2.5 mA/V	5 mA/V

r_o in parallel with the controlled-current source $g_m v_{gs}$. The two diodes shown are the junction diodes created by the diffusion of the source and drain regions into the substrate.

The values of the capacitances in the models of Fig. 5-11*a* and *b* can be calculated from the manufacturers' data sheets, which give the input capacitance with the output short-circuited (C_{iss}) and the output capacitance with the input short-circuited (C_{rss}). From the model we observe that

$$C_{rss} = C_{GD}$$

and

$$C_{iss} = C_{GD} + C_{GS}$$

In a junction FET, C_{GS} and C_{GD} are junction capacitances which vary inversely with V_{DS}. In a MOSFET, the capacitances are independent of

V_{DS} and are from one-half to one order of magnitude smaller than in a junction device. There also exists a small drain-to-substrate capacitance, which can be lumped with any external load capacitance C_L. For simplicity, we shall usually assume constant values for all capacitors.

Both the junction FET and the MOSFET have additional leakage resistances, series ohmic drops, and stray capacitances. Since in most cases of circuit analysis these elements may be neglected, they have been omitted from the simplified models of Fig. 5-11*a* and *b*.

5-4 Biasing and Small-signal Applications

In order to use the FET as an amplifier, we must be able to set the quiescent operating point wherever we wish within the active region. For a junction FET operating in the depletion mode, this means that the gate-to-channel junction must be back-biased. Inserting a bias resistor in the source load, as shown in Fig. 5-12*a*, is a convenient means of obtaining the requisite bias voltage. The gate-to-source voltage is

$$V_{GS} = -I_G R_G - I_D R_S \tag{5-30}$$

Normally, $I_G \approx 0$ and the drop due to $I_G R_G$ can be neglected. The resultant equation for V_{GS} is a straight line through the origin. This is plotted on the transfer characteristics of Fig. 5-12*b*. The intersection at $I_D = -0.61$ mA and $V_{GS} = -0.61$ V is the quiescent operation point. The same result could have been obtained analytically by substituting Eq. (5-30) into the square-law expression for I_D [Eq. (5-27)] and solving for I_D.

The advantage of self-biasing, as contrasted to fixed biasing, is that self-biasing, to some extent, is self-regulating. FET device characteristics normally vary over the wide range shown in Fig. 5-12*c*. With a -0.61-V fixed bias, the low I_{DSS} devices would be operating at Q_2, which is practically at pinchoff. With self-bias, the intercept with the bias line at Q_3 is still near the center of the transfer curve.

Once we have found the operating point, we can then evaluate the small-signal parameters needed in the circuit of the model shown in Fig. 5-13.

We might note that the tangent to the square-law transfer characteristic from the zero bias point, I_{DSS}, intersects the voltage axis at $V_p/2$. Thus

$$g_{mss} = \frac{2I_{DSS}}{V_p}$$

and since g_m varies as the square root of the drain current, at the actual operating point

$$g_m = \frac{2I_{DSS}}{V_p}\sqrt{\frac{I_D}{I_{DSS}}} = 1.03 \times 10^{-3} \text{ mho}$$

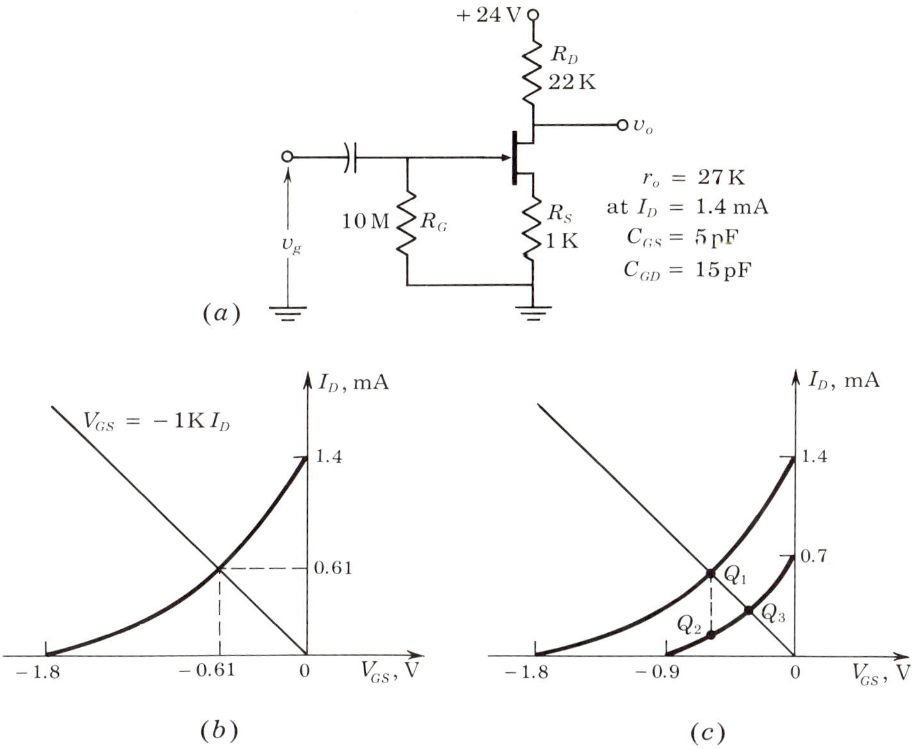

Fig. 5-12 (*a*) Junction-FET circuit; (*b*) junction transfer characteristics showing bias line and *Q* point; (*c*) illustrating shift in *Q* point because of typical spread of device characteristics.

Assuming that the variation of r_o is approximately inversely proportional to I_D,

$$r_o \approx r_{oss} \frac{I_{DSS}}{I_D} = 61 \text{ K}$$

The voltage-amplification factor at this operating point is

$$\mu = g_m r_o = 63$$

Figure 5-13*a* and *b* show the two possible circuit models for the FET. In the first, we have a current source shunted by the output impedance; the second is the Thévenin equivalent circuit, where the output impedance is in series with a voltage source. The circuit of Fig. 5-12*a* is simpler to analyze if we use the model of Fig. 5-13*b*. From this model we note that

$$v_{gs} = v_g - i_d R_S$$

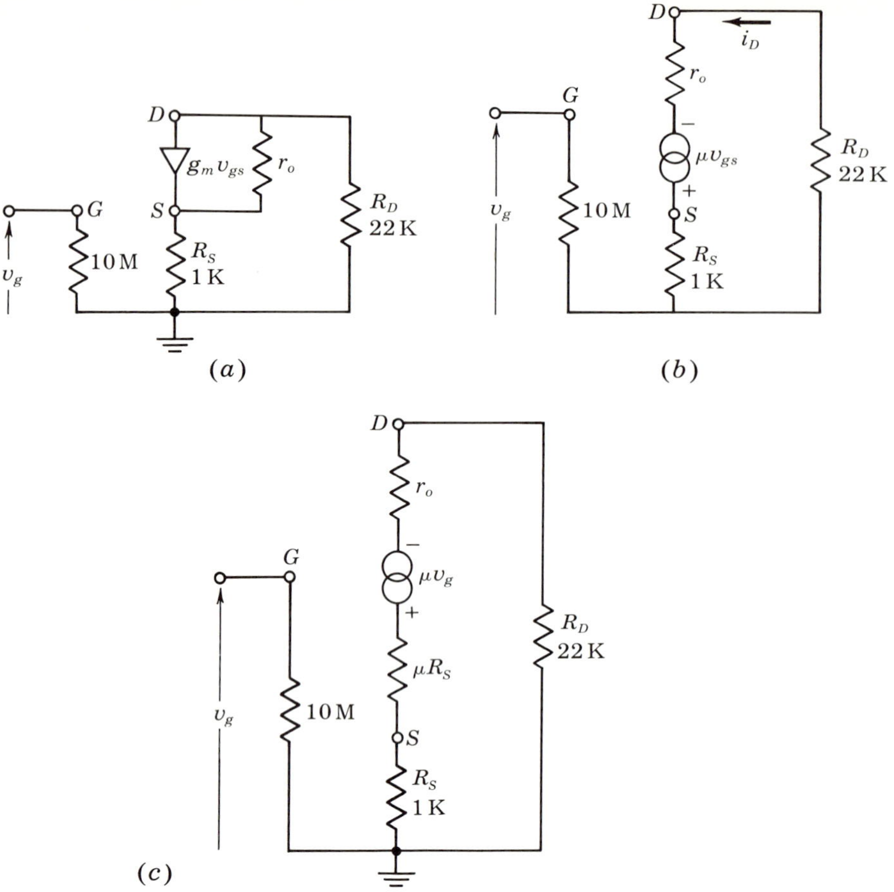

Fig. 5-13 (*a*) Small-signal model for the circuit of Fig. 5-12*a*; (*b*) Thévenin equivalent model for the drain-source branch; (*c*) voltage generator converted from dependence on v_{gs} to dependence on v_g.

Substituting back into the controlled source, we obtain the equivalent voltage generator

$$-\mu v_{gs} = -\mu v_g + \mu R_S i_d \tag{5-31}$$

Since the second term of Eq. (5-31) is proportional to the current flowing through the voltage generator, we can replace it by the equivalent resistance μR_S. From the resulting model of Fig. 5-13*c*, we find that the small-signal voltage amplification from the gate to the drain is

$$A = \frac{-\mu R_D}{r_o + R_D + (1 + \mu) R_S} = -9.4 \tag{5-32}$$

If R_S is bypassed, the midfrequency gain becomes

$$A = \frac{-\mu R_D}{r_o + R_D} = -16.7$$

Because of the FET's extreme nonlinearity, the amplifier's large-signal transfer characteristic must be obtained by a point-by-point plot from the drain volt-ampere characteristics.

MOSFET operation in the enhancement mode must be biased so that the gate voltage is of the same polarity as the drain voltage. In addition, to ensure that the operating point is in the saturation region, $|V_{GS} - V_T| \leq |V_{DS}|$. If we bias the device, as shown in Fig. 5-14*a*, then

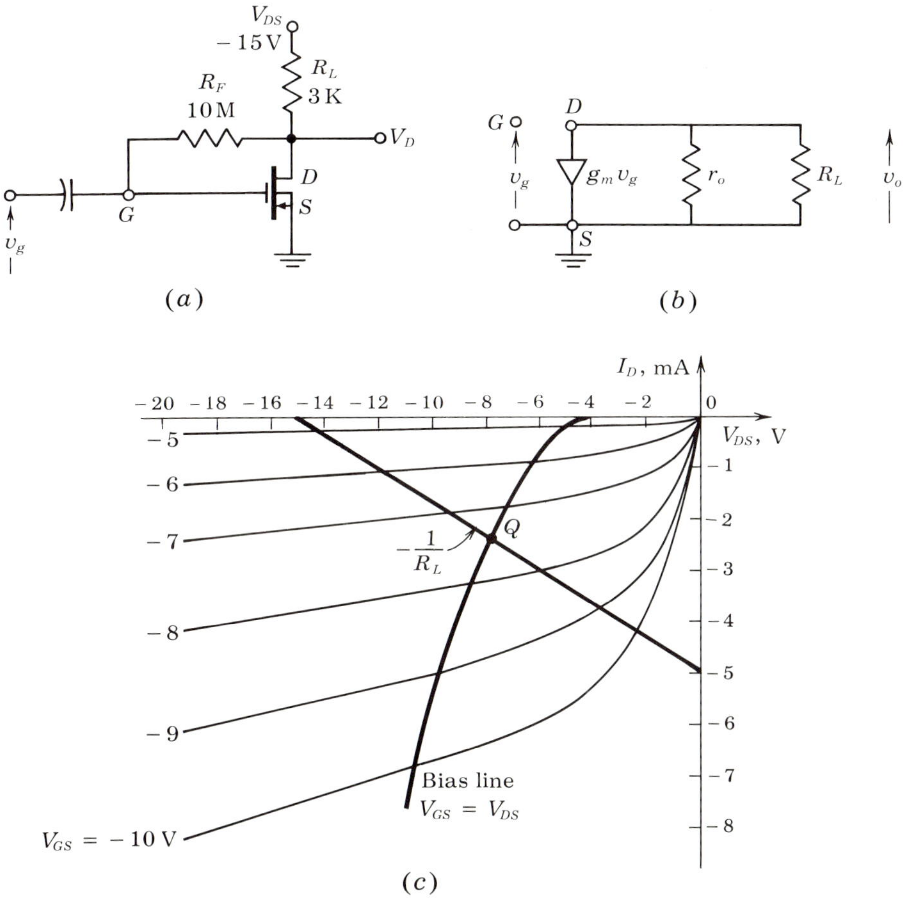

Fig. 5-14 (*a*) MOSFET amplifier circuit; (*b*) small-signal model; (*c*) volt-ampere characteristics showing bias line and load line.

since $I_G \approx 0$, $V_{GS} = V_{DS}$ and the condition for saturation-region operation is satisfied for every value of V_{DS}. The intersection of the dc load line with the bias locus, $V_{GS} = V_{DS}$, on the volt-ampere characteristic of Fig. 5-14*c* is the quiescent operating point.

In order to find the small-signal gain, we first calculate the values of g_m and r_o at the operating point. Using the model of Fig. 5-14*b*, if $r_o \gg R_L$, we see that the gain becomes

$$A = -g_m R_L \tag{5-33}$$

5-5 MOSFET Used as a Load Resistor

Because of the high resistance of a saturated MOSFET, it can be used as the load resistor in the drain circuit of another amplifier FET. By doing this, we save a tremendous area on the integrated-circuit chip. For example, if the base diffusion of 200 Ω/□ were used as the load resistance, a 20-K resistor would require a ½-mil line 50 mils long. This is 25 mils^2 before including the necessary isolation-region area. With a MOSFET, we can obtain an equivalent load in an area of less than 1 mil^2. In addition, the nonlinearity of the load compensates for the nonlinearity of the amplifier MOS volt-ampere characteristics, thus linearizing the overall response.

As a purely illustrative example, we shall consider the simple circuit of Fig. 5-15*a*, having the geometry shown in Fig. 5-15*b*. We shall first examine the operation of the load MOSFET. For it to be saturated,

$$|V_{GS} - V_T| \leq |V_{DS}|$$

If the load gate is connected directly to the drain, then $V_{GG} = V_{DD}$ and $V_{GS} = V_{DS}$. We see that the saturation condition is always satisfied. The locus of equal gate and drain voltages can be plotted on the volt-ampere characteristics, as shown in Fig. 5-16*a*. The locus of points separating the ohmic and saturation regions is also plotted on the same characteristic.

At any point along the locus of equal drain and source voltages, the incremental conductance is

$$g_{DSL} = \left.\frac{\partial I_D}{\partial V_{GS}}\right|_{V_{GD}=\text{const}} + \left.\frac{\partial I_D}{\partial V_{DS}}\right|_{V_{GS}=\text{const}}$$

$$= g_m + g_o \tag{5-34}$$

Thus the effective load resistance is the reciprocal of the two conductances in parallel. The result of Eq. (5-34) can also be found from the model of the load MOSFET shown in Fig. 5-15*c*. Because the controlled-current

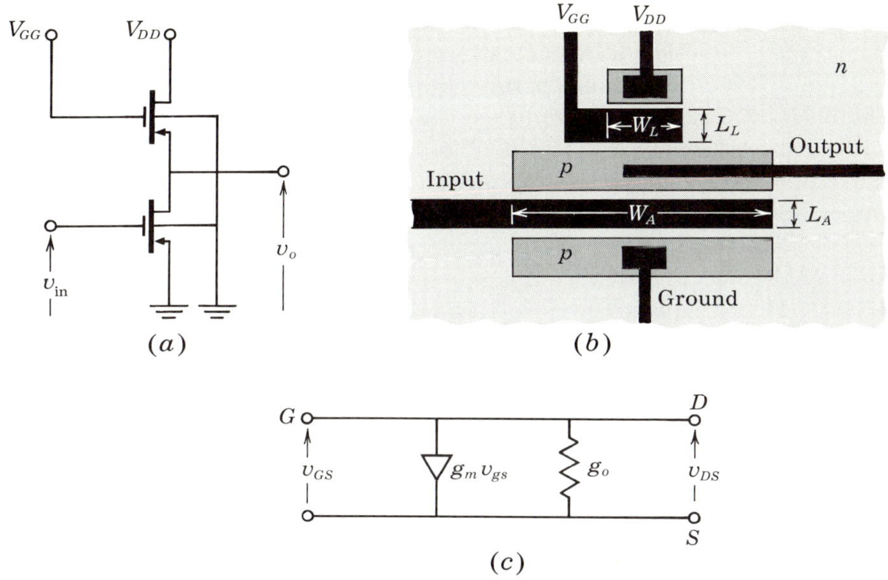

Fig. 5-15 (*a*) MOSFET amplifier using a second MOSFET as a load; (*b*) amplifier geometry; (*c*) small-signal model of load with $V_{GS} = V_{DS}$.

source is proportional to the voltage across it, it can be replaced by the equivalent conductance g_m in parallel with the output conductance g_o.

In all MOS devices operating at low values of I_D, $g_m > g_o$. It follows that the active drain-to-source resistance depends on the reciprocal of the g_m of the load device. We conclude that in order to maximize the gain, a FET having a high g_m should be used as the amplifier while a low-g_m FET would act as the high impedance needed for the load.

In an integrated circuit, since both devices are formed on the same chip, the g_m's have identical dependency on the geometry. Neglecting for the moment the effects on the gain of the r_o of either device, we have

$$A = -g_{mA} r_{DSL} = -\frac{g_{mA}}{g_{mL}}$$

Both g_m's must be evaluated at the same value of I_D. From Eqs. (5-21) and (5-22), we see that the g_m's are determined by the square root of the ratio of channel width to length:

$$A = -\sqrt{\frac{K_A}{K_L}} = -\sqrt{\frac{W_A}{L_A}\frac{L_L}{W_L}} \tag{5-35}$$

Equation (5-35) shows that, to a very good approximation, the gain of the stage is a function solely of the channel geometry. In the particular example of Figs. 5-15 and 5-16, we selected a geometric ratio $K_A/K_L = 4$. The transfer function of Fig. 5-16*c* verifies that the voltage gain is indeed -2.

Layout constraints and permissible node capacity usually determine the W/L ratio of the load, and then the geometry of each region is adjusted for the desired gain. Typical W/L ratios for the amplifier and load are 10 and 0.3, respectively, yielding a stage gain of 5.7. A gain of 10 is usually the upper limit. The thinness of the line which can be deposited for the gate is often the limiting factor. Certainly, for any required W/L ratio, the thinner the line, the smaller the gate area and the smaller the gate capacitance. High-frequency performance is degraded by a large L.

So long as the amplifier remains in the saturation region, to a first-order approximation the nonlinear load compensates exactly for the nonlinearity of the amplifier MOS volt-ampere characteristic. Since r_o varies roughly inversely with I_D, only at the higher drain currents will the output conductance of both the load and the amplifier MOS become important. The load output impedance shunts g_{mL} and reduces the effective load resistance. The amplifier r_{oA} shunts the load MOS, also resulting in a lower overall gain. If, however, we use a low g_m load for maximum gain, then it will limit the maximum current flow through the amplifier MOSFET and the effect of the shunting by r_{oA} will be negligible.

When the amplifier is driven into the ohmic region, the change in characteristics causes a rapid falloff in gain. The nonlinearity at higher drain currents is clearly shown by the change in the slope of the transfer characteristics of Fig. 5-16*c*.

Because the load and amplifier FETs are in series, the same drain current will flow through each device. We can obtain the path of operation by plotting the drop across the load, at a given I_D, on the amplifier volt-ampere characteristic. The load-path equation is

$$V_{D,\text{amp}} = V_{DD} - V_{DS,\text{load}} \tag{5-36}$$

Figure 5-16*b* shows the locus of the load-MOSFET operation (Fig. 5-16*a*) plotted on the amplifier characteristics. By a point-by-point plot, we finally obtain the transfer characteristic of Fig. 5-16*c*.

One limit of amplifier operation is the OFF condition. For input voltages below the threshold of the amplifier MOSFET, only a small leakage current flows through the load. The load is always biased within the saturation region. Therefore the minimum drop across the load must be equal to the threshold voltage V_T. The output under OFF conditions becomes our logical-1 output level for logic operations:

$$V_1 = V_{\text{OFF}} = V_{DD} - V_T \tag{5-37}$$

The second limit occurs at the higher currents, where the amplifier V_{DS} drops and the device enters the ohmic region. The change of slope of the transfer curve corresponds to the intersection of the load path of operation and the boundary between the saturation and ohmic regions. In Fig. 5-16*b* and *c*, these points are indicated by β_1 and β_2. The gain certainly starts to drop off beyond β_1 and β_2.

The final bottoming voltage, V_0, can be found by the simultaneous solution of the saturation-region equation for the load device and the ohmic-region equation for the amplifier MOSFET. Boundary conditions are, for the amplifier, $V_{GS} = V_1$ and $V_{DS} = V_0$, and for the load, $V_{GS} = V_{DD} - V_0$ (load line 1). Substituting into Eqs. (5-16) and (5-15), we have

$$I_{DL} = \frac{K_L}{2}(V_{DD} - V_0 - V_T)^2 = \frac{K_L}{2}(V_1 - V_0)^2 \qquad (5\text{-}38a)$$

$$I_{DA} = \frac{K_A}{2}[2(V_1 - V_T)V_0 - V_0^2] \qquad (5\text{-}38b)$$

I_D must always be the same in both devices. Solving Eqs. (5-38*a*) and (5-38*b*) for the intersection, we find that the V_0 output is given by

$$V_0 = V_{\mathrm{ON}} \approx \frac{V_1^2}{2A^2(V_1 - V_T)} \qquad (5\text{-}39)$$

The approximation holds when $A^2 \gg 1$. A is defined by Eq. (5-35). V_0 is nominally about $0.05V_{DD}$.

We can also introduce a simple linear model for the ohmic region. The bottomed amplifier MOSFET is replaced by the low resistance

$$R_{\mathrm{ON}} = \frac{1}{g_m} = \frac{1}{K_A(V_1 - V_T)}$$

The load device acts as a constant-current source. If we approximate $V_0 \approx 0$, then

$$I_{DL} \approx \frac{K_L}{2}(V_{DD} - V_T)^2$$

Forming the product $I_{DL}R_{\mathrm{ON}}$, we arrive at

$$V_0 \approx \frac{K_L V_1^2}{2K_A(V_1 - V_T)}$$

which is identical to Eq. (5-39). The substitution of R_{ON} for the ohmic device is particularly useful when the load is a fixed resistor. V_0 is then found from the R_L, R_{ON} voltage divider.

When the MOS amplifier is used as an inverter in a digital system, it is driven between the extremes of OFF and ON, yielding the logical-1 and logical-0 output states given, respectively, by Eqs. (5-37) and (5-39).

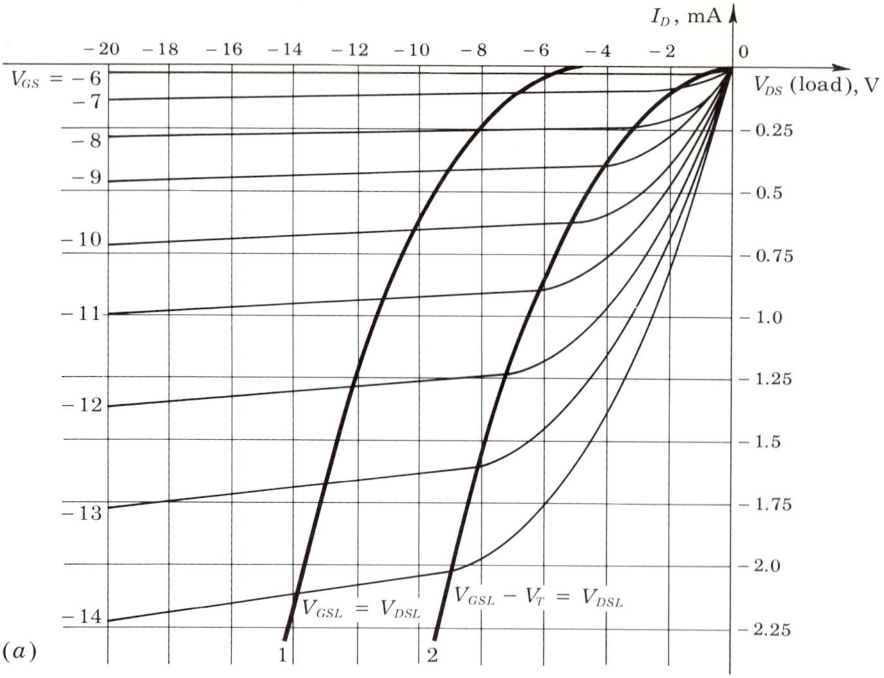

I_D, mA
−20 −18 −16 −14 −12 −10 −8 −6 −4 −2 0
V_{DS} (load), V
V_{GS} = −6
−7
−8
−9
−10
−11
−12
−13
−14
−0.25
−0.5
−0.75
−1.0
−1.25
−1.5
−1.75
−2.0
−2.25
$V_{GSL} = V_{DSL}$
$V_{GSL} - V_T = V_{DSL}$
1
2

(a)

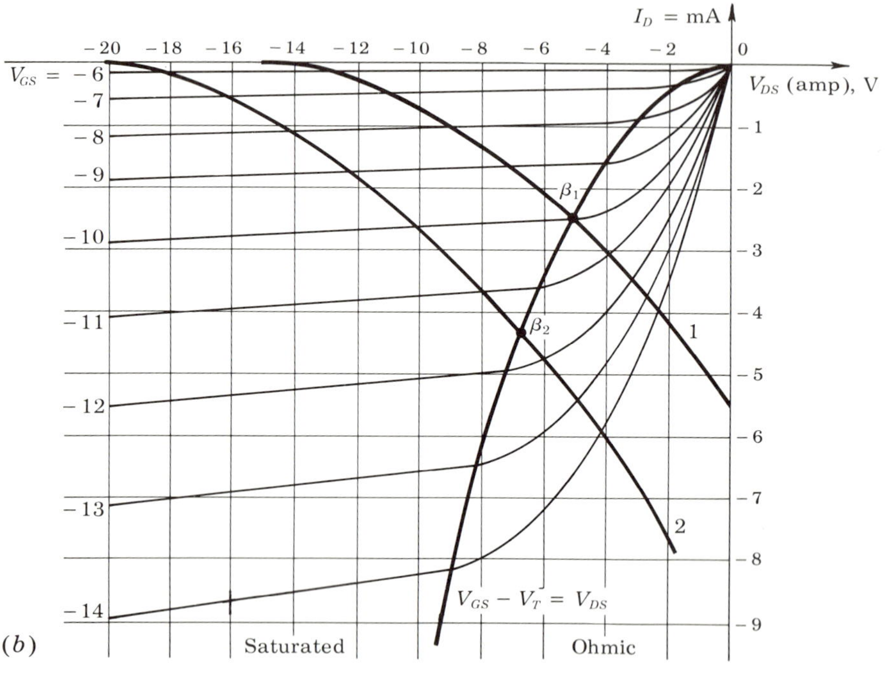

I_D = mA
−20 −18 −16 −14 −12 −10 −8 −6 −4 −2 0
V_{DS} (amp), V
V_{GS} = −6
−7
−8
−9
−10
−11
−12
−13
−14
−1
−2
−3
−4
−5
−6
−7
−8
−9
β_1
β_2
1
2
$V_{GS} - V_T = V_{DS}$
Saturated
Ohmic

(b)

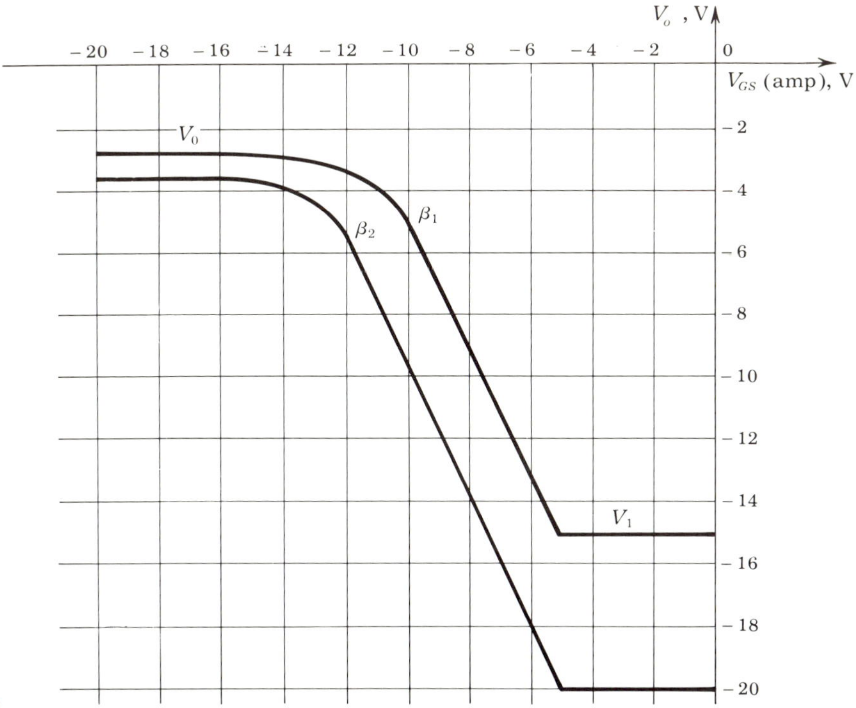

(c)

Fig. 5-16 (a) Load MOSFET volt-ampere characteristics showing loci of $V_{GS} = V_{DS}$ and loci separating ohmic and saturation regions; (b) paths of operation on amplifier MOSFET volt-ampere characteristics; (c) transfer characteristics corresponding to operating paths of (b).

If we make the load gate voltage more negative than the drain supply, the path of operation of the load shifts toward the ohmic region. At $V_{GG} = V_{DD} + V_T$, the load is at the border between ohmic and saturation operation. For higher values of V_{GG}, the load path is always in the ohmic region. Because of the shift in load path toward the origin, we can obtain much greater swings from the same amplifier. Figure 5-16b, curve 2, is a second path of load operation located at the border between ohmic and saturation operation. It clearly shows the increase in the permissible dynamic swings.

Since the drain and bias voltages are not only of the same polarity but also of the same order of magnitude, the output of one amplifier stage using an enhancement-mode MOS may be direct-coupled to the gate of the next stage. This is not true of the depletion-type junction-FET amplifiers, where the polarity of the gate bias is opposite to that of the drain voltage. Coupling, in an integrated circuit, is accomplished during the metallization step. Figure 5-17 shows a three-stage amplifier using only MOSFETs. The input and output capacitors and the 20-M feedback biasing resistors

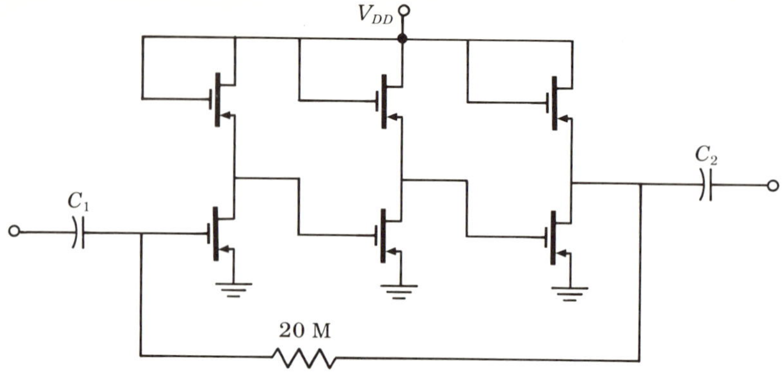

Fig. 5-17 Three-stage direct-coupled MOSFET amplifier.

are lumped external elements. Because of the symmetry, each stage will be biased at the same value of V_{GS} and I_D.

5-6 Inverter Transient Response—Part I, Piecewise-linear Approximation

A FET has a capacitive input (C_{GS}) and capacitive coupling to the output (C_{GD}) and works into a capacitive load (C_L). Impedances are usually high, and therefore any transient analysis must consider several important time constants. To simplify the analysis of the multiple energy-storage-element system, we shall look for the most significant time constant for each portion of the response. Any solution, consequently, will be only approximate.

We shall now investigate the transient response of the bottoming inverter of Fig. 5-18*a* excited by an input pulse switching between the logical-0 and logical-1 levels. The circuit model contains three energy-storage elements and is difficult to solve in general terms. Let us instead attempt to pose a problem which can be solved in a simple manner. We do this by ensuring that the poles are widely separated. Once this is true, then we can first solve for the fast leading portion of the response and assume that it is complete before the start of the slow trailing portion. To separate the significant poles, we assume that $R_D \gg R_G$ and $R_D \gg R_{ON}$. These assumptions are not stringent since a small R_G implies that the gate is driven by a square-wave voltage source and we would normally expect a large R_D to obtain a reasonable gain. After we solve the simplified problem, we can then consider the effect of bringing the poles closer together.

Let us start the analysis at $t = 0^-$ with the FET off (Fig. 5-18*b*). This means that the gate voltage is beyond pinchoff, $|V_{GS}| > |V_p|$, in a junction-

FET device (an enhancement-type MOS requires $|V_{GS}| < |V_T|$). The channel is an open circuit. The output capacitor (C_L) is charged to the supply voltage V_{DD}, and the input capacitor (C_{GS}) is charged to the logical-0 level V_0.

At $t = 0$, the gate voltage abruptly switches from V_0 to V_1. We shall assume $V_1 = 0$ for a junction FET ($|V_1| \gg |V_T|$ for a MOS). Because C_{GS} was previously charged to V_{OFF}, the gate voltage and hence the OFF-channel state do not change immediately with the switching-voltage change. Thus the first portion of the turn-on transient can be found from the model of Fig. 5-18*b*, with V_0 replaced by V_1.

The gate charges toward V_{ON} with the fast time constant

$$\tau_1 \approx R_G\left(C_{GS} + \frac{C_{GD}C_L}{C_{GD} + C_L}\right) \tag{5-40}$$

In $2.2\tau_1$, the gate voltage rises from 10 to 90 percent of the final value. After the initial portion of the transient, we shall assume that the gate has reached $V_{GS} = 0$, the final value. The FET channel has turned on.

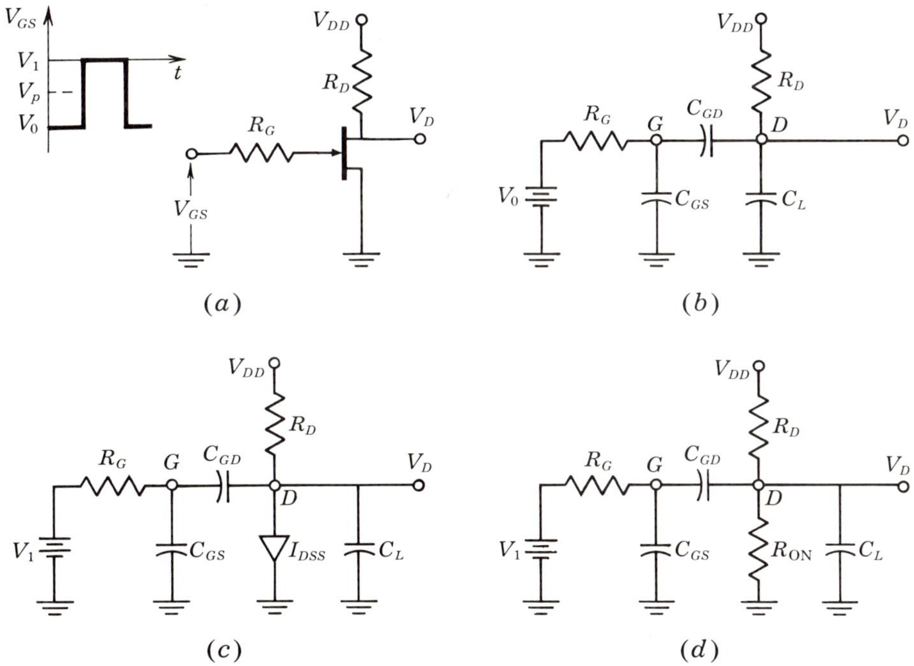

Fig. 5-18 Models for the piecewise-linear calculation of the approximate transient response of an inverter stage. (*a*) Circuit; (*b*) OFF model; (*c*) saturation-region ON model; (*d*) steady-state ohmic region ON model.

While C_{GS} is charging, capacitors C_{GD} and C_L act as a capacitive divider (normally, $C_L > C_{GD}$). Any voltage change at the gate will divide inversely proportionately across C_{GD} and C_L. We can justify this argument because the drain-time constant $R_D C_L$ is so very much longer than the input-time constant τ_1 that the current through R_D will not contribute to the charging of C_L during the initial period of interest. The output jumps up from V_{DD} by

$$\Delta V_{D,\max} = \frac{C_{GD}}{C_{GD} + C_L} \Delta V_{GS} \tag{5-41}$$

where $\Delta V_{GS} = V_1 - V_0$ is the magnitude of the positive gate transition.

Because of the initial charge stored in and the voltage rise across C_L, the drain-to-source voltage remains large enough to keep the FET in the pinchoff region even after the gate voltage reaches 0. The drain current increases to I_{DSS}. From Fig. 5-18*c*, we see that the Thévenin equivalent steady-state voltage toward which the drain charges is

$$V_{D,\mathrm{Th}} = V_{DD} - I_{DSS} R_D \tag{5-42}$$

The FET switches into the ohmic region once the output drops to

$$V_{D\Omega} = V_{DS} = -V_p + V_1 \tag{5-43}$$

and with $V_1 = 0$, $V_{D\Omega} = -V_p$.

Assuming that the input capacitor is fully charged and that the input-time constant is fast enough to follow all changes at the output, the output charging-time constant during the drain-voltage rundown is

$$\tau_2 \approx R_D(C_L + C_{GD})$$

We now know the initial [Eq. (5-41)], final [Eq. (5-43)], and steady-state [Eq. (5-42)] voltages. The time required for the drain-to-source voltage to run down from the initial value to the boundary of the ohmic region is, therefore,

$$t_2 = \tau_2 \ln \frac{V_{D,\mathrm{Th}} - V_{D,\max}}{V_{D,\mathrm{Th}} + V_p} \tag{5-44}$$

Once in the ohmic region, we can use the linear model of Fig. 5-18*d* to approximate the final portion of the rundown. The steady-state voltage changes to the bottoming value

$$V_B = \frac{R_{\mathrm{ON}}}{R_{\mathrm{ON}} + R_D} V_{DD}$$

and the new time constant is much smaller because of the shunting of C_L by R_{ON}:

$$\tau_3 \approx R_D \parallel R_{\mathrm{ON}}(C_L + C_{GD}) \approx R_{\mathrm{ON}}(C_L + C_{GD})$$

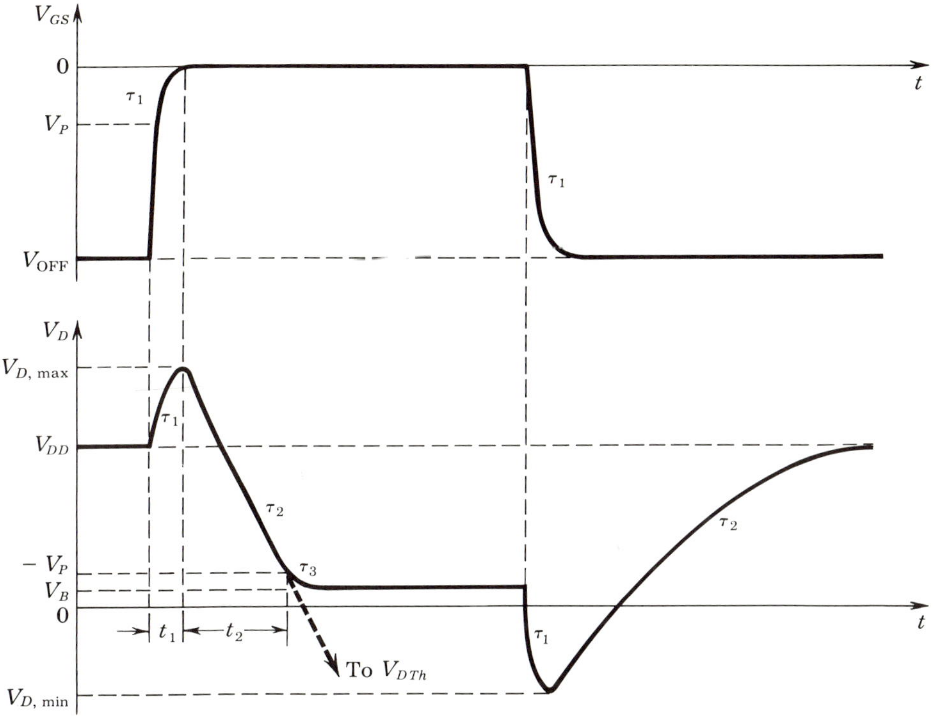

Fig. 5-19 Transient response of a FET inverter.

In the complete waveshape shown in Fig. 5-19, we can clearly see the three portions of the turn-on response: the fast rise and overshoot during the gate charging, the almost linear rundown during the time the channel remains pinched off, and the final decay to the bottoming voltage.

To turn off the inverter, we switch the gate drive from V_1 to V_0. The FET starts changing state from the bottoming point in the ohmic region. Consequently, during the initial portion of the turn-off transient, the drain-time constant, determined by R_{ON}, is small. The drain follows the fall in the gate voltage. However, only a small change in the gate voltage will bring the FET into the pinchoff region, where the drain resistance becomes large. Any further change at the gate is then capacitively coupled to the output. We can approximate the total initial portion of the turn-off period by the single gate-time constant used during the beginning of the turn-on period (τ_1). At the output we observe a negative jump from V_B down to

$$V_{D,\min} = V_B - \frac{C_{GD}}{C_{GD} + C_L}(V_1 - V_0)$$

Since the gate is beyond pinchoff, the drain recovers to the final steady-state voltage V_{DD}, with the long drain-time constant τ_2. Figure 5-19 shows the overshoot present during the turn-off period and the extremely slow recovery time.

The overshoots may be reduced in amplitude or eliminated completely by driving the gate with signals having the sloping leading and trailing edges that might be expected from the output of a similar inverter. A large value of C_L will also reduce the amplitude of the jumps by increasing the attenuation of the capacitive divider.

The gate input time constant τ_1 increases linearly with the switching-signal source impedance R_G. For large values of R_G, τ_1 will increase until it is equal to or even larger than the drain-time constant τ_2. When this happens, we can no longer solve separately for the initial and trailing edges of the turn-on and turn-off transients. We observe, however, that a longer input-time constant means that the output decay starts while the input is still rising. The leading edge of the output waveshape is slower, and the overshoots will be smeared out and reduced in amplitude. Turn-on and turn-off times both increase.

The case of the large R_G is exactly the situation faced in solving for the turn-on transient of a logic element fed from the output of another FET stage. Since the logical 1 appears at the drain of an OFF FET, the output impedance is the drain load R_D. When directly coupling to the next state, we turn it on through this high impedance.

The logical 0 is produced by the bottoming of the FET. In this case, the output impedance is the small R_{ON} of the FET and the transient solution follows that outlined above.

5-7 Inverter Transient Response—Part II, Exact Response

The piecewise-linear techniques discussed in Sec. 5-6 are a reasonable first approximation to the analysis of an inverter having a resistive load. Even though the FET is itself highly nonlinear, it controls only the final portion of the turn-on transient.

Different techniques are necessary for the analysis of the inverter of Fig. 5-21*a*. Here the MOS load is extremely nonlinear, decreasing from a very high impedance at low currents to a much smaller value when the amplifier device bottoms in the ohmic region. But because both the amplifier and the load characteristics are explicitly known, we can solve exactly for the response. All turn-on and turn-off delays and overshoots are found as described above. For simplicity, in the following discussion we shall assume extremely fast switching with no spikes. We shall concentrate our attention on the charge and discharge of the output-node capacitor.

Instead of evaluating the voltage response, we shall solve directly for the interval necessary for the capacitor to charge from a specified initial voltage to a known final value. Circuit stable conditions and switching points establish the boundaries of each region. Our analysis starts from the capacitor-current-defining equation, where the current is expressed as a function of the capacitor voltage:

$$C \frac{dv_q}{dt} = i_c = f(v_q)$$

By separating variables and integrating, we obtain the time interval in the form of the definite integral

$$t_1 = C \int_{V_i}^{V_f} \frac{dv_q}{f(v_q)} \tag{5-45}$$

For purposes of illustration, let us now consider the resistive charging of a capacitor (Fig. 5-20). Here, once the switch is opened,

$$i_c = \frac{V_{DD} - v_q}{R} = \frac{V_{DD}(1 - x)}{R}$$

where x is the normalized voltage ratio v_q/V_{DD}. As the capacitor charges, x increases from 0 to 1. The same normalization yields $dv_q = V_{DD}\,dx$. Substituting into Eq. (5-45), we have

$$\begin{aligned} t_R &= CR \int_{x_i}^{x_f} \frac{dx}{1 - x} = -\tau_R \ln (1 - x) \Big|_{x_i}^{x_f} \\ &= \tau_R \ln \frac{1 - x_i}{1 - x_f} \end{aligned} \tag{5-46}$$

Equation (5-46) can be identified as the identical exponential charging-time equation found in Eq. (1-9). If in the integral of Eq. (5-46) we first used the indefinite upper limit of x and then solved for $x(t)$, we would obtain the known exponential time response. We see that this form of solution presents the desired time intervals directly and the voltage response only indirectly.

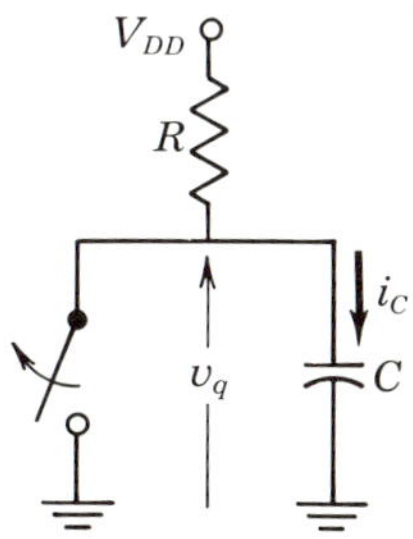

Fig. 5-20 *RC* capacitor charging.

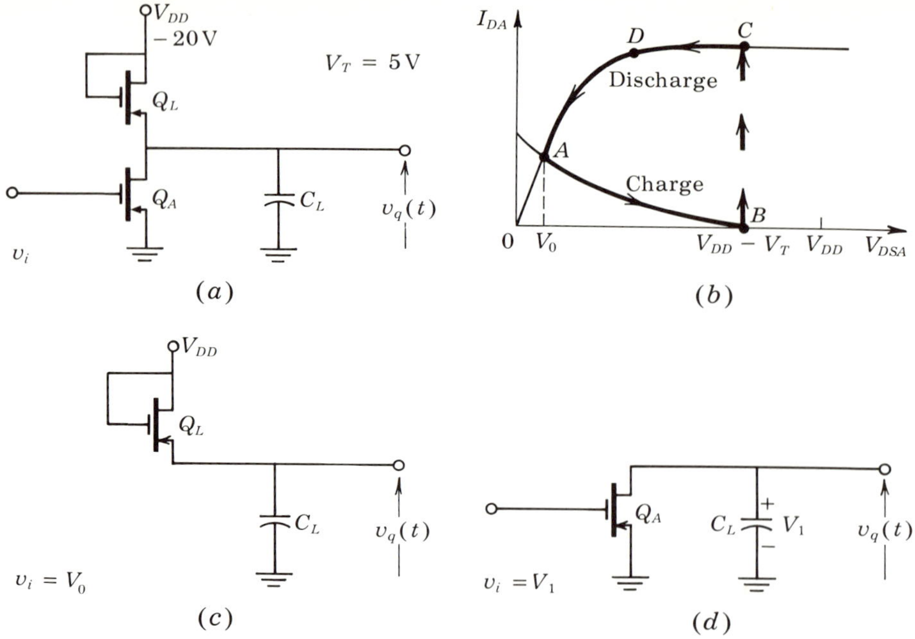

Fig. 5-21 MOSFET inverter charging and discharging a capacitive load. (*a*) Basic circuit; (*b*) amplifier volt-ampere characteristic, showing charge and discharge paths; (*c*) reduced circuit during charging; (*d*) reduced circuit during discharge.

The rise time of the RC circuit, i.e., the time needed to charge from 10 to 90 percent of the final value, is given by Eq. (5-46) as

$$t_R = \tau_R \ln \frac{1 - 0.1}{1 - 0.9} = 2.2\tau_R$$

Capacitor charging

Now let us approach the problems posed by the switching of the inverter stage of Fig. 5-21*a*. Our starting point will be with the inverter on, just being switched off. Figure 5-21*b* indicates the path of operation on the amplifier volt-ampere characteristics. The logical-0 starting point of A is at the intersection of the $V_{GS} = V_1$ characteristic of the amplifier MOS and the superimposed load characteristic. After the amplifier is switched off (by applying a logical 0 to the gate), the load capacitor charges through the load, along path AB, to the logical-1 output of $V_{DD} - V_T$. Figure 5-21*c* shows the appropriate model of Q_A off and Q_L on. Since the load MOSFET is always operating in the saturation region, the charging current is given by

$$I_c = I_{DL} = \frac{K_L}{2}(V_{GS} - V_T)^2$$

But $V_{GS} = V_{DD} - v_q$. Substituting for V_{GS} and normalizing, we have

$$I_c = \frac{K_L}{2}(V_{DD} - V_T)^2(1 - x)^2 \tag{5-47}$$

where $x = v_q/(V_{DD} - V_T) = v_q/V_1$ and $dv_q = V_1\,dx$.

Substituting Eq. (5-47) into Eq. (5-45), the resultant charging-time integral becomes

$$t = \tau_c \int_0^x \frac{dx}{(1 - x)^2} \tag{5-48}$$

In Eq. (5-48) the proportionality constant τ_c has the units of time:

$$\tau_c = \frac{2C}{K_L V_1} \tag{5-49a}$$

At $x = 0$, $V_{GS} = V_{DD}$ and $I_{DL} = I_{DL,\max} = (K_L/2)V_1^2$. From Eqs. (5-21) and (5-26), we find that $2R_{ONL,\min} = V_1/I_{DL,\max}$. By solving for K_L and substituting into Eq. (5-49a), we obtain a second form for the proportionality constant:

$$\tau_c = \frac{CV_1}{I_{DL,\max}} = 2CR_{ONL,\min} \tag{5-49b}$$

Because we can readily measure R_{ONL}, this second form of τ_c is particularly useful in verifying experimental results.

For simplicity, we assume that the initial point of the charging path was at $v_q = 0$. Equation (5-48) thus has the solution

$$t = \tau_c \frac{x}{1 - x} \tag{5-50}$$

It takes $8.9\tau_c$ for the capacitor voltage to rise from $x = 0.1$ to $x = 0.9$. This poor rise time is the direct result of the increase in the effective value of the load resistance with the decrease in charging current.

For absolute values of the rise time, we must evaluate τ_c. Assuming a load capacity of 3 pF, a logical-1 level of 15 V, and a W_L/L_L ratio of 2, and from K given by Eq. (5-17) $(6.4 \times 10^{-6}W/L)$, we find $\tau_c = 47$ ns. Thus the total rise time is $8.9 \times 47 = 418$ ns.

Equation (5-50) can be solved for the normalized time response:

$$x(t) = \frac{t/\tau_c}{1 + t/\tau_c} \tag{5-51}$$

Figure 5-22 shows the normalized response of Eq. (5-51) together with the solutions of the other types of capacitor charging discussed in this section. By comparing the results, we can easily determine which response is preferable. The plot of Fig. 5-23 is the actual solution for the capacitive charge and discharge. Since the relative time constants vary from region to region, this plot is needed to place the results in proper perspective.

The load MOSFET could be considered to be a source follower driven by the logical-1 input of V_{DD}. Thus we can expect any source follower to be a poor current source for charging a capacitor. The transient response at the source will be given by Eqs. (5-50) and (5-51). This is in contrast to the fast charging obtained from the bipolar emitter follower.

Capacitor discharge

When the amplifier MOS is switched on hard by a logical-1 signal from a similar inverter, we can, as a reasonable approximation, neglect the charging current supplied by the high-impedance MOS load. The charged output-node capacitor will supply all the current drained by the amplifier device. Figure 5-21*d* shows the model defining the full discharge interval.

The path of operation starts with a current jump from point B to C on the amplifier characteristic of Fig. 5-21*b*. This is followed by the constant-current discharge to the boundary of the ohmic region (point D) and, finally, the discharge through the ohmic region to the logical-0 output back at point A. For the first portion of the decay (C to D), the amplifier MOSFET is in the saturation region. I_c is a constant given by

$$I_c = I_{DA} = \frac{K_A}{2}(V_{GS} - V_T)^2 = \frac{K_A}{2} V_1^2(1 - x_k)^2$$

where $V_{GS} = V_1$ and $x_k = V_T/V_1$.

From Eq. (5-45), setting the lower limit of the integral at $x = 1$ (because we are discharging a charged capacitor C), we obtain

$$t = \tau_{D1}\frac{1 - x}{(1 - x_k)^2} \qquad \tau_{D1} = \frac{2C}{K_A V_1} \tag{5-52}$$

Equation (5-52) shows the linear relationship between t and x expected of the linear rundown constant-current discharge. $x(t)$ either can be written directly or can be found from Eq. (5-52):

$$x(t) = 1 - (1 - x_k)^2 \frac{t}{\tau_{D1}}$$

The upper bound of the ohmic region occurs when $v_{q\Omega} = V_{DS} = V_{GS} - V_T$. But with $V_{GS} = V_1$ at the edge of the ohmic region, we have

$$v_{q\Omega} = V_1\left(1 - \frac{V_T}{V_1}\right) \qquad \text{or} \qquad x_\Omega = (1 - x_k)$$

When x_Ω is substituted into Eq. (5-52), we obtain the time interval of the linear portion of the discharge:

$$t_{D1} = \tau_{D1} \frac{x_K}{(1 - x_K)^2}$$

x_K is usually on the order of 0.15 to 0.3. Even taking the $x_K = 0.33$ in the circuit used for illustration, $t_{D1} = 0.75\tau_{D1}$, which is still less than one time constant. Since we knew that a constant-current discharge is among the fastest ways of discharging a capacitor, we had expected this interval to be very short.

If we now compare τ_{D1} with τ_c of Eq. (5-49), we find that they are proportional to the ratio of the device constants, K_A/K_L. In Eq. (5-35) we saw that the ratio of constants determines the inverter's gain:

$$\tau_c = \frac{K_A}{K_L} \tau_{D1} = A^2 \tau_{D1}$$

Using the previously found τ_c and with $A = 5$, $\tau_{D1} = \tau_c/25 = 1.9$ ns. Thus

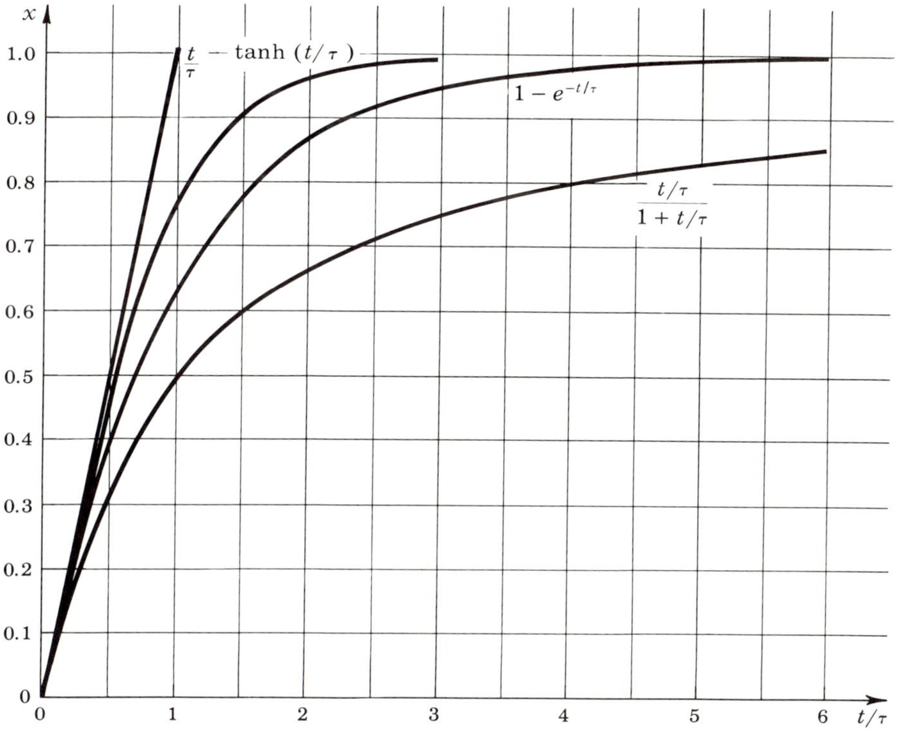

Fig. 5-22 Normalized time responses possible in a FET circuit.

the linear portion of the rundown from 15 to 10 V will last for only $t_{D1} = 1.43$ ns.

The longer portion of the decay will be governed by the ohmic-region behavior of the MOSFET. From Eq. (5-15) we now have for the path DA of Fig. 5-21b

$$I_{DA} = K_A \left[(V_{GS} - V_T)V_{DS} - \frac{V_{DS}^2}{2} \right]$$

Substituting $V_{GS} = V_1$, $V_{DS} = v_q$ and normalizing anew with

$$x = \frac{v_q}{V_1 - V_T}$$

I_{DA} is given by

$$I_{DA} = \frac{K_A}{2} (V_1 - V_T)^2(2x - x^2)$$

As a result of treating this portion of the response as an independent problem, the limits on the discharge will be $x = 1$ to $x = 0$. The integral defining the time interval becomes

$$t = \tau_{D2} \int_1^x \frac{dx}{2x - x^2} \qquad \tau_{D2} = \frac{2C}{K_A(V_1 - V_T)} \tag{5-53}$$

To evaluate the integral of Eq. (5-53) in the most useful form, we first made the linear transformation $x = 1 - y$, resulting in

$$t = \tau_{D2} \int_0^{1-x} \frac{-dy}{1 - y^2} = \tau_{D2} \tanh^{-1} (1 - x) \tag{5-54}$$

The time response for the decay interval defined by Eq. (5-54) is the inverse function

$$x(t) = 1 - \tanh \frac{t}{\tau_{D2}} \tag{5-55}$$

The time required to sweep through 90 percent of the ohmic region, i.e., to fall to within 10 percent of the final value (approximated at 0 V), is found to be

$$t_{D2} = \tau_{D2} \tanh^{-1} (1 - 0.1) = 1.47\tau_{D2} \tag{5-56}$$

τ_{D2} will be on the same order of magnitude as τ_{D1}. Comparing Eq. (5-53) with Eq. (5-52), we find that

$$\tau_{D2} = \frac{V_1}{V_1 - V_T} \tau_{D1}$$

or $\tau_{D2} = 2.85$ ns. From Eq. (5-56) this yields the final decay interval of $t_{D2} = 4.2$ ns.

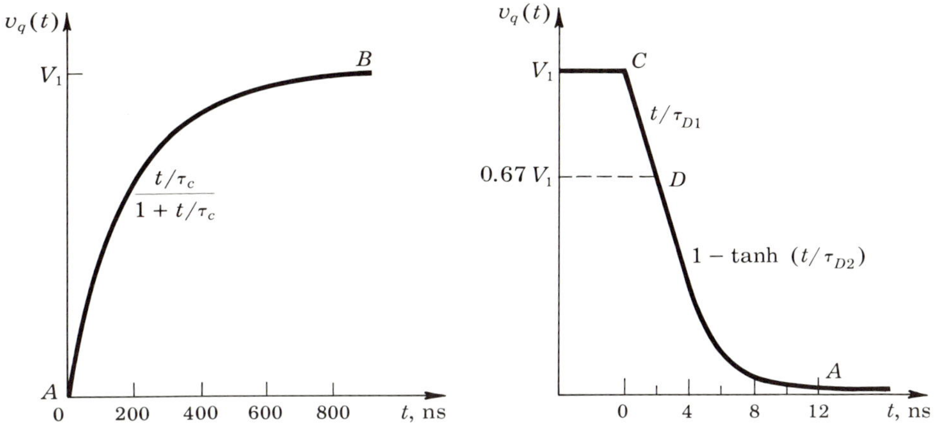

Fig. 5-23 Charge and discharge waveshapes for the circuit of Fig. 5-21*a*. Note that the time scales are different for the two portions of the response.

If, instead of discharging C, we charged C through the ohmic region of a MOSFET, the form of the time solution would become

$$t = \tau_{D2} \tanh^{-1} x \tag{5-57}$$

where x now increases from 0 to 1. Equations (5-57) and (5-54) are just two forms of the same equation. We can go from one to the other by a simple linear transformation.

In a junction FET turned on with the logical-1 level of 0 V (Sec. 5-6), the amplifier-device constant turns out to be

$$K_A = \frac{2I_{DSS}}{V_p^{\,2}}$$

Consequently, the significant proportionality factor becomes

$$\tau_{D2} = \frac{CV_p}{I_{DSS}}$$

But $g_{mss} = 2I_{DSS}/V_p$, and the bottoming resistor $R_{ON} = 1/g_{mss}$. Our proportionality constant may be rewritten as

$$\tau_{D2} = 2R_{ON}C \tag{5-58}$$

A similar interpretion may be applied to τ_{D2} of the MOS amplifier device. In that case, R_{ON} would be the reciprocal of the g_m determined by the gate excitation V_1. [See also Eq. (5-49*b*).]

The above analysis leads us to conclude that a MOSFET is a very good choice as a current sink for discharging a capacitor. But when connected as a source follower, it is a poor charging path: the high impedance results in excessively long rise times. The poor ratio of charge to discharge times is inherent in the design of the stage. To achieve a reasonable amplification, we require a large ratio of K_A/K_L. But this ratio also controls the relative time constants. Furthermore, the nature of the charging itself results in a longer relative rise time. Both phenomena work against fast rises at the output.

5-8 The FET as a Logic Element

The FET inverter pair of Figs. 5-15 and 5-21 is, in many respects, the ideal logic element. Logical-0 and logical-1 levels are separated and hence easily distinguishable. The built-in threshold results in inherently good noise immunity. In the logical-1 output state, the device is off and no power is consumed. Even in the logical-0 bottomed state, only a small amount of power will be dissipated in the high-impedance load. MOSFETs are extremely small and are ideally suited for large-scale integration. Circuits containing over 4,000 devices have been fabricated on a single 90- by 100-mil^2 chip.

The capacitive gate input of the MOSFET (leakage resistance $> 10^{12}\ \Omega$) may also be used for short-time information storage (see Sec. 14-2). Current flow in the induced drain-to-source channel does not disturb the stored gate charge.

Fan-out capabilities must be traded off against switching speed. The essentially capacitive gate input does not impose restrictions on the dc fan-out, but excessive capacitance at an output node will result in very poor rise and fall times. Expected gate propagation delays are about 100 ns. We conclude that MOSFETs are most suitable for cheap, large-scale, relatively slow logic arrays.

Figure 5-24 shows a MOSFET NOR gate. It is exactly the same circuit configuration as that used for the DCTL of Sec. 4-4. So long as one of the input devices is biased on hard with a logical-1 input, it conducts and the output is pulled down to the logical-0 level defined by Eqs. (5-38). When more than one input gate is turned on, the several devices conduct in parallel, effectively increasing the K_A/K_L ratio and resulting in a reduced V_0.

Only when all inputs are simultaneously back-biased by logical-0 signals will the output node rise to the logical-1 level of $V_{DD} - V_T$. Thus the logic statement is $Y = \overline{A + B + C}$.

A cascode arrangement of MOSFET devices (Fig. 5-25) yields a NAND gate. So long as at least one gate element is in or beyond the saturation region, only a small leakage current flows and, as in the NOR gate, the

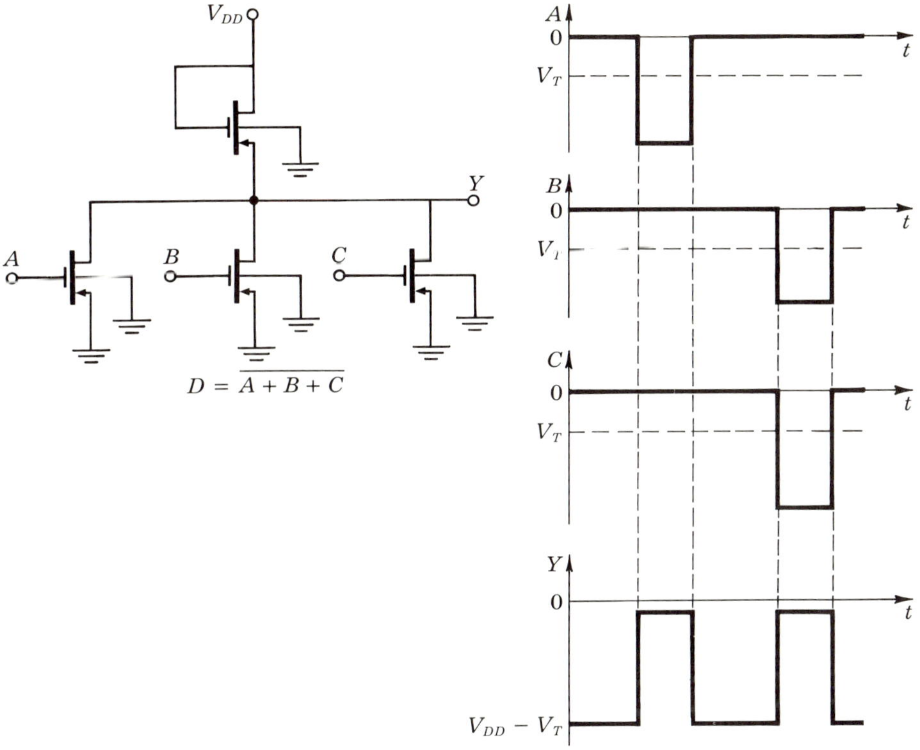

Fig. 5-24 MOSFET three-input NOR gate.

output will be at the logical-1 level. Only when all inputs are at the logical-1 level will the three devices be in the ohmic region and the output bottom at the logical-0 level. Because the individual gates are in series, the logical-0 output will be due to three ohmic drops. To minimize this voltage, the NAND-gate channels are made wider, making R_{ON} proportionally smaller. By adjusting the circuit geometry, the outputs of all types of gates can be kept approximately equal.

NAND and NOR functions can be combined in a single gate through the wired AND output connection at the common drain point. (See Sec. 4-9.) Figure 5-26 illustrates one possible configuration. When either branch is driven into the ohmic region, the common-mode voltage becomes approximately zero. One branch is a NOR, and the other is a NAND combination. To ensure a constant logical-0 output level, the drop across each of the series gates B and C must be one-half the drop of the A input. The geometry of the devices can be modified to satisfy this condition.

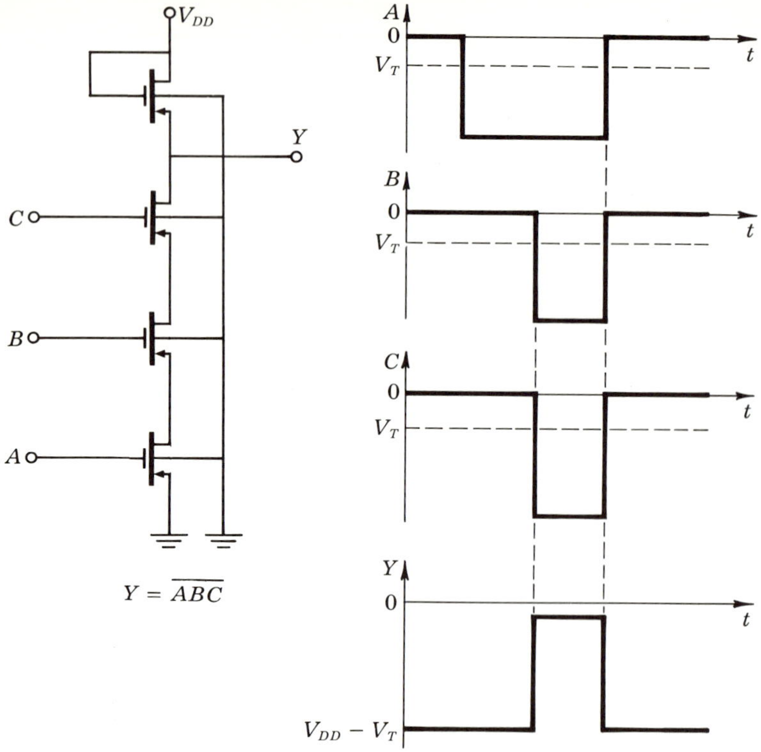

Fig. 5-25 A three-input MOSFET NAND gate.

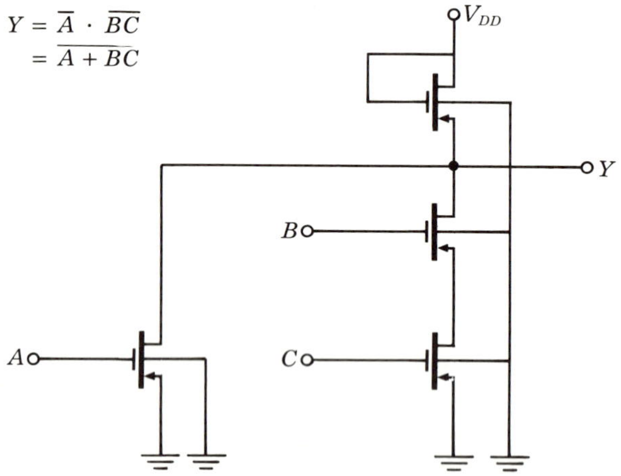

Fig. 5-26 Combination MOSFET gate.

Noise immunity

For a logical-0 input, the voltage at which the gate begins to turn on is V_T. The noise immunity becomes

$$NI_0 = \overline{V}_0 - \underline{V_T} \tag{5-59}$$

where V_0 is given by Eq. (5-39). For $V_0 = -2.75$ V and $V_T = -5$ V, $NI_0 = 2.25$ V. In NAND gates, the threshold of the upper devices must be modified to account for substrate effects [Eq. (5-39)]. For the logical-1 condition,

$$NI_1 = \underline{V_1} - \overline{V}_{1\Omega} \tag{5-60}$$

where $V_{1\Omega}$ is the edge of the ohmic region. If we assume that the input is the worst-case logical 0, that is, V_T, then from the definition of the boundary between the ohmic and saturation regions ($V_{DS} = V_{GS} - V_T$), we obtain the highest $V_{1\Omega}$:

$$V_{1\Omega} = 2V_T$$

Substituting back into Eq. (5-60), the logical-1 noise immunity is given by

$$NI_1 = V_1 - 2V_T$$

Using the numbers of the circuit of Fig. 5-15, where $V_1 = -15$ V, $NI_1 = 5$ V.

Clocked inverter logics

Most logic systems operate with clock-timed pulse intervals. Only during the pulse periods will logic inputs be present, and only during this time will the logic gates have to function. The intervening dead intervals are usually of the same duration as the pulse periods. Use can be made of the dead time either to improve switching speed or to reduce overall power consumption.

Section 5-7 clearly showed that the slow charging of the output-node capacity through the MOSFET load is the limiting factor in logic switching speed. The logics of Figs. 5-24 to 5-26 all suffer from this limitation. In the circuit of Fig. 5-27*a*, the rise-time delay is avoided by charging C only during the dead time. When the extra MOSFET, Q_2, is clocked on during every single clock pulse, it connects the output of the logic (indicated by the single FET, Q_1) to the output node.

During the dead space between successive clock pulses, Q_2 is turned off and C is *primed;* that is, C is charged to the logical-1 level through the load Q_L. This slow charging process is completed during the dead time. It sets the minimum interval between successive clock pulses. Once the clock pulse turns on Q_2, it completes the logic path from the drain of Q_1

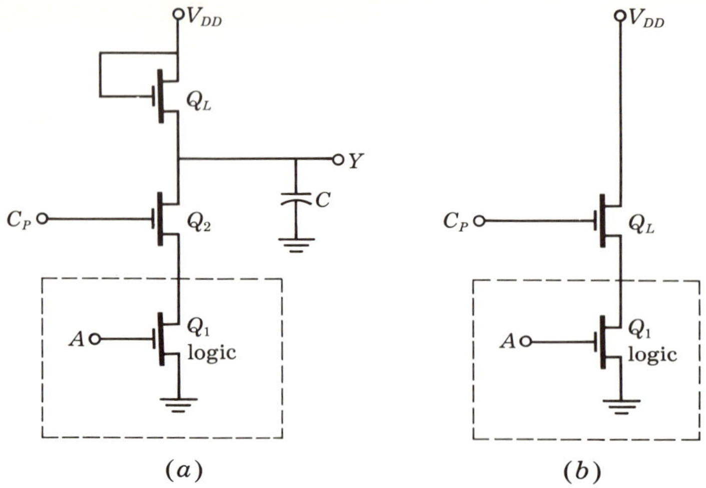

Fig. 5-27 Clocked MOSFET logic circuitry. (*a*) Speed improvement through control over charging interval; (*b*) dissipation reduction through clocking load MOSFET on and off.

to the output node. If the *sensed* required output is a logical 1, then the capacitor remains in its primed state. If, however, the sensed logic output is at the logical-0 state, C must discharge from the primed state through the Q_2-Q_1 NAND gate. The discharge is of the form tanh (t/τ_D), and as we saw in Sec. 5-6, it is very fast. Thus slow charging paths are reserved for the intervals when the gates are off. Only the fast discharge must be accounted for within the ON interval.

Figure 5-27*b* illustrates a technique used to reduce the power consumption of a logic system. By simultaneously triggering all gates of the MOSFET loads between 0 and V_{DD} (or above), we can activate the complete logic only during the timed intervals when logic inputs are actually present. During the dead time, the 0 V on the gate of the load device keeps it back-biased and no current flows. If the inverters are kept off for half the time, we cut the power input in half. When hundreds of inverters are fabricated on a single chip, the reduction in power will be significant.

Complementary-symmetry logic

The complementary inverter of Fig. 5-28*a* combines the fast switching speed of an ON MOSFET with an extremely low power consumption. Let us start our analysis of this circuit by examining the two limiting cases. A logical-1 signal of V_{DD} applied to the gates of both the n- and p-channel MOSFETs turns on the n channel and drives it into the ohmic region.

The p-channel device has its gate and source at the same potential. Since $|V_{GS}| < |V_T|$, no channel is induced. Only the small reverse current of the back-biased n-substrate p-drain junction will flow. This constant-current source acts as the load for the n-channel device (Fig. 5-28b).

The same steady-state current must always flow through both devices. Therefore the logical-0 output corresponds to the intersection of the $V_{GSn} = V_{DD}$ characteristic and the I_{RSp} constant-current line. This point is labeled 0 in Fig. 5-28b. The intersection is extremely close to 0 V.

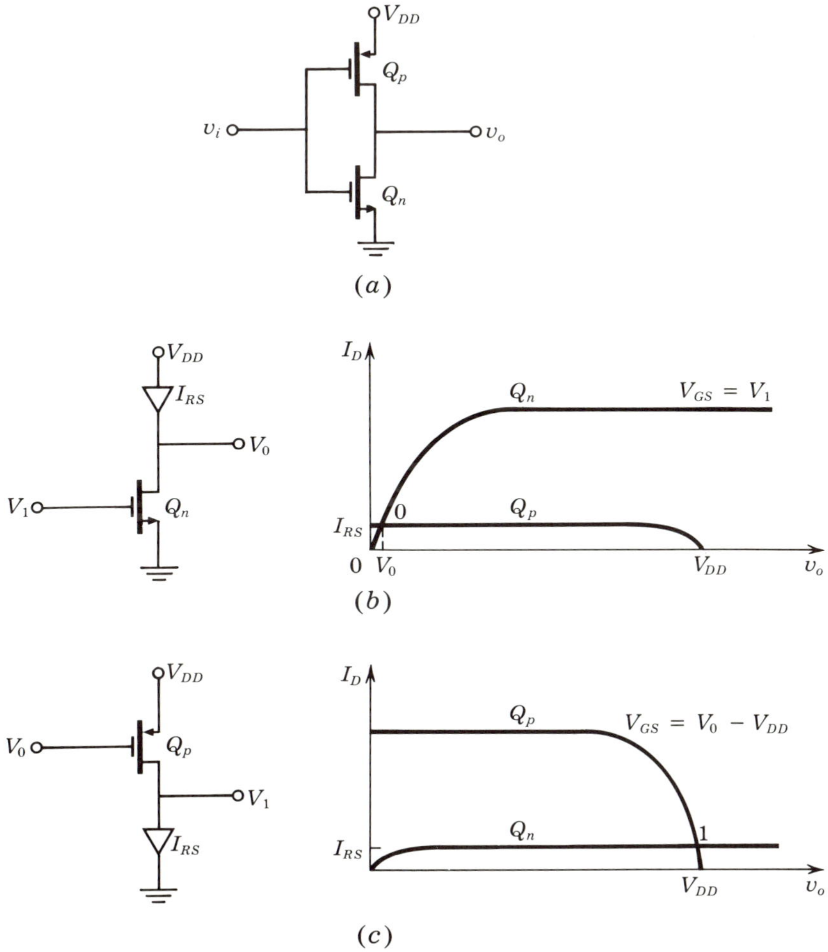

Fig. 5-28 Complementary-symmetry logic. (a) Basic complementary-pair inverter; (b) reduced model under logical-1 excitation—stable logical-0 operating point shown on the volt-ampere plane; (c) reduced model under logical-0 excitation—stable logical-1 operating point shown on the volt-ampere plane.

Logical-0 inputs of 0 V applied to both gates back bias the n-channel MOSFET. Now Q_n may be replaced by the small reverse current, as shown in Fig. 5-28c. For the p-channel device, we have $V_{GSp} = -V_{DD}$, and hence the device is biased on in the ohmic region. Because of the particular circuit configuration of Fig. 5-28a, the p-device volt-ampere characteristics may be plotted on the common $V_D I_D$ plane by inverting them and shifting them to the right by V_{DD}. Figure 5-28c shows the single characteristic for $V_{GS} = -V_{DD}$. Its intersection with I_{RSn} is the logical-1 output (point 1), which we see lies close to V_{DD}.

Suppose that the output node is capacitively loaded. Under a logical-1 input, C sees the ON MOSFET Q_n. A charged capacitor will discharge through Q_n, tracing out the path $BCDA$ described in Fig. 5-21b. A short interval of a constant-current discharge will be followed by the fast tanh (t/τ_{D2}) response.

For logical-0 inputs, the capacitor sees the same type of charging path through the ON p-channel MOSFET. C will now charge to V_{DD}, following the identical sequence: first the linear ramp through the saturation region, followed by the tanh (t/τ_{D2}) charging. Per-stage rise and fall times of a few nanoseconds become possible.

The complete transfer characteristics for the complementary pair can be plotted from the two families of volt-ampere curves. Since the same current must flow through both devices, we look for the intersections of corresponding characteristics; for the n device, $V_{GSn} = v_i$, and for the p device, $V_{GSp} = v_i - V_{DD}$. Figure 5-29 shows the resultant transfer function.

Note that at the midpoint of the active region, where both devices are

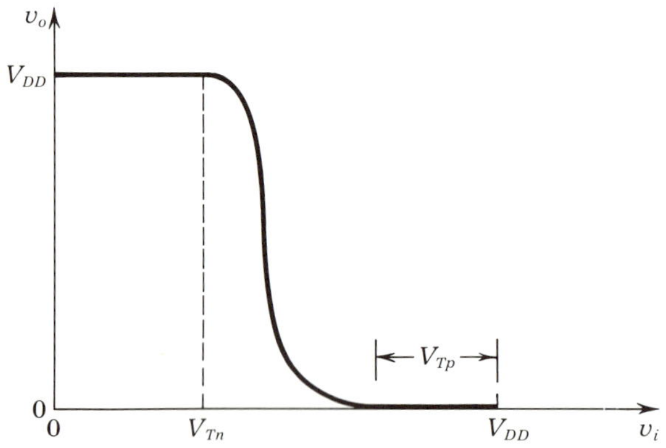

Fig. 5-29 Transfer characteristic of the complementary MOSFET inverter. Ideal devices exhibit infinite gain in the active region.

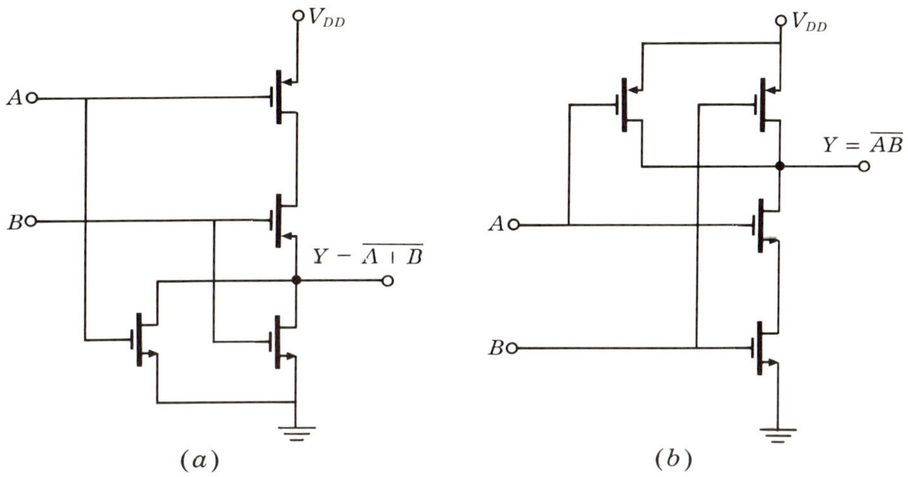

Fig. 5-30 Two complementary MOSFET logics. (*a*) NOR gate; (*b*) NAND gate.

in the saturation region, the gain approaches infinity. It is limited only by the presence of the channel-spreading resistance r_o. Since the high gain is independent of channel geometry, very small-area, almost equal-area, MOSFETs may be employed for both the *p*- and *n*-channel devices. This was not true in the MOS inverter of Fig. 5-15, where the gain was directly dependent on device geometry.

Figure 5-30*a* and *b* show typical complementary NOR and NAND gates. Each *n*-channel input requires its own separate *p*-channel load. Combined logic functions can also be synthesized, with appreciable savings in the numbers of active elements.

The major disadvantage of the complementary-pair logic is the increased complexity of processing both *n*- and *p*-channel MOSFETs within a single *n* substrate. Separate *p*-region islands must be diffused to serve as the *n*-channel substrate. Since the *n* and *p* devices require different diffusions, additional masks are needed. Clearly, the overall process is not that much different from the fabrication of transistors and is certainly well within the state of the art.

5-9 Analog Switching Applications—Transmission Gates

The FET's gate, isolated from the channel, turns on or off the conduction path between the drain and source. Because the resistance changes from a very low value in the ON condition (20 to 500 Ω) to a very high value in the OFF condition (10^9 to 10^{14} Ω), the channel serves as a controlled

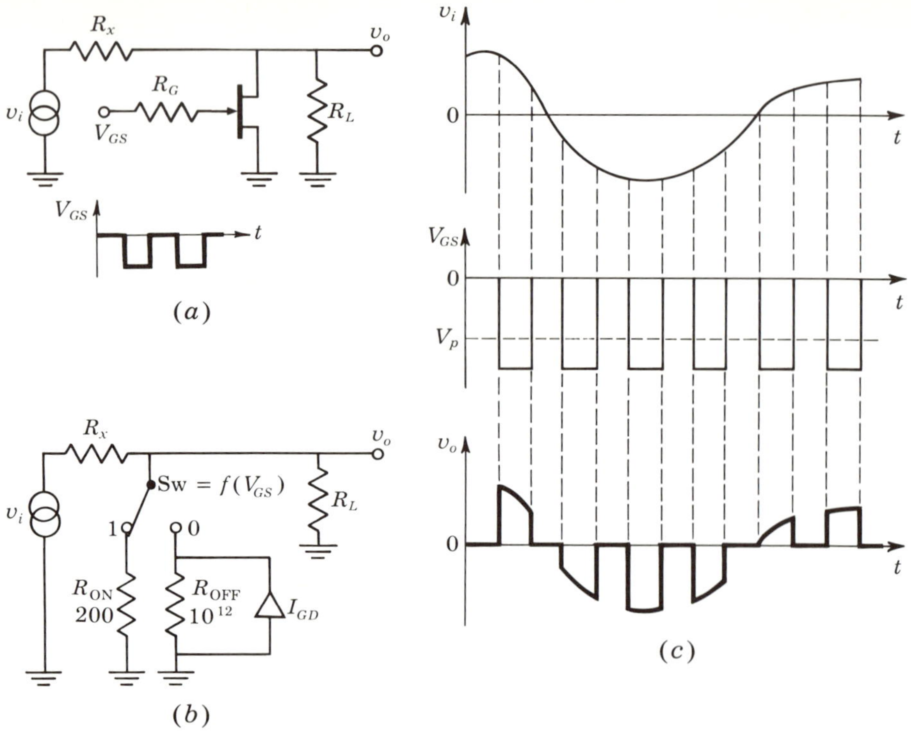

Fig. 5-31 (*a*) Shunt chopper; (*b*) model; (*c*) input, gating, and chopped-output waveshapes.

switch establishing or interrupting a transmission path, chopping the analog signal at the control rate.

Figure 5-31*a* shows a shunt chopper. When on, the channel shunts the load with its very low resistance, thus reducing the output to near zero. When off, the channel, effectively an open circuit, does not affect the signal appearing across the load. For optimum response, $R_X \gg R_L$ and $R_{ON} \ll R_L$.

To establish the ON state of a junction FET, $V_{GS} \approx 0$. From the model of Fig. 5-31*b*, with the controlled switch in position 1, we have

$$v_{o_1} \approx \frac{R_{ON}}{R_{ON} + R_X} v_{in} \tag{5-61}$$

When the gate voltage changes to where $|V_{GS}| < |V_p|$, the FET channel is pinched off. The drain-to-source leakage resistance is extremely high and may be neglected. Because the drain-to-gate junction is back-biased, we also observe the small reverse saturation current of the junction, I_{GD}, at the drain. We obtain the output from the model of Fig. 5-31*b* (the switch is in position 0):

$$v_{o_0} \approx \frac{R_L}{R_X + R_L} v_{\text{in}} + I_{GD} R_L \| R_X \tag{5-62}$$

I_{GD} is very small. For $R_L = 100$ K, the dc offset voltage, $I_{GD}(R_L \| R_X)$, is normally less than a few microvolts.

An ideal shunt chopper should give zero output in the ON state. In the OFF state the slope of the transfer characteristic should be 1. To simultaneously minimize Eq. (5-61) and maximize Eq. (5-62) we must satisfy the contradictory conditions

$$R_X \gg R_{\text{ON}} \qquad \text{and} \qquad R_X \ll R_L$$

These conditions are identical to those derived in Chaps. 2 and 4 for the diode clippers and for the diode bridge. Thus we choose

$$R_X = \sqrt{R_{\text{ON}} R_L}$$

as a compromise solution.

The smaller the ON resistance, the more nearly the chopper approximates the response of the ideal switch. Since the electron mobility is about twice the hole mobility in the FET, an n-channel FET which will have one-half the resistance of a p-channel device is the better choice for the analog chopper.

A controlled switch, when placed in series with the input, will transmit information in the ON condition and will interrupt the transmission path when turned off. Figure 5-32 shows the series chopper together with the

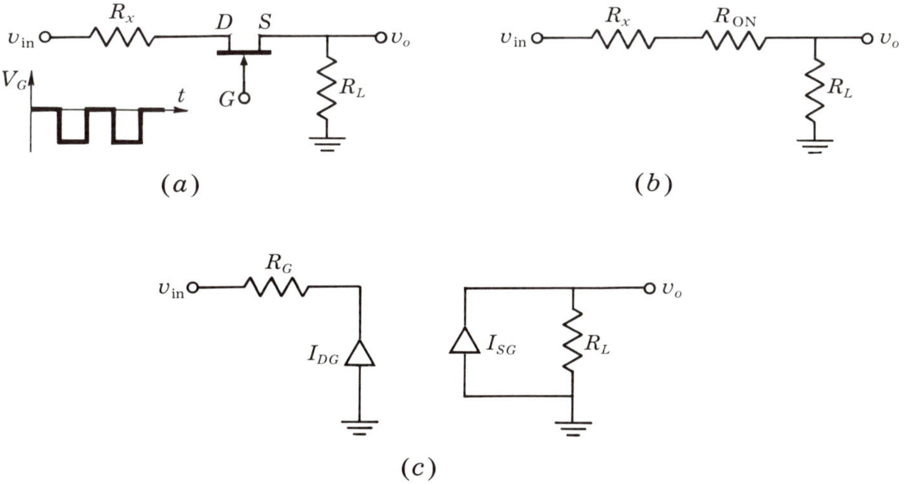

Fig. 5-32 Series chopper and models. (*a*) Series chopper; (*b*) ON model; (*c*) OFF model.

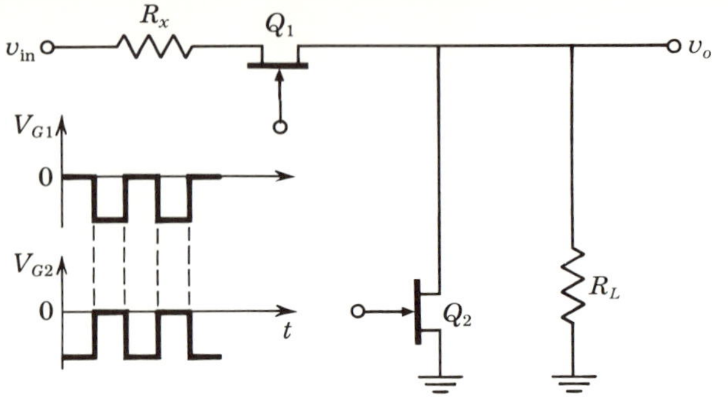

Fig. 5-33 Series-shunt chopper.

ON and OFF models. In the ON state,

$$v_{o_1} = \frac{R_L}{R_L + R_{ON} + R_X} v_{in}$$

In the OFF state,

$$v_{o_0} \approx I_{SG}R_L \tag{5-63}$$

where, in the model, I_{DG} and I_{SG} are the back-biased drain-to-gate and source-to-gate junction currents, respectively. Since these currents are very small, in the OFF state v_{o_0} will be close to zero.

A series chopper very nearly approximates the ideal. We only have to maximize the ON output. This requires as large an R_L as is consistent with the permissible dc offset in the OFF state.

The dc offset of Eq. (5-63) may be reduced through shunting R_L with the ON resistance of a FET. Figure 5-33 shows the resulting series-shunt chopper. Q_1 is the series switch, which, when on, closes the transmission path to the output. So that it will not shunt R_L by its low resistance, Q_2 must be off when Q_1 is on. The two gates must be driven out of phase.

When Q_1 is off, the transmission path is open. The output will be the small offset voltage due to the reverse junction current. But with Q_2 now turned on, R_1 is shunted by the low ON resistance of Q_2. The OFF output becomes

$$v_{o_0} = I_{DG}R_{ON} \tag{5-64}$$

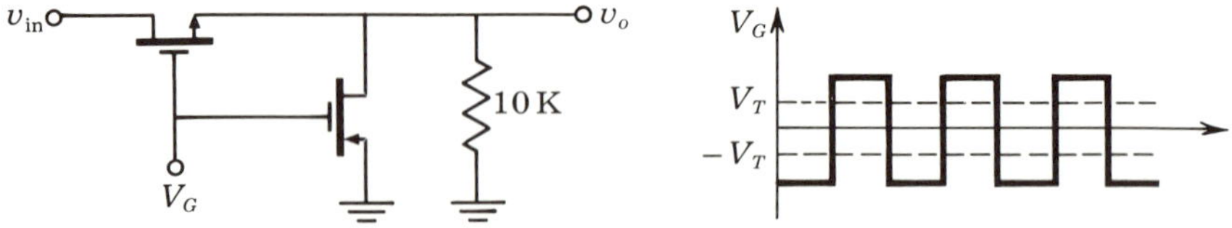

Fig. 5-34 MOSFET complementary series-shunt chopper.

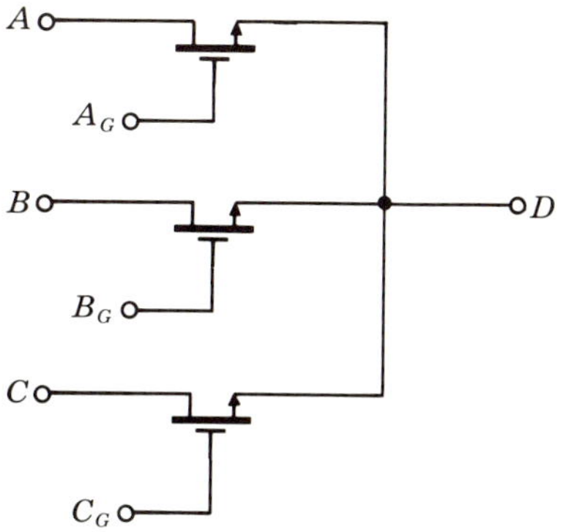

Fig. 5-35 MOSFET commutation.

which is several orders of magnitude smaller than the OFF output of the series chopper [Eq. (5-63)]. The lower output impedance helps to discharge the output capacitance. It also speeds up the chopper switching time.

FET characteristics are symmetrical about the origin (see Fig. 5-38), and consequently the choppers will switch positive or negative signals with equal linearity. Unlike bipolar transistors, there is no junction offset voltage to overcome and thus there is no zero-region dead zone. Very small signals can be chopped with acceptable linearity.

Figure 5-34 shows a MOSFET series-shunt chopper which uses complementary devices for the series and shunt elements. Since the induced *p*-channel MOSFET is turned on by a negative gate voltage and off by a positive gate voltage and the induced *n* channel is turned on by a positive gate signal and off by a negative signal, the two gates may be simultaneously driven by a single square wave. This avoids the difficulty of generating two perfectly complementary high-frequency square waves.

The chopper circuitry is also useful for analog commutation. As we apply a logical-1 signal in the circuit of Fig. 5-35, the appropriate series FET is driven on, connecting point A (or B or C) to point Y. Transmission paths from A or B or C can be established in sequence. The sequence will be determined by the outputs of NAND or NOR logic arrays feeding the appropriate series gate. Combinations of series MOSFET gates can be used to interconnect large numbers of input and output paths.

5-10 Composite Bipolar-Unipolar and Other Specialized FET Devices

Field-effect and bipolar transistors may be interconnected, as shown in Fig. 5-36, to form a composite device which affords the advantages of the high input impedance of the FET and the high g_m associated with the

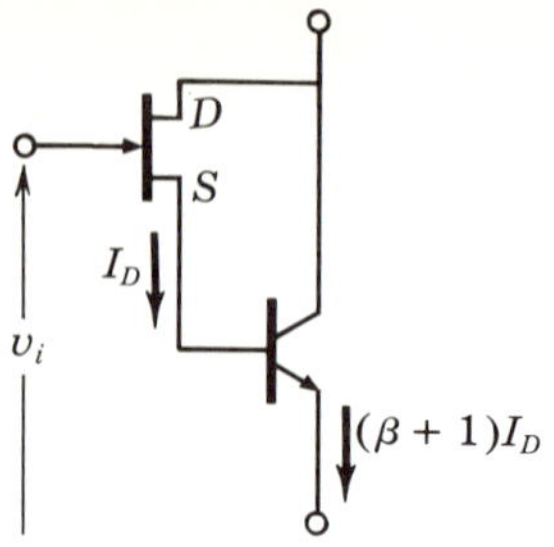

Fig. 5-36 Composite field-effect–bipolar transistor.

bipolar transistor. The drain current of the FET is the base input current of the bipolar. At the collector, the amplified current βI_D is combined with the drain current of the FET, yielding the composite current

$$I_X = I_D(\beta + 1)$$

For small signals, the incremental $i_D = g_m v_1$, and therefore

$$i_x = g_m(\beta + 1)v_1 \tag{5-65}$$

We find the same current and the same effective g_m in the emitter lead. A FET operating at a low drain current, where $g_m = 250$ μmhos, combined with a bipolar transistor having $\beta = 49$ results in a composite $g_{m_c} = 12{,}500$ μmhos. But since the r_o of the FET serves as a feedback resistor from the collector to the base of the bipolar, the negative feedback reduces the overall output impedance.

The composite device has a g_m even higher than might be expected from a bipolar transistor, as well as the high input impedance associated with the FET. It is particularly suitable for use as the input stage of any amplifier requiring very high input impedances.

A current-limiting diode

A junction FET with the gate connected directly to the source becomes a two-terminal device having a path of operation always along the single characteristic curve $V_{GS} = 0$ (Fig. 5-37). So long as $V_{DS} > -V_p$, the device is biased on in the saturation region at the fixed current I_{DSS}. Over the V_{DS} voltage range from V_p to reverse breakdown, the device may be modeled as a constant-current source shunted by the channel-length modulation resistance r_o. The terminal current of a 6-mA device will change by about 30 μA over the voltage range of 10 to 50 V. By adjusting the cross-sectional area of the channel, any reasonable I_{DSS} can be obtained.

From the terminal characteristics of Fig. 5-37, we can conceive of the current-limiter diode as the dual of the bipolar junction zener diode; and

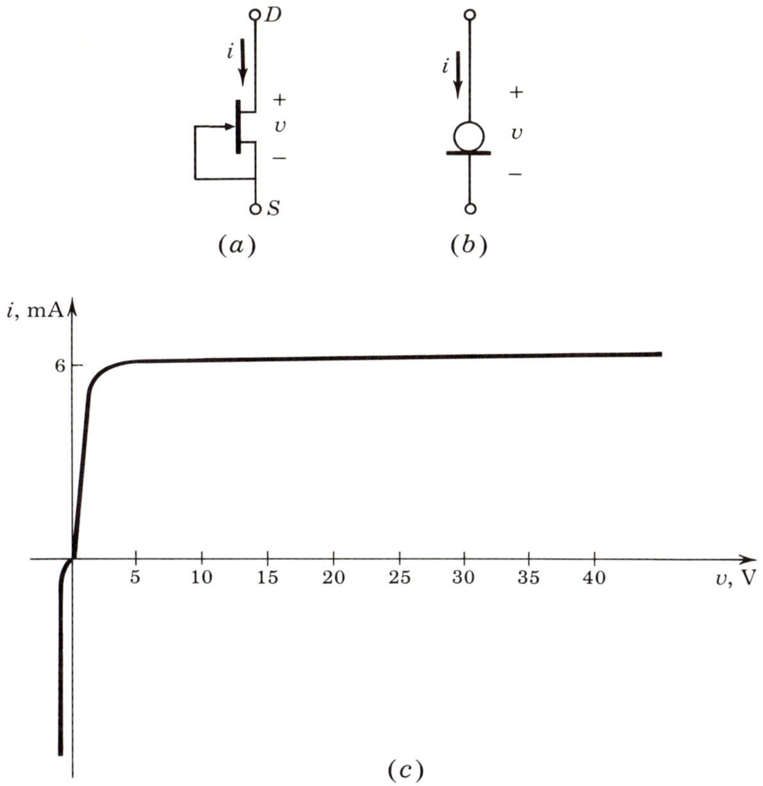

Fig. 5-37 Current limiter. (*a*) Device; (*b*) symbol; (*c*) volt-ampere characteristic (junction breakdown at ~100 V).

to shift the voltage region down to or below zero, a zener diode may be inserted in series with the current limiter.

A voltage-controlled resistor

If we carefully examine the symmetrical volt-ampere characteristics of a junction FET in the vicinity of $V_{DS} = 0$ (Fig. 5-38), we see that they are a family of straight lines passing directly through the origin. The slope of these lines is an inverse function of V_{GS}. Thus we conclude that the drain-to-source characteristics are exactly those of a voltage-controlled resistor (VCR). The characteristics of Fig. 5-38 were drawn for an *n*-channel junction FET. With due allowance for the conductivity enhancement of MOS, in place of the conductivity depletion of the junction device, a MOSFET will also look like a VCR.

In the vicinity of the origin, we have, from Eq. (5-19),

$$R_{\mathrm{ON}} = \frac{1}{K(V_{GS} - V_p)}$$

where K is a function of the geometry of the channel.

The controlling signal V_{GS} is isolated from the controlled element by either the back-biased junction of the junction FET or the SiO_2 layer of the MOSFET. V_{GS} will not therefore be dc-coupled into any networks containing our VCR. The capacitive coupling still exists, and it will pose problems at high frequencies.

One application of a VCR is in a voltage-controlled attenuator (Fig. 5-39*a*). Attenuation of the input signal is a direct function of the applied control signal. Several networks can be cascaded to improve the sensitivity.

A second example is in the tunable network of Fig. 5-39*b*, where the pole locations now depend on the control voltage V_{GS}. When combined with the voltage-controlled capacitor of a back-biased junction, we achieve great flexibility in the design of voltage-variable networks. Many complex feedback problems become much easier to implement. For example, for frequency stabilization, a derived dc control voltage can establish pole

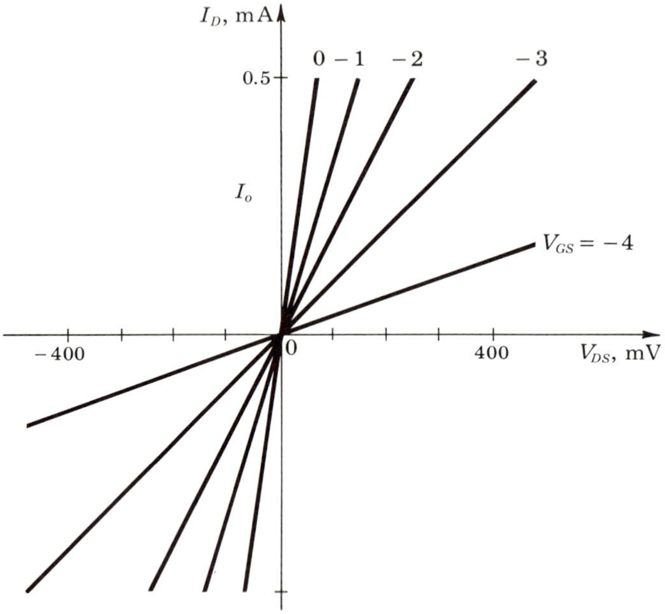

Fig. 5-38 Expanded drain-source volt-ampere characteristics of a junction FET in the vicinity of the origin.

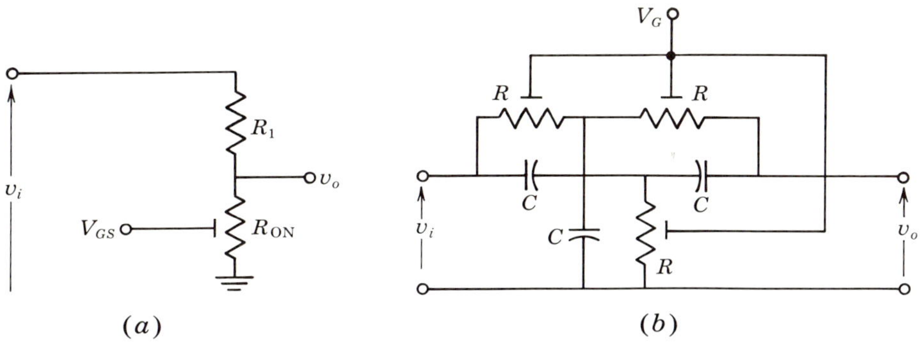

Fig. 5-39 Examples of voltage-controlled networks. (*a*) Variable attenuator (see also Fig. 16-34); (*b*) tuned twin-T network (see also Fig. 16-9).

locations; pole migrations will change the dc level and retune the network. Sweep signals can also be applied to the controlled elements to scan the network's full response.

Of course, in both applications unless the terminal voltage is kept small, the device will be driven out of its linear resistance region.

Problems

5-1 (*a*) Calculate the channel resistance of a junction silicon *p*-channel FET at $V_{DS} \approx 0$. The device has a channel height of 1 μm at $V_{GS} = 0$, a length of 10 μm, and a width of 25 μm ($\mu_p = 300$ cm^2/V-s).

(*b*) Repeat your calculations for an identical *n*-channel device ($\mu_n = \frac{3}{2}\mu_p$).

(*c*) Express the variation of channel height as a function of V_{GS} in the vicinity of $V_{DS} = 0$. The *n* channel pinches off at $V_{GS} = -5$ V.

5-2 An induced-*n*-channel MOSFET has a measured current of 10 mA at $V_{DS} = 20$ V and $V_{GS} = 10$ V. The threshold is at 5 V.

(*a*) Plot the saturation-region g_m as a function of V_{GS}.

(*b*) Plot the g_m in the ohmic region as a function of V_{GS} ($V_{DS} \approx 0$).

(*c*) Plot R_{ON} as a function of V_{GS} in the vicinity of $V_{DS} \approx 0$. Over what range of V_{DS} will R_{ON} be within 5 percent of the $V_{DS} = 0$ value?

5-3 Two induced-*p*-channel MOSFETs are fabricated on the same substrate. They are identical except for geometry. Q_1 has a channel width of 50 μm and a length of 10 μm. Q_2 has a channel width of 10 μm and a length of 20 μm. Both devices have the same threshold of -3 V.

(*a*) The two MOSFETS are connected in parallel, drain to drain and source to source. What is the effective g_m in the saturation region of the combination? Express your answer in terms of the g_{m1} and g_{m2} of the individual devices.

(*b*) Repeat part *a* when the two devices are connected in series. The gates remain connected together.

5-4 Show that the composite saturation-region behavior of the two devices of Prob. 5-3 can be predicted from the parallel and series combination of their ohmic-region resistances ($V_{DS} \approx 0$).

5-5 As a simple model, let us assume that the pinchoff region extends along the channel in the same manner as the depletion region of a *pn* junction broadens when operating under back-bias conditions. The pinchoff-region length is $L_p = K_D/\sqrt{V_D}$, where V_D is the drop across the pinchoff region. The drop across the remaining ohmic portion of the channel remains constant at $V_{GS} - V_T$. This is the drop that determines the current flow.

On the basis of the above reasoning, calculate the variation of I_D with V_{DS} within the saturation region. Show that when $L \gg L_p$, I_D is effectively constant.

5-6 In designing a single inverter stage, such as that shown in Fig. 5-12 (with the bias resistor bypassed with a large C), we have the choice of three FETs. Each has the same V_p, but each has a different channel width, making $I_{DSS} = 1.4$ mA, 7 mA, and 14 mA, respectively. In each, r_o is inversely proportional to I_{DSS}. (The 1.4-mA device parameters are given in Fig. 5-12.)

(*a*) If you use each device successively in the circuit of Fig. 5-12 (bypassed resistor), calculate the new Q point and the gain.

(*b*) Find the small-signal gain when we readjust the bias to set the Q point of each FET at 0.61 mA. Can you draw any conclusions as to which device should be used in this circuit? Are there any disadvantages? Explain.

5-7 In the circuit of Fig. 5-40, the junction FET has $V_p = -2$ V, $I_{DSS} = 2$ mA,

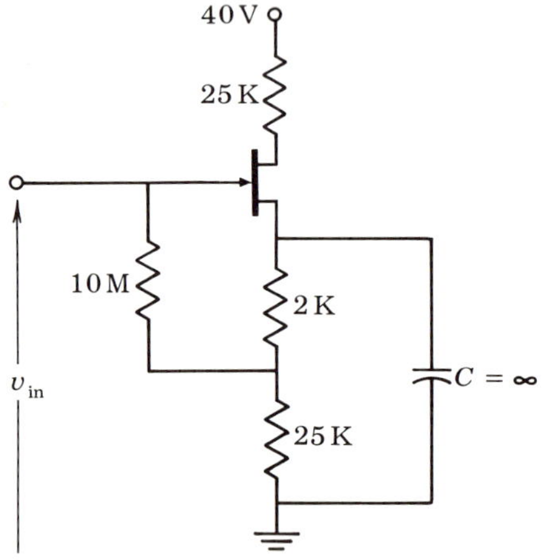

Fig. 5-40

and $r_o = 200$ K at I_{DSS}. Find the Q point. Draw the small-signal model holding at the Q point. Calculate the gain and the midband input and output impedances.

5-8 The bypass capacitor in Fig. 5-40 is removed. Calculate the gain from gate to source, the input impedance, and the output impedance at the source.

5-9 Plot the dc-voltage transfer characteristic of the circuit of Fig. 5-12. The input is applied directly to the gate, and the output appears at the drain.

5-10 (*a*) Plot the voltage transfer function of the circuit shown in Fig. 5-41. The MOSFET has $V_T = -3$ V, and $I_D = 5$ mA at $V_{GS} = -4$ V (saturation region).

(*b*) Draw the small-signal model holding at $I_D = 2$ mA. From this model, calculate the gain and the input and output impedances.

(*c*) What amplitude constitutes small signal? (A small signal is defined as an input signal large enough to cause a gain variation of 5 percent over the limits.)

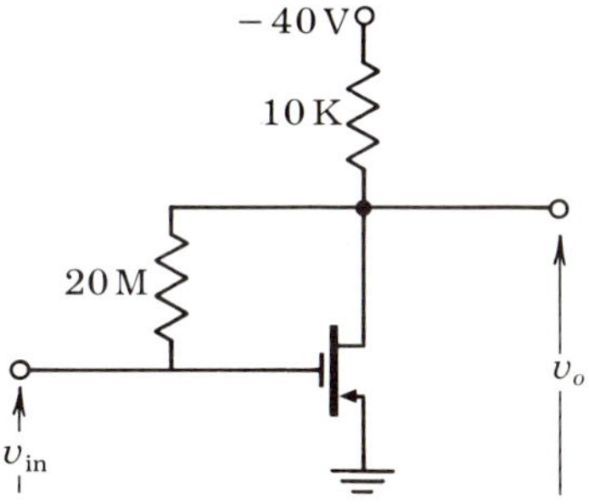

Fig. 5-41

5-11 Repeat Prob. 5-10 when an unbypassed 1-K resistor is inserted in series with the source in the circuit of Fig. 5-41.

5-12 We wish to make an amplifier using a MOSFET having the characteristics of Fig. 5-16*b* as the amplifier device. The load is a MOSFET having the same channel length but only one-tenth the width. Plot load lines for $V_{DD} = -15$ V and $V_{GS_L} = V_{DS_L}$, $V_{GS_L} = V_{DS_L} + V_T$, $V_{GS_L} = V_{DS_L} + 2V_T$ on the amplifier volt-ampere characteristics. From these curves, plot the transfer characteristic. Calculate the gain of the stage.

5-13 The switch in the circuit of Fig. 5-42 is closed at $t = 0$ and opened 1 ms

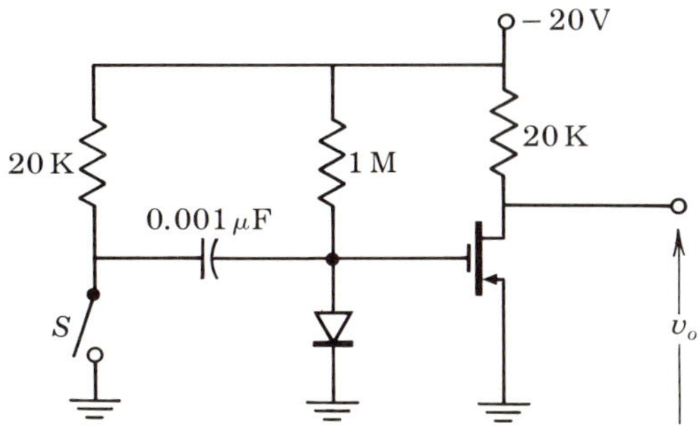

Fig. 5-42

later ($V_T = -3$ V, $I_D = 5$ mA at $V_{GS} = -4$ V). Sketch and label the gate and output waveshapes. Specify all significant break voltages, times, and time constants.

5-14 Repeat Prob. 5-13 when the MOSFET is replaced by a p-channel junction FET. The diode is removed. ($V_p = 3$ V, $I_{DSS} = 5$ mA.)

5-15 In the circuit of Fig. 5-43, the switch is opened at $t = 0$ and closed 25 μs later. The MOSFET has $V_T = -3$ V and $I_D = 10$ mA at $V_{GS} = -4$ V. Plot the output waveshape to scale, labeling it with all significant voltage values, times, and time constants. Make all reasonable approximations.

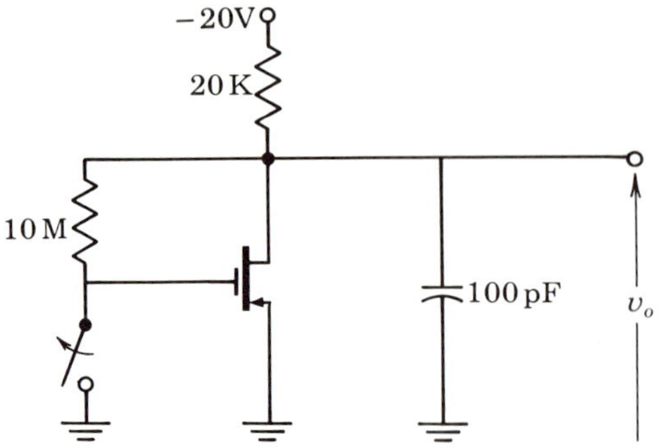

Fig. 5-43

5-16 Repeat Prob. 5-15 when the load resistor and capacitor are transferred into the source lead. This converts the inverter into a source follower driving the load capacitor. Make all reasonable approximations in your calculations.

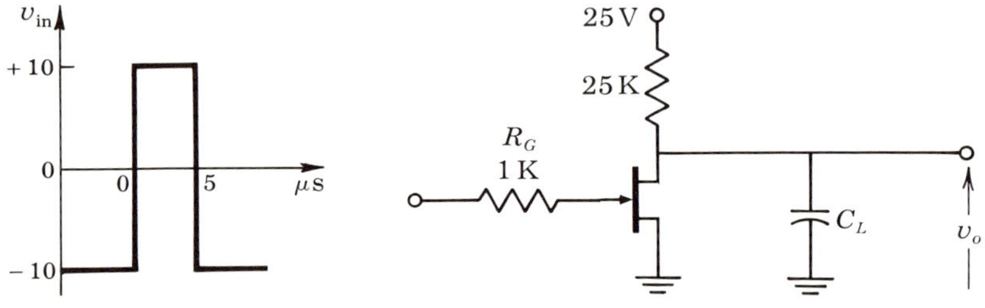

Fig. 5-44

5-17 The inverter of Fig. 5-44 is switched on with a 5-μs pulse. Sketch and label completely a reasonable approximation of the time response at the output.

Specify any delay times and the rise and fall times. Repeat when the gate resistance is increased to 50 K. ($V_p = -4$ V, $I_{DSS} = 25$ mA, $r_o = 200$ K at I_{DSS}, $C_{GD} = 5$ pF, $C_{GS} = 10$ pF, $C_L = 100$ pF.)

5-18 The inverter of Fig. 5-15*a*, having the characteristics of Fig. 5-16 ($V_{GG} = V_{DD} = -20$ V), is pulsed on with a single 20-μs 20-V negative pulse. The output is loaded by N subsequent logic stages, each input of which may be modeled by a 10-pF capacitor shunted by a 25-M leakage resistance.

(*a*) Plot the output response for $N = 1$ and $N = 10$ on the same graph. Make all reasonable approximations.

(*b*) What is the maximum loading N if the worst-case rise (or fall) time is not to exceed 1 μs?

5-19 Assume that each of the transistors of the complementary-symmetry inverter of Fig. 5-28 has the volt-ampere characteristics of Fig. 5-16*b*, with, however, due consideration given to polarity. A 10-μs 15-V pulse is used to turn on the inverter, which is loaded with a 100-pF capacitor. (For each device, $I_{rs} = 10^{-15}$ A, $V_{DD} = 25$ V.)

(*a*) Plot the path of operation for charging and discharging the load capacitor.

(*b*) Sketch and label completely the output-voltage waveshape, giving all voltages and the rise and fall times.

5-20 By plotting the transfer function, prove that the gain of the complementary-symmetry inverter of Fig. 5-28 is limited solely by the r_o of the devices. Each device has the characteristic given in Fig. 5-16*a*, and $V_{DD} = 20$ V. Specify the location of the knees of the curve. From this information and the bottoming voltages, give the logical-0 and logical-1 noise immunity.

5-21 Exercise your ingenuity in designing MOSFET arrays satisfying the following logic statements. Draw the geometry of the logic. Make sure that the stage gain is at least 3 (see Prob. 5-3).

(*a*) $Y = \overline{AB + CDE}$

(*b*) $Y = \overline{AB} + CD$

(*c*) $Y = AB + \overline{C}D$

5-22 Repeat Prob. 5-21*b* and *c* for a complementary-symmetry logic.

5-23 Design an interface circuit that will invert and translate the TTL positive-logic levels to the negative-logic levels required by the MOSFET logic (logical $0 > -2$ V, logical $1 < -20$ V). You have available MOSFETs, diodes, silicon transistors with $\beta = 50$, and resistors. The two supply voltages are +4.5 and −40 V.

5-24 Design an interface circuit that translates the MOSFET negative logic to the positive-logic levels required by the ECL. See Prob. 5-23. The power supplies available are the −40 V used with the MOSFET logic and the −5.2 V used with the ECL.

5-25 The shunt chopper of Fig. 5-31, with $R_x = R_L = 10$ K and $C_L = 10$ pF, makes use of the junction FET of Prob. 5-14. It is switched on and off by a 10-V peak-to-peak 20-μs square wave having a −5-V dc level. Suppose we set $v_{in} = 0$; what is the output waveshape? What is the adverse effect of this output? (*Hint:* Consider the transient-response approximation of the circuit of Fig. 5-18.)

5-26 In the circuit of Fig. 5-45, $V_p = -2$ V, $I_{DSS} = 5$ mA, $r_o = 200$ K at I_{DSS}, while $\beta = 50$ and $V_{BE} = 750$ mV. Find the quiescent operating point. At this

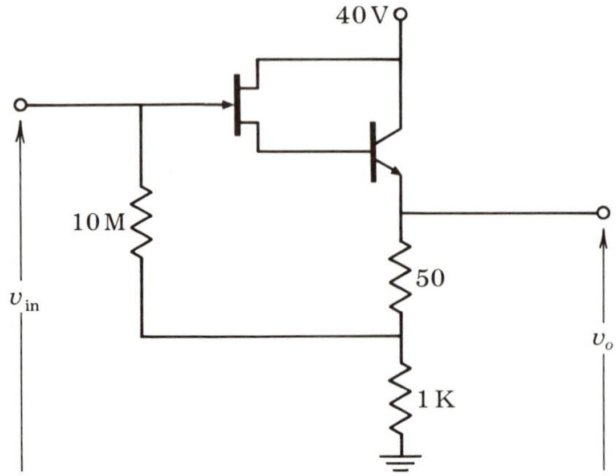

Fig. 5-45

point, calculate the composite g_m, the overall gain, the input impedance, and the output impedance.

5-27 Repeat Prob. 5-26 for the inverter configuration shown in Fig. 5-46.

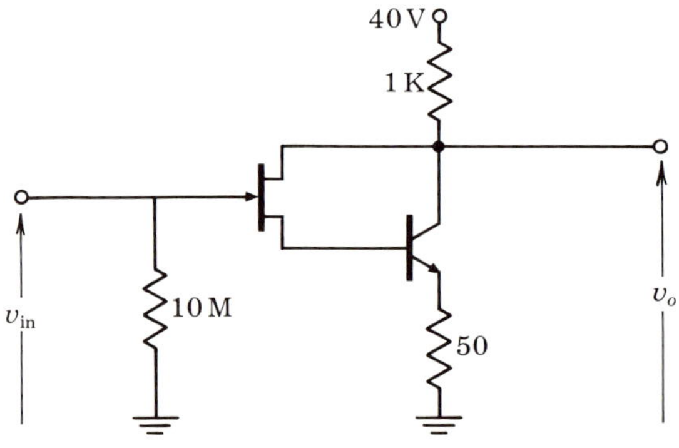

Fig. 5-46

5-28 Figure 5-47 shows a combination of an induced-channel MOSFET (Q_1) and a junction FET (Q_2). Q_1 has a threshold of -3 V and a measured current in the pinchoff region of 5 mA at $V_{GS} = -6$ V. Q_2 has a pinchoff voltage of -3 V and $I_{DSS} = 5$ mA.

(*a*) Plot the volt-ampere transfer characteristic of the composite device. Give the voltage bounds for pinchoff-region operation.

(*b*) The composite device is fixed-biased at its null point. A 100-kHz 2-V peak-to-peak sinusoidal voltage is used as the input. The output voltage is developed across a 10-K resistor in the composite drain (V_{DD} = 100 V). What will the output voltage be?

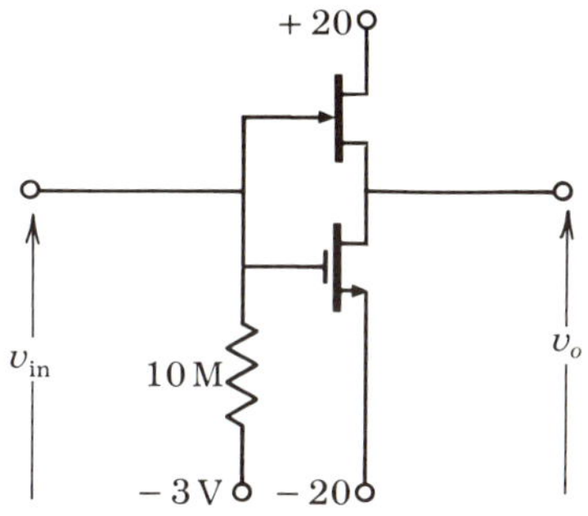

Fig. 5-47

Bibliography

Burns, J. R.: Switching Response of Complementary Symmetry MOS Transistor Circuits, *RCA Rev.*, vol. 25, no. 4, pp. 627–661, December, 1964.

Crawford, R. H.: "MOSFET in Circuit Design," McGraw-Hill Book Company, New York, 1967.

Hofstein, S. R., and F. P. Heiman: The Silicon Insulated-gate Field-effect Transistor, *Proc. IEEE*, vol. 51, pp. 1190–1202, September, 1963.

Research Triangle Institute: Integrated Silicon Device Technology, VI, *Unipolar Transistors* (*AD*613951), Clearinghouse for Federal Scientific and Technical Information, Washington.

Sah, C. T.: Characteristics of the Metal-oxide-semiconductor Transistors, *IEEE Trans. Electron. Devices*, vol. ED-11, pp. 324–345, July, 1964.

Sevin, L. J.: "Field-effect Transistors," McGraw-Hill Book Company, New York, 1965.

Shockley, W.: A Unipolar Field-effect Transistor, *Proc. IRE*, vol. 40, pp. 1365–1376, November, 1952.

Vadasez, L.: The Use of MOS Structure for the Design of High Value Resistors in Monolithic Integrated Circuits, *IEEE Trans. Electron. Devices*, vol. ED-13, pp. 459–465, May, 1966.

Wallmark, J. T., and H. Johnson: "Field Effect Transistors," Prentice-Hall, Inc., Englewood Cliffs, N.J., 1966.

PART 2

timing

6.
linear voltage sweeps

The preceding chapters were concerned with the shaping of voltage and current signals by various configurations of linear and nonlinear elements, e.g., differentiation, integration, amplification, clamping, clipping, and logical-switching functions. No consideration was given to the origin of the signals applied, but as we now begin this study, we should realize that many of the analytic techniques previously developed are applicable to our paramount problem, the generation of simple waveshapes.

In this chapter we shall concentrate our attention on some characteristics of one particular waveshape, the linear voltage sweep. Ideally, the output voltage increases linearly with time until it reaches a predetermined final value, instantaneously returns to zero, and starts increasing again when the cycle is reinitiated. This waveshape finds widespread application: in the horizontal-deflection circuit of an oscilloscope so that time-varying signals may be displayed against a linear time scale; in television transmitters and receivers as a means of generating the raster; and in various control systems where it is used to time preprogrammed functions.

Each of the three sweep essentials—rise, recovery, and switching—will be separately treated with a view toward optimizing the overall performance. After first defining the characteristics of an ideal voltage sweep, we shall use active circuit elements to establish an almost ideal linear charging path. Three methods of linearization, all involving feedback, are in common use: (1) negative-current feedback for constant-current charging of the capacitor, (2) the effective multiplication of the time constant and charging voltage through the use of negative-voltage feedback, and (3) positive-voltage feedback, wherein the needed correction term is first developed and then applied.

At the same time that we examine each one of these methods of linearization, we shall also consider the improvement of the sweep recovery time and the nature of the switching process.

6-1 Basic Voltage Sweeps

All linear-voltage-sweep circuits have a common basic method of operation: some form of exponential charging of a capacitor followed by a change in the circuit to discharge the capacitor at the proper point of the cycle. Differences in sweep circuits usually concern techniques of using active elements to linearize the sweep and methods of rapidly discharging the capacitor. We can gain some understanding of these problems by examining the very simple sweep circuit of Fig. 6-1a, which generates the waveshape shown in Fig. 6-1b. The sweep capacitor is charged from the current source. Its voltage builds up from the initial value V_i toward a steady-state value:

$$V_{ss} = IR \tag{6-1}$$

The resistance R is the parallel combination of the current source's internal impedance and any external load resistance. At the instant the voltage

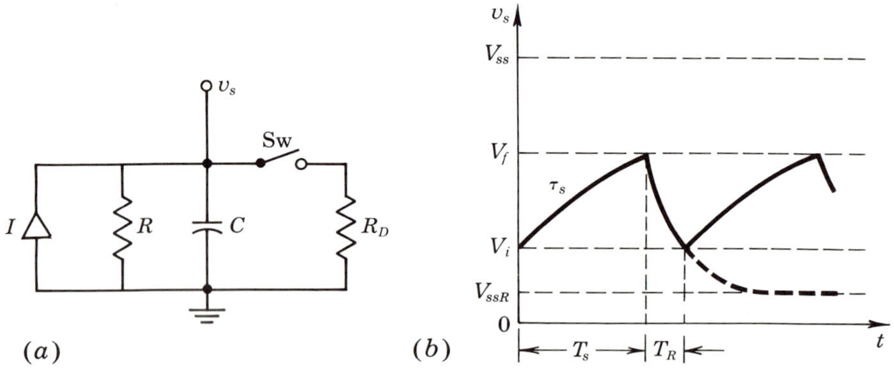

Fig. 6-1 (a) Basic voltage-sweep circuit; (b) periodic waveshape produced by periodically opening and closing the switch.

reaches some predetermined final value V_f, we close the switch and connect the discharge circuit. When the sweep voltage falls to its initial value, the switch is again opened, and the cycle repeats.

If there were no source or load impedance, we would have the ideal linear sweep produced by the constant-current charging of a capacitor:

$$v_s = \frac{1}{C}\int I\,dt = \frac{It}{C} \tag{6-2}$$

Of prime importance is the recognition that we are actually generating not a linear sweep, but an exponential charging curve, and only through restricting the sweep amplitude to a relatively small portion of the total possible curve will the capacitor voltage even approximate a linear rise. Thus our first problem is to approximate the ideal constant-current charging of the timing capacitor.

During the capacitor-charging interval,

$$v_s(t) = V_{ss} - (V_{ss} - V_i)e^{-t/\tau_s} \tag{6-3}$$

At the termination of the sweep, $v_s(T_s) = V_f$. Substituting into Eq. (6-3) and solving for the sweep time, we find that to sweep from the initial voltage V_i to the switching value V_f takes

$$T_s = \tau_s \ln \frac{V_{ss} - V_i}{V_{ss} - V_f} \tag{6-4}$$

where $\tau_s = RC$.

When we restrict the total sweep amplitude

$$V_s = V_f - V_i \tag{6-5}$$

to a small fraction of the total possible swing, $V_{ss} - V_i$, then it is reasonable to approximate the sweep by a linear rise having a slope equal to the slope of the exponential at the origin. Finding the slope from Eq. (6-3) and using the initial condition, we obtain the linear-sweep equation

$$v_s(t) \approx V_i + \frac{V_{ss} - V_i}{\tau_s} t \tag{6-6}$$

Substituting the final voltage V_f at $t = T_s$, we find that the sweep time may be approximated by

$$T_s \approx \tau_s \frac{V_f - V_i}{V_{ss} - V_i} = \tau_s \frac{V_s}{V_{ss} - V_i} \tag{6-7}$$

If we further restrict $V_s \leq 0.08(V_{ss} - V_i)$, the approximate value of the sweep duration T_s will be less than 2.5 percent shorter than the exact answer found from Eq. (6-4).

At time T_s, the switch is closed and the capacitor discharges toward the new steady-state value

$$V_{ssR} = IR \parallel R_D \tag{6-8}$$

with the new time constant $\tau_R = CR \parallel R_D$. The initial value during recovery is the final value during the sweep interval, and vice versa. Thus, after

$$T_R = \tau_R \ln \frac{V_{ssR} - V_f}{V_{ssR} - V_i} \tag{6-9}$$

the circuit has recovered back to its original state.

We note that the retrace time T_R is a direct function of R_D and that T_R becomes insignificant only when R_D is very small. Making R_D zero obviously reduces the discharge time to zero, but other circuit constraints may negate this choice. Thus our second problem will be to minimize the recovery interval.

The third question is: How do we throw the switch at the correct time? This problem will be considered later in the discussion of specific practical voltage sweeps.

6-2 A Simple Periodic-voltage-sweep Generator

An extremely simple sweep voltage generator (Fig. 6-2*a*) makes use of the double-valued characteristics of a silicon-controlled rectifier (SCR) to periodically discharge the sweep capacitor. The mechanisms giving rise to these characteristics are discussed in some detail in Sec. 11-8. For now, it is sufficient simply to use the volt-ampere characteristic of Fig. 6-2*b*. We see that so long as the SCR anode voltage remains below a breakdown ignition potential that we can call V_f, the SCR supports only a very limited current flow; for all practical purposes, it is an open circuit. However, once the voltage rises to V_f and the SCR fires, the terminal voltage drops to a much lower, almost constant value V_o. V_f is a function of the Miller breakdown voltage of the junction and also depends on the very small gate bias current. $I_G = 0$ sets the maximum possible anode firing voltage V_{fm}. Increasing I_G decreases the ignition voltage.

After breakdown, the SCR anode current (i_a) is limited solely by the external circuit. The gate no longer has any control over the device so long as it remains within the breakdown region. The SCR turns off once the anode current falls below a minimum maintaining current I_x.

The total sweep operation is defined by the two models of Fig. 6-2*c* and *d*. Since we know the firing voltage but do not know the initial sweep

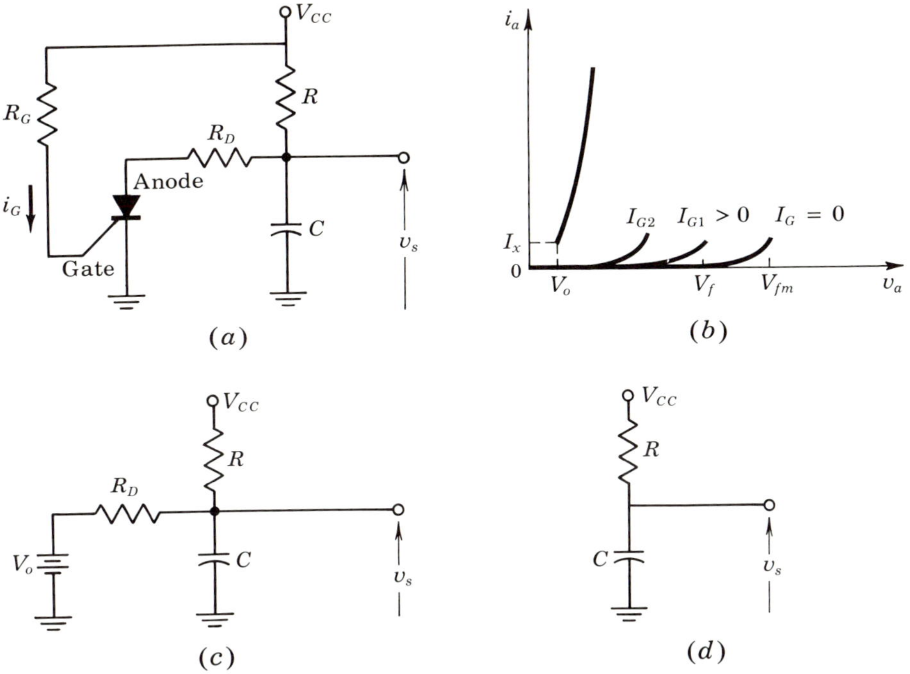

Fig. 6-2 (*a*) SCR periodic-sweep circuit; (*b*) device volt-ampere characteristics; (*c*) capacitor-discharge model (SCR conducting); (*d*) capacitor-charge model (SCR off).

voltage, it is easier to start our analysis with the discharge model. The capacitor voltage at the instant of firing is V_f. Immediately after the SCR conducts, C starts discharging toward the Thévenin equivalent voltage across its terminals:

$$V_{\mathrm{Th}} = \frac{R}{R + R_D} V_o + \frac{R_D}{R + R_D} V_{CC} \tag{6-10}$$

Extinction occurs when the SCR current decays to I_x, which corresponds to a sweep capacitor voltage of

$$v_s(T_R) = V_i = V_o + I_x R_D$$

Since we now know the initial, final, and steady-state voltages of the decay region, substituting into the sweep-time equation (6-4) we obtain

$$T_R = \tau_R \ln \frac{V_{\mathrm{Th}} - V_f}{V_{\mathrm{Th}} - (V_o + I_x R_D)} \tag{6-11}$$

where $\qquad \tau_R = (R \parallel R_D)C \approx R_D C$

Normally, R_D serves to limit the maximum SCR current to a safe value and is much smaller than the charging resistor R. Reexamination of Eq. (6-10) indicates that under these circumstances the Thévenin equivalent voltage will be only slightly larger than V_o. The steady-state discharge voltage, the initial voltage of the sweep, and the SCR extinction voltage are so close together that we might as well approximate them by the single value V_o. We conclude that the discharge curve will constitute virtually the complete exponential, requiring roughly four time constants for completion.

Once the SCR extinguishes, the capacitor charges from V_o back toward V_{CC} with the much longer time constant $\tau_s = RC$. This is exactly the situation described in deriving Eqs. (6-3) and (6-4), and thus the sweep time T_s, shown in Fig. 6-3, is given by

$$T_s = \tau_s \ln \frac{V_{CC} - V_o}{V_{CC} - V_f}$$

A glance at Fig. 6-2*c* discloses that during conduction a portion of the SCR anode current is contributed by V_{CC}. If this current component is greater than I_x, the SCR will never extinguish. To prevent continuous conduction, the inequality of Eq. (6-12) must be satisfied:

$$I_a = \frac{V_{CC} - V_o}{R + R_D} < I_x \tag{6-12}$$

Equation (6-12) sets the lower limit of R for guaranteed repetitive sweep operation. Since $R \gg R_D$,

$$R > \frac{V_{CC} - V_o}{I_x}$$

Limits must also be imposed on R_D; if it is too large, the discharge time becomes excessive, and if it is too small, the peak current may exceed the

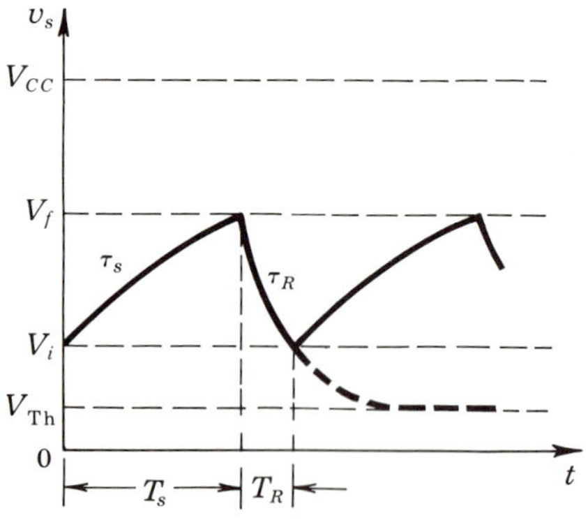

Fig. 6-3 Periodic voltage waveshape generated by the SCR circuit of Fig. 6-2.

device rating. Therefore, we have

$$R_D \geq \frac{V_{fm} - V_o}{I_{a,\max}}$$

where $I_{a,\max}$ is the maximum allowable SCR current.

6-3 Sweep Linearity

Since our announced objective is the generation of a linear sweep, we should have a means of expressing the linearity, or departure from linearity, as a measure of the sweep quality. This entails making a comparison between an ideal sweep and the actual sweep, with any divergence representing the nonlinearity. Two possible choices are presented in Fig. 6-4. The first is a straight line drawn tangent to the sweep at the origin (Fig. 6-4*a*). This choice exaggerates the timing inaccuracy of the less-than-ideal sweep. The second choice is a line which connects the end points of the sweep (Fig. 6-4*b*) and illustrates the amplitude distortion of the actual sweep.

Depending on the particular application of the sweep, either choice will be eminently satisfactory as a basis for comparison. Whenever the sweep-time duration becomes important, the sweep is usually synchronized by an external reference signal which constrains its end points. Under these conditions it seems reasonable to calculate the amplitude deviation from an ideal sweep connecting the two restricted ends. Referring to Fig. 6-4*b*, the deviation is

$$\delta(t) = v_e(t) - v_l(t) \tag{6-13}$$

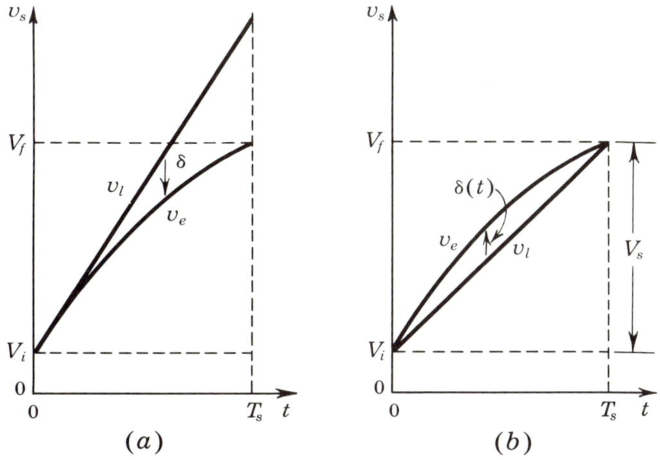

Fig. 6-4 Two methods of defining sweep nonlinearity.

The ratio of maximum deviation to sweep amplitude, expressed as a percentage, is defined as the sweep nonlinearity:

$$NL \equiv \frac{\delta_{\max}}{V_s} 100\%$$

where $V_s = V_f - V_i$.

By shifting the axis and writing all equations with respect to V_i, the equation of the linear sweep of Fig. 6-4*b* becomes

$$v_l(t) = kt = \frac{V_s}{T_s} t \tag{6-14}$$

and that of the exponential sweep

$$v_e(t) = (V_{ss} - V_i)(1 - e^{-t/\tau_s}) \tag{6-15}$$

When the limits of V_s and T_s are substituted into Eq. (6-15), the coefficient can be expressed in terms of the known sweep quantities:

$$V_{ss} - V_i = \frac{V_s}{1 - e^{-T_s/\tau_s}} \tag{6-16}$$

The first two terms of the power-series expansion of the exponential [Eq. (6-17)] represent its linear approximation, and the remaining higher-order terms, the deviation from linearity. Because we are interested in an almost linear sweep, the contribution of the higher-order terms must be kept to a minimum by using only a small portion of the total exponential curve. Therefore t/τ_s will always be much less than unity, and we can assume that the complete nonlinearity is due to the square term of Eq. (6-17):

$$e^{-t/\tau_s} = 1 - \frac{t}{\tau_s} + \frac{1}{2!}\left(\frac{t}{\tau_s}\right)^2 - \frac{1}{3!}\left(\frac{t}{\tau_s}\right)^3 + \cdots \tag{6-17}$$

Substituting Eq. (6-16) into Eq. (6-15) and using the first three terms of the expansion of Eq. (6-17) in both numerator and denominator, the equation of the exponential sweep becomes

$$v_e(t) \approx \frac{\tau_s V_s}{T_s(1 - T_s/2\tau_s)} \left[\frac{t}{\tau_s} - \frac{1}{2}\left(\frac{t}{\tau_s}\right)^2\right]$$

and since $T_s/\tau_s \ll 1$,

$$v_e(t) \approx \frac{\tau_s}{T_s} V_s \left(1 + \frac{T_s}{2\tau_s}\right)\left[\frac{t}{\tau_s} - \frac{1}{2}\left(\frac{t}{\tau_s}\right)^2\right] \tag{6-18}$$

The one missing piece of information, the time location of $\delta_{\max}$, will be found from geometric considerations. Any curve having only slight curvature, such as the exponential sweep of Fig. 6-4*b*, may be approximated by the arc of some circle with large radius. The linear sweep becomes a chord of this circle, and the maximum distance from a chord to the arc, measured at any fixed angle, occurs at the center of the chord, at the point corresponding to $T_s/2$. We evaluate Eqs. (6-14) and (6-18) at this time and substitute the answers into Eq. (6-13), with the result

$$\begin{aligned}\delta_{\max} &= \delta\left(\frac{T_s}{2}\right) = \left[\frac{T_s}{8\tau_s} - \frac{1}{16}\left(\frac{T_s}{\tau_s}\right)^2\right] V_s \\ &\approx \frac{T_s}{8\tau_s} V_s \end{aligned} \tag{6-19}$$

Consequently, the percentage of nonlinearity becomes

$$NL = \frac{T_s}{8\tau_s} 100\% \tag{6-20}$$

Substitution of the exact period, expressed in terms of the circuit voltages [Eq. (6-4)], provides us with an alternative form for the nonlinearity, one which brings to the forefront the dependency on the voltage limits imposed:

$$NL = \frac{100\%}{8} \ln \frac{V_{ss} - V_i}{V_{ss} - V_f} \tag{6-21}$$

A third expression for the nonlinearity is possible when the sweep starts close to zero and constitutes only a small fraction of the total exponential. If $V_{ss} \gg V_i$ and $V_{ss} \gg V_f$, then by using only the first term of the series expansion of Eq. (6-21), we obtain

$$NL \approx 12.5\% \frac{V_s}{V_{ss}} \tag{6-22}$$

Equations (6-20) to (6-22) verify our previous contention that good sweep linearity can be realized only by operating over a small portion of the total exponential. For less than 1 percent nonlinearity, the sweep time must be restricted to 8 percent of the time constant and its amplitude to 8 percent of the total possible sweep voltage.

6-4 The Transistor as a Current Source

One of the simplest voltage sweeps makes use of the transistor as the current source for charging the sweep capacitor (Fig. 6-5). So long as the switch is closed, the transistor is cut off and the capacitor charges to V_{CC}. At $t = 0$, the switch is opened and the transistor turns on. The base current becomes

$$I_b = \frac{V_{CC} - V_{BE}}{R_B} = \frac{10 - 0.7}{10\text{ K}} = 0.93\text{ mA} \tag{6-23}$$

and for $\beta = 50$, $I_c = \beta I_b = 46.5$ mA. The charged sweep capacitor at the collector ensures that the transistor becomes active upon switching. C can supply all the collector current required by the given base-current drive.

From the model of Fig. 6-5*b*, we note that the collector voltage heads from V_{CC} toward the equivalent steady-state value

$$V_{ss1} = V_{CC} - \beta I_b R_L = -455 \text{ V} \tag{6-24}$$

with the time constant $\tau_s = R_L C$. The collector voltage runs down almost linearly until it reaches the final value, which is determined by the transistor saturating at $V_{CE,\text{sat}}$. Assuming that the bottoming voltage is 150 mV, the sweep amplitude is

$$V_s = V_i - V_f = 10 - 0.150 = 9.85 \text{ V}$$

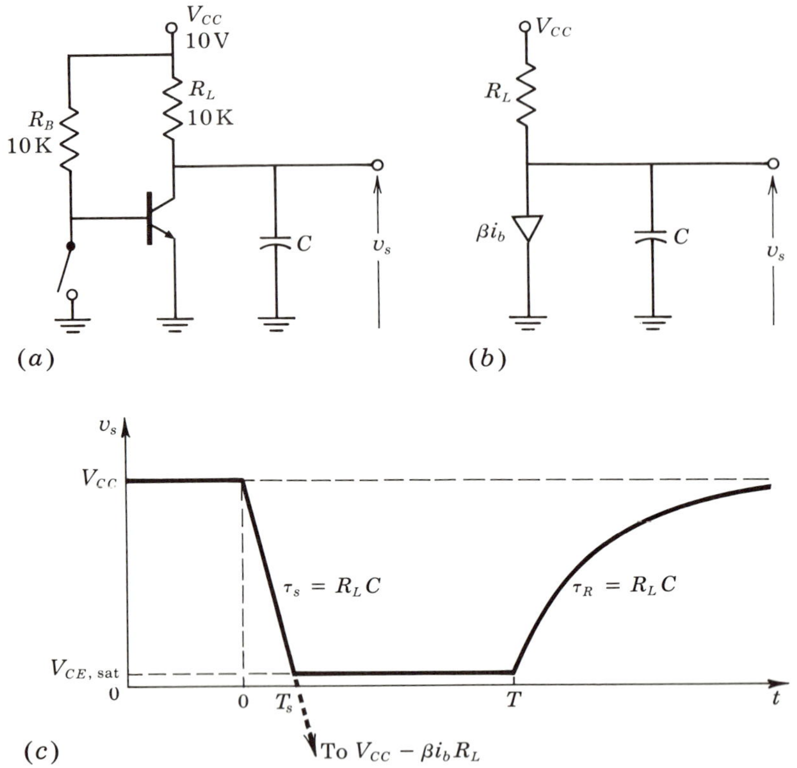

Fig. 6-5 (*a*) Simple transistor voltage sweep; (*b*) model holding during the linear rundown; (*c*) waveshape produced. For the example in the text, $\beta = 50$, $V_{BE} = 700$ mV, and $V_{CE,\text{sat}} = 150$ mV.

Since the maximum output swing is only a small fraction of the total exponential of $V_{ss1} - V_i = -465$ V, it is valid to use the linear approximation for the rundown time. Substituting into Eq. (6-7), we obtain

$$T_s = \tau_s \frac{V_{CE,\text{sat}} - V_{CC}}{V_{CC} - \beta I_b R_L - V_{CC}} = 0.0212\tau_s \tag{6-25}$$

The actual sweep is only 2 percent of the total effective exponential, and as a result, the nonlinearity is only 0.265 percent.

Equation (6-25) points up the dependency of the rundown time on β and hence the necessity of recalibration upon the replacement of the individual transistor. We also note that the sweep rundown time T_s can be easily controlled in the base circuit: decreasing I_b increases the sweep time.

A short time after saturation, we close the switch and turn the transistor off. C now recovers to the initial voltage V_{CC}, with a time constant $\tau_R = R_L C$. We assume recovery is complete within four time constants.

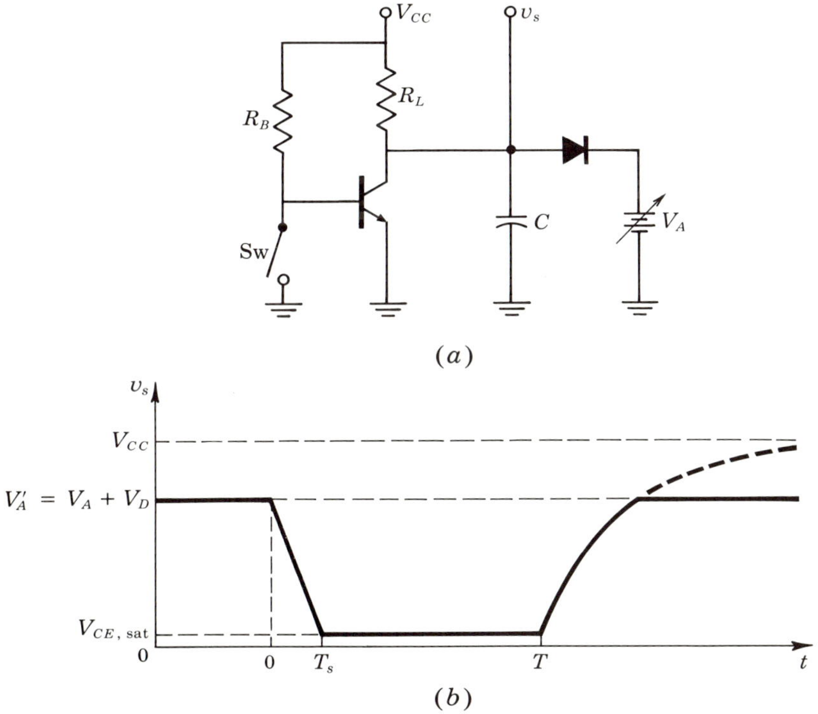

Fig. 6-6 External voltage control over sweep time and output waveshape. Note the great improvement in recovery time.

Since the circuit has identical rise and fall time constants, the recovery interval will be very long compared with the linear rundown

$$T_R = 4R_LC \gg T_s$$

Voltage control over sweep time

External voltage control of the sweep rundown time may be incorporated by coupling the collector of the sweep through a catching diode to the control voltage V_A (Fig. 6-6). When the circuit is in the normal, unexcited state (switch closed), the diode conducts through R_L, setting the collector voltage to

$$V_A' = V_A + V_D \tag{6-26}$$

Just after the switch is opened and the transistor conducts, the initial portion of the rundown at the collector will back bias the catching diode. Collector rundown proceeds as before, starting, however, from V_A' instead of from V_{CC}. Thus, by replacing V_{CC} with V_A' in Eq. (6-25), we note that the sweep time

$$T_s = \tau_s \frac{V_A' - V_{CE,\text{sat}}}{\beta I_b R_L} \tag{6-27}$$

becomes a linear function of the control voltage V_A.

By eliminating the long trailing end of the recovery exponential, the catching diode also shortens the recovery interval. Recovery toward V_{CC} terminates once the diode "catches" the collector voltage at V_A'. The recovery time is now given by

$$T_R = \tau_R \ln \frac{V_{CC} - V_{CE,\text{sat}}}{V_{CC} - V_A'} \tag{6-28}$$

If we set $V_A' = 0.5V_{CC}$, the recovery time is reduced almost sixfold—from $4\tau_R$ down to $0.693\tau_R$.

Recovery-time improvement

We know that the constant-current charging during rundown results in a sweep whose duration is only a very small fraction of the total exponential. We therefore ask: Can recovery be improved by using a second transistor to supply the necessary recharging current? Figure 6-7 illustrates this technique. Transistor Q_1 is the original sweep transistor, and transistor Q_2 is used to speed recovery.

The saturation of Q_2 (with the switch open) determines the initial conditions for the rundown of C. After the switch is closed, the collector rundown of Q_1 starts from $V_{CC} - V_{CE,\text{sat}}$ and continues to the bottoming

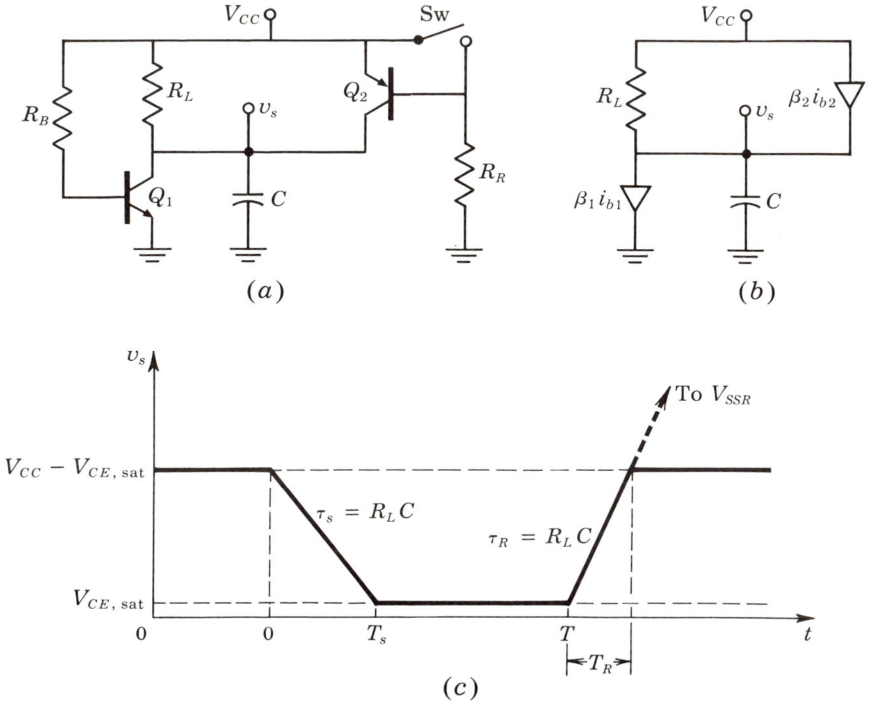

Fig. 6-7 (*a*) Complementary-transistor sweep circuit to speed recovery; (*b*) mode holding when both transistors are active (switch open); (*c*) waveshape produced.

value of $V_{CE,\text{sat}}$. After the switch is opened, both transistors become active and C charges toward the recovery Thévenin equivalent voltage

$$V_{SSR} = V_{CC} + R_L(\beta_2 i_{b2} - \beta_1 i_{b1}) \tag{6-29}$$

where $i_{b2} = (V_{CC} - V_{BE})/R_R > i_{b1}$. Recovery proceeds until Q_2 saturates at $V_{CE,\text{sat}}$. Through the choice of R_R, we can easily control the value of V_{SSR} and consequently the recovery time.

Figure 6-7*c* shows the waveshape produced. Either the linear rundown or the linear runup can be used as the sweep. Simply interchanging the switching will change the roles of Q_1 and Q_2. If the switch is normally closed and is opened for the sweep, the runup will occur before the rundown. As an alternative, the switch can be placed from the base of Q_1 to ground and can sequentially turn Q_1 on and off.

A second approach to improving recovery time is to recharge the capacitor through a low impedance such as an emitter follower. (Also see Sec. 6-5 below.) Figure 6-8 illustrates the resultant circuit. When the switch at the base of Q_2 is closed, the emitter follower is cut off and the sweep rundown

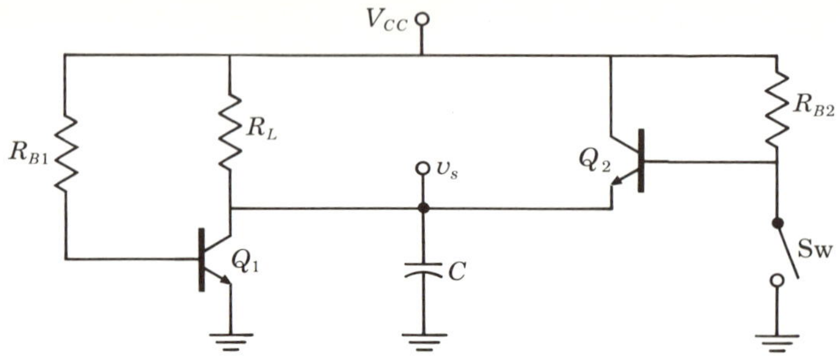

Fig. 6-8 Sweep circuit having an emitter follower (Q_2) for fast recovery.

is as before. At the end of the sweep, the switch is opened. Looking into the emitter of the emitter follower, the capacitance sees the low impedance $R_{B2}(1 - \alpha)$. Thus recovery proceeds with the very small time constant

$$\tau_R = [R_{B2}(1 - \alpha) \parallel R_L]C$$

It is left as an exercise (Prob. 6-10) to find the final recovery voltage (which will be the initial value for the sweep rundown) and to compare the recovery time of the emitter-follower recovery circuit with the time required in the sweep recovery circuit of Fig. 6-7.

6-5 Current-feedback Linearization—Miller Sweep

The simple circuit of Sec. 6-4 generates a sweep of acceptable linearity, but because the sweep duration is a function of β, it is not suitable for use as a precision ramp generator. Negative feedback can be used to minimize the sweep-time dependence on the β of the transistor. In the circuit of Fig. 6-9, which is a current-feedback version of a Miller sweep circuit, the charging current is fed back directly into the transistor's base. With the switch open, the base diode is conducting and the capacitor charges to $V_{CC} - V_{BE}$. Closing the switch connects the charged capacitor to the collector, and the transistor becomes active. From the model of Fig. 6-9*b*, we see that the base current may now be expressed as

$$i_b = I_1 + i_q$$

where $I_1 = (V_{CC} - V_{BE})/R$. The controlled source current

$$\beta i_b = \beta I_1 + \beta i_q \tag{6-30}$$

contains a term directly dependent on the capacitor branch current i_q. The controlled current βi_q may be combined with i_q by replacing C with the equivalent capacitor $C(\beta + 1)$, as shown in Fig. 6-9*c*. By doing this, we have reflected the sweep capacitor into the output circuit and reduced the active-region model to the simple sweep of Fig. 6-5.

The circuit of Fig. 6-9 differs from that of Fig. 6-5 only in that the run-down-time constant is now dependent on β. From Fig. 6-9*c*,

$$\tau_s = R_L C(\beta + 1) \tag{6-31}$$

By adjusting R, the equivalent steady-state voltage,

$$V_{ss} = V_{CC} - \beta I_1 R_L$$

can be made large enough to ensure acceptable linearity. Using the linear approximation of Eq. (6-7), the time required for the collector voltage to run down from V_{CC} to the bottoming voltage $V_{CE,\text{sat}}$ is

$$T_s = \tau_s \frac{V_s}{V_{ss} - V_i} = R_L C(\beta + 1) \frac{V_{CC} - V_{CE,\text{sat}}}{\beta I_1 R_L} \tag{6-32}$$

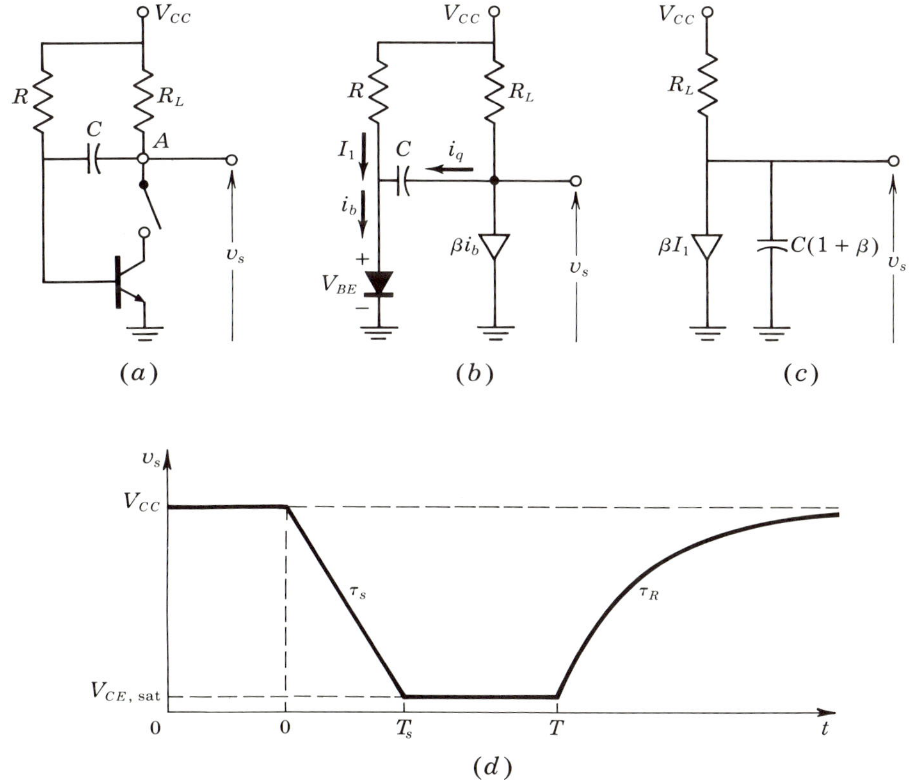

Fig. 6-9 (*a*) Transistor Miller-sweep circuit; (*b*) model holding during rundown (transistor active); (*c*) reduced-circuit model (equivalent sweep capacitor shown at output node); (*d*) sweep waveshape.

For a high-β transistor, $\beta + 1 \approx \beta$ and

$$T_s \approx C \frac{V_{CC} - V_{CE,\text{sat}}}{I_1} \tag{6-33}$$

If $V_{CC} \gg V_{CE,\text{sat}}$ and when $V_{CC} \gg V_{BE}$, then $I_1 \approx V_{CC}/R$ and Eq. (6-33) may be further reduced to

$$T_s \approx RC \tag{6-34}$$

RC is *not* the time constant of the circuit; it is only the approximate time of the linear rundown.

Equation (6-34) shows that we have succeeded in virtually eliminating the sweep-time dependency on the transistor β. In addition, if the supply voltage is large enough for the approximations to hold, the sweep time becomes virtually independent of the supply voltage. The total sweep stability depends only on the two physical elements, the current-determining resistor R and the timing capacitor C.

Upon opening the switch, after bottoming, the capacitor recharges through the conducting base-emitter diode and R_L. The output voltage recovers to V_{CC} with the time constant $\tau_R = R_L C$.

Linear runup

In a current-feedback circuit, the sweep limits are determined by the bounds of the active region of the transistor (or other active device). The linear rundown at the collector starts with the active device cut off (or disconnected) and terminates with the saturation of the transistor. It is just as easy to design the sweep output to run up through the active region from the edge of saturation to cutoff. To do this, the base resistor must be switched from a positive supply establishing saturation to a negative supply which will ensure eventual cutoff. Figure 6-10 shows one possible switching arrangement together with the models holding just before and just after switching.

At $t = 0^-$, the switch is open and the current through R_1 keeps the transistor saturated. The capacitor is charged to the difference between V_{BE} and $V_{CE,\text{sat}}$, or about 600 mV. We shall assume that the closing of the switch causes the transistor to become active. The circuit model changes to that of Fig. 6-10*c*. The capacitor prevents the collector voltage from changing instantaneously. Thus, from Fig. 6-10*c*,

$$I_2(0^+) = I_2(0^-) = \frac{V_{CC} - V_{CE,\text{sat}}}{R_L} = \frac{10 - 0.15}{1\text{ K}} = 9.85\text{ mA}$$

and $$I_1(0^+) = \frac{V_{\text{Th}} - V_{BE}}{R_1 \parallel R_2} = \frac{-10 - 0.7}{5\text{ K}} = -2.14\text{ mA}$$

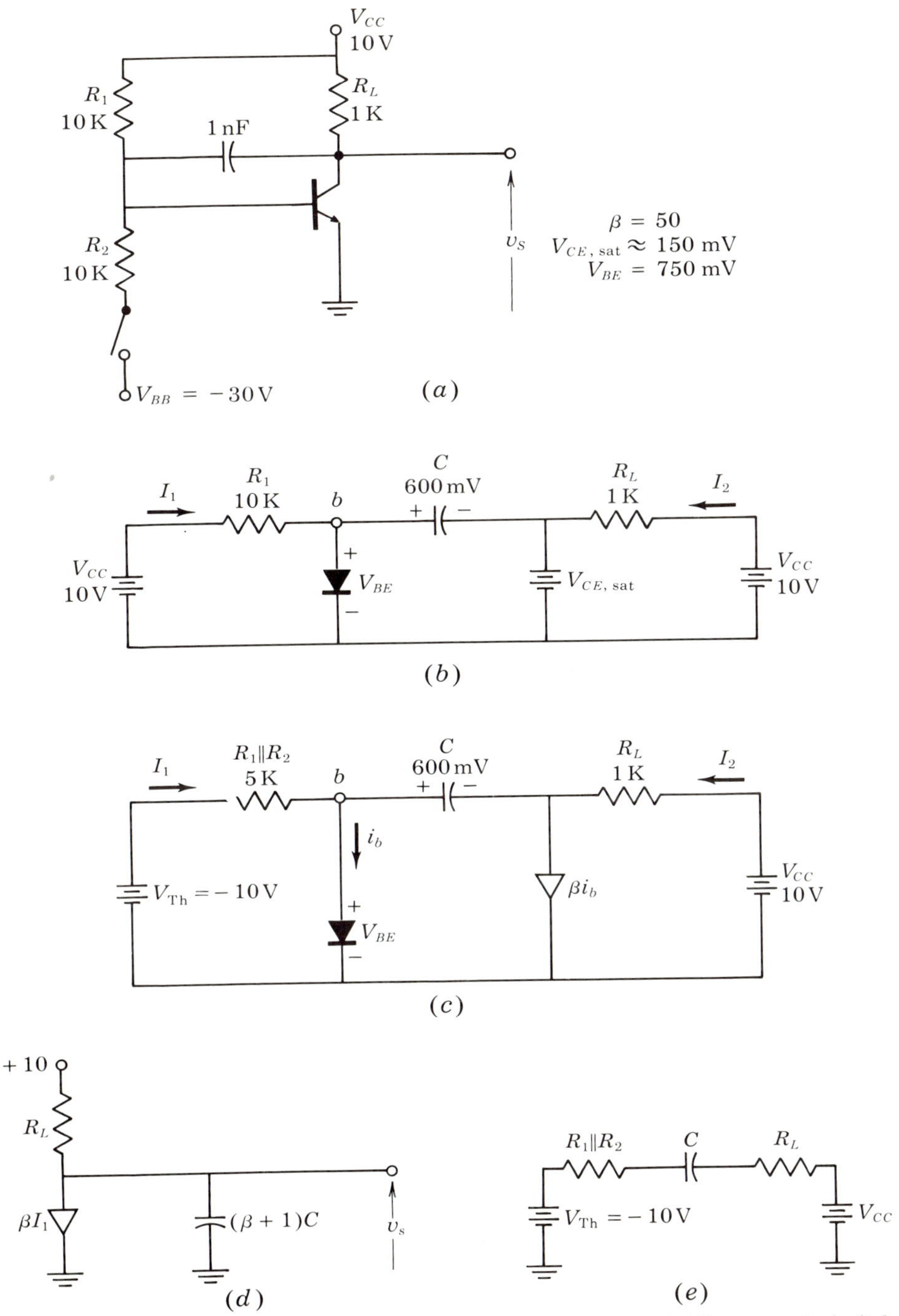

Fig. 6-10 (*a*) Miller sweep connected for a linear runup; (*b*) model holding at $t = 0^-$ (switch open); (*c*) active-region model holding during linear runup (switch closed); (*d*) reduced active-region model, showing equivalence to the circuit of Fig. 6-9; (*e*) model holding after transistor cuts off (switch closed).

By summing the branch currents, we obtain

$$I_2 - \beta i_b = -I_1 + i_b$$

or

$$i_b = \frac{I_2 + I_1}{\beta + 1} \tag{6-35}$$

At $t = 0^+$, $i_b = 151\ \mu\text{A}$.

We must check to see if our assumption that the transistor becomes active at switching is correct. For active-region operation,

$$\beta i_b \leq I_2$$

Substituting into Eq. (6-35), we see that the transistor will be active so long as

$$I_1 < \frac{I_2}{\beta}$$

and since, in this circuit, $I_1 < 0$ and $I_2 > 0$, certainly the required condition is satisfied and we have justified our assumption as to the state of the transistor.

This sweep circuit is identical to the sweep of Fig. 6-9 except that now a negative value of I_1 replaces the previously positive I_1. Thus the same active-region model (Fig. 6-10*d*) and equations hold. Because of the negative I_1, the collector runs up from $V_{CE,\text{sat}}$ toward a steady-state voltage

$$V_{ss1} = V_{CC} - \beta I_1 R_L = 10 + 107 = 117\ \text{V}$$

with the time constant $\tau_s = R_L C(\beta + 1) = 51\ \mu\text{s}$.

The linear-sweep runup continues until the transistor becomes cut off once i_b drops to zero. From Eq. (6-35) we have, at cutoff,

$$I_2 = -I_1$$

Thus at the edge of cutoff, V_{BE} has dropped to approximately 700 mV and

$$v_s(T_s) = V_{CC} - I_1 R_L = 7.86\ \text{V}$$

Substituting the initial, the steady-state, and the limiting voltages at the output into Eq. (6-7), we obtain the time required for the sweep to run up through the active region,

$$T_s = 51\ \mu\text{s}\ \frac{7.86 - 0.15}{117 - 0.15} = 3.36\ \mu\text{s}$$

We now must use the model of Fig. 6-10*e*, which holds after the transistor is cut off. From this, we see that charging continues with the collector

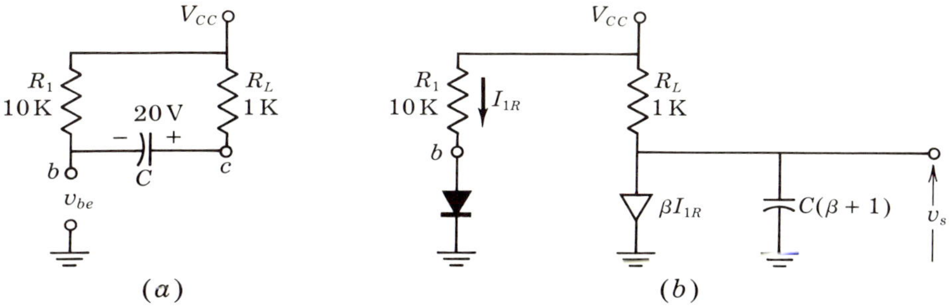

Fig. 6-11 Recovery-region models for the linear-runup Miller sweep of Fig. 6-10. (*a*) Model holding just after switch is opened (transistor still off); (*b*) transistor active, model holding for linear rundown.

heading toward 10 V and the base toward −10 V. The new time constant is

$$\tau_2 = C(R_1 \parallel R_2 + R_L) = 6\ \mu\text{s}$$

We assume charging is complete in $4\tau_2$. The waveshape generated appears in Fig. 6-12.

In calculating the sweep recovery, we shall start by assuming that immediately after the switch is opened the transistor remains cut off. From the model of Fig. 6-11*a*, we find that the base jumps from −10 to −8.2 V, and since the base is still negative, this satisfies our initial assumption. This jump is coupled by the capacitor to the collector, raising its voltage to 11.8 V. The base charges toward V_{CC} with the time constant

$$\tau_3 = (R_L + R_1)C = 11\ \mu\text{s}$$

and the transistor turns on when $V_{BE} = 700$ mV. The time lapse before entering the active region is

$$t_3 = 11 \ln \frac{10 - (-8.2)}{10 - 0.7} = 7.4\ \mu\text{s}$$

and during this time the collector voltage has decayed to 10.92 V.

After the transistor becomes active, the rundown follows the standard Miller-sweep pattern. The collector voltage heads toward

$$V_{ss4} = V_{CC} - \beta I_{1R} R_L = -36.5\ \text{V}$$

with the time constant $\tau_s = R_L C(\beta + 1)$. Since the total rundown from 10.92 V to the bottoming point at $V_{CE,\text{sat}}$ is an appreciable portion of the total sweep amplitude, we must use the exponential form to find the time:

$$T_R = \tau_s \ln \frac{-36.5 - 10.92}{-36.5 - 0.15} = 13.2\ \mu\text{s}$$

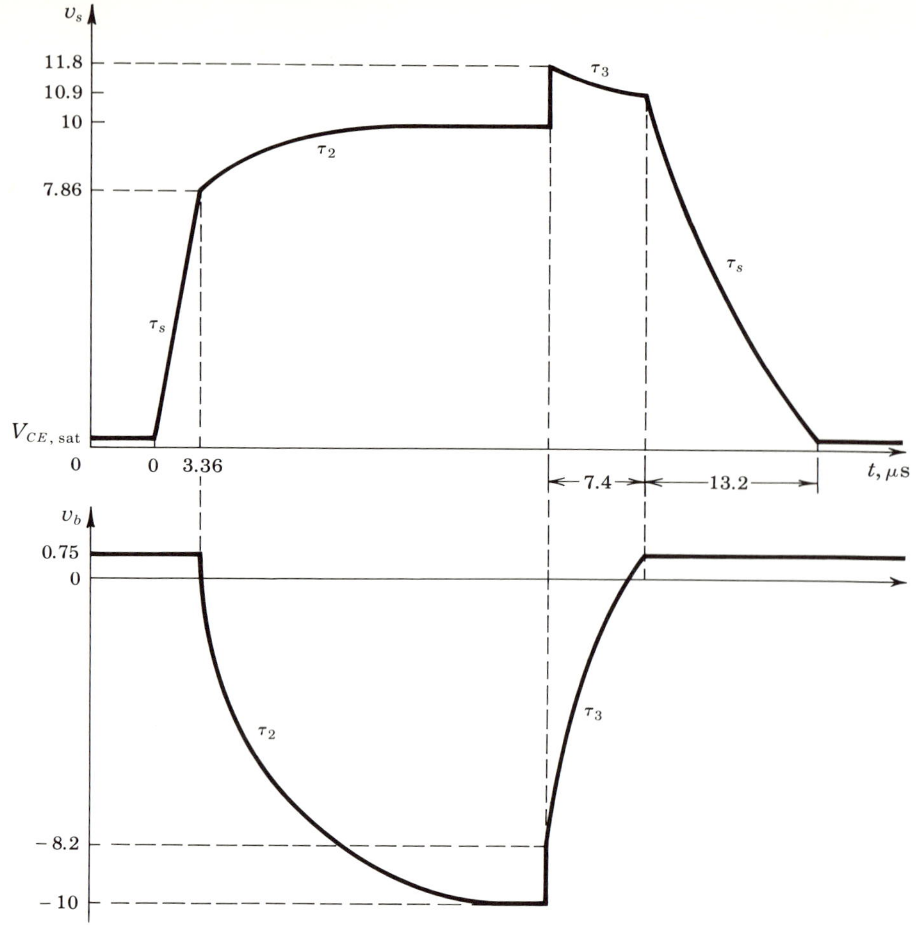

Fig. 6-12 Waveshapes present in the circuit of Fig. 6-10.

In this particular circuit, total recovery time is appreciably longer than the linear-sweep runup. The techniques discussed in Sec. 6-4 can be used to shorten the recovery time.

6-6 Voltage-feedback Linearization—Miller Sweep

A Miller sweep which utilizes negative-voltage feedback for sweep linearization generates a sweep virtually independent of the amplifier gain. Consider the basic circuit of Fig. 6-13*a*, where we use an integrated-circuit operational amplifier as the voltage amplifier. A typical integrated-circuit amplifier will have a gain A of $-1{,}500$, a dc input resistance (R_i) greater than 20 K, and output impedance (R_o) of less than 200 Ω.

To obtain the active-region model of Fig. 6-13*b*, we first convert the input resistive network of V_1, R_1, and R_i to the Thévenin equivalent, where

$$V = V_1 \frac{R_i}{R_1 + R_i} \qquad \text{and} \qquad R = R_i \parallel R_1$$

Usually $R_i \gg R_1$, and therefore $V \approx V_1$ and $R \approx R_1$. Next, we look at the feedback path. By writing the input-node-current equation,

$$i_i = \frac{v_i - Av_i}{R_o + 1/sC}$$

and solving for the input impedance,

$$Z_{\text{in}} = \frac{v_i}{i_1} = \frac{R_o}{1 - A} + \frac{1}{sC(1 - A)} \tag{6-36}$$

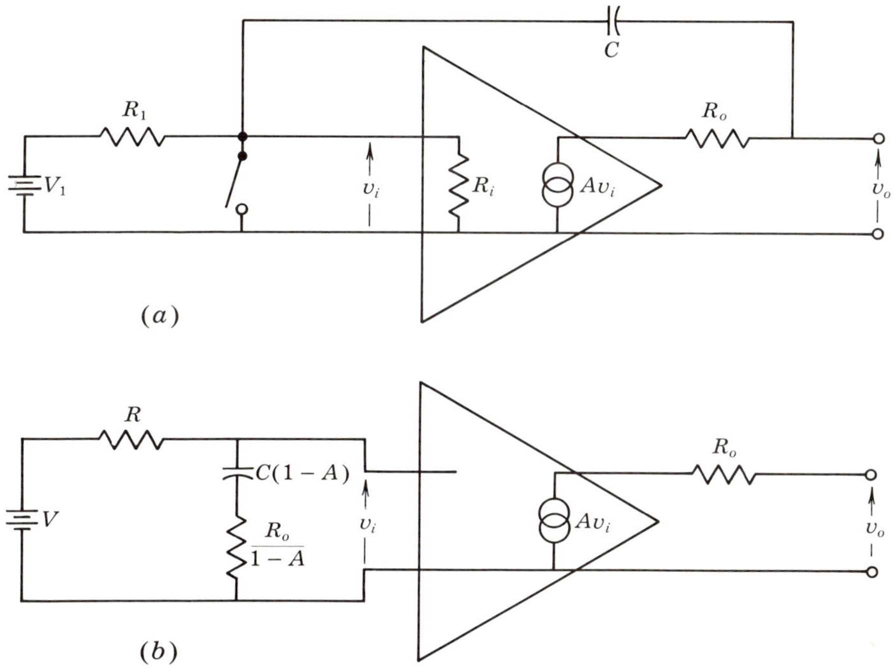

Fig. 6-13 (*a*) Basic voltage-feedback Miller voltage sweep; (*b*) equivalent active-region model.

we find that the equivalent input circuit may be represented as shown in Fig. 6-13*b*. Equation (6-36), the Miller effect, expresses the effective multiplication of the feedback capacity and reduction in resistance as we reflect the complete feedback network into the input node. With current feedback, the multiplied capacitance appeared at the amplifier output. (See Fig. 6-9*c*.)

If the amplifier output impedance is small enough, we can neglect the loading effect of the feedback current on the output voltage. We have thus effectively isolated the input and output circuits, allowing an independent solution for the input-node voltage. It may then be reflected into the output by multiplying by the circuit gain A.

Since we are interested in generating a linear sweep, charging must be limited to a small portion of the total exponential. The input circuit will be treated as an integrator, with the reflected resistance $R_o/(1 - A)$ contributing an additional small constant term:

$$v_i = \frac{R_o}{R_o + R(1 - A)} V + \frac{1}{[R + R_o/(1 - A)]C(1 - A)} \int_0^t V\,dt$$

and the amplified input voltage appears at the output:

$$v_o = Av_i = \frac{AR_oV}{R_o + R(1 - A)} + \frac{AV}{[R + R_o/(1 - A)]C(1 - A)} t \tag{6-37}$$

If the amplifier gain is a very large negative number, then $|1 - A| = |A|$ and the magnitude of A cancels out in the sweep term of Eq. (6-37). This necessary condition must be satisfied in order to make the sweep relatively independent of the amplifier parameters. As a further result of the large gain, $R \gg R_o/(1 - A)$ and

$$v_o \approx -\frac{R_o}{R} V - \frac{V}{RC} t \tag{6-38}$$

Usually the first term of Eq. (6-38) is extremely small.

With an integrated-circuit amplifier, we can obtain either a negative- or a positive-going sweep. The polarity of V will dictate the slope. The sweep will continue until the amplifier gain drops to zero because of either the saturation or the cutoff of a stage in the amplifier.

An amplifier with a gain of $-1{,}000$ and a source voltage V of 10 V can generate an output sweep of 10 V while the input changes by $1/A$ times the output, or by 0.01 V. Since this is only a very small portion of the possible 10-V input charging curve, the sweep will be extremely linear. Nonlinearity of better than 0.1 percent is easily achieved. With a large enough gain, second-order effects and not the exponential charging usually account for the sweep nonlinearity.

Junction-FET Miller sweep

The procedure used in calculating the circuit waveshapes and in subsequent verification of sweep linearity is outlined below for the single-stage junction-FET Miller sweep of Fig. 6-14. The solution follows in time sequence the behavior of the circuit. In each region we draw the model holding and use it to evaluate the initial condition, the steady-state voltage, and the time constant.

Before closing the switch at $t = 0$, the gate-substrate diode, returned through R to V_{DD}, is conducting. The gate is at the small diode-voltage drop, which is about 700 mV. The capacitor is charged to

$$v_q(0^-) = V_{DD} - 0.7 \tag{6-39}$$

Once the switch closes, the first question of interest is: What are the new values of gate and drain voltages? Our qualitative argument as to the circuit behavior follows. The voltage across the capacitor cannot change instantaneously; therefore any change in the drain voltage must also appear as an equal change at the gate. When the switch closes, the gate voltage either rises, falls, or remains constant at its slightly positive value. We shall first assume that it remains unchanged. The drain current corresponding to the positive gate voltage is very large, causing a large voltage drop at the drain, which, coupled by the capacitor, drives the gate beyond pinchoff. Hence a contradiction arises, and the original assumption must have been incorrect.

This contradiction may be resolved by observing that only a small change in drain voltage may be tolerated without pinching off the FET. The current flow after switching must be small, and we conclude that the FET gate is driven almost to pinchoff.

The exact drop may be calculated from the active-region model which holds immediately after switching, Fig. 6-15a. We first convert the

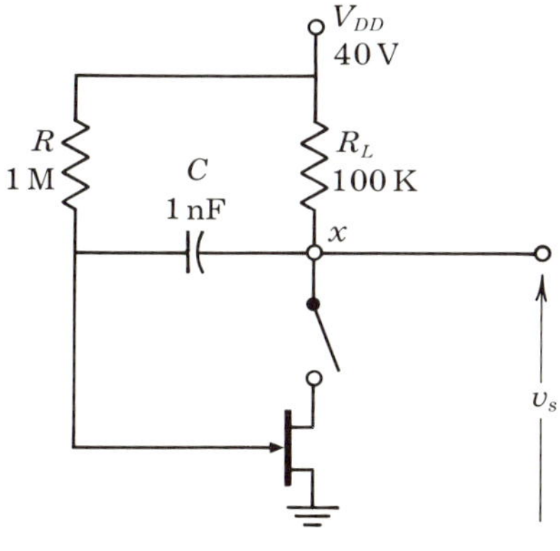

Fig. 6-14 A junction-FET voltage sweep. For our example, we shall assume the following FET parameters: $V_p = -4$, $I_{DSS} = 10$ mA, and $r_o = 100$ K at I_{DSS}.

circuit at the drain to the Thévenin equivalent, resulting in the circuit of Fig. 6-15*b*. Next, by writing the loop equation in terms of I_D, we obtain

$$\begin{aligned} V_G &= V_{\text{eq}} - I_D R_L \parallel R \\ &= 4 - I_D 91\text{ K} \end{aligned} \tag{6-40}$$

where
$$V_{\text{eq}} = V_{DD} - \frac{v_q(0^+)R}{R + R_L}$$

$v_q(0^+)$ is the charge in the capacitor at $t = 0^+$. Equation (6-40) is the equation of a load line which can be plotted on the FET transfer characteristic, as shown in Fig. 6-15*c*. The intercept at x is the initial operating point. We note that x is very close to the pinchoff voltage of -4 V. Approximating V_G by V_p in Eq. (6-40), we find $I_D = 95\ \mu$A.

We could also solve for V_G analytically by the simultaneous solution of Eq. (6-40) and the square-law I_D-vs.-V_{GS} transfer characteristic for the FET. [A more exact answer would be $V_G(0^+) = -3.64$ V, which is very close to the approximate value $V_p = -4$ V.]

A MOSFET sweep would operate in exactly the same manner, and we would follow the same procedure in solving for the initial conditions. With normal circuit-resistor values, the initial gate voltage turns out to be close to but slightly beyond the threshold. To obtain the same self-limiting inherent in the gate-substrate diode of the junction FET, a diode connected to conduct outside the active region can be shunted from the gate to the drain.

At low drain currents, r_o becomes large enough to be neglected and g_m becomes small compared with the values at I_{DSS}. In this case, $g_m = 465\ \mu$mhos at 85 μA. The gain becomes

$$A = -g_m R_L = -46.5$$

The gate voltage now begins rising from the initial value of -4 V toward 40 V, with the time constant

$$\tau_s = C(1 - A)\left(R + \frac{R_L}{1 - A}\right) \approx RC(1 - A) = 47.5\text{ ms} \tag{6-41}$$

Using the linear approximation [Eq. (6-6)], the gate voltage may be expressed as

$$v_G \approx V_p + \frac{V_{DD} - V_p}{\tau_s} t = -4 + \frac{44}{\tau_s} t \tag{6-42}$$

The 4.7-V drop at the gate during switching is coupled by C to the drain. Here the rundown is A times as fast as the gate rise:

$$v_D = V_{DD} - \Delta v_G(0) + \frac{A(V_{DD} - V_p)}{(1 - A)RC} t \tag{6-43}$$

and since $|A| \gg 1$,

$$v_D \approx 35.3 - \frac{44}{RC} t$$

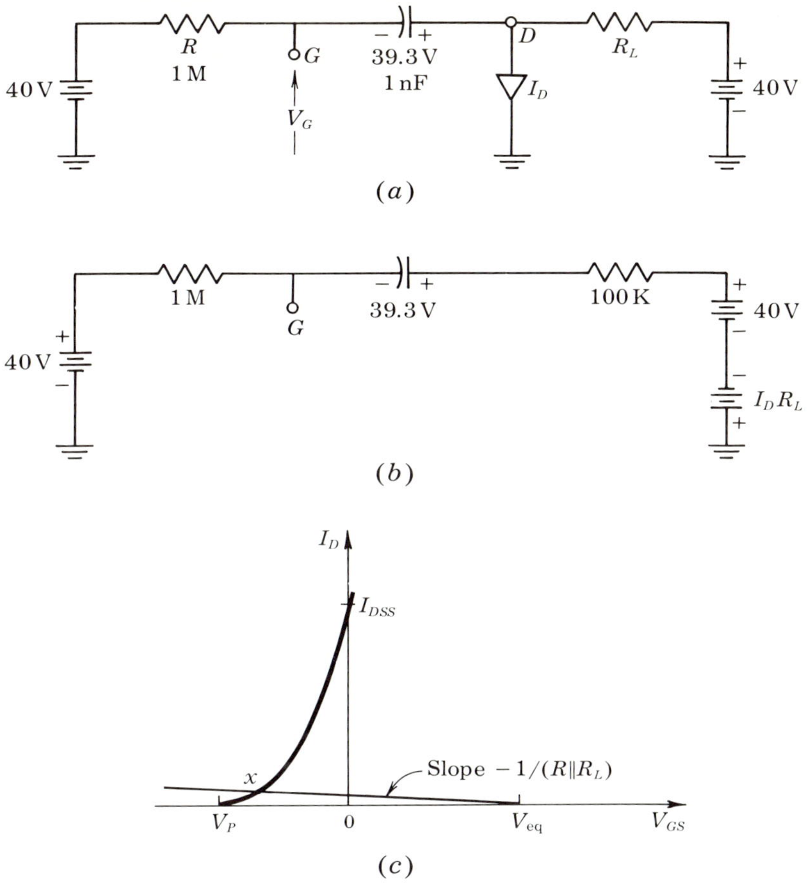

Fig. 6-15 *(a)* Active-region model of the circuit of Fig. 6-14; *(b)* equivalent model holding just after switching ($t = 0^+$); *(c)* effective bias load line from *(b)* plotted on the FET transfer characteristic to obtain the initial operating point.

The sweep will terminate either when the gate voltage becomes positive and the gate-to-substrate diode conducts or when the drain rundown causes the drain-source channel to become ohmic at some low drain-to-source voltage. Since the gain is -46.5, the gate rises by only 0.75 V while the drain runs down almost 35 V. Thus the drain bottoms first. The bottoming voltage can be found from the bounds of the ohmic region:

$$V_\Omega = V_{DS} = V_{GS} - V_p$$

Assuming the final value of $V_{GS} = -3$ V, the boundary of the ohmic region will be at $V_\Omega = 1$ V. Substituting into Eq. (6-43), we find that the linear rundown takes

$$T_s = RC\,\frac{34.3}{44} = 785\ \mu\text{s}$$

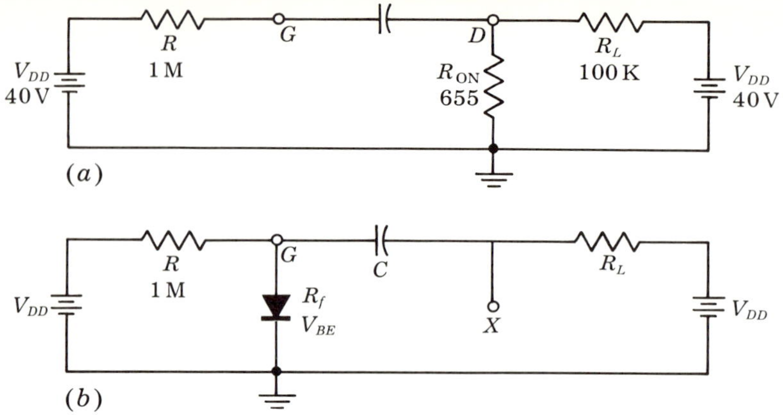

Fig. 6-16 (*a*) Model holding for bottoming in the Miller sweep of Fig. 6-14; (*b*) recovery-region model.

After the drain enters the ohmic region, we can approximate the FET by the bottoming resistance $R_{ON} = 1/g_m$. The model used is the one shown in Fig. 6-16*a*. We see that the gate continues rising toward 40 V and that the drain voltage continues the rundown, although with a different time constant and toward a new steady-state voltage. Because both the gate and the drain voltages are changing, R_{ON} decreases from its initial value (which will be determined by V_{GS} at the bottoming point of the linear gate runup) to the much lower value when $V_{GS} = 0$.

The inital operating point is close to the quiescent point x in Fig. 6-15. At x, $g_{mx} = 465$ μmhos and $R_{ONx} = 1/g_{mx} = 2{,}150\ \Omega$. At $V_{GS} = 0$, $g_m = 5{,}000$ μmhos and $R_{ON} = 200\ \Omega$.

We shall select the root-mean-square value as a suitable compromise constant resistance to approximate the actual response in the bottoming region. Thus, with $R_{ON} = 655\ \Omega$ in the model of Fig. 6-16*a*, we find that the drain heads toward 260 mV with the time constant

$$\tau_2 = C(R + R_{ON}) \approx CR = 1 \text{ ms}$$

After the gate reaches 0.7 V, the gate diode conducts, terminating the gate rise. The drain continues falling toward 260 mV with the much faster time constant

$$\tau_3 = CR_{ON} = 650 \text{ ns}$$

Figure 6-17 shows the total output waveshape.

The recovery portion of the waveshape can be calculated from the model holding after the switch is opened. From Fig. 6-16*b* we see that point x will rise toward V_{DD} with the time constant

$$\tau_R = R_L C = 100\ \mu\text{s} \tag{6-44}$$

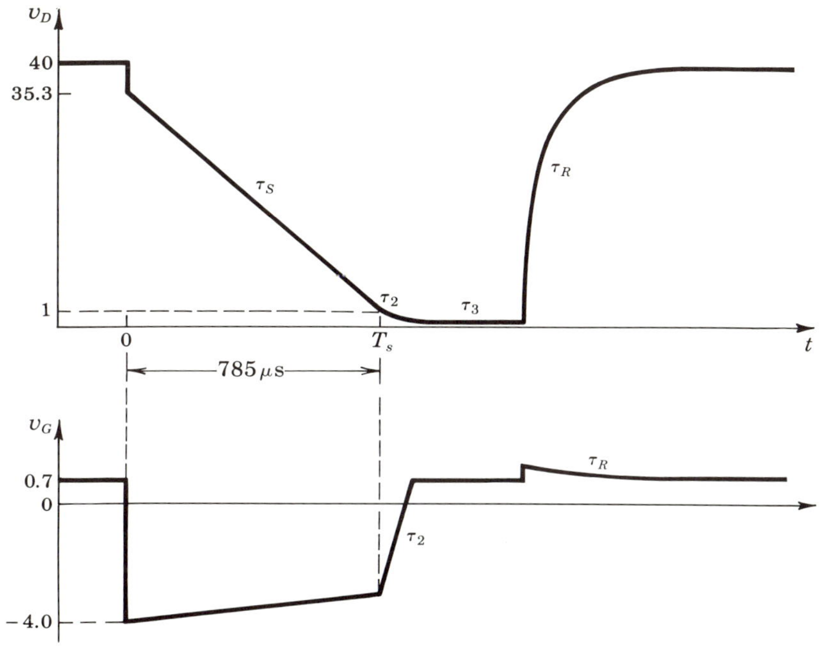

Fig. 6-17 Waveshapes present in the junction-FET Miller sweep of Fig. 6-14.

In the circuit of Fig. 6-18, we have added voltage control over the rundown time previously discussed with respect to the circuit of Fig. 6-6. Rundown will proceed as before but from a drain starting point of $V_A + 700$ mV instead of from V_{DD}. The lower we make V_A, the lower the starting point and the shorter the sweep. The size of the initial jump and the time of the rundown can be found from the previously derived equations. By substituting the new drain voltage for V_{DD} in Eqs. (6-39) and (6-40), we find the new initial gate drop. Runup and rundown proceed from the new starting point to the same final voltage.

A second FET, Q_2, is also incorporated in this circuit as a switch. It is a voltage-controlled series switch which is normally kept open by the large negative gate voltage (below pinchoff). A positive pulse switches Q_2 into the low-resistance ohmic region connecting the amplifier FET, Q_1, to point x, and the sweep begins.

Recovery-time improvement

The long recovery time is primarily a function of the large drain load of the Miller-sweep FET. If, in the capacitor charge and discharge paths, we could somehow replace this large source impedance by a much smaller

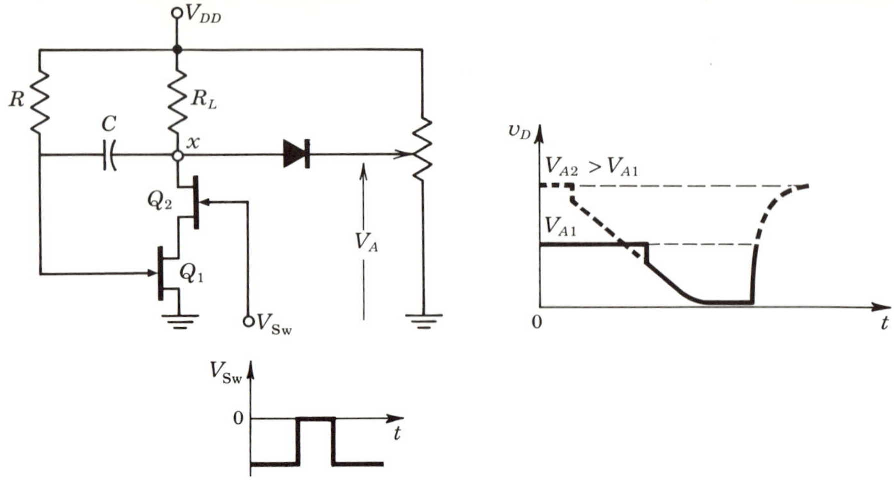

Fig. 6-18 A FET Miller sweep incorporating voltage control over sweep time. A second FET is used for series gating.

value without otherwise affecting the circuit, then the recovery time would be reduced proportionately. One convenient method is to isolate the capacitor from the drain circuit by means of an intervening source follower (Fig. 6-19).

A source follower's gain, especially when using a high-g_m FET and a large value of R_S, is almost unity. Since $A_{SF}v_i \approx v_i$, the source follower couples everything happening at the drain of Q_1 to the capacitor through its low

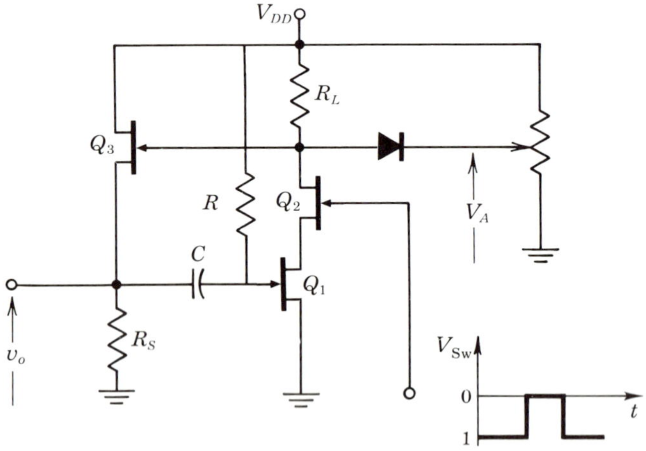

Fig. 6-19 Miller sweep incorporating a source follower for fast recovery.

output impedance. The overall circuit is identical with that previously discussed with respect to Figs. 6-14 and 6-18 except that the effective source resistance must be changed from R_L to

$$R_{S3} = \frac{r_{o3}}{\mu_3 + 1} \parallel R_S \approx \frac{1}{g_{m3}} = R_{\text{ON}3} \qquad (6\text{-}45)$$

R_{ON} is not constant; it varies inversely with the square root of the drain current [Eq. (5-21)]. For a piecewise-linear approximation to the actual response, it is often convenient to assume a constant average value for R_{ON}. A reasonable choice is three to six times the expected minimum value. In a typical FET, the minimum R_{ON} would be between 50 and 2,000 Ω. Conditions during the initial voltage jump and the linear rundown remain as before, except for the introduction of R_{S3} in place of R_L in all pertinent equations.

The isolation of C from the drain of the Miller-sweep FET allows extremely rapid recovery of the drain, limited only by any stray circuit capacity from the drain of the switch Q_2 to ground. This jump brings the source follower's gate voltage back to V_{DD}.

The recharging of C through the source follower follows the path AB described in Fig. 5-21*b*. With respect to that path, we found the form of the rise to be [Eq. (5-51)]

$$v(t) = V_{ss} \frac{t/\tau_c}{1 + t/\tau_c}$$

Assuming for the source follower that $R_{\text{ON,min}} = 300$ Ω, we have $\tau_c = 2R_{\text{ON,min}}C = 600$ ns. The rise time during recovery becomes

$$t_{R_{\text{SF}}} = 8.9\tau_c = 5.3 \ \mu\text{s}$$

This should be compared with the rise time of the exponential response when charging through the large R_L. From Eq. (6-44) we had a rise time of

$$t_R = 2.2\tau_R = 220 \ \mu\text{s}$$

The source follower has reduced the recovery time by a factor of 40.

The large initial charging current from the source follower will cause a small voltage jump at the gate upon switching into the recovery region. This jump can be found from the model of Fig. 6-20:

$$V_G(0^+) = 0.7 + I_{D3}(0^+)R_{G1} \approx 1 \text{ V}$$

where R_{G1} is the piecewise resistance of the conducting gate-to-substrate diode. The approximation of a 1-V peak output is not unreasonable since, for an increasing $I_D(0^+)$, the effective diode resistance decreases.

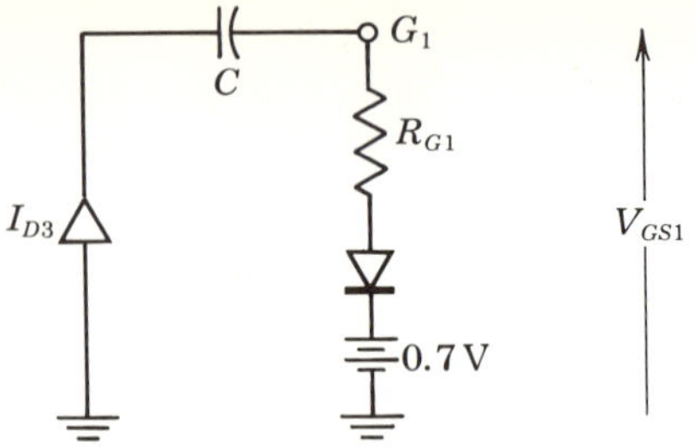

Fig. 6-20 Piecewise model used to calculate jumps in voltage during recovery.

The source follower also serves the important secondary function of making the sweep voltage available at a low impedance point for coupling to the external circuit.

6-7 Bootstrap Sweep

Bootstrapping, a form of positive feedback, linearizes the sweep by first determining what voltage-correction term is needed and then deriving it through acting on an already existing voltage. The necessary circuit modification usually follows directly from the analysis of the original problem posed. For example, the nonlinear charging in the RC circuit of Fig. 6-21 is a consequence of the dependency of the charging current upon the capacitor voltage:

$$i = \frac{V' - v_s}{R} \tag{6-46}$$

If, somehow, we could cancel the effects of v_s, through the adjustment or replacement of one of the circuit components under our control, then the desired linear charging would be ensured. One obvious course is to replace V' by $V + v_s$. This required drive can be developed quite easily by amplifying the capacitor voltage and connecting the charging resistor to the amplifier output instead of to the battery. Because of the special

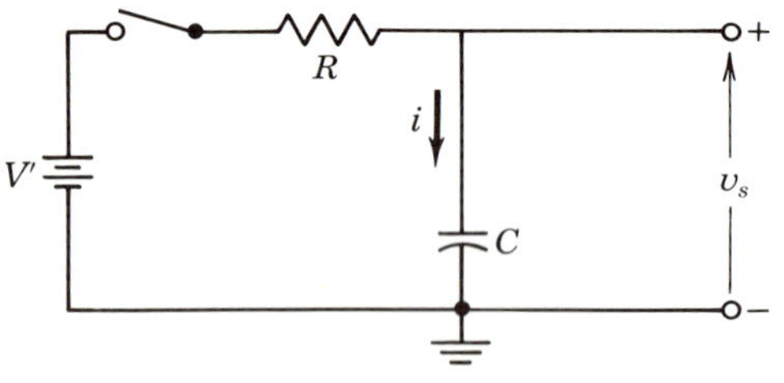

Fig. 6-21 Simple RC sweep.

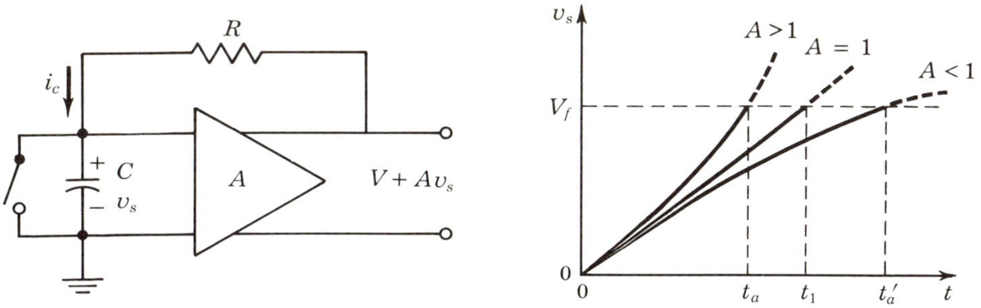

Fig. 6-22 General bootstrap sweep and possible output waveshapes.

charging requirements, the output must have a dc level of V. Thus it follows that Fig. 6-22 represents the general form of the bootstrap sweep. In this case, positive-voltage feedback is employed to increase the effective charging source voltage at the same rate as the increase in the sweep voltage v_s.

Constant-current charging of C is contingent on the amplifier gain A remaining unity. Any variation is immediately reflected as a change in the output waveshape and sweep period (Fig. 6-22). This dependency on A may, if serious, prevent the use of the bootstrap circuit for all but noncritical applications.

The sensitivity of sweep duration with respect to the voltage amplification may be interpreted quantitatively by starting from the equivalent circuit controlling the charging current flow. Referring to Fig. 6-22 and writing the input-node equation yield

$$i_c = \frac{V + Av_s - v_s}{R} = \frac{V}{R} - \frac{(1 - A)v_s}{R} \tag{6-47}$$

The right-hand side of Eq. (6-47) represents the circuit as seen by the capacitor: it consists of a constant-current generator in parallel with its own internal admittance (Fig. 6-23).

We might observe, from Fig. 6-23, that when $A \neq 1$, the charging of C is exponential rather than linear. The final steady-state voltage, determined by setting $i_c = 0$, is $V/(1 - A)$, and by inspection of the circuit, we see that the charging-time constant is $CR/(1 - A)$. Therefore the sweep voltage may be expressed as

$$v_s(t) = \frac{V}{1 - A}\left(1 - e^{-[(1-A)/RC]t}\right) \tag{6-48}$$

The output will rise A times as fast as the input. However, when $A = 1$, G_o becomes zero and C is charged directly from the ideal constant-current source.

$$v_s = \frac{V}{RC}t \tag{6-49}$$

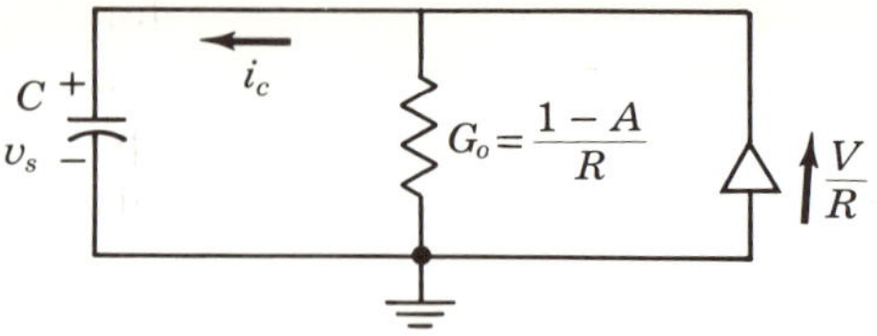

Fig. 6-23 Bootstrap equivalent circuit.

Suppose that a change in either the active device or the circuit parameters makes A slightly greater than unity. Under these circumstances, G_o becomes negative and the terminal voltage increases exponentially toward infinity. The equation defining this region of operation is the same as Eq. (6-48). But $A > 1$ means that the time constant is now negative and that the system's single pole has moved into the right half plane. Since a physical device cannot handle infinite-amplitude signals, any slight disturbance immediately drives the amplifier toward saturation.

In order to see how the sweep time depends on A, we first calculate the sweep duration from Eq. (6-48):

$$t'_a = \frac{RC}{1 - A} \ln \frac{V}{V - (1 - A)V_f}$$

No generality is lost by choosing a convenient termination point, one which will keep A in the forefront and at the same time eliminate superfluous voltage terms. The greatest simplification occurs by setting $V_f = V$:

$$t_a = \frac{RC}{1 - A} \ln \frac{1}{A} \tag{6-50}$$

As a basis for comparison, the duration of the linear sweep of the same amplitude is evaluated from Eq. (6-49):

$$t_l = RC \tag{6-51}$$

and from Eqs. (6-50) and (6-51),

$$\frac{t_a}{t_l} = \frac{1}{1 - A} \ln \frac{1}{A} \tag{6-52}$$

For $A = 0.95$, 5 percent below the optimum value of unity,

$$\frac{t_a}{t_l} = 1.082$$

or the sweep time is now 8 percent longer than the perfectly linear sweep. When $A = 1.05$, the sweep duration is reduced to $0.976t_l$. The separation from the linear sweep with respect to changes in A becomes even more pronounced as the sweep amplitude increases. From the above discussion, the only possible conclusion drawn is that the gain of the bootstrap amplifier must be highly stabilized if we are to avoid sweep-duration instability. But even with the high degree of sensitivity of the sweep period to changes in A, the sweep linearity and maximum sweep amplitude are still superior to what it is possible to achieve with a simple RC network.

A bipolar-transistor bootstrap sweep

The bootstrap sweep of Fig. 6-24a uses an emitter follower as the voltage amplifier. In the active region, the gain from base to emitter is

$$A_{EF} = \frac{(\beta + 1)R_E}{(\beta + 1)R_E + h_{ie}}$$

With a high-β transistor, we can reasonably expect a gain of 0.95 to 0.99. The series battery in the feedback branch supplies the necessary charging current for the sweep capacitor.

To avoid delays in starting the sweep, we want the transistor to turn on immediately upon opening the switch. With the switch closed, we use the drop across a conducting diode to store the necessary turn-on voltage V_D in the sweep capacitor. The small resistor R_1 controls the current division between the transistor and diode, ensuring that the transistor is barely on. When the switch is opened, the battery starts charging the capacitor, and because the transistor is active, we have the bootstrap equivalent circuit shown in Fig. 6-24b.

The capacitor charges from V_D toward the large steady-state voltage

$$V_{ss} = \frac{V}{1 - A}$$

with the time constant

$$\tau_s = \frac{CR}{1 - A}$$

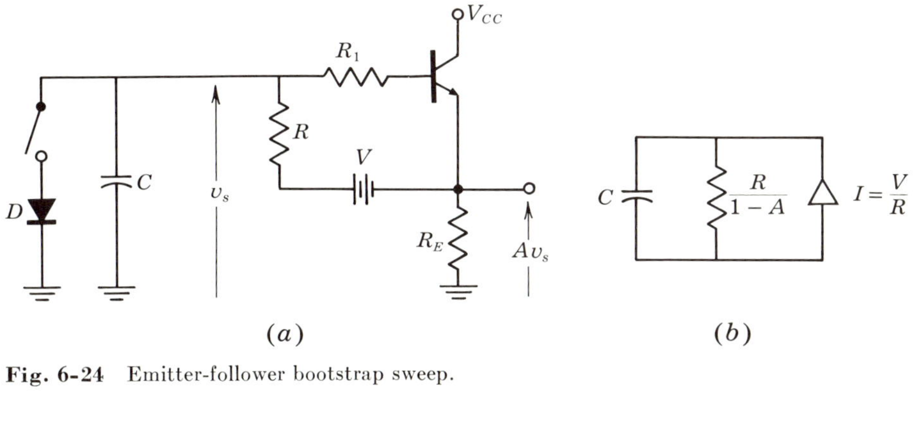

Fig. 6-24 Emitter-follower bootstrap sweep.

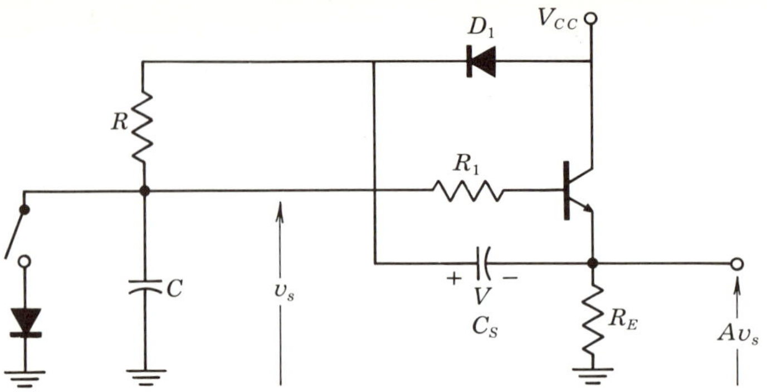

Fig. 6-25 A bootstrap sweep containing a self-charging supply (C_S).

Again using the linear approximation, we have the sweep defined by

$$v_s(t) = V_D + \frac{V_{ss} - V_D}{\tau_s} t$$

The emitter rises from zero as

$$v_e(t) = A \frac{V_{ss} - V_D}{\tau_s} t$$

Because the sweep ends with the transistor saturating when

$$v_e(T_s) = V_{CC} - V_{CE,\text{sat}}$$

we obtain a runup time of

$$T_s = \frac{CR}{1 - A} \frac{V_{CC} - V_{CE,\text{sat}}}{[A/(1 - A)]V - V_D} \tag{6-53}$$

Equation (6-53) points up the sweep-time dependency on the amplifier gain.

Capacitive power supply

Any circuit such as the one of Fig. 6-24 requiring a separate, isolated power supply is quite unsatisfactory for general use. However, in this particular sweep, it is possible to replace the battery by a charged capacitor (C_S in Fig. 6-25), and provided that we allow only a slight discharge of C_S over any sweep interval, the basic sweep operation will be unaffected. Of course, some provision must be incorporated for automatic recharging. Examination of Fig. 6-25 will show how this may be accomplished.

Under the normal operating condition of this circuit, the switch across C is closed. This permits the energy-storage capacitor C_S to charge through R_E and the conducting diode to $V_{CC} - V_D$. Upon opening the switch, the sweep begins. As the base starts rising, the emitter voltage follows it. Since C_S remains almost fully charged over the complete sweep cycle, the voltage at the bottom of the diode D_1 becomes

$$v_x = V_{CC} - V_D + Av_s(t)$$

where the gain A is very close to unity.

Hence the slightest increase of the sweep voltage automatically back-biases the charging diode D_1, and it may be removed from our consideration during the linear-sweep interval. At the end of the sweep, the switch is closed, C discharges, and the diode again conducts, finally allowing C_S to recover toward V_{CC}.

If we neglect the small base current which is also supplied from the charge stored in C_S, then during the sweep, charge is simply transferred from C_S to C. Since the total charge in the circuit must remain constant, it follows that

$$\Delta Q = C(V_{CC} - V_D) = C_S \, \Delta V \tag{6-54}$$

where ΔV is the drop in voltage across C_S over the complete sweep interval. For best linearity, ΔV should be as small as possible, leading us to conclude that C_S must be very large compared with C. An adverse effect introduced by the large storage capacity is the long time constant during circuit recovery, $R_E C_S$.

Double bootstrapping

The less-than-unity gain of an emitter follower or source follower results in a less than perfectly linear sweep. By applying a second correction term, we can more nearly approach the ideal. Figure 6-26a shows a FET

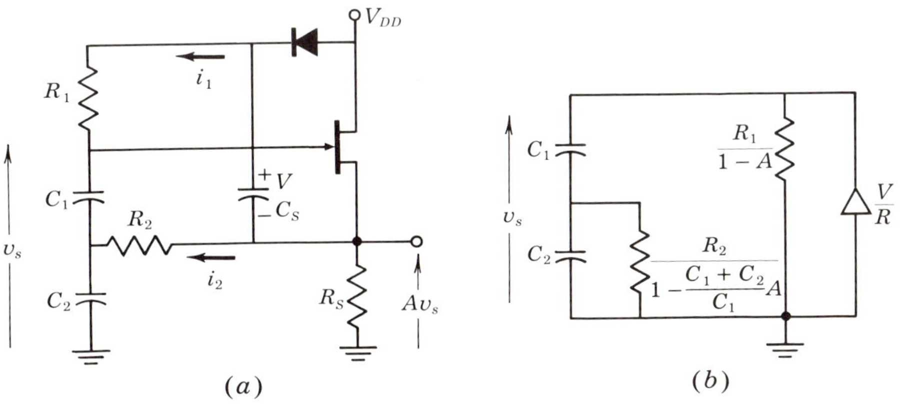

Fig. 6-26 (a) Junction-FET bootstrap with additional feedback path for sweep linearization; (b) equivalent active-region model.

bootstrap where we split the sweep capacitor into two parts and incorporate two separate bootstrap paths. The main bootstrap current i_1 charges both capacitors, and the extra current i_2 adds additional charge to capacitor C_2:

$$i_1 = \frac{V - v_s(1 - A)}{R_1} \tag{6-55}$$

$$i_2 = \frac{1}{R_2}\left(Av_s - \frac{C_1}{C_1 + C_2} v_s\right) \tag{6-56}$$

where V is the voltage across the storage capacitor C_S.

Since each current equation contains a term proportional to the particular node voltage, we can represent these terms by equivalent resistances. The total resultant model is shown in Fig. 6-26*b*.

The current term i_2 is equivalent to placing across C_2 a resistance

$$R_{eq} = \frac{R_2}{1 - [(C_1 + C_2)/C_1]A} \tag{6-57}$$

If the very simple condition $A > C_1/(C_1 + C_2)$ is satisfied, R_{eq} becomes a negative resistance. The negative resistance implies a positive pole and a rising exponential associated with C_2. This positive exponential term applies the correction factor to compensate for the sweep nonlinearity due to the nonideal charging of the timing capacitor.

The total response can be approximated by noting that the positive pole is used only to add a small correction term to the regular bootstrap-circuit response. If, as a first-order approximation, we assume that the two poles are independent, then

$$v_s \approx \frac{V}{1 - A}\left[(1 - e^{-t/\tau_1}) + \frac{C_1}{C_1 + C_2}(e^{-t/\tau_2} - 1)\right] \tag{6-58}$$

where $$\tau_1 = \frac{R_1}{1 - A}\frac{C_1C_2}{C_1 + C_2} \qquad \text{and} \qquad \tau_2 = R_{eq}C_2 \tag{6-59}$$

τ_1 is the time constant of the regular bootstrap sweep, and τ_2 is the extra pole introduced because of the effective negative resistance shunting C_2.

By choosing R_2 and the ratio C_1/C_2, we can add a correction term which will compensate for the exponential falloff of the normal sweep. The final adjustment is usually made experimentally.

6-8 Switching

Most linear sweeps are turned on and off by signals derived from input information. The input signals are amplified and clipped square and may then be differentiated and/or applied directly to gate the sweep. Many

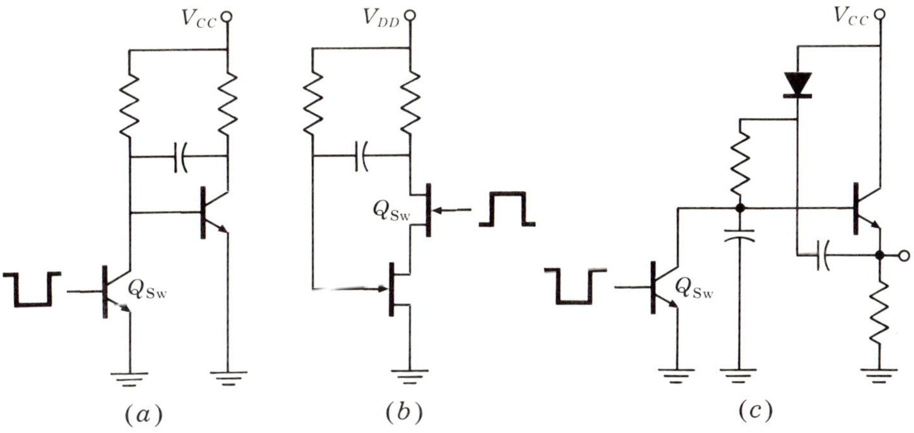

Fig. 6-27 Three possible switching networks. (*a*) Shunt switching by a saturated transistor; (*b*) series switching by an ohmic FET; (*c*) constant-current discharge by an active transistor.

of the logic gates of Chap. 4 are suitable to turn a sweep on or off. Some more obvious switches shown in Fig. 6-27 are:

1. A bipolar transistor driven into saturation or a FET driven into the ohmic region used to connect one of the circuit nodes to ground. Or it may be cut off and may thus disconnect the mode from ground.
2. Series gating, using an active device or a series gate, as a switch to connect the amplifier into the feedback path. This is analogous to the analog chopping described in Sec. 5-7.
3. Sweep capacitor discharge through a current source. This is equivalent to the current-controlled sweeps of Sec. 6-4.
4. The voltage-controlled change of state of certain negative-resistance devices. This is exactly what was used in the SCR periodic sweep of Sec. 6-2. (See below, Chap. 11.)
5. The voltage comparators of Chap. 10. Since the sweep capacitor is to be discharged at a fixed voltage point, these are suitable for the automatic periodic trigger generation necessary when we wish to generate a periodic sweep.

6-9 Synchronization

The period of a sweep depends on many factors: on the circuit time constant, on the voltage at which the capacitor discharge commences and at which it terminates, and finally, on the steady-state supply voltage. These parameters are never completely constant but vary slightly because of ambient-temperature change, line-voltage variation, aging of components, replacement of the active device, noise, etc. As a direct consequence of the random circuit changes, sequential sweep cycles will not have identical duration but will jitter about some average time. When this sweep is used for oscilloscope display, a corresponding jitter appears in the pattern seen on the screen. If the sweep is used for timing, there will be uncertainty as to its exact time duration. However, by injecting an external control signal, the sweep time can be locked to some multiple of the synchronizing signal period, and when this input is derived from the display signal, the oscilloscope pattern will remain stationary.

Synchronization of the sweep is accomplished by forcing the free-running sweep period t_1 to some integral multiple of the signal period t_s, that is, by changing t_1 to nt_s. In effecting the change of period, the sweep amplitude may also be affected, but as the amplitude can always be adjusted elsewhere in the system, this is relatively unimportant.

The sweep period is controlled by varying either the initial voltage or the discharge point. For example, in the SCR sweep of Fig. 6-2, the sync signal is injected at the gate, with its fluctuation about the quiescent gate-current point changing the instantaneous firing voltage. Figure 6-28 indicates the firing voltage in the presence of a sinusoidal sync signal. We observe that because of the assumed linearity of the gate-anode transfer

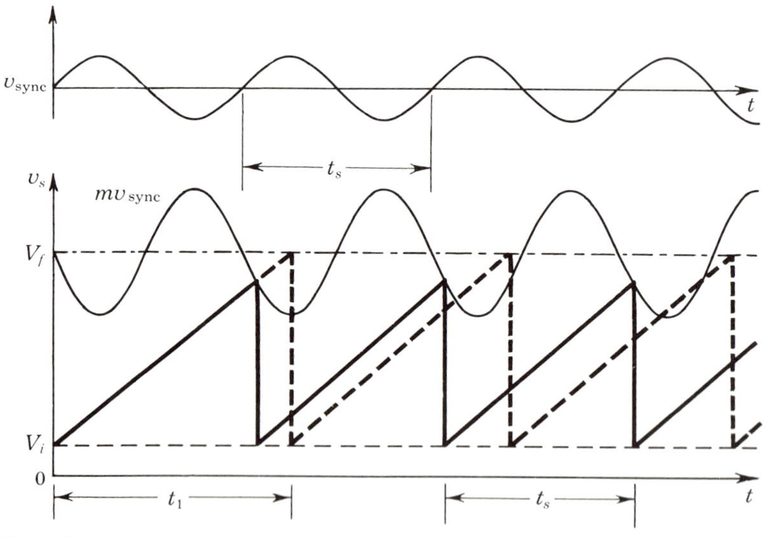

Fig. 6-28 Applied synchronizing signal and its effect on the free-running sweep duration.

characteristics, the gate-current waveshape is simply multiplied by the gate-current-anode firing-voltage transfer factor, reflected into the anode, and plotted about the quiescent firing line. The sweep ends earlier because it intersects the firing curve at a lower value. As a consequence, it fires at the same point of each cycle, whereas the free-running sweep (dashed line of Fig. 6-28) terminates randomly with respect to the applied sinusoidal signal.

Synchronization information can also be used to modify the starting point of the active-device sweeps. For example, the synchronization signal may be injected at the collector (or the drain) through a diode, and together with dc level, it will set the initial value of the rundown. This is exactly what was done in the voltage-controlled sweeps of Figs. 6-6 and 6-18. The rundown or runup time is dependent on the total voltage swing available; it does not really matter whether the amplitude restriction is placed on the beginning or on the end of the sweep.

If we use an ideal free-running sweep instead of an actual SCR or voltage-controlled sweep, the quantitative treatment below will be greatly simplified. In addition, the lack of identification with any particular circuit increases the generality of our discussion. This perfectly linear sweep starts rising from zero at a rate of k V/s, ending at the firing voltage at time t_1.

The introduction of a sync signal of proper amplitude and period pulls the sweep into synchronization, as shown in Fig. 6-29. This buildup may take one or several cycles, but eventually the sweep will start and end at the same point of the sync cycle.

One obvious question facing us is: Why does synchronization increase the stability of the sweep with respect to random circuit changes? Our argument might begin with the observation that these changes result in sweep-time instability, and since the effect produced is the same, any

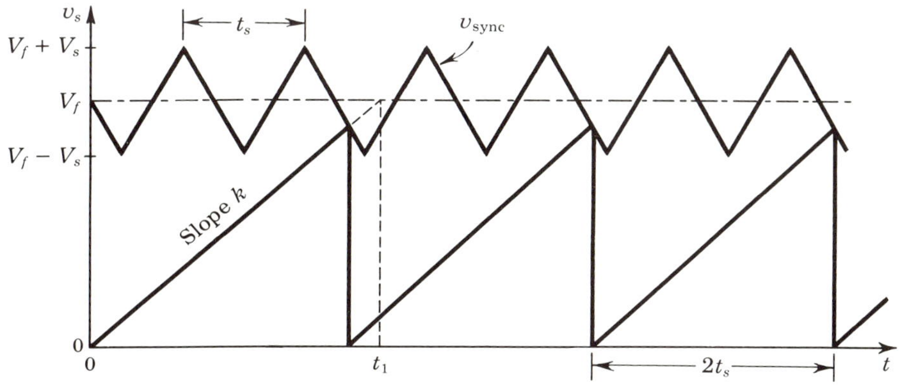

Fig. 6-29 Ideal sweep synchronized by a triangular wave.

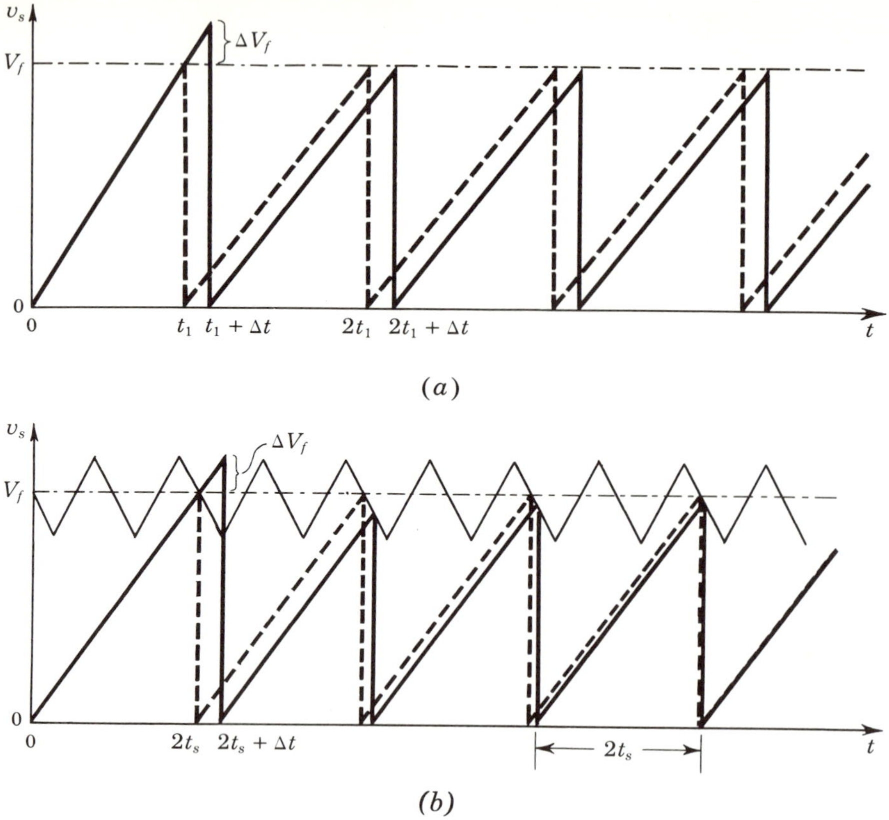

Fig. 6-30 (*a*) Sweep perturbation, no sync signal; (*b*) sweep perturbation, sync signal present.

circuit variation may be represented as an equivalent small perturbation of firing voltage. In Fig. 6-30*a*, the slight increase of V_f by ΔV_f increases a single free-running sweep period by Δt. The subsequent cycles are back to the normal period, but each is delayed by Δt. However, once the sweep is synchronized by terminating it at a point on the sync signal that has a slope of opposite sign to the sweep slope (in this case, on a point of negative slope), then the time perturbation rapidly damps out. Figure 6-30*b* shows how the sweep returns to its original firing point within several cycles of the original disturbance.

The reason for the rapid damping can easily be seen when we expand two adjacent cycles of the sweep for a closer look at the end point (Fig. 6-31). At the first firing point, the voltage jitter ΔV_f produces an increase in the sweep period of Δt. Thus the second cycle starts Δt later, but because of the geometry at the intersection of the two straight lines, this

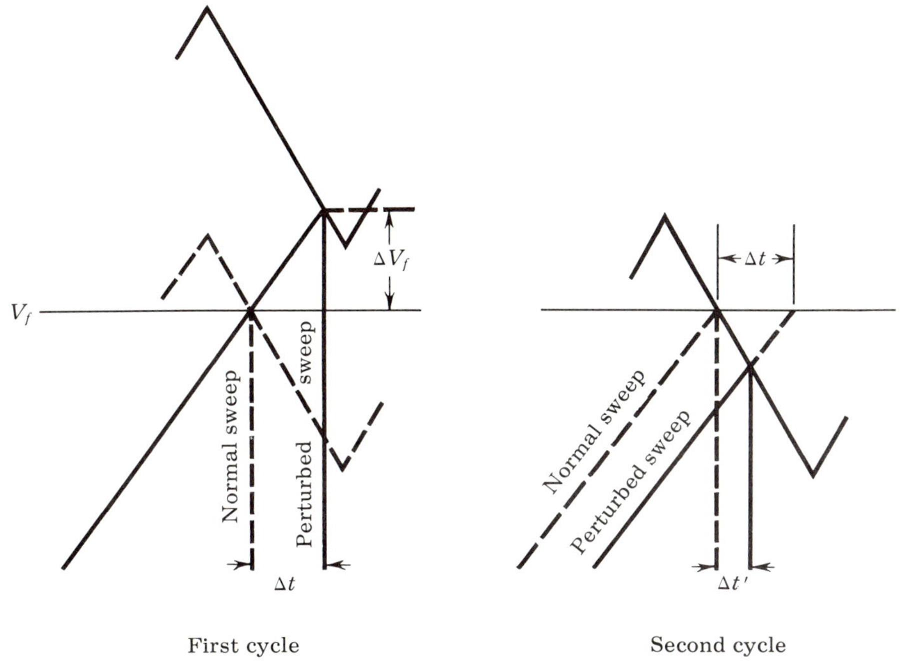

Fig. 6-31 Expanded-sweep firing point, showing stable synchronization.

cycle ends only $\Delta t'$ later. The time variation is reduced by a factor δ, at the instant of firing, where

$$\delta = \frac{\Delta t'}{\Delta t} \tag{6-60}$$

When the two intersecting curves have slopes of opposite sign, as in Fig. 6-31, δ will always be less than unity. The third sweep starts $\Delta t'$ later, and at the instant of firing, this is also reduced by the factor δ. After n sweeps, the total time displacement from the normal time will be $\delta^n \Delta t$ and

$$\lim_{n \to \text{large}} \delta^n \Delta t = 0$$

Thus the sweep eventually returns to its original firing point with respect to the sync signal. We see from Fig. 6-31 that the steeper the slope of the synchronizing signal at the point of sweep intersection, the smaller the value of δ and the fewer the number of cycles required for sweep-time recovery. We further conclude that the optimum point to effect synchronization (with any input signal) is where the slope is the steepest, such as the crossover point of a sine wave. If a choice presents itself, the optimum sync signal would be a square wave or pulse train where the sweep can terminate on a point of infinite slope.

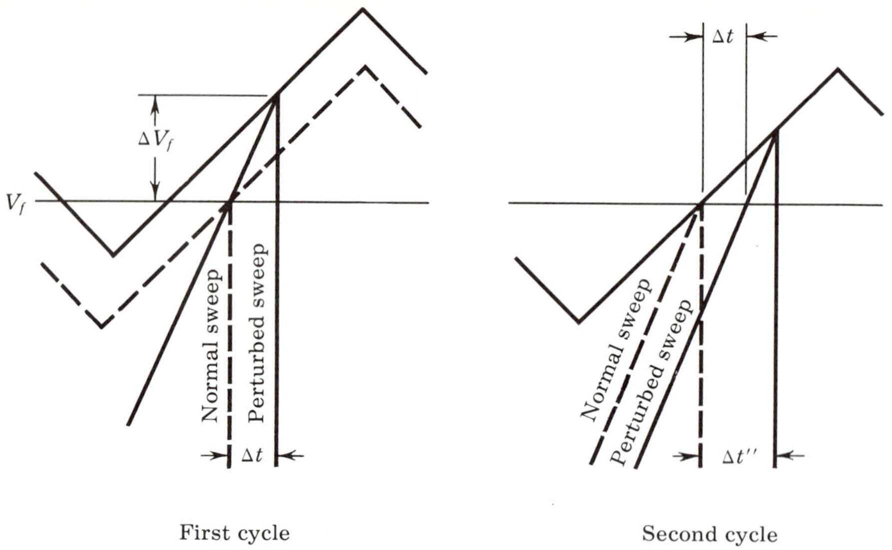

Fig. 6-32 Unstable sweep synchronization—expanded firing region.

Suppose that the sweep intersects the sync signal at a point that has the same sign of slope; i.e., both are positive (Fig. 6-32). Under these circumstances any slight time perturbation Δt increases at the point of firing by the factor δ' to $\Delta t''$:

$$\delta' = \frac{\Delta t''}{\Delta t} > 1 \tag{6-61}$$

Instead of damping out, the change in sweep time now builds up geometrically as $(\delta')^n$ over n cycles. Eventually, as the sweep shifts its relative position, it intersects the sync voltage at a point of negative slope. If the sync signal has the proper time and amplitude relationship, with respect to the sweep, synchronization will now be effected.

6-10 Regions of Synchronization

Synchronization is not automatically guaranteed upon the injection of just any external signal. The sweep will not be pulled in unless the sync amplitude and period happen to fall within regions well defined in terms of the sweep constants—regions which, we shall see below, are also a function of the sync signal waveshape. As an initial example, refer to Fig. 6-33, where a symmetrical square wave of variable amplitude is used as the control signal. The first sweep fires when it terminates on the square wave

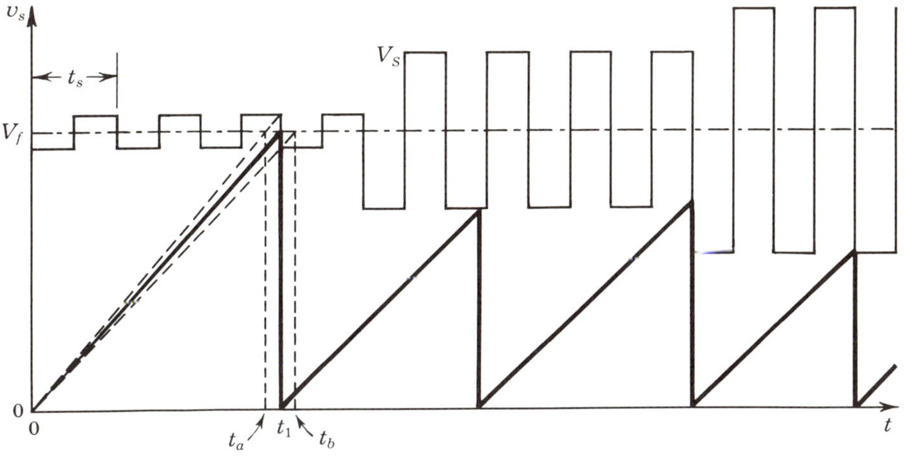

Fig. 6-33 Square-wave synchronization.

at $3t_s$, and if the input remains unchanged, the sweep will continue to fire once every three cycles. However, the increased square wave causes the second and third sweeps to end at nonintegral multiples of the sync signal period, and therefore they are unsynchronized. The fourth cycle is again synchronized, this time at the next smaller integral multiple of the sync period, at $2t_s$. We might also note that any free-running sweep bounded by the dashed lines shown in the first sweep cycle, that is, $t_a < t_1 < t_b$, intersects the changed firing voltage along the same portion of the square wave and also fires at $3t_s$.

Let us now assume that the control signal is a square wave having a constant amplitude of V_s peak value. Then if its period falls anywhere within the two extremes given in Fig. 6-34*a* and *b*, the fixed free-running sweep will fire over the same integral number of sync cycles, over n cycles. At the one limit shown in Fig. 6-34*a*, the sweep period becomes foreshortened because a portion of the sync signal extends below the original firing line. Figure 6-34*b* illustrates the possible lengthening of the sweep as a consequence of the effective increase in the firing voltage.

We shall first concern ourselves with expressing the lower usable limit of the sync period (Fig. 6-34*a*). Since the two sweep triangles $0BC$ and $0DE$ are similar, the ratios of their equivalent sides must be equal:

$$\frac{0E}{0C} = \frac{DE}{BC} \qquad \frac{nt_{s,\min}}{t_1} = \frac{V_f - V_s}{V_f}$$

Hence one limit becomes

$$\frac{t_{s,\min}}{t_1} = \frac{1}{n}\left(1 - \frac{V_s}{V_f}\right) \tag{6-62}$$

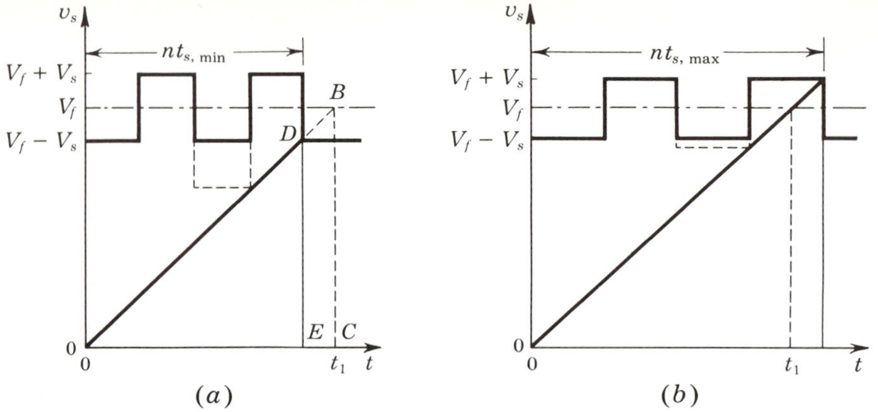

Fig. 6-34 *(a)* Minimum sync period producing synchronization; *(b)* maximum sync period producing synchronization.

The solution of Fig. 6-34*b*, also by similar triangles, yields a second limit,

$$\frac{t_{s,\max}}{t_1} = \frac{1}{n}\left(1 + \frac{V_s}{V_f}\right) \tag{6-63}$$

If the sync amplitude is now increased, the sweep will stay in synchronization until the corner of the preceding half cycle intersects the rising sweep, causing premature firing. This condition is indicated, for both cases, by the dashed lines of Fig. 6-34. Consequently, for any value of nt_s, the following condition on the maximum allowable sync amplitude may be written:

$$\frac{nt_s - \frac{1}{2}t_s}{t_1} = \frac{V_f - V_{s,\max}}{V_f}$$

or

$$\frac{V_{s,\max}}{V_f} = 1 - \frac{2n-1}{2}\frac{t_s}{t_1} \tag{6-64}$$

Equations (6-62) to (6-64) are all straight lines in terms of the normalized coordinates t_s/t_1 and V_s/V_f. Each delineates one set of the boundary values which limit possible synchronization. By plotting these equations with n as a parameter, we define the regions of synchronization for a square-wave input (Fig. 6-35). Any combination of periods and voltages lying within a region ensures proper synchronization over n cycles of the square wave; and to allow for possible fluctuation of sync and sweep voltage amplitude, and sweep period, the best combination of parameters lies at the center of any one region. Furthermore, the area of each region decreases extremely rapidly as n increases, with the consequence that it becomes very difficult to ensure synchronization over ratios greater than

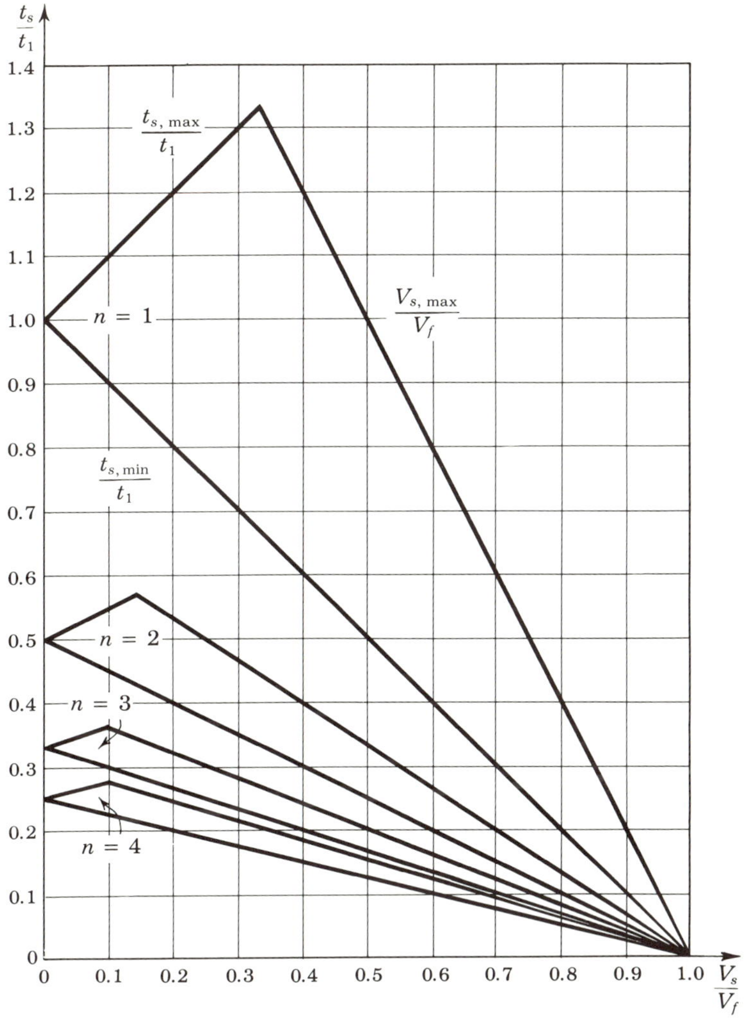

Fig. 6-35 Regions of synchronization—square-wave input signal.

4 or 5. The expected component variations would easily shift the operating point out of the small synchronization regions holding for large n.

All sync signals that effectively increase and decrease the firing voltage by equal amounts, e.g., sine waves, triangles, sawtooths, will have two limits of their synchronization regions defined by Eqs. (6-62) and (6-63). Equivalent limits for nonsymmetrical signals may be calculated quite easily from their geometry.

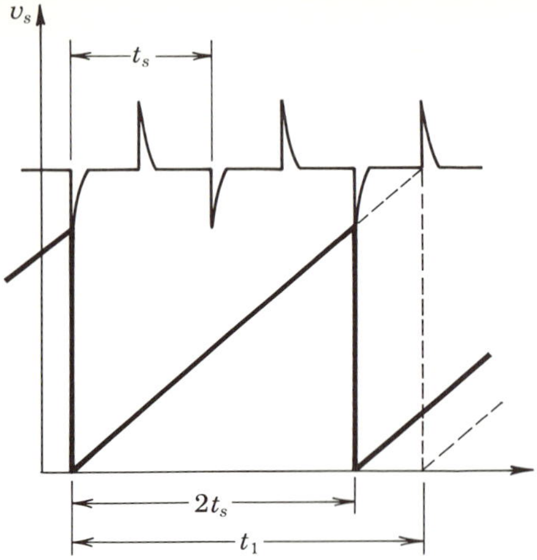

Fig. 6-36 Synchronization with pulses.

For example, if the square wave used above is differentiated and the train of equal-amplitude positive and negative pulses is applied, then only the negative pulse will influence the timing. The sweep cannot be lengthened since it will always terminate on the negative pulse or along the original firing line, as in Fig. 6-36. In this case, Eq. (6-63) reduces to

$$\frac{t_{s,\max}}{t_1} = \frac{1}{n}$$

Equation (6-62) will still hold as a second limit. A new equation must be found to delineate the third boundary (Prob. 6-22).

In many sweeps, a significant portion of the overall period must be devoted to the recovery interval. We can treat these cases by interpreting the recovery time as a fixed delay between each linear sweep segment. From Fig. 6-37 we see that the linear portion of each cycle will now start at t_r instead of at 0. The sync interval of $nt_s - t_r$ must now be compared with the free-running linear segment t_1. Equations (6-62) to (6-64) become

$$\frac{t_{s,\min}}{T_1} = \frac{1}{n}\left(1 - \frac{T_1}{t_1}\frac{V_s}{V_f}\right) \tag{6-65a}$$

$$\frac{t_{s,\max}}{T_1} = \frac{1}{n}\left(1 + \frac{t_1}{T_1}\frac{V_s}{V_f}\right) \tag{6-65b}$$

$$\frac{V_{s,\max}}{V_f} = \frac{T_1}{t_1}\left(1 - \frac{2n-1}{2}\frac{t_s}{T_1}\right) \tag{6-65c}$$

where $T_1 = t_1 + t_r$ is the overall sweep period.

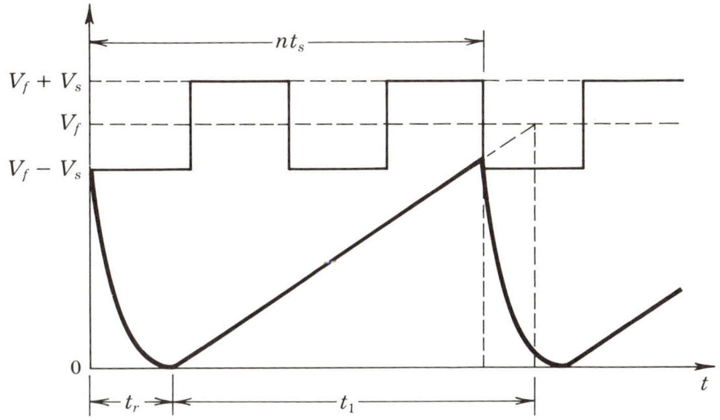

Fig. 6-37 Illustrating delay caused by the recovery interval (total sweep period $T_1 = t_r + t_1$).

Equations (6-62) and (6-63) may also be applied to nonsymmetrical waves by properly interpreting the various voltage terms they contain. We rewrite these equations:

$$\frac{t_{s,\min}}{t_1} = \frac{1}{n}\left(1 - \frac{V_{sn}}{V_f}\right) \tag{6-66a}$$

$$\frac{t_{s,\max}}{t_1} = \frac{1}{n}\left(1 + \frac{V_{sp}}{V_f}\right) \tag{6-66b}$$

where V_{sn} is the negative peak of the sync signal and V_{sp} is the effective positive peak, both measured from the original firing line. Of course, as with the pulse, the sweep must be able to terminate at these voltage values.

The difficult problem in finding the regions of synchronization is the evaluation of the third boundary,

$$\frac{V_{s,\max}}{V_f} = f\left(\frac{t_s}{t_1}\right)$$

For signals having simple geometric waveshapes, it is possible to find an explicit analytic solution, but for more complex signals we would have to turn to a tedious graphical construction.

We can appreciate the increase in complexity by treating the case of triangular-wave synchronization (Fig. 6-38). Two limits are known [Eqs. (6-62) and (6-63)], and the third will be found from Fig. 6-38, which illustrates the conditions existing at the maximum possible value of the sync signal. If we assume that the sweep is still properly synchronized but on the verge of limiting,

$$\frac{nt_s}{t_1} = \frac{V_f + KV_{s,\max}}{V_f} \tag{6-67}$$

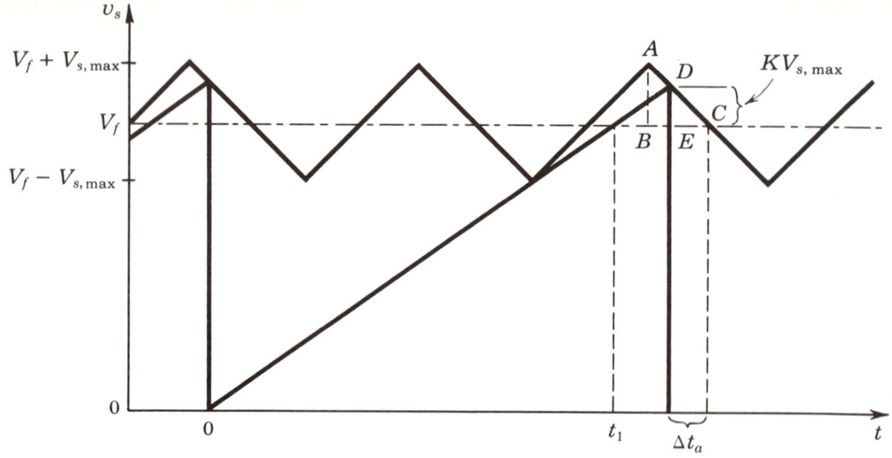

Fig. 6-38 Triangular-wave synchronization—maximum value of sync signal.

where K is a positive or negative constant relating the point of intersection to the peak value of the triangular wave. The sweep starts and ends when it intersects the sync signal Δt_a before the triangular wave crosses the quiescent firing line. From the similar triangles ABC and DEC,

$$\Delta t_a = \frac{Kt_s}{4} \tag{6-68}$$

When the sync voltage becomes infinitesimally larger than the value assumed for Eq. (6-67), the sweep no longer clears the negative peak. It now fires prematurely upon intersecting the sync signal at $3t_s/4 - \Delta t_a$ before the original firing point. Thus the following ratio now holds:

$$\frac{nt_s - (3t_s/4 - \Delta t_a)}{t_1} = \frac{V_f - V_{s,\max}}{V_f} \tag{6-69}$$

Substitution of Eq. (6-68) into Eq. (6-69) provides us with a second equation relating t_s/t_1 and $V_{s,\max}/V_f$, one which, however, also includes the third variable K. The simultaneous solution of Eqs. (6-67) and (6-69), eliminating K, completes the solution, with the final answer given in Eq. (6-70):

$$ny^2 - y + (4n - 3)xy - 4x + 4x^2 = 0 \tag{6-70}$$

where $$x = \frac{V_{s,\max}}{V_f} \qquad y = \frac{t_s}{t_1}$$

Equation (6-70), the equation of a conic section, defines the third boundary of the synchronization regions plotted in Fig. 6-39. The larger areas indicate that synchronization will be maintained in face of wider-latitude parameter variations than can be tolerated for the square wave (compare

Figs. 6-35 and 6-39). However, a square wave or pulse is vastly preferable because of the fast recovery from sweep perturbations.

The two waveshapes above were presented primarily in an effort to establish graphically the interlocking conditions which must be evaluated and satisfied for precise and accurate output control. We made no pretense of covering the multitudinous possible sync signals, such as pulses and sine waves. When order-of-magnitude or approximate boundaries will serve our purposes, the replacement, for calculation, of complex

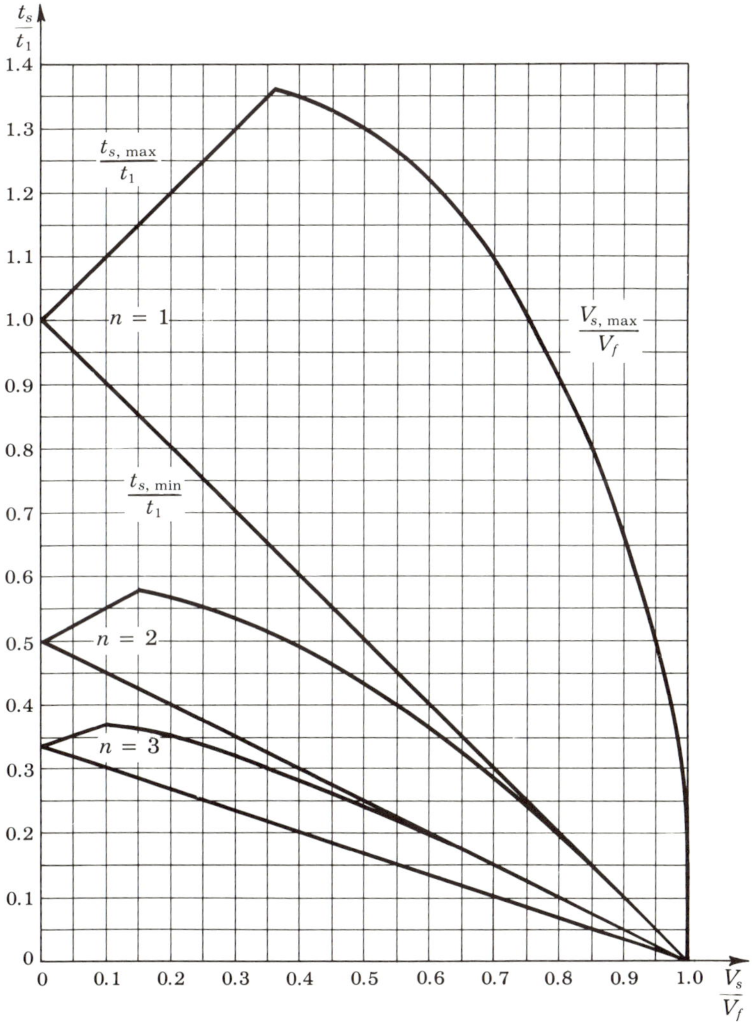

Fig. 6-39 Synchronization regions—triangular-wave sync signal.

waveshapes by simpler ones suffices. For example, a sinusoidal signal might be approximated by a truncated triangular wave, leading to regions slightly smaller than those of the triangle and slightly larger than those of the square wave. If we are interested in only one parameter's variation, then evaluation from a sketch gives the allowable limits directly.

Synchronization is employed in many circuits other than the linear sweep. In complex systems, often all waveform generators are controlled by one master, which is made extremely stable with respect to time changes. The process of synchronization imparts this same stability to the remaining circuits. Each free-running generator, oscillator, multivibrator, etc., has its own mode of synchronization contingent on its individual control characteristics. Consequently, the individual regions will be well defined but will differ from one circuit to the other. When needed, they may be calculated by application of the methods discussed in this chapter, modified as necessary to suit the particular problem.

Example 6-1

A signal having a highly complex structure with many jagged peaks must be used to synchronize a sweep over a ratio of 2:1. In its original form, the input is not suitable for synchronization, but by amplifying and clipping, it can be converted into a square wave; and, if needed, we can differentiate it to obtain pulses or integrate it to obtain a triangular wave. The questions of interest are:

(*a*) Which signal is best for synchronization?
(*b*) At what point should synchronization be effected?

Solution For optimum operation, the sweep should be synchronized so that the nominal operating point is in the center of the appropriate region. By this we mean that it should be located as far as possible from the boundaries, allowing the time and voltage to vary over the widest possible limits before the sweep becomes erratic. In general, the operating point which permits the greatest variation in the time ratio will not be the optimum one with respect to voltage changes. However, from Figs. 6-35 and 6-39 we see that the time bounds are much narrower, and therefore we shall usually design the circuit for the best response with respect to time variations.

1. For square-wave synchronization, the plot of Fig. 6-35 indicates that the widest changes in t_s/t_1 occur along a line through the apex of the region $n = 2$. The bounds are

$$0.425 \leq \frac{t_s}{t_1} \leq 0.575$$

and the nominal free-running sweep should be set at the center, i.e., at $t_s = 0.5t_1$. Assuming a constant sweep time, the sync signal can increase from $0.15V_f$ to $0.25V_f$ or can decrease to zero before the sweep becomes unstable.

2. With the triangular-wave input, the maximum permissible change in t_1 does not occur along a line through the apex. In Fig. 6-39 we see that at $V_s/V_f = 0.3$,

$$0.35 \leq \frac{t_s}{t_1} \leq 0.53$$

The free-running sweep should be adjusted to the center value $t_1 = 2.27t_s$. This establishes the following nominal limits on the sync amplitude from the operating point given above:

$$0.12V_f \leq V_s \leq 0.5V_f$$

3. The regions which might be drawn for pulse synchronization (Prob. 6-22) would indicate that the optimum operating point should lie at

$$t_1 = \frac{t_s}{0.375} \qquad V_s = 0.5V_f$$

Permissible time and voltage variations are between the limits

$$0.25 \leq \frac{t_s}{t_1} \leq 0.5 \qquad 0.25 \leq \frac{V_s}{V_f} \leq 0.625$$

To summarize, the percentage variations of free-running sweep time and sync voltage, as related to the nominal value of t_1 and V_f, are:

	PERCENTAGE VARIATION		
	SQUARE	TRIANGLE	PULSE
$\Delta t/t_1$	± 15	± 20	± 33
$\Delta V_s/V_f$	$-15, +10$	$-18, +20$	$-25, +12.5$

We conclude that pulses are best for synchronization purposes. Besides allowing the widest tolerances in time, their steep edges force recovery of any perturbation within one, or at most two, cycles. The triangular wave does permit wider latitude to the sync amplitude, but the difficulties involved in its generation as well as the longer recovery time would usually preclude its use; pulses, or even the square wave, are to be preferred.

Problems

6-1 An SCR is used in the simple free-running sweep of Fig. 6-2*a*. It will fire at $V_{fm} = 50$ V when $I_G = 0$. After firing, the SCR may be approximated by a constant drop of 1 V. The extinction current $I_x = 50\ \mu$A, and $I_{\max} = 50$ mA. Design a free-running sweep, linear to within 5 percent and having a period of 100 μs. Specify the resistor, capacitor, and power-supply values.

6-2 Repeat Prob. 6-1 when we replace the sweep resistor R with the constant-current diode of Fig. 5-37. This diode supports a current flow of 5 mA over the range of 5 to 100 V. It has an internal impedance of 100 K. What advantages does this circuit offer over the circuit of Prob. 6-1? Recheck the circuit design and, if necessary, modify the circuit.

6-3 We wish to use the SCR of Prob. 6-1 to generate an approximately triangular waveshape (the rise and fall times are identical) having a period of 100 μs. A 250-V power supply is available. Design this circuit, specifying all component values. Give the nonlinearity of each portion of the waveshape.

6-4 The sweep of Fig. 6-40 uses the SCR of Prob. 6-1. Its firing voltage may be expressed as $V_f \approx 50 - 0.5 \times 10^6 I_G$. Plot the output waveshape as a function of time. ($V_D = 0.7$ V.)

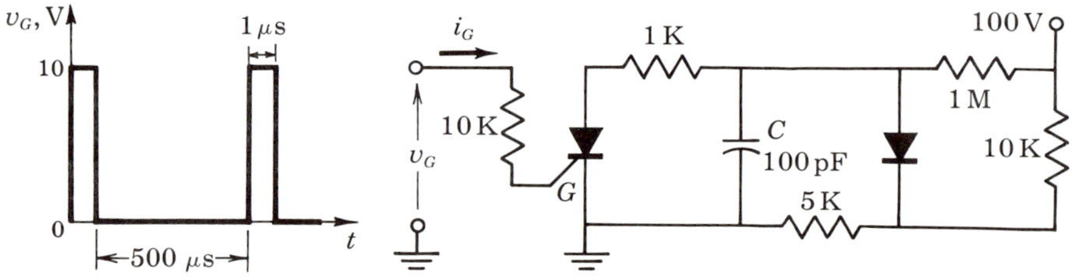

Fig. 6-40

6-5 Derive an expression for the percentage of sweep nonlinearity, starting from the time-distortion sweep representation of Fig. 6-4*a*. Express the answer in terms of the sweep period and time constant. Also express the nonlinearity in terms of the critical sweep voltages.

6-6 The sweep circuit of Fig. 6-41 employs the constant-current source of Prob. 3-8 to supply the capacitor charging current. The *pnp* transistor Q_2 is used as a switch to discharge C. Q_2 is normally saturated. A 10-V 100-μs switching pulse turns Q_2 off, and this permits the generation of a sweep. Find the value of C that will generate a 100-μs 10-V ramp. Sketch and label the output voltage. For both transistors, $\beta_F = 50$, $\beta_R = 5$, $V_{BE} = 750$ mV, and $V_{CE,\text{sat}} = 150$ mV. $V_D = 750$ mV.

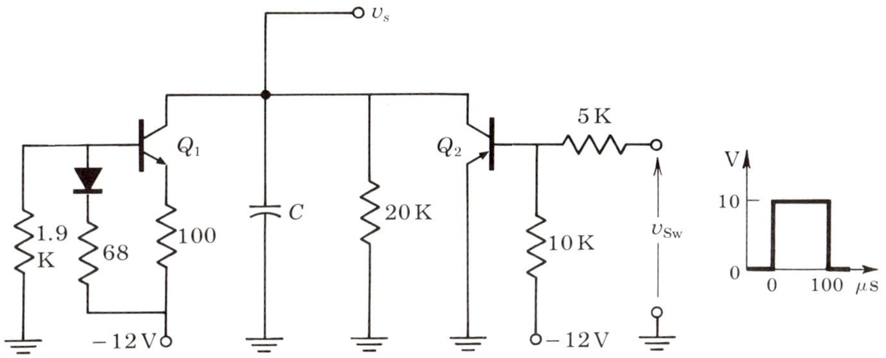

Fig. 6-41

6-7 (a) Plot the waveshape appearing at the collector of the circuit of Fig. 6-42 if the switch is opened at $t = 0$ and closed as soon as the transistor saturates. Specify the sweep time. ($V_A = 10$ V, $\beta = 100$, $V_{BE} = 700$ mV, $V_D = 700$ mV.)

(b) Find an equation for the rundown time as a function of V_A.

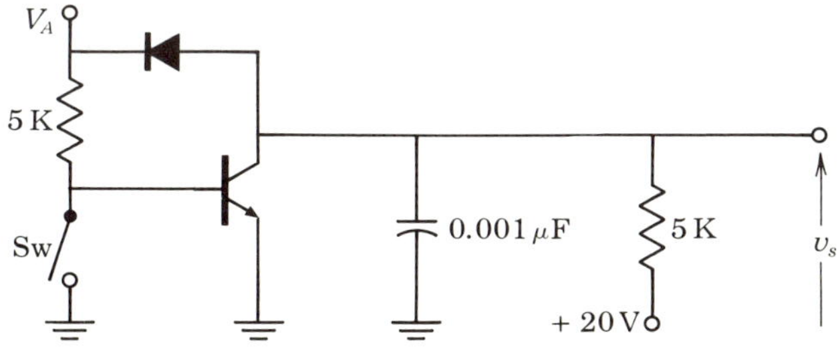

Fig. 6-42

6-8 The circuit of Fig. 6-43 is used to generate a specific nonlinear sweep. Plot both the base current and the collector voltage as a function of time. What is the equation of the sweep? (The switch is thrown from ground to V_{CC} at $t = 0$ and is

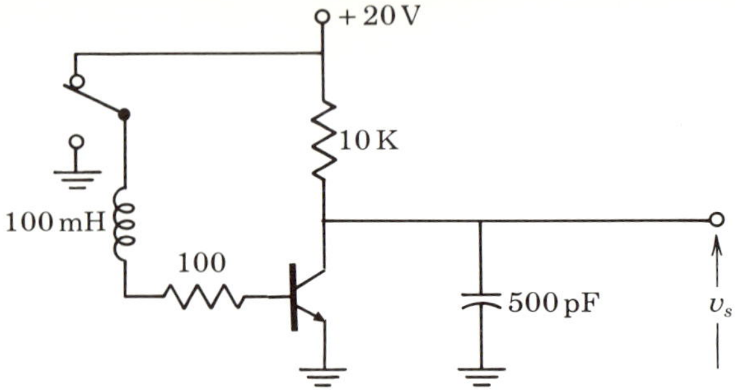

Fig. 6-43

returned to ground as soon as the transistor saturates. $\beta = 100$, $V_{BE} = 700$ mV.) Make all reasonable approximations in your calculations.

6-9 In the circuit of Fig. 6-7, the switch is closed at $t = 0$ and opened as soon as the sweep voltage runs down by 10 V. The circuit values are $R_B = 10$ K, $R_L = 2$ K, $R_R = 5$ K, $C = 1{,}000$ pF, $V_{CC} = 12$ V. $\beta_1 = \beta_2 = 100$, and $V_{BE} = 700$ mV.

(*a*) Sketch the output waveshape and label all significant times and time constants.

(*b*) Repeat part *a* when the values of R_B and R_R are interchanged.

6-10 Repeat Prob. 6-9 for the circuit of Fig. 6-8. All the component values are as given in Prob. 6-9 ($R_{B2} = R_R$).

6-11 The circuit of Fig. 6-44 is a Miller sweep which generates a linear runup. The switch is opened at $t = 0$ and is closed when the sweep reaches 10 V. Sketch and label the base-current and collector-voltage waveshapes. What advantage does this mode of switching offer over that used in Figs. 6-9 and 6-10? ($\beta_F = 100$, $\beta_R = 10$, $V_{BE} = 700$ mV.)

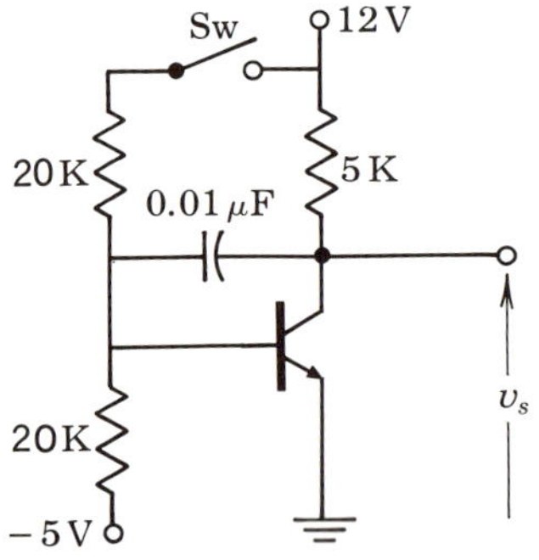

Fig. 6-44

6-12 The circuit of Fig. 6-45 represents a bipolar version of a voltage-feedback Miller sweep. By incorporating the resistor in the emitter circuit, we change the base input to the high impedance required of the voltage-feedback Miller sweep.

The gain, however, is not large enough for all the approximations to hold. The bottoming diode prevents transistor saturation.

(a) Sketch and label the collector and base voltage waveshapes for this circuit. Calculate the sweep linearity. (The switch is opened at $t = 0$ and is closed shortly after the rundown is complete. $\beta = 100$, $V_{BE} = 700$ mV, $V_D = 700$ mV.)

(b) Is the sweep at the collector improved or is it degraded when the emitter resistor is removed? Explain.

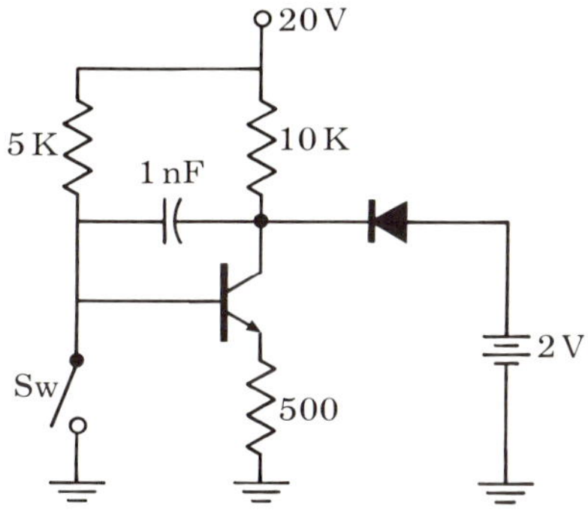

Fig. 6-45

6-13 We have available an integrated-circuit operational amplifier having a gain of $-1{,}200$, an output impedance of 1 K, and an input impedance of 50 K. The maximum output swing is ± 10 V (power supplies of ± 12 V). We wish to use this amplifier in a Miller-sweep circuit to generate an ideal trapezoidal sweep at the output. Under steady state, the output is at -10 V and the sweep consists of a jump to -5 V followed by a linear runup to $+10$ V in 10 ms. To generate the pedestal, we may have to add resistance in series with C, thus increasing the first term of Eq. (6-38).

Draw the circuit, showing how the switching would be incorporated. Specify sweep linearity. Give values for all voltages, resistors, and capacitors.

6-14 The circuit of Fig. 6-14 is converted into the voltage-controlled sweep of Fig. 6-18. When Q_2 is triggered on into the ohmic region, it can be considered as 200 Ω in series with the drain of Q_1. $V_A = 19.3$ V, $V_D = 0.7$ V, and all other values remain as given in Fig. 6-14.

Sketch and label completely the waveshape at the drain. Pay particular attention to the initial jump.

6-15 Convert the Miller sweep of Fig. 6-14 to a sweep that will generate a linear runup. Incorporate a bottoming diode so that your sweep will start at 5 V. It should run up to 15 V at the rate of 10^6 V/s. Draw the circuit, giving all values and showing the switching. Sketch the output waveshape, including the recovery region, indicating all significant voltages, times, and time constants.

6-16 Figure 6-46 shows a Miller sweep with a second MOSFET used as a switch. For both MOSFETs, $V_T = -3$ V and $I_D = 20$ mA at $V_{GS} = -4$ V. The bottoming of Q_2 sets $V_{GS_1} = -1$ V. Sketch and label both the gate and the drain waveshapes of Q_1. Give the values of all jumps, times, and time constants. You may make all reasonable approximations.

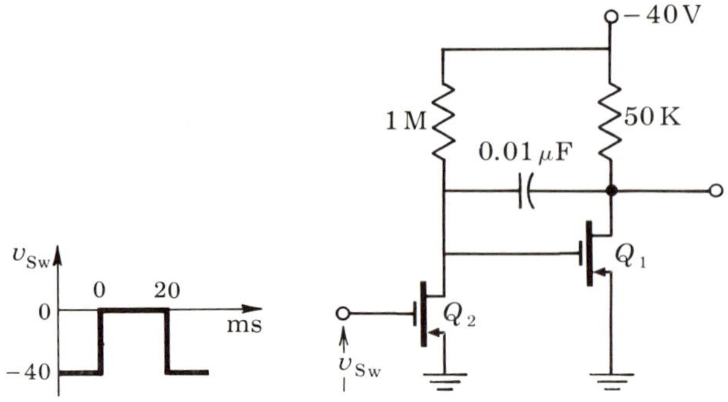

Fig. 6-46

6-17 Repeat Prob. 6-16 when we add a source follower to Fig. 6-46 to improve the recovery. The source resistor is 20 K.

6-18 The switch in the circuit of Fig. 6-47 is opened at $t = 0$ and is closed just after the sweep terminates. Plot the gate and drain voltages to scale from just before $t = 0$ until the circuit recovers completely. Pay particular attention to initial jumps. ($V_T = 3$ V and $I_D = 30$ mA at $V_{GS} = 4$ V.)

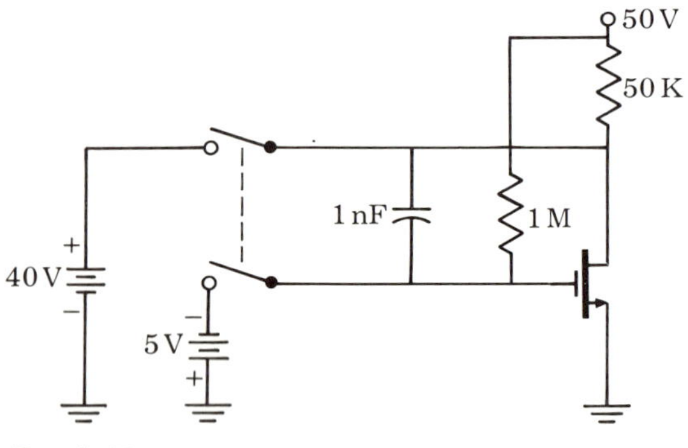

Fig. 6-47

6-19 Plot the base-current and output-voltage waveshapes to scale for the circuit of Fig. 6-24. The switch is opened at $t = 0$ and is closed as soon as the output reaches 10 V. $\beta_F = 100$, $\beta_R = 10$, $R_E = 5$ K, $R = 5$ K, $R_1 = 100\ \Omega$, $C = 1$ nF, $V_{CC} = 12$ V, $V = 10$ V, $V_D = V_{BE} = 700$ mV at $I_e = I_D = 1$ mA, $V_{CE,\text{sat}} =$ 100 mV.

6-20 Repeat Prob. 6-19 for the circuit of Fig. 6-25. $C_S = 0.1\ \mu$F. What is the drop in voltage across C_S over the sweep interval?

6-21 Figure 6-48 shows an alternative version of a bootstrap sweep. For this circuit, sketch and label completely the output voltage if the switch is opened at $t = 0$ and is closed once the output drops to 1 V. ($\beta_F = 100$, $V_{BE} = V_D = 700$ mV, $V_{CE,\text{sat}} = 100$ mV.)

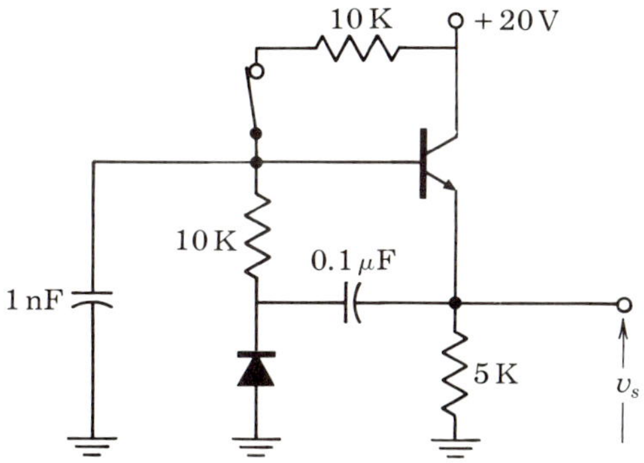

Fig. 6-48

6-22 Plot the regions of synchronization for a train of equally spaced positive and negative pulses. Assume that the pulses are extremely narrow. Explain your answer.

6-23 Repeat Prob. 6-22 for a free-running sweep whose recovery time is 50 percent of the total period.

6-24 Plot the regions of synchronization for a train of equally spaced positive and negative pulses for $n = \frac{1}{2}, \frac{1}{3}, \frac{1}{4}$, etc. Can this be called synchronization?

6-25 A perfectly linear sweep has a normal period of 1.0 ms. This can be expected to vary by as much as ± 10 percent because of the variations in the firing voltage. In the absence of synchronization, the sweep starts at 0 V and terminates at the nominal value of $V_f = 100$ V. The retrace time is zero. It is required that the sweep be synchronized so that its period is exactly 1.0 ms despite the instability in amplitude.

Three synchronization waveshapes are shown in Fig. 6-49 plotted at the firing line of sweep. In each, t_s is 1.0 ms or some integral submultiple.

(*a*) Which among these will provide the desired synchronization?

(*b*) When the sweep is synchronized under the conditions specified, what is the possible number of cycles of the synchronizing wave over which synchronization can take place?

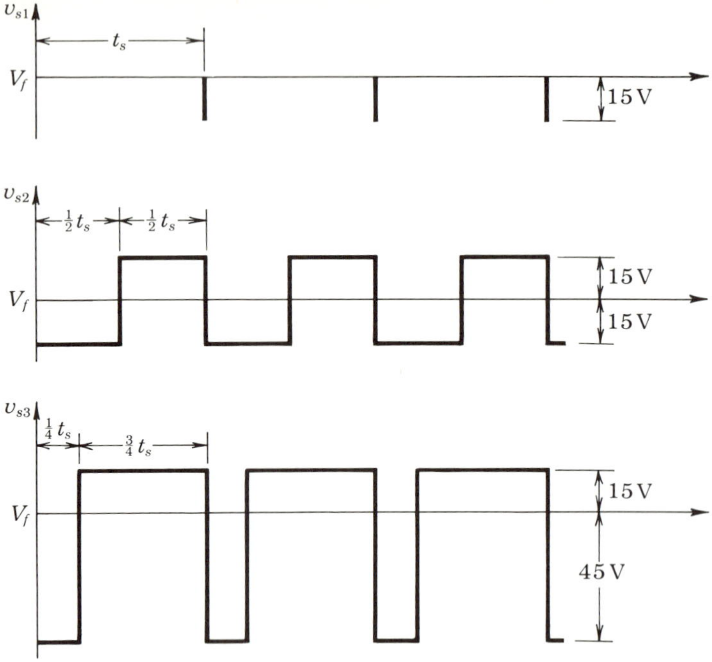

Fig. 6-49

6-26 (*a*) Verify Eq. (6-70) by direct solution from the conditions of triangular-wave synchronization.

(*b*) Calculate the slope of Eq. (6-70) at $x = 1$ as a function of n and compare it with the slope of the $t_{s,\min}/t_1$ line of the $n - 1$ region.

Bibliography

Briggs, B. H.: The Miller Integrator, *Electron. Eng.*, vol. 20, pp. 243–247, August, 1948; pp. 279–284, September, 1948; pp. 325–330, October, 1948.

Chance, B., et al.: "Waveforms," M.I.T. Radiation Laboratory Series, vol. 19, McGraw-Hill Book Company, New York, 1949.

MacLean, W. R.: The Synchronization of Oscilloscope Sweep Circuits, *Communications*, March, 1943.

Nambiar, K. P. P., and A. R. Boothroyd: Junction Transistor Bootstrap Linear Sweep Circuits, *Proc. IEE (London)*, pt. B, vol. 104, pp. 293–306, 1957.

Puckle, O. S.: "Time Bases," 2d ed., John Wiley & Sons, Inc., New York, 1951.

7.
linear current sweeps

The electric deflection of a high-energy electron beam imposes a severe strain on the circuit designer in that it requires an excessively large sweep amplitude. For this reason, magnetic deflection, which is produced by a linearly increasing current in the deflection coil, is used instead. The most common example is a television receiver, where the raster is generated by the magnetic deflection of the beam, a single horizontal scan taking about 53 μs and a vertical scan about 16 ms. Radar display scopes and some electromechanical systems employ similar deflection circuits. Furthermore, a coil excited with a linear current may be mechanically rotated, thus generating a spiral sweep.

Our treatment of current sweeps in this chapter will closely parallel the development of the various voltage sweeps, with, however, the deflection-coil inductance replacing the capacitor as the basic timing element. We should expect our thinking to be influenced by the previous discussion and should feel free to adapt any existing circuit to our current need. For purposes of comparison and in order to gain perspective, the reader might occasionally refer back to the appropriate sections of Chap. 6.

7-1 Basic Current Sweeps

If the ideal inductance did exist, as does the almost ideal capacitor, then the problem of producing a linearly increasing current would become trivial;

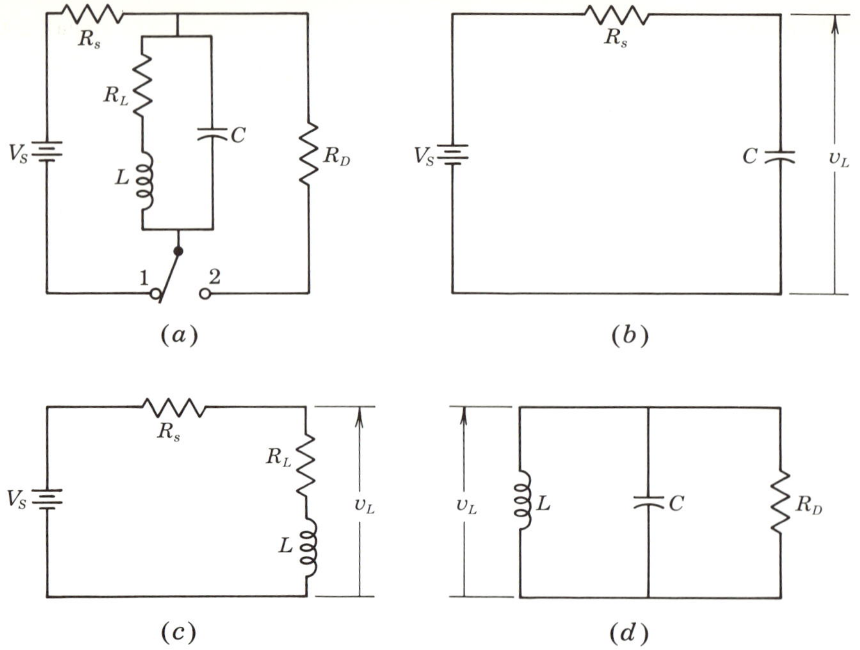

Fig. 7-1 (*a*) Basic current sweep; (*b*) approximate model holding during initial charging; (*c*) model representing the current-sweep region; (*d*) recovery-region model.

simply switching a constant voltage across the coil would suffice. Because $V = L\,di/dt$, the current would immediately become

$$i = \frac{V}{L}t \tag{7-1}$$

Unfortunately all coils have distributed winding resistance and interwinding and stray capacity; a typical iron-core deflection coil of 50 mH may have a resistance of 70 Ω and an effective capacity of 200 pF. These parasitic elements are generally represented by the lumped parameters (R_L, C) shown in the basic sweep circuit of Fig. 7-1*a*, and their effects on the sweep waveshape must be accounted for in any analysis.

Since we are interested in a linear sweep, the portion of the cycle devoted to the initial charging of C must be much smaller than the time required for the coil current to build up to its final value. Immediately upon closing the switch in Fig. 7-1*a*, the complete circuit current flows into C. Thus the beginning of the sweep may be represented by the model of Fig. 7-1*b*, which implicitly assumes that the start of the inductive-current buildup will be delayed until the terminal voltage reaches V_S. The time constant of this circuit,

$$\tau_1 = R_s C \tag{7-2}$$

is quite small, and the delay is proportional ($t_1 = 4\tau_1$); even when charging through 50 Ω, the 200-pF capacity introduces a delay of less than 50 ns (Fig. 7-2).

Sweep region

We can now assume that the circuit enters into its sweep region, which is defined by the model given in Fig. 7-1*c*. Neglecting the small current change occurring during the charging of C, the time response becomes

$$i_L = \frac{V_S}{R_s + R_L}(1 - e^{-t/\tau_2}) \qquad \tau_2 = \frac{L}{R_L + R_s} \tag{7-3}$$

To justify our separating the response into these two segments, the sweep-time constant τ_2 must be much longer than τ_1. In the case considered above ($L = 50$ mH, $R_L = 70$ Ω, and $R_s = 50$ Ω), $\tau_2 = 420$ μs, and the assumption made above is certainly valid. Moreover, in order to ensure good sweep linearity, the charging interval t_2 must be limited to some small fraction of τ_2.

Suppose that we employ this coil in a deflection circuit where the current must reach a peak of 200 mA in 53 μs. From Eq. (7-3), the necessary supply voltage is found to be

$$V_S = \frac{120 I_m}{1 - e^{-0.126}} = 202 \text{ V}$$

The inductive component of the voltage decays to 178 V over the sweep interval. An additional 14 V (200 mA through 70 Ω) appears across the resistive component, raising the final coil voltage to 192 V.

The waveshapes produced during this portion of the cycle are sketched in Fig. 7-2. Note the delay in starting the sweep and the voltage change appearing across the coil during the sweep interval.

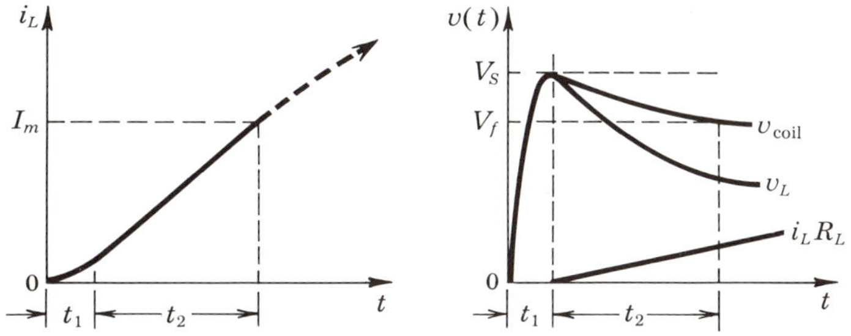

Fig. 7-2 Sweep current and voltage waveshapes for the basic sweep of Fig. 7-1 (switch in position 1).

Sweep recovery

At the end of the required sweep interval, the switch is thrown to position 2. The response of the coil with its parasitic capacity and external damping, R_D, is found from the single node equation of the circuit in Fig. 7-1*d*. We assume, in writing Eq. (7-4), that the series coil resistance will have but little influence during the portion of the recovery time with which we shall be concerned.

$$i_c + i_L + i_R = C\frac{dv_L}{dt} + \frac{1}{L}\int v_L\,dt + \frac{v_L}{R_D} = 0 \tag{7-4}$$

The roots of Eq. (7-4) are

$$s_{1,2} = -\frac{1}{2R_DC} \pm \sqrt{\left(\frac{1}{2R_DC}\right)^2 - \frac{1}{LC}} \tag{7-5}$$

Three bounding possibilities serve to delineate the response characteristics of the network. In one case, $(1/2R_DC)^2 < 1/LC$ and the circuit exhibits a damped sinusoidal response. Here the poles are located at

$$s_{1,2} = -\alpha \pm j\beta$$

where $\alpha = \dfrac{1}{2R_DC} \qquad \beta = \sqrt{\omega_0{}^2 - \alpha^2} \qquad \omega_0{}^2 = \dfrac{1}{LC}$

The complete solution of Eq. (7-4) may be expressed as

$$i_L = Ae^{-\alpha t}\cos(\beta t + \phi)$$

If the system is but slightly damped, $\alpha \ll \beta$. By including the initial conditions, which are the final voltage and current reached during the linear sweep (V_f and I_m of Fig. 7-2),

$$i_L = I_m e^{-\alpha t}\cos\omega_0 t \tag{7-6a}$$

$$v_L \approx -\sqrt{\frac{L}{C}}\,I_m e^{-\alpha t}\sin\omega_0 t \tag{7-6b}$$

Since the damping factor α is so very small, the peak coil voltage occurs when $\sin\omega_0 t = 1$ (in ¼ cycle of f_0 or at $t = \pi\sqrt{LC}/2$):

$$V_p \approx -\sqrt{\frac{L}{C}}\,I_m \tag{7-7}$$

For the coil considered above, this peak reaches almost −3,200 V only 4.98 μs after recovery begins. Figure 7-3*a* illustrates the rapid oscillation and slow decay of the recovery portion of the sweep waveshape.

This enormous voltage peak can be reduced through proper choice of C and/or R_D. Anything that acts to reduce the peak does so at the expense of the time which must be allotted to the recovery interval. Increasing C from 200 to 2,000 pF reduces the peak voltage from 3,200 to 1,000 V.

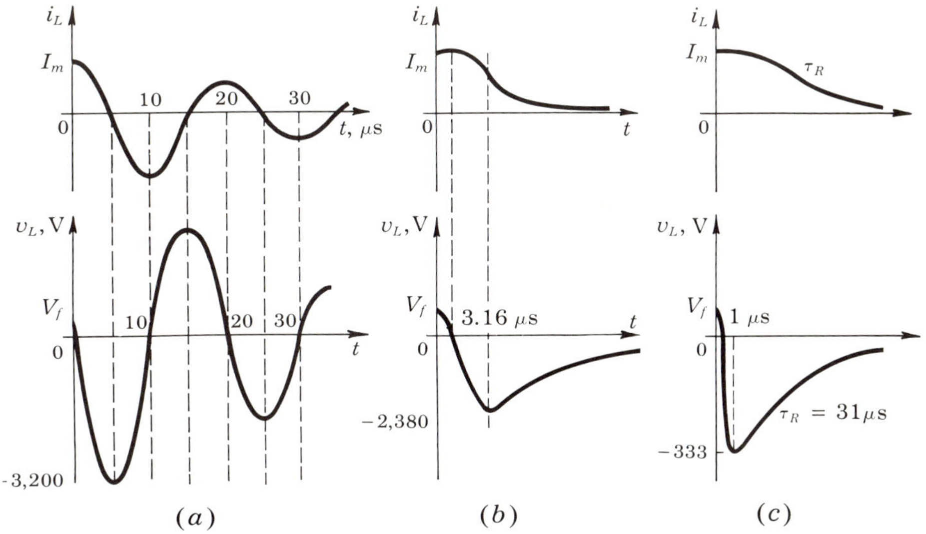

Fig. 7-3 (*a*) Recovery waveshape when lightly damped; (*b*) recovery waveshape when critically damped; (*c*) recovery when overdamped (not drawn to the same scale).

The peak occurs much later, at 15 μs instead of at 5 μs. The tenfold increase in C also causes a proportionate increase in the sweep turn-on delay, from 50 to 500 ns.

A second bounding location of the roots of Eq. (7-4) occurs when R_D is adjusted to critically damp the response ($R_D = \sqrt{L/4C}$). The two roots are now identical, both lying on the real axis at

$$s_{1,2} = -\frac{1}{\sqrt{LC}}$$

When the initial conditions are included, the solutions for the critically damped case become

$$i_L = \left[I_m + \left(\frac{V_f}{L} + \frac{I_m}{\sqrt{LC}}\right)t\right]\exp\frac{-t}{\sqrt{LC}} \tag{7-8a}$$

$$v_L = \left[V_f - \left(\frac{V_f}{\sqrt{LC}} + \frac{I_m}{C}\right)t\right]\exp\frac{-t}{\sqrt{LC}} \tag{7-8b}$$

The peak voltage is much smaller, and it occurs sooner; for the same coil considered above, it reaches −2,380 V at t = 3.16 μs. Figure 7-3*b* shows this portion of the complete waveshape.

In comparing Fig. 7-3*a* and *b*, we conclude that the fastest possible recovery corresponds to a single half cycle in the completely undamped case. For periodic sweeps, the "switch" is returned to position 1 at the time that the coil current reaches its minimum point and the capacitor voltage returns to zero. The full energy stored in the capacitor is recovered and applied for the next sweep cycle. Current will build up from approximately $-I_m$, instead of from zero, with the sweep amplitude and duration increasing accordingly.

A further reduction in R_D causes the poles to separate from coalescence at $-1/\sqrt{LC}$ along the negative real axis. For $R_D \ll \sqrt{L/4C}$, the poles are far enough apart so that the effect of each pole may be considered separately. The total response is found by the technique of time-sequential solution discussed in Sec. 1-8.

The far-out pole, which is located at

$$s_D \approx -\frac{1}{R_D C} \qquad \tau_D \approx R_D C$$

controls the voltage transition across L as the voltage builds up from V_{CC} to the negative peak V_p. We can conceive of this transition as introducing a delay lasting for $t_D = 4\tau_D$ at the start of the recovery interval (see Fig. 7-3c).

Only the dominant pole is significant during the final recovery interval. It is located at

$$s_R \approx -\frac{R_D}{L} \qquad \tau_R = \frac{L}{R_D}$$

The approximate decay interval is defined by the single exponential response

$$i_L \approx I_m e^{-t/\tau_R} \tag{7-9a}$$

$$v_L = V_p e^{-t/\tau_R} \tag{7-9b}$$

where $$V_p = -R_D I_m$$

Figure 7-3c illustrates this type of recovery for $R_D = 0.1\sqrt{L/4C} = 1.67$ K ($\tau_D = 333$ ns and $\tau_R = 31.6$ μs). The short transition region lasts for slightly over 1 μs, compared with the 150 μs required for final recovery. In this particular example, the total recovery requires three times as long as the time devoted to the sweep. Recovery in the overdamped case will always last longer than in the critically damped case of Fig. 7-3b.

We can define a figure of merit which will show the trade-off between the peak voltage and the recovery time. Forming the product $\tau_R V_p$ for the overdamped case, we obtain

$$\tau_R V_p = |L I_m| \tag{7-10}$$

In the critically damped case, the figure of merit is given approximately by

$$\tau_R V_p \approx |L I_m e^{-1}|$$

which is an improvement of almost 3:1 over the overdamped response. In the damped sinusoidal case, the product of the peak voltage and the location of the peak will also result in the LI_m product as a figure of merit. We conclude that any reduction in peak voltage will always be accompanied by an increased recovery time. Critical damping results in a significant improvement over the overdamped case whenever a nonoscillatory recovery is desired. Sufficient shunt C and R_D should be used to control the peak voltage and the form of the recovery waveshape. Of course, the only way to simultaneously minimize the peak flyback voltage and the recovery time is to use the minimum inductance consistent wiht the sweep or switching constraints.

7-2 Switched Inductive Loads

Figure 7-4 illustrates the application of a transistor to the basic inductive sweep of Fig. 7-1. We shall initially assume that the circuit is inert, i.e., that it contains no stored energy. When an input pulse is applied to the base, the transistor switches on. At the collector, it sees the uncharged capacitor C_s. In Fig. 7-5*a*, we have the model holding during the initial portion of the turn-on delay. The turn-on path starts with a jump from Q to point A, followed by the constant-current charging of C_s to point B, where the transistor saturates.

Finally, the transistor bottoms near the origin as C_s completes its charging through the very small r_{sc}. We can estimate the turn-on delay caused by a small C_s by the linear voltage rundown at the collector from V_{CC} to $V_{CE,\text{sat}}$:

$$v_c(t) = V_{CC} - \frac{\beta I_b}{C_s} t \qquad (7\text{-}11)$$

Taking $V_{CE,\text{sat}} \approx 0$, the delay in starting the exponential current buildup is quite small:

$$t_D = C_s \frac{V_{CC}}{\beta I_b} \qquad (7\text{-}12)$$

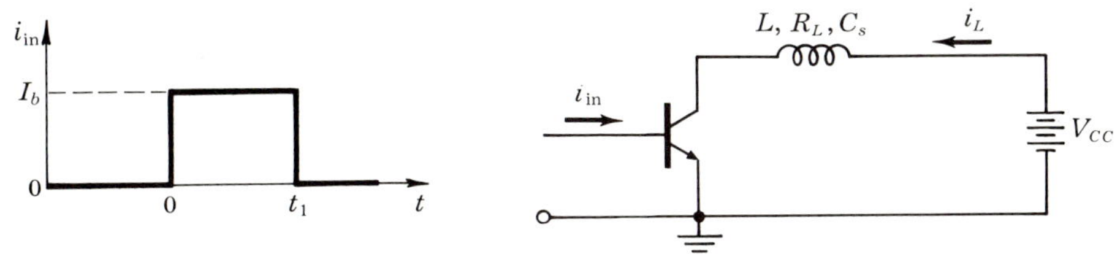

Fig. 7-4 A switched current sweep—the transistor is driven from cutoff to saturation.

After the turn-on delay interval of Eq. (7-12), the inductive current starts increasing from 0 toward the intersection with the steady-state load line of slope $-1/R_L$ and the saturation-region characteristic (point D in Fig. 7-5b). Because the charged C_s is shunted by the very small collector saturation resistance, the $C_s r_{sc}$ time constant is extremely small. It seems

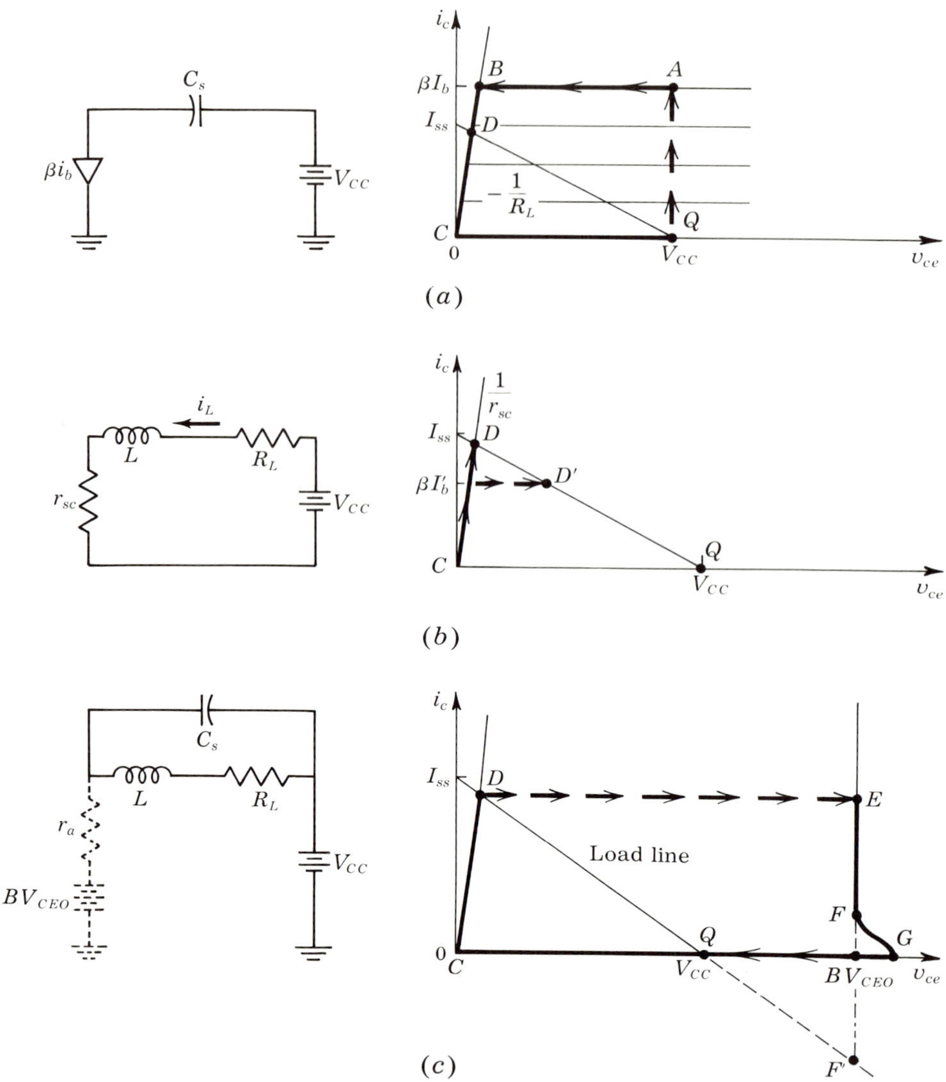

Fig. 7-5 Models for the sweep of Fig. 7-4. (a) Initial model indicating turn-on delay and the path of operation on the volt-ampere plane; (b) sweep-region operation along saturation-region characteristics; (c) recovery-region model showing avalanche-region operation.

reasonable to neglect C_s during the inductive-current charging. We are left with the model of Fig. 7-5*b*. From this reduced model, we have

$$I_{ss} = \frac{V_{CC}}{R_L + r_{sc}} \qquad \tau_s = \frac{L}{R_L + r_{sc}} \tag{7-13}$$

In general, $R_L \gg r_{sc}$, and both terms of Eq. (7-13) are determined primarily by the coil and the power supply; they are almost completely independent of the transistor. The transistor simply acts as a switch connecting the coil to the power supply. Current builds up exponentially in the coil, reaching I_{ss} in five time constants.

If $\beta I_b < I_{ss}$, the transistor will again become active before reaching point D and the operating point will stabilize at point D' of Fig. 7-5*b*. When $\beta I_b \ll I_{ss}$, only a small portion of the exponential exists and it may be approximated by a linear sweep.

After the input pulse is removed, the transistor becomes back-biased. Depending on the damping present, the output waveshape will be similar to one of the cases shown in Fig. 7-3.

As was pointed out in Sec. 7-1, the flyback voltage across the inductance can reach extremely large values. If, during flyback, the voltage across the transistor exceeds the reverse breakdown voltage, the retrace will follow the avalanche characteristic, as shown in Fig. 7-5*c*. But since the avalanche region is a low-resistance, almost constant voltage drop, the retrace-time constant will be extremely slow. It becomes

$$\tau_R = \frac{L}{R_L + r_a}$$

where r_a is the avalanche-region resistance.

The recovery path starts with a fast oscillatory transition from the peak-current point at D to point E on the avalanche characteristic. This is followed by a slow exponential rundown along path EF toward the intersection of the extended load line $(-1/R_L)$ and the extended piecewise-linear avalanche characteristic at F'. From F to G, the path follows the avalanche characteristic back to the zero-current intercept. The final portion of the recovery is the fast voltage decay back to the OFF quiescent point at Q.

Breakdown-region operation during recovery has two adverse effects. First, the long time constant means that an excessive amount of time will have to be allotted for the recovery interval. Second, since the slow recovery is accompanied by a very high collector voltage as well as high collector currents, we shall have a large power dissipation. The power can even exceed the transistor limitations, and the transistor can then be permanently damaged.

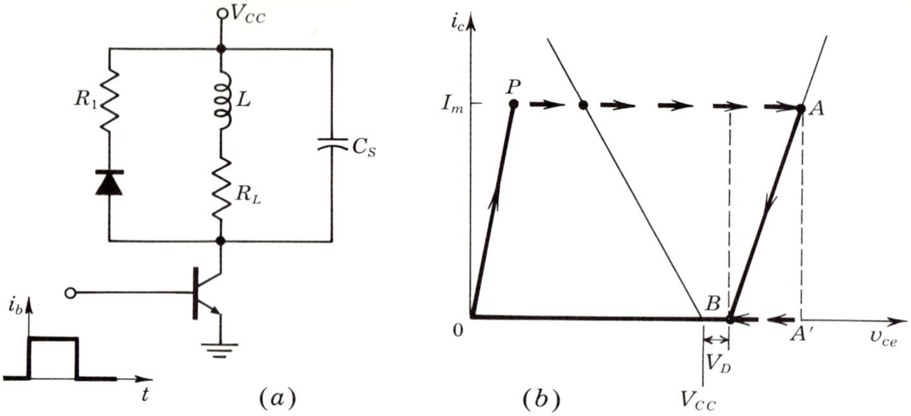

Fig. 7-6 (a) Diode-resistor voltage limiting of recovery-region voltage; (b) path of operation on the transistor volt-ampere plane.

In Fig. 7-6a, we prevent avalanche-region operation of the switching transistor by shunting the coil with a diode and series resistor. During the exponential charging, the diode is back-biased and R_1 is out of the circuit. During recovery, the diode conducts and L discharges through R_1. The peak collector voltage is limited to

$$V_{pc} = V_{CC} + V_D + I_m R_1 \tag{7-14}$$

R_1 is chosen to limit V_{pc} to less than the collector-emitter breakdown voltage. The larger the R_1, the smaller the recovery-time constant. For fastest recovery, C_s and R_1 can be adjusted for critical damping as well as for an acceptable peak flyback voltage.

In the characteristics of Fig. 7-6b, the switching transistor is off during the complete retrace interval. The current indicated along the path segment AB actually flows through the shunt diode. The transistor current path is a jump from A to A' and then along the path $A'B$.

7-3 Periodic Switched Sweeps

In the special case of a periodic sweep where the switching transistor must be periodically turned on and off, a sinusoidal response within the recovery region enables us to recover some of the energy stored in the inductance. Figure 7-7 shows the trajectory of a periodic sweep drawn on the volt-ampere plane of the switching transistor. The initial portion of the exponential buildup follows the path $0A$ along the saturation characteristics. To achieve acceptable sweep linearity, only a small portion of the total exponential is used as the sweep. At point A, which is far below the effective steady-state current given by Eqs. (7-13), the transistor is switched off. In the oscillatory region, the coil current and collector voltage follow the elliptical trajectory $ABCD$ of Fig. 7-7. The coil must be shunted with sufficient capacitance to set the peak of the oscillatory response at B below the collector breakdown voltage.

At point A, the energy of the complete system,

$$W_L = \tfrac{1}{2}LI_m^2 \tag{7-15}$$

is stored in the magnetic field of the coil. Along the path AB, the stored energy is transferred to the electric field of the capacitor. As the coil voltage reverses and builds up to the negative peak of V_p, the coil current decreases to zero. At point B,

$$W_C = \tfrac{1}{2}CV_p^2 \tag{7-16}$$

From Eqs. (7-15) and (7-16),

$$V_p = \sqrt{\frac{L}{C}}\, I_m$$

which is identical with the result given in Eq. (7-7).

Along the second portion of the elliptic trajectory BD, the energy is retransferred from C to the magnetic field of the deflection coil. The

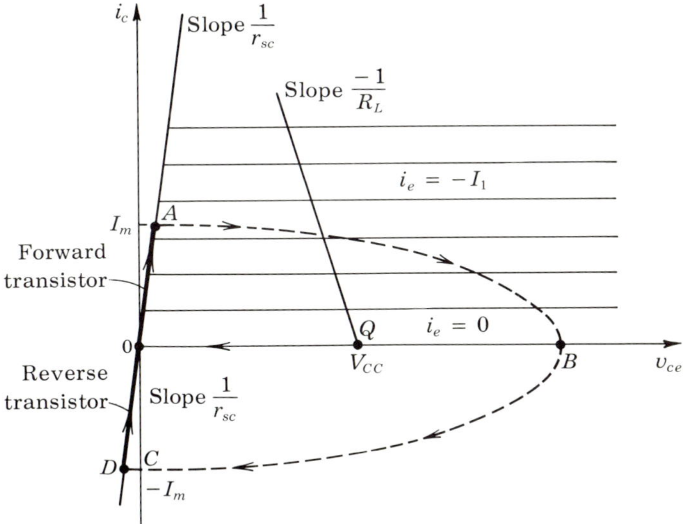

Fig. 7-7 Trajectory of the collector voltage and coil current for a periodic switched sweep of the type of Fig. 7-4. The dashed portion is the path taken in the oscillatory recovery region, and the solid-line segments represent the path followed in the linear-sweep region.

voltage now decreases, and the current builds up to $-I_m$. The dashed portion is the path taken in the oscillatory recovery region, and the solid-line segments represent the path in the linear-sweep region.

Slightly after one half cycle of the oscillation (Figs. 7-7 and 7-8), the positive voltage developed across the coil exceeds the negative supply voltage and the transistor is biased on in the reverse direction. Current now builds up from the negative peak of $-I_m$ back toward the steady-state value given by Eq. (7-13). The charging path is now along the reverse transistor saturation characteristic, line segment $D0$ in Fig. 7-7.

To generate a periodic sweep, the base must be periodically switched from off to on. The time of the application of the input gating pulse is not extremely critical provided that it occurs before the coil voltage again rises to V_{CC}. Under these conditions, the transistor changes state smoothly from its operation as a saturated reverse transistor to a saturated forward transistor. Buildup continues along the path $D0A$ to I_m. Along this line segment, the charging equation is

$$i_L(t) = I_{ss} - (I_{ss} + I_m)e^{-t/\tau_s} \tag{7-17}$$

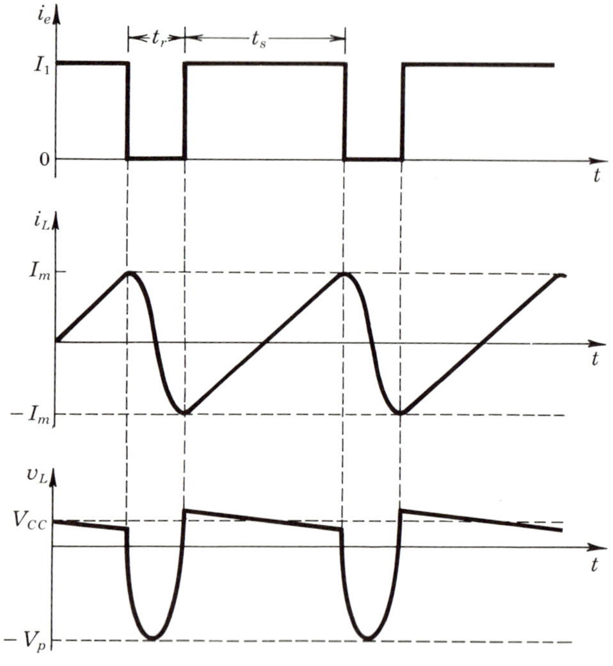

Fig. 7-8 Periodic sweep produced by the circuit of Fig. 7-4 upon periodic pulsed-current excitation.

and from Eq. (7-17), the time required for the current sweep to build up from $-I_m$ to I_m is

$$t_s = \tau_s \ln \frac{I_{ss} + I_m}{I_{ss} - I_m} \tag{7-18}$$

where τ_s is given by Eqs. (7-13).

At the end of the linear-sweep interval, the excitation is removed and the circuit recovers as discussed above. The portion of the period devoted to the flyback constitutes only one half cycle of the oscillatory wave, or

$$t_r = \frac{1}{2f_0} = \frac{\pi}{\omega_0} = \pi\sqrt{LC} \tag{7-19}$$

We should note that this periodic sweep circuit is extremely efficient. During the half cycle that the forward transistor is saturated, energy is transferred from V_{CC} to the coil. During the half cycle when the reverse current flows through the saturated reverse transistor, the stored energy is returned to the power supply. Of course, some power is dissipated over each half cycle in the circuit resistance.

Satisfactory switching may be obtained from current pulses, coupled into the base through an emitter follower. The sweep produced is shown in Fig. 7-8.

Example 7-1

We wish to see whether the sweep of Fig. 7-4 can be used in the horizontal-deflection circuit in a portable television receiver. Of the 63-μs total period, the linear sweep requires 55 μs, with 8 μs allowed for the sweep recovery. The deflection yoke, which must be excited by a peak-to-peak current of 1.0 A, has the following parameters: $L = 0.5$ mH, $R_L = 4.5\ \Omega$, and $C_s = 50$ pF.

(*a*) With this coil, what supply voltage must be used?
(*b*) How can the peak reverse voltage be minimized?
(*c*) What is the sweep nonlinearity?

Solution Assume that the saturation resistance of the transistor employed is 1 Ω. The time constant of the sweep interval is $\tau_1 = 91\ \mu$s. Solving Eqs. (7-13) with $I_m = 0.5$ A yields

$$I_{ss} = 1.7 \text{ A}$$

Hence $V_{CC} = 1.7(r_{sc} + R_L) = 9.35$ V.

As C is increased, the peak reverse voltage decreases. The upper limit of C is that value which increases the half period of the recovery sinusoid to 8 μs. From Eq. (7-19),

$$C_m = \frac{1}{L}\left(\frac{t_r}{\pi}\right)^2 = 0.013\ \mu\text{F}$$

Thus the peak voltage becomes

$$V_p = 0.5\sqrt{\frac{0.5 \times 10^{-3}}{1.3 \times 10^{-8}}} = 106\ \text{V}$$

Sweep nonlinearity is given by Eq. (6-20):

$$\text{NL} = 12.5\% \times {}^{55}\!/_{91} = 7.5\%$$

For acceptable linearity, the coil inductance must be increased or the total resistance decreased.

7-4 Current-sweep Linearization

The constant-voltage charging of Sec. 7-1 is not a very satisfactory sweep for critical applications because of its delay in starting and its inherent nonlinearity. We now propose to reexamine the basic sweep of Fig. 7-10*b* and ask: What applied waveshape will ensure a linearly increasing current flow?

In order to minimize the sweep delay, the initial portion of the excitation waveshape should be a voltage impulse. The impulse, shown in Fig. 7-9, is a pulse whose amplitude increases toward infinity as its duration decreases toward zero. Its area remains constant during the limiting process. It follows that the unit impulse $\delta(t)$ may be defined by the area integral

$$\int_0^{0^+} \delta(t)\, dt = \int_0^{\Delta T} \frac{1}{\Delta T}\, dt = 1 \tag{7-20}$$

where, in the limit, $\Delta T = 0^+ - 0$.

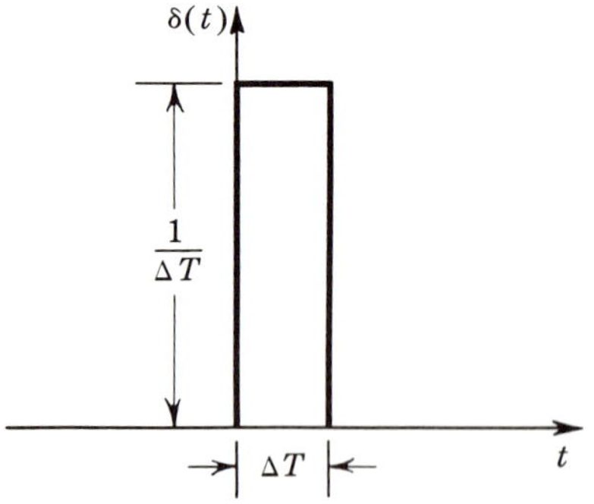

Fig. 7-9 Defining the unit impulse.

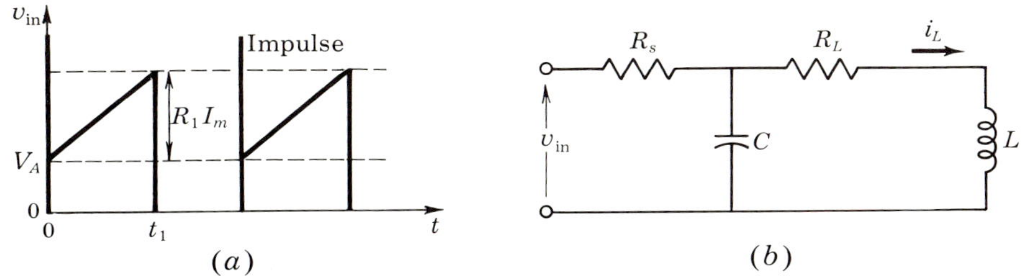

Fig. 7-10 (*a*) Required drive for the ideal current sweep; (*b*) equivalent circuit of the sweep.

If we excite the coil through the source impedance R_s with a voltage impulse of area K, then the stray capacity would be charged by the current

$$i = \frac{K\delta(t)}{R_s}$$

From Eq. (7-20), the voltage across C is given by

$$V_C(0^+) = \frac{1}{C}\int_0^t dt = \frac{K}{CR_s}\int_0^{0^+} \delta(t)\, dt = \frac{K}{CR_s}$$

In order for this voltage to equal V_A at $t = 0^+$, the weight of the impulse must be

$$K = CR_sV_A$$

It is impossible to generate the ideal impulse, but it is possible to inject a finite-amplitude pulse at the institution of the sweep to speed the charging of C. Any large-amplitude pulse will reduce the sweep starting time by an appreciable factor.

To find the required drive voltage over the sweep region, we assume that the desired linearly increasing current is already flowing:

$$i = kt = \frac{I_m}{t_1}t \tag{7-21}$$

Substitution of this linear term into the circuit loop equation results in

$$v_{\text{in}} = L\frac{di}{dt} + R_1 i = Lk + R_1 kt \tag{7-22}$$

where $R_1 = R_L + R_s$. Equation (7-22) defines the trapezoidal driving voltage shown in Fig. 7-10*a*. The first term in the above equation ($Lk = V$) is the magnitude of the pedestal; i.e., it is the constant voltage which must be developed across an ideal inductance if the current is to increase linearly with time. The second term represents the linearly increasing voltage drop appearing across the resistive components of the circuit due to the current sweep produced in the coil.

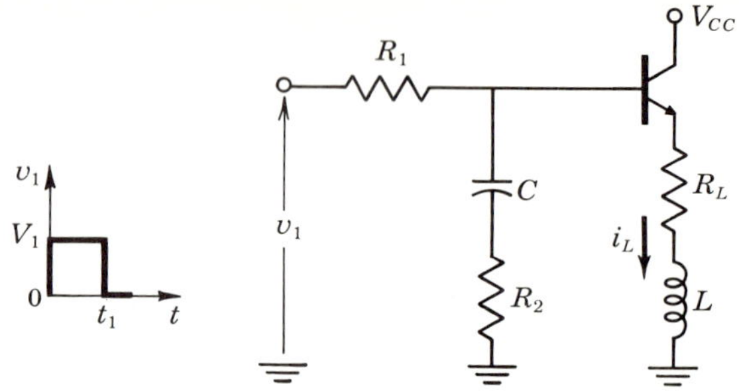

Fig. 7-11 Emitter-follower drive circuit for a current sweep.

The individual components of the trapezoidal drive voltage necessary for a linear sweep may be generated separately, added together, and used to excite the coil. For example, by integrating a rectangular voltage wave in the input circuit of the emitter follower of Fig. 7-11, or alternatively in the gate circuit of a source follower, the voltage impressed across the coil would be of the proper trapezoidal form. The low source impedance of the driver stage limits the peak excursion of the output voltage to a reasonable value. A starting impulse must be separately injected. Many other modes of signal generation and coupling are also possible—through a transformer from a high-impedance source, by capacitive coupling to eliminate the dc component in the coil, or directly from a push-pull circuit.

Example 7-2

The emitter follower used in the sweep of Fig. 7-11 employs as its active element the composite transistor of Sec. 3-6. Each unit has $\alpha = 0.98$. The 200-mH 30-Ω deflection coil must be excited by a linear-sweep current which reaches a maximum of 0.5 A in 20 ms. In generating the input signal, the end of R_1 is connected directly to the 50-V supply. What must R_1, R_2, and C be to satisfy the necessary sweep conditions?

Solution With the composite transistor, the large value of $\alpha_c = 0.9996$ given by Eq. (3-51) means that the overall gain is extremely close to unity. Consequently, the reflected input impedance of the emitter load (R_L and L) back into the base will be large enough so that its loading of the input RC network will be negligible. From Eq. (7-22), the height of the pedestal must be

$$V_A = Lk = 200 \times 10^{-3} \times \frac{0.5}{20 \times 10^{-3}} = 5 \text{ V}$$

At the base we must have an additional 750 mV to allow for V_{BE}. The maximum height of the triangle, neglecting the small emitter resistance, is

$$V_{Tm} = R_L I_m = 30 \times 0.5 = 15 \text{ V}$$

Assume that the initial current through the RC network is 1.0 mA. To obtain a 5.75-V jump at the base input upon excitation,

$$R_2 = \frac{5.75 \text{ V}}{1 \text{ mA}} = 5.75 \text{ K}$$

The remaining 44.25 V must be developed across the other input resistor:

$$R_1 = \frac{44.25 \text{ V}}{1 \text{ mA}} = 44.25 \text{ K}$$

The base must charge from 5.75 to 20.75 V in the 20-ms sweep duration. From the exponential charging equation,

$$\tau_s = \frac{t_s}{\ln [(50 - 5.75)/(50 - 20.75)]} = 48.6 \text{ ms}$$

Since $R_1 + R_2 = 50$ K, $C = 0.97$ μF. We should note that the sweep duration is independent of the 6.7-ms coil-time constant; the reasonably linear sweep is now three times as long.

7-5 A Transistor Bootstrap Sweep

For a second method of linearizing the current sweep, we might consider a circuit which is the dual of the RC voltage sweep of Fig. 6-21, i.e., the parallel combination of a coil and a conductance charged from a constant-current source (Fig. 7-12). This circuit's node equation may be written as

$$v_L = \frac{I' - i_L}{G} \tag{7-23}$$

At this point, Eq. (7-23) should be compared with the equation written for the voltage sweep [Eq. (6-46)] and the argument employed with respect to that circuit reread.

We know that bootstrapping generates an extremely linear voltage sweep. Since we now have at our disposal a current-controlled active circuit element, the transistor, we should consider the possibility of developing the dual circuit, a current bootstrap, as our current sweep. Choosing Eq. (7-23) for the starting point, we see that the replacement of I' by

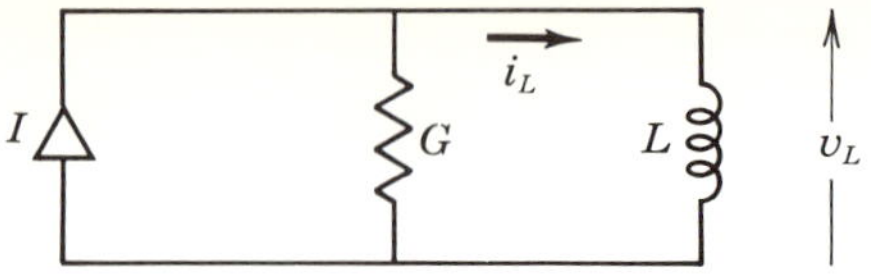

Fig. 7-12 Simple *LG* current sweep.

$I + A_c i_L$ provides the correction term which is so essential. When this is done,

$$v_L = \frac{I + A_c i_L - i_L}{G} \tag{7-24}$$

where A_c is the current gain of the feedback amplifier. Figure 7-13*a* and *b* illustrate the current bootstrapping process defined in Eq. (7-24). It should be noted that the role of the current amplifier is analogous to that performed by the voltage amplifier in the voltage bootstrap. Both supply the additional energy made necessary by the dissipative elements of the external circuit. By this means the system maintains either constant-current or constant-voltage charging of the energy-storage element. Since under these circumstances the dissipative energy remains constant (I^2R or V^2G), the system must supply an ever-increasing amount of energy for storage in the capacitor or inductance. Unless an active circuit element is present to cancel the effects of the circuit resistance (which limits the available storage energy), it is impossible to achieve an absolutely linear sweep.

The role of the amplifier becomes somewhat clearer if we solve for the Thévenin equivalent of the current generator and its internal conductance in Fig. 7-13*b*. One of the two voltage generators resulting is of the form

$$v_a = RA_c i_L$$

where $R = 1/G$. But since this voltage rise is proportional to the series current flow, it may be replaced by a negative resistance $-RA_c$, as shown in Fig. 7-13*c*. The amplifier's action serves to reduce the overall series resistance and to increase the circuit time constant. The current sweep is now defined by

$$i_L = \frac{V}{R_L + R(1 - A_c)}(1 - e^{-t/\tau_s}) \qquad \tau_s = \frac{L}{R_L + R(1 - A_c)}$$

It is immediately apparent that there will be no improvement over the first circuit discussed unless A_c is positive and greater than unity. When $A_c < 1$, there will actually be deterioration of the sweep quality as a result of the decrease in the steady-state current and the time constant. Since both terms change in the same direction, the slope at the origin remains invariant with respect to any instability of A_c. Any improvement in the sweep permits a larger-amplitude output before the nonlinearity becomes excessive.

Ideal constant-voltage charging occurs when there is complete cancellation of the resistance. From Fig. 7-13*c* we see that the necessary amplifier gain is

$$A_c = 1 + \frac{R_L}{R} \tag{7-25}$$

Usually $R \gg R_L$, and consequently the current gain required for linear charging will be but slightly greater than unity. Since we are dependent on the cancellation of two terms of equal magnitude for sweep linearization, any instability in A_c will have pronounced repercussions on the current waveshape, particularly on the time required to reach a predetermined final value. Generally, the amplifier is adjusted so that the net resistance will be small but positive and so that any expected gain instability will never result in a net negative circuit resistance. The high degree of stabilization necessary, if the sweep time is to be very long, sets extremely stringent requirements on the current amplifier.

Previously, we proved that a trapezoidal voltage drive must appear across the physical inductance when a linear-sweep current flows through it [Eq. (7-22)]. Substituting the assumed ideal current $i_L = kt$ into the bootstrap equation (7-24) and equating the result to the known drive signal affords another method of determining the essential system conditions:

$$v_L = R_L kt + Lk = \frac{I + (A_c - 1)kt}{G}$$

Comparing similar terms on both sides of the above equation yields two relationships:

$$A_c - 1 = R_L G = \frac{R_L}{R} \tag{7-26}$$

$$I = LkG \tag{7-27}$$

The first of the above equations establishes the necessary system gain, and as might be expected, it is identical with Eq. (7-25). The second

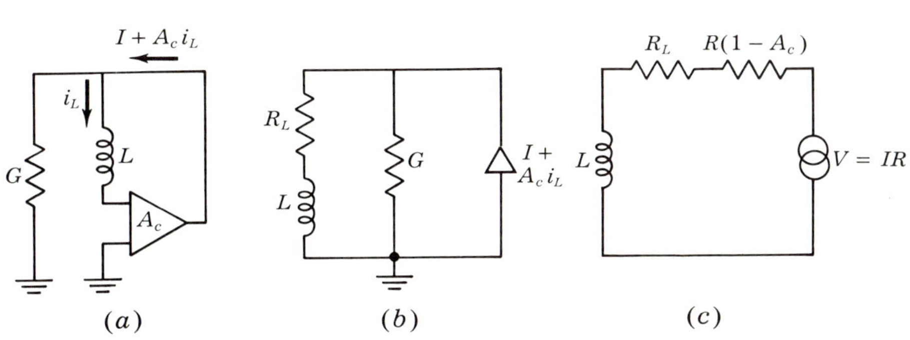

Fig. 7-13 Basic current bootstrap circuit. (*a*) Circuit connection; (*b*) schematic representation; (*c*) Thévenin equivalent circuit.

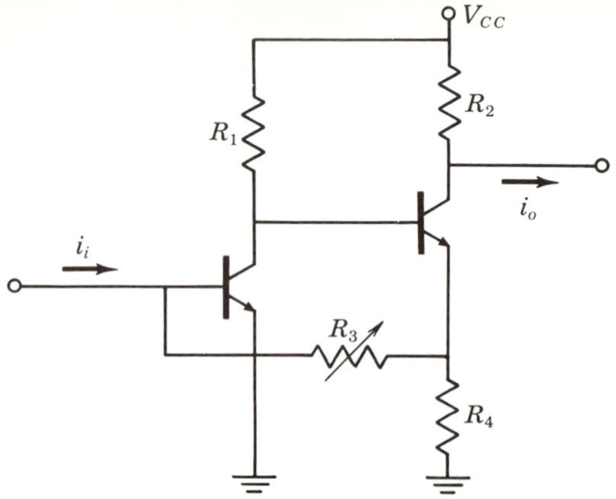

Fig. 7-14 Possible bipolar amplifier for a current bootstrap circuit.

condition [Eq. (7-27)] expresses the constant-voltage charging of the inductive component of the external circuit. By rewriting this in the form

$$V = IR = L\frac{di}{dt} = Lk$$

it also offers a convenient design criterion for the choice of the parallel resistance.

The positive current gain of a single transistor is necessarily limited to somewhat less than α, and consequently a multiple-stage amplifier must be used in this bootstrap sweep. Figure 7-14 illustrates an appropriate circuit, which employs current feedback both to increase the gain stability and as a means of adjusting the gain to the optimum value. This will generally be somewhat less than that given by Eq. (7-25), thus allowing a margin for any remaining amplifier instability. Since the coil resistance will not be completely canceled, if originally R_L was small, little or nothing will be gained by bootstrapping and the switched sweep of Fig. 7-4 might prove to be just as satisfactory.

The current sweep discussed above is the dual of the bootstrap voltage sweep treated in Chap. 6. By analogy it would seem likely that a current-controlled Miller sweep, depending for its timing on the inductance, might also be practical. However, the impossibility of isolating the coil resistance causes any operation on the inductance to be accompanied by an equal effect on the associated coil resistance. Multiplying or dividing both L and R_L by the same factor leaves the time constant unchanged. Because there is no improvement, this method of linearization is never used. It is worth noting that not all forms of voltage-operated systems can, in fact, be

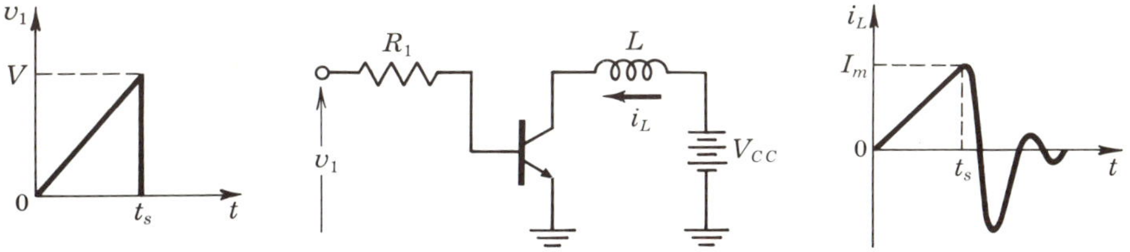

Fig. 7-15 Current sweep generated from a constant-current source.

converted to their duals. Often, as in this case, the unavailability of the needed ideal circuit element (pure inductance) will be the decisive factor.

7-6 Constant-current-source Current Sweep

A contrasting line of thought actually obviates the generation of a current sweep. The transistor is utilized solely as a voltage-to-current converter. A highly accurate voltage sweep is generated elsewhere and applied as a drive: in Fig. 7-15 the constant-current output of the transistor furnishes a driving current proportional to this voltage. We thus bypass the need to develop separately the trapezoidal voltage across the coil, the negative resistance in series with the coil, or any other correction term necessitated by the nonideal inductance. Moreover, by restricting the sweep generation to voltage waveshapes, we are able to employ an almost ideal energy-storage element (i.e., a capacitor), with consequent case of linearization.

The current flow through L, found from the model holding for the active region, is a linear function of the driving voltage:

$$i_L = \beta \frac{v_1(t) - V_{BE}}{R_1} \tag{7-28}$$

At the end of the sweep cycle, the input voltage drops to zero. The turn-off process is identical with the removal of the excitation in the switched sweeps of Sec. 7-2. Here also the recovery would be controlled either by an external damping diode and resistance or by a recovery diode for recurrent sweeps.

Problems

7-1 A coil having an inductance of 100 mH, a resistance of 20 Ω, and stray capacity of 200 pF is charged through a source impedance of 1,000 Ω (Fig. 7-1). The sweep current must reach 150 mA in 50 μs.

(*a*) Specify the necessary charging voltage and calculate the sweep linearity.

(*b*) Sketch the current and voltage waveshapes when the circuit is critically damped.

(c) Repeat part b when the damping resistance is reduced to 1,000 Ω. Specify the peak voltage and the time at which it occurs. Make all reasonable approximations. (*Hint:* Treat in the same way as the double-energy circuit of Chap. 1.)

7-2 The coil of Prob. 7-1 is shunted with additional capacity so that one half cycle of the oscillatory recovery lasts for 60 μs ($R_D = \infty$). Calculate the peak voltage and then compare with the value found for the coil without additional shunting capacity. What does the additional capacity do to the initial sweep delay?

7-3 Prove that the critically damped case has the optimum figure of merit, $\tau_R V_p$.

7-4 The transistor used in the sweep of Fig. 7-4 has $\beta_F = 50$, $\beta_R = 5$, $V_{BE} =$ 700 mV, $V_{CE,\text{sat}} \approx 0$, $r_{sc} = 5\ \Omega$, $BV_{BEO} = 50$ V, $BV_{CEO} = 100$ V, $r_a = 5\ \Omega$. The circuit is excited by a 1-ms current pulse. Coil parameters are $L = 50$ mH, $R_L =$ 150 Ω, $C_s = 5$ nF. $V_{CC} = 24$ V, and the coil is shunted with sufficient resistance to critically damp the recovery response.

(a) What is the minimum input-pulse amplitude that will ensure that the transistor remains saturated over the complete 1-ms sweep interval?

(b) Sketch the current and voltage waveshapes, giving all significant values.

(c) Calculate the current-sweep nonlinearity.

7-5 Design a periodic sweep using the transistor and coil of Prob. 7-4. The sweep interval should be 1 ms. The transistor is used in a grounded-base configuration with the periodic current waveshape injected into the emitter. By this means, we take advantage of the higher $BV_{CBO} = 150$ V. The coil is shunted by a resistor that will keep the peak reverse voltage below BV_{CBO}.

(a) What is the maximum-amplitude sweep possible if the recovery time is one-half of the total period?

(b) Sketch the voltage and current waveshapes and the path of operation in the volt-ampere plane.

(c) We wish to reduce the sweep period to half and to double the sweep amplitude. Is this possible without exceeding the transistor ratings?

7-6 As an alternative to the switched sweep of Fig. 7-4, a voltage excitation is used with an emitter follower, as shown in Fig. 7-16. Use the transistor of Prob. 7-4

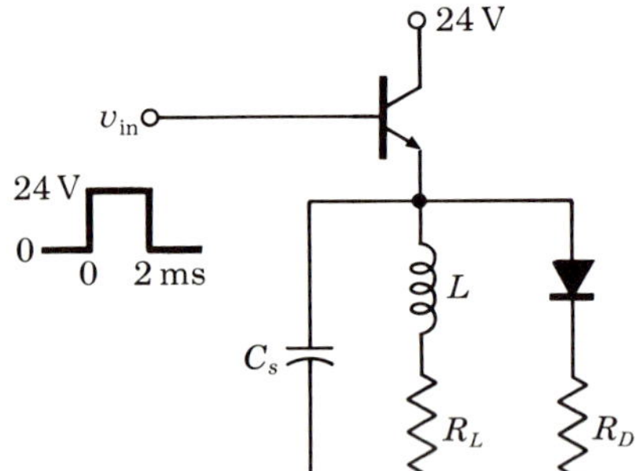

Fig. 7-16

with a coil having $L = 100$ mH, $R_L = 100\ \Omega$, $C_s = 500$ pF. The coil is shunted with a resistor to keep the peak reverse voltage below the junction breakdown voltage.

(*a*) Sketch and label completely the coil-current and emitter-voltage waveshapes.

(*b*) Plot the path of operation on the volt-ampere plane.

7-7 The transistor of Prob. 7-4 is used to switch the state of a relay in its collector circuit (Fig. 7-17). The relay pulls in at a coil current of 10 mA and drops out at a current of 1 mA. When unenergized, the relay looks like an inductance of 100 mH in series with a resistance of 500 Ω. After it is energized and the air gap is closed, the inductance rises to 500 mH. The relay is shunted by an *RC* diode damping network. *C* is selected for critical damping. In the following calculations, make all reasonable approximations.

(*a*) Sketch the waveshape of the coil current, indicating the delay time for relay engagement and the delay time for relay disengagement.

(*b*) Plot the path of operation on the volt-ampere characteristic. Pay particular attention to the retrace path.

(*c*) What will happen if the recovery network is removed?

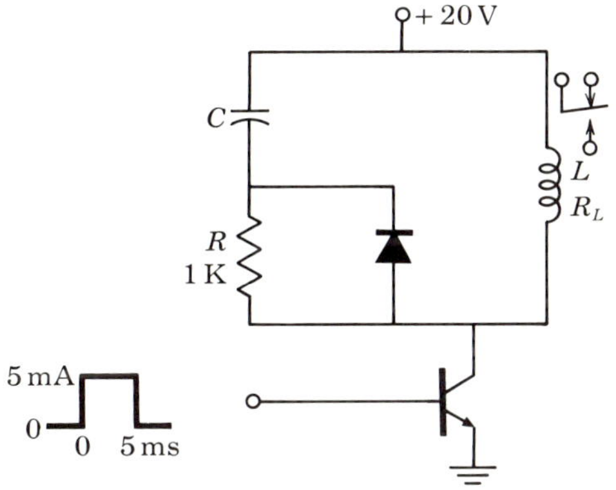

Fig. 7-17

7-8 Repeat Prob. 7-7 with the diode removed. What role does the diode play?

7-9 Repeat Prob. 7-7 with the relay and its recovery network shifted into the emitter of an emitter follower. The input excitation is a 5-ms 20-V pulse.

7-10 Show the approximate path of operation on the transistor volt-ampere characteristic for the circuit of Fig. 7-17 under the following two cases:

(*a*) $C \approx 0$

(*b*) $C \approx \infty$

7-11 Repeat Prob. 7-10 with the diode connected from the collector directly to V_{CC}.

7-12 Prove that the current bootstrap cricuit employing only a single transistor, such as the one shown in Fig. 7-18, produces a sweep of poorer quality than the simple switched coil of Fig. 7-4. Use the coil and transistor of Prob. 7-4. The coil is critically damped.

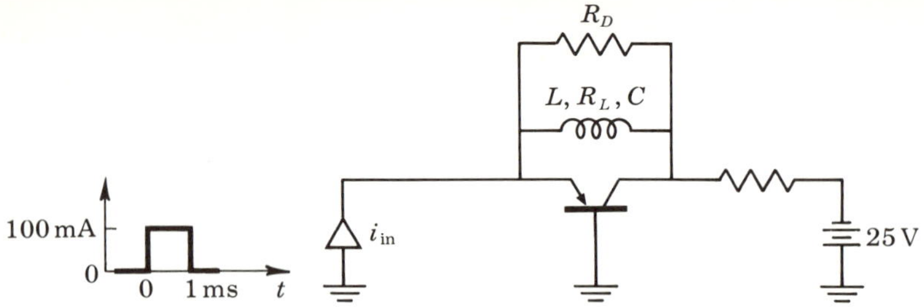

Fig. 7-18

7-13 In the base amplifier of Fig. 7-14, the two transistors are identical, each having $\beta = 50$ and $V_{BE} = 700$ mV. The power supply $V_{CC} = 30$ V, and $R_1 = R_2 = R_4 = 1$ K. Plot the current gain as a function of R_3.

7-14 We wish to use the circuit of Fig. 7-19*a* to produce a perfectly linear current sweep $i = kt$ after the switch is closed at $t = 0$.

(*a*) For this condition, sketch the volt-ampere characteristic of N, labeling all slopes.

(*b*) If $i = kt$ with N properly adjusted, what is the value of k?

(*c*) When N is represented by a two-stage current amplifier with feedback (Fig. 7-19*b*), specify the necessary current gain β in terms of R_1, R_S, and R_L.

(*d*) How does the time constant of the circuit vary as a function of β, for small variations in β, in the vicinity of the optimum value?

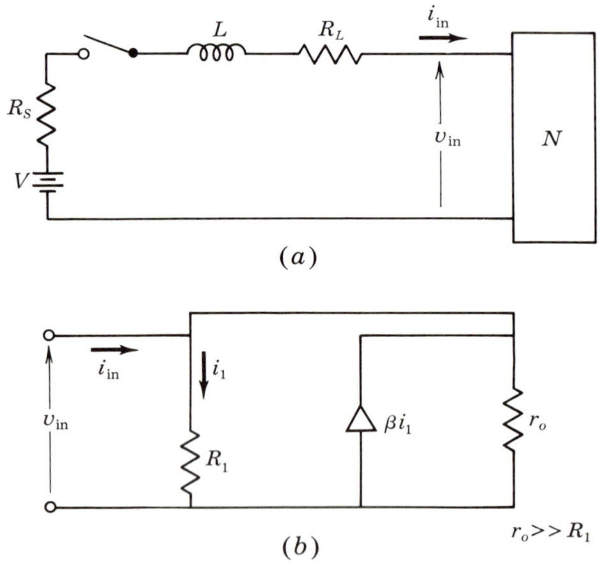

Fig. 7-19

7-15 In the current sweep of Fig. 7-15, the driving voltage increases linearly from 0 to 10 V in 100 μs. The circuit components are $R_1 = 1$ K, $L = 500$ mH, $R_L = 50\ \Omega$, $C = 0.01\ \mu$F, $V_{CC} = 20$ V, $\alpha = 0.98$, $r_{sc} = 20\ \Omega$, and $V_{BE} = 700$ mV.

(*a*) Sketch the coil current and voltage to scale, specifying all values. Make all reasonable approximations.

(*b*) Calculate the value of the damping resistance which should be included across the coil if the maximum collector voltage must be limited to 100 V. Repeat part *a*, first when this resistor is directly across the coil and then when it is in series with a damping diode.

7-16 A series *RL* circuit is excited by a linear-sweep voltage from a low-impedance source such as an emitter follower. The total series resistance is 100 Ω, and the coil inductance is 1 H. Evaluate and sketch the current and voltage waveshapes appearing across L:

(*a*) When τ is long compared with the time duration of a single applied drive signal

(*b*) When τ is short compared with the time duration of the drive signal

7-17 (*a*) Prove that the coil current in the sweep of Fig. 7-15 can be expressed as

$$i = Ae^{-t/\tau} + Bt + C$$

where A, B, C, and τ are constants of the circuit.

(*b*) Evaluate the above constants in terms of the circuit and transistor parameters.

(*c*) Is it possible to generate a perfectly linear sweep? What conditions must be satisfied if the sweep nonlinearity is to be minimized?

Bibliography

Chance, B., et al.: "Waveforms," M.I.T. Radiation Laboratory Series, vol. 19, McGraw-Hill Book Company, New York, 1949.

Goodrich, H. C.: A Transistor Horizontal Deflection System, *RCA Rev.*, vol. 18, pp. 293–321, September, 1957.

Nillesen, A. H.: Field Deflection Circuit for Tiny-vision Receivers, *Electron. Appl.*, vol. 27, no. 1, pp. 12–29, 1966–1967.

Soller, T. M., A. Starr, and G. E. Valley, Jr.: "Cathode-ray Tube Displays," M.I.T. Radiation Laboratory Series, vol. 22, McGraw-Hill Book Company, New York, 1948.

Sziklai, G. C., R. O. Lohman, and G. B. Herzog: A Study of Transistor Circuits for Television, *Proc. IRE*, vol. 41, no. 6, pp. 708–717, June, 1953.

Texas Instruments Inc.: "Application Note; Switching Inductive Loads with Transistors," Dallas, Tex., 1963.

PART 3
switching

8.
collector-base and drain-gate coupled multivibrators

8-1 Basic Multivibrator Considerations

Any closed-loop regenerative system having two or more clearly defined stable or quasi-stable states, each of which is maintained without recourse to external forcing, might well be categorized as a multivibrator. The application of a low-energy-content stimulus, a trigger, starts the switching action, which drives the operating point away from one stable state and toward the next. But the important attribute of the multivibrator which differentiates it from other multistate systems such as a switch or relay is that once the trigger brings the circuit into its regenerative region, the additional energy necessary to complete the transition is supplied by the system itself.

The necessary and sufficient conditions for any system to operate as a multivibrator may best be established by considering the functional form of the multivibrator. In the block diagram of Fig. 8-1, A_1 and A_2 are the ideal amplifiers containing all of the circuit's active elements; K_1 and K_2 are completely passive coupling networks.

Upon the application of any input stimulus v_{in} (even noise), the signal returned through the loop is

$$v_o = A_1 A_2 K_1 K_2 v_{\text{in}} = A_L v_{\text{in}} \tag{8-1}$$

When the loop gain $A_L > 1$, the returned signal is larger than the applied stimulus, and once the switch is closed, it will reinforce the original signal. The closed-loop system is unstable, with buildup continuing until the balance

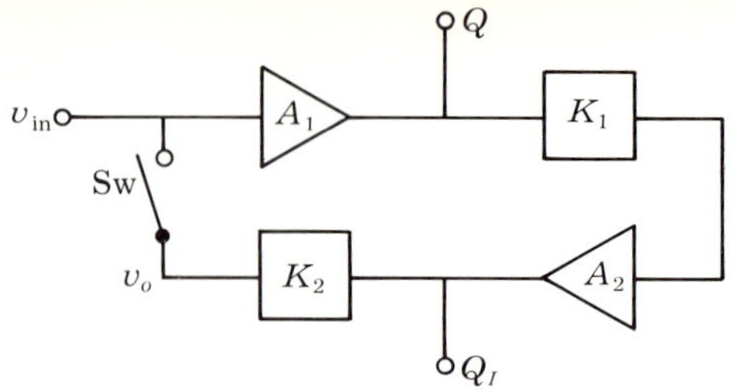

Fig. 8-1 Basic representation of a closed-loop regenerative system—outputs at Q and Q_I.

of the returned signal and the input is achieved, either through a decrease in the amplifier gain (saturation) or through the opening of the transmission path (active-element cutoff).

In order to satisfy the switching condition, that is,

$$A_L > 1$$

the two amplifiers must have the same sign. Usually an inverter which has a negative gain is employed for each amplifier stage. The inverter may range from a single bipolar or MOS transistor stage in a lumped-element multivibrator to the four-transistor totem-pole output stage of the DTL or TTL logic discussed in Chap. 4 (Fig. 4-28). Integrated-circuit operational or differential amplifiers may also be cross-connected to form a multivibrator.

Balance conditions require that when the switch is closed,

$$v_o' = v_{in} \tag{8-2}$$

The switching action of the bistable multivibrator is illustrated by the circuit of Fig. 8-2, where two saturating RTL inverters are directly coupled (see Fig. 4-16). The coupling constants K_1 and K_2 in Eq. (8-1) are both unity. Let us assume that Q_1 is the ON transistor. Its logical-0 output of $V_{CE,\text{sat}} \approx 200$ mV coupled to Q_2 turns Q_2 OFF. The output of Q_2 is at the high logical-1 level, and consequently, it drives Q_1 into saturation, setting Q_1's output to the originally assumed logical-0 level. Thus our initial assumption is proved to be a possible circuit state.

Figure 8-2*b* shows the transfer function of a single RTL inverter stage under the conditions where its output is loaded by a second inverter. In calculating this response, the loop was opened at point x. We shall assume that when the transistor saturates, $V_{BE} = 700$ mV. Our linearized model further assumes that a transistor starts becoming active approximately 100 mV below the saturation base-emitter drop. Thus the edge of the active region, which occurs at $i_b = 0$, corresponds to $v_x = 600$ mV. At saturation,

$$v_x = I_{b1,\text{sat}}R_1 + 0.700 = 0.760 \text{ V}$$

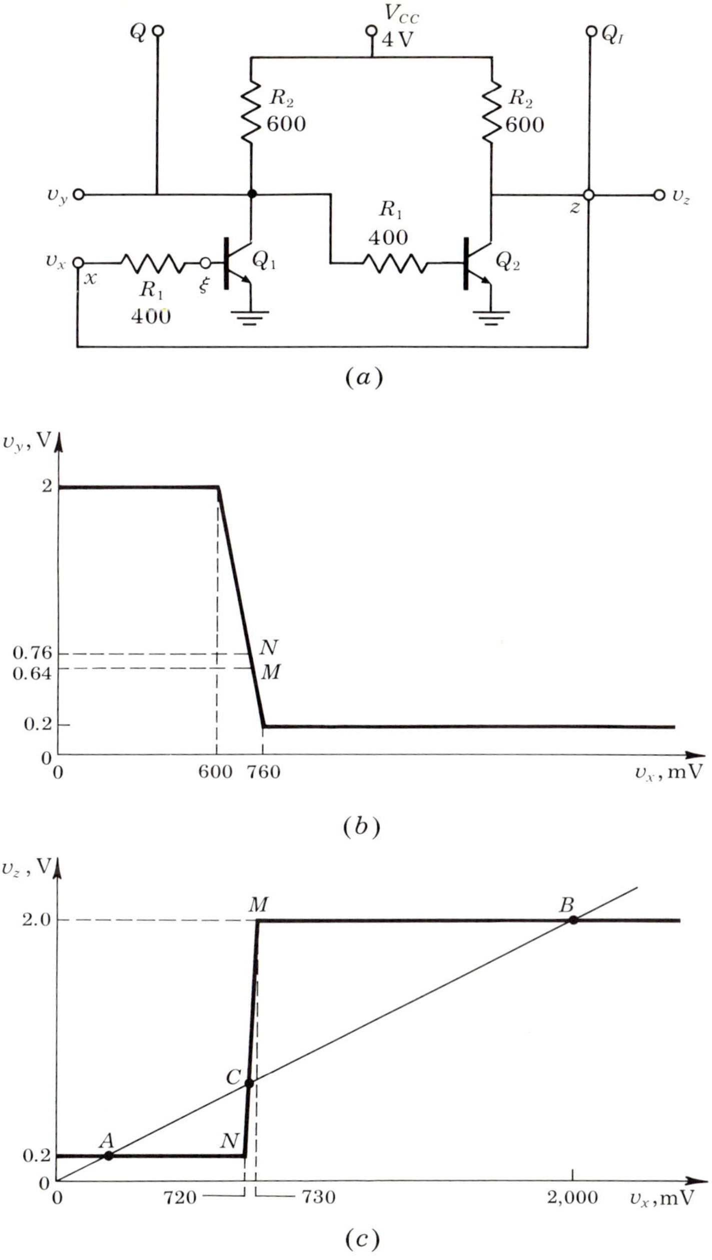

Fig. 8-2 (*a*) Bistable circuit composed of direct-coupled RTL inverter stages; (*b*) transfer characteristic of a single stage (Q_1); (*c*) overall transfer characteristic, showing balance conditions at *A*, *C*, and *B*. Points corresponding to the knees of the curve at *N* and *M* (respectively, the saturation and cutoff of Q_2) are marked on the single-stage characteristic of Q_1 in (*b*).

Figure 8-2c shows the overall open-loop transfer characteristic from point x to the output Q_I at z. In plotting the overall transfer function, the output voltages shown in Fig. 8-2b are applied as the input to a second identical stage. Superimposed on this plot is the unity gain locus ($v_x = v_z$), and we see that three points on the characteristic (A, C, B) satisfy the closed-loop balance condition.

Satisfaction of the balance criterion is a necessary but insufficient condition for stable operation. Consider point C, where the unity gain intersects the loop transfer characteristic in the region where both amplifiers are active. The slope of the transfer characteristic, evaluated at a particular point, is simply the value of the corresponding loop gain A_L. At point C, $A_L > 1$. Any slight perturbation which might momentarily shift the operating point to the right will be amplified and returned as a much larger disturbance. The operating point travels up the transfer characteristic from C to point B, where the balance condition is again satisfied. At B, the output inverter is cut off and the input inverter is saturated. Here the incremental gain A_L is less than 1; it is, in fact, zero. Any small disturbance away from B immediately damps out, and the operating point returns to B, one of the two stable points. Similar reasoning shows that the other stable point is A, where the output inverter is saturated and the input inverter is cut off.

We conclude, from the previous discussion, that the second necessary condition for stable switching is that $A_L > 1$ in the active region and that $A_L < 1$ at the stable point. It also follows that there must always be a point of unstable equilibrium between each set of stable points. As an exercise, the reader may furnish the proof of this statement.

The initial starting from any unstable point, such as C, would be caused by circuit noise or some other small perturbation. However, in order to switch from one absolutely stable point to the other, we must disturb the system at least enough to shift the operating point into the regenerative region, where $A_L > 1$. This region is easily delineated by marking its boundary points ($A_L = 1$). We simply find the two points on the characteristic with unity slope (M, N). These points appear at the knees of the piecewise-linear transfer characteristic of Fig. 8-2c. Thus the trigger amplitude required to switch from A to B becomes

$$v_T > v_N - v_A$$

To shift back from B, we must introduce a positive disturbance greater than the distance BM. Switching techniques are treated in Sec. 8-3 below, as well as in Chap. 13.

In any particular circuit, the design conditions depend on the input and output characteristics of the two amplifiers. We start by drawing the models. For the simple inverter bistable of Fig. 8-2a, the models holding

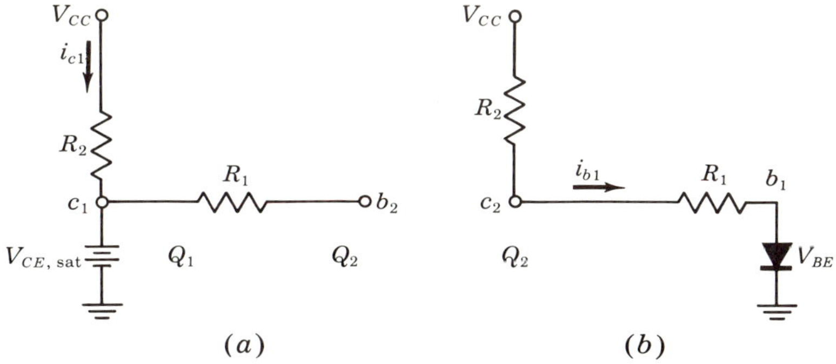

Fig. 8-3 Models of the transistors for each state of the bistable. (*a*) Coupling from the ON Q_1 to the OFF Q_2; (*b*) coupling from the OFF Q_2 to the ON Q_1.

in each region are shown in Fig. 8-3. To ensure the saturation of Q_1, I_{b1} must satisfy the condition

$$I_{b1} = \frac{V_{CC} - V_{BE}}{R_1 + R_2} \geq \frac{V_{CC} - V_{CE,\text{sat}}}{\beta R_2} \tag{8-3}$$

If we approximate the solution of Eq. (8-3) by setting $V_{BE} \approx 0$ and

$$V_{CE,\text{sat}} \approx 0$$

the maximum value of R_1 is given by

$$R_1 \leq (\beta - 1)R_2 \tag{8-4}$$

By treating this circuit as a current amplifier, we shall now demonstrate that the condition described above is the very one necessary for guaranteed regenerative switching. Suppose that the loop is broken at the base of Q_1 and that point ξ in Fig. 8-2 is shorted to ground. Then the loop current gain, from an input at Q_1 to point ξ, is (assuming an ideal input, where $h_{ie} \ll R_1$)

$$\frac{i_\xi}{i_{b1}} = A_c \approx \left(\frac{R_2}{R_1 + R_2}\right)^2 \beta^2 \tag{8-5}$$

For regenerative operation, $A_c > 1$, and thus

$$R_1 < (\beta - 1)R_2$$

which is identically the condition previously given by Eq. (8-4).

We can draw some further conclusions from the transfer characteristics of Fig. 8-2*b* and *c*. The flat portions of the characteristic arise as a direct result of driving one of the active elements into saturation or cutoff. Under balance conditions, the two identical amplifiers must be in different states. When A_1 is saturated, A_2 is off, and vice versa. Near the knee of the curve even when transistor Q_1 is active, its low collector voltage keeps

Q_2 off. In some circuits, Q_2 may also become saturated while Q_1 remains in the active region. Since both active elements cannot remain in the same condition, the only permissible combinations of inverter states are those tabulated below. We assume that A_1 is in the ON condition (low output) and A_2 is OFF (high output).

Permissible inverter states

A_1	A_2
Saturated	Active
Saturated	Cutoff
Active	Cutoff

When energy-storage elements are included in the coupling networks, the multivibrator can operate in a toggling mode, where the exponential decay of the energy stored will eventually bring the circuit back from one of the limiting regions to the regenerative region. As a consequence, the multivibrator switches between the points satisfying the balance conditions without resorting to external triggers. But since the system remains at each particular state for a well-defined interval, this only changes a stable point to a quasi-stable one.

Transition from state to state takes a small finite time interval, which will be evaluated in Sec. 8-4. The transistor on-off and minority-carrier-storage times will further affect the switching speed. These delays are not germane to the system's basic behavior, and we shall therefore ignore them in the following discussion.

8-2 Bistable Multivibrators

A logic-based direct-coupled bistable multivibrator operates by using the logical-0 output of one inverter as the input to the other inverter, and vice versa. Interconnections are direct because logic elements are designed to provide standard logical-0 and logical-1 outputs from standard logical-1 and logical-0 inputs.

A MOSFET bistable

Devices which have high input impedances under all conditions can be directly interconnected to form a bistable circuit. One input will not load the preceding output. In Fig. 8-4, each of the induced-n-channel MOSFET inverters uses another MOSFET operating in the high-resistance saturation

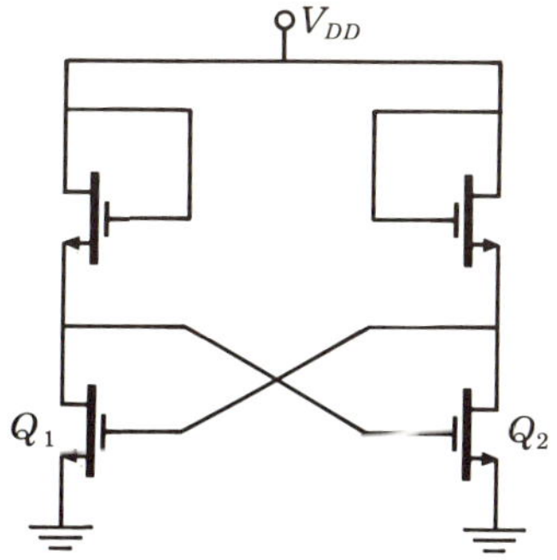

Fig. 8-4 A MOSFET directly coupled bistable circuit (induced-n-channel devices).

region as a load (see Sec. 5-5). The logical-0 output of the assumed bottomed Q_1 is

$$V_{D0} \approx 0$$

This is well above the threshold V_T of Q_2, and consequently Q_2 is cut off. Q_2 has a logical-1 output,

$$V_{D1} = V_{DD} - V_T$$

which must be well beyond V_T to ensure that Q_1 is actually in the low-resistance ohmic region. Triggering introduced at the gate can turn the OFF device on (or the ON device off), thus switching the bistable's state.

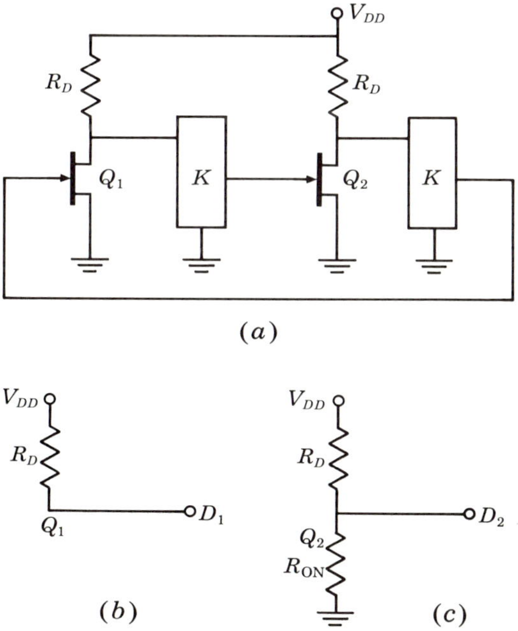

Fig. 8-5 (*a*) A junction-FET bistable; (*b*) Q_1 OFF model; (*c*) Q_2 ON model.

A junction-FET bistable

Let us now consider the more general case illustrated in Fig. 8-5. The junction-FET inverter used as the base amplifier is driven between the two limiting states of cutoff and bottoming. The necessary input gate voltage and corresponding output drain voltage for each stage are given in the table below. We have assumed that there is no loading of R_D by coupling networks K_1 and K_2.

FET	STATE	INPUT	CONDITION	OUTPUT
Q_1	OFF	Logical 0	$V_G < V_p$	Logical 1 $V_{D1} = V_{DD}$
Q_2	ON (ohmic region)	Logical 1	$V_G > 0$	Logical 0 $V_{D0} = V_{ON} \approx V_{DD} \dfrac{R_{ON}}{R_{ON} + R_D}$

In the above table we see that the logical-0 output of Q_2, which must be coupled to the gate of Q_1, is well above the input 0 threshold of V_p. If we directly couple the two stages, then the positive output of Q_2 will also drive Q_1 into the ohmic region and the two devices will not be in different states.

Once a problem is stated properly, the very statement often defines the nature of the solution. Proper multivibrator operation is predicated on establishing the correct input (gate) voltages in each of the two circuit states. Since these depend on the constants of the coupling networks, we should determine their interrelationship, at least at the limits, as a guide in circuit design.

We see from the table of device conditions that the coupling network must shift the high positive-output levels down to the logical-0 negative-input levels required at the gate of the next stage. Two simple forms of level shifting are shown in Fig. 8-6. In Fig. 8-6*a*, a series zener diode results in a subtractive voltage shift. The two defining conditions for the gate voltage at the output of the level shifter become

Logical 1: $$V_{G1} = V_{DD} - V_Z > 0 \tag{8-6a}$$

Logical 0: $$V_{G0} = V_{ON} - V_Z < V_p \tag{8-6b}$$

where V_{ON} is the bottoming voltage of a FET driven into the ohmic region. The series R and the negative bias V_{GG} establish a bias current through

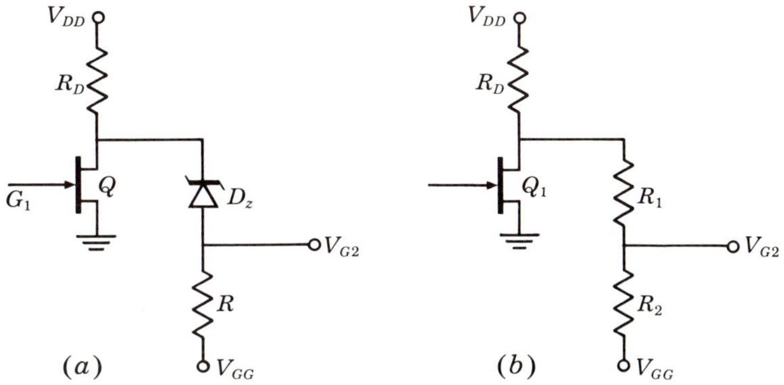

Fig. 8-6 Level-shifting networks. (*a*) Subtractive level shifter using a zener diode; (*b*) resistive divider level shifter.

the zener diode and permit the shifting down to the negative voltage required to cut off the FET.

In writing Eqs. (8-6*a*) and (8-6*b*) we assumed that the current through the level-shifting network is very small and that it will not change the unloaded output logical-1 and logical-0 levels. If this assumption is unjustified in any specific circuit, then the voltages on the left side of the inequality must be modified accordingly.

Figure 8-6*b* illustrates a purely resistive level shifter. Again assuming no loading of R_D by R_1 and R_2 and using a coupling factor,

$$K = \frac{R_2}{R_1 + R_2}$$

The two defining equations become

Logical 1: $$V_{G1} = KV_{DD} + (1 - K)V_{GG} > 0 \tag{8-7a}$$

Logical 0: $$V_{G0} = KV_{\text{ON}} + (1 - K)V_{GG} < V_p \tag{8-7b}$$

Since $V_{\text{ON}} \approx 0$, the inequality of Eq. (8-7*b*) may be approximated by

$$(1 - K)V_{GG} < V_p \tag{8-8}$$

Gate-to-substrate diode conduction will limit the gate voltage to one diode drop regardless of the solution of Eq. (8-7*a*) or (8-6*a*). Equations (8-7*a*) and (8-7*b*) can be simultaneously satisfied by choosing the two constants V_{GG} and K. But since K is also an attenuation factor, a third constraint must be considered. This comes from the loop gain condition $A_L > 1$. Each identical stage must satisfy the gain requirement

$$|g_m R_D K| > 1$$

Usually the supply voltages V_{DD} and V_{GG} are constraints imposed on the circuit designer by the available power supplies. The two conditions of Eqs. (8-7*a*) and (8-8) reduce to

$$1 - \frac{V_p}{V_{GG}} > K > -\frac{V_{GG}}{V_{DD} - V_{GG}}$$

The mean of the inequalities is often chosen as a reasonable compromise design condition.

Figure 8-7 shows the complete FET bistable coupled by a resistive level-shifting network. The voltage levels indicated take into account the loading of R_D by the coupling networks. The capacitor C_1 shunting R_1 in the coupling network is called a *speedup capacitor*. During the switching interval, it couples the full change in voltage at the drain to the gate, speeding up the transition time. The input impedance at the gate of the FET is primarily capacitive, and it is shunted by the coupling network resistance R_2. In Sec. 1-8 it was shown that a compensated attenuator has a much faster rise time than an uncompensated attenuator.

Example 8-1

In this problem we shall investigate the role performed by the speedup capacitor C_1 in the bistable multivibrator of Fig. 8-7. The circuit states

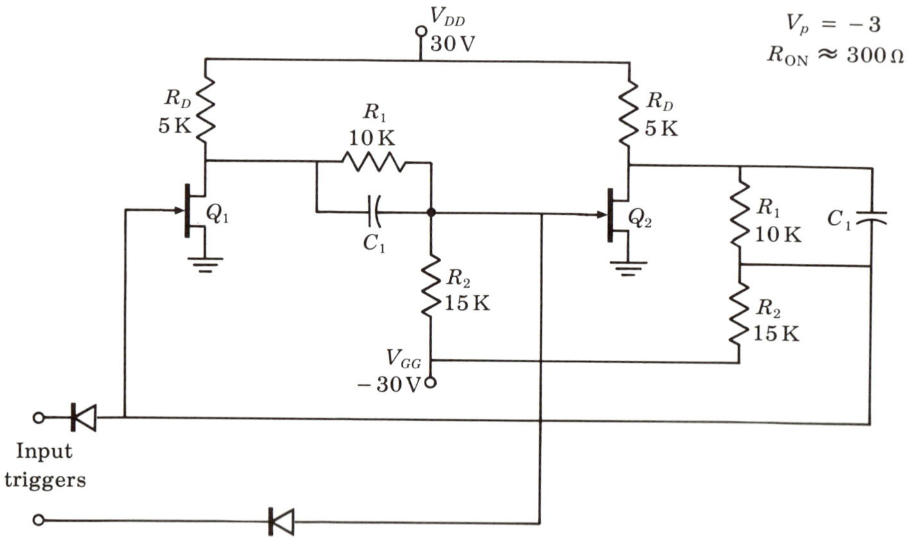

Fig. 8-7 Complete FET bistable, showing compensated-attenuator coupling and diode trigger injection.

are triggered by positive pulses injected at both gates through capacitors. The FET used has $V_p = -3$ V, $r_o = 200$ K, $I_{DSS} = 6$ mA, $C_{GS} = 5$ pF, and $C_{GD} = 5$ pF. We see that the drain voltage will swing from 40 V when the FET is cut off to approximately 0 V when the gate is driven slightly positive. To prevent false triggering, the gate of the OFF FET is maintained at -12 V, which is well beyond the pinchoff voltage.

After the external pulse is injected, the cutoff gate of Q_2 will charge from -12 V toward the new value established by the input trigger, with the time constant

$$\tau_a = (R_D + R_1) \parallel R_2 C_{GS} = 37.5 \text{ ns}$$

Eventually Q_2 will become active, and regeneration will complete the switching. If the input pulse is extremely narrow, the net change of the gate voltage may be insufficient to maintain the switched states after the trigger is removed; the multivibrator will subsequently return to its original condition.

By shunting R_1 with C_1, the coupling network is changed into the compensated attenuator discussed in Sec. 1-8. Usually, for the fastest possible switching, it is adjusted to be slightly overcompensated. However, for the purpose of this discussion, assume that $C_1 R_1 = R_2 C_{GS}$, or that

$$C_1 = 7.5 \text{ pF}$$

This capacity is far from critical, and in an actual circuit it may vary from 5 to 10 pF. The switching-time constant now becomes, from Eq. (1-35),

$$\tau_a' \approx \frac{C_1 C_{GS}}{C_1 + C_{GS}} R_D = 15 \text{ ns}$$

We see that the presence of C_1 reduces the critical time constant τ_a by a factor of 2.5.

The above argument attempted to show, in a relatively qualitative manner, that C_1 speeds up the switching process and permits operation with quite narrow trigger pulses. Many factors are changing during the interval considered: one FET is turning off; the other one becomes active, then its gate becomes positive, and then it saturates. Therefore no attempt will be made to define exactly the required pulse width. We can, however, say that it should be somewhat longer than the calculated time constant τ_a' but much narrower than the pulse needed in the uncompensated circuit.

A junction-transistor bistable with level shifting

The same technique of level shifting can be used with a junction transistor to speed up the turn-off process of the saturated transistor. In the symmetrical circuit of Fig. 8-8, the coupling network is returned to a negative supply. As before in analyzing this circuit, we first draw the models holding under each circuit state. From the models of Fig. 8-8*b* and *c*, we can write the defining equations.

With Q_2 saturated, to ensure that Q_1 is off,

$$KV_{CE,\text{sat}} + (1 - K)V_{BB} < 0 \tag{8-9a}$$

For Q_2 saturated,

$$I_{b2} = \frac{V_{CC} - V_{BE}}{R_L + R_1} + \frac{V_{BB} - V_{BE}}{R_2} > \frac{V_{CC} - V_{CE,\text{sat}}}{\beta R_L} \tag{8-9b}$$

where K is defined, as before, to be $K = R_2/(R_1 + R_2)$. When V_{BB} is negative, the second term of Eq. (8-9*b*) is a negative current which shifts the

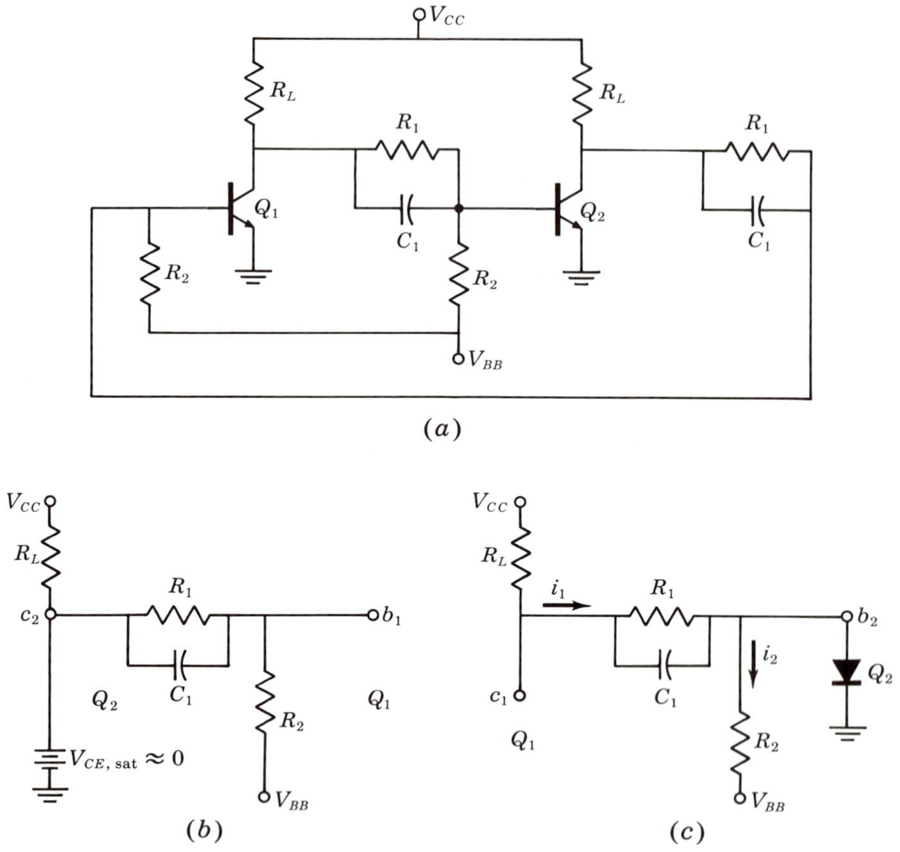

Fig. 8-8 (*a*) Junction-transistor bistable with level shifting; (*b*) coupling network from saturated Q_2 to OFF Q_1; (*c*) coupling network from OFF Q_1 to saturated Q_2.

current levels down. The first term is the positive base current, which flows through the load resistor at the collector of the OFF transistor. Once Q_1 turns on, this current term drops to zero. But Q_2 will not begin to turn off until the total excess base charge is removed, either through recombination or through the use of a negative base current. The second term of Eq. (8-9*b*) is exactly this negative current, and, as was pointed out in Chap. 3, when negative base currents are possible, the storage-delay time will be greatly reduced.

The charge stored in the shunting capacitor C_1 also speeds up the switching process through supplying negative charge to the base. If we set

$$R_1C_1 = \tau_B$$

where τ_B is the base recombination-time constant, then the charge stored in C_1 will be exactly equal and opposite to the charge stored in the base. Once Q_1 saturates and returns the left side of C_1 to ground, the capacitively stored negative charge will supply the carriers needed to neutralize the minority carriers stored in the base. For this reason, speedup capacitors are normally used in all coupling networks, even in the one shown in Fig. 8-2.

8-3 Triggering of Bistables

The bistable multivibrators of Secs. 8-1 and 8-2 can be set into any desired state by individually triggering each base. A positive pulse coupled into the OFF transistor will turn it on, or a negative pulse into the ON transistor will turn it off. Obviously, sequences of alternating positive and negative triggers capacitively coupled into either base will program the bistable states. The input triggers may also be introduced sequentially to alternative inputs by satisfying the conditions of a logic network at each half of the bistable. Chapter 13 is devoted to a detailed discussion of logic triggering.

It is much more convenient to use a train of positive (or negative) pulses applied simultaneously to both bases for triggering. The pulse must be injected so that it will affect only one-half of the bistable; i.e., a positive pulse must turn only the OFF transistor on, a negative pulse must trigger only the ON transistor off. Two possibilities of simultaneously coupling the pulses into both bases are shown in Fig. 8-9.

We shall first consider triggering with positive pulses, as shown in Fig. 8-9*a*, starting our analysis at the instant the pulse is applied. Looking into the common node N, each path has two diodes to ground, the coupling diode in series with the base-emitter diode of the bistable transistors. The ON transistor's base is at V_{BE}. This voltage is established by the

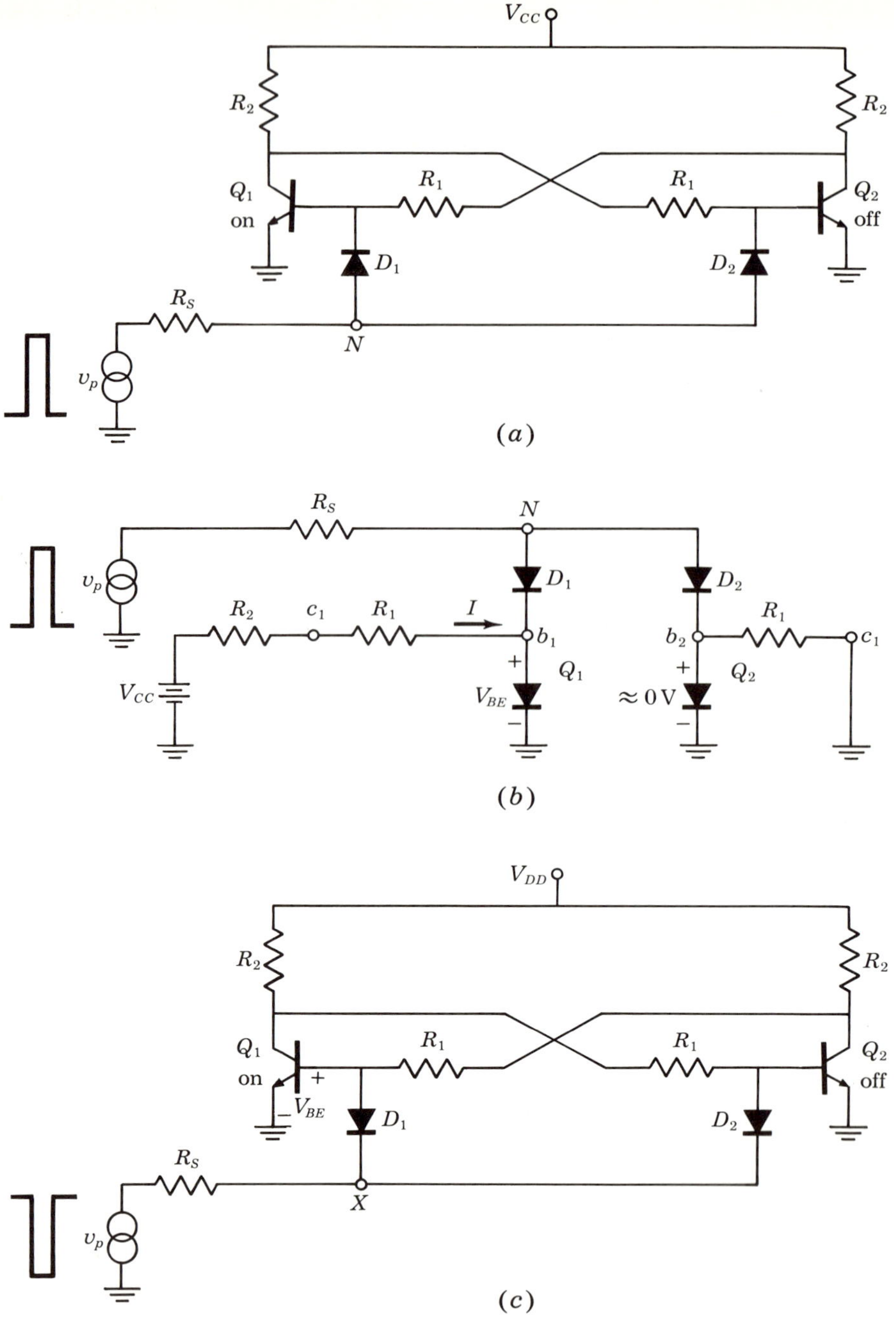

Fig. 8-9 Diode triggering networks for bistables. (*a*) Positive-pulse trigger circuit; (*b*) equivalent input circuit at the bases; (*c*) negative-pulse triggering circuit.

current flow through the coupling network. From the model of Fig. 8-9*b*, we can write

$$V_N = V_{BE1} + V_{D1} = V_{D2} + V_{BE2} \qquad (8\text{-}10)$$

where $$V_{BE1} = 700 \text{ mV} \qquad \text{and} \qquad V_{BE2} \approx 0$$

Substituting into Eq. (8-10) and solving, we find that

$$V_{D1} \approx 0 \qquad V_{D2} \approx 700 \text{ mV}$$

This result shows that the drop across the conducting transistor's base back biases the trigger diode D_1 and forward biases the trigger diode D_2. D_2 is the diode coupling the applied pulse to the OFF transistor. It supports a current flow

$$I = I_{rs}e^{V_{D2}/V_T}$$

into the OFF transistor, turning it on. We conclude that the current supplied by the positive trigger serves solely to turn on the OFF transistor. It has little or no effect on the ON transistor.

After the OFF transistor is driven into the active region, regeneration completes the transition by turning off the ON transistor.

If the input trigger duration is too long, the circuit may retrigger. When both bases are at the same potential (because of equal transistor currents), the input current will divide equally and may drive both toward saturation. Erratic triggering can be prevented by limiting the area of the pulse. External charge supplied should start the switching but should not be sufficient to saturate both transistors.

Negative pulses may also be employed, as shown in Fig. 8-9*c*. The diode connected to the base of the ON transistor (D_1) conducts and, by introducing an auxiliary current path, shunts the base current that normally flows in the base of Q_1. Q_1 turns off. Diode D_2 is back-biased until Q_2 turns on. Again the pulse area must be controlled within narrow limits or erratic triggering will occur.

The diodes, back to back with respect to the transistor bases, have no effect on circuit operation in the absence of pulses.

Triggers may also be simultaneously injected into both bases through capacitors. The results will be similar.

8-4 Multivibrator Rise Time

After the external trigger brings the circuit into the regenerative region, we face the question: How much time is required for the transition between states?

Switching between the stable states of the bistable, or between the quasi-stable states of the astable (Sec. 8-7), is actually a transition through the entire active region of the loop amplifier. Within this region the gain response is characterized by the feedback equation

$$A = \frac{A_L}{1 - A_L}$$

where A_L is the gain of the two cascaded inverter stages.

For regenerative behavior, at least one pole of the multivibrator must lie in the right half plane. This pole (which is a zero of $1 - A_L$) will determine the nature of the transient response during the self-switching process.

Once we assume that the high-frequency response of each identical inverter is primarily determined by a single dominant pole located at ω_c, we can write the inverter gain in the form

$$A_I = -\frac{A_o}{1 + s/\omega_c}$$

where A_o is the midband gain. Substituting $A_I^2 = A_L$ into $1 - A_L = 0$, we obtain the polynomial

$$s^2 + 2\omega_c s + (1 - A_o^2)\omega_c^2 = 0 \tag{8-11}$$

Equation (8-11) has two roots; the significant one is located at

$$s_p = (A_o - 1)\omega_c \tag{8-12}$$

We see from Eq. (8-12) that only if $A_o > 1$ will this pole lie in the right half plane and only then will the multivibrator be regenerative.

In the active region, the transient response is of the form

$$v(t) \approx A + Be^{t/\tau_p} \tag{8-13}$$

where $\tau_p = 1/s_p$ from Eq. (8-12). The constants A and B must be determined from the initial circuit conditions, which depend upon the amplitude of the applied trigger. Because, in general, this is not known, we instead define the rise (and fall) time of a pulse. This is the time required for the output to increase from the 10 to the 90 percent point. The 9:1 increase in amplitude is a function solely of the positive exponential term in Eq. (8-13). Solving for $e^{t_R/\tau_p} = 9$, we obtain a rise time of

$$t_R = 2.2\tau_p \tag{8-14}$$

which is identical to that given in Eq. (1-32) for a negative exponential response.

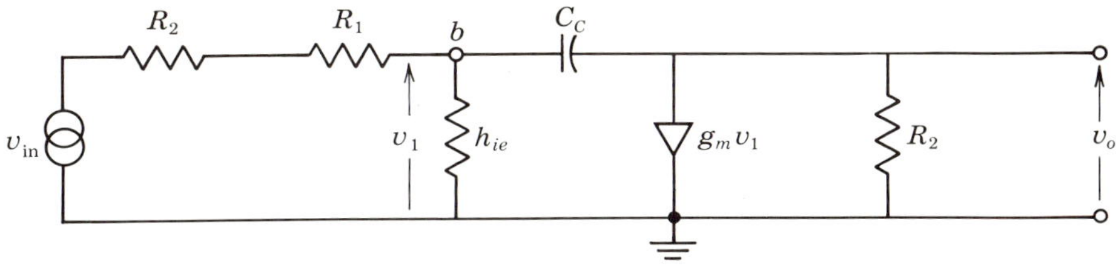

Fig. 8-10 Simplified high-frequency model of a single-stage RTL inverter used in the bistable of Fig. 8-2.

In normal circumstances, the high-frequency break (ω_c) is a function of the gain of the inverter stage. For example, let us examine the simplified high-frequency small-signal model of the RTL inverter stage used in the bistable of Fig. 8-2. From Fig. 8-10 we find that ω_c is given by

$$\omega_c = \frac{1}{[(R_1 + R_2) \parallel h_{ie}]C_C(1 - A_V)} \tag{8-15}$$

In Eq. (8-15), $C_C(1 - A_V)$ is the input capacity of the stage seen at the base. $A_V = -g_m R_2$ is the voltage gain from the base to the collector.

The magnitude of the midband voltage gain can also be found from the model of Fig. 8-10:

$$|A_o| = \frac{h_{ie}}{h_{ie} + R_1 + R_2} g_m R_2$$

If we have a high-gain stage where $|A_V| \gg 1$ and $|A_o| \gg 1$, then when we substitute into Eq. (8-12), we find that the positive pole is located at

$$s_p = \frac{1}{C_C(R_1 + R_2)}$$

It follows that fast rise times are only possible with a high-frequency transistor (small C_C) operating in a low-impedance circuit. In the circuit of Fig. 8-2, with $C_C = 7.5$ pF,

$$s_p = 1.33 \times 10^8$$

and the rise time of the multivibrator becomes

$$t_R = 16.5 \text{ ns}$$

8-5 Monostable Multivibrators—Bipolar Transistor

The introduction of a single energy-storage element into one of the coupling networks converts a bistable multivibrator into a circuit having one stable and one quasi-stable state. If we examine the nature of the new network

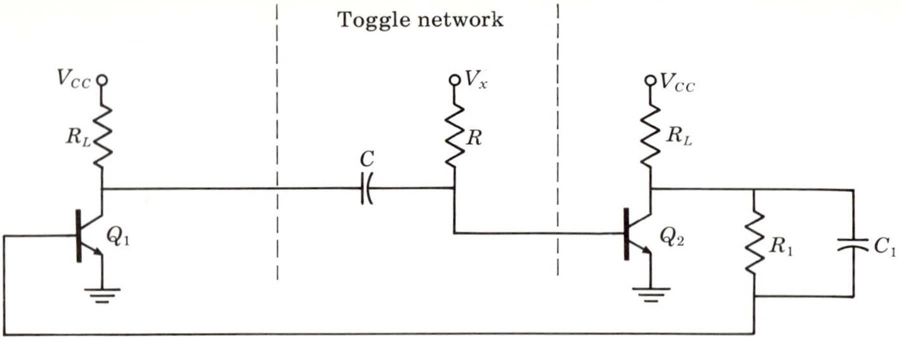

Fig. 8-11 A bipolar-transistor monostable multivibrator showing the *toggle network* coupling between stages.

in Fig. 8-11, which we can call a *toggling network*, we observe that the capacitor effectively opens the loop at zero frequency, reducing the steady-state gain to zero. The multivibrator's single stable state depends on the voltage to which we return the resistor R. If it is returned to a logical-1 level ($V_x = V_{CC}$), then Q_2 will be saturated. (Of course, the base-emitter diode limits the voltage at the base side of the capacitor to one diode drop.) The collector 0-level voltage coupled by the direct-coupling network turns Q_1 off. If, on the other hand, R is returned to a voltage at or below the logical-0 level, Q_2 will be off and Q_1 will be on.

Case 1—keep-off timing, Q_2 normally on

Let us first consider the operation of the monostable with Q_2 normally on. We establish this condition by setting $V_x = V_{CC}$. The quasi-stable state comes into existence after an external positive pulse drives Q_1 on from its normal OFF state. The collector of Q_1 drops from V_{CC} to $V_{CE,\text{sat}}$. Since the voltage across the capacitor cannot change instantaneously, C couples the full negative transition to the base of Q_2, where it drives the base voltage down and turns (or toggles) Q_2 off. The positive jump at the collector of Q_2 coupled into the base of Q_1 keeps Q_1 on after the termination of the triggering pulse.

As the timing capacitor C charges through R toward V_{CC}, the base of Q_2 eventually reaches the 700-mV base-emitter drop necessary to turn Q_2 on. Q_1 turns off, and the multivibrator is back in its monostable state.

From our initial discussion of the circuit operation, we conclude that the duration of the quasi-stable state is controlled by the RC toggle network. External triggering and internal regeneration just cause Q_1 to switch from off to on and then back to off. Once we replace Q_1 by a controlled switch, the full monostable can be described by the single model of Fig. 8-12*a*.

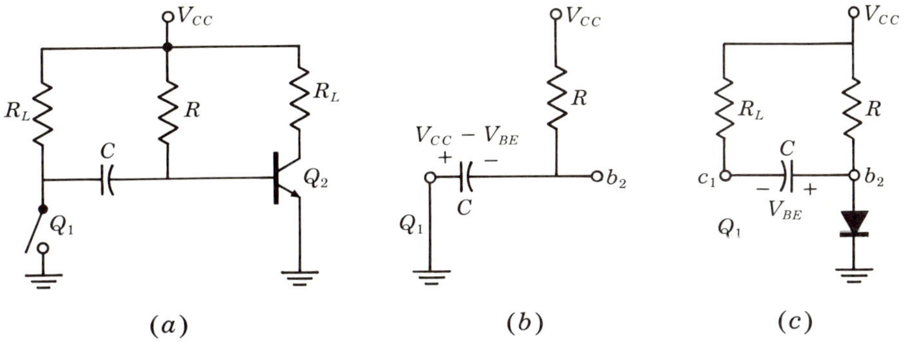

Fig. 8-12 (*a*) Monostable multivibrator of Fig. 8-11 shown as a switched circuit; (*b*) model holding during the pulse-timing interval (Q_1 saturated); (*c*) model holding during recovery (Q_1 off).

Starting when Q_1 is off (switch open), we obtain the initial conditions at $t = 0^-$:

$$v_{c1}(0^-) = V_{CC} \tag{8-16a}$$

$$v_{b2}(0^-) = V_{BE} \tag{8-16b}$$

$$v_{c2}(0^-) = V_{CE,\text{sat}} \approx 0 \tag{8-16c}$$

$$v_q(0^-) = V_{CC} - V_{BE} \tag{8-16d}$$

$v_q(0^-)$ is the steady-state voltage on the timing capacitor. Regenerative triggering closes the switch. The model reduces to the RC circuit of Fig. 8-12*b*. At $t = 0^+$,

$$v_{c1}(0^+) = V_{CE,\text{sat}} \approx 0 \tag{8-17a}$$

$$v_{b2}(0^+) = V_{BE} - V_{CC} \tag{8-17b}$$

$$v_{c2}(0^+) = V_{CC} \tag{8-17c}$$

After the initial negative drop of $-V_{CC}$, the base of Q_2 recovers toward V_{CC} with the sweep-time constant $\tau_S = RC$:

$$v_{b2}(t) = V_{CC} - (2V_{CC} - V_{BE})e^{-t/\tau_S}$$

When v_{b2} reaches 700 mV, Q_2 turns back on and regenerative action reswitches the state. Substituting the known initial, final, and steady-state values of the base voltage during the quasi-stable state into the timing equation, we obtain

$$T_S = \tau_S \ln \frac{V_{CC} + (V_{CC} - V_{BE})}{V_{CC} - V_{BE}} \tag{8-18}$$

When the supply voltage V_{CC} is much larger than the 700-mV diode drop, Eq. (8-18) may be approximated by

$$T_S \approx \tau_S \ln 2 = 0.693\tau_S \tag{8-19}$$

Equation (8-19) proves that to a very good first approximation the pulse duration is almost completely independent of V_{CC}.

Figure 8-13 shows the exponential timing waveshape appearing at the base of Q_2. The collector voltage of Q_2 follows the change of state of the base. Because only a small stray capacity (and the small speedup capacity)

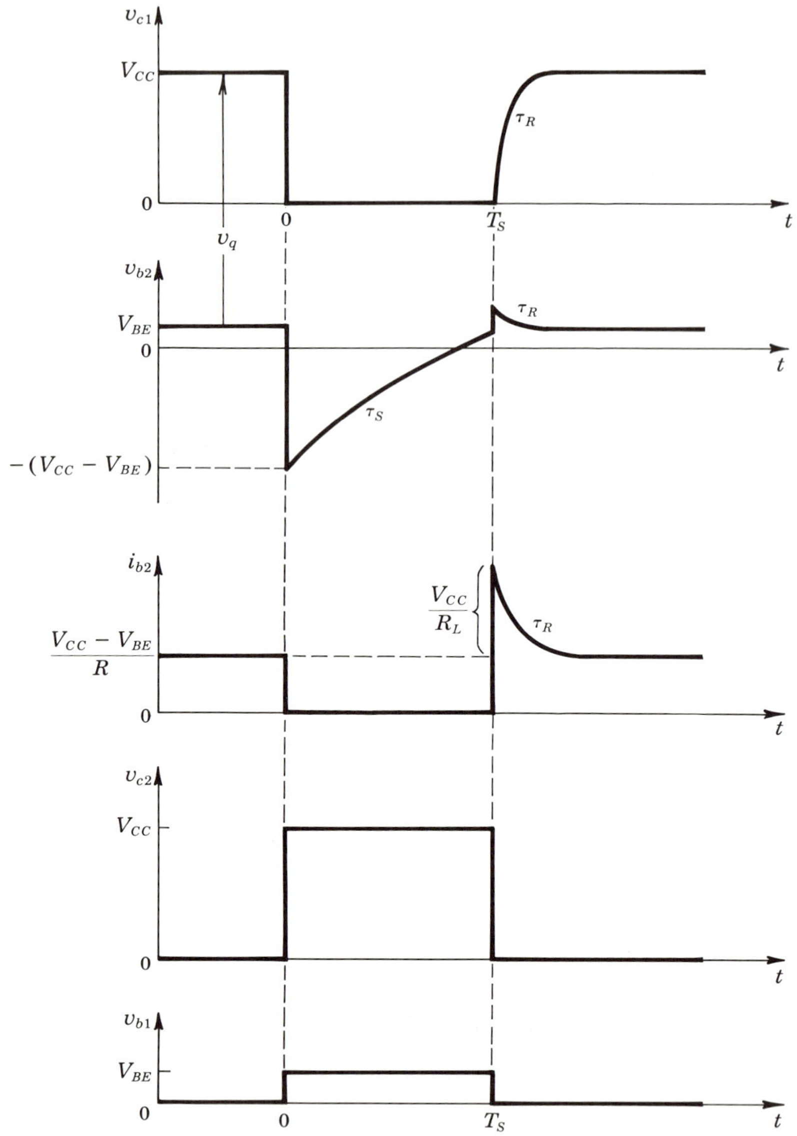

Fig. 8-13 Waveshapes appearing in the monostable multivibrator of Fig. 8-11 under keep-off timing conditions.

is present at this collector, the monostable multivibrator generates a positive pulse having fast rise and fall times. Of course, the storage-delay times and the transistor turn-on and turn-off times will affect the pulse shape when generating extremely narrow pulses.

Recovery region—case 1

The model for the recovery region is given in Fig. 8-12c. We observe that once the switch reopens (Q_1 off), C starts recharging toward its original condition. The recharging current flows into the base of Q_2 and drives it even harder into saturation. At the new $t = 0^+$, the total base current is

$$I_{b2m} \approx \frac{V_{CC} - V_{BE}}{R} + \frac{V_{CC}}{R_L} \tag{8-20}$$

The first term of Eq. (8-20) is the normal ON bias current of Q_2, and the second term is the excess base current due to the recharging of C. We know from the exponential diode volt-ampere response that the additional current will be accompanied by an increased voltage drop (60 mV for every 10:1 increase in the total emitter-diode current). To calculate the overshoot we would have to solve the saturated Ebers-Moll model under the conditions of the known peak base current. We conclude that the recovery transient starts with a small voltage jump at the base of Q_2.

The collector voltage of Q_1 must also jump by this same small amount. It will then recover toward the steady-state V_{CC} with the recovery time constant

$$\tau_R = R_L C \tag{8-21}$$

As the charging current drops to zero, the base voltage of Q_2 recovers back to its lower normal diode drop. Complete recovery requires approximately four time constants.

Because we must wait for complete recovery before retriggering, an excessively long recovery time is undesirable. We usually restrict the time constants so that the recovery interval is shorter than the pulse duration. From Eqs. (8-19) and (8-21) we can write the necessary resistor inequality,

$$4R_L < 0.693R \tag{8-22}$$

The jumps and charging regions during recovery are clearly indicated in the waveshapes of Fig. 8-13.

In the interests of simplicity, we have been assuming that the base-emitter turn-on voltage and the final quiescent diode drop of 700 mV are the same. Since the actual turn-on voltage may be 50 or 100 mV lower (see Fig. 8-2), the magnitudes of both base and collector jumps will be slightly higher.

Base-emitter-junction breakdown

The above discussion of the pulse timing implicitly assumed that the initial drop at the base of Q_2 did not exceed the reverse breakdown voltage of the base-emitter junction (BV_{BEO}). When, however, the supply voltage V_{CC} does exceed BV_{BEO}, the drop at the collector of Q_1 coupled to the base of Q_2 causes the base-emitter junction of Q_2 to break down. We find that a constant avalanche drop of $-BV_{BEO}$ appears between the base of Q_2 and ground (Fig. 8-14a). This drop is in series with the charged timing capaci-

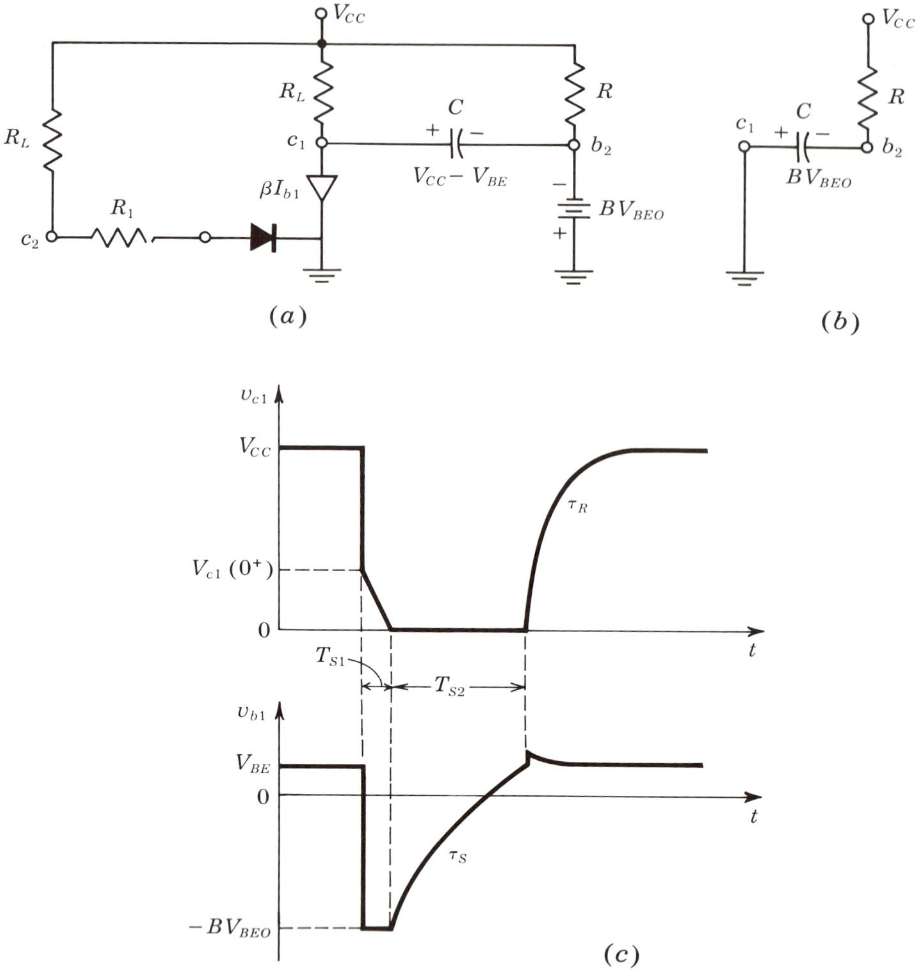

Fig. 8-14 Monostable operation with base-emitter junction breakdown. (a) Breakdown region model defining the initial portion of the pulse timing interval; (b) model defining the final portion of the pulse timing interval; (c) significant collector and base voltage waveshapes.

tor, resulting in an initial voltage at the collector of Q_1 of

$$v_{c1}(0^+) = V_{CC} - V_{BE} - BV_{BEO}$$

Since $v_{c1}(0^+) > 0$, Q_1 is not driven into saturation.

The first portion of the charging of C is evaluated from the model of Fig. 8-14*a*, which is identical to the current sweep of Sec. 6-4. If we neglect the current flow through R_L, then C is charged by the constant-current source βI_{b1}, where

$$I_{b1} = \frac{V_{CC} - V_{BE}}{R_L + R_1}$$

This portion of the timing ends once the collector voltage of Q_1 falls to $V_{CE,\text{sat}} \approx 0$ and Q_1 saturates. The time required for the collector rundown from $v_{c1}(0^+)$ to 0 volts is

$$T_{S1} = C(R_L + R_1)\,\frac{v_{c1}(0^+)}{\beta(V_{CC} - V_{BE})}$$

During this entire interval, the base voltage of Q_2 remains constant at $-BV_{BEO}$ (Fig. 8-14*c*). Once Q_1 saturates at T_{S1}, the net reverse base current of Q_2 drops to zero and the emitter-base junction comes out of breakdown.

The second portion of the charging is a normal exponential (Fig. 8-14*b*). The base voltage of Q_2 heads from $-BV_{BEO}$ toward V_{CC}. Since the pulse terminates once the base voltage rises to V_{BE}, this interval is defined by

$$T_{S2} = \tau_S \ln \frac{V_{CC} - BV_{BEO}}{V_{CC} - V_{BE}}$$

T_{S2} is usually much longer than T_{S1}, and it determines the pulse width.

Figure 8-14*c* shows the waveshapes at the collector of Q_1 and at the base of Q_2. These should be compared with the waveshapes shown in Fig. 8-13.

Case 2—keep-on timing, Q_2 normally off

This second case (Fig. 8-15*a*) corresponds to interchanging the roles of the recovery and timing regions described under case 1. Since Q_2 is normally biased off, by returning the resistor R to a negative voltage ($V_{BB} \leq 0$), the quasi-stable state comes into existence when the capacitor-charging current drives Q_2 into saturation. Timing is controlled by the base current flow. The significant voltage and current waveshapes appearing in this circuit are drawn in Fig. 8-16.

From the model of Fig. 8-15*b*, we see that after triggering Q_2 on, the base current may be written in the form

$$i_{b2} = \frac{V_{BB} - V_{BE}}{R} + \frac{V_{CC} - v_q - V_{BE}}{R_L}$$

$$= \frac{-3 - 0.7}{10\text{ K}} + \frac{6 - v_q - 0.7}{1\text{ K}} \tag{8-23}$$

where v_q is the voltage across the capacitor. Just before, and therefore just after, triggering into the quasi-stable state, the capacitor voltage is equal to the -3-V bias supply (Fig. 8-15*a*).

We can recognize the first term of Eq. (8-23) as the steady-state level-shifting current (I_{bss}) through R, which is required to ensure that Q_2 is

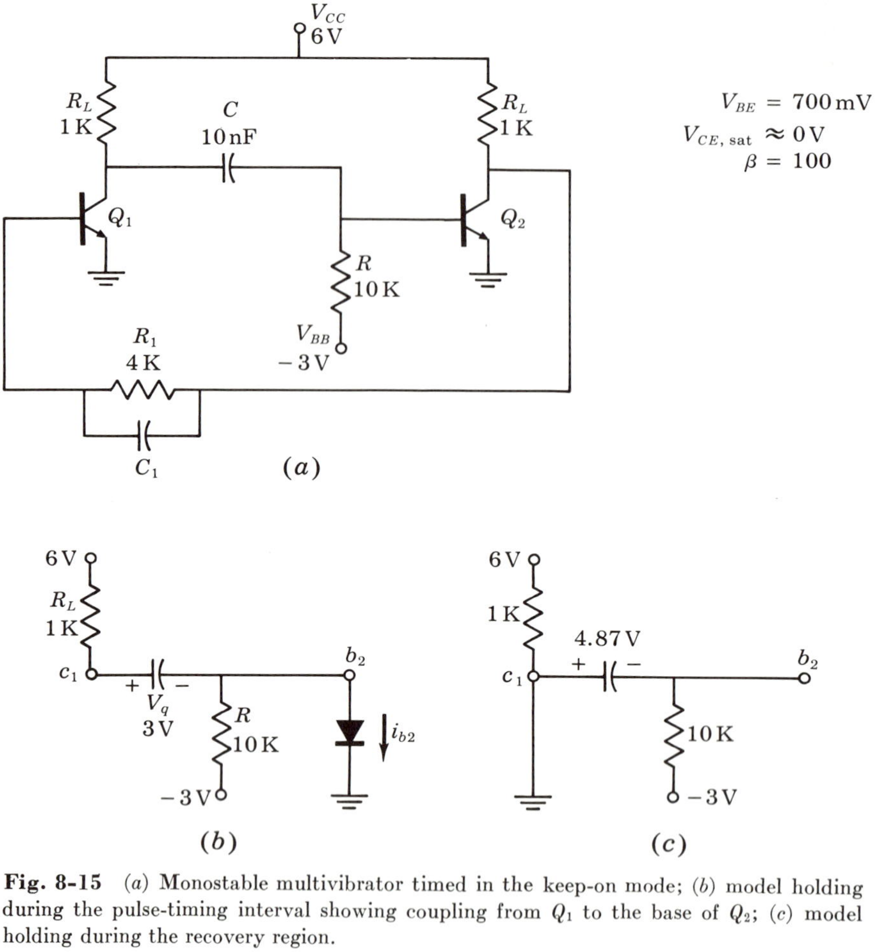

Fig. 8-15 (*a*) Monostable multivibrator timed in the keep-on mode; (*b*) model holding during the pulse-timing interval showing coupling from Q_1 to the base of Q_2; (*c*) model holding during the recovery region.

normally in the OFF state. Of course, a net negative current never actually flows in the base. The second term, the capacitor-charging current, decays toward 0 as C charges. Solving Eq. (8-23) at $t = 0^+$, the initial base current becomes

$$i_{b2}(0^+) = -0.37 + 2.3 = 1.93 \text{ mA}$$

where $I_{bss} = -0.37$ mA. But the saturation of Q_2 requires only

$$I_{b2,\text{sat}} \geq \frac{V_{CC}}{\beta R_L} = 60\ \mu\text{A}$$

The current-decay-time constant, which determines the pulse duration, is

$$\tau_S = R_L C = 10\ \mu\text{s}$$

τ_S is the same as the recovery-region time constant of the case-1 circuit. The width of the pulse generated can now be found by substituting the current values into the timing equation:

$$T_S = \tau_S \ln \frac{I_{bss} - i_{b2}(0^+)}{I_{bss} - I_{b2,\text{sat}}}$$

$$= 10 \ln \frac{-0.37 - 1.93}{-0.37 - 0.06} = 17\ \mu\text{s} \qquad (8\text{-}24)$$

The duration of the pulse will to some extent depend on the particular transistor chosen. The $I_{b2,\text{sat}}$ term in the denominator of Eq. (8-24) is an inverse function of β. If the β is very large, this term becomes relatively unimportant and, as before, the pulse duration depends primarily on the passive circuit elements.

Recovery region—case 2

The final conditions of the timing region determine the initial conditions for recovery. At the instant Q_2 becomes active (from saturation), the capacitor-charging current must supply both the base and the level-shifting currents.

$$i_q(T_S) = I_{bss} + I_{b2,\text{sat}} = 0.37 + 0.06 = 0.43 \text{ mA}$$

This current flows through R_L, resulting in a collector voltage at Q_1 of

$$V_{c1} = V_{CC} - i_q(T_S)R_L = 5.57 \text{ V}$$

The voltage on the left-hand side of C is 700 mV.

Our model for the recovery region appears in Fig. 8-15*c*. As Q_1 saturates, the negative transition is transmitted by C to the base of Q_1, driving it down to -4.87 V. The base recovers to the steady state -3 V, with the long time constant

$$\tau_R = RC$$

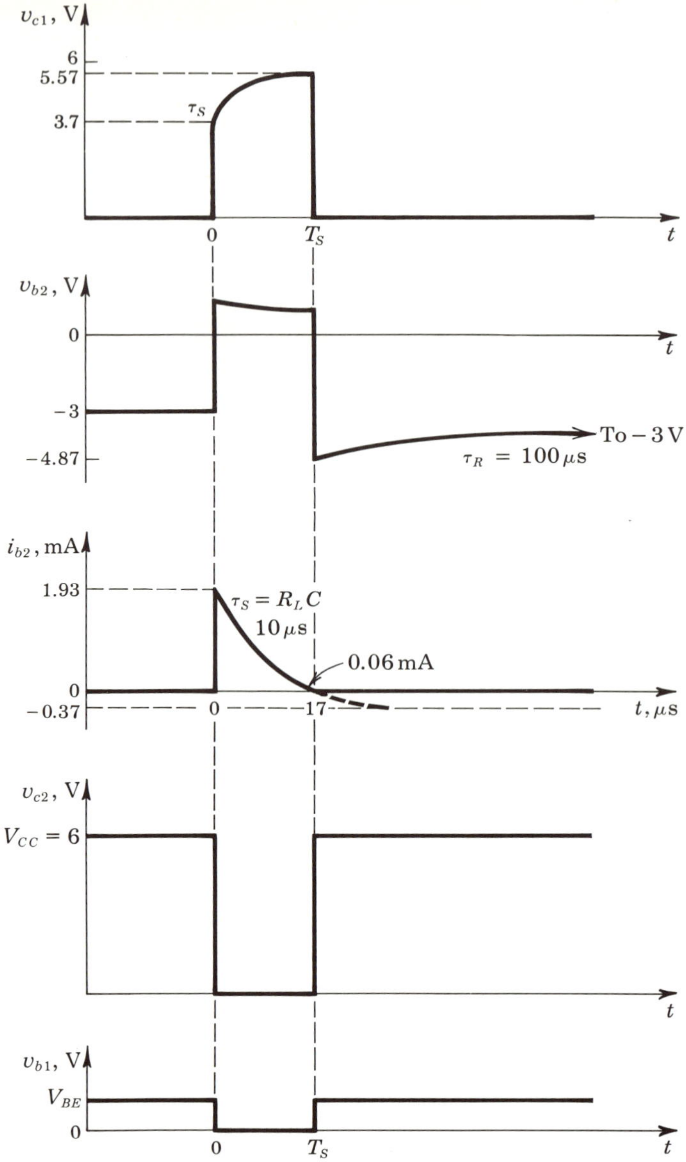

Fig. 8-16 Waveshapes appearing in the monostable multivibrator of Fig. 8-15 (keep-on timing).

Several disadvantages of operation under keep-on timing conditions become apparent. First, the recovery time is very much longer than the pulse duration, necessitating long delays between successive triggerings. Second, the circuit requires two power supplies. Third, the pulse duration is, to some slight extent, dependent on the active-element parameters. [See Eq. (8-24).]

Inductively timed toggling network

The timing operation in which a transistor is forced into saturation for a controlled interval is current-controlled. Therefore, it seems reasonable to see if other current-switching operations can replace the capacitive coupling of Fig. 8-15.

Figure 8-17 shows an inductively timed circuit. The steady-state collector current of Q_1 is stored in the inductance of the toggling network. Upon the external triggering of Q_1 into the cutoff state, the current flow is transferred to the base of Q_2, where it turns Q_2 on. With an appropriate choice of R_2, the new steady-state base current can be set at a level less than the saturation current, and once the current decay reaches the edge

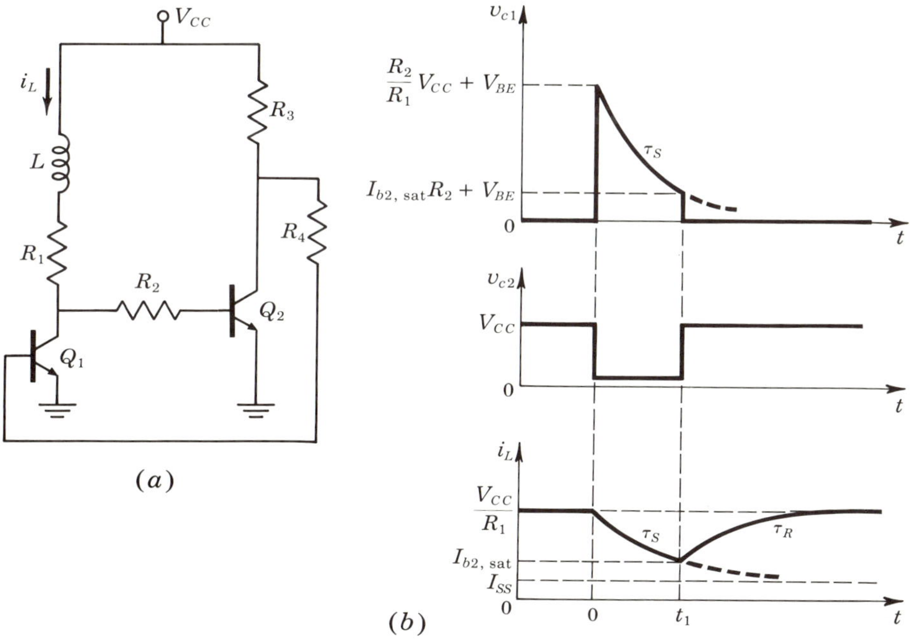

Fig. 8-17 (*a*) An inductively timed monostable multivibrator; (*b*) waveshapes present in the circuit.

of saturation, regenerative switching will reset the circuit into its original state.

As before, the complete operation can be delineated by finding the significant currents. During the timing interval, we have

$$i_b(0^+) = \frac{V_{CC}}{R_1} \tag{8-25a}$$

$$i_b(\infty) = \frac{V_{CC} - V_{BE}}{R_1 + R_2} \tag{8-25b}$$

$$I_{b,\text{sat}} = \frac{V_{CC}}{\beta R_3} \tag{8-25c}$$

For regenerative switching, the circuit must enter the active region by itself. This requires that $i_b(\infty) < I_{b,\text{sat}}$. The decay-time constant is

$$\tau_S = \frac{L}{R_1 + R_2} \tag{8-26}$$

During the recovery interval, where the inductance recharges to its original steady-state current,

$$i_{c1}(0^+) = I_{b,\text{sat}} \tag{8-27a}$$

$$i_{c1}(\infty) = \frac{V_{CC}}{R_1} \tag{8-27b}$$

and the recovery-time constant is

$$\tau_R = \frac{L}{R_1} \tag{8-28}$$

We should note that here, as with the previous case of keep-on timing, the recovery-time constant is longer than the sweep-time constant.

8-6 Monostable Multivibrators—FET Active Element

Every active element or amplifier imposes specific boundary conditions on the monostable coupling network. Besides the obvious switching-point conditions, the device used influences the energy storage in the timing capacitor, the size of the overshoots, and the rate of final recovery.

In a MOSFET, the input constraints are almost exactly opposite to the conditions set by a bipolar transistor. The MOSFET gate is always a high impedance, while in the ON state the bipolar transistor's input impedance is very low. By comparing the results of the analysis of the following circuit with the monostable of Sec. 8-5, we gain understanding of the effects of the amplifier input impedance on the waveshapes produced.

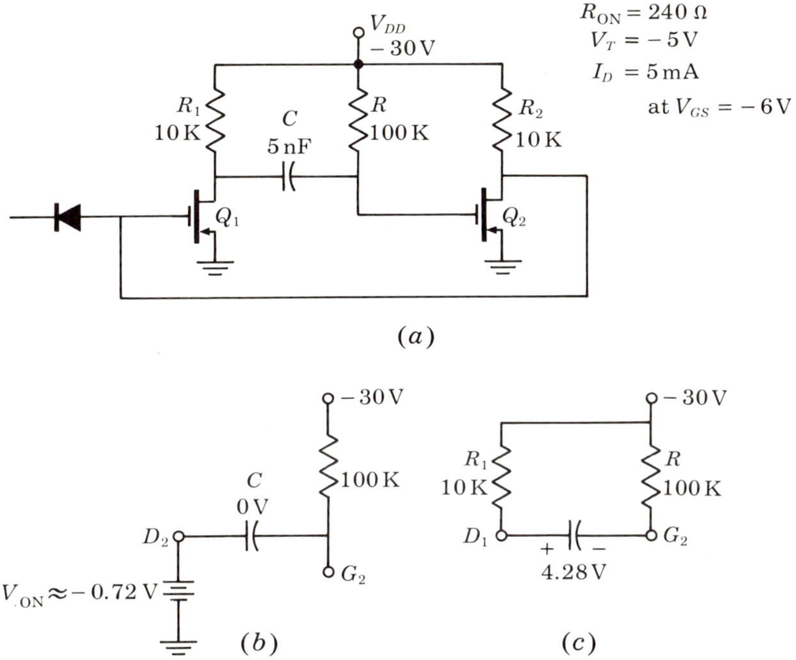

Fig. 8-18 (a) A MOSFET monostable multivibrator; (b) model holding in the quasi-stable region, showing initial conditions; (c) model holding in the recovery region, showing initial conditions.

The general operation of the monostable of Fig. 8-18a, which generates the waveshapes shown in Fig. 8-19, is quite similar to the bipolar circuit previously discussed. In Fig. 8-18a, Q_2 is the ON device. Since the gate is returned to V_{DD}, Q_2 has bottomed. The steady-state conditions are

$$V_{G2}(0^-) = V_{DD} = -30 \text{ V}$$
$$V_{G1}(0^-) = V_{D2} = V_{ON}$$
$$V_{D1}(0^-) = V_{DD} = -30 \text{ V}$$

where $$V_{ON} \approx \frac{R_{ON}}{R_{ON} + R_2} V_{DD} \approx -0.72 \text{ V} > V_T$$

Since both sides of C are returned to the same voltage, the capacitor is uncharged.

Immediately after triggering into the quasi-stable state, the voltage at the drain of Q_1 changes from -30 V to

$$V_{D1}(0^+) = V_{ON} = -0.72 \text{ V}$$

The full positive jump, coupled by the capacitor, raises the gate voltage of Q_2 to

$$V_{G2}(0^+) = V_{\text{ON}}$$

which is above the threshold, proving that Q_2 is off and that the assumed regenerative switching has, in fact, occurred (see Fig. 8-18*b*). With the direct coupling from the drain of Q_2 to Q_1, we have

$$V_{G1}(0^+) = -30 \text{ V}$$

Q_1 is solidly on.

The steady-state voltage at the gate of Q_2 remains invariant across all transition regions. As a consequence of the positive jump, the gate must recover from V_{ON} toward

$$V_{G2}(\infty) = V_{DD} = -30 \text{ V}$$

When the gate reaches the threshold voltage of -5 V, a channel will be induced, the loop gain will reach unity, and regeneration will cause switching of states.

At the end of the quasi-stable state, the voltage across the capacitor will be

$$v_q(T_S) = V_{\text{ON}} - V_T = 4.28 \text{ V}$$

Because the initial (V_{ON}) and final (V_T) voltages are close together, we expect that the pulse generated will be much less than a time constant. Substituting the appropriate conditions, the width of the pulse generated is given by

$$T_S = \tau_S \ln \frac{V_{DD} - V_{\text{ON}}}{V_{DD} - V_T} = 500 \ln \frac{-30 - (-0.72)}{-30 - (-5)} = 79.3\ \mu\text{s} \qquad (8\text{-}29)$$

where

$$\tau_S \approx RC \qquad (8\text{-}30)$$

It must be noted that since the V_T of the MOS may vary over a ratio of 2:1 or more, the duration of the quasi-stable state will vary somewhat with a change of device.

To obtain the recovery waveshapes, we must start from the model of Fig. 8-18*c*. Upon switching, the previously stored capacitor voltage (4.28 V) divides proportionately across the two resistors R_1 and R. The gate voltage of Q_2 jumps from -5 V down to

$$V_{G2}(0^+) = -33.89 \text{ V}$$

At the same time, the drain of Q_1 jumps by the same amount from -720 mV to

$$V_{D1}(0^+) = -30.39 \text{ V}$$

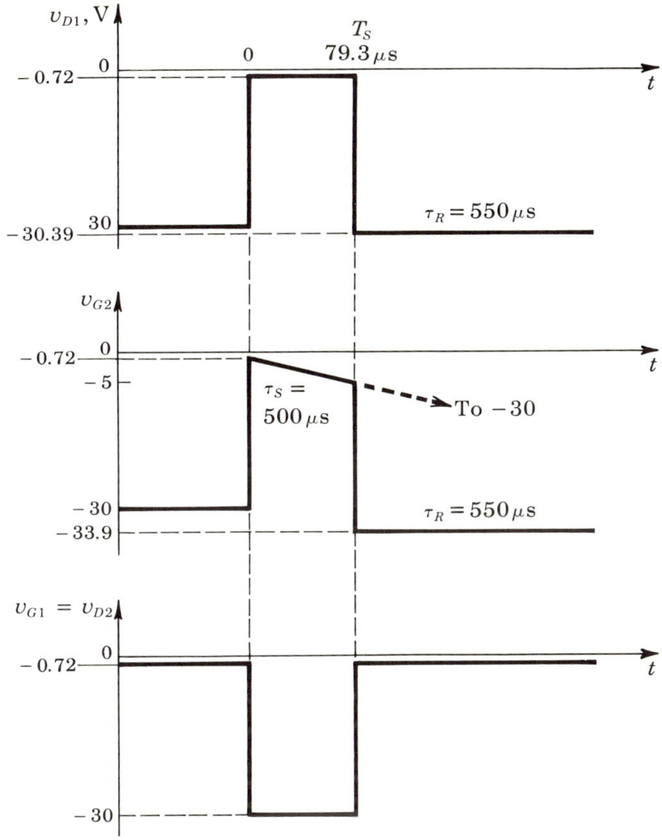

Fig. 8-19 Voltage waveshapes appearing in the MOSFET monostable multivibrator of Fig. 8-18*a*.

Recovery proceeds at both the drain and the gate with the single time constant

$$\tau_R = C(R_1 + R) = 550\ \mu s \tag{8-31}$$

Here the full exponential is needed for recovery. But in Eq. (8-29) only a small fraction of almost the same time-constant exponential defined the pulse width. Again we have a circuit with an excessively poor ratio of pulse width to recovery time.

The extremely long recovery time is a direct consequence of the lack of a low-impedance region in the gate input characteristics. During recovery in the bipolar-transistor circuit, the base-emitter diode effectively shorted the timing resistor R, greatly reducing the time constant. The same result can be obtained in the MOSFET circuit of Fig. 8-18*a* by shunting R with a diode that will conduct only during the recovery portion of the

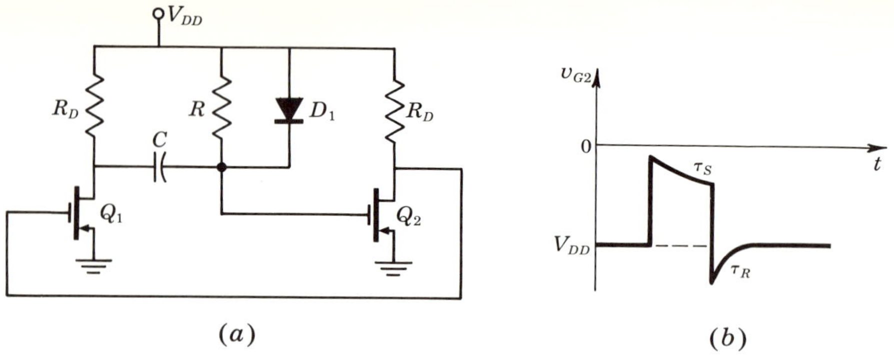

Fig. 8-20 (*a*) Circuit of Fig. 8-18 modified with a diode to shorten the recovery interval; (*b*) gate waveshape of Q_2, showing improvement in recovery time.

cycle. This is easy to do because we note from the waveshapes of Fig. 8-19 that the gate voltage drops below the -30-V V_{DD} level only during the recovery period.

Figure 8-20 shows the resultant circuit and waveshapes. During recovery, D_1 conducts, limiting the overshoot to V_D. The recovery-time constant is now

$$\tau_R = CR_1 = 50\ \mu\text{s} \tag{8-32}$$

Equation (8-32) should be compared with the preceding, much longer time constant given in Eq. (8-31).

Junction-FET monostable multivibrators

Junction-FET monostables (Fig. 8-21) have the same diode limiting at the conducting gate junction as does the bipolar-transistor base-emitter junction.

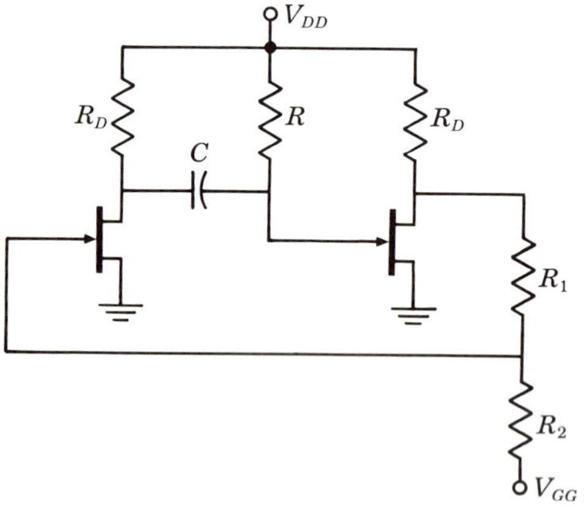

Fig. 8-21 A junction-FET monostable multivibrator timed in the keep-off mode.

Switching behavior, waveshapes, and pulse durations are similar. Level-shifting coupling is required for the dc path for the reasons given in discussing the bistable. The pulse generated will be shorter than that in a bipolar circuit because the transition occurs earlier—at V_p instead of at V_{BE}.

The junction-FET monostable may be timed in the keep-turned-off mode shown (where the negative transition turns off Q_2), or by changing the voltage to which R is returned, the pulse can be timed in the keep-on mode (where the gate diode current flow keeps the FET bottomed in the ohmic region). The keep-turned-off mode is preferable because here the recovery-time constant is much shorter than the duration of the timing waveshape.

8-7 Astable Multivibrators—Sequential Toggle Mode

The insertion of a second capacitor in the transmission loop, i.e., capacitive toggle-mode coupling from each inverter output to the other inverter's input, makes both circuit states quasi-stable. Figure 8-22a shows the circuit of a bipolar-transistor astable multivibrator where both transistors are normally biased in the keep-on mode because only in this mode is the recovery time less than the pulse time. Timing is in the keep-off mode.

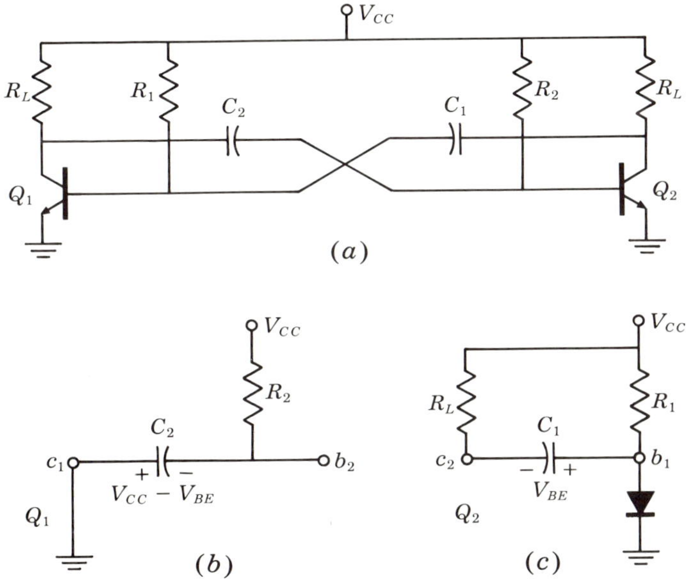

Fig. 8-22 (a) A bipolar transistor astable multivibrator; (b) model showing the coupling from the collector of the ON Q_1 to the base of the OFF Q_2; (c) model showing the coupling from the collector of the OFF Q_2 to the base of the ON Q_1.

(Section 8-9 below discusses the complications arising from incomplete recovery.) If both transistors are assumed to be in their active regions, regeneration will rapidly drive one into cutoff and the other to saturation. This state is unstable, lasting only until the particular capacitor that is keeping the transistor cut off discharges, at which point the switching action reverses the circuit states (the OFF device goes on, and the ON device goes off). The astable multivibrator is self-starting and free-running.

We might note that the timing network coupling each output to the next input is exactly the same as the single timing circuit of the monostable multivibrator (Fig. 8-11). The astable multivibrator may well be considered as two monostable circuits operating sequentially, one triggered off whenever the other one turns on. Waveshapes produced (Fig. 8-23) will be basically those of the monostable (Fig. 8-13) with alternate transistor states displaced by half of the total period. The previous techniques of analysis are readily applicable, and by starting anywhere, we shall, within no more than two cycles, reach the final periodic solution.

Since the exponential timing after the change of circuit state depends upon the conditions existing immediately before switching, probably the best starting point is where Q_2 is full on, just going off, and Q_1 is cut off, just going on. At this instant, the model holding is identical with the one defining the stable state of the monostable multivibrator (Fig. 8-12*c*). In our initial analysis, in order not to obscure or conceal the basic simplicity of the behavior of this circuit beneath mathematical manipulations, we shall ignore the effects of any base-voltage jumps above the normal V_{BE}. The initial circuit voltages are those given in Eqs. (8-16*a*) to (8-16*d*). From the initial conditions and the assumed starting point, we can determine the charge in each capacitor at the instant before switching into the new state. These are

$$v_{q1}(0^-) = V_{BE}$$

$$v_{q2}(0^-) = V_{CC} - V_{BE}$$

The voltages on the capacitors remain constant across the change of transistor states. These known charges enable us to calculate the voltages at each transistor terminal after the circuit switches states. The new models appear in Fig. 8-22*b* and *c*.

As the circuit changes state, the collector voltage of the formerly OFF Q_1 drops from V_{CC} to 0. This negative transition is coupled by C_2 to the base of Q_2 and cuts off Q_2. Figure 8-22*b* shows the timing circuit which defines the unstable interval. The base of Q_2 recovers toward V_{CC}, with the transistor turning back on when it reaches V_{BE}. The time required is

$$T_2 = \tau_2 \ln \frac{V_{CC} + (V_{CC} - V_{BE})}{V_{CC} - V_{BE}} \qquad \tau_2 = R_2C_2 \tag{8-33}$$

When the multivibrator reswitches, the sequence of events repeats in the network, coupling the collector of Q_2 to the base of Q_1. The duration of the other half cycle is given by

$$T_1 = \tau_1 \ln \frac{V_{CC} + (V_{CC} - V_{BE})}{V_{CC} - V_{BE}} \qquad \tau_1 \approx R_1C_1 \tag{8-34}$$

The two waveshapes generated are displaced in time from one another, one starting when the other ends (Fig. 8-23). Thus the total period is the sum of the times given by Eqs. (8-33) and (8-34). If a square-wave output is desired, the circuit is made completely symmetrical and $T_1 = T_2$.

Up to this point we have concerned ourselves only with the circuit behavior during the OFF period and we have ignored anything that happened at the base of the ON transistor. When Q_2 switches off (Fig. 8-22*c*), the capacitor's charging current drives Q_1 deeper into saturation and causes a small voltage overshoot at the base. The same jump appears at the collector of Q_2. Recovery back to the steady-state conditions, shown in the model, is with the time constant

$$\tau_{R2} = C_1R_L$$

Proper astable operation is predicated on complete collector recovery at the OFF transistor before the circuit reswitches. In a symmetrical astable multivibrator, $C_1 = C_2 = C$ and $R_1 = R_2 = R$. The half period may be approximated by Eq. (8-19), and we obtain the inequality

$$T_1 = RC \ln 2 > 4R_LC$$

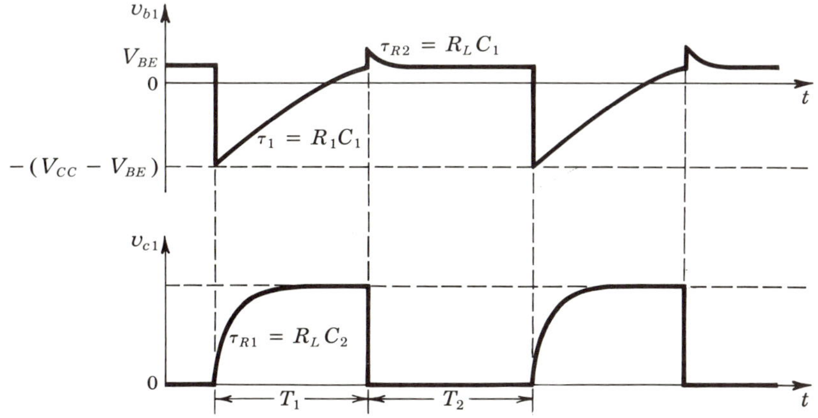

Fig. 8-23 Waveshapes of the transistor astable multivibrator at transistor Q_1 Those at the other transistor are identical but displaced by a half period.

If, in addition, we insist that the ON transistor remain saturated over the complete half cycle, we must also satisfy the additional approximate inequality

$$R < \beta R_L$$

Combining the two limiting conditions, we see that R must be restricted to the range

$$\beta R_L > R > 5.8R_L \tag{8-35}$$

Since C is the same in both the timing and the recovery charging paths, the larger the size of R consistent with Eq. (8-35), the smaller the percentage of the half period devoted to the collector recovery exponential. This means that high-β transistors must be used if we are to generate a relatively square pulse at the collector.

Astable multivibrators may occasionally start with both transistors saturated. Because of the capacitive coupling from stage to stage, this is a permissible circuit state (see Fig. 8-22). The transistors will not alternately switch between cutoff and saturation as described above. Usually this particular problem arises when the inverters are designed with quiescent base currents very much higher than needed to saturate the transistor. When the power supply is switched on, the transistors never enter the active region and the regenerative switching necessary to initiate the astable operation never occurs. Reducing the base current by increasing R or increasing the collector saturation current by reducing R_L will usually resolve this design problem. R and R_L should be chosen closer to the upper limit of Eq. (8-35).

Example 8-2

In this example we shall consider the response of a symmetrical astable multivibrator nominally biased on in the active region. This mode of operation is somewhat different from that previously considered for the transistor, but since the basic behavior remains the same, we may use the standard method of analysis. We shall assign component values in the circuit of Fig. 8-22*a* of $V_{CC} = 10$ V, $R_L = 1$ K, $C_1 = C_2 = 100$ pF, and $\beta = 50$. By setting $R_1 = R_2 = R = 75$ K, Eq. (8-35) is not satisfied and the transistors switch between their cutoff and active regions.

As our arbitrary starting point, we shall assume that Q_1 is off, on the verge of switching on, and that Q_2 is in the opposite state. In the initial discussion, the contribution of the charging current through C_1 and the base of Q_1 will be neglected, and we shall further assume that the total

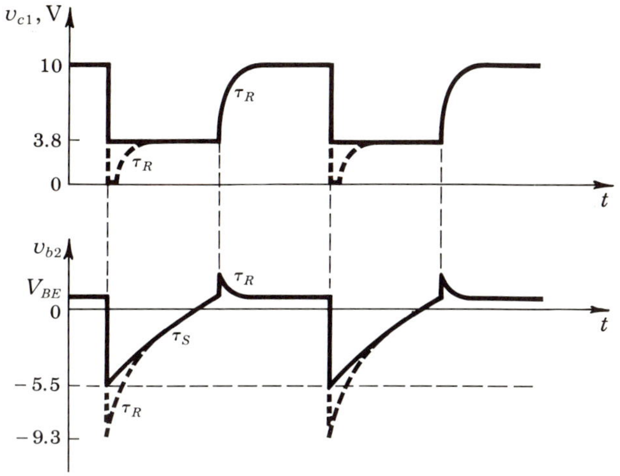

Fig. 8-24 Waveshapes at the collector of Q_1 and the base of Q_2 for the multivibrator of Example 8-2. The other collector and base waveshapes are identical but displaced by a half period.

bias current is $(V_{CC} - V_{BE})/R$. When Q_1 turns on, its collector falls from 10 V to

$$v_{c1}(0^+) = V_{CC} - \beta i_b R_L = 10 - 50\,\frac{10 - 0.7}{75\text{ K}}\,1\text{ K} = 3.8\text{ V}$$

The base of Q_2 falls by the same amount and immediately starts charging from -5.5 toward 10 V with the time constant

$$\tau_S = (R_L + R)C = 76\ \mu\text{s}$$

Since the turn-on point is 700 mV, all the information necessary to calculate the switching time is now known. The half period is

$$T_S = 76 \ln \frac{10 + 5.5}{10 - 0.7} \approx 38.9\ \mu\text{s}$$

In the previous calculations we neglected the component of the base current of Q_1 contributed by the recharge of C_1. This current flows through the 1-K collector load resistor, starting from a peak of

$$I_m = \frac{10 - 3.8}{R_L} = 6.2\text{ mA}$$

and decaying toward zero with the fast time constant

$$\tau_R = R_L C = 100\text{ ns}$$

The extra current flow drives the transistor into saturation for a portion of the cycle.

In order to obtain a fast rise time, $\tau_R \ll \tau_S$. As the recovery portion of the half cycle will be complete long before the end of the timing interval, there will be but slight modification needed as to the actual duration of the OFF period. It seems that a reasonable approach is to simply superimpose the effects of a recovery waveshape on the sweep waveshapes that were found by ignoring the base jump. This is indicated by the dashed lines in Fig. 8-24. We note that the ON transistor remains in saturation for a small portion of the half period.

Under the conditions of this problem, the initial voltage drop, and hence the half period, are dependent on the value of β. A 10 percent reduction, from 50 to 45, reduces the size of the collector-voltage drop from 6.2 to 5.6 V and the sweep time to 35.4 μs. If the transistor switches between cutoff and saturation, as it does when generating the waveshapes of Fig. 8-23, then the period is determined solely by the RC network and not by the parameters of the active circuit element.

A junction-FET astable multivibrator

The circuit of a symmetrical junction-FET astable multivibrator appears in Fig. 8-25*a*. Each FET sequentially switches between bottoming and cutoff in a manner similar to the single unstable state of the monostable multivibrator. Operation is also identical to that of the bipolar-transistor circuit—with, however, a different turn-on voltage. The FET turns back on once the gate rises to V_p. From the circuit of Fig. 8-25*a*, we find that upon switching on, the drain voltage drops from V_{DD} to V_{ON}. The gate drops by the same amount from its starting point of $V_{G\text{-sub}}$ and then recovers toward V_{DD}. The duration of the half period is given by

$$T_1 = \tau_S \ln \frac{V_{DD} + (V_{DD} - V_{\text{ON}}) - V_{G\text{-sub}}}{V_{DD} - V_p} \tag{8-36}$$

where

$$\tau_S = CR$$

The beginning of the recovery interval is delineated by the gate jump from V_p to and slightly above $V_{G\text{-sub}}$. The drain of the other FET (now going off) must jump by the same amount. Because the gate-substrate diode is now conducting, the recovery-time constant is the much smaller

$$\tau_R = CR_D$$

Waveshapes at one drain and at the coupled gate are shown in Fig. 8-25*b*. Any overshoots during recovery appear at the drain of the ON FET and hence at the gate of the OFF FET. They would be superimposed on the idealized single exponential response.

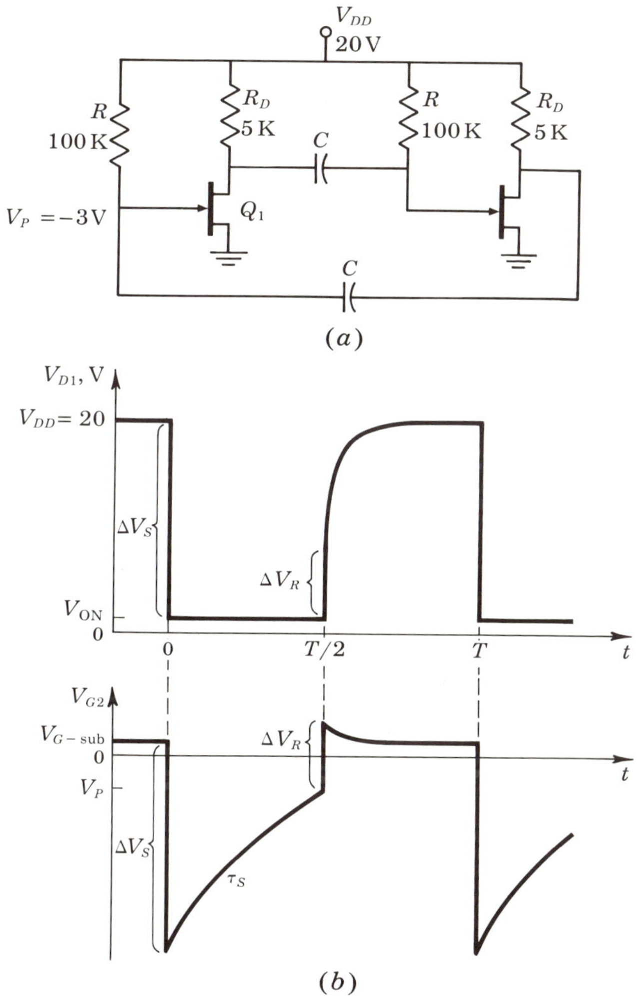

Fig. 8-25 (a) A junction-FET astable multivibrator; (b) waveshapes at a drain and the coupled gate.

A MOSFET astable multivibrator

The MOSFET monostable multivibrator of Fig. 8-18a cannot be converted into an astable multivibrator by the addition of a second coupling capacitor. In that circuit, the recovery interval greatly exceeded the sweep time. Consequently, we cannot postulate complete recovery before reswitching. This case is treated below, in Sec. 8-9. If diodes are added to control the recovery interval (Fig. 8-20), then astable operation becomes possible, and, as with all astables, the analysis would simply be that of two monostables sequentially switched from on to off.

8-8 Integrated-circuit Monostable and Astable Multivibrators

Monostable and astable multivibrators require energy-storage elements for timing the duration of the quasi-stable state. But the size of a capacitor which can be formed on a chip is directly proportional to the surface area available for capacitance ($\sim$0.35 pF/mil^2). Even when the monostable is used to generate a narrow pulse, the capacitance required would have to occupy the major portion of the chip or even an area greater than the total chip. Therefore, the timing network must be external to the inverter circuits. By using lumped elements for the timing network, we also achieve the capability of adjusting the pulse duration.

Bistable multivibrators are either directly coupled or coupled through a resistor network. Along with the associated triggering and logic circuitry, they can be conveniently integrated on a portion of a single silicon chip. Chapter 13 discusses these devices.

In Fig. 8-26, two TTL NAND gates are externally coupled to form a monostable multivibrator. Note that the output of gate B is the direct input to gate A. This is the dc or bistable coupling. The output of gate A is coupled to the input of gate B through a lumped-element toggling network.

The particular mode of timing depends on the input characteristics of the inverter. Because the TTL input is essentially an AND gate, keep-off timing is most suitable. Timing the OFF interval is accomplished by coupling a positive $0 \rightarrow 1$ transition into the gate-B input through the RC network. This positive pulse turns off the input gate. To establish the quiescent state of gate B, we therefore return the bias resistor R to a negative voltage. The diode limits the negative ON level at the B input to one diode drop. Thus the stable-state B output is at the logical-1 level and the A output is at the logical-0 level.

When a logical-0 trigger is applied at one of the inputs to gate A, the A output jumps from a logical 0 to a logical 1. The capacitor couples the positive transition to gate B and switches its state. The associated input emitter of the TTL stage is effectively disconnected. The loop regeneration keeps the output of gate A at the logical-1 level. C now charges toward the negative bias level of $-|\text{logical } 1|$. The logic gate turns back on at the edge of the transition region. Ignoring any diode drops and approximating the transition edge at zero, we obtain

$$\begin{aligned} T_S &\approx \tau_S \ln \frac{-1 - |-1|}{-1} \\ &= \tau_S \ln 2 \end{aligned}$$

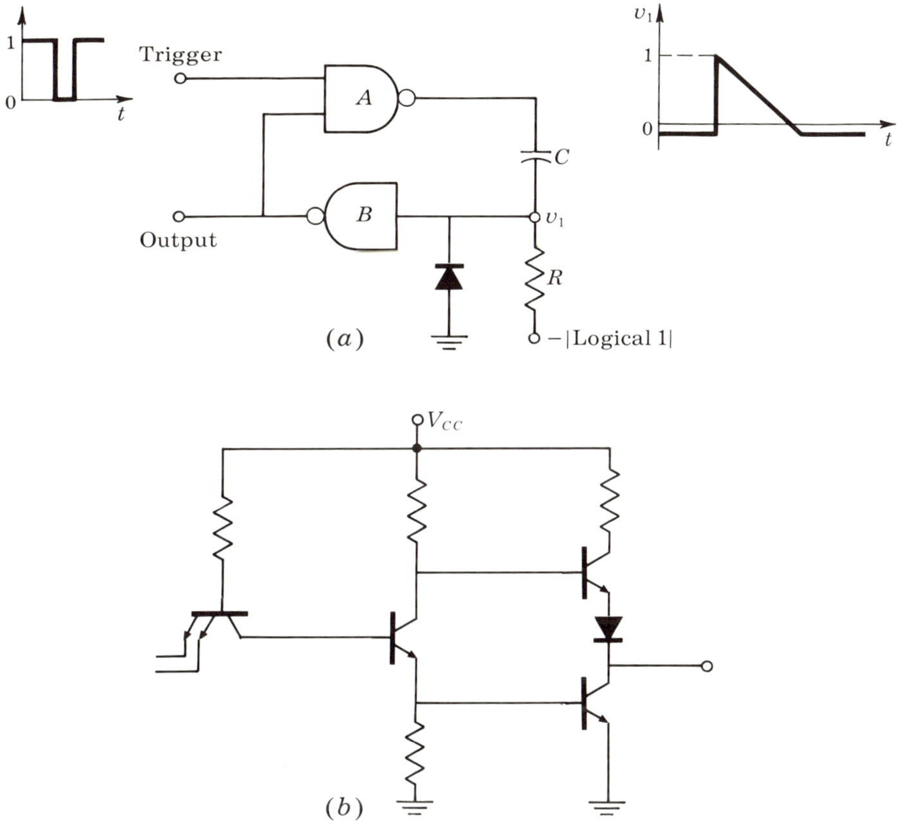

Fig. 8-26 (*a*) A monostable multivibrator constructed out of NAND gates; (*b*) the circuit of a single TTL gate used in the monostable. From two to four gates are usually available on a single chip. The diode and *RC* network are lumped external elements.

where $\tau_S = RC$ and where the 1's represent the logic levels. Because of the actual diode drops of the ON transistor, the pulse duration will differ slightly from that defined above. T_S is, however, a reasonable first approximation. After switching back into the stable region, final circuit recovery is relatively fast. The capacitor-charging current depends on the internal base current and reverse β of the TTL input stage.

Astable multivibrators can also be synthesized by using two toggling networks.

Other logic elements may also be cross-coupled to form multivibrators. We must, however, carefully examine the input circuitry of the basic gate to determine whether keep-on or keep-off timing would be more suitable. Just as in the above TTL circuit, diode limiting of voltage levels is usually necessary.

8-9 Incomplete Circuit Recovery

We now propose to examine the additional complexity introduced in calculating the duration of the half period if the collector recovery is not complete when the multivibrator switches states. The symmetrical junction-transistor astable of Fig. 8-22 is relatively easy to treat, once it builds up to periodicity, since the waveshapes of both transistors are identical. The base of one transistor is driven off by the drop in the collector voltage of the other transistor as it switches full on. At the end of the unknown half period t_1, the collector voltage has risen only from zero to the value found from the recovery exponential [Eq. (8-37)]. (Also see the waveshape in Fig. 8-27.)

$$v_c(t_1) = V_{CC}(1 - e^{-t_1/\tau_R}) \qquad \tau_R = R_L C \tag{8-37}$$

Therefore the negative-transition ON switching is $-v_c(t_1)$ instead of $-V_{CC}$. The equation defining the rise of the base voltage becomes

$$v_b(t) = V_{CC} - [V_{CC} + v_c(t_1)]e^{-t/\tau_S} \qquad \tau_S = RC \tag{8-38}$$

A change of state again occurs when $v_b(t_1) = V_{BE}$. But the elapsed time depends on the initial drop [Eq. (8-37)], which itself is determined by the unknown half period. The interrelationships of the two equations are illustrated in Fig. 8-27, especially by comparing it with Figs. 8-23 and 8-24.

The above argument has furnished us with two transcendental equations [(8-37) and (8-38)], which must be simultaneously solved for the unknown half period t_1. One standard method of solution is to graph v_{c1} as a function of various values of t_1, as individually found from each equation. The intersection represents the unique solution for both v_{c1} and t_1.

In the special case of a symmetrical transistor multivibrator, by substituting Eq. (8-37) into Eq. (8-38), we find that the half period t_1 corresponds to the solution of

$$\left(1 - \frac{V_{BE}}{V_{CC}}\right) - 2e^{-t_1/\tau_S} + e^{-t_1/\tau_S}e^{-t_1/\tau_R} = 0 \tag{8-39}$$

Since normally $V_{BE}/V_{CC} \ll 1$, by setting $V_{BE}/V_{CC} = 0$ and by making the following additional substitutions,

$$\tau_S = \lambda\tau_R \qquad x = e^{-t_1/\tau_S}$$

Eq. (8-39) reduces to

$$x^{\lambda+1} - 2x + 1 = 0 \qquad x \leq 1 \tag{8-40}$$

The restriction on x must be imposed so that the solution of Eq. (8-40) for the half period will correspond to real values of time.

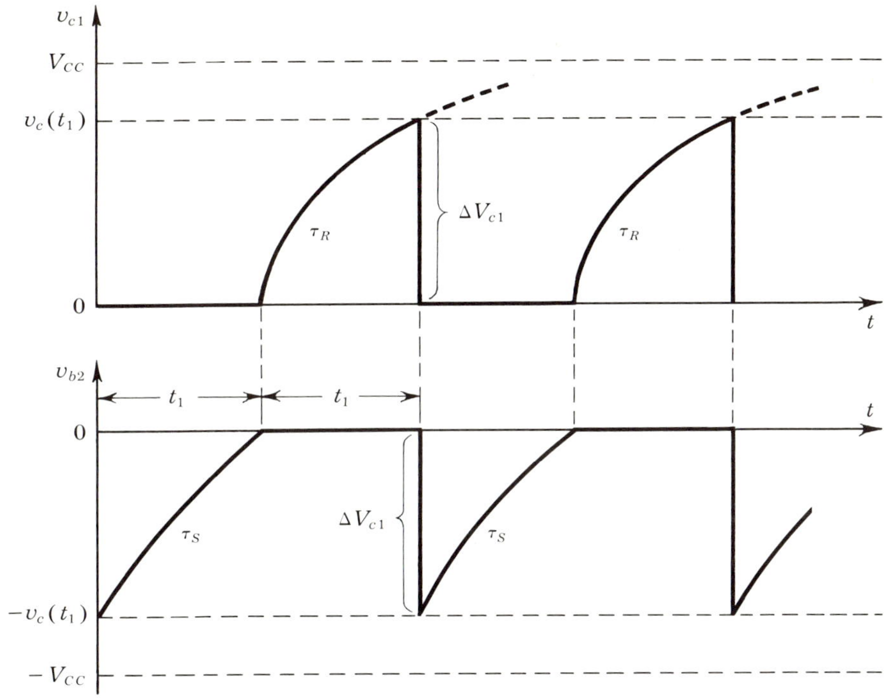

Fig. 8-27 Idealized collector and base waveshapes due to incomplete recovery (symmetrical circuit).

In general, λ is not an integer and Eq. (8-40) would have to be solved by numerical or graphical methods for the single root lying within

$$0 < x_1 < 1$$

Finally, since $x_1 = \exp[-(t_1/\tau_S)]$, the half period is

$$t_1 = \tau_S \ln \frac{1}{x_1} \tag{8-41}$$

For λ large, the recovery-time constant τ_R is very much smaller than that of the rise, and the root of Eq. (8-40) is located at

$$x_{1,\min} = \tfrac{1}{2}$$

This simply yields the half period for the complete recovery waveshape, i.e., the maximum possible sweep duration of

$$t_1 = \tau_S \ln 2$$

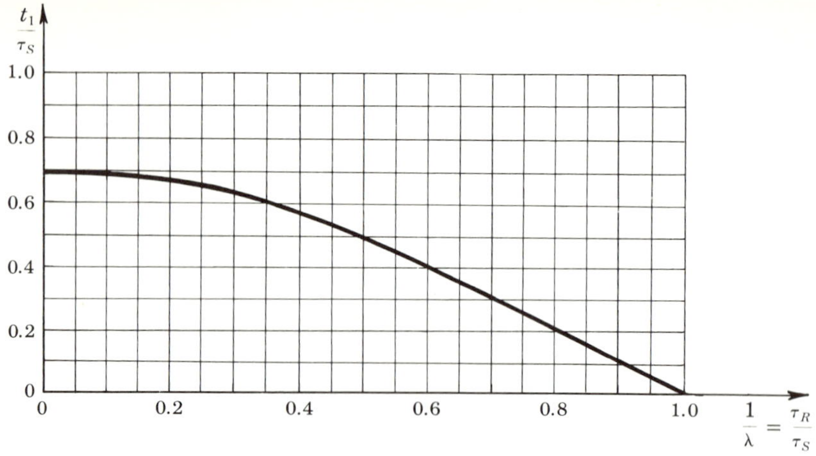

Fig. 8-28 Normalized half period as a function of the ratio of time constants [Eq. (8-41)].

which was given in Eq. (8-19). The second limiting location of the timing root appears where $\tau_S = \tau_R$ (i.e., when $\lambda = 1$). Under this condition, Eq. (8-40) may be expressed as

$$(x - 1)^2 = 0$$

and the half period is reduced to zero.

A plot of the solution of Eq. (8-41) using the root of Eq. (8-40) appears in Fig. 8-28. We can see that the recovery exponential has almost no effect on the sweep period when $\tau_R < 0.2\tau_S$, but as τ_R approaches the sweep-time constant τ_S, the sweep time rapidly decreases toward zero. Operation in this region is usually avoided since the slightest variation in circuit parameters has pronounced repercussions on the waveshape produced. There are multivibrators, however, which depend on the incomplete recovery behavior to generate extremely fast pulses. In this case, the transistor would never be allowed to saturate and the analysis would be slightly more complicated than that given above.

If the multivibrator is asymmetrical, we shall be faced with four transcendental equations, two for each half cycle. The evaluation of the two unknown time intervals and the two starting voltages would clearly be most tedious.

The preceding discussion proves that it is impossible to make an astable multivibrator when the recovery-time constant is equal to or exceeds the sweep-time constant. Thus we conclude that the MOSFET monostable of Fig. 8-18 and a bipolar monostable timed in the keep-turned-on mode (Figs. 8-15 and 8-17) cannot be converted into an astable by introducing a second energy-storage element.

The astable circuits we have discussed so far had a separate timing capacitor in each toggle network. These capacitors independently kept the active device off during each sequential-timing period.

Every active device may, however, be timed in either a keep-off or a keep-on mode. If we adjust the biasing of the basic amplifier so that it normally operates in the active region (open loop), then a single energy-storage element can be used to sequentially control the switching characteristics of a single active element, maintaining it first in the OFF state for one timing interval and then in the ON state for the other timing interval. This particular type of astable operation is illustrated in the MOSFET multivibrator of Fig. 8-29 generating the waveshapes of Fig. 8-31.

Under open-loop conditions (C open-circuited), both gates of the MOSFET are biased at the same potential, $V_{GS} = V_{DS}$, which satisfies the requirements for pinchoff-region operation (see Sec. 5-4).

A convenient starting point for the analysis is the instant before the gate of Q_2 reaches the threshold and switches Q_2 on. The significant circuit voltages are

$$v_{G2}(0^-) = V_T \qquad (8\text{-}42a)$$

$$v_{D1}(0^-) = V_{\text{ON}} \qquad (8\text{-}42b)$$

$$v_{D2}(0^-) = v_{G1}(0^-) = V_{DD} \qquad (8\text{-}42c)$$

The voltage across the capacitor before switching states is

$$v_q(0^-) = v_{G2} - v_{D1} = V_T - V_{\text{ON}} \qquad (8\text{-}43)$$

This voltage remains invariant across the circuit transition to Q_2 ON. The model holding is given in Fig. 8-30*a*.

From the model of Fig. 8-30*a* we can calculate the voltages just after Q_1 switches into the ohmic region and verify that bottoming has, in fact,

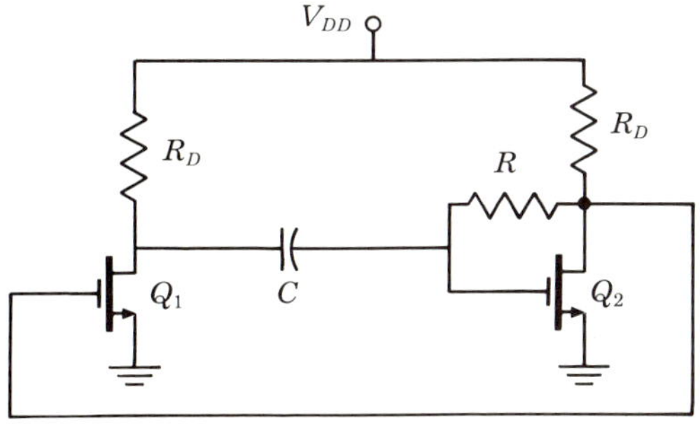

Fig. 8-29 A MOSFET astable multivibrator—normally biased in the active region.

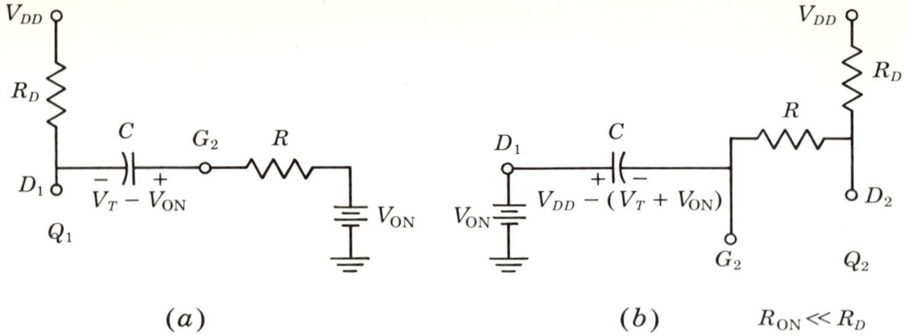

Fig. 8-30 Models holding in the two timing regions of the MOSFET astable multivibrator of Fig. 8-29. (*a*) Keep-on timing from cutoff Q_1 to ON Q_2; (*b*) keep-off timing from bottomed Q_1 to OFF Q_2. (In drawing these models, the small $R_{ON} \parallel R_D$ in series with V_{ON} was omitted in the interests of simplicity.)

occurred. Since $R_D \ll R$, almost the full circuit voltage will appear across R:

$$v_{G2}(0^+) \approx V_{DD} + V_T + V_{ON} \tag{8-44a}$$

$$v_{D1}(0^+) \approx V_{DD} \tag{8-44b}$$

$$v_{D2}(0^+) = v_{G1}(0^+) = V_{ON} \tag{8-44c}$$

The gate of Q_2 is certainly at a high enough voltage to ensure bottoming. The gate recovers toward V_{ON} with the time constant

$$\tau_R = (R + R_D)C \tag{8-45}$$

When the gate voltage falls enough so that the MOSFET enters its active region and the loop gain increases to unity, regeneration will again cause switching.

The bounds of the ohmic region are defined by

$$V_{DS} \le V_{GS} - V_T$$

But in the keep-on region, $V_{DS} = V_{ON}$; therefore, for

$$V_{GS} \ge V_{ON} + V_T \tag{8-46}$$

we consider that the MOSFET has bottomed. Of course, since V_{GS} falls as V_{ON} increases, the boundary is not as clearly defined as indicated in Eq. (8-46). We shall, however, use the lower limit of Eq. (8-46) as a fixed-voltage switching point in the timing equation.

The duration of the keep-on interval is given by

$$T_R = \tau_R \ln \frac{V_{ON} - (V_{DD} + V_T + V_{ON})}{V_{ON} - (V_T + V_{ON})}$$

$$= \tau_R \ln \left(1 + \frac{V_{DD}}{V_T}\right) \tag{8-47}$$

Usually $V_{DD} \gg V_T$ and the keep-on interval corresponds to almost the full exponential, lasting for $4\tau_R$. At the end of this interval, the voltage across the capacitor is

$$v_q(T_R) = V_{DD} - (V_T + V_{ON})$$

This voltage carries across the switching transient into the keep-off timing region. From the model now holding (Fig. 8-30*b*), we obtain the new initial conditions

$$v_{G2}(0^+) = 2V_{ON} + V_T - V_{DD} \tag{8-48a}$$

$$v_{D1}(0^+) = V_{ON} \tag{8-48b}$$

The gate of Q_2 charges from the negative voltage given by Eq. (8-48*a*) back toward V_{DD}, with switching again occurring when the gate voltage reaches V_T. The duration of this interval is

$$T_S = \tau_S \ln \frac{V_{DD} - (2V_{ON} + V_T - V_{DD})}{V_{DD} - V_T} \tag{8-49}$$

where $\tau_S = (R + R_D)C = \tau_R$

For large V_{DD}, we can approximate Eq. (8-49) by

$$T_S \approx \tau_S \ln 2$$

Bipolar-transistor and FET astables can be timed in exactly the same way. Here the keep-on period depends on the base-current flow keeping

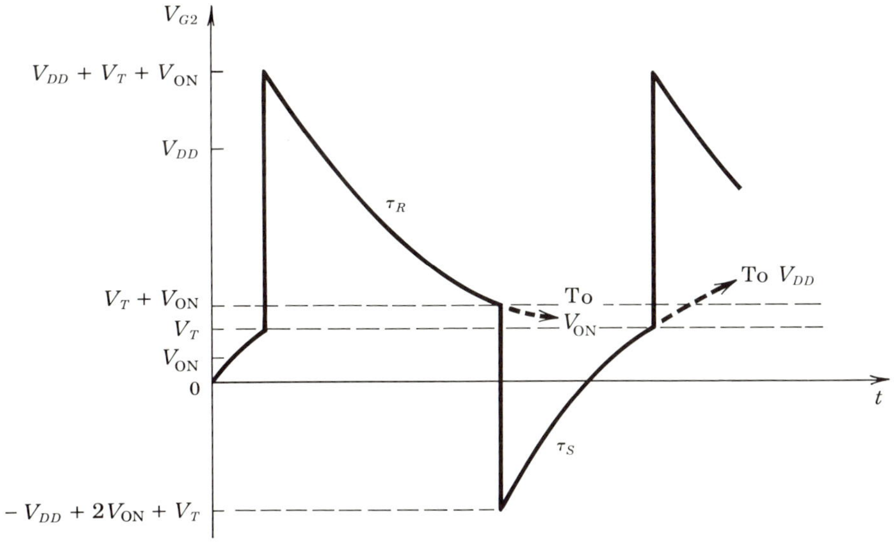

Fig. 8-31 Gate-voltage waveshape at Q_2 in the astable multivibrator of Fig. 8-29. The circuit is timed sequentially in the keep-off and keep-on modes.

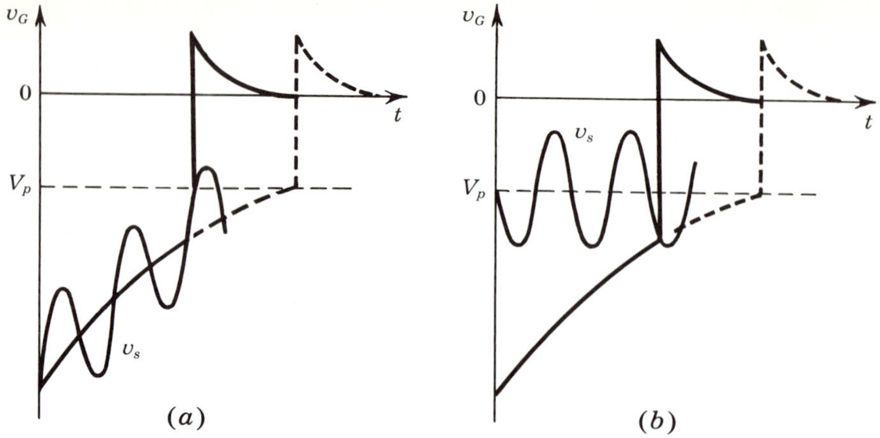

Fig. 8-32 Sine-wave synchronization of a FET astable multivibrator.

the transistor on and the keep-off period depends on back biasing the base. This type of astable operation may be considered as the sequential timing of the keep-on and keep-off cases of Sec. 8-5.

8-11 Multivibrator Synchronization

A synchronizing signal may be injected into one or both bases (or gates) of an astable multivibrator to lock its free-running period to some submultiple of the sync frequency. The charging curve would no longer be a simple exponential but would have the external signal superimposed. The FET conducts when the total gate voltage rises to pinchoff (Fig. 8-32*a*), and the transistor conducts when the base voltage rises to V_{BE}; and as was done in Chap. 6, the sync signal might be considered as effectively changing the FET's pinchoff voltage (Fig. 8-32*b*). If we approximate the exponential sweep by a straight line, then all the results of Chap. 6 dealing with synchronization may be applied to the astable multivibrator. In Fig. 8-32, the broken lines indicate the normal free-running waveshapes while the solid curves are the ones due to synchronization by the sinusoidal signal v_s.

If the sync signal is simultaneously applied to both gates (or bases) of a symmetrical astable multivibrator, then each half cycle will separately satisfy the synchronization conditions of Sec. 6-10. Provided that we redefine the free-running sweep period $t_1 = T/2$, where T is the astable period, the regions of multivibrator synchronization will be identical to those previously found.

When the sync signal is applied to only one base or gate, the unaffected half cycle may be considered a recovery interval. Equations (6-65*a*) to (6-65*c*) are now used to define the bounds of the regions of synchronization.

8-12 Summary

In retrospect, we might well enumerate the particular mode of analysis followed in the multivibrator circuits considered. First, the circuit is stripped to its very essentials, those necessary to treat only the important attributes of the system. We should especially try to remove any components which might change single energy-storage-element loops into multiple energy-storage systems. Of course, where this simplification drastically changes the characteristics of the multivibrator, the final answer will at best be an approximation. Second, the remaining circuit is linearized and the models which define its various possible regions are drawn. Finally, any logical place is chosen as our starting point and the calculations are begun. We now follow in time sequence the multivibrator's behavior, drawing new models whenever the circuit changes state. Any contradiction which may arise must be resolved by starting the solution again, from a new assumed condition. Jumping in blindly will eventually lead to a correct answer, but only if all circuit ramifications are followed to their logical conclusions. Our approach must remain consistent throughout the complete analysis of the circuit.

Differences between calculated and experimental results can usually be traced back to the approximations made in drawing the models. One example is that the threshold of a MOSFET always differs from the assumed value used in the model. The physical device turns on at a different time, and the pulse generated may be either shorter or longer than expected. If any other major "errors" are found, the original treatment must be modified in a "second look" at the circuit.

Problems

8-1 (*a*) Draw the loop transfer characteristic for the active network of Fig. 8-33. Assume that all amplifiers and feedback networks are ideal over the complete operating range. What are the coordinates of the stable points? How large must the input pulses be to guarantee triggering?

(*b*) Repeat part *a* when K_1 is reduced to 0.01.

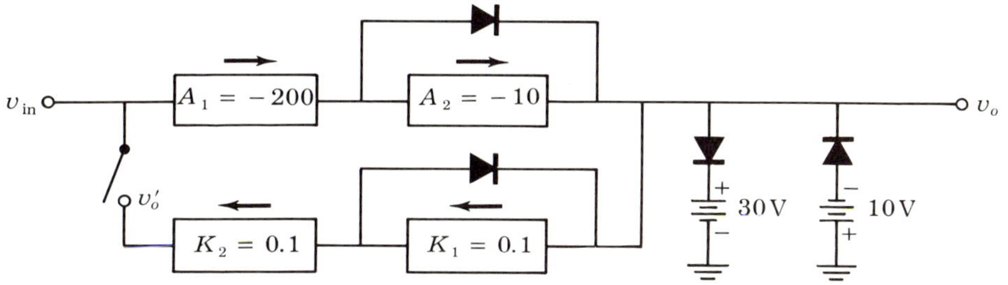

Fig. 8-33

8-2 Find the stable conditions of a bistable multivibrator constructed from the two DTL NAND gates of Fig. 4-9. Give the minimum-amplitude input pulse required to switch states.

8-3 Repeat Prob. 8-2 when the basic inverters are the ECL stages of Fig. 4-29.

8-4 Show that, to a very good first approximation, the stable conditions of the silicon bipolar direct-coupled bistable of Fig. 8-34 are independent of R_C and depend solely on V_{CC} and β. If $\beta = 10$, what is the minimum V_{CC} for a 1-V separation between the 0 and 1 output conditions?

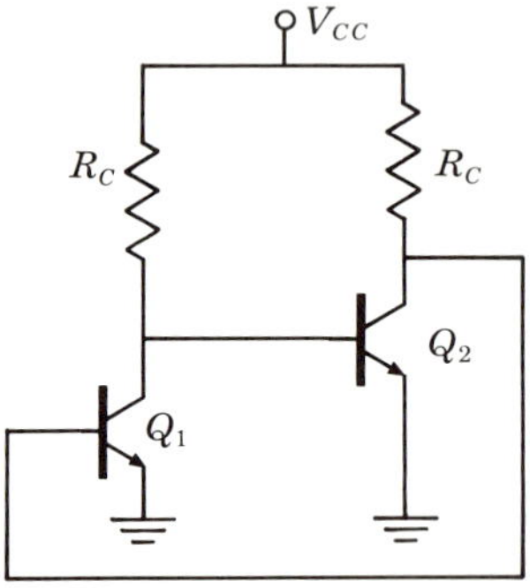

Fig. 8-34

8-5 The MOSFET bistable of Fig. 8-4 makes use of the inverter of Fig. 5-16. ($V_{DD} = 20$ V.)

(*a*) Give the output levels.

(*b*) If $C_{GS} = 3$ pF, find the switching speed for the transition between 0 and 1 and between 1 and 0.

8-6 Figure 8-35 shows a complementary-symmetry MOSFET bistable. This circuit switches extremely fast. Each device has $|V_T| = 3$ V, $|I_D| = 10$ mA at $|V_{GS}| = 4$ V, $I_{rs} = 10^{-12}$ A, and $C_{G\text{-sub}} = 1$ pF.

(*a*) What are the stable operating points for this circuit?

(*b*) Calculate the approximate switching speed of one inverter for the $0 \to 1$ and $1 \to 0$ transitions.

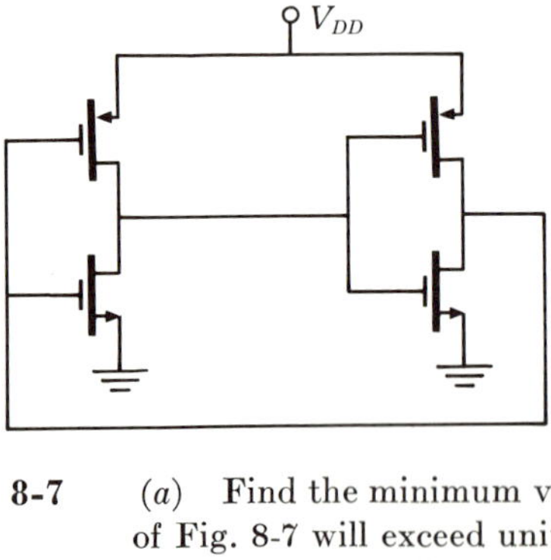

Fig. 8-35

8-7 (*a*) Find the minimum value of I_{DSS} for which the loop gain of the bistable of Fig. 8-7 will exceed unity.

(*b*) FETs having an I_{DSS} equal to ten times the value found in part *a* are used for the bistable. Calculate the stable operating conditions.

(c) Under the conditions of part b, what is the minimum trigger needed for switching?

8-8 By coupling together the sources of the two FETs of a bistable, as shown in Fig. 8-36, the need for a second power supply is eliminated. Calculate the values of R_D and R_s that will ensure that the gate of one FET is 2 V below pinchoff and the gate of the other FET is at +2 V (neglecting diode limiting). What role is served by C_s? Should it be very large or relatively small?

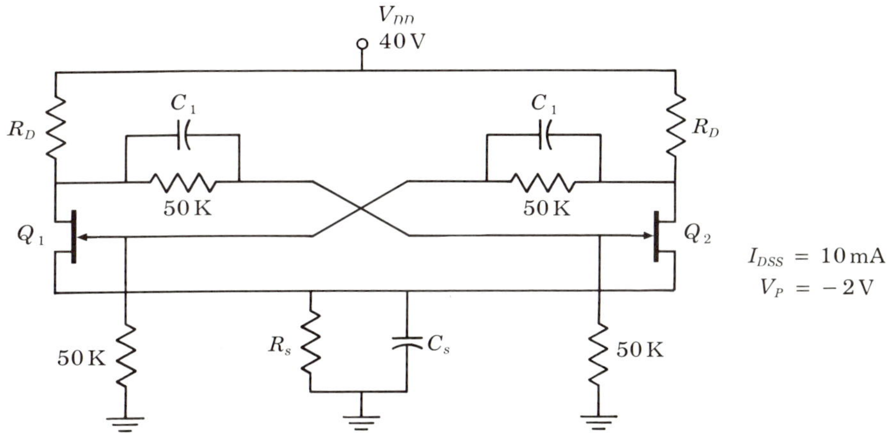

Fig. 8-36

8-9 The circuit of Fig. 8-8 uses silicon transistors with $\beta = 30$, $V_{BE} = 700$ mV, and $V_{CE,sat} = 150$ mV. $R_L = 1$ K, and $V_{CC} = 10$ V.

(a) Find the stable conditions if $R_1 = R_2 = 100$ K and $V_{BB} = -10$ V. Show that this condition corresponds to less-than-unity loop gain.

(b) What are the bounding values of $R_1 = R_2$ for proper bistable operation?

8-10 (a) Show that in a bistable designed to operate well within the saturation region, the negative real-axis pole will have little effect on the transition time.

(b) Justify the rise time given in Eq. (8-14) by direct solution from Eq. (8-13).

8-11 In a junction bistable of the type of Fig. 8-2, we shunt R_1 with a speedup capacitor such that $R_1C_1 = \tau_B$ (see Sec. 3-7).

(a) Under the above conditions, what is the ratio of the charge stored in C_1 to the charge stored in the base of the saturated transistor?

(b) Show how the charge stored in C_1 will affect the switching time. Which portion of cycle does it improve the most?

8-12 Figure 8-37 shows a chain of four inverters. Each transistor has $\beta = 100$, $V_{BE} = 700$ mV, and $V_{CE,sat} = 100$ mV. The turn-on delay of each stage is 10 ns, and the base-charge-storage turn-off delay is 20 ns.

(a) Find the voltages at all bases and collectors under steady-state conditions. Give the states of each transistor. What is the minimum negative disturbance at a collector needed for switching?

(*b*) Sketch the periodic waveshape appearing at any collector when the chain is increased to five stages. (*Hint:* Start with one set of assumed conditions and follow these through the loop.)

(*c*) Repeat part *b* for nine stages.

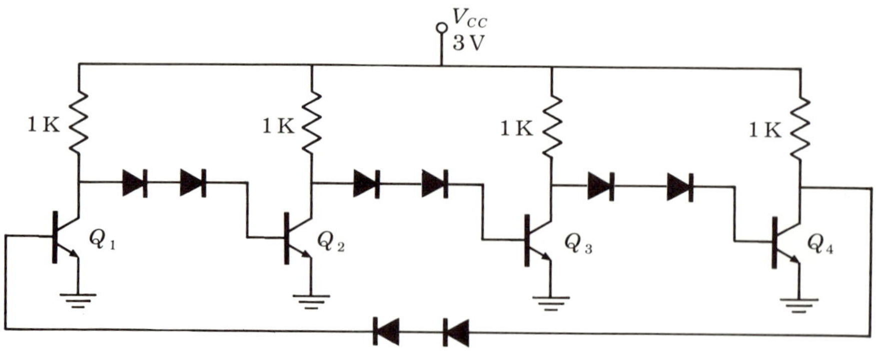

Fig. 8-37

8-13 The components in the circuit of Fig. 8-11 are $V_{CC} = 12$ V, $R_L = 2$ K, $R = 40$ K, and $C = 10$ nF. The transistors have $\beta = 100$, $V_{BE} = 750$ mV at $I_e = 1$ mA, $V_{CE,\text{sat}} = 100$ mV, and $r_{sc} = 25\ \Omega$. A trigger pulse switches the circuit into its quasi-stable state at $t = 0$. Sketch and label completely the waveshapes of the base and collector voltages and base current of Q_2. ($V_x = 5$ V.)

8-14 Using the values of Prob. 8-13, make a plot of the duration of the pulse interval of the circuit of Fig. 8-11 as a function of V_x over the range $+20\text{ V} \leq V_x \leq -20$ V.

8-15 Repeat Prob. 8-13 when the transistor has a base-emitter-junction breakdown voltage of 7.5 V. The avalanche-region resistance may be taken as a constant 25 Ω.

8-16 Figure 8-38 illustrates an alternative configuration for an inductively timed multivibrator. Show that in the normal state, Q_1 is on and Q_2 is cut off.

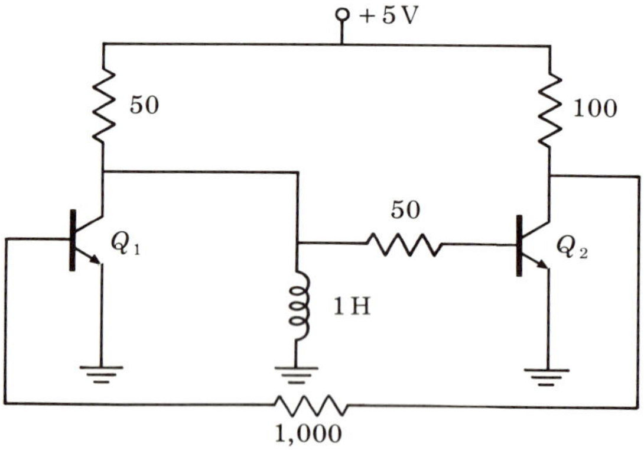

Fig. 8-38

$\beta = 30$, $V_{BE} = 700$ mV. Sketch the collector-voltage and base-current waveshapes of Q_2 and the coil-current waveshape. Give all times, time constants, and voltage and current values at the break points. What are the advantages or disadvantages of this circuit compared with the circuit of Fig. 8-17?

8-17 The RTL logic bistable of Fig. 8-2 is to be converted into a monostable. Draw the circuit and specify an *RC* toggle network for keep-off timing of a 10-μs pulse. Sketch and label the waveshapes appearing at the input (not the base) and output of both inverters.

8-18 The bistable of Fig. 8-7 is to be converted into a monostable multivibrator by replacing one level-shifting network by a toggle network. The timing resistor is 1 M.

(*a*) Specify the capacitor needed to generate a 100-μs pulse. Use keep-off timing. Sketch and label the waveshapes appearing at both gates and drains.

(*b*) Repeat part *a* for keep-on timing. Which is to be preferred? Why?

8-19 (*a*) We wish to check on the pulse-duration stability of a monostable timed in the keep-off mode. The timing resistor is returned to V_{CC}. How does the pulse duration vary with a 20 percent change in V_{CC} for a bipolar circuit ($V_{CC} \gg V_{BE}$)? For a FET circuit ($V_{DD} \gg V_p$)?

(*b*) Repeat part *a* when the pulse is timed in the keep-on mode. The timing resistor is returned to a keep-off voltage derived from the same power supply as V_{CC} or V_{DD}.

8-20 Sketch and label the gate- and drain-voltage waveshapes of Q_2 for the circuit of Fig. 8-20. All component values are given in Fig. 8-18.

8-21 Repeat Prob. 8-20 for keep-on timing. Show the modified circuit. Compare the results with those of Prob. 8-20.

8-22 Figure 8-39 shows a bipolar complementary-symmetry monostable. Each transistor has $\beta = 100$, $V_{BE} = 700$ mV at $I_e = 1$ mA, and $V_{CE,\text{sat}} = 100$ mV.

(*a*) Calculate the quiescent conditions.

(*b*) Sketch and label completely the waveshapes at the base of Q_1 and Q_2 and at the collector of Q_2. The circuit is triggered into the unstable mode at $t = 0$.

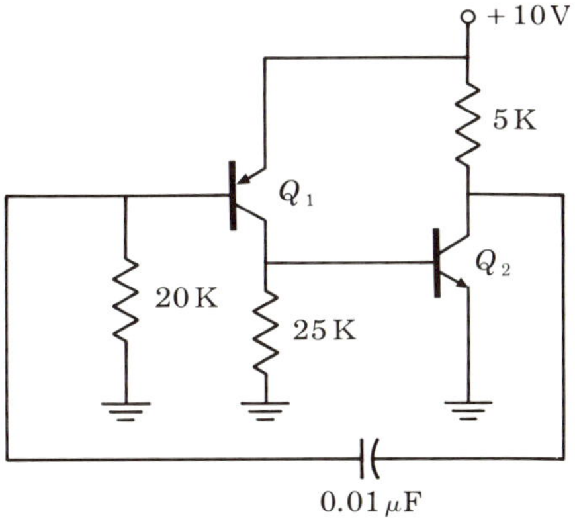

Fig. 8-39

8-23 Design an astable multivibrator that will generate a pulse whose duration is 10 percent of the total period of 200 μs. Use the transistors of Prob. 8-22, a power supply of 12 V, and collector resistors of 500 Ω. Make allowance in your design for a 20 percent variation in β; check to see if recovery is complete before the circuit reswitches.

8-24 Design a square-wave generator using junction FETs. The peak-to-peak amplitude should be 30 V, and the period should be adjustable by a potentiometer over the range of 50 to 500 μs. Use a FET having $V_p = -3$ V and $I_{DSS} = 10$ mA. Show the circuit, giving all component values. If necessary, incorporate diodes to square the output waveshape. Draw all significant waveshapes at a setting of 200 μs.

8-25 Interconnect the TTL inverters of Fig. 4-28*a* to make a symmetrical astable multivibrator having a 5-μs period. Show the circuit, giving values of all external components. Draw the waveshapes existing at the input and output of one inverter.

8-26 Repeat Prob. 8-25 using the ECL inverter of Fig. 4-29.

8-27 Figure 8-40 shows an astable multivibrator used as a linear frequency-modulation generator. Each capacitor-charging resistor is replaced by the constant-current source I, whose current is a linear function of the input driving voltage (for example, the collector current of a transistor depends on the base drive).

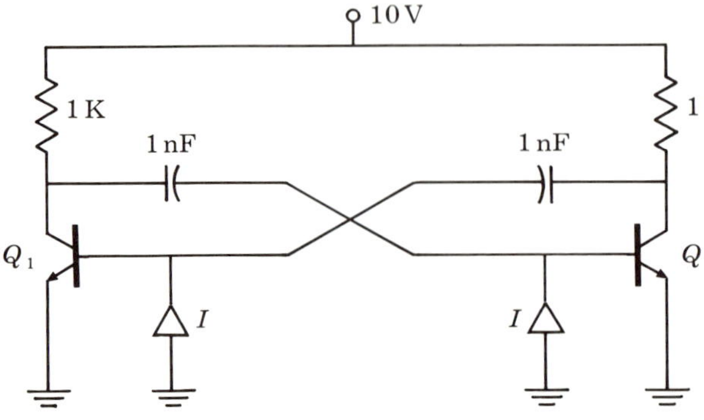

Fig. 8-40

(*a*) Find the frequency of oscillation as a function of I, assuming saturated operation and complete recovery within each cycle.

(*b*) Over what range of frequencies are the requirements of saturated operation and complete recovery satisfied?

8-28 The circuit of Fig. 8-11 becomes astable when the following components are employed: $R_L = 1$ K, $R = R_1 = 100$ K, $C = 0.01$ μF, $V_x = V_{CC} = 10$ V, $\beta = 50$, $V_{BE} = 700$ mV at $I_e = 1$ mA. Draw the base-current, base-voltage, and collector-voltage waveshapes of Q_2, giving all significant values.

8-29 (*a*) Derive Eq. (8-40).

(*b*) Show that the complete restriction on x in Eq. (8-40) must be

$$0.5 \leq x \leq 1.0$$

(*c*) For the case of $\tau_S = \tau_R$, what are the stable states of the two transistors?

(*d*) Under the symmetrical conditions of part *c*, calculate the loop current gain of the circuit. Explain the significance of your answer.

8-30 A symmetrical multivibrator is synchronized by injecting positive pulses into both bases (gates) through diodes. Because of the loading by the ON device, we can assume that these pulses will affect only the OFF device and that they will act to shorten the OFF time. For simplicity, we shall further assume that the charging curve is absolutely linear, starting at $-V_{CC}$ $(-V_{DD})$ and taking t_1 to rise to the idealized firing point at zero.

(*a*) Calculate and plot the regions of synchronization.

(*b*) Repeat part *a* if the pulses are injected into only one active device.

Bibliography

Abraham, H., and E. Bloch: Le Multivibrateur, *Ann. Phys. (Paris)*, vol. 12, p. 237, 1919.

Eccles, W. H., and F. W. Jordan: A Trigger Relay Utilizing Three Electrode Thermionic Vacuum Tubes, *Radio Rev.*, vol. 1, no. 3, pp. 143–146, 1919.

Linvill, J. G.: Nonsaturating Pulse Circuits Using Two Junction Transistors, *Proc. IRE*, vol. 43, pp. 826–834, 1955.

Millman, J., and H. Taub: "Pulse, Digital, and Switching Waveforms," McGraw-Hill Book Company, New York, 1965.

9.
emitter-coupled multivibrators

9-1 Transistor Emitter-coupled Multivibrator—Monostable Operation

The emitter-coupled multivibrator is a circuit worthy of more than a cursory glance, illustrating as it does the possibility of one circuit operating in any one of several modes. Selection of the particular mode and a measure of control over the performance within this mode are functions of the control voltage V_x, which is labeled in the emitter-coupled circuit of Fig. 9-1.

Because a general treatment often obscures the detailed functioning of the switching circuit, we shall, as an introduction, explicitly state one operating point and analyze the resultant circuit. By doing this, we may make a straightforward study of the time response without, at the same time, being forced to account for the effects of a varying control voltage. In Sec. 9-2 the possible modes will be defined, and any required modification of our original treatment may then be considered.

We shall assume that the circuit of Fig. 9-1 is adjusted so that, in its normal state, Q_2 is saturated and Q_1 is cut off. The application of a positive trigger at the base of Q_1 switches the multivibrator into the single quasi-stable state: Q_1 is driven into its active region, and its collector voltage drop, coupled through C, turns Q_2 off. Transistor Q_1, now operating as an emitter follower, stabilizes with an emitter-to-ground voltage one diode drop below V_x. As C begins charging, the base-to-ground voltage of Q_2 rises toward V_{CC}. Once it exceeds the new emitter voltage by 700

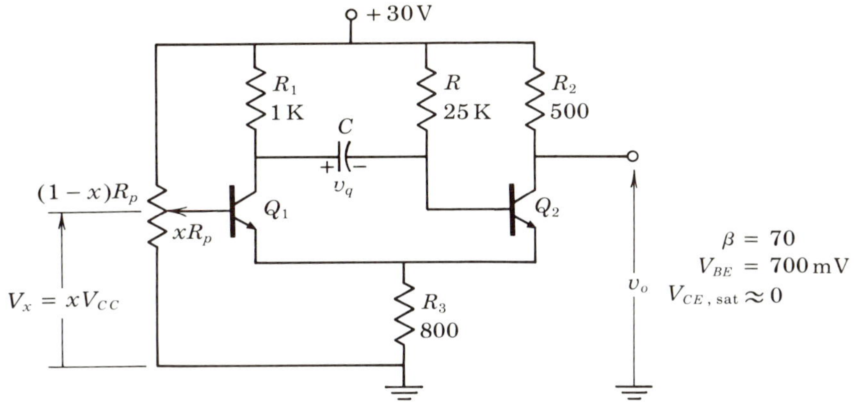

Fig. 9-1 Emitter-coupled multivibrator—component values shown for Examples 9-1 and 9-2.

mV, Q_2 will turn back on. The increased current in R_3 raises the common-emitter voltage above V_x, and it follows that Q_1 must be driven off. The positive transition at the collector of Q_1 is coupled to Q_2, driving it back into saturation.

Note that the coupling from Q_1 to Q_2 is through both the common-emitter resistance R_3 and the collector-to-base network of C and R. But the coupling from Q_2 back to Q_1 is only through the emitter resistor, which leaves R_2 relatively isolated from the transmission path. The collector of Q_2 therefore serves as a convenient place at which to take the output. At this point, a small amount of capacitive loading may be tolerated without seriously affecting the pulse duration.

In order to simplify the quantitative analysis of the multivibrator of Fig. 9-1, we shall make the following further assumptions:

1. β is very large, allowing $R \gg R_1$ and $R \gg R_2$.
2. The Thévenin equivalent impedance looking from the base of Q_1 into the control-voltage potentiometer is so very small that it also will be neglected.
3. $V_{CE,\text{sat}} \approx 0$.

We shall first define the ratio

$$x_B \equiv \frac{V_{BE}}{V_{CC}} \tag{9-1}$$

x_B is a normalization factor representing the relative importance of the junction drop compared with the supply. When x_B is very small, it is convenient to set $x_B = 0$; by thus removing V_{BE}, we simplify the circuit calculations.

Preceding switching, the values of significant voltages, measured with respect to ground, are

$$v_{c1}(0^-) = V_{CC} \tag{9-2a}$$

$$v_e(0^-) = v_{c2}(0^-) = \frac{R_3}{R_2 + R_3} V_{CC} \tag{9-2b}$$

$$v_{b2}(0^-) = v_e(0^-) + V_{BE} = \left(\frac{R_3}{R_2 + R_3} + x_B\right) V_{CC} \tag{9-2c}$$

$$\begin{aligned} v_q(0^-) &= V_{CC} - \left(\frac{R_3}{R_2 + R_3} V_{CC} + V_{BE}\right) \\ &= \left(\frac{R_2}{R_2 + R_3} - x_B\right) V_{CC} \end{aligned} \tag{9-2d}$$

where $v_q(0^-)$ is the initial charge on the capacitor.

Immediately after switching, the models of Fig. 9-2 represent the multivibrator's state and they must be used for the calculation of the system's time response. From Fig. 9-2*a*, we can see that the base voltage of Q_1 is always constant at xV_{CC}. Whenever Q_1 conducts, the common-emitter voltage will be one diode drop lower:

$$\begin{aligned} V_e &= xV_{CC} - V_{BE} \\ &= V_{CC}(x - x_B) \end{aligned} \tag{9-3}$$

Since we know the drop across R_3, we can easily find the emitter current of Q_1:

$$I_{e1} = \frac{V_{CC}(x - x_B)}{R_3} \tag{9-4}$$

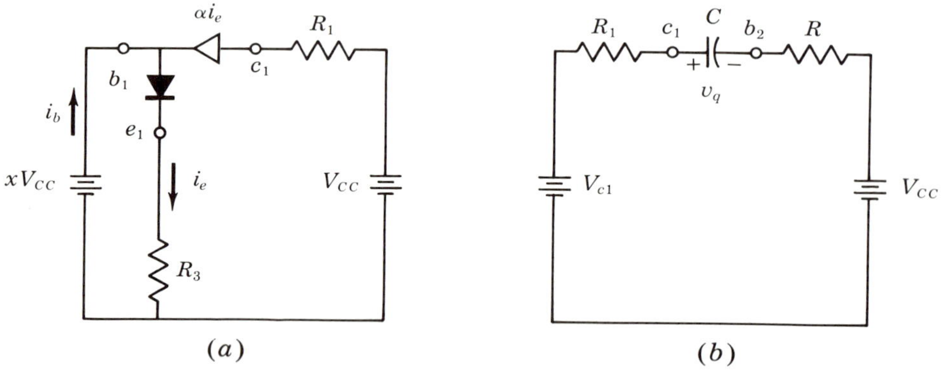

Fig. 9-2 Emitter-coupled multivibrator—models holding for the quasi-stable state. (*a*) Active-region model for Q_1; (*b*) Thévenin equivalent circuit coupling the collector of Q_1 to the base of Q_2.

With a high-β transistor, the collector current of Q_1 will be essentially the same as I_{e1}. The new collector voltage of Q_1 (neglecting capacitor-charging current) is

$$v_{c1}(0^+) \approx V_{CC} - I_{e1}R_1 = V_{CC}\left[1 - \frac{R_1}{R_3}(x - x_B)\right] \tag{9-5}$$

We can identify the whole second term of Eq. (9-5) as simply the magnitude of the negative transition at the collector of Q_1, once the circuit enters the quasi-stable region. Because our original assumption ($R \gg R_1$) allows us to ignore the loading of R_1 by R, the voltage fall at the base of Q_2 will be identically the drop of Eq. (9-5). We may immediately write

$$v_{b2}(0^+) = \frac{R_3}{R_2 + R_3} V_{CC} + V_{BE} - V_{CC}(x - x_B)\frac{R_1}{R_3} \tag{9-6}$$

Recovery from this sharp change of voltage is toward V_{CC}, with the time constant

$$\tau_S = (R_1 + R)C \approx RC$$

Once $v_{be2} = V_{BE}$ or, from Eq. (9-3), once the base-to-ground voltage of Q_2 reaches

$$v_{b2}(T_S) = xV_{CC}$$

Q_2 turns back on. We now know the initial, the final, and the steady-state voltages at the base of Q_2; their substitution into, and the solution of, the exponential response equation yields

$$T_S = \tau_S \ln \frac{V_{CC}\left[1 - \dfrac{R_3}{R_2 + R_3} + \dfrac{xR_1}{R_3} - x_B\left(1 + \dfrac{R_1}{R_3}\right)\right]}{V_{CC}(1 - x)} \tag{9-7}$$

for the pulse duration. V_{CC} in the numerator and denominator cancel. All terms except x_B are independent of the active element and of the power supply. x_B does depend on the power supply, but this dependence becomes significant only when the multivibrator is operated with a low supply voltage.

The multivibrator reenters the regenerative transition region with the base of Q_2 at a voltage lower than its steady-state voltage [Eq. (9-2*c*)]. We suspect that there will be a voltage jump, with a high probability that this jump will be larger than necessary to return the base to its original point. The new base, emitter, and collector voltages can easily be calculated by drawing the model holding after the jump, i.e., during recovery (Fig. 9-3).

Just before the multivibrator switches, the final voltage across C is

$$v_q(T_S) = V_{c1} - xV_{CC} = V_{CC}\left[1 - x\left(1 + \frac{R_1}{R_3}\right) + \frac{R_1}{R_3}x_B\right] \tag{9-8}$$

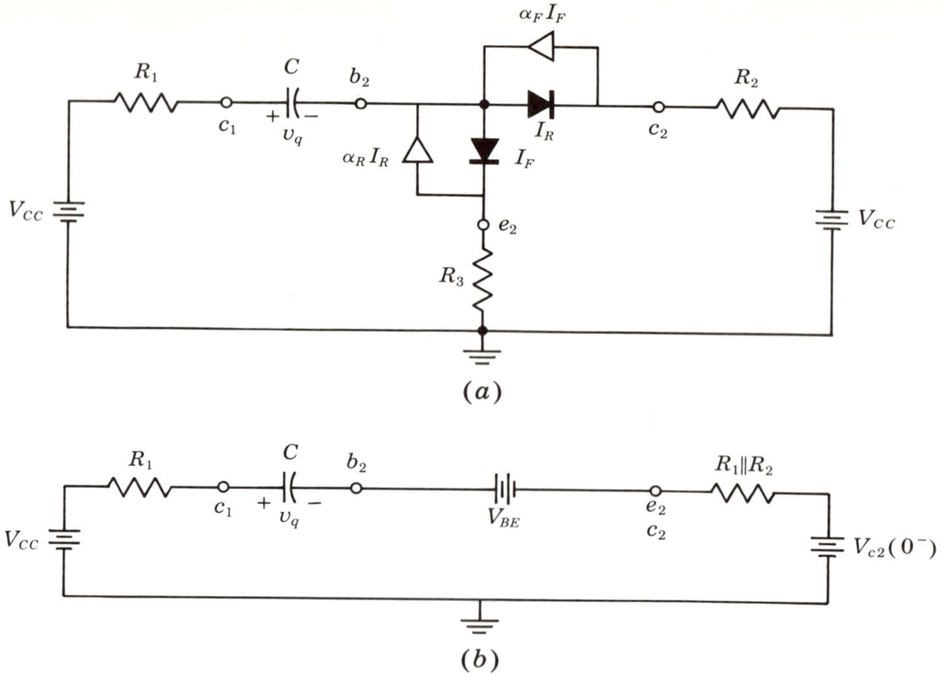

Fig. 9-3 *(a)* Recovery-region model of the emitter-coupled multivibrator (Q_2-saturated Ebers-Moll model shown); *(b)* Thévenin equivalent model of the circuit of *(a)*.

By substituting this value into Fig. 9-3, we arrive at the new operating point of the transistor. The algebra is much simpler when we can use numbers, and the solution will therefore be left to the reader as a part of his analysis of specific problems and to the calculations performed in Example 9-1.

After the final jump, the complete circuit recovers toward the original steady-state conditions, with the time constant

$$\tau_R = C[R_1 + (R_2 \parallel R_3)]$$

Waveshapes appear in Fig. 9-4, with all jumps and time constants indicated. Observe that the base, emitter, and collector of Q_2 all have identical recovery waveshapes (the saturated transistor is a short circuit).

Example 9-1

By setting $x = 0.4$, the monostable multivibrator of Fig. 9-1 is adjusted to operate in the mode described above, where Q_1 switches between its

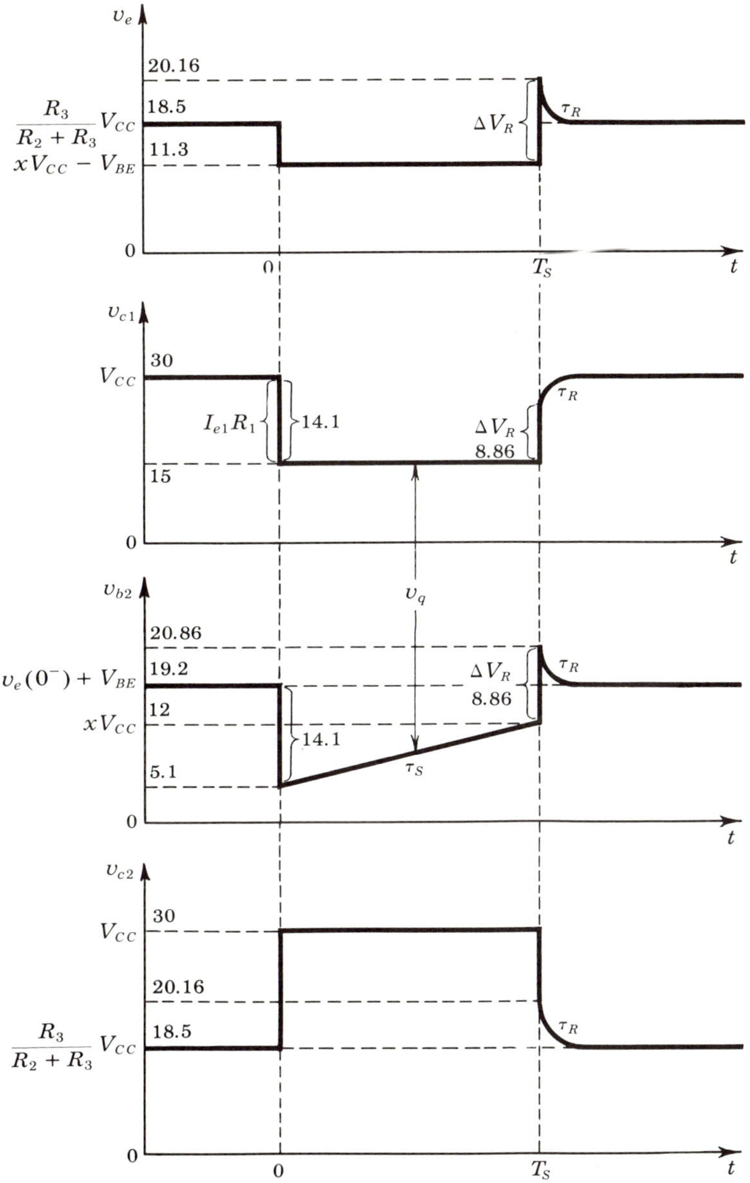

Fig. 9-4 Emitter-coupled multivibrator waveshapes—monostable operation-mode 12. Values given are for the solution of Example 9-1.

cutoff and active regions. All circuit components are as given in Fig. 9-1. We do not need to specify C for the following analysis since it would be selected on the basis of the required pulse width.

The duration of the unstable state, which is of primary interest, is controlled by the charge in C, maintaining Q_2 cut off. Only two unknown voltages must be found in order to evaluate this interval, the initial value at the base of Q_2 as the circuit enters its unstable region and the voltage at which Q_2 again turns on.

From Eq. (9-2c), the stable base voltage of Q_2 is

$$v_{b2}(0^-) = \left(\frac{800}{1{,}300} + 0.023\right)30 = 19.2 \text{ V}$$

where $x_B = V_{BE}/V_{CC} = 0.023$. With $x = 0.4$, the voltage at the base of Q_1 remains fixed at

$$v_{b1} = xV_{CC} = 12 \text{ V}$$

When the trigger is injected, Q_1 turns on and the common-emitter voltage drops from 18.5 to $xV_{CC} - V_{BE} = 11.3$ V.

If the small base-current flow is neglected, the voltage drop across R_1 will be proportional to the 11.3-V drop across R_3:

$$\frac{R_1}{R_3} = \frac{\Delta V_{c1}}{V_e}$$

or $\Delta V_{c1} = 14.1$ V. C couples this transition to the base of Q_2, driving it from its original value of 19.2 down to 5.1 V. C immediately starts charging, as shown in Fig. 9-2b, toward 30 V with the time constant $\tau_S = 1{,}000C$. When $v_{be2} = V_{BE}$ or when $v_{b2} = xV_{CC} = 12$ V, Q_2 turns back on. Consequently,

$$T_S = \tau_S \ln \frac{30 - 5.1}{30 - 12} = 0.324\tau_3$$

At the end of the sweep interval, the voltage across C is only 3.8 V. Substituting this value into the model of Fig. 9-3b, we find that the initial current after switching into the recovery region is

$$i_R(0^+) = \frac{V_{CC} - v_q(T_S) - V_{BE} - V_{c2}(0^-)}{R_1 + (R_2 \parallel R_3)}$$

This current gives rise to a voltage jump at the base of Q_2 from 12 to 20.86 V. The same-magnitude jump will appear both at the Q_1 collector side of the coupling capacitor and at the emitter of the now ON Q_2. Since Q_2 is saturated, its collector voltage must be the same as at the emitter.

The circuit returns to the original state with the recovery-time constant

$$\tau_R = 1.310C$$

The voltages calculated above are used to label the typical circuit waveshapes of Fig. 9-4.

9-2 Modes of Operation of the Emitter-coupled Multivibrator

The common-element (emitter) coupling permits this multivibrator to operate in a possible multiplicity of modes in each of its two classes. Since Q_2 is always conducting in the normal state, these classes might well be defined in terms of the quiescent condition of this transistor—class 1 existing when the bias current is adjusted so that Q_2 is placed in its active region and class 2 when it is saturated. The particular class depends solely on the relationship of R to R_2; for Q_2 saturated (class 2), the approximate relationship becomes

$$R < \beta R_2$$

We can prove the sufficiency of this condition by considering the transistor model holding under saturation. [A more exact condition is $R < \beta(1 - x_B)R_2$.]

Class-2 operation is to be preferred over class 1; here both the initial conditions and the sweep duration are relatively independent of the transistor parameters. However, the saturated transistor has a long recovery time and we would be forced to incorporate antibottoming diodes. We thus transfer the minority-carrier-storage problem from the transistor to the diodes, where it is more readily handled.

The various possible modes within each class are determined by the switching conditions of Q_1, that is, by its starting and ending states. Modes will be classified by two numbers, the first describing the state of Q_1, as listed below, and the second the class (state of Q_2).

States of Q_1

	BEFORE SWITCHING	AFTER SWITCHING
0	Cut off	Cut off
1	Cut off	Part on
2	Cut off	Full on
3	Part on	Full on
4	Full on	Full on

Since the state of Q_1 is a direct function of V_x, investigation of the multivibrator's performance will proceed, following the potentiometer setting, from $x = 0$ to $x = 1$. Each mode will be delineated by establishing the boundary values (of x) in terms of the circuit parameters. At any particular setting of x, detailed time-response calculations would simply follow the analysis outlined in Sec. 9-1.

We shall mainly concern ourselves with modes $n2$, representing as they do the more stable operation of the emitter-coupled multivibrator. In order to complete the discussion, one astable mode of group $n1$ is briefly treated in Sec. 9-4. All approximations and assumptions made in Sec. 9-1 are also applicable in the following arguments.

Mode 02

In this mode, Q_1 is normally off. Application of a trigger turns Q_1 part on, but not far enough on so that its collector-voltage change can drive Q_2 into cutoff. Upon removal of the trigger, Q_2 is still in the active region, and it immediately switches back to its normal state, full on. The output pulse generated is of small amplitude and of exactly the same duration as the input trigger.

In order for Q_1 to turn on at all,

$$xV_{CC} \geq V_{BE}$$

which leads to the lower bound of x for operation in any mode:

$$x \geq x_B \tag{9-9}$$

By assuming that Q_2 has turned full off and that Q_1 is still part on, the model drawn is much simpler than the one required if both transistors are taken in their active regions (the model for Q_1 is given in Fig. 9-2a). The region's boundary remains the same regardless of the direction from which we approach it. After the circuit switches state, the initial value of Q_2's base voltage is given by Eq. (9-6). Whenever Q_1 is on, the drop across the emitter resistor is always $xV_{CC} - V_{BE}$. In order for Q_2 to be part on, its base must be roughly V_{BE} higher. The boundary condition is found from

$$v_{b2}(0^+) = \frac{R_3}{R_2 + R_3} V_{CC} + V_{BE} - V_{CC}(x - x_B) \frac{R_1}{R_3} \geq xV_{CC} \tag{9-10}$$

Solving Eq. (9-10) for the limit of x satisfying the inequality stated, we find that the limits of mode 02 are

$$x_B < x < \frac{R_3}{R_1 + R_3} \frac{R_3}{R_2 + R_3} + x_B \tag{9-11}$$

This calculation serves mainly to define the lower limit of mode 12, the first useful range.

Mode 12

An input trigger will again switch Q_1 part on, but now the drop in its collector voltage is sufficient to drive Q_2 into cutoff. The upper limit of x is at that particular potentiometer setting that would allow Q_1 to switch from cutoff to saturation upon being triggered. To find this setting, we should draw the model holding for Q_1 full on (Fig. 9-5) and solve for the corresponding emitter-to-ground voltage:

$$v_e = xV_{CC} - V_{BE} \approx \frac{R_3}{R_1 + R_3} V_{CC} \tag{9-12}$$

The limits of x for operation in mode 12 are from the upper limit of mode 02 to the value found from Eq. (9-12):

$$x_B + \frac{R_3}{R_1 + R_3} \frac{R_3}{R_2 + R_3} < x < \frac{R_3}{R_1 + R_3} + x_B \tag{9-13}$$

For the multivibrator of Example 9-1, the limits are $0.296 < x < 0.468$.

Mode 12 was the one treated in detail in Sec. 9-1. Its pulse width, defined by Eq. (9-7), is a function of the potentiometer setting x. Increasing the setting increases both the drop in the base voltage of Q_2 and the value of the emitter voltage after switching into the unstable state. As a consequence of the greater separation between the initial and final values of the base voltage, the pulse duration increases with potentiometer rotation.

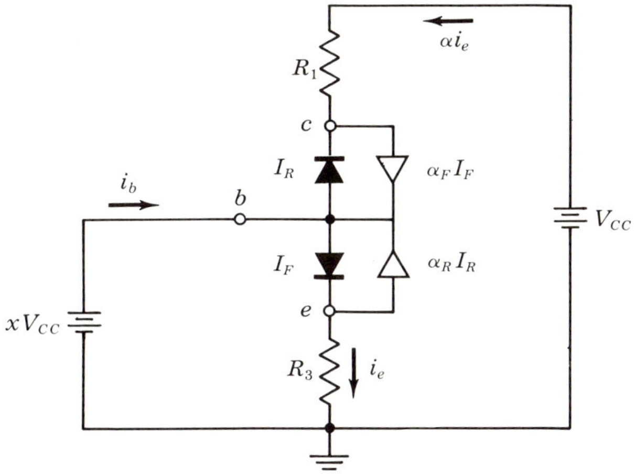

Fig. 9-5 Model for the upper limit of mode 12 (Q_1 saturated).

Substituting the limits of x into Eq. (9-7) gives the approximate permissible range of output-pulse width as

$$0 < T_{S1} < \tau_S \ln \left[1 + \frac{R_2(R_1 + R_3)}{R_1(R_2 + R_3)}\right] \tag{9-14}$$

In finding the limits of Eq. (9-14), we have made the approximation that at the upper bound $x_{12,\max} > x_B$. This approximation is reasonable when $V_{CC} \gg V_{BE}$. For the circuit of Fig. 9-1 we have found that $x_{12,\max} = 0.468$, which is certainly much larger than $x_B = 0.023$. When the circuit is symmetrical ($R_1 = R_2$), the maximum pulse width generated is $T_{S,\max} \approx \tau_S \ln 2$.

Mode 22

In this mode, Q_1 switches from cutoff to full on. The next possible circuit change occurs when the normal state of Q_1 changes from cutoff to part on. But with Q_1 cut off, Eq. (9-2*b*) holds; by solving for $v_{be1} = V_{BE}$, we find the upper limit of mode 22:

$$v_{be1} = xV_{CC} - \frac{R_3}{R_2 + R_3} V_{CC} \leq V_{BE} \tag{9-15}$$

The limits of x for this mode become

$$x_B + \frac{R_3}{R_1 + R_3} < x < \frac{R_3}{R_2 + R_3} + x_B \tag{9-16}$$

and in Example 9-1, the upper limit is located at 0.638.

Note that unless $R_1 > R_2$, mode 22 will not exist and the transition will be directly from mode 12 to mode 32. Furthermore, if $R_1 = R_2$, the range of voltage settings for mode-22 operation degenerates into a single point.

When Q_1 goes full on, the drop across R_3 is $xV_{CC} - V_{BE}$. The remaining voltage must appear across R_1; therefore the drop in the collector voltage of Q_1 will now be

$$\Delta v_{c1} = (1 - x + x_B)V_{CC} \tag{9-17}$$

This same drop appears at the base of Q_2, turning it off. The base of Q_2 charges from its initial voltage, found from Eqs. (9-2*b*) and (9-17),

$$v_{b2}(0^+) = \frac{R_3 V_{CC}}{R_2 + R_3} + V_{BE} - (1 - x + x_B)V_{CC} \tag{9-18}$$

toward V_{CC} with the time constant τ_S. Q_2 finally conducts when

$$v_{b2}(T_{S2}) = xV_{CC}$$

Consequently, in mode 22, the pulse width becomes

$$T_{S2} = \tau_S \ln \frac{2 - R_3/(R_2 + R_3) - x}{1 - x} \tag{9-19}$$

Equation (9-19) indicates that in mode 22 the pulse width depends only on the resistor ratios and the potentiometer setting x; it is independent of the power supply and of x_B. This is the mode to use when generating a fixed-width pulse.

Mode 32

Under the circuit conditions existing within this mode of operation, Q_1 always conducts, maintaining the emitter voltage fixed at $xV_{CC} - V_{BE}$ across the complete transition region. Limits, on x, are from the upper limit of mode 22 to that value which puts both transistors into permanent saturation (Fig. 9-6).

The upper limit, found from the model of Fig. 9-6, is

$$x < \frac{R_3}{(R_1 \parallel R_2) + R_3} + x_B$$

Thus the range of x-defining mode 32 becomes

$$x_B + \frac{R_3}{R_2 + R_3} < x < \frac{R_3}{(R_1 \parallel R_2) + R_3} + x_B \tag{9-20}$$

When R_3 is very large compared with R_1 and R_2, the limits of x for which this mode holds shrink almost to the vanishing point. The circuit conditions before switching are determined from the appropriate model (Q_1 part on and Q_2 saturated). Since both transistors are conducting, their base voltages are almost identical. At $t = 0^-$,

$$v_{b2}(0^-) = v_{b1}(0^-) = xV_{CC}$$

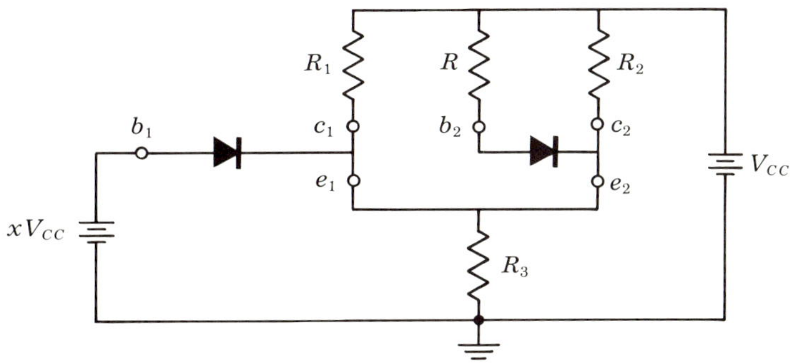

Fig. 9-6 Both transistors in saturation.

The total current through the common-emitter resistor is always determined by the control voltage V_x:

$$I_{R_3} = \frac{xV_{CC} - V_{BE}}{R_3} \tag{9-21}$$

Since Q_2 is saturated at $t = 0^-$, its emitter current is given by

$$i_{e2}(0^-) \approx \frac{V_{CC} - (xV_{CC} - V_{BE})}{R_2} \tag{9-22}$$

Thus the current through Q_1 will be determined from

$$i_{e1}(0^-) = I_{R_3} - i_{e2}(0^-) \tag{9-23}$$

Substituting and solving, we find the quiescent collector voltage of Q_1 to be

$$\begin{aligned} v_{c1}(0^-) &= V_{CC} - i_{e1}(0^-)R_1 \\ &= V_{CC}\left[1 + \frac{R_1}{R_2} - \left(\frac{R_1}{R_2} + \frac{R_1}{R_3}\right)(x - x_B)\right] \end{aligned} \tag{9-24}$$

But after switching, Q_1 saturates and

$$v_{c1}(0^+) = v_e = xV_{CC} - V_{BE} \tag{9-25}$$

The net change in the collector voltage of Q_1 [Eqs. (9-24) and (9-25)] is coupled to the base of Q_2. As the multivibrator enters the quasi-stable region, Δv_{c1} determines the initial value of v_{b2}:

$$v_{b2}(0^+) = xV_{CC}\left[1 + \frac{R_1}{R_2} - \left(1 + \frac{R_1}{R_2} + \frac{R_1}{R_3}\right)(x - x_B)\right] \tag{9-26}$$

As in all the other modes, the base voltage rises toward V_{CC}, causing switching at xV_{CC}. The pulse duration, still a function of x, turns out to be

$$T_{S3} = \tau_S \ln \frac{2 - x + (R_1/R_2) - [1 + (R_1/R_3) + (R_1/R_2)](x - x_B)}{1 - x} \tag{9-27}$$

In the three modes of operation that generate a usable pulse, its duration, to the first approximation, is independent of V_{CC}, resulting in extremely good pulse-width stability.

Mode 42

This mode exists when both transistors are always saturated. Thus an input trigger will have absolutely no effect. The limits are

$$x_B + \frac{R_3}{R_1 \parallel R_2 + R_3} < x < 1 \tag{9-28}$$

It is included only to complete the range of settings of the potentiometer and, like mode 02, is of no practical use.

A summary of the results is presented in Tables 9-1 and 9-2. Table 9-1 shows the operating regions. Within each mode, the normalized pulse duration is given by

$$\frac{T_n}{\tau_S} = \ln \frac{b + ax}{1 - x} \tag{9-29}$$

The constants a and b are listed in Table 9-2.

In the interests of simplicity, for the remaining discussion we shall assume $x_B = 0$. This is not a particularly gross assumption since $x_B \ll x$ over most of the operating range whenever $V_{CC} \approx V_{BE}$.

Table 9-1 Possible transistor states, emitter-coupled multivibrator ($x_B = 0$)

MODE	UPPER LIMIT OF x	Q_1	Q_2
02	$\frac{R_3}{R_1 + R_3} \frac{R_3}{R_2 + R_3}$	Off	Full on
12	$\frac{R_3}{R_1 + R_3}$	Off ↓ Part on	Full on ↓ Cut off
22	$\frac{R_3}{R_2 + R_3}$	Off ↓ Full on	Full on ↓ Cut off
32	$\frac{R_3}{(R_1 \parallel R_2) + R_3}$	Part on ↓ Full on	Full on ↓ Cut off
42	1	Full on	Full on

To summarize further the behavior of the multivibrator in terms of the particular modes, the complete normalized period is plotted as a function of x for the circuit of Example 9-1. The solid curve 1 of Fig. 9-7 illustrates the varying pulse duration as a function of x when the impedance of the control potentiometer is negligible. This corresponds to the mode boundaries discussed above. Both the demarcation between the individual modes and the different rates of pulse-width variation with x within each mode are quite clear. It might be noted that, in this special case, the pulse duration appears to vary almost linearly with x within mode 12.

Table 9-2 Pulse duration, emitter-coupled multivibrator—modes $n2$ ($x_B = 0$)

MODE	b*	a*
12	$1 - \dfrac{R_3}{R_2 + R_3}$	$\dfrac{R_1}{R_3}$
22	$2 - \dfrac{R_3}{R_2 + R_3}$	-1
32	$2 + \dfrac{R_1}{R_2}$	$-\left(2 + \dfrac{R_1}{R_3} + \dfrac{R_1}{R_2}\right)$

* Constants for Eq. (9-29).

Effects of finite potentiometer impedance

When the resistance of the control potentiometer is appreciable, we must include its equivalent source impedance in all calculations involving the actual base and emitter voltages of Q_1. The dashed curve 2 of Fig. 9-7 shows the effect of a 10-K potentiometer on the pulse duration in the various modes of operation. This plot is the result of inserting the Thévenin equivalent impedance of

$$R_s = x(1 - x)R_p \tag{9-30}$$

in series with xV_{CC}.

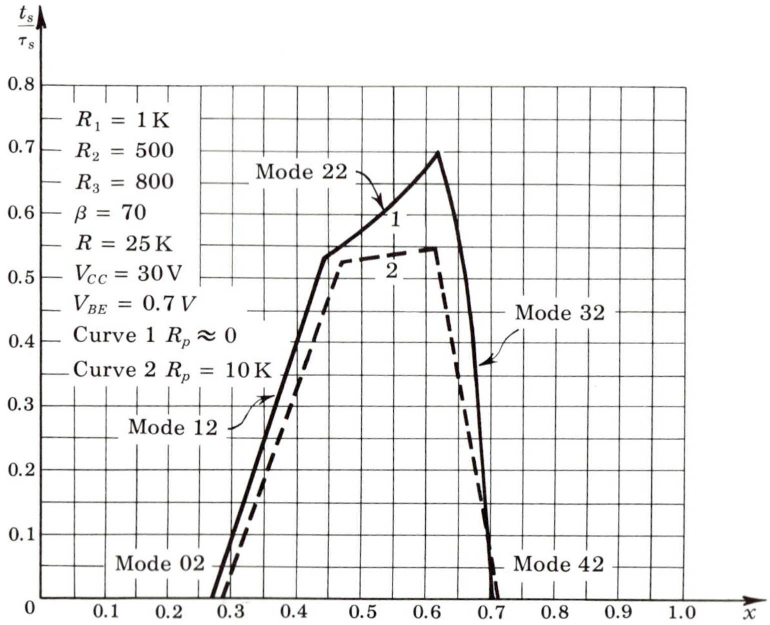

Fig. 9-7 Pulse duration as a function of potentiometer setting.

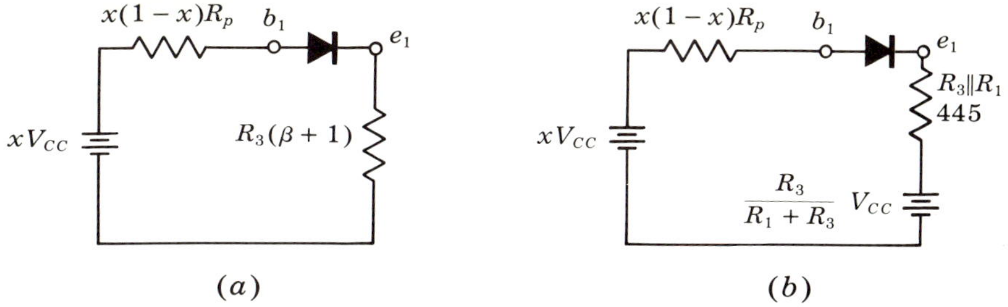

Fig. 9-8 Equivalent input circuit of Q_1, including the potentiometer source impedance. (*a*) Mode-12 operation; (*b*) Q_1 saturated (mode 22).

For mode 12, the input model is shown in Fig. 9-8*a*. Here, for any setting of x, the source impedance is much less than the active-region input impedance $(\beta + 1)R_3$ of the transistor. With the specific circuit values of Example 9-1, the input impedance is 56 K, compared with the maximum value of R_s of 2.5 K. The loading of the potentiometer is negligible, and the base voltage is but slightly less than the open-circuit voltage xV_{CC}. To generate a given-duration pulse, the value of x must be set slightly higher than previously found.

Once the multivibrator begins operating in mode 22, the input model must be changed to the one shown in Fig. 9-8*b*, which is obtained by replacing the saturated transistor by the Thévenin equivalent circuit seen looking into its base. The input impedance is reduced from 56 K to the parallel combination of R_1 and R_3 (445 Ω). Furthermore, the potentiometer varies about the midpoint of its setting, and its impedance remains close to the maximum possible value (R_s = 2,500 Ω). Because most of the voltage will be developed across R_s, any change in the control voltage will have but small effect on the base voltage of Q_1. The common-emitter voltage will not vary much, and neither will the pulse duration. This segment of the curve in Fig. 9-7 becomes much more horizontal. In Example 9-2 we shall consider one specific setting of x, lying in this mode, to illustrate the role of R_s in determining the circuit voltages and timing.

Mode-32 operation is restricted to such a small fraction of the range that, in general, it is of little interest. R_s would be treated as above, by including it in the appropriate models. The reader may perform this substitution and calculation himself.

Example 9-2

The multivibrator of Fig. 9-1 uses a 10-K potentiometer for its timing control. When it is set at $x = 0.55$, the circuit operates near the center of mode 22. All other circuit components are the same as in Example 9-1, and the results of the problem may be used in this solution.

The initial voltage at the base of Q_2, before triggering, is 19.2 V (from Example 9-1). After the trigger turns Q_1 on, the saturated transistor voltage may be found from the model of Fig. 9-8*b*. Here $xV_{CC} = 16.5$ V and $R_s = 2{,}475\ \Omega$. From Fig. 9-8*b*,

$$v_{c1}(0^+) = v_e(0^+) = 13.33 + \frac{445}{2{,}920} \times (16.5 - 13.33 - 0.7)$$
$$= 13.71 \text{ V}$$

If R_s were 0, this voltage would be $xV_{CC} = 16.5$ V. The loading of the potentiometer reduces the transistor voltage to slightly above its saturation value. Thus, as Q_1 turns on, the drop at its collector is only

$$\Delta v_{c1} = 30 - 13.7 = 16.3 \text{ V}$$

This drop is coupled by C, driving the base of Q_2 down to

$$v_{b2}(0^+) = 19.2 - 16.3 = 2.9 \text{ V}$$

Recovery is toward 30 V, with the circuit reswitching at one diode drop above the common-emitter voltage of 13.71 V. The pulse generated lasts for

$$T'_S = \tau_S \ln \frac{30 - 2.9}{30 - 14.4} = 0.554\tau_S$$

When the source impedance is small enough to be neglected, the base of Q_1 will be at xV_{CC} (16.5 V). In this case the change in collector voltage is only 14.2 V and the base of Q_2 is driven only to 5 V. The reswitching value is also higher, 16.5 V. Since a greater percentage of the total exponential is used, the sweep period becomes somewhat longer:

$$T_S = \tau_S \ln \frac{30 - 5}{30 - 16.5} = 0.615\tau_S$$

9-3 Monostable Pulse Variation

The question confronting us in this section is: What circuit relationships must be satisfied to achieve some physically permissible variation of the pulse duration with respect to x? In modes 12 and 32, the function might well be a linear one, and in mode 22 it might be either constant or linear. One commonly used method of attacking this problem is to first expand the general time equation (9-29) into some series and then, by examining the individual terms, attempt to eliminate those that contribute the nonlinearity. But since x takes on a relatively large range of values, there is no possible series that will converge fast enough so that all the higher-order terms in x can be eliminated. A solution of this type is quite tedious, and in addition, the design requirements are often obscured beneath the algebra needed in handling the existent higher-order terms.

A second, more fruitful approach is through an investigation of the properties of the slope of the curve rather than of the curve itself. When we differentiate Eq. (9-29) with respect to x, the resultant expression for the slope is no longer an exponential function of x, and consequently it is of simpler form than is the original equation:

$$S = \frac{d(T_S/\tau_S)}{dx} = \frac{-(b + a)}{ax^2 - (a - b)x - b} \tag{9-31}$$

In order for the pulse-width variation to be linear with respect to changes in x, the slope must be independent of x; it should be a constant. But this requires that the denominator of Eq. (9-31),

$$y = ax^2 - (a - b)x - b \tag{9-32}$$

also be a constant for all values of x within any one mode. Equation (9-32) is the equation of a parabola, and therefore it is impossible to satisfy the required condition; we can never expect a completely linear relationship. A good compromise will keep the change in y small for the range of x within the mode under investigation.

We would conclude, from scrutinizing the properties of a parabola, that it is most nearly constant with respect to small variations of x in the vicinity of its vertex. The best design would adjust the circuit parameters so that the point of inflection of Eq. (9-32) lies in the middle of the mode in which the linear function is desired. Differentiating Eq. (9-32) and equating to zero locate the vertex at

$$x_m = \frac{a - b}{2a} \tag{9-33}$$

For mode 12 of the multivibrator that was treated in making the plot of Fig. 9-7 (zero control-source impedance), x_{m1} is at 0.345, compared with the lower and higher limits of $x_{l1} = 0.272$ and $x_{h1} = 0.444$. It is quite close to the optimum placement of the vertex at $x = 0.358$. But in mode 22 of the same circuit, x_{m2} is at 1.19, well outside the mode limits of $x_{l2} = 0.444$ and $x_{h2} = 0.614$. The difference in linearity is quite apparent. If x_B were taken into account, the limits would shift up by 0.023. This shift would not significantly affect the vertex location.

The second condition for good linearity also comes from the examination of the properties of the parabola. If the focus is close to the vertex, then the parabola is wide open and will be more nearly constant in the vicinity of the vertex, further improving the linearity. This distance, for the parabola of Eq. (9-32), is

$$D = \frac{1}{4a}$$

Generally it is not possible to make D small and still satisfy Eq. (9-33). Setting $D = 0$ in mode 12 is a trivial case, requiring a zero value for R_3, which would make the multivibrator inoperative. But even in this case, making R_3 as small as possible compared with R_1 minimizes D within the other design constraints that now determine the lower limit of R_3. The more important condition is the one given by Eq. (9-33), and this is the one which must be satisfied for best linearity.

Under some other circumstances, it may be desirable to make the pulse width relatively independent of x. From Eq. (9-31) it can be seen that this requires $a = -b$. All modes give ridiculous answers in satisfying this condition—either a pulse duration of zero or negative values of circuit resistors.

By including the source impedance of the control path in the derivation of the timing waveshape, we would find it possible to maintain the pulse width almost constant with respect to small changes in x in mode 22. The resulting equation would be in the same form as Eq. (9-29), with, of course, different constants. Either from this equation or by considering the effect on the base voltage of Q_1 (as in Example 9-2), the slope of the time-vs.-x curve can be minimized by maximizing R_s. Any changes in the transistor parameters would now have but little influence on the timing, and the sweep stability would be greatly improved.

9-4 Emitter-coupled Astable Multivibrator

Astable operation of an emitter-coupled multivibrator must depend on the single coupling capacitor alternately timing transistor Q_2 first in the keep-off mode and second in the keep-on mode. The operation will be similar to the MOSFET astable analyzed in Sec. 8-10. For this type of switching behavior, we must start with both transistors biased in their active regions.

For Q_2 active: $\quad R > \beta R_2$

For Q_1 active:

$$x_B + \frac{(1+\beta)R_3(1-x_B)}{R+(1+\beta)R_3} < x < \frac{R_3[R+(1+\beta)R_1] + x_B R(R_3+R_1)}{R(R_1+R_3)+(1+\beta)R_3R_1} \tag{9-34}$$

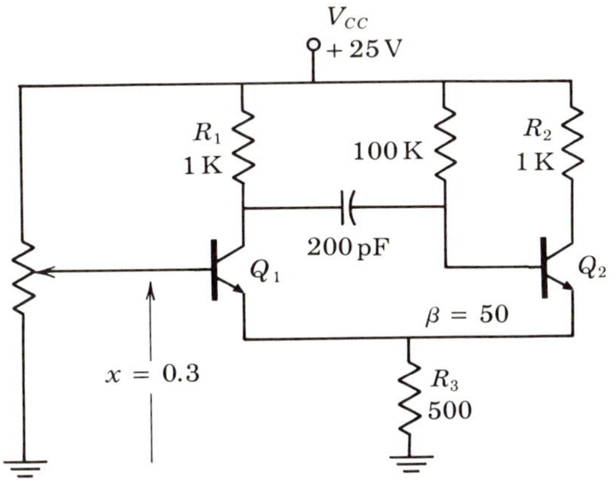

Fig. 9-9 An emitter-coupled astable multivibrator.

The complete limits of astable operation expressed by Eq. (9-34) are readily derived from the appropriate models, one with Q_1 cut off and Q_2 active and the other for Q_1 saturated while Q_2 remains active.

Once the feedback loop is again closed, the regenerative action of the multivibrator rapidly drives Q_2 into cutoff. As C recovers toward V_{CC}, the base of Q_2 eventually reaches the switching voltage xV_{CC}, at which point Q_2 reenters its active region. The increased current through R_3 now raises v_e above $xV_{CC} - V_{BE}$ and thus drives Q_1 off. The resultant jump in the collector voltage of Q_1 forces Q_2 toward, or even into, full conduction, the capacitor-charging current contributing the necessary additional base-current flow. However, C will again recover. In the process, the total emitter current drops; the base voltage of Q_2 and the common-emitter voltage exponentially decay. Finally, v_e recrosses the $xV_{CC} - V_{BE}$ voltage. At this instant, Q_1 turns back on, and the cycle repeats. We see that the single capacitor performs both timing functions; first it maintains Q_2 cut off, and next it supplies the additional current that keeps Q_2 turned on and/or Q_1 cut off.

As a convenient starting point for the purpose of computation, we might well look at the conditions of Fig. 9-9 while Q_2 is on, just going off, and Q_1 is off, on the verge of turning on. The initial voltages at $t = 0^-$, neglecting any charging current which might be flowing, are

$$v_{c1}(0^-) = V_{CC} = 25 \text{ V}$$
$$v_{b1} = v_{b2}(0^-) = xV_{CC} = 7.5 \text{ V}$$
$$v_q(0^-) = v_{c1}(0^-) - v_{b2}(0^-) = 17.5 \text{ V}$$

Immediately after the circuit changes state, at $t = 0^+$, Q_2 becomes cut off and Q_1 active. Therefore the model with which we must be concerned is the one shown in Fig. 9-10a. Since the 100-K base bias resistor of Q_2 is so very much larger than the 1-K collector resistor of Q_1, we may ignore the additional loading introduced by R. From Fig. 9-10a, the conditions at $t = 0^+$ are

$$v_{b1}(0^+) = 7.5 \text{ V}$$

$$I_{e1} = \frac{v_{b1} - V_{BE}}{R_3} = \frac{6.8}{500} = 13.6 \text{ mA}$$

$$v_{c1}(0^+) \approx V_{CC} - I_{e1}R_1 = 25 - 13.6 = 11.4 \text{ V}$$

The collector voltage of Q_1 drops 13.6 V, and the base of Q_2 drops the same amount, from 7.5 to -6.1 V. It immediately starts charging toward V_{CC} with the time constant

$$\tau_S = RC = 20 \ \mu\text{s}$$

But Q_2 turns back on once $v_{b2}(T_S) = 7.5$ V, and thus the first portion of the cycle lasts for

$$T_S = \tau_S \ln \frac{V_{CC} - v_{b2}(0^+)}{V_{CC} - xV_{CC}} = 20 \ln \frac{25 - (-6.1)}{25 - 7.5} = 11.5 \ \mu\text{s}$$

The final charge in C, which we shall need as a boundary condition in the next operating region, is

$$v_q(T_S) = v_{c1}(T_S) - v_{b2}(T_S) = 3.9 \text{ V}$$

After Q_2 turns back on, the jump in the collector voltage of Q_1, as it turns off, coupled through C, forces Q_2 into saturation (Fig. 9-10b). Before

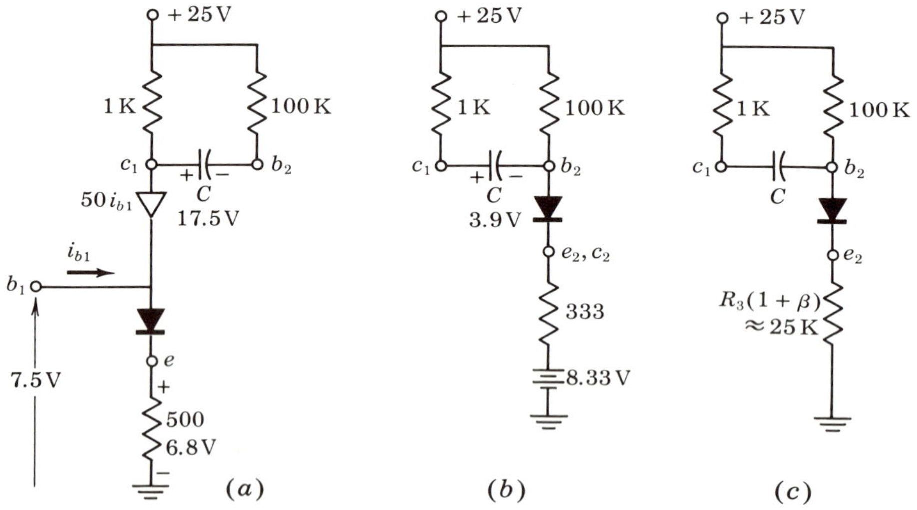

Fig. 9-10 Transistor models used for the calculation of astable operation. (a) Q_1 on, Q_2 off; (b) Q_1 off, Q_2 saturated; (c) Q_1 off, Q_2 active.

we can calculate the time response, we must verify the correctness of this assumption. From the model drawn for the saturated transistor, we see that the equivalent-emitter voltage is

$$\frac{V_{CC}R_3}{R_1 + R_3} = 8.33 \text{ V}$$

The additional capacitor-charging current flowing through the 333-Ω source impedance ($R_3 \parallel R_1$) raises the voltage at the emitter of Q_2 to 11.33 V, a value well above the saturation-threshold voltage. The initial current flow through C and the base of Q_2 becomes

$$i_{b2}(0^+) = \frac{25 - 3.9 - 0.7 - 8.33}{1{,}333} = 9.05 \text{ mA}$$

However, the base current necessary to sustain saturation of Q_2 is merely

$$i_{b2,\text{sat}} = \frac{v_e}{(\beta + 1)R_3} = \frac{8.33}{25 \text{ K}} = 0.333 \text{ mA}$$

and of this, the amount contributed by the normal bias-current flow through R is

$$i_{b2R} = \frac{V_{CC} - V_{BE} - v_e}{R} = \frac{25 - 0.7 - 8.33}{100 \text{ K}} = 0.16 \text{ mA}$$

Thus the equation of the base-current component due to the capacitor-charging current is

$$i_{b2C}(t) = 9.05e^{-t/\tau_R}$$

where $$\tau_R = [R_1 + (R_2 \parallel R_3)]C = 0.266 \text{ } \mu\text{s}$$

Q_2 finally enters its active region (from saturation) when

$$i_{b2C}(T_R) = 0.173 \text{ mA}$$

As a consequence of the large initial and small final values of current flow, the duration of the first portion of the recovery period takes virtually the complete exponential:

$$T_R = \tau_R \ln \frac{9.05}{0.173} \approx 4\tau_R = 1.06 \text{ } \mu\text{s}$$

At the end of this interval, a new model is needed for Q_2, one representing the active region (Fig. 9-10*c*). The transistor input resistance, measured from the base of Q_2 to ground, is $(\beta + 1)R_3 \approx 25$ K, and the equivalent base open-circuit voltage is

$$V_{\text{Th}_b} = \frac{(\beta + 1)R_3(V_{CC} - V_{BE})}{R + (\beta + 1)R_3} + V_{BE} = \frac{25 \text{ K}(25 - 0.7)}{100 \text{ K} + 25 \text{ K}} + 0.7 = 5.56 \text{ V}$$

Moreover, since the transition from the saturation to the active region will be smooth (without any voltage jump), there is no need to calculate the

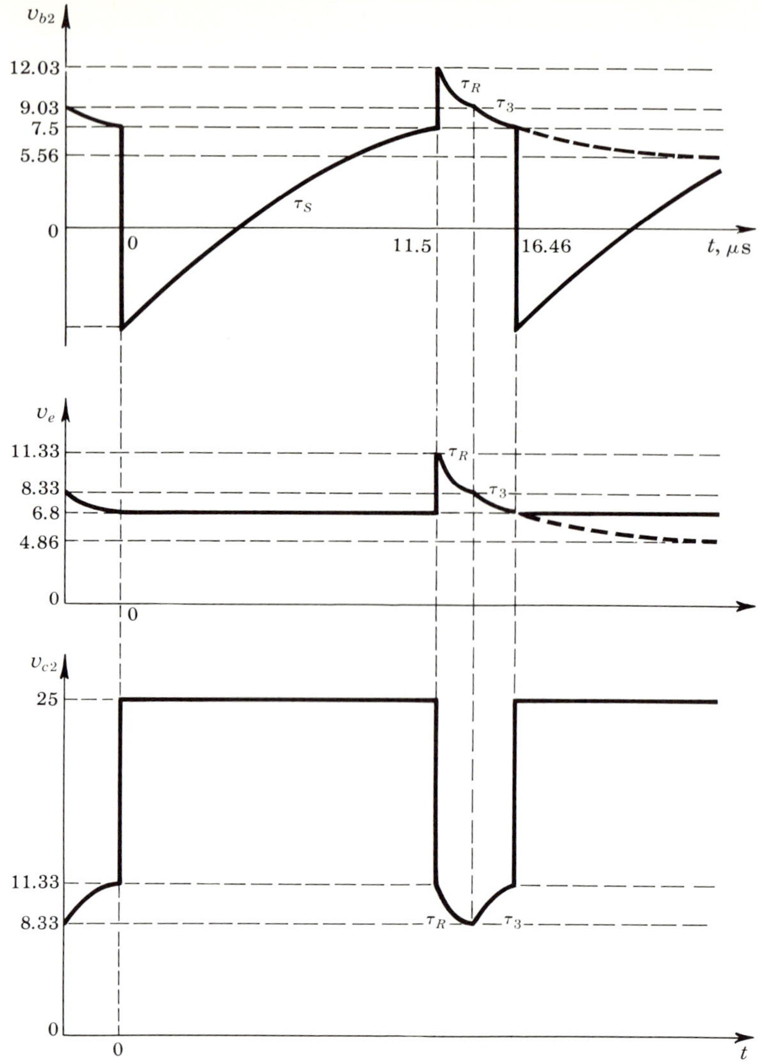

Fig. 9-11 Controlling waveshapes of the emitter-coupled astable multivibrator of Fig. 9-9.

initial charge on C. Charging continues, but now toward 5.56 V with a new time constant $\tau_3 = [(\beta + 1)R_3 \parallel R]C = 6.7\ \mu s$. Once the base of Q_2 falls to 7.5 V and the common-emitter voltage to 6.8 V, Q_1 again conducts, the circuit reswitches, and the cycle repeats.

$$T_3 = \tau_3 \ln \frac{5.56 - 9.03}{5.56 - 7.5} = 3.9\ \mu s$$

During this final interval, Q_2 is in its active rather than its saturated state. As its base current continues to decrease, the collector voltage now starts rising. The generated waveshapes at the base of Q_2 and at the common-emitter junction appear in Fig. 9-11.

Problems

9-1 The emitter-coupled circuit of Fig. 9-1 employs the following components: $R_1 = R_2 = 2$ K, $R_3 = 1$ K, $R = 50$ K, $C = 1\ \mu$F, $\beta = 50$, $V_{BE} = 750$ mV, and $V_{CC} = 5$ V.

(*a*) For $xV_{CC} = 1.5$ V, plot to scale the waveshapes seen at both collectors, at the base of Q_2, and at the common emitter.

(*b*) Repeat part *a* for $xV_{CC} = 2.5$ V. Superimpose these plots on the waveshapes drawn for part *a*.

9-2 We wish to adjust the multivibrator of Prob. 9-1 so that it will generate a 2-V 10-ms pulse at the collector of Q_2. This should be the maximum possible pulse width which can be produced by this circuit.

(*a*) Specify the new values of R_3, x, and C needed to satisfy these conditions.

(*b*) Sketch the waveshape appearing at the base of Q_2.

9-3 (*a*) Calculate the limits on x, separating the various modes of operation for the multivibrator of Fig. 9-12. Select a value of x that lies halfway between the extremes of mode 12, and calculate the pulse duration and amplitude at the collector of Q_2 (assume R_d small).

(*b*) Solve for the setting of x that will produce a pulse of the same duration but in mode 32. Plot the collector waveshape to scale on the same graph as in part *a*.

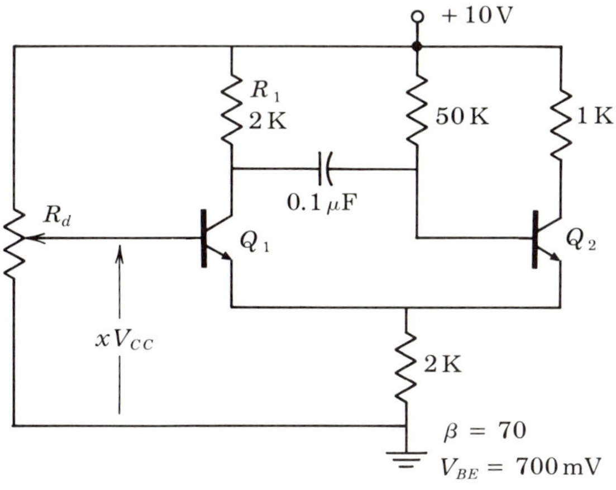

Fig. 9-12

9-4 Design an emitter-coupled multivibrator that will generate, at its maximum setting, a 10-ms pulse. This circuit should switch directly from mode 12 to mode 32 and must operate in mode 12 over the widest possible range of x. Base your design on a transistor having $\beta = 50$, $V_{BE} = 700$ mV, and $V_{CC} = 12$ V. Set $R_3 = 1$ K. Specify the range of x for both operating modes. Sketch the various circuit waveshapes produced when the multivibrator generates its 10-ms pulse.

9-5 (*a*) The multivibrator of Fig. 9-12 is controlled by a 6-V battery in place of the potentiometer. V_{CC} suddenly drops by 5 percent, from 10 to 9.5 V. What effect will this have on the pulse duration?

(*b*) If this same control voltage is derived by setting the potentiometer at $x = 0.6$, by how much will the pulse width change when V_{CC} is reduced by 5 percent?

(*c*) Repeat parts *a* and *b* if a 2-K resistor is inserted in series with the base of Q_1.

9-6 (*a*) Show three methods of triggering the multivibrator of Fig. 9-12. Discuss the required source impedance of the signal generator and the means of coupling the pulse into the circuit.

(*b*) If a pulse is applied through a diode to the common-emitter terminal, calculate the required amplitude at $x = 0.2, 0.4, 0.6$.

9-7 (*a*) The emitter-coupled multivibrator shown in Fig. 9-12 must be adjusted, by changing R_1, to generate the longest possible pulse. If $x = 0.4$, find the required value of R_1 and the width of the pulse produced.

(*b*) With $R_1 = 1$ K and with x set to give the maximum possible swing at the collector of Q_1, sketch and label fully the waveshape at the collector of Q_1.

(*c*) Sketch and label the waveshapes at both collectors when $x = 0.33$ and when $R_1 = 1$ K.

9-8 (*a*) Plot the collector-voltage swing of Q_2 as a function of the potentiometer setting for the multivibrator described in the text in Example 9-1. Label all modes of operation.

(*b*) Repeat the plot, on the same graph, for the collector voltage of Q_1.

9-9 (*a*) Verify the bounds of each mode for the multivibrator discussed in the text in Example 9-2. Assume that R_s remains constant at 2,500 Ω. Is this approximation justified?

(*b*) Prove that the x boundary between mode 22 and mode 32 is independent of the potentiometer employed for control.

9-10 In the circuit of Fig. 9-12, R_d is a 5-K control. Plot the waveshapes at each transistor element, to scale, when $x = 0.5$. How does the pulse duration compare with the one generated when R_d is very small?

9-11 The equation

$$y = \ln \frac{4 - 5x}{1 - x} \qquad 0 < x < 0.75$$

defines the operating path of a circuit.

(*a*) For what value of x is y most nearly a linear function of x?

(*b*) If x is varied by ± 0.05 about this point, what is the nonlinearity of y?

(*c*) What is the NL of y if x varies by ± 0.05 about the point $x = \frac{1}{2}$? (Define the NL of y as the maximum variation in y from a straight line drawn between its bounded end points divided by the change in y between the same two end points.)

9-12 (*a*) Derive the limits of Eq. (9-34).

(*b*) Find the numerical values of x that delineate the modes of operation of the multivibrator discussed in Sec. 9-4. Tabulate, under each mode, the normal and the switched state of each transistor. Are any modes of operation monostable? Explain.

9-13 Repeat the calculations for the sample problem given in Sec. 9-4 when x is increased to 0.5. Make all reasonable approximations.

9-14 Figure 9-13 shows a junction-FET source-coupled multivibrator. The operation of this circuit is quite similar to that of the bipolar emitter-coupled multivibrator, with, however, the switching conditions depending on the bounds of the FET. Plot the gate- and drain-voltage waveshapes of Q_2 from just before the circuit is triggered into the unstable state until final recovery. Give all significant times, time constants, and voltage values. The FET has $V_p = -2$ V and $I_{DSS} = 10$ mA; set $x = 0.25$.

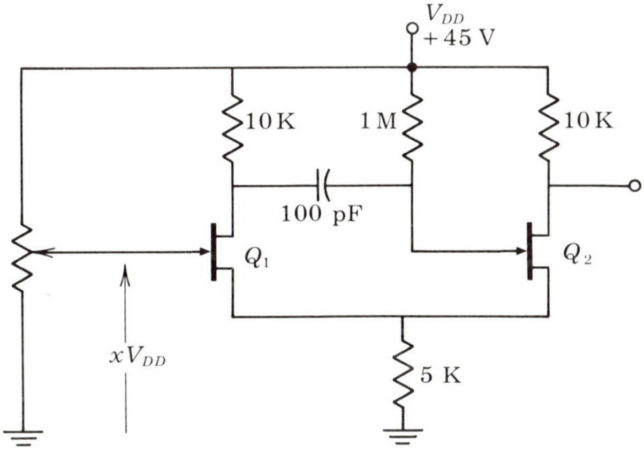

Fig. 9-13

Bibliography

Chance, B., M. H. Johnson, and R. H. Phillip: Precision Delay Multivibrator for Range Measurements, *M.I.T. Rad. Lab. Rept.* 63-2.

Glegg, K.: Cathode-coupled Multivibrator Operation, *Proc. IRE*, vol. 38, no. 6, pp. 655–657, June, 1950.

Schmitt, O. H.: A Thermionic Trigger, *J. Sci. Instr.*, vol. 15, p. 24, 1938.

10.
voltage comparators

Voltage comparators are used to compare an input signal with a fixed or variable reference voltage. The comparator output changes from a logical 0 when the input is below the reference to a logical 1 once the input-signal amplitude exceeds the reference (or vice versa). This amplitude-discrimination property may be used to quantize analog signals in an analog-to-digital converter or as a pulse-height discriminator. By setting the reference level at 0 V, the comparator functions as a sense (or polarity) detector.

10-1 Difference-amplifier Comparator

The basic voltage comparator (Fig. 10-1) uses a differential-amplifier pair where the input signal, applied at one input, is compared with the reference signal applied at the other input. In order to make the logical-0 output levels independent of the reference signal, a constant-current sink is used as the common-emitter load. This current supply also aids in the temperature stabilization of the circuit.

The common-emitter current I of the differential pair Q_1 and Q_2 is established by the biasing of Q_3. Transistor Q_4 is diode-connected by short-circuiting the collector and base and is used in the biasing network for temperature compensation. Neglecting the base current of Q_3 and assum-

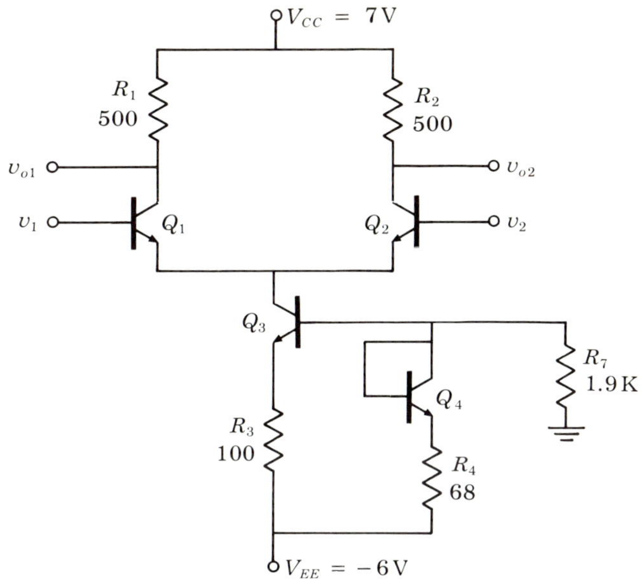

Fig. 10-1 Basic differential amplifier.

ing initially a diode drop of 750 mV across Q_4, we obtain

$$I_d = \frac{-V_{EE} - V_D}{R_4 + R_7} = \frac{5.25}{1.97\ \text{K}} = 2.56\ \text{mA} \tag{10-1}$$

The diode characteristics are represented by

$$I_d = I_{rs}e^{V_D/V_T} \tag{10-2}$$

where $I_{rs} = 3 \times 10^{-16}$ A is a typical value of the reverse saturation current for the integrated-circuit transistors used in the above circuit. At room temperature, $V_T \approx 26$ mV. From Eq. (10-2), with a diode current of 2.56 mA, we obtain a diode drop of

$$V_D = V_T \ln \frac{I_d}{I_{rs}} = 26 \ln \frac{2.56 \times 10^{-3}}{3 \times 10^{-16}} = 775\ \text{mV} \tag{10-3}$$

which is very close to the initial assumed value. If it were not, we would have to recalculate, starting with an assumed V_D equal to the trial solution of Eq. (10-3).

Comparing the drops in the base and emitter circuits of Q_3, we have

$$V_{BE3} + I_{e3}R_3 = I_dR_4 + V_D$$

and therefore

$$I_{e3} = \frac{(V_D - V_{BE3}) + I_dR_4}{R_3} \tag{10-4}$$

Assuming equal diode and junction voltage drops, we obtain

$$I_{e3} = I_d \frac{R_4}{R_3} = 1.88 \text{ mA}$$

Substituting this current into the diode equation (10-2), we find that $V_{BE} = 767$ mV, which is only 8 mV less than V_D.

Both V_D and V_{BE} vary in the same manner with temperature. Thus the cancellation shown in the numerator of Eq. (10-4) indicates that the biasing circuit is relatively insensitive to temperature changes.

For $\alpha = 0.97$, we have a collector current into Q_3 of

$$I = \alpha I_{e3} = 1.83 \text{ mA}$$

This is the constant-current drain for the differential pair Q_1 and Q_2.

The output voltages at the collectors of the differential pair will depend on the distribution of the constant common-emitter current I between Q_1 and Q_2. From the model of Fig. 10-2, we can write

$$v_1 - V_{BE1} = v_2 - V_{BE2} \tag{10-5}$$

and

$$I_{e1} + I_{e2} = I \tag{10-6}$$

Substituting the exponential volt-ampere relationship for the emitter-current terms in Eq. (10-6),

$$I_{es}e^{V_{BE1}/V_T} + I_{es}e^{V_{BE2}/V_T} = I$$

By factoring out I_{e1},

$$I_{e1}(1 + e^{(V_{BE2}-V_{BE1})/V_T}) = I \tag{10-7}$$

From Eq. (10-5) we note that $V_{BE2} - V_{BE1} = v_2 - v_1$. Substituting into Eq. (10-7), we obtain

$$I_{e1} = \frac{I}{1 + e^{(v_2-v_1)/V_T}} \tag{10-8}$$

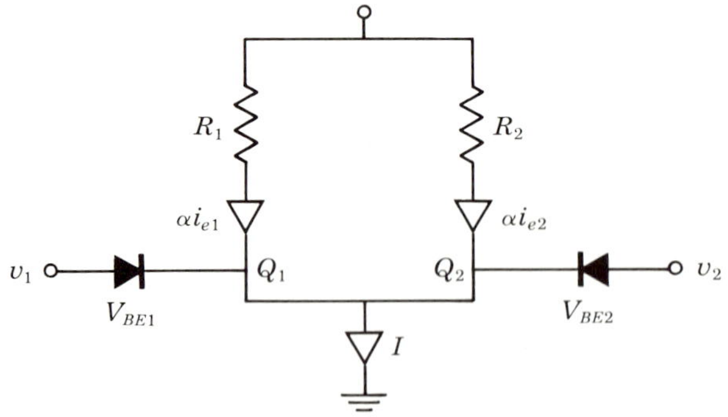

Fig. 10-2 Equivalent circuit of differential amplifier.

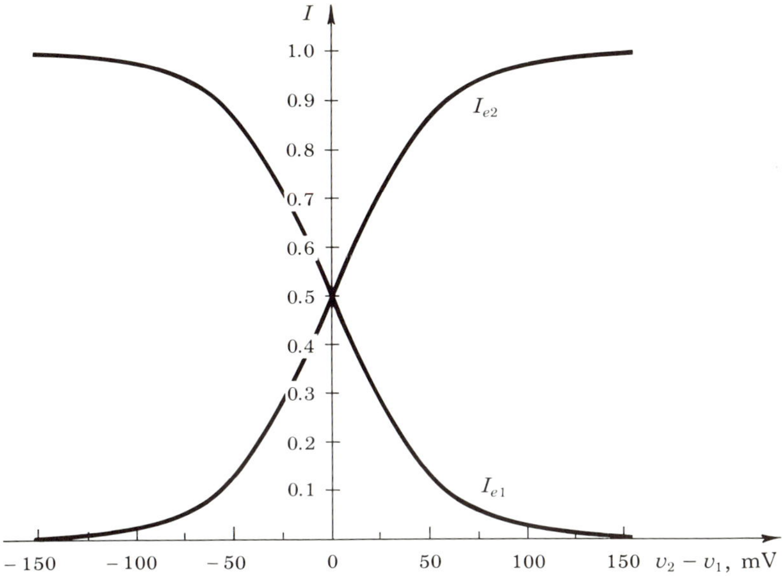

Fig. 10-3 Emitter-current distribution in a differential-amplifier pair as a function of the differential input voltage.

In a similar manner, solving for I_{e2},

$$I_{e2} = \frac{Ie^{(v_2-v_1)/V_T}}{1 + e^{(v_2-v_1)/V_T}} \tag{10-9}$$

We can see from the plots of Fig. 10-3 that these functions are symmetrical about the crossover value $I_{e1} = I_{e2} = \frac{1}{2}I$ occurring at $v_1 = v_2$.

If $v_1 = v_2$, the circuit is symmetrically biased; half the sink current flows in each transistor. When v_1 is 58.5 mV below v_2,

$$I_{e1} = 0.1I \qquad \text{and} \qquad I_{e2} = 0.9I$$

When v_1 is 58.5 mV above v_2,

$$I_{e1} = 0.9I \qquad \text{and} \qquad I_{e2} = 0.1I$$

The transition region, defined as the differential input necessary to drive the current from 0.1 to 0.9 of the final value, is 117 mV wide. This region is centered about the common mode (or dc level) of either v_1 or v_2, whichever input we choose to use as the reference level.

The output voltage v_{o2} changes from the logical-0 value of

$$V_{o2_0} \leq V_{CC} - \alpha 0.9IR_2 = 6.2 \text{ V} \tag{10-10}$$

when v_1 is 58.5 mV or more below v_2, to the logical-1 level of

$$V_{o2_1} \leq V_{CC} - \alpha 0.1 I R_2 = 6.9 \text{ V} \tag{10-11}$$

when $v_2 - v_1 \geq 58.5$ mV.

The change of a single transistor's collector current with respect to the push-pull input is defined as the transconductance of the differential pair. From Eq. (10-8) or (10-9) we find that when $v_2 = v_1$,

$$g_m = \frac{dI_{c2}}{d(v_2 - v_1)} = \frac{\alpha I}{4V_T} = 17{,}100\ \mu\text{mhos} \tag{10-12}$$

The voltage gain from the push-pull input to the output at the collector of Q_2 becomes

$$A_2 = \frac{\Delta v_{o2}}{v_2 - v_1} = -g_m R_2 = -8.55 \tag{10-13}$$

The push-pull output gain in a symmetrical circuit ($R_1 = R_2$) will be twice as large.

So long as all transistors remain active, the output levels are independent of the reference voltage; they depend only on the change in the IR_2 drop as the current is transferred from one transistor to the other. The permissible range of input-signal levels can be found from the dc circuit conditions. The lower limit will be determined by the saturation of Q_3. At the edge of the saturation of Q_3, the differential pair's common-emitter voltage is

$$v_e = V_{EE} + I_{e3}R_3 + V_{CE,\text{sat}} = -5.6 \text{ V} \tag{10-14}$$

where we assume $V_{CE,\text{sat}} = 200$ mV. The corresponding common-mode voltage, $v_1 = v_2$, is one base-emitter drop higher:

$$v_{2,\text{min}} = -4.85 \text{ V} \tag{10-15}$$

The upper limit of v_2 is that voltage which will just saturate Q_2 when the full sink current flows through Q_2. Under saturation conditions, the collector voltage of Q_2 becomes

$$v_{o2} = V_{CC} - \alpha I R_2 = 6.10 \text{ V} \tag{10-16}$$

The emitter voltage of the saturated transistor will be 200 mV lower. Since the base voltage is 750 mV above the emitter, we have

$$v_{2,\text{max}} = 6.10 - 0.2 + 0.75 = 6.65 \text{ V} \tag{10-17}$$

The single differential-amplifier pair can compare an input signal v_1 with a reference voltage v_2 over the range -4.85 to $+6.65$ V. We can determine the point of coincidence only to within ± 58.5 mV centered about a crossover point equal to the reference voltage. Because of the balanced operation, the crossover voltage and the output levels are quite insensitive to temperature changes.

10-2 High-gain Differential Amplifier

The wide transition region is the major disadvantage of the differential-amplifier pair used as a voltage comparator. But since the crossover point ($I_c = I/2$) is temperature-insensitive and is independent of the reference voltage v_2, we can always cascade additional stages of voltage gain and use only a small portion of the input transition region to drive the output stage from the logical-0 to the logical-1 level. By doing this, the effective input transition region can be reduced from 117 mV to a few millivolts centered about v_2.

Figure 10-4 shows the complete comparator. The input differential amplifier is followed by a second stage of gain, which then drives an emitter-

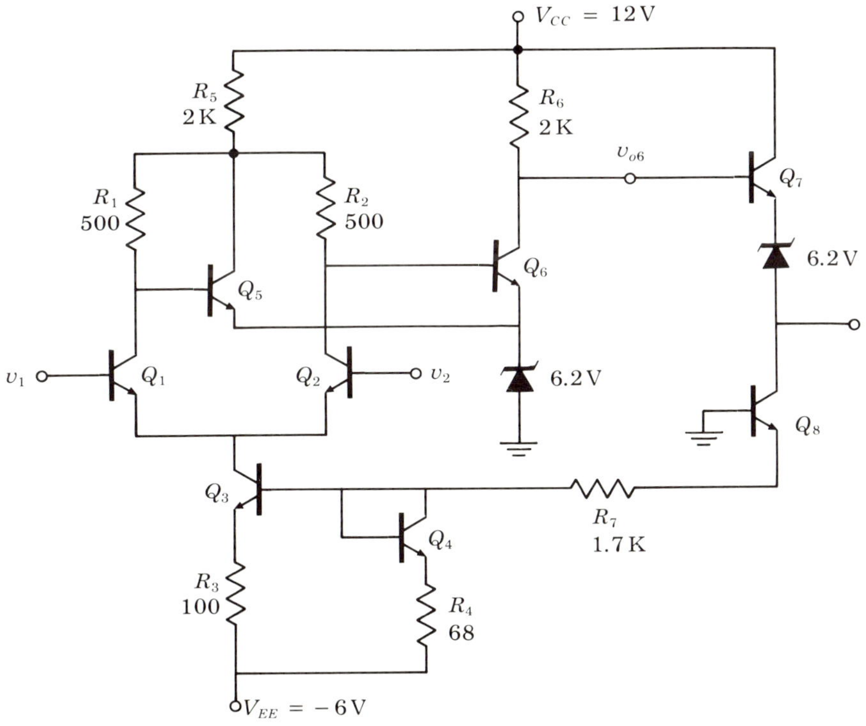

Fig. 10-4 Full differential amplifier based on the type-710 amplifier. Q_5 and Q_6 form the middle amplifier stage. Q_8 is a constant-current load in the emitter-follower (Q_7) output stage.

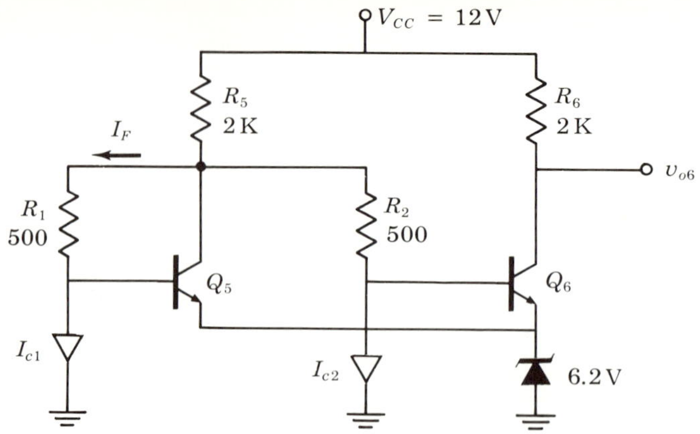

Fig. 10-5 Middle amplifier stage of amplifier of Fig. 10-4. The input differential amplifier is represented by the two collector-current sources I_{c1} and I_{c2}.

follower output stage. In Fig. 10-5, the balanced-gain stage, driven by the collector currents of the differential pair, is shown separately. This stage combines the out-of-phase outputs of the two halves of the differential input stage, amplifies them, and yields a single-ended output which is the input to the emitter-follower output stage. The zener diode simply establishes the dc levels at a value which allows true differential operation over a wide range of common-mode inputs.

Biasing calculations are performed with $v_1 = v_2$, which means that

$$I_{c1} = I_{c2} = \tfrac{1}{2}\alpha I = 0.89 \text{ mA}$$

Because transistors Q_5 and Q_6 are both biased from the collector of Q_5 through equal resistors, they will have identical quiescent currents. Neglecting the base-current flow, we can write the base voltage of Q_5 as

$$v_{b5} = V_{CC} - (I_{e5} + I_{c1} + I_{c2})R_6 - I_{c1}R_1 = 6.2 + V_{BE5} \tag{10-18}$$

Solving Eq. (10-18) yields

$$I_{e5} \approx 1 \text{ mA} \tag{10-19a}$$

and by symmetry,

$$I_{e6} \approx 1 \text{ mA} \tag{10-19b}$$

The first stage, Q_5, effectively inverts the current I_{c1}, and the inverted current is combined with I_{c2} at the input to Q_6. Q_6 then supplies the total stage amplification. To illustrate this, by looking at Fig. 10-6a we see that the base current of Q_5 may be written as

$$I_{b5} = I_F - I_{c1} \tag{10-20}$$

where I_F is the feedback current through R_1. The output model of Q_5 will appear as shown in Fig. 10-6*b*. Since the βI_F-controlled current source appears in parallel with the controlling current I_F, by reducing R_1 to $R_1/(\beta + 1)$, we may remove the controlled source without changing the total input current. We obtain the simplified model of Fig. 10-6*c*. If $R_5 \gg R_1/(\beta + 1)$ and if $\beta + 1 \approx \beta$, then the Thévenin equivalent output circuit will appear as shown in Fig. 10-6*d*.

Figure 10-7 shows how this output, used as the input to Q_6, combines with I_{c2}. The small-signal base current of Q_6 may be written (for $R_2 = R_1$) as

$$i_{b6} = (I_{c1} - I_{c2}) \frac{R_2}{R_2 + h_{ie}}$$

The output voltage becomes

$$v_{o6} = -\beta(I_{c1} - I_{c2}) \frac{R_2}{R_2 + h_{ie}} R_6 \tag{10-21}$$

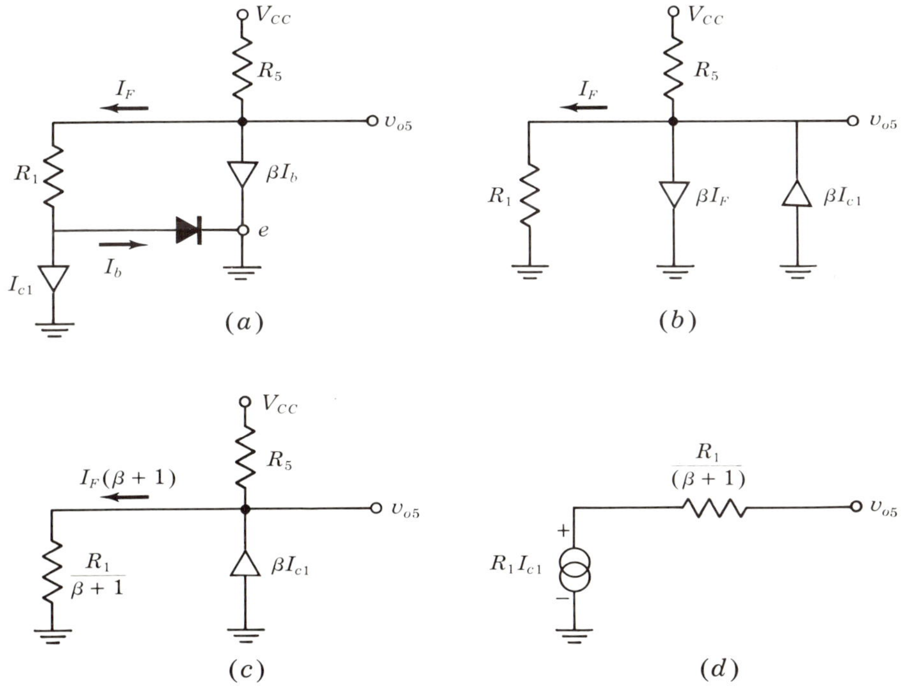

Fig. 10-6 Illustrating the simplification of the input amplifier stage (Q_5) of Fig. 10-5. (*a*) Active-region model; (*b*) equivalent-output model; (*c*) reduced-output model; (*d*) Thévenin equivalent seen at the output of Q_5 [$R_5 \gg R_1(\beta + 1)$].

and by expressing I_{c1} and I_{c2} in terms of g_m [Eq. (10-12)] and the differential input $(v_1 - v_2)$, taking into account the signal polarities,

$$v_{o6} = -2g_m\beta \frac{R_2}{R_2 + h_{ie}} R_6(v_1 - v_2) \tag{10-22}$$

For $I_{e6} = 1$ mA, $\beta = 33$, $h_{ie} \approx 1{,}100\ \Omega$. The gain from the push-pull input to the amplifier output becomes

$$A = \frac{v_{o6}}{v_1 - v_2} = -680 \tag{10-23}$$

The output voltage swing is determined by the saturation and cutoff of Q_6 (Q_5 remains active). For Q_6 cut off, the output is at the logical-1 level:

$$V_{o6_1} = V_{CC} = 12\ \text{V} \tag{10-24}$$

For Q_6 saturated, we obtain the logical-0 level of

$$V_{o6_0} = 6.2 + V_{CE,\text{sat}} = 6.4\ \text{V} \tag{10-25}$$

The logical-1 and logical-0 levels at the collector of Q_6 are well above the levels generally used in logic circuits. Some form of level translation is needed to set the logical-0 output at 0 V. Ideally, whatever is done should not reduce the overall gain. For maximum fan-out capabilities, we should combine a low impedance output with the level shifting.

The output stage (Fig. 10-4) is a straightforward emitter follower using a constant-current source for the emitter load. A 6.2-V zener diode is used to level-shift the output voltage down so that the logical-0 output will be at approximately 0 V. The emitter current of Q_8 is the diode current

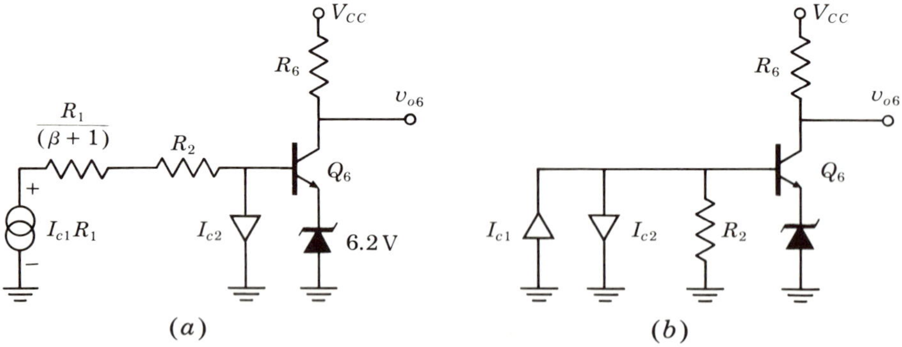

Fig. 10-7 Circuit of Q_6, showing how output of Q_5 (Fig. 10-6d) converts to the current subtraction $I_{c1} - I_{c2}$ at the base. ($R_1 = R_2$, and $\beta \gg 1$.)

of Q_4. For the values given in Fig. 10-4, using the diode drops previously calculated,

$$I_{e8} = \frac{6 - 0.775 - 0.775}{1.77} = 2.50 \text{ mA}$$

The output levels will be one base-emitter drop plus the output zener-diode drop below the input. From Eqs. (10-24) and (10-25),

$$V_{o1} = 12 - 6.97 = 5.03 \text{ V} \tag{10-26}$$

$$V_{o0} = 6.4 - 6.97 = -0.57 \text{ V} \tag{10-27}$$

Since the overall gain is 680, the 5.6-V output swing corresponds to an input transition width of ± 4.2 mV. This circuit has a greater output swing and a much smaller transition region than does a single differential stage. By changes in design, we can easily trade off output swing for a narrower transition region. Increased gain will also reduce the width of the input switching region.

The second-stage configuration discussed above is widely used for several reasons. First, the subtraction of the differential-amplifier currents at the base of Q_6 further reduces the circuit's dependence on the input-voltage levels. Since, at the balance point $v_1 = v_2$, equal currents flow in each of the differential-amplifier transistors, any changes in the absolute values of these currents cancel at the input to Q_6. A second advantage is that the quiescent operating point is relatively insensitive to variations of the positive power supply.

The differential amplifier is also widely used as an operational amplifier. For this application, it is necessary that the output be zero for a 0 input. We had previously found that for $v_1 = v_2$, $I_{e6} = 1$ mA [Eq. (10-19*b*)]. The collector voltage of Q_6 becomes

$$V_{c6} = V_{CC} - \alpha I_{e6} R_6 = 10.06 \text{ V}$$

and to obtain a 0 output, -10.06 V of level shifting is necessary. Since the emitter follower has a V_{BE} drop of 775 mV, a 9.3-V zener diode will bring the output to zero. The biasing of the current source in the emitter must also be changed; the base of Q_8 now has to be tied to a negative potential instead of to ground.

Figure 10-8 shows the type-710 integrated-circuit comparator, which is basically the same as the simpler circuit described above. It differs in only two respects:

1. The diode-connected transistor Q_{10}, from the collector of Q_6 to the collector load of Q_5, is used to control the output logical-1 level. When Q_6 is cut off, Q_{10} conducts and establishes a logical-1 output level of 3.1 V. The logical-0 level is, as before, at -0.57 V.

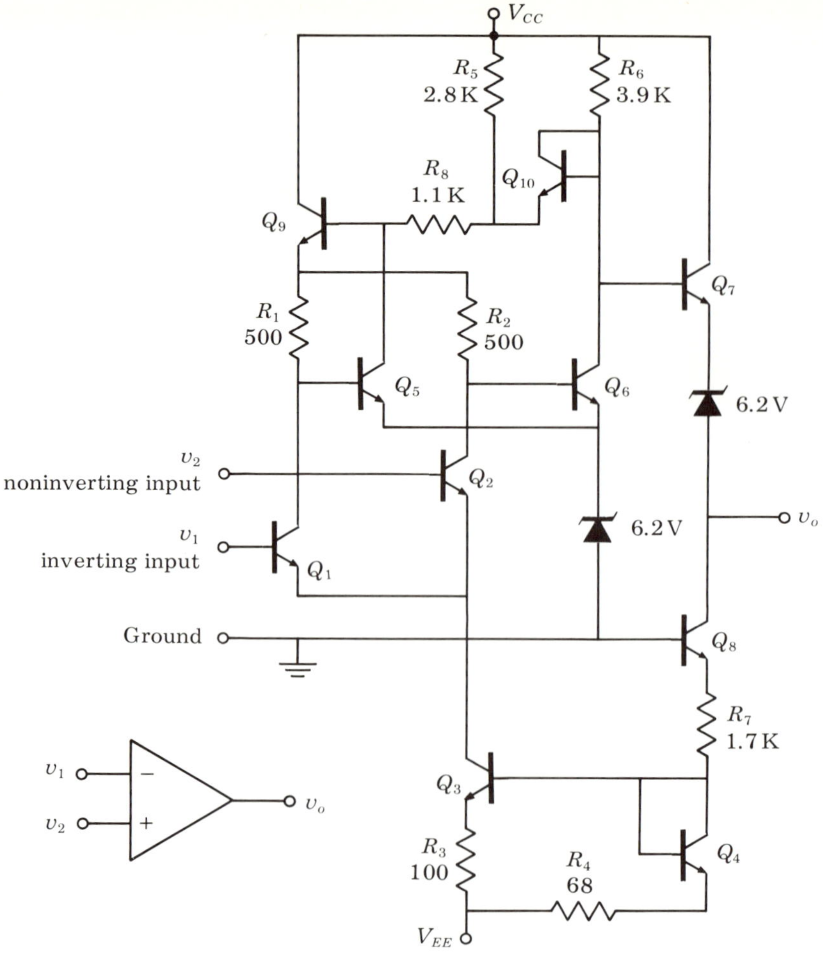

Fig. 10-8 Type-710 integrated-circuit comparator—overall gain, 1,500; logical output levels, 3.1 and −0.5 V.

2. The collector of Q_6 is coupled back to the bases of Q_5 and Q_6 through an emitter follower Q_9. This permits use of a larger collector load resistance for Q_5 and Q_6, with the resultant increased gain.

Both changes result in a narrower input transition region. The typical gain of 1,500 corresponds to a ±1-mV input switching region.

In any differential comparator, it really does not matter to which input the reference signal is applied; the differen ial input stage is symmetrical. If a low output voltage is required for a low signal input voltage, the inverting input v_1 would be used for the reference voltage (when $v_1 - v_2 > 0$, the output is low). Contrariwise, applying the reference signal at the noninverting input v_2 results in a high output voltage for a low input voltage and a low level output for a high level input.

10-3 Regenerative Comparators

The differential amplifier has a high positive overall gain from the reference-signal input v_2 to the output. Through the use of positive feedback from the output back to the v_2 input, we can convert the straight-through comparator into a regenerative circuit. Figure 10-9 shows the resulting circuit. (This circuit is also known as a *Schmitt trigger*.)

To improve switching behavior, R_1 is usually shunted with the small speedup capacitor C_1. During the switching interval, the capacitor couples the full output change back to the amplifier input. This serves the same function as the multivibrator speedup capacitor described in Chap. 8.

The overall gain of the amplifier with the positive feedback factor K can be written as

$$A = \frac{-A_o}{1 - KA_o} \tag{10-28}$$

where $-A_o$ is the gain of the base amplifier from $(v_1 - v_2)$ to the output.

If we assume that the base amplifier has a very small output impedance, then the feedback factor is given by

$$K = \frac{R_2}{R_1 + R_2}$$

For regenerative operation, the loop-gain equation must blow up. This requires

$$K > \frac{1}{A_o} \tag{10-29}$$

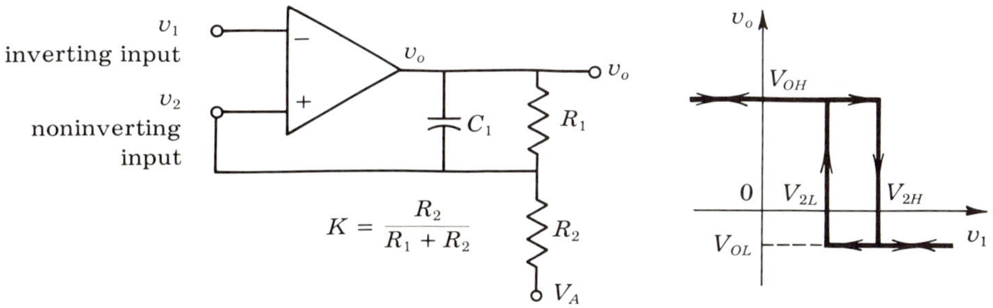

Fig. 10-9 Regenerative-comparator arrangement of the differential amplifier (Schmitt trigger) and transfer characteristic.

When Eq. (10-29) is satisfied, the comparator operates as the voltage-controlled bistable described below.

If Eq. (10-29) is not satisfied, the positive feedback merely acts to increase the overall gain of the amplifier. But this in itself is an advantage since any increase in gain causes a corresponding decrease in the width of the input transition region.

Now let us consider the case of regenerative operation. The input at v_2 is derived by combining a fraction of the output with a fixed voltage V_A. For low values of v_1, the output is at the high logical-1 level and the resultant reference signal will be high. By superposition, we can write

$$V_{2H} = V_A(1 - K) + KV_{oH} \tag{10-30}$$

where V_{oH} is the logical-1 output voltage. K, as defined above, is the fraction of the output fed back to the noninverting input terminal.

Once v_1 rises to the active region of the amplifier, the output starts falling toward the logical-0 level. With a high-gain amplifier, the width of the active region, referred to the input, is extremely narrow (only a few millivolts). We can reasonably approximate the response by assuming that the amplifier becomes active when $v_1 = v_2$. Thus, on a rising input, the edge of the active region occurs at

$$v_{1H} = V_{2H}$$

The change at the output is coupled by R_1 and R_2 to the v_2 input, where it acts to reduce the reference signal. It consequently increases the positive input difference voltage $v_1 - v_2$. The regenerative action described drives the output to the logical-0 level.

The new steady-state voltage at v_2 is now determined by the low logical-0 output signal:

$$V_{2L} = V_A(1 - K) + KV_{oL} \tag{10-31}$$

Since the regenerative switching establishes a new reference signal which is lower than the original voltage, the output will remain latched at the logical-0 level.

On a falling v_1, the signal must decrease to the new reference signal V_{2L} before regeneration starts. The rise at the output, coupled back to the v_2 input, increases the negative difference $v_1 - v_2$. The output latches at the logical-1 level.

We note that the regenerative operation has introduced hysteresis into the comparator's transfer function. The width of the hysteresis loop is given by

$$\Delta V_i = V_{2H} - V_{2L} = K(V_{oH} - V_{oL}) = K\,\Delta V_o \tag{10-32}$$

which is directly proportional to the feedback factor.

The feedback factor K may also be expressed as

$$K = \frac{\Delta V_i}{\Delta V_o}$$

which we can interpret as the aspect ratio of the hysteresis loop, i.e., the ratio of the minor to the major dimension.

If we use a feedback factor of $\frac{1}{50}$ with the basic differential amplifier of Sec. 10-1 ($A_o = -680$), the width of the hysteresis loop becomes

$$\Delta V_i = \tfrac{1}{50}(V_{oH} - V_{oL}) = \tfrac{1}{50}(5.05 + 0.63) = 113 \text{ mV}$$

From Eqs. (10-30) and (10-31), we find that the hysteresis loop is centered at

$$V_c = V_A(1 - K) + \frac{K}{2}(V_{oH} - V_{oL}) \qquad (10\text{-}33)$$

We can solve Eq. (10-33) for the supply voltage V_A necessary to center the loop at the desired point:

$$V_A = \frac{2V_c - K(V_{oH} - V_{oL})}{2(1 - K)}$$

The major advantage of the regenerative comparator is the inherent self-latching action of the hysteresis loop. Since the down and up switching levels are separated, the circuit has a built-in noise immunity equal to the width of the hysteresis loop. Once the comparator registers an output, small perturbations will not cause reswitching and consequent erratic indications. We pay for this by the broadening of the input transition zone and the inability to narrowly define a single voltage-comparator level.

10-4 The Differential Amplifier as an Astable Multivibrator

We have seen that the regenerative switching of a positive-feedback comparator creates a hysteresis transfer characteristic (Fig. 10-9). The two different switching levels of the inverting input are caused by the change in the reference voltage applied at the noninverting input. If we incorporate a sweep circuit at the noninverting input (Fig. 10-10), the circuit can be adjusted to operate as an astable multivibrator timed first in the keep-on and then in the keep-off mode (Sec. 8-9).

Let us start by assuming that the output of the circuit of Fig. 10-10 is in the high state. The reference voltage at v_2 is

$$V_{2H} = KV_{oH}$$

The capacitor from the inverting input to ground is charging toward the higher voltage V_{oH}. Thus v_1 will eventually reach V_{2H}, and the circuit

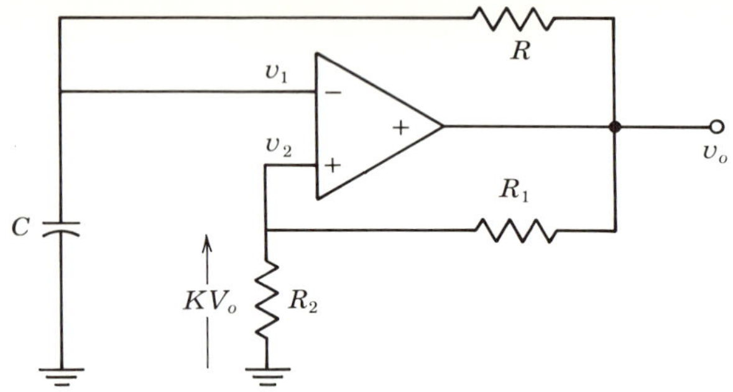

(a)

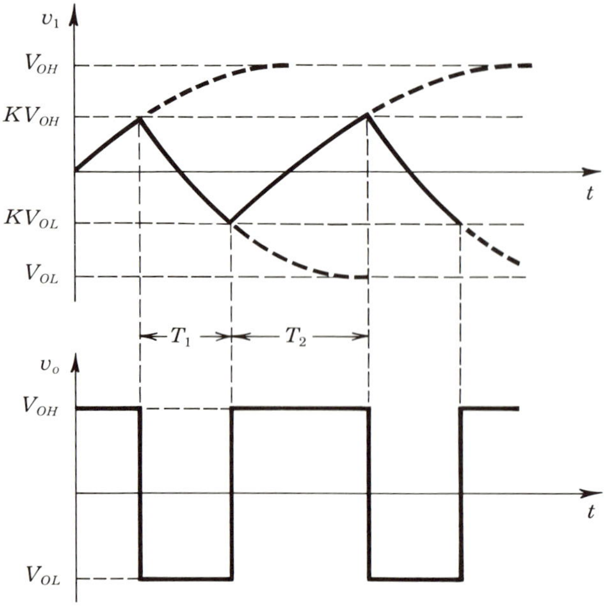

(b)

Fig. 10-10 (a) Differential amplifier used as an astable multivibrator; (b) waveshapes present in the circuit.

will switch state. At the instant before and the instant after switching state,

$$v_1(0^-) = v_1(0^+) = V_{2H} = KV_{oH} \tag{10-34}$$

The rising v_1 has caused an output down transition, from V_{oH} to V_{oL}. As a result of this transition, the timing capacitor now charges toward V_{oL}. The new reference voltage for the up transition is

$$V_{2L} = KV_{oL}$$

Now if

$$KV_{oL} > V_{oL} \tag{10-35}$$

then the switching voltage will lie between the initial voltage and the new steady-state voltage. Once the voltage across C decays to KV_{oL}, we shall observe another state transition. But since $1 > K > 0$, Eq. (10-35) implies that the necessary condition is

$$V_{oL} < 0$$

We now know the initial, steady-state, and transition voltages. The duration of the rundown portion of the cycle lasts for

$$t_1 = \tau_S \ln \frac{V_{oL} - KV_{oH}}{V_{oL} - KV_{oL}} \tag{10-36}$$

where $\tau_S = RC$.

At t_1, the output jumps to V_{oH} and the voltage across C starts rising. The next transition occurs at $v_1 = KV_{oL}$. By symmetry, we see that the duration of the rising portion of the cycle is defined by

$$t_2 = \tau_S \ln \frac{V_{oH} - KV_{oL}}{V_{oH} - KV_{oH}} \tag{10-37}$$

If we compare Eqs. (10-36) and (10-37), we note that when $V_{oL} = -V_{oH}$, the waveshape generated will be symmetrical.

10-5 Schmitt-trigger Circuits

The discrete-component regenerative voltage comparator is the bistable version of the emitter-coupled multivibrator discussed in Chap. 9. The simple Schmitt trigger circuit of Fig. 10-11, which predates the integrated-circuit version, has been widely used in systems made from individual components. But since a single lumped component is much more expensive than a single circuit element on the integrated-circuit chip, the integrated comparator has almost completely displaced the lumped circuit. Besides being as cheap or cheaper, the higher gain of the multistage integrated circuit results in much better performance.

The lumped-element comparator is still quite suitable for many applications which only require crude voltage quantization and for systems where a wide hysteresis loop is required.

In the circuit of Fig. 10-11a, positive feedback is obtained by coupling the output of transistor Q_1 to the base of Q_2. The common-emitter resistor couples the signal back to Q_1. If the loop gain is greater than unity, the circuit will switch into a limiting state where one of the transistors

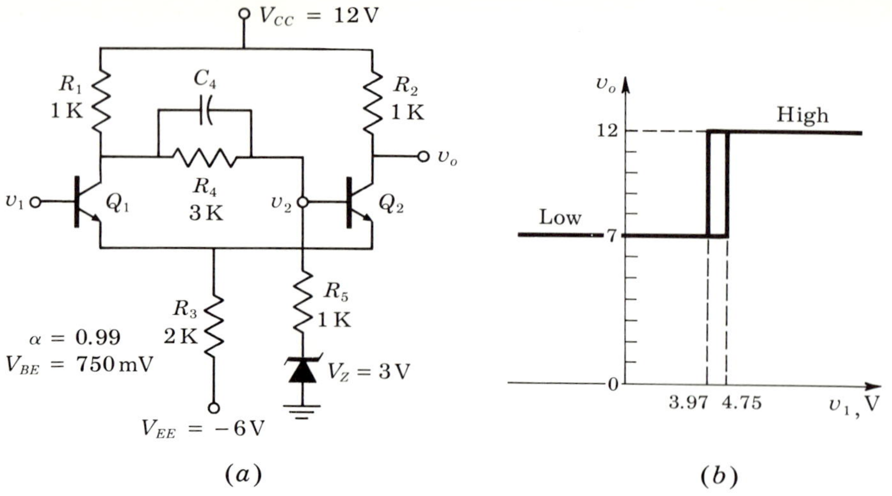

Fig. 10-11 (*a*) Discrete-component regenerative comparator (Schmitt trigger); (*b*) transfer characteristic, showing hysteresis.

is either cut off or saturated. The specific states of each transistor depend upon the values of the circuit components. Possible modes will be treated below, in Sec. 10-6. But in order to switch states, the two transistors must first be brought into the active region; this is a function of the level of the input signal v_1.

For convenience, the output is taken at the collector of Q_2. Since this is at a point removed from the feedback path, changes in external loading will not affect the circuit operation. The output follows the input: a low-level input produces a low-level output, and a high input signal results in a latched high-level output.

In analyzing this circuit, we shall start at a low value of v_1, which ensures that the trigger is in its low state—Q_1 off and Q_2 on. The base voltage of Q_2 is at the high value, V_{2H}. Neglecting base-current flow, by superposition we find the base voltage of Q_2 to be

$$V_{2H} = (1 - K)V_Z + KV_{CC}$$
$$= (1 - 0.2)3 + 0.2 \times 12 = 4.8 \text{ V} \qquad (10\text{-}38)$$

where $K = R_5/(R_1 + R_4 + R_5) = 0.2$. The common-emitter voltage is one diode drop lower.

We find the emitter current of Q_2 from the known drop across R_3:

$$I_{e2} = \frac{-V_{EE} + V_{2H} - V_{BE}}{R_3} = 5.02 \text{ mA} \qquad (10\text{-}39)$$

With this emitter current, the low-level output voltage at the collector of Q_2 becomes

$$V_{oL} = V_{CC} - \alpha I_e R_2 = 7 \text{ V}$$

Since $v_{ce2} = 7 - 4.05 > 0$, transistor Q_2 is not saturated. If Q_2 turned out to be saturated, we would have to recalculate the voltage at the base of Q_2 from the saturated model.

Up-switching occurs once Q_1 becomes active. Assuming a total input transition region of 200 mV (see Fig. 10-3), the up-switching point will occur when v_1 is 100 mV below V_{2H}:

$$V_{1H} = V_{2H} - 0.1 = 4.7 \text{ V}$$

Once Q_1 turns on, the drop at its collector, coupled by the speedup capacitor to the base of Q_2, turns Q_2 off. As before, we find I_{e1} from the known drop across R_3. From Fig. 10-12,

$$I_{e1} = \frac{-V_{EE} + v_1 - V_{BE}}{R_3} \tag{10-40}$$

At the switching point of $v_1 = V_{1H}$, $I_{e1} = 4.98$ mA. For all practical purposes, this is the same current as that which previously flowed through Q_2.

The drop $\alpha I_{e1} R_1$ at the collector of Q_1, attenuated by the constant K, appears at the base of Q_2:

$$v_2 = V_{2H} - K\alpha I_{e1} R_1 \tag{10-41}$$

With $I_{e1} = 4.98$ mA, $v_2 = 3.8$ V, which is certainly low enough to ensure that Q_2 is cut off.

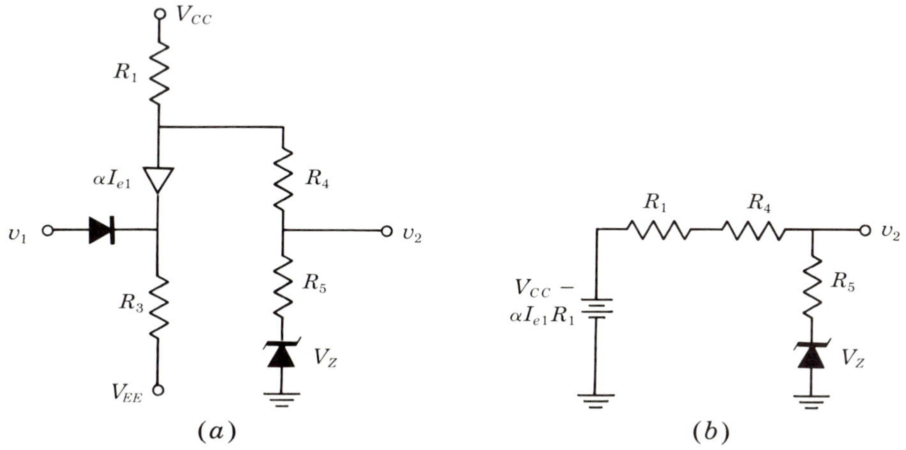

Fig. 10-12 (*a*) Model for circuit of Fig. 10-11*a* when Q_1 is on and Q_2 is off; (*b*) Thévenin equivalent.

The output rises to the high level

$$V_{oH} = V_{CC}$$

To find the down-switching voltage, we must first solve for v_2 in terms of v_1. Combining Eqs. (10-40) and (10-41), we obtain

$$v_2 = V_{2H} + K\frac{\alpha R_1}{R_3}(V_{EE} + V_{BE}) - K\frac{\alpha R_1}{R_3}v_1 \qquad (10\text{-}42)$$

Substituting numbers,

$$v_2 \approx 4.27 - 0.1v_1$$

Transistor Q_2 will again become active once its base voltage rises to within 100 mV of v_1. Substituting the down-switching condition into Eq. (10-42),

$$V_{1L} = \frac{V_{2H} + 0.1 + K(\alpha R_1/R_3)(V_{EE} + V_{BE})}{1 + K(\alpha R_1/R_3)} = 3.97\text{ V} \qquad (10\text{-}43)$$

V_{1L}, and thus the width of the hysteresis loop of Fig. 10-11*b*, depends on the term $K\alpha R_1/R_3$. For narrow hysteresis loops, we want

$$K\frac{\alpha R_1}{R_3} \ll 1$$

We can recognize the term $\alpha R_1/R_3$ as the gain of the first stage of the base amplifier (Q_1) when the second stage (Q_2) is cut off. K is the attenuation factor to the base of Q_2. As we reduce the term $K\alpha R_1/R_3$, we also reduce the loop gain. For very small values, the overall loop gain will become less than unity and the regenerative switching action will cease.

A current-coupled logic trigger

The emitter-coupled logic (ECL) discussed in Sec. 4-8 registered the logical-0 or logical-1 output by transferring the current from one transistor of the differential pair to the other transistor. But this is just what happens in the Schmitt trigger of Fig. 10-11. By adding an external

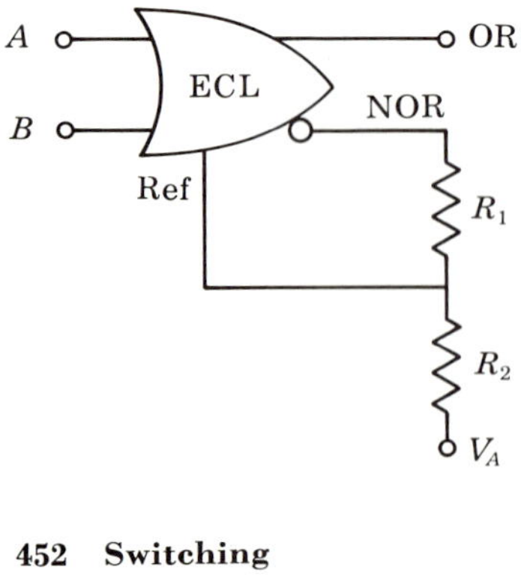

Fig. 10-13 Emitter-coupled logic (ECL) connected to form a regenerative trigger.

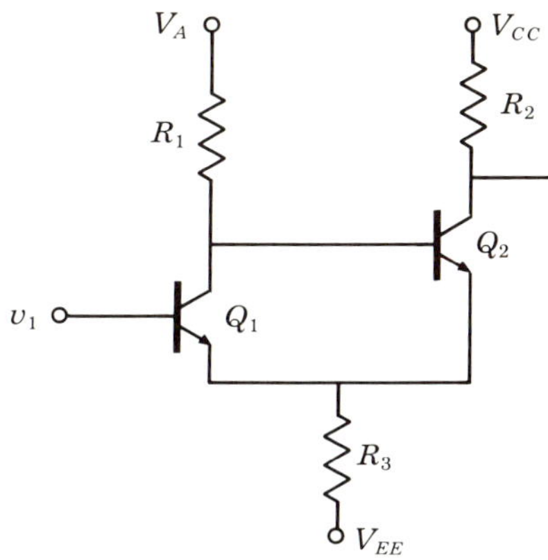

Fig. 10-14 Direct-coupled Schmitt trigger.

positive-feedback path and level voltage from the ECL NOR output back to the reference-voltage input, we create the regenerative triggering described above. Figure 10-13 shows the circuit. The multiple inputs A and B add an OR logic stage at the comparator input. Triggering levels and the width of the hystersis loop can be calculated as above, using the appropriate value of K.

A direct-coupled trigger circuit

Figure 10-14 shows a direct-coupled version of the circuit of Fig. 10-11. Usually the supply voltage V_A and the effective load resistance R_1 of transistor Q_1 are obtained by connecting the collector to a tap on a resistive divider connected from V_{CC} to ground or from V_{CC} to V_{EE}. As a result of the direct coupling from Q_1 to the base of Q_2, the coupling-time constant is very much smaller, resulting in much faster switching times.

The direct coupling means that the feedback factor $K = 1$. Using this value in the equations previously derived, we obtain the following up- and down-switching voltages. Since

$$V_{2H} = V_A$$

the up-trigger level becomes

$$V_{1H} = V_A - 0.1$$

The down-switching levels are

$$V_{2L} = V_A - I_{e1}R_1$$

$$V_{1L} = \frac{V_A + 0.1 + (\alpha R_1/R_3)(V_{EE} + V_{BE})}{1 + (\alpha R_1/R_3)}$$

We might note that in order to obtain a narrow hysteresis loop, the first stage must have a small gain ($\alpha R_1/R_3 \ll 1$).

A composite FET–junction-transistor comparator

The input impedance of both the discrete and the integrated junction-transistor comparators is finite. It changes from a relatively low value during the switching interval, when both transistors are active, to a much larger value when Q_1 is active and Q_2 is off. Of course, in the low state (Q_1 off), the input impedance is the very high impedance of the back-biased junction. If the input signal has a high source impedance, then the IR drop due to the voltage-comparator loading will result in triggering at an incorrect voltage level. The change in loading occurring during the switching interval may even cause erratic triggering.

If either a junction FET or a MOSFET is used as the input stage of a composite comparator (Fig. 10-15), then we always see a very high input impedance. This circuit operates in exactly the same manner as the junction trigger circuit; only the switching levels will change to satisfy the new boundary conditions of the FET.

When v_1 is low, Q_1 is off and Q_2 is on. As before,

$$V_{2H} = V_Z(1 - K) + KV_{DD}$$

The emitter voltage is one diode drop lower, resulting in an emitter current in Q_2 of

$$I_{e2} = \frac{V_{2H} - V_{BE}}{R_3}$$

Thus the low-level output voltage at the collector of Q_2 is

$$V_{oL} = V_{DD} - \alpha I_{e2} R_2 \tag{10-44}$$

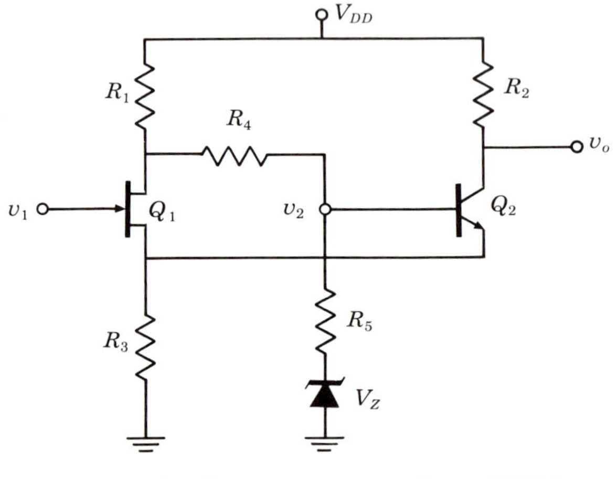

Fig. 10-15 A voltage comparator using a FET input stage.

Up-switching occurs when the FET turns on. This requires that the gate voltage be within the active region, $V_{GS} > V_p$; in fact we must be far enough within the active region for the FET to have a transconductance that will yield a greater-than-unity loop gain. Assuming, for simplicity, that the turn-on point is $V_{GS} = V_p$, the up-switching voltage will be given by

$$V_{1H} \approx V_{2H} - V_{BE} + V_p \tag{10-45}$$

After switching, the output at the collector of the OFF Q_2 will be

$$V_{oH} = V_{DD}$$

Down-switching occurs when the base-to-emitter voltage of Q_2 rises on a falling v_1. Assuming that $r_o \gg R_1$, the model looking into the drain of the FET will be the same as the model shown in Fig. 10-12. The voltage at the base of Q_2 will depend on the drain current of Q_1:

$$v_2 = V_{2H} - I_D R_1 K \tag{10-46}$$

The emitter voltage is also proportional to I_D:

$$v_e = I_D R_3$$

Switching occurs once $V_{BE2} = 650$ mV. Setting $v_2 = V_{2L}$, we have

$$V_{BE} = 0.65 = V_{2L} - I_{DL} R_3 \tag{10-47}$$

Substituting v_2 from Eq. (10-46) and solving for the drain current at the edge of down-triggering yield

$$I_{DL} = \frac{V_{2H} - 0.65}{KR_1 + R_3} \tag{10-48}$$

We can find the V_{GS} corresponding to this drain current from the square-law transfer characteristic,

$$I_D = I_{DSS}\left(1 - \frac{V_{GS}}{V_p}\right)^2$$

Solving for V_{GSL}, we obtain

$$V_{GSL} = V_p\left(1 - \sqrt{\frac{I_{DL}}{I_{DSS}}}\right) \tag{10-49}$$

and finally, the down-switching voltage is found by summing the drops

$$V_{1L} = V_{GSL} + I_{DL} R_3 \tag{10-50}$$

In order to obtain comparison levels to and below zero, the bottom of R_3 must be returned to a negative supply.

10-6 Modes of Operation—Schmitt Trigger

The trigger circuits of Sec. 10-5 can operate in several different modes. Following the definitions given in Sec. 9-2, the possible switching conditions are:

MODE	Q_1	Q_2
11	Off → active	Active → off
21	Off → sat	Active → off
12	Off → active	Sat → off
22	Off → sat	Sat → off

Preferred operation is mode 11, the mode analyzed in detail above (Sec. 10-5). Since neither device ever saturates, the switching time does not include any storage-delay time. Saturated operation is also used because it has the major advantage that the logical-0 output level is independent of transistor parameters; it depends only on the resistor ratios.

Mode boundaries depend on the biasing, on the circuit configuration, and even on the level of the up-switching voltage. A circuit operating in the active mode will change to saturated operation as the switching level is raised.

For simplicity, we shall determine mode boundaries only for the simplified trigger circuit of Fig. 10-16*a*. Assuming that Q_2 is saturated when Q_1 is off, the circuit reduces to the very simple model of Fig. 10-16*b*. The condition for saturation is found by setting the base voltage at the divider network above the base voltage determined from the saturated-transistor model:

$$KV_{CC} > \frac{R_3}{R_3 + R_1} V_{CC} + V_{BE} \tag{10-51}$$

where, as before, $K = R_5/(R_1 + R_4 + R_5)$. Thus the conditions on the operation of Q_2 are

Modes $n1$: $$\frac{R_3}{R_3 + R_1} > K - \frac{V_{BE}}{V_{CC}} = K' \tag{10-52}$$

Modes $n2$: $$\frac{R_3}{R_3 + R_1} \leq K' \tag{10-53}$$

K' is the coupling factor K modified to also account for the V_{BE} drop. So long as Q_2 is not saturated (modes $n1$), the up-switching voltage V_{1H} will be almost exactly equal to the base voltage of Q_2, KV_{CC}.

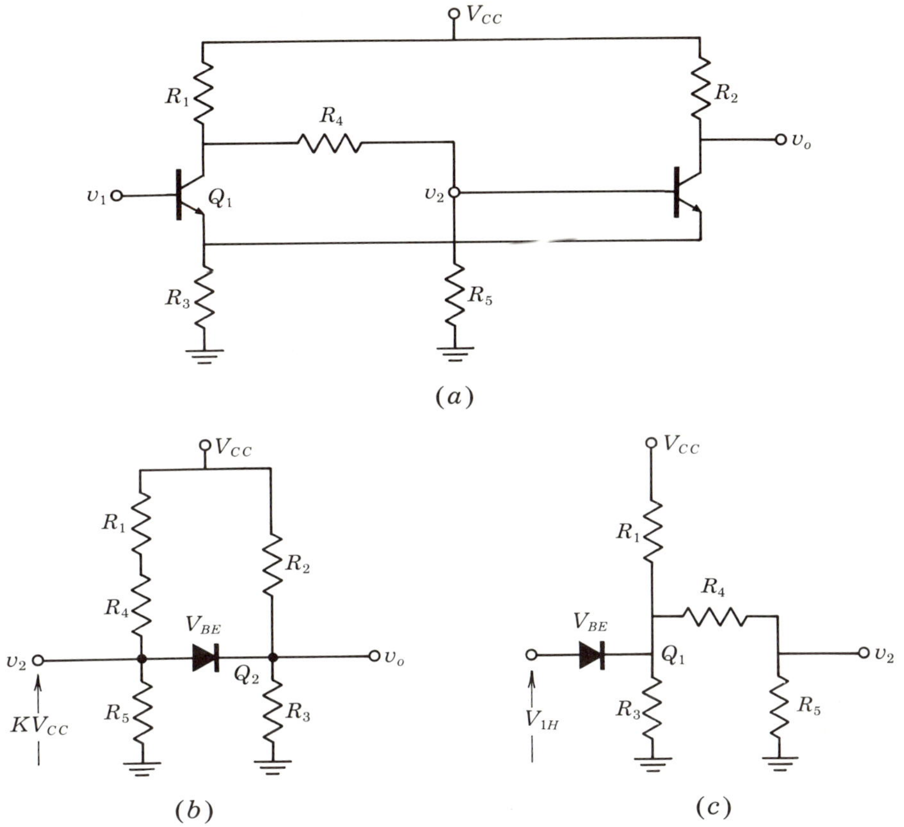

Fig. 10-16 (*a*) Trigger circuit for mode investigation; (*b*) model with Q_2 saturated and Q_1 off; (*c*) model with Q_1 saturated and Q_2 off.

The second set of conditions is found from the model of Fig. 10-16*c*. When $R_1 \gg R_4 + R_5$, we obtain

Mode 11: $$\frac{R_3}{R_1 + R_3} \geq K' \tag{10-54}$$

Mode 21: $$\frac{R_3}{R_1 + R_3} < K' \tag{10-55}$$

When Q_2 is saturated (modes $n2$), its base voltage is determined by the saturation emitter current found from Fig. 10-16*b*:

$$V_{2H} = V_{BE} + \frac{R_3}{R_3 + R_2} V_{CC} \tag{10-56}$$

The up-switching voltage is now independent of the R_4, R_5 coupling network.

In order for Q_1 to switch from cutoff directly into saturation (where, just before switching, we had $V_e = V_{2H} - V_{BE}$), from Fig. 10-16*c* we find that

$$V_{1H} = V_{2H} \geq V_{BE} + \frac{R_3}{R_3 + R_1} V_{CC} \qquad (10\text{-}57)$$

Equating (10-56) to (10-57) and solving, the second set of conditions for modes 12 and 22 becomes

Mode 12: $$R_1 < R_2 \qquad (10\text{-}58)$$

Mode 22: $$R_2 \geq R_1 \qquad (10\text{-}59)$$

Analysis of the circuit within all modes follows the technique of Sec. 10-4, using, for the saturated modes, the up-switching voltages given above. Any remaining calculations are straightforward.

10-7 Some Comparator Applications

The comparator of Fig. 10-17 is a sense or zero-crossing detector. Whenever the input exceeds the zero reference signal, the output registers a logical-1 output. The sense detector thus generates a positive pulse whose duration is equal to the duration of the positive-polarity signal.

By differentiating the output signal, we obtain a positive pulse for each 0-to-1 transition and a negative pulse for every 1-to-0 transition. Counting the number of positive pulses over a given time interval gives the frequency of a periodic input waveshape.

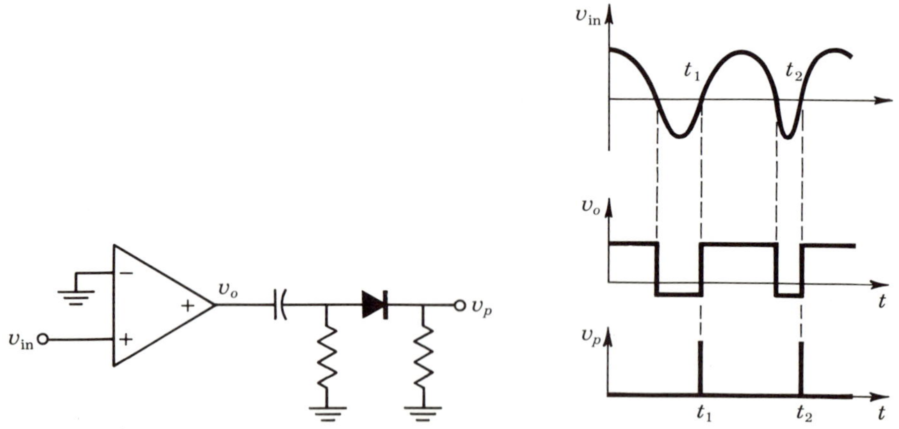

Fig. 10-17 Sense or zero-crossing detector.

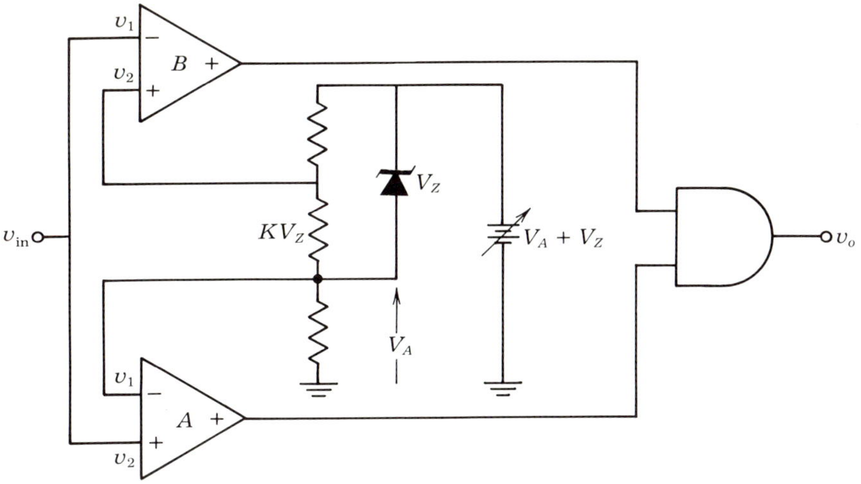

Fig. 10-18 Pulse-height analyzer. v_0 = logical 1 for $V_A + KV_Z \geq v_{in} \geq V_A$. ($v_1$ is the inverting input, and v_2 is the noninverting input.)

The time interval between successive positive pulses may be measured by using the first positive pulse to trigger a counter on and cause it to count pulses from its own internal clock. The next positive pulse turns the counter off. The number of pulses registered is proportional to the interval between successive positive zero crossings.

Figure 10-18 illustrates a pulse-height analyzer. The same input signal is applied to two comparators. In comparator A, the input is compared with a reference signal V_A in such a way that a logical-1 output is generated whenever the input exceeds V_A. Comparator B, inversely connected, compares the same input with another reference signal $V_A + KV_Z$. This comparator has a logical-1 output only if the input is less than $V_A + KV_Z$. The two comparators are ANDed, yielding a logical 1 when both are at the logical-1 level. But this occurs only for inputs lying between

$$V_A + KV_Z \geq v_{in} \geq V_A \tag{10-60}$$

We have created a window that will indicate, by a logical-1 output, the presence of pulses falling within the required height interval. The full range of pulse heights can be examined by slowly scanning V_A while still maintaining the separation.

In the analog-to-digital converter of Fig. 10-19, the input signal is compared with a linear ramp. So long as the input is above the ramp voltage, the comparator output will be at the logical-1 level. As the ramp rises linearly, the two signals eventually become equal, the output dropping

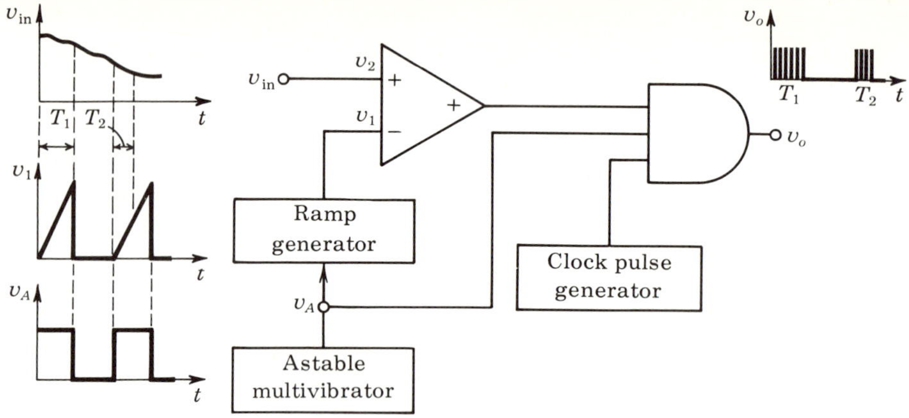

Fig. 10-19 Analog-voltage–to–digital-pulse converter.

to the logical-0 level. Thus the duration of the logical-1 signal depends on the amplitude of the input signal. The comparator has converted amplitude into a proportional time interval.

At the instant the ramp starts rising, the astable multivibrator output is applied to an AND gate. The coincidence of the comparator output and the astable output (or ramp) opens the AND gate and permits the transmission of a train of positive clock pulses. The number of pulses transmitted by the AND gate is directly proportional to the time that the gate is open, and this time is proportional to the input-voltage amplitude. The discrete output pulses can be counted and stored.

The sampling time depends on the astable-multivibrator period. The ON time must be long enough for the ramp to rise from zero to the highest required voltage. The OFF time, which determines the sampling interval, is determined by the rate of change of the input voltage. Rapidly varying signals require a higher sampling rate. When the sampled signal varies very rapidly, the analog-to-digital converter may not be able to register enough points to define the input properly.

Problems

10-1 Figure 10-20 is the basic circuit of a differential pair having a fixed-resistor-emitter load. Each transistor has $\beta = 100$, $V_{BE} = 700$ mV, and $r_b = 200\ \Omega$.

(*a*) Calculate the differential small-signal gain to each output at the Q point $v_1 \approx v_2 = 0$.

(*b*) Plot the in-phase transfer function of this stage, that is, v_{o2} vs. $v_1 = v_2$. Calculate the in-phase gain.

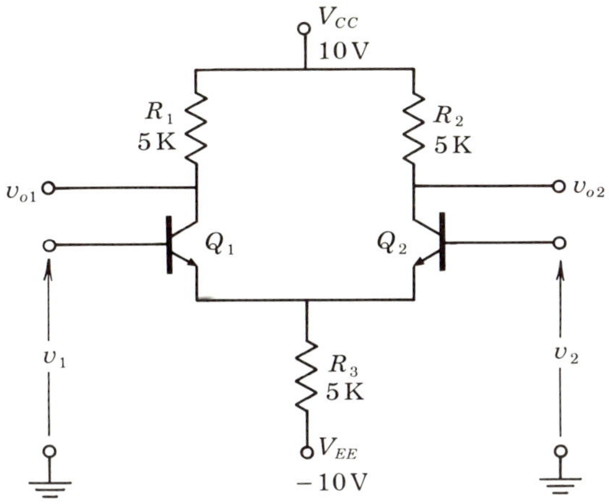

Fig. 10-20

(c) Will changes in the absolute value of the in-phase signal level, $v_1 \approx v_2$, affect the differential gain? Explain.

10-2 (a) Repeat part b of Prob. 10-1 for the circuit of Fig. 10-1.

(b) Repeat the plot of part a when the two transistors are not identical. $\beta_1 = 100$ and $\beta_2 = 75$.

10-3 Calculate the input impedance in each of the regions of operation of the differential-amplifier input stage of Fig. 10-1. Both inputs are connected to low-impedance sources.

10-4 Evaluate the output impedance of the simplified type-710 amplifier of Fig. 10-4. What are the output-current limitations when used as a source? As a sink?

10-5 The type-μA702 amplifier of Fig. 3-23 uses a more complex output stage to obtain a higher overall gain as well as level shifting. This stage (Q_6, Q_7, and Q_8) employs positive feedback to obtain the required gain. Since Q_7 has approximately 2.5 times the emitter area, its I_{es} is 2.5 times the value used for the other transistors in the text in Secs. 10-1 and 10-2.

(a) Calculate the differential gain and dc conditions from the inputs to the base of Q_6.

(b) Find the gain and level shifting introduced by the output stage. Give the amplifier's overall gain.

10-6 Calculate the output impedance of the μA702. What are the maximum current capabilities when used as a source? As a sink? Compare these results with those found for the amplifier of Fig. 10-4.

10-7 Figure 10-21 is a junction-FET differential-amplifier pair employing a third FET as a constant-current element in the common source. Each FET has the parameters $I_{DSS} = 10$ mA, $V_p = -2$ V, $r_o = 100$ K at $I_D = I_{DSS}$.

(a) Plot the drain-current distribution of Q_1 and Q_2 as a function of the differential input voltage (neglect r_o).

(b) What is the differential-input-voltage range for a 10 to 90 percent change in the current in a single FET of the differential pair?

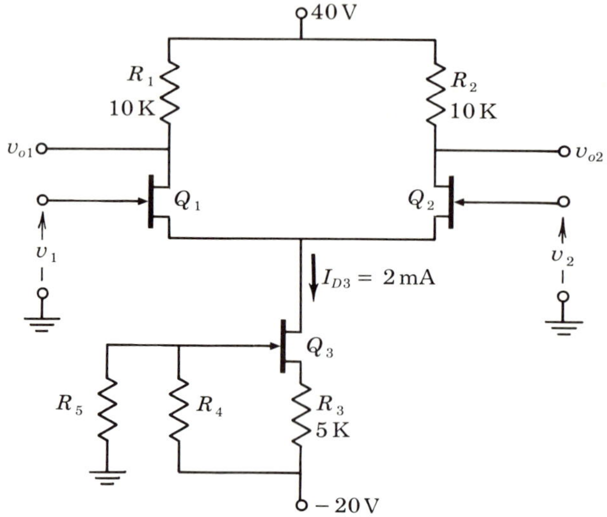

Fig. 10-21

10-8 (a) Specify R_4 and R_5 in Fig. 10-21 to set $I_{D3} = 2$ mA.

(b) Calculate the small-signal differential gain.

(c) Over what range of $v_1 = v_2$ will this circuit operate as a differential amplifier?

(d) What is the common-mode gain? (v_{o2}/v_1 when $v_1 = v_2$.) Take r_o of Q_3 into account.

10-9 The amplifier of Fig. 10-4 is employed in the feedback comparator of Fig. 10-9. Plot the transfer functions for the following values of K on the same graph. Pay particular attention to the location of the knees of the transfer curve. Use a piecewise-linear approximation for simplicity.

(a) $K = 1/5A_o$

(b) $K = 1/A_o$

(c) $K = 5/A_o$

A_o is the gain of the differential amplifier.

10-10 Make use of the amplifier of Fig. 10-4 to create a monostable circuit. This circuit is to generate a 100-ms pulse at the output after triggering. Show the external feedback and biasing, giving all component values. Show where and how to inject a triggering pulse. Plot the significant input- and output-voltage waveshapes. Give all times, time constants, and voltage values.

10-11 Plot the transfer function of the Schmitt trigger of Fig. 10-22 for the two cases given below. ($\beta = 100$, $V_{BE} = 750$ mV.)

(a) $V_z = 4$ V

(b) $V_z = 8$ V

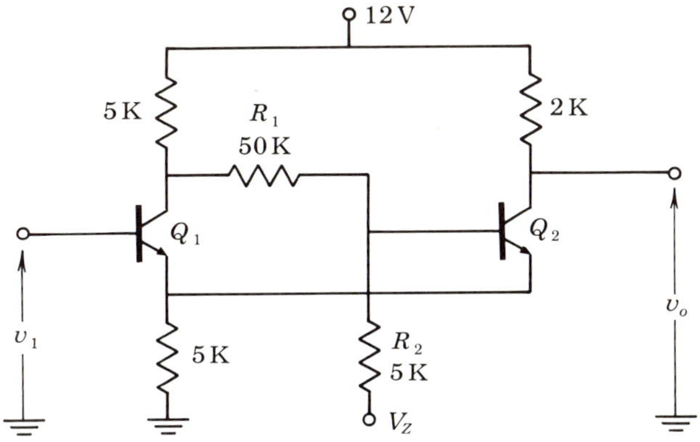

Fig. 10-22

10-12 Plot the transfer function of the circuit of Fig. 10-14 when Q_1 is replaced by a junction FET. $R_1 = 10$ K, $R_2 = 2$ K, $R_3 = 5$ K, $V_{CC} = 20$ V, $V_{EE} = -10$ V, $V_A = 10$ V. For Q_1: $V_p = -2$ V, $I_{DSS} = 5$ mA. For Q_2: $\beta = 100$, $V_{BE} = 700$ mV. Give the width of the hysteresis region.

10-13 Design a Schmitt comparator based on the ECL logic of Fig. 4-29. The center of the switching region is to be located at -3 V, and the width of the hysteresis loop must be less than 400 mV. Specify all external resistors and voltage supplies. Draw the piecewise-linear transfer characteristic, giving all voltage levels.

10-14 Convert the FET-differential-amplifier stage of Fig. 10-21 to a Schmitt comparator. The up-switching point is to be at 20 V, with down-switching occurring 1 V lower. Plot the transfer characteristic, paying particular attention to the location of the corners of the curve.

10-15 Design a comparator switching at -10 ± 0.5 V using MOSFETs for the active devices. The devices available have $V_T = -3$ V and $I_D = 5$ mA at $V_{GS} = -4$ V. $V_{DD} = -30$ V. Give all component values. Draw the transfer characteristic.

10-16 Define the mode conditions in terms of the resistor values for the direct-coupled Schmitt trigger of Fig. 10-14.

Bibliography

Giles, J. N.: "Fairchild Semiconductor Linear Integrated Circuits Applications Handbook," Fairchild Semiconductor Division, Mountain View, Calif., 1967.

Middebrook, R. D.: "Differential Amplifiers," John Wiley & Sons, Inc., New York, 1963.

Schmitt, O. H.: A Thermionic Trigger, *J. Sci. Instr.*, vol. 15, pp. 24–26, January, 1938.

Widlar, R. J.: A Unique Circuit Design for a High Performance Operational Amplifier Especially Suited to Monolithic Construction, *NEC Proc.*, vol. 21, pp. 85–89, 1965.

11.
negative-resistance switching circuits

A two-terminal active network which exhibits a negative-resistance driving-point impedance over a portion of its range may be used, together with an energy-storage element, as a switching circuit. Our initial discussion concerns a postulated ideal device in various circuit configurations. After considering several available active elements that are used for this type of switching, we shall modify the original argument to account for the frequency dependence of the nonideal active elements. A physical device differs markedly from an ideal one, and some of the later results may differ from the earlier conclusions. If the reader considers only the problem posed in each individual network, the arguments employed will be found to be consistent.

11-1 Basic Circuit Considerations

The regenerative circuits treated in Chaps. 8 and 9—bistable, monostable, and astable multivibrators—depended for their proper self-switching action on the charge stored in a capacitor (the single energy-storage element of the circuit) maintaining the transistor or FET cut off. Only after a sufficient amount of the stored energy had been dissipated did the active element reenter its active region; then, during recovery, the original stored energy was replenished through a positive-source impedance. But in order to transfer the capacitor from a region of low potential energy (off) to one of high energy (recovery), the switch itself must contain an energy source. Thus a switching circuit might well be characterized as a device having two stable or quasi-stable states, i.e., dissipative regions, and a single transition zone.

Any two-terminal device with these attributes may, when combined with a capacitor or inductance, function as a multivibrator. Figure 11-1 presents the idealized volt-ampere characteristic of one class of such devices. In regions I and III, the curve has a positive slope with a corresponding positive incremental resistance. These are the dissipative segments. Region II is the negative-resistance portion of the characteristic, and it is here that the switching of circuit states takes place.

Negative-resistance regions are created through the use of internal or external positive-feedback mechanisms operating on the controlled source of an active device. If we restrict our observation to a device or circuit which does not itself contain an internal power supply, then the terminal response must be completely dissipative. This implies that the terminal volt-ampere characteristics must lie completely in the first and/or third quadrants of the volt-ampere plane. At every point in these two quadrants, the VI product is positive, implying that power must be supplied from an external source to set a particular bias point. A segment of these volt-ampere characteristics may have a negative slope. Over the small finite range of negative resistance, energy is transferred, by the device, from the external power supply to the circuit switching operations.

Yet no physical device can be expected to have an unlimited range of negative resistance; that would imply an infinite availability of energy.

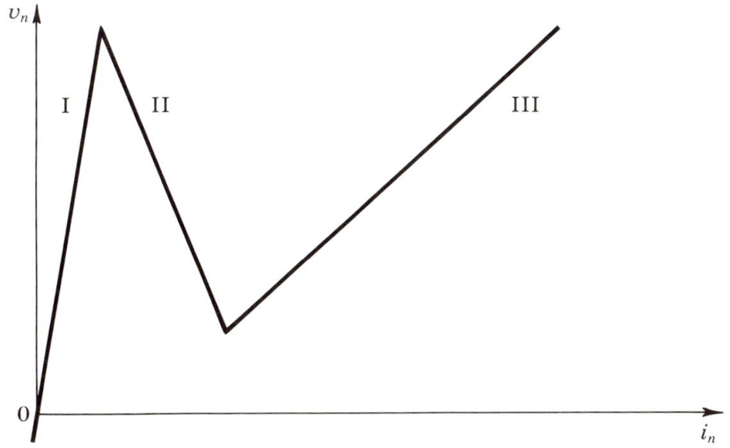

Fig. 11-1 Idealized negative-resistance characteristic (current-controlled).

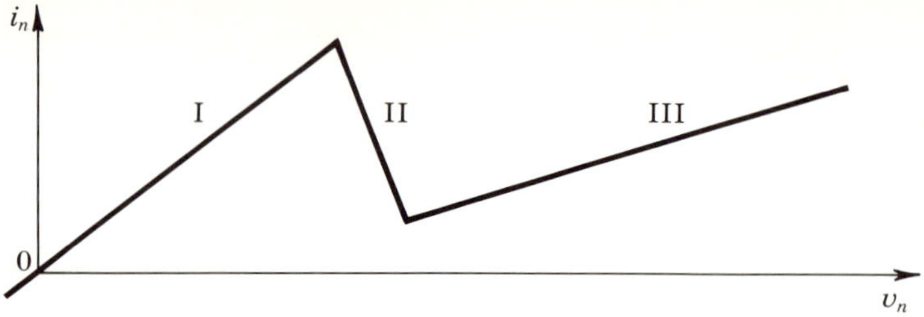

Fig. 11-2 Voltage-controlled negative-resistance characteristic.

Therefore each negative slope will always be flanked by two regions of positive resistance.

Because the terminal voltage is a single-valued function of current, we classify the characteristic of Fig. 11-1 as a current-controlled nonlinear resistance (CNR). On the other hand, any one of three values of current becomes possible upon excitation by an input voltage. The particular value that will flow is determined by the constraints imposed by the external circuit, with this very multivaluedness permitting the several different circuit states. Since only the current is unrestricted, in our switching circuit we must utilize an energy-storage element allowing instantaneous current changes during the switching process, i.e., a capacitor.

An alternative form of the nonlinear resistance, the dual of the above, also exists. Its voltage-controlled characteristic (VNR) is illustrated by the sketch of Fig. 11-2. To complete the duality, an inductance would be used instead of the controlling capacitor to permit rapid terminal-voltage changes.

Negative-resistance regions are exhibited by many devices. Current-controlled characteristics appear in unijunction transistors, avalanche diodes, silicon-controlled rectifiers, and gas tubes. For examples of the voltage-controlled nonlinear resistance (VNR), we may refer to the characteristics or the tunnel diode and the *pnpn* transistor. In addition, the use of feedback allows the production of almost any desired form of NR. Generally voltage feedback furnishes the voltage-controlled characteristics and current feedback establishes those which are current-controlled.

Behavior of the two classes of NR devices is similar enough so that any conclusions derived from the examination of one class may be applied to the other. Of course, caution as to the duality relationship must be observed while interpreting the results.

11-2 Basic Switching Circuits

The basic form of the switching circuit with which we shall be concerned is shown in Fig. 11-3. For the negative-resistance element, we shall use the current-controlled characteristic previously given. In this circuit we are first interested in the necessary conditions for stable operation at any particular operating point. We shall start the analysis by writing the single-node equation:

$$i_r = i_n + i_c = i_n + C\frac{dv_n}{dt} \tag{11-1}$$

and

$$V_{nn} - i_r R = v_n \tag{11-2}$$

Solving Eq. (11-2) for i_r and substituting the result into Eq. (11-1) yield

$$(V_{nn} - i_n R) - v_n = RC\frac{dv_n}{dt} \tag{11-3}$$

Let

$$\mathbf{v}_t \equiv RC\frac{dv_n}{dt} \tag{11-4}$$

However, $V_{nn} - i_n R$ is simply the static (dc) load line which we can superimpose on the volt-ampere characteristic. The distance measured from the known curve to the load line must be $\mathbf{v}_t$, and as shown in Fig. 11-4, it may be treated as a vector. $\mathbf{v}_t$ is a measure of the time rate of change of the terminal voltage v_n. A positive $\mathbf{v}_t$ indicates a voltage increasing with time, while a negative $\mathbf{v}_t$ indicates a terminal voltage decreasing with time. Thus, by examining the nature of $\mathbf{v}_t$, we can obtain some information as to the response of the circuit without bothering to find the complete solution.

The steady-state operating point, the intersection of the load line and the characteristic curve, is stable if, and only if, any momentary perturbation will create the conditions forcing a return to the original point. Since the volt-ampere plot represents the locus of all possible behavior of the NR, any disturbance will shift the operating point along it—say, from A to A'. Equation (11-4) defines the vector $\mathbf{v}_t$, drawn from any point on the characteristics to the load line, as being proportional to the time rate of change of v_n. At point A', this vector is negative, indicating that the

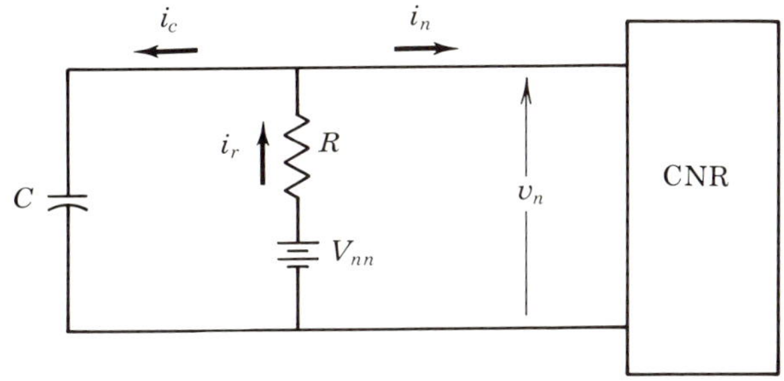

Fig. 11-3 Basic CNR switching circuit.

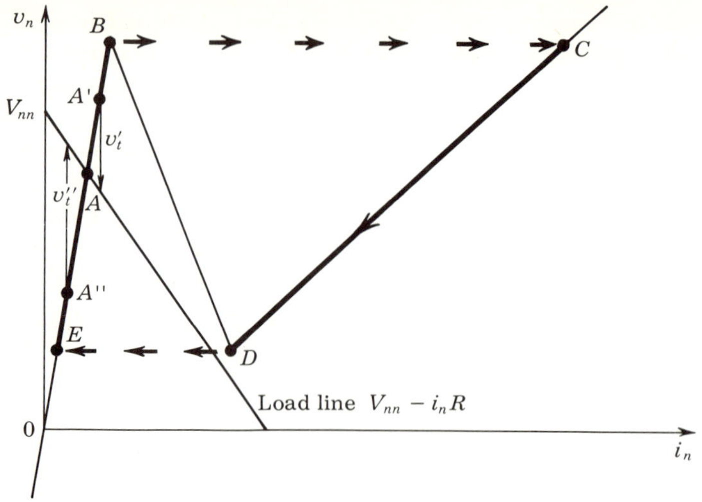

Fig. 11-4 Graphical solution of the circuit of Fig. 11-3.

terminal voltage across the capacitor must decrease with time. If, instead, the initial disturbance were from A to A'', the vector would be positive, with the terminal voltage increasing with time toward point A. Only at A itself is $\mathbf{v}_t$ zero, and only here do we have a stable operating point.

Operation as a monostable switching circuit

A voltage pulse, applied in series with the bias supply, momentarily shifts the load line parallel to itself to the position indicated by the dashed line of Fig. 11-5. While the pulse is present, the shifted load line establishes a new stable point at F. But $\mathbf{v}_t$, as drawn from the original operating point A to the shifted load line, is positive, and therefore the terminal voltage increases toward point B' lying at the intersection of the extended segment EB and the shifted load line. Here $\mathbf{v}_t$ is zero. But at the peak point B, $\mathbf{v}_t$ is still positive, forbidding travel down the line segment BD. Operation must be restricted to the actual characteristics and must stabilize at point F. It can do so only if the operating point can transfer from segment EB to CD. The voltage across the capacitor may not change instantaneously, but the current can and does; it jumps from B to C.

Along line segment CF, $\mathbf{v}_t$ becomes negative. After the jump, the direction of the operating path is down toward point F.

Removal of the input trigger permits the load line to return to its original position. Even after it does so, the terminal voltage, constrained by the capacitor to change slowly, remains on segment CD and travels toward the $\mathbf{v}_t = 0$ intersection at point D'. At the valley point D, $\mathbf{v}_t$ is still negative, preventing the traverse up the line segment DB. A second jump in current is indicated, this time from D to E. Following this jump, $\mathbf{v}_t$ becomes positive, and the operating path is finally back toward the original stable point A.

In order to repeat the operating cycle, another trigger pulse must be injected. The minimum-size pulse, for guaranteed triggering, is that value necessary to shift the load line so that it just clears the peak point B.

As an alternative method of triggering, we can inject a current impulse, having an area of K A-s, into the common node of Fig. 11-3. $K\delta(t)$ changes the voltage across the capacitor from the initial value of V_A to

$$v(0^+) = V_A + \frac{K}{C}$$

If $v(0^+)$ is greater than the peak voltage V_B, the operating point will switch to the appropriate voltage coordinate along the extension of line segment CD. Recovery proceeds as before, from the new initial value established by the applied impulse.

Since this system has one, and only one, stable condition, any disturbance, no matter how large, eventually results in the circuit returning to this point; it functions as a monostable multivibrator.

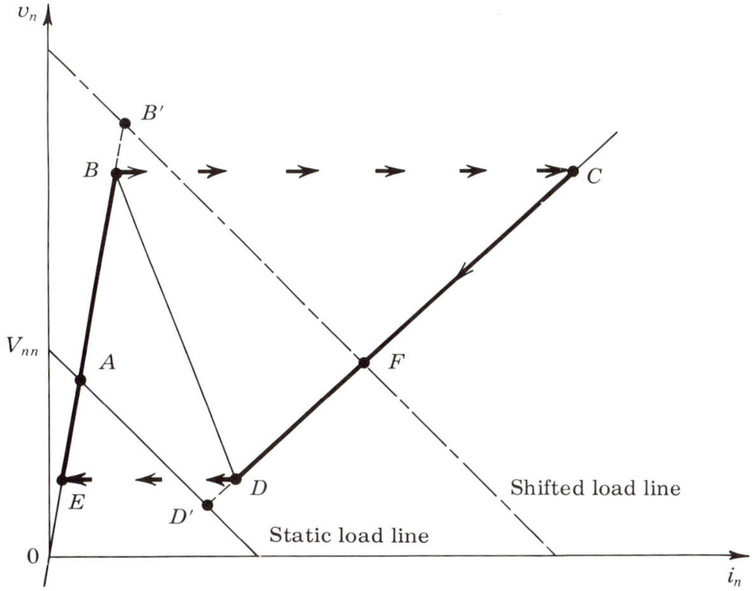

Fig. 11-5 Monostable operation—load line and trajectory.

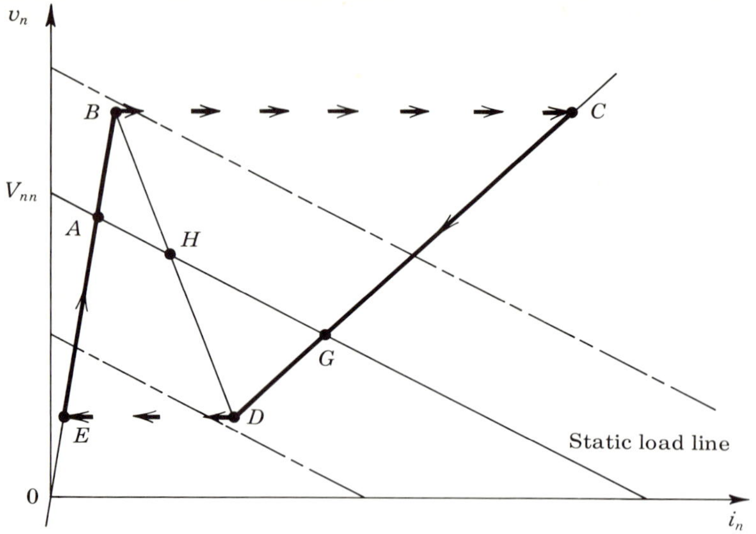

Fig. 11-6 Bistable operation—load line and trajectory.

Bistable operation

For operation as a bistable, the values of V_{nn} and the load resistance will have to be changed so that the circuit has the two stable points (A, G) indicated in Fig. 11-6. Upon first examination, H appears to be a third stable point, but once switching begins, the transition will always be from A to G for positive triggers and from G to A when a negative trigger is applied. The operating path is in the direction of the arrows. Regardless of how or where the triggers are injected, it is impossible to establish a transition path to point H. This is clearly seen when we examine the switching paths established by the two dashed shifted load lines of Fig. 11-6.

In effecting the change of state, the external trigger forces the operating point to move from its stable position to the peak of the curve, i.e., from A to B or from G to D. In sweeping through this distance, power proportional to the product of the changes in voltage and current will be dissipated, and this power must be supplied from the trigger source. We conclude that if the stable points are established closer to the peaks, the switching-power requirement imposed on the trigger generator will be much less severe.

The sequence of the multivibrator's behavior in switching from one stable point to the other is identical with that discussed for the monostable circuit.

Astable operation

When the load line is chosen so that it intersects the volt-ampere characteristic only on the negative slope portion of the NR, no stable point of operation exists. This case is illustrated in Fig. 11-7. In the vicinity of the point of intersection J, the total current available for storage in an external energy-storage element is

$$i_c = i_r - i_n = \frac{v_n}{R} - \frac{v_n}{R_n} \tag{11-5}$$

where $-R_n$ is the slope of the negative-resistance region. But since for astable operation $R > |R_n|$, the net current is negative, indicating that current will flow into the capacitor and cause a voltage buildup toward either point B or point D.

Once the peak is reached, the path of operation becomes $BCDE$, with the instantaneous current jumps taking place from B to C and D to E. We might observe that at all points on the characteristic, the value of $\mathbf{v}_t$ is such as to direct the time rate of voltage change toward the peak or valley.

An alternative way of viewing the buildup phenomenon is to recognize that in region II the negative resistance $-R_n$ and the load resistor R are

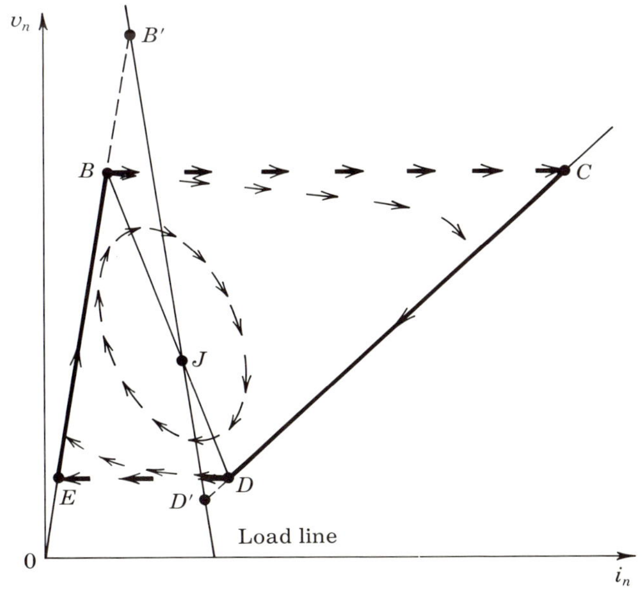

Fig. 11-7 Astable operation—load line and trajectory.

in parallel. Surplus energy means that the parallel combination is still a negative resistance, with the system's pole located at

$$s = -\frac{1}{C(R \parallel -R_n)}$$

This pole lies on the real axis in the right half of the complex plane. Obviously, the circuit is unstable, having an increasing exponential for its transient response. Terminal voltage builds up until limiting occurs when the operating point enters the positive-resistance region.

Instantaneous current jumps are forbidden in any physical CNR circuit by the inductive component of the driving-point impedance. This modifies the path of operation, introducing the curvature shown by the dashed lines of Fig. 11-7. For very-low-frequency operation, the transition time is small enough so that the path approaches the ideal. But as the generated-pulse duration decreases, the transition time becomes an appreciable portion of the total cycle and the locus constricts to the dashed curve sketched. At very high frequencies, the path may even turn into the small ellipse in Fig. 11-7.

Stability of ideal CNR switching circuits

For absolute stability in the presence of an energy-storage element, i.e., the system's pole remaining on the left-hand half of the real axis, the parallel combination of the ideal negative resistance and the external load resistor must be positive:

$$R_p = \frac{R(-R_n)}{R + (-R_n)} > 0$$

In the particular circuit of Fig. 11-3, where the graphical solution is given by Fig. 11-6, this condition can be satisfied only when

$$R < |R_n| \tag{11-6}$$

If the load is a short circuit, then regardless of the value of R_n, the parallel combination will always be positive. In addition, the now horizontal load line intersects the positive-resistance, as well as the negative-resistance, regions of the CNR characteristic and consequently establishes at least two stable operating points. Alternatively, an infinite-resistance load could intersect only the negative-resistance portion, and it would cause astable behavior in the presence of any external capacity. Other considerations will cause us to modify this conclusion for a physical CNR, and reference should be made to Sec. 11-9, where the topic of stability is treated in much greater detail.

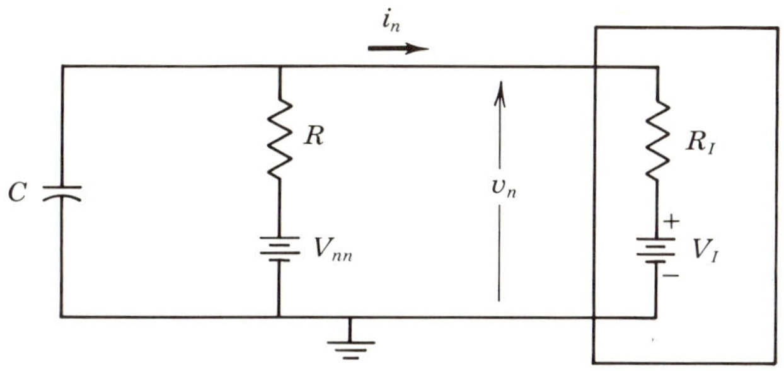

Fig. 11-8 Piecewise-linear circuit used for the calculation of waveshapes.

11-3 Calculation of Waveshapes

The calculation of the waveshapes of an ideal switching circuit (no parasitic energy-storage elements present) may be greatly simplified by representing each linear region of the NR by a resistor in series with a battery. If we extend the straight lines of the NR characteristic until they reach the voltage axis, the point of intersection is the battery voltage and the slope of the line segment is the resistance value. Figure 11-8 shows the resulting equivalent circuit for region I. Since the instantaneous current jump carries the operating point through the negative-resistance region, only regions I and III need be considered.

Within each region the exponential terminal response has, as its single boundary condition, voltage continuity across the current jumps of the circuit. The problem resolves itself into finding the boundary values and the time constants holding within regions I and III.

At the instant that the operating point enters region I, the voltage across the capacitor is that of point E in Fig. 11-5, 11-6, or 11-7. It rises with a time constant τ_1 toward the steady-state voltage of the Thévenin equivalent circuit across its terminals V_1:

$$\tau_1 = C(R \parallel R_{\text{I}}) \tag{11-7}$$

$$V_1 = \frac{RV_{\text{I}} + R_{\text{I}}V_{nn}}{R_I + R} \tag{11-8}$$

This voltage [Eq. (11-8)] is simply the value of the intersection of the load line with the extension of the line segment defining region I. In the astable circuit, V_1 is the voltage at point B' in Fig. 11-7. V_1 must always be larger than V_B; otherwise the output will never reach the voltage peak B. However, if the circuit is adjusted for bistable or monostable operation and if it has a stable point along the segment EB, then V_1 will be the voltage at the stable point. Once this value is reached, sweep action ceases.

At the peak, $v_n = V_B$ and the operating point flips from B to C. Upon entering the new region (III), the circuit of Fig. 11-8 changes; R_{III} and V_{III} replace R_I and V_I. The terminal voltage immediately starts charging toward the new Thévenin equivalent voltage with a time constant τ_3:

$$V_3 = \frac{RV_{III} + R_{III}V_{nn}}{R_{III} + R} \qquad \tau_3 = C(R \parallel R_{III}) \tag{11-9}$$

Equation (11-9) defines the final value of any circuit having a stable point lying on line segment CD. For astable operation, $V_3 < V_D$ and V_3 will be the voltage at D' in Fig. 11-7. After reaching the valley point D, the operating path jumps to point E, and the cycle repeats.

The current values at the switching points are simply those given by the NR characteristic at points B, C, D, and E. In order to find the steady-state values, we must consider the circuit when the capacitor is fully charged and the full current flows into the NR terminals. In region I,

$$I_1 = \frac{V_{nn} - V_I}{R_I + R} \tag{11-10}$$

and in region III,

$$I_3 = \frac{V_{nn} - V_{III}}{R_{III} + R} \tag{11-11}$$

These values also correspond to the points of intersection B' and D' of the extended segments of the volt-ampere characteristic and the load line.

From the characteristic and from Eqs. (11-7) to (11-11), we have all the voltage and current values necessary to sketch the terminal waveshapes. Figure 11-9 shows the current and voltage waveshapes for an astable configuration, and Fig. 11-10*b* shows those appearing in the monostable circuit of Example 11-1.

The durations of the two portions of the generated waveshape, evaluated by substituting the appropriate voltage values into the general time equation, are

$$t_1 = \tau_1 \ln \frac{V_1 - V_E}{V_1 - V_B} \tag{11-12}$$

and

$$t_3 = \tau_3 \ln \frac{V_3 - V_B}{V_3 - V_E} \tag{11-13}$$

In some NR devices we can exercise a measure of control over the shape of the volt-ampere characteristic. Adjusting the ratio of R_I to R_{III} determines the ratio of pulse width to pulse spacing. Furthermore, if R_I is made very much larger than R_{III} and if V_1 is very much larger than V_B, then the circuit becomes an almost linear sweep generator. Its fast recovery time is mainly dependent on R_{III}.

We see that the greater the degree of control over the characteristic, the more versatile the system. Not all devices allow for wide latitude of

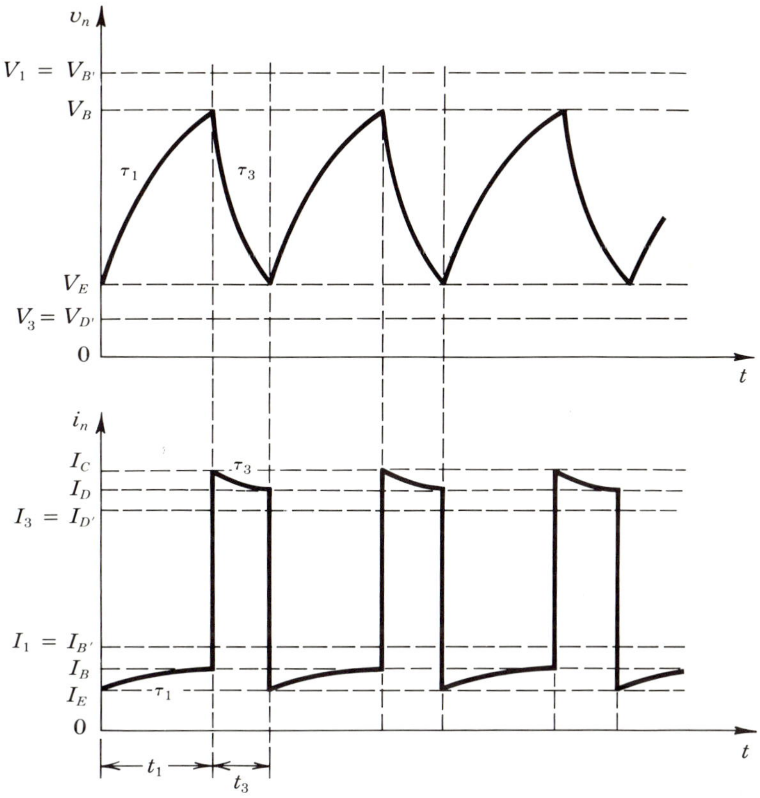

Fig. 11-9 Waveshapes produced in a CNR switching circuit (astable operation).

adjustment, but even in those which do not, the characteristics might be modified by external diodes biased for conduction at the desired voltage levels (Chap. 2).

Example 11-1

A CNR having the volt-ampere characteristics given in Fig. 11-10*a* is used as a monostable pulse generator. Its external load consists of a 1-nF capacitor and a 250-Ω resistor in series with 10-V source, as shown in Fig. 11-3. The circuit is triggered by a 15-V 500-ns pulse applied in series with the 10-V supply. The trigger pulse momentarily shifts the load line to the position indicated by the dashed line.

Only two sets of coordinates are unknown: (1) the location of the intersection of the load line with the line segment defining region I, that is,

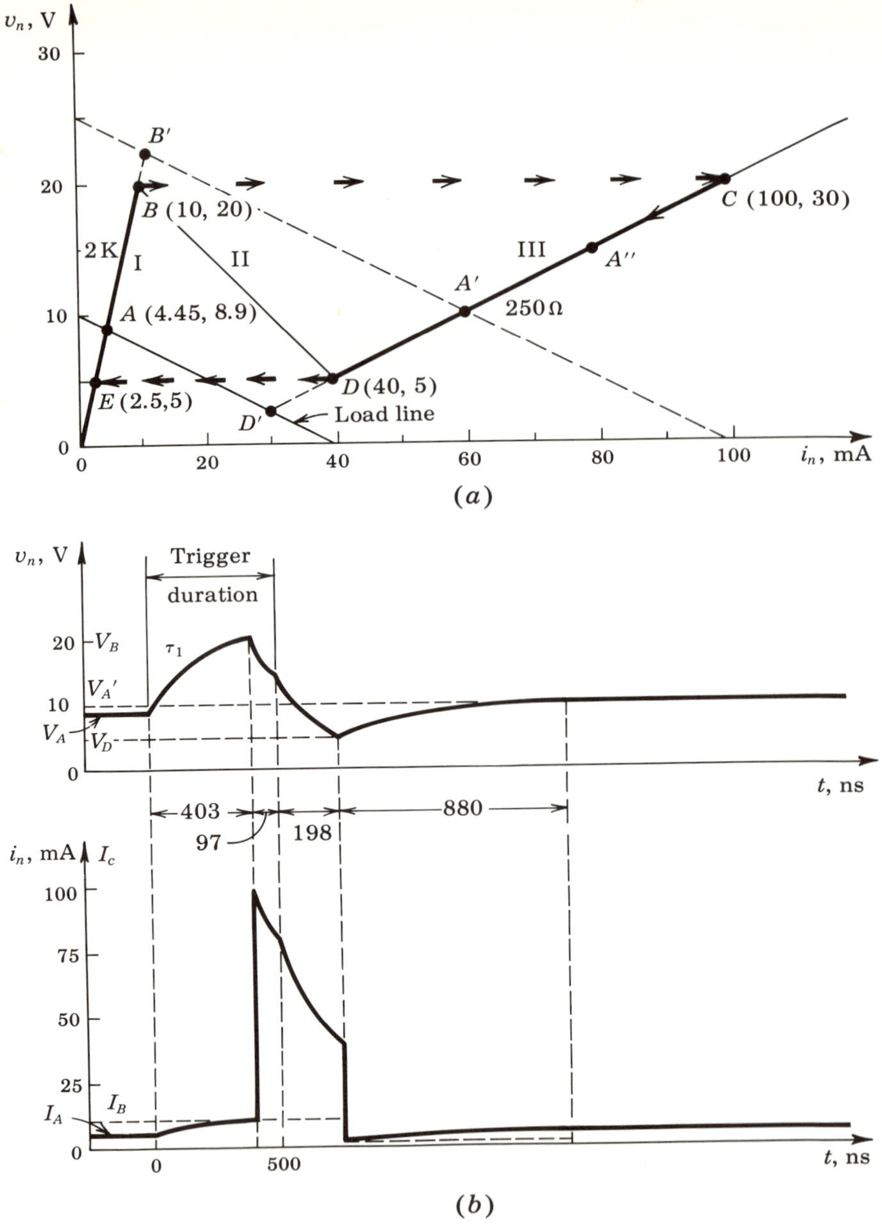

Fig. 11-10 Monostable operation. (*a*) CNR characteristic and graphical construction for Example 11-1; (*b*) current and voltage waveshapes produced after triggering.

the single stable point A; (2) the intersection of the load line with the extension of segment I, that is, $V_{B'}$ and $I_{B'}$. In order to find these points, we first solve the graph for the three defining equations

$$\begin{aligned} &\text{Region I:} && v_{n\mathrm{I}} = (2\ \mathrm{K})i_n \\ &\text{Region III:} && v_{n\mathrm{III}} = -5 + 250i_n \\ &\text{Load line:} && v_{nL} = 10 - 250i_n \end{aligned}$$

By equating $v_{n\mathrm{I}}$ to v_{nL}, we find that the stable point is located at

$$I_A = 4.45\ \mathrm{mA} \qquad \text{and} \qquad V_A = 8.9\ \mathrm{V}$$

Upon applying the 500-ns trigger, the load line shifts to the position indicated by the dashed line. The new equation becomes

$$v_{nL} = 25 - 250i_n$$

The shift of the load line makes point A an unstable point, and the operating point moves toward the intercept B'. The coordinates of this point are

$$I_{B'} = 11.1\ \mathrm{mA} \qquad V_{B'} = 22.2\ \mathrm{V}$$

The time required to reach the peak point B is

$$t_1 = 220 \times 10^{-9} \ln \frac{22.2 - 8.9}{22.2 - 20} = 403\ \mathrm{ns}$$

where $\tau_1 = (2\ \mathrm{K} \parallel 250)1\ \mathrm{nF} = 220\ \mathrm{ns}$.

Once the operating point reaches the peak, we observe a current jump to point C and then a decay toward the intercept at A':

$$I_{A'} = 60\ \mathrm{mA} \qquad V_{A'} = 10\ \mathrm{V}$$

The new decay-time constant is $\tau_3 = (R_{\mathrm{III}} \parallel R_L)C = 125\ \mathrm{ns}$. In the 97 ns remaining out of the 500-ns pulse, the voltage decays from the 20-V peak to point A'', where

$$V_{A''} = 10 - (10 - 20)e^{-97/125} = 14.6\ \mathrm{V}$$

$$I_{A''} = 78.5\ \mathrm{mA}$$

At the end of the trigger pulse, the load line goes back to its original position. The path of operation is now down toward point D' with the time constant τ_3:

$$I_{D'} = 30\ \mathrm{mA} \qquad V_{D'} = 2.5\ \mathrm{V}$$

After the operating point runs down to the valley point D, we observe another current jump to E. The time required for the rundown is

$$t_3 = 125 \ln \frac{2.5 - 14.6}{2.5 - 5} = 198\ \mathrm{ns}$$

After the final jump, recovery is back toward the stable point A with the time constant τ_1.

Figure 11-10*b* shows the current and voltage waveshapes generated in this circuit. Note that the initial trigger must be wide enough to permit the operating point to transfer from region I to region III. Triggers less than 403 ns wide will not cause triggering.

We might further note that the ratio of the recovery time to the pulse duration depends on the relative slopes of the two positive-resistance regions. For a fast pulse and a long recovery, the single stable point should be established in the high-resistance region, as it is in the above example. For a relatively long sweep and a fast recovery, the stable point should be shifted into the low-resistance region by increasing the bias voltage.

11-4 Composite Terminal Characteristics

Let us now examine the composite terminal response of the NR device and the external biased load resistor. By summing the load and device current over each region, at the same terminal voltage, we obtain the composite characteristic:

$$i_c = \frac{v_n - V_{nn}}{R} + i_n \tag{11-14}$$

The first term of Eq. (11-14) is the biased load current, and the second term is the device current.

Figure 11-11 shows how the two currents may be summed graphically. In Fig. 11-11*b*, the parallel bias path is drawn as both a load line and a parallel biased resistor. At every value of v_n, we can sum the path currents. For example, at V_A the current into the CNR is positive and exactly equal in magnitude to the negative current through R. The net composite current is zero. The new region I′ line intersects the voltage axis at V_A. In a similar manner, the currents in the CNR and R can be summed to yield the composite characteristic of Fig. 11-11*c*. Since these are the volt-ampere characteristics seen by the external energy-storage element, they will determine the circuit's stability as well as the nature of the switching response.

To obtain the bistable configuration shown in Fig. 11-11, the external load line must have three intersections with the CNR characteristic. This requires that the external load line have a slope less than the slope of the negative-resistance region, which implies that

$$R < |R_n|$$

When this condition is satisfied, the parallel combination of R and R_n will be positive, as shown in region II′ of the composite characteristic. Because all three regions now have a positive slope, the composite characteristic is double-valued in both voltage and current. The stable points are the three intersections with the voltage axis. The middle region is, however, inaccessible. On a rising voltage, we traverse the path AE, jump to F,

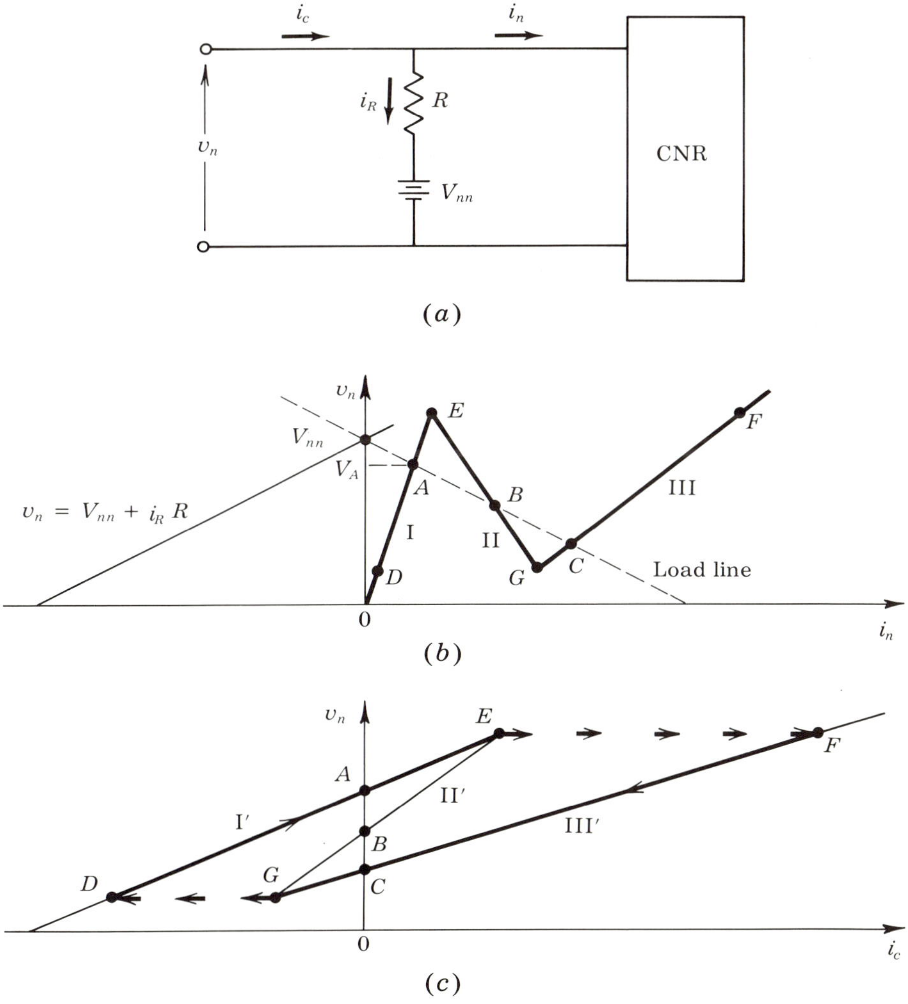

Fig. 11-11 Composite characteristic of a bistable circuit.

and then go down segment III′ to the stable point C. On a falling voltage, the operating path is CG, a jump to D, and then up along segment I′. There is no way of traversing segment II′.

The hysteresis inherent in a characteristic such as the one of Fig. 11-11 is typical of all bistable circuits. The fact that all regions have positive slopes implies that all of the energy needed to switch from one accessible stable point to the other must come from the external trigger. The capacitor is not essential for bistable operation; it only enables us to control the transition-time constants in regions I′ and III′.

An astable circuit's composite characteristic is shown in Fig. 11-12. Here the single intersection with the voltage axis lies along the negative-resistance region. When the external capacitor is present, the path of operation is $BCDE$. Within regions I′ and II′, the Thévenin equivalent voltages are found under the $i_c = 0$ condition.

Time constants are the products of the timing capacitor and the slopes of the composite characteristics.

More complex composite characteristics can be obtained by paralleling two identical CNR devices which we shall call X and Y (Fig. 11-13). As the current increases, the operating points of both devices travel up region I toward the peak. Suppose device X peaks first at a slightly lower voltage than Y. A further increase of the drive current causes its operating point to move down along region II. The decrease in terminal voltage forces the operating point of device Y back down region I. Device X traces out regions II and III up to the peak voltage of device Y, while Y again moves up along region I. After the peak is reached, device X remains in region III while Y traces out regions II and III. On the composite curve of Fig. 11-13b, the corresponding regions of devices X and Y are indicated.

The composite characteristic of two negative-resistance devices exhibits two negative-resistance regions bounded by three positive-resistance regions.

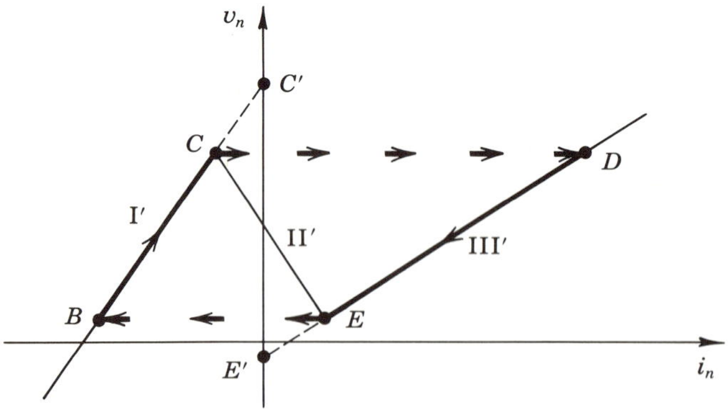

Fig. 11-12 Astable composite characteristic.

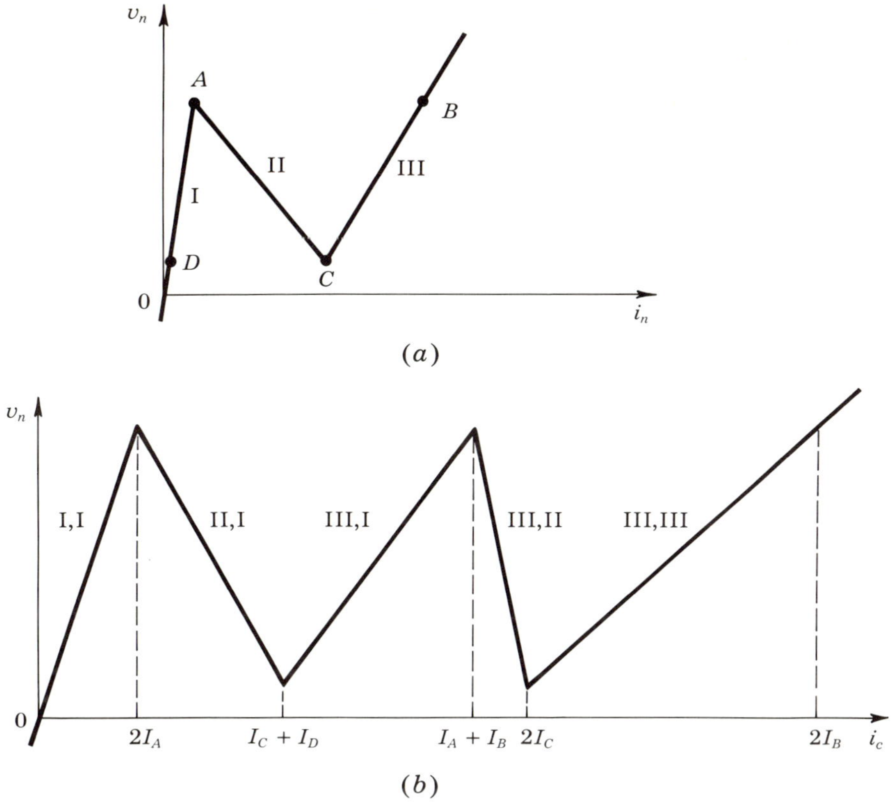

Fig. 11-13 Composite of two negative-resistance devices. (*a*) Volt-ampere characteristic of a single device; (*b*) parallel combination of two identical devices, indicating the regions of operation of each device and the currents of all peaks and valleys.

With the appropriate load and with controlled-amplitude triggers, *tristable operation* becomes possible. Each additional parallel device will introduce another possible region of operation. Eventually the shunting of a negative-resistance region by all the parallel positive resistances will cause the appearance of a region multivalued in both voltage and current (identical to that of Fig. 11-11).

11-5 Voltage-controlled NR Switching Circuits

The basic switching circuit associated with the VNR is simply the dual of the one discussed in Sec. 11-2. Figure 11-14 illustrates the resultant series switching circuit together with the graphical construction holding when it is adjusted for monostable operation.

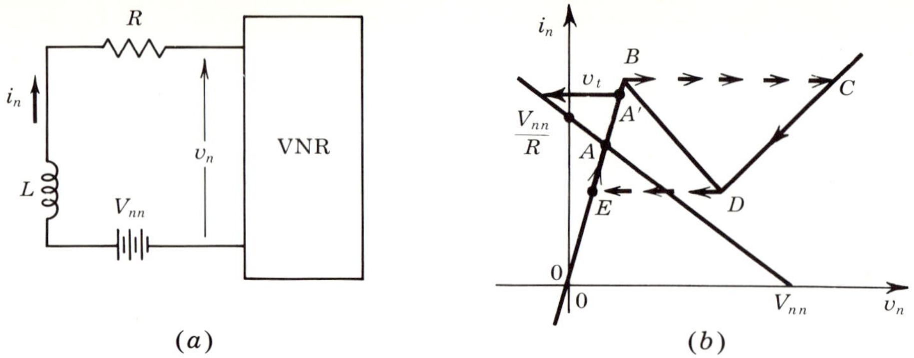

Fig. 11-14 (*a*) Basic VNR switching circuit; (*b*) monostable path of operation.

By properly arranging the terms, we can write the input-loop equation in a form that will represent the steady-state solution as well as indicate the permissible time variation:

$$(V_{nn} - i_n R) - v_n = L\frac{di_n}{dt} = \mathbf{v}_t \tag{11-15}$$

It follows from Eq. (11-15) that the intersection of the load line $(V_{nn} - i_n R)$ with the device characteristics at A is the only possible stable operating point $(di_n/dt = 0)$. Any perturbation from this point creates a time rate of change of current that forces the operating point back to the original stable point (point A in Fig. 11-14*b*). Thus the path of operation, the location of the stable points, the calculation of the minimum trigger amplitude, and the calculation of circuit waveshapes are similar to those discussed for the CNR. The major difference is that now the current is constrained to change slowly, but this only supports the duality relationship, i.e., the interchange of the circuit's voltage and current responses.

The second important difference in the ideal VNR concerns the condition which must be satisfied for absolute system stability. Here the external load appears in series with any internal incremental resistance. In order to prevent the system pole from migrating into the right half plane, the total series combination must remain positive:

$$R_s = R + (-R_n) > 0$$

or

$$R > R_n \tag{11-16}$$

This stability condition holds only for the ideal VNR circuit. A practical device differs in important respects from the ideal, and as with the CNR, the stability criterion will be reexamined in Sec. 11-9.

The composite characteristic of a VNR and the external load is found by summing the voltage drops $(v_c = i_n R + v_n - V_{nn})$ at the same terminal current. The shape of the resultant curve will be the same as that found for the CNR. Because an inductance is used for switching, the current axis becomes the load line ($v_c = 0$).

11-6 NR Characteristics—Collector-Base-coupled Monostable Multivibrator

As an example of how all single-energy-storage-element switching circuits may be treated from the viewpoint of their NR characteristics, we shall now reexamine the collector-base-coupled monostable multivibrator of Chap. 8. Since the logical place to look for the negative resistance is across the energy-storage-element terminals, Fig. 11-15*a* shows the basic circuit with the timing capacitor omitted. In our analysis we shall make use of the equivalent circuit drawn in Fig. 11-15*b*. Both transistors are shown in their active region, and as each is driven into either cutoff or saturation, the model must be accordingly modified.

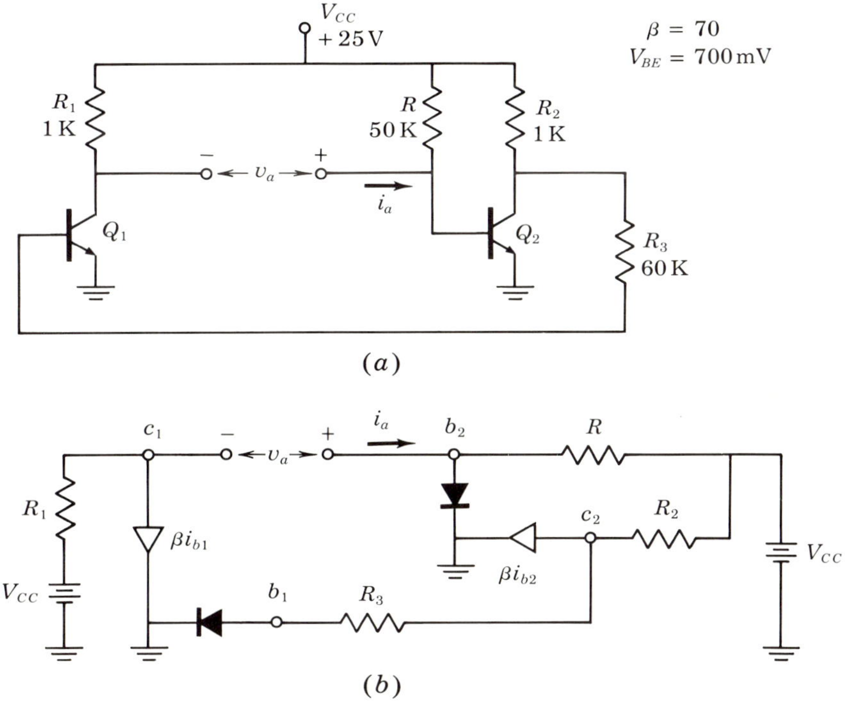

Fig. 11-15 (*a*) Monostable multivibrator; (*b*) equivalent circuit. The timing capacitor is normally connected across the driving-point terminals.

The circuit is initially adjusted so that in its normal state Q_2 is fully on and Q_1 is cut off. Even though the storage time of the saturated transistor limits the switching time, this configuration is considered because of its inherent simplicity. Further simplification arises from the following conditions:

1. β is very large; therefore $\beta + 1 \approx \beta$.
2. $R \gg R_2$, and $R_3 \gg R_2$.
3. $V_{CE,\text{sat}} \approx 0$.

Generally, three or four line segments are sufficient to define the complete voltage-ampere characteristic. In finding these segments, attention should be focused on the controlling current i_a. By starting from a high positive value of input current and considering the permissible changes of state as the current decreases to, and below, zero, the possible operating regions are:

Region 1. Q_2 saturated and Q_1 off
Region 2. Both transistors in their active region
Region 3. Q_2 active and Q_1 saturated
Region 4. Q_2 off and Q_1 active
Region 5. Q_2 off and Q_1 saturated

At most, two of the last three regions can exist. Once Q_1 saturates, the decreasing input current may still drive Q_2 through its active to its cutoff region. Thus there is the possibility of operating over regions 3 and 5. On the other hand, if Q_2 is driven into cutoff while Q_1 is still active, the final bounding region will be region 4.

In each region, we shall solve for the equation of v_a as a function of i_a, in the form

$$v_a = V_i + R_i i_a$$

V_i is the voltage intercept found at $i_a = 0$, just as if the defined region remained invariant. R_i is the incremental driving-point resistance in the region under examination. A sequence of such straight lines presents, in a piecewise manner, the complete volt-ampere characteristic. In order to find the boundaries of the individual segments, we solve for the coordinates of the intersection of the appropriate two lines. We might also consider the physical limitations imposed by the active elements employed in establishing these bounds.

Region 1—Q_2 saturated and Q_1 off

Here $i_{b1} = 0$ and $v_{c2} = 0$, and hence

$$v_a = (V_{BE} - V_{CC}) + i_a R_1 \tag{11-17}$$

The lower limit of this region, where both Q_1 and Q_2 become active, occurs once

$$v_{c2} = V_{CC} - \beta i_{b2} R_2 \approx 0 \tag{11-18}$$

where

$$i_{b2} = \frac{V_{CC} - V_{BE}}{R} + i_a \tag{11-19}$$

From Eqs. (11-18) and (11-19), we find that this zone is limited to

$$i_a > \frac{V_{CC}}{\beta R_2} - \frac{V_{CC} - V_{BE}}{R}$$

and since $R < \beta R_2$, region 1 holds down to some negative value of i_a.

If, on the other hand, the circuit were designed so that Q_2 were active ($R < \beta R_2$), region 1 would end at a positive i_a and the negative-resistance region would intersect the voltage axis. The circuit would operate as an *astable* instead of as a monostable multivibrator.

Region 2—Q_1 and Q_2 active

From the solution of the circuit's loop equation, when i_{b2} is given by Eq. (11-19) and $i_{b1} = (v_{c2} - V_{BE})/R_3$,

$$v_a = (V_{BE} - V_{CC})\left[1 - \frac{\beta R_1}{R_3}\left(1 - \frac{\beta R_2}{R}\right)\right] - \left(\beta^2 \frac{R_1 R_2}{R_3} - R_1\right) i_a \tag{11-20}$$

Note that once a transmission path involving the gain of both transistors exists, the slope of the volt-ampere characteristic becomes negative. Here the positive feedback provides the required energy source. This negative-resistance region is bounded when Q_2 turns off at $i_{b2} = 0$. The corresponding constraints on i_a are

$$\frac{V_{CC}}{\beta R_2} - \frac{V_{CC} - V_{BE}}{R} \geq i_a \geq -\frac{V_{CC} - V_{BE}}{R} \tag{11-21}$$

Region 3—Q_2 active and Q_1 saturated

The conducting emitter-base diode of Q_2 and the saturated Q_1 now short-circuit the input terminals. The input equation becomes

$$v_a = V_{BE} \tag{11-22}$$

Region 4—Q_2 off and Q_1 active

Under the conditions holding in this region,

$$i_{b1} \approx \frac{V_{CC} - V_{BE}}{R_3}$$

and therefore

$$v_a = \beta(V_{CC} - V_{BE})\frac{R_1}{R_3} + i_a(R + R_1) \tag{11-23}$$

Q_1 finally becomes saturated at

$$i_{a,\text{sat}} = -\beta \frac{V_{CC} - V_{BE}}{R_3} + \frac{V_{CC}}{R_1}$$

Region 5—Q_2 off and Q_1 saturated

Q_2 is now an open circuit, and Q_1 is a short circuit. The defining equation becomes

$$v_a = V_{CC} + i_a R \tag{11-24}$$

Example 11-2

In this example we shall calculate and plot the characteristic for the typical transistor monostable multivibrator of Fig. 11-15*a*.

As the first step, the defining equation for each region is evaluated:

Region 1	$v_a = -24.3 + (1\text{ K})i_a$	Eq. (11-17)
Region 2	$v_a \approx -35.6 - (81\text{ K})i_a$	Eq. (11-20)
Region 3	$v_a = 0.7$	Eq. (11-22)
Region 4	$v_a = 28.4 + (51\text{ K})i_a$	Eq. (11-23)
Region 5	$v_a = 25 + (50\text{ K})i_a$	Eq. (11-24)

By plotting the five straight lines, as in Fig. 11-16, we can determine the permissible circuit states. As the current decreases, we note that the segments traversed are 1, 2, 3, and 5. There is no possible way to reach line segment 4, and it therefore represents a forbidden mode of operation for this particular circuit.

One of the two remaining coordinates of interest is the location of the valley point. To find the intersection of lines 1 and 2, we equate (11-17) to (11-20), with the result

$$I_v = -\frac{35.6 - 24.3}{81\text{ K} + 1\text{ K}} = -0.138\text{ mA}$$

$$V_v = -24.44\text{ V}$$

The current intercept of regions 5 and 3 is found from Eq. (11-24) by setting $v_a = V_{BE}$:

$$I_p = -\frac{25 - V_{BE}}{50\text{ K}} = -0.486\text{ mA}$$

All the resistors have been included in the base amplifier, and consequently the voltage axis is also the steady-state load line. The circuit remains at its single stable point ($i_a = 0$, $v_a = 24.3$ V) until a trigger forces it into its unstable state. Transition is directly from region 1 to region 5. Here the slope is simply the timing resistance (50 K), and it is here that the output pulse is generated. When the circuit switches back to segment 1, it enters the low-resistance (1-K) recovery portion of the cycle.

Since the load line is the voltage axis, the time constants are the product of the slope of the line segment and the timing capacitor. Moreover, the steady-state charging voltage is the voltage intercept; the steady-state current must be zero. All the information necessary to calculate the voltage across, and the current through, C is contained in the volt-ampere characteristic of Fig. 11-16.

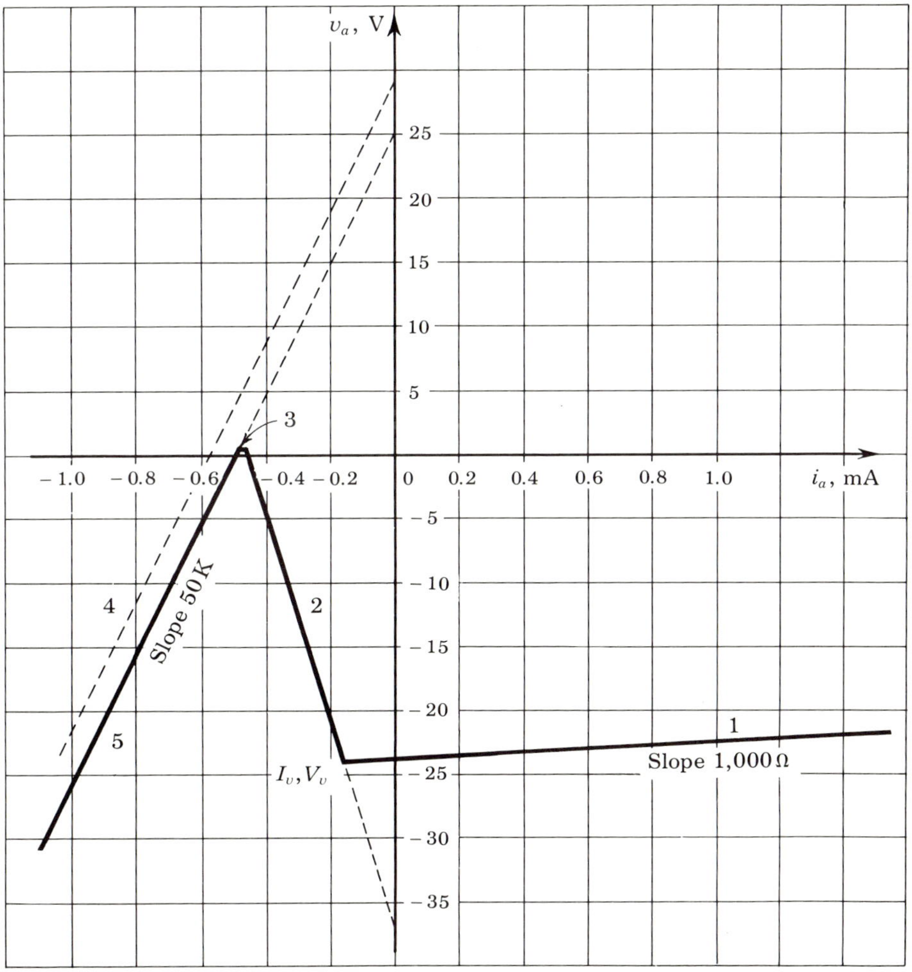

Fig. 11-16 CNR volt-ampere characteristic of the collector-base-coupled monostable multivibrator of Example 11-2.

11-7 Negative-impedance Converters

All negative-impedance converters (NIC) use positive feedback to develop a driving-point impedance that is the negative of one of the passive elements used in the feedback loop. If the feedback reference element is a resistor, the two-terminal characteristic is a negative resistance. A negative input capacitance or negative inductance results when we replace the feedback resistor by a capacitor or an inductance.

Because they are the most useful, the discussion below will be limited to the production of negative-resistance characteristics.

Either a voltage or a current amplifier may be used as the active device. Positive-current feedback creates the current-controlled negative-driving-point impedance, while voltage feedback is used to obtain a voltage-controlled negative impedance.

CNR characteristics

Figure 11-17 illustrates how positive-current feedback is used with an ideal current amplifier to produce a negative input impedance. The total input current flows through the short-circuit input of the current amplifier and through the feedback element Z. The output current also flows in Z. The input voltage may be expressed in terms of the input controlling current as

$$v_1 = (i_1 - A_c i_1)Z = (1 - A_c)Zi_1$$

and we recognize the driving-point impedance to be

$$Z_{\text{in}} = (1 - A_c)Z$$

If the current amplification within the active region is positive and greater than unity, the input impedance becomes negative. Under the particular circumstances where A_c is set equal to $+2$, we obtain the very convenient result

$$Z_{\text{in}} = -Z$$

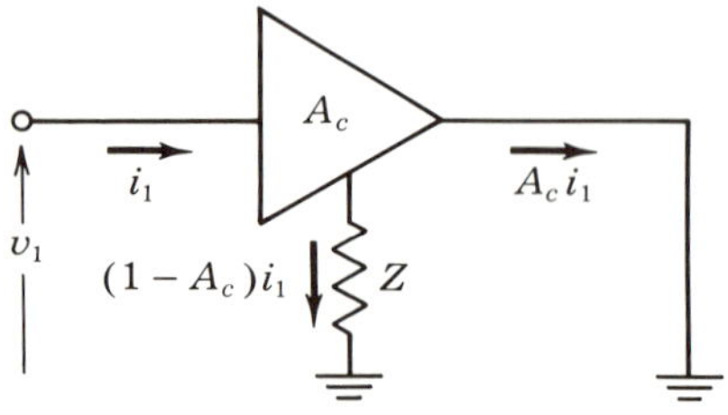

Fig. 11-17 Basic current-feedback negative-impedance converter (NIC).

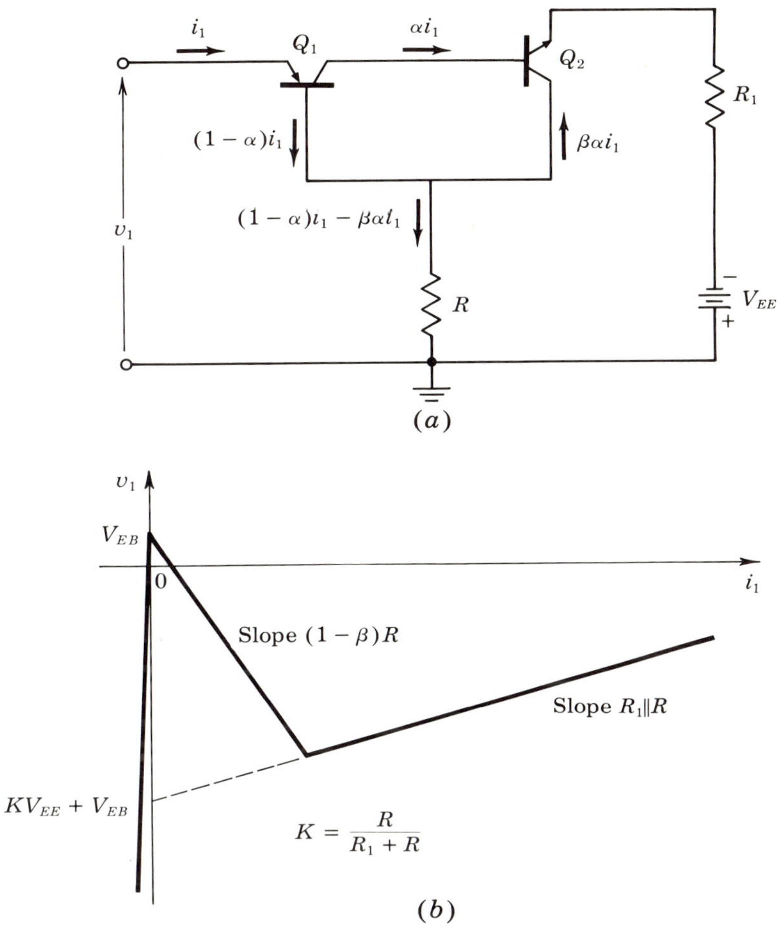

Fig. 11-18 (*a*) Practical current-feedback NIC using a *pnp-npn* pair for the amplifier; (*b*) input volt-ampere characteristic showing the CNR region.

Figure 11-18*a* shows a practical way of realizing the circuit of Fig. 11-17. For convenience in biasing, the current amplifier uses a *pnp* input stage followed by an *npn* output stage. For negative currents, the input transistor Q_2 is cut off. The reverse-biased emitter-base junction can support the small reverse current only for voltages less than the avalanche breakdown voltage. A positive input current turns on transistor Q_1. Using the active-region currents shown in Fig. 11-18*a*, the input voltage becomes

$$v_1 = V_{EB} + (1 - \beta)Ri_1 \tag{11-25}$$

Eventually both transistors saturate, terminating the negative-resistance region. By drawing the model holding, we can see that in the saturation region the volt-ampere characteristic is defined by

$$v_1 = \frac{R}{R_1 + R} V_{EE} + V_{EB} + i_1(R_1 \parallel R) \tag{11-26}$$

The current and voltage coordinates of the valley point can be found from the simultaneous solution of Eqs. (11-25) and (11-26). The total input volt-ampere characteristic is shown in Fig. 11-18*b*.

The value of the negative resistance developed by the preceding circuit is directly dependent on the gain of the current amplifier. This undesirable situation can be corrected by modifying the required defining equation. When we change it from

$$v_n = (1 - A_c)Ri_n$$

to

$$v_n = \frac{1 - A_c}{A_c} Ri_n \approx -Ri_n \tag{11-27}$$

the magnitude of the negative resistance becomes independent of the active element. All we require is a high-gain current amplifier to convert any positive impedance to an equal-magnitude negative-impedance element.

But how do we begin to synthesize the function given in Eq. (11-27)? By first multiplying out Eq. (11-27) and rearranging terms, we eventually obtain

$$\frac{i_n}{A_c} = \frac{v_n}{R} + i_n \tag{11-28}$$

If we now identify the left-hand side of Eq. (11-28) as the input current to the amplifier, then the terms on the right-hand side are the current components which must be fed back from the output to the input to create the negative impedance. Since, in this case, v_n and i_n are the output voltage and current, the negative impedance generated will appear as the amplifier's output impedance.

We may not be able to synthesize a circuit in exactly the manner defined by Eq. (11-28), but from this equation we can see the form of the required feedback terms. One of the input currents must be proportional to the driving-point current, and the other current must be proportional to the developed terminal voltage.

Figure 11-19*a* shows a practical circuit where this combination of voltage and current feedback is used to develop a CNR. The positive-current feedback through resistor R_2 determines the nature of the negative resistance, while the negative feedback through resistor R_1 makes the characteristic independent of the amplifier gain. In analyzing this circuit,

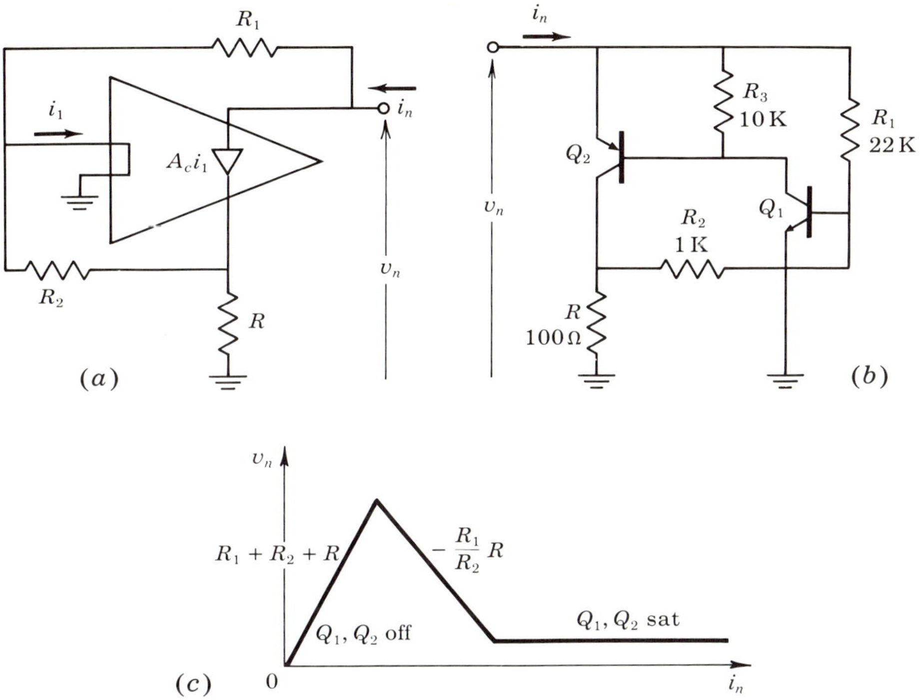

Fig. 11-19 (*a*) Basic CNR-NIC feedback circuit; (*b*) practical circuit, base-amplifier current gain $\sim\beta^2$ (R_3 is used only for circuit stabilization); (*c*) volt-ampere characteristic.

we can make the simplifying assumptions

$$i_n \approx A_c i_1 \qquad R_2 \gg R$$

Then, in the active region,

$$i_1 = \frac{v_n}{R_1} + \frac{A_c i_1 R}{R_2}$$

yielding

$$v_n = \frac{1 - A_c(R_1/R_2)}{A_c} R i_n \tag{11-29}$$

When we also insist that $1 \ll A_c R_1/R_2$, Eq. (11-29) reduces to

$$v_n = -\frac{R_1}{R_2} R i_n$$

So long as the base-amplifier gain is very high, the effect of the negative-voltage feedback through R_1 is to eliminate the circuit dependency on the amplifier gain. The value of the negative resistance is determined solely by the resistor values.

CNR–voltage-amplifier synthesis

CNR characteristics can also be synthesized by using current feedback in combination with a voltage amplifier (Fig. 11-20). In the active region, the output voltage is given by

$$v_n = A_v v_1 + R i_n$$

Since $v_1 = -R i_n$, we obtain

$$v_n = R(1 - A_v) i_n \tag{11-30}$$

which we recognize as the equation of a current-controlled negative resistance.

Any two-stage amplifier having a positive gain greater than 1 can be used. Figure 11-20 illustrates a suitable two-stage MOSFET amplifier. The resistive divider R_1 and R_2 sets the quiescent operating condition.

Voltage-controlled negative resistance

To obtain a voltage-controlled negative resistance, we turn to voltage feedback. The basic amplifier may be either a current or a voltage amplifier. In the case of a current amplifier, the feedback results in a negative output impedance (Fig. 11-21*a*). From the basic circuit we note that the feedback current is proportional to the output voltage v_n:

$$i_1 = \frac{v_n}{R}$$

The driving-point current i_n may be written as

$$i_n = i_1 - A_c i_1 = (1 - A_c) \frac{v_n}{R} \tag{11-31}$$

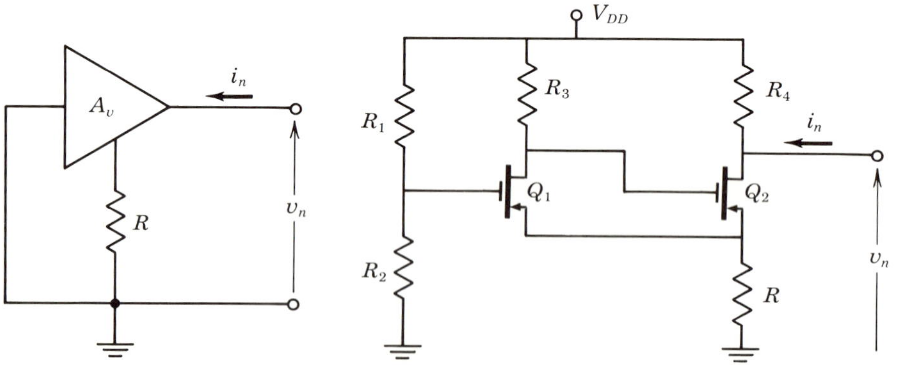

Fig. 11-20 Basic CNR-NIC using a voltage amplifier for the active element and a possible MOSFET circuit.

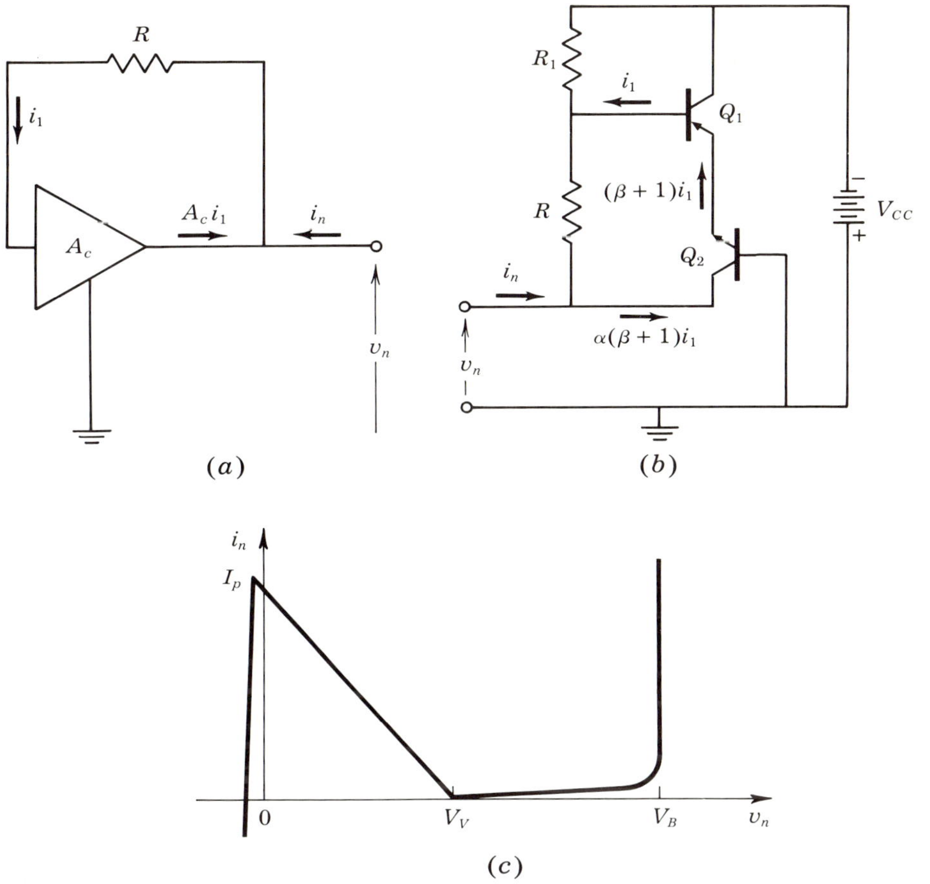

Fig. 11-21 Negative-impedance converter—VNR. (*a*) Basic voltage-feedback circuit; (*b*) practical circuit; (*c*) VNR volt-ampere characteristic.

indicating the presence of a voltage-controlled negative conductance when $A_c > 1$:

$$G_n = G(1 - A_c)$$

Figure 11-21*b* shows a practical circuit for obtaining the VNR characteristics of Fig. 11-21*c*. This circuit is a variation of the basic current amplifier of Fig. 11-18*a*. In order to obtain the ideal current-source output seen at a collector, we reverse the sequence of amplification by the two transistors. Here the first stage supplies the current gain of $\beta + 1$, while the second stage has only the gain of α. The overall current gain remains the same: β.

For a negative terminal voltage, Q_2 is saturated. The current flows through the conducting base-collector diode and is limited only by the external circuit. At $v_n = -V_{BE}$, the output transistor starts becoming active. The base current of Q_1 is given by

$$i_1 = \frac{V_{CC} - 2V_{BE}}{R_1} - \frac{v_n + 2V_{BE}}{R} \tag{11-32}$$

In the active region, the driving-point current may be expressed as

$$i_n = \beta i_1 + \frac{v_n + 2V_{BE}}{R} \tag{11-33}$$

Combining Eqs. (11-32) and (11-33), we obtain

$$i_n \approx \beta\left(\frac{V_{CC}}{R_1} - \frac{2V_{BE}}{R \parallel R_1}\right) - \frac{\beta v_n}{R} \tag{11-34}$$

The first term of Eq. (11-34), the current intercept, is very close to the peak value. The second term shows the voltage-controlled negative conductance.

At the valley point, both transistors are cut off. Setting $i_n = 0$ in Eq. (11-34),

$$V_v = R\left(\frac{V_{CC}}{R_1} - \frac{2V_{BE}}{R_1 \parallel R}\right)$$

For voltages above the valley point, the current remains approximately zero. At a high enough voltage, the back-biased collector-base junction of Q_2 breaks down. Avalanche multiplication causes the current to increase until limited by the external circuit.

The second basic technique for obtaining the VNR is illustrated in Fig. 11-22. Voltage feedback is applied around a voltage amplifier, yielding the driving-point equation

$$i_n = \frac{1 - A_v}{R} v_n \tag{11-35}$$

Any two-stage amplifier having a gain greater than 1 will create a negative conductance. One possibility is the MOSFET amplifier used in the circuit of Fig. 11-20. The feedback resistor would be connected from the drain

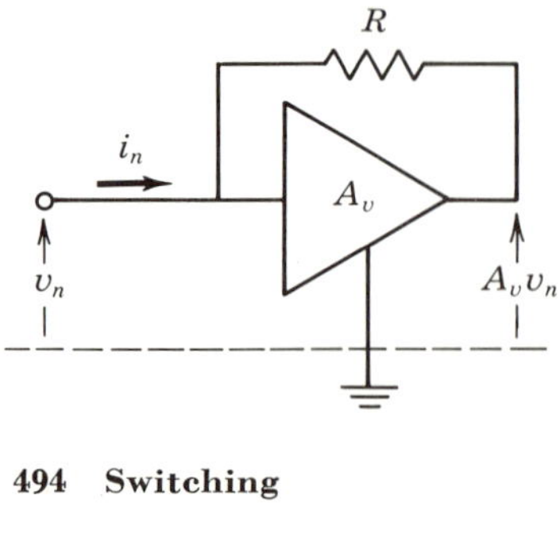

Fig. 11-22 VNR input characteristic obtained through the use of voltage feedback.

of Q_2 back to the gate of Q_1, and the sources of both FETs would be grounded.

FET–bipolar VNR

The interesting FET–bipolar-transistor combination circuit of Fig. 11-23 has the necessary voltage feedback inherent in the gate voltage control over the drain current and hence over the collector current of the bipolar transistor. For positive voltages, the collector-base junction becomes forward-biased. Current is limited solely by the external circuit. With a negative voltage, the bipolar transistor becomes active. The FET gate-to-source voltage is

$$V_{GS} = v_n - V_{EB} \tag{11-36}$$

The drain current and consequently the collector current will follow the FET square-law response. The terminal current becomes

$$i_n = -\alpha I_D = -\alpha I_{DSS}\left(1 - \frac{v_n - V_{EB}}{V_p}\right)^2 \tag{11-37}$$

I_D is a maximum at $v_n = V_{EB}$ and decreases to zero as V_{GS} approaches pinchoff. The volt-ampere characteristic has a negative slope.

Beyond pinchoff, the FET is essentially an open circuit. The terminal current drops to the very small reverse current of the collector-base junction.

The negative conductance will be multiplied by β when the emitter and base terminals are interchanged. The peak current increases to βI_{DSS}.

Feedback-stabilized VNR

Through the use of a combination of positive-voltage and negative-current feedback, the VNR output characteristics may be made independent of the amplifier gain. This is the dual of the stabilization technique

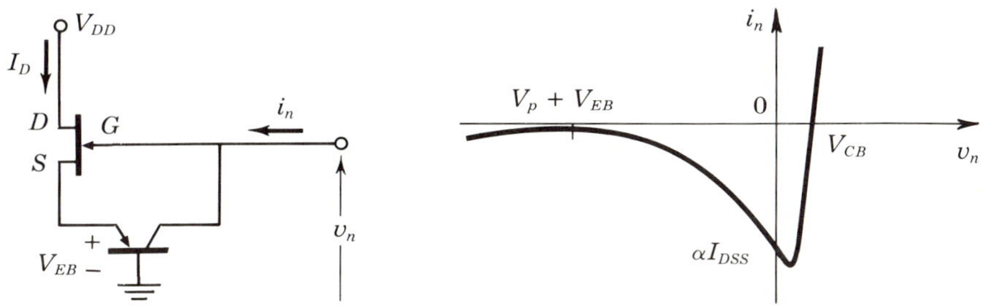

Fig. 11-23 Composite FET–bipolar-transistor voltage-feedback amplifier and VNR input characteristic.

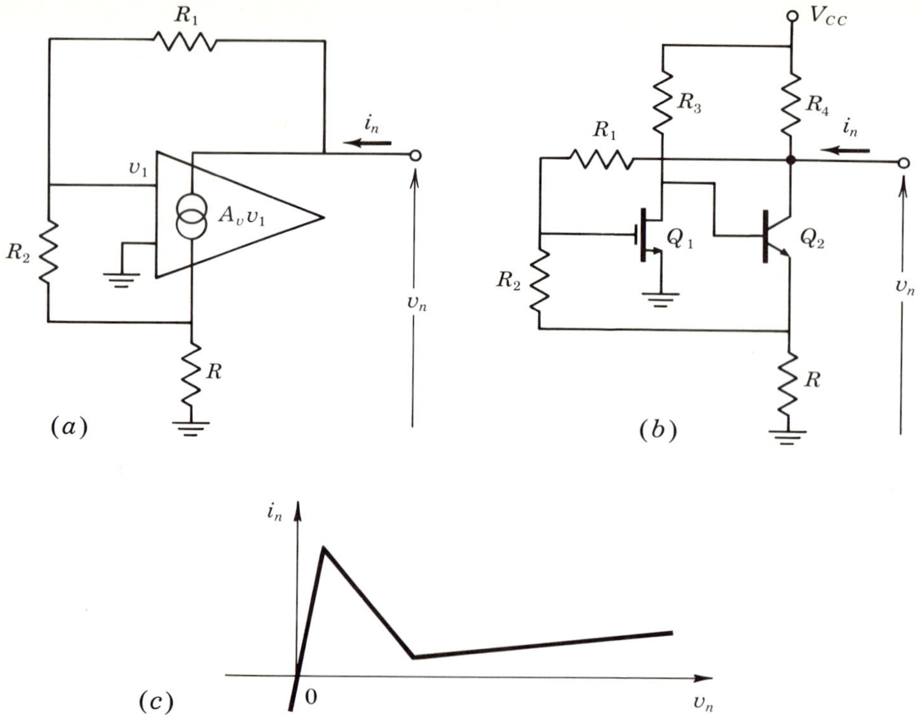

Fig. 11-24 (*a*) Basic circuit for obtaining a negative-feedback-stabilized VNR characteristic; (*b*) possible MOSFET-bipolar circuit; (*c*) VNR volt-ampere characteristic.

previously described for the CNR. From the circuit of Fig. 11-24*a*,

$$v_1 = Kv_n + (1 - K)i_n R \tag{11-38}$$

where for $R_1 + R_2 \gg R$, $K = R_2/(R_1 + R_2)$, and

$$v_n \approx A_v v_1 + i_n R$$

Substituting into Eq. (11-38) and solving,

$$i_n = \frac{1 - A_v K}{1 + A_v(1 - K)} G v_n \tag{11-39}$$

where $G = 1/R$. For a high-gain amplifier where $A_v K \gg 1$, Eq. (11-39) reduces to

$$i_n \approx -\frac{R_2}{R_1} G v_n$$

which proves that the negative input conductance is relatively independent of the amplifier characteristics.

11-8 Devices Possessing NR Characteristics

Many semiconductor devices exploit internal-feedback mechanisms to produce negative-resistance regions. The internal feedback acts in a manner similar to the external feedback used in the negative-impedance converters of Sec. 11-7. In fact, the terminal characteristics can often be explained by multitransistor models. The inherent feedback mechanisms which operate within the devices that we shall consider include the avalanche current multiplication of a back-biased junction, conductivity modulation by the injection of carriers, and tunneling through a junction.

pnpn four-layer diode

A three-junction diode having the *pnpn* structure of Fig. 11-25*a* has an extremely sharp jump transition from the high-resistance OFF condition to the very low resistance of a normal diode when in the ON state. The internal mechanisms giving rise to the current-controlled characteristic of Fig. 11-25*c* can best be understood from the two-transistor equivalent model of Fig. 11-25*b*. In replacing the diode by the coupled *pnp-npn* transistor pair, the middle *n* and *p* layers are split in two. (Compare this equivalent circuit with the NIC of Fig. 11-18*a*. Except for the feedback resistor R and the internal power supply, they are identical.)

Since the four-layer diode is a two-terminal device, the net current through any junction must equal the driving-point current i_n. With a positive terminal voltage, junctions J_1 and J_3 are forward-biased and junction J_2 is back-biased. Summing the currents at J_2,

$$i_n = \alpha_1 i_n + \alpha_2 i_n + I_{co} \tag{11-40}$$

Solving for the terminal current,

$$i_n = \frac{I_{co}}{1 - (\alpha_1 + \alpha_2)} \tag{11-41}$$

We see that as $\alpha_1 + \alpha_2$ approaches unity, the terminal current increases without limit. This current multiplication is equivalent to internal current amplification.

Let us now examine how α varies with current and voltage (Fig. 11-26). At low emitter currents, α is quite small, increasing rapidly with I_e to its normal value. The increase of α with I_e is related to the reduction in importance of surface and space-charge recombination currents in the total base current. At higher current densities, there also exists a longitudinal electric-field component which helps the transport of minority carriers through the base.

By using at least one wide-base region, we can ensure that at a low input current, the sum $\alpha_1 + \alpha_2 \ll 1$. This bounds i_n.

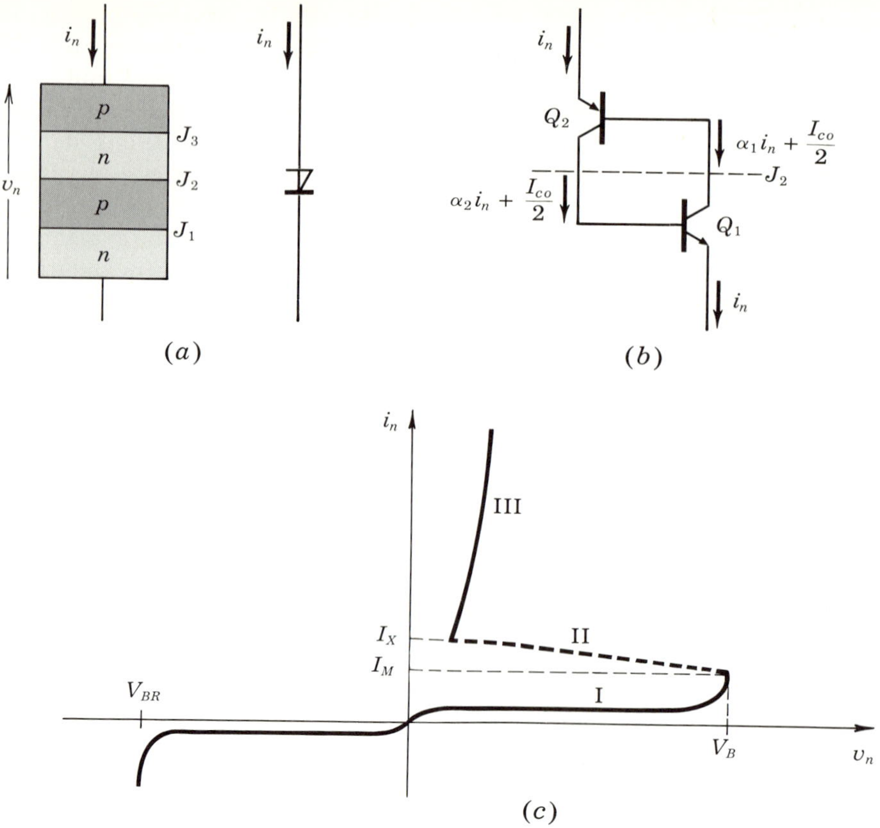

Fig. 11-25 (a) *pnpn* diode shown schematically and its symbol; (b) *pnp-npn* two-transistor equivalent model; (c) volt-ampere characteristic, indicating the current-controlled NR region. Region I is the high-resistance forward-blocking region, and region III is the low-resistance conducting-diode region.

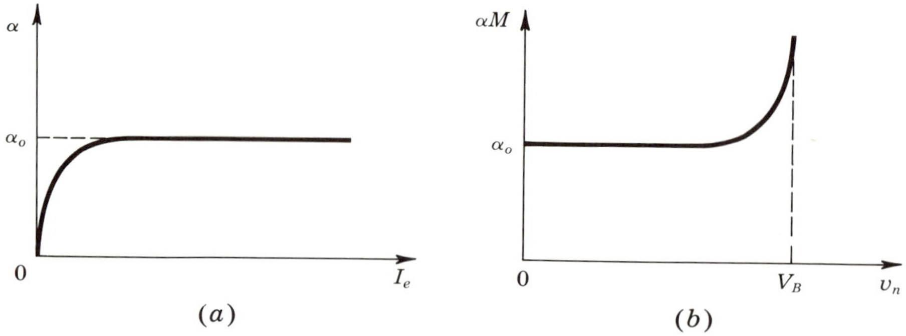

Fig. 11-26 (a) Variation of α with emitter current; (b) variation of α with back-biased junction voltage.

Any increase in the terminal voltage appears across the single back-biased junction J_2. The avalanche multiplication, which accounts for the increase in collector current at high fields, begins to become important. On being accelerated across the junction, the carriers gain enough energy to produce M additional electron-hole pairs upon colliding with the lattice. To account for this, the current i_n crossing the junction J_2 is multiplied by

$$M = \frac{1}{(1 - V/V_B)^\gamma}$$

M is 1 at low reverse-junction voltages and increases very rapidly toward ∞ in the vicinity of V_B.

V_B is the avalanche breakdown voltage. At this voltage, the collision-produced electron-hole pairs are sufficient to cause a self-sustained carrier flow. γ is a constant of the junction; it takes on values from 1.2 to 8, depending on the particular material and doping levels.

If we now multiply the right-hand side of Eq. (11-40) by M and solve, we obtain

$$i_n = \frac{MI_{co}}{1 - M(\alpha_1 + \alpha_2)} \tag{11-42}$$

At low voltages where $M = 1$ and $\alpha_1 + \alpha_2 < 1$, the current is quite low. Once the terminal voltage rises to near the junction breakdown voltage V_B, the increase in M causes the current function, Eq. (11-42), to become singular. The current through junction J_2 increases until it is limited by the external circuit. But the current increase causes both α_1 and α_2 to increase (Fig. 11-26*a*). Even after the terminal voltage drops and M is again unity, the rise in α maintains the singular condition of Eq. (11-42) and the high-current operation.

On switching into the high-current region, both transistors saturate and all the junctions become forward-biased. The drop across the *np* junction J_2 will almost cancel one of the *pn*-junction drops. The terminal response is of the exponential form expected from a single forward-biased silicon junction. Forward-conduction drops are approximately 700 to 1,000 mV.

The four-layer diode switches out of the low-resistance, high-current region when the current drops below a maintaining current I_x. This is the current at which $\alpha_1 + \alpha_2 \leq 1$.

Because the two α multiplication mechanisms are different, it is possible for the current at the peak to exceed the maintaining current I_x. For example, local heating due to operating in the high-current region can change the transistor parameters. In the case where $I_p > I_x$, the characteristic exhibits a hysteresis region; the negative-resistance region disappears.

In the reverse direction, we see two back-biased *pn* junctions in series. The characteristics are essentially the same as those of any other back-biased silicon diode.

Gas discharge tubes, such as a helium-neon laser or a neon bulb, have similar volt-ampere characteristics. The blocking voltage depends on the gas pressure and is above the ionization potential of the gas. After ionization, so long as the current flow is sufficient to maintain conduction, the voltage drop across the gas discharge remains constant at the ionization potential.

Silicon-controlled rectifier (SCR)

The SCR (also called a Thyristor) is a *pnpn* structure identical to that discussed above, with, however, an external connection to the base of the transistor. We call this base the gate terminal. A current injected into the gate flows through the forward-biased junction J_1 to the cathode. The net current through the middle junction J_2 must still remain equal to the anode current i_n. Summing the currents and solving, we find, as a result of the injection of I_G, that Eq. (11-42) is modified to

$$i_n = \frac{MI_{co} + M\alpha_1 I_G}{1 - M(\alpha_1 + \alpha_2)} \tag{11-43}$$

The injected gate current increases the emitter current of Q_1, and as a consequence, α_1 increases. The increase in the sum $\alpha_1 + \alpha_2$ means that the current singular point will be reached at a lower blocking voltage. The injection of a large gate current may even make $\alpha_1 + \alpha_2 > 1$ in a region where M is still unity. This will cause an immediate change of state into the conduction region.

Two modes of SCR operation are possible. In the first, the gate current is used as a fixed bias, establishing the desired breakpoint. When the anode voltage rises to the preset blocking voltage, the SCR "fires" and the current becomes limited by the external circuit. The second mode is where the blocking voltage is above the maximum expected anode voltage. The SCR remains in the blocked condition until an injected current pulse switches the circuit state into the low-resistance conduction region. The SCR will remain conducting even after the end of the trigger. A gate current of 50 mA is sufficient to switch 100 A in the anode circuit. In low-power SCRs, the maximum gate current needed for triggering will be only a few milliamperes. An SCR's turn-on time can be as short as 100 ns.

To turn off an SCR, the anode current must be reduced below the sustaining current I_x. Usually the anode circuit is opened or, with ac operation, the anode voltage reverses. Alternatively, the SCR can be momentarily

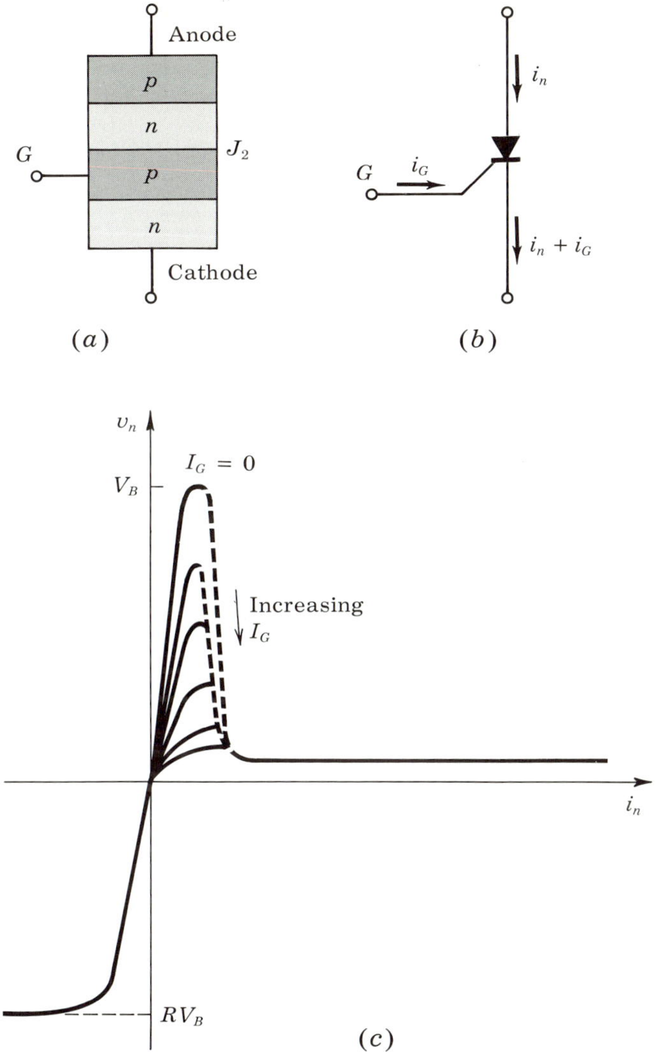

Fig. 11-27 (*a*) Schematic representation of an SCR; (*b*) SCR symbol; (*c*) volt-ampere characteristic.

shunted with a very low resistance and the current will transfer into the shunt path. Provided that the gate is back-biased, once the SCR turns off, it will stay off. The shunt path can then be opened.

Certain SCRs afford gate control over the turn-off process. A negative gate current reduces the magnitude of the numerator of Eq. (11-43), and even small negative currents can reduce it to zero. Any reduction in current will have a secondary effect in reducing α_1 of the transistor. Once the term

$M(\alpha_1 + \alpha_2)$ in the denominator of Eq. (11-43) becomes less than unity, the regenerative switching turns the SCR off. Negative gate currents will improve the turn-off of any SCR, but unless the SCR is specially designed, the turn-off gate current can be very large. When $\alpha_1 \gg \alpha_2$, turn-off current gains of 10 or less are possible, compared with turn-on current gains of 1,000 or more.

Triac

A symmetrical SCR, called a Triac or a double-ended Thyristor, is available for ac control applications. These devices conduct current equally well in both directions. They trigger symmetrically. The injection of a positive current triggers the device on when the anode voltage is positive. A negative trigger current flows in the gate when the anode voltage reverses.

Fabrication of a Triac results in the five-layer semiconductor sandwich shown in Fig. 11-28. We note that in one direction (reading down), we have a *pnpn* structure while in the other direction (reading up), we have a similar *pnpn* structure (using a different final *n* region). Turn-on and turn-off times are somewhat longer than in the nonsymmetrical SCR.

The unijunction transistor

In its simplest form, the unijunction transistor (UJT) consists of a bar of *n*-type silicon with a single *p* emitter junction located somewhat above the center (at about the 50 to 80 percent point). Figure 11-29 shows the device and its characteristic.

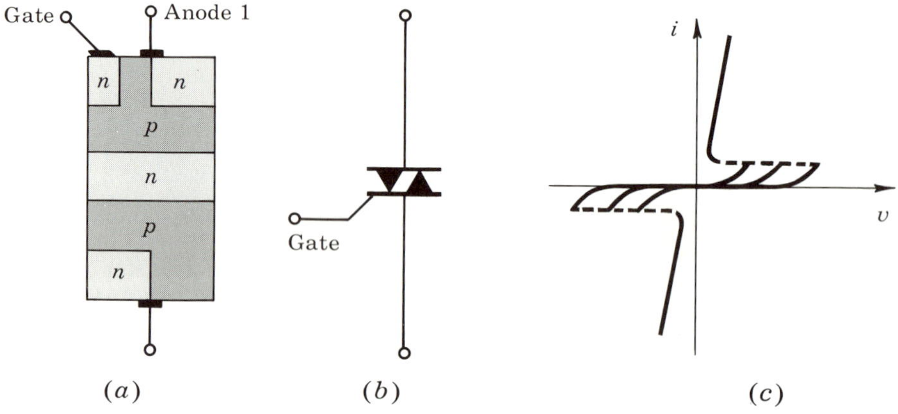

Fig. 11-28 A Triac (double-ended SCR). (*a*) Structure; (*b*) symbol; (*c*) volt-ampere characteristic.

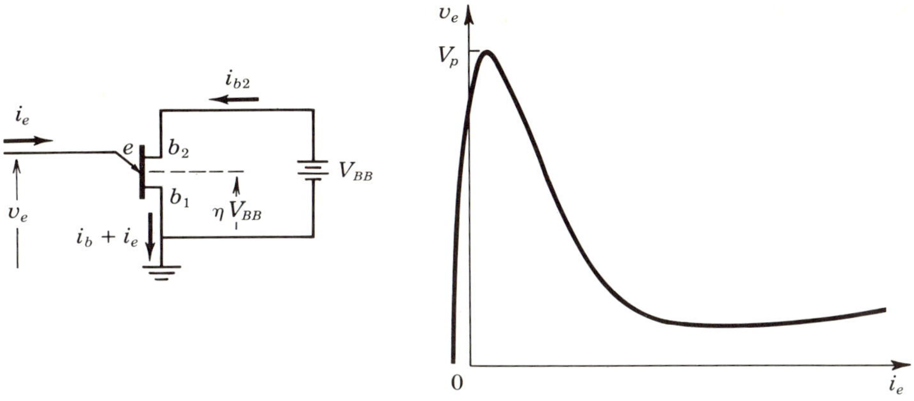

Fig. 11-29 Unijunction-transistor circuit and *vi* characteristic.

Under cutoff conditions, a reverse junction current of a few picoamperes flows out of the emitter. The conductive current flow from base 2 to base 1, through the high-resistivity n bar, establishes a back bias at the emitter proportional to the location of the junction. The emitter remains back-biased until the driving-point voltage exceeds the conductive drop.

At the peak voltage of

$$V_p = V_D + \eta V_{BB}$$

the emitter diode conducts. η is the intrinsic standoff ratio

$$\eta = \frac{r_{b1}}{r_{b1} + r_{b2}}$$

where r_{b2} and r_{b1} are the resistances of the top and bottom portions of the silicon bar. The emitter current at the conduction point is very small (less than 1 μA). The corresponding diode drop will be about 400 mV.

Holes injected into the n bar from the emitter increase the minority-carrier concentration in the region between the emitter and base 1. Since the conductivity is proportional to the number of available charge carriers, the resistance r_{b1} decreases with increasing current. We have entered the negative-resistance region. The increase in emitter current is accompanied by a decreasing terminal voltage.

The negative-resistance region is bounded when the hole concentration is approximately 10^{16} holes per cm^3. At this level, the hole mobility begins to decrease, with a further increase in the carrier concentration. Base-1 conductivity remains almost constant.

We can represent the base-1 resistance by two components, a small saturation resistance R_s in series with a controlled negative resistance. The terminal impedance is written as

$$R_{\text{in}} = R_s + (1 - \gamma)R_n \tag{11-44}$$

The factor γ is the conductivity modulation term and is a function of the mobility. It varies from about 3 or 4 at low carrier concentrations to unity at the limiting saturation condition.

A large measure of control over the slopes and intercepts of the individual regions is afforded by the insertion of additional resistance in series with b_1 and b_2. Any resistance we add to b_1 has the most pronounced effect on the saturation region, while that included at b_2 mainly increases the magnitude of the negative input resistance.

Currently available unijunction transistors are limited to switching applications below 2 MHz. The correspondingly large input inductive component of the input impedance (see Sec. 11-9) will adversely affect the switching path even at low frequencies. Rather than the expected large instantaneous current jumps from the peak and valley, the path of operation becomes almost elliptic.

Avalanche-region devices

We have already discussed avalanche current multiplication at a single back-biased junction. This is exactly the situation in a transistor operating in the common-base configuration. The collector avalanche breakdown voltage, BV_{CBO}, is defined under the condition $I_e = 0$. The total collector current

$$I_c = M\alpha I_e + MI_{co} \tag{11-45}$$

increases very rapidly in the vicinity of BV_{CBO}. But since the increase in current is not accompanied by a decrease in voltage, this is not a negative-resistance region.

Let us now examine what happens when the transistor is operated in the common-emitter configuration. The collector current is still given by Eq. (11-45). Substituting $I_e = I_b + I_c$ and solving,

$$I_c = \frac{M\alpha I_b}{I - M\alpha} + \frac{I_{co}}{I - M\alpha} \tag{11-46}$$

The first consequence of the common-emitter configuration is that the collector-emitter singular voltage (called BV_{CEO}) is appreciably lower than the avalanche breakdown, BV_{CBO}. Equation (11-46) defines a family of characteristics, each breaking sharply near BV_{CEO} (see Fig. 11-30).

Suppose, however, that the external base circuit can sustain a negative base current. When $I_b = -I_c$, I_e must be zero. The transistor is effectively operating as a common-base device. As we see in Fig. 11-30, collector current will now break sharply upward at BV_{CBO}.

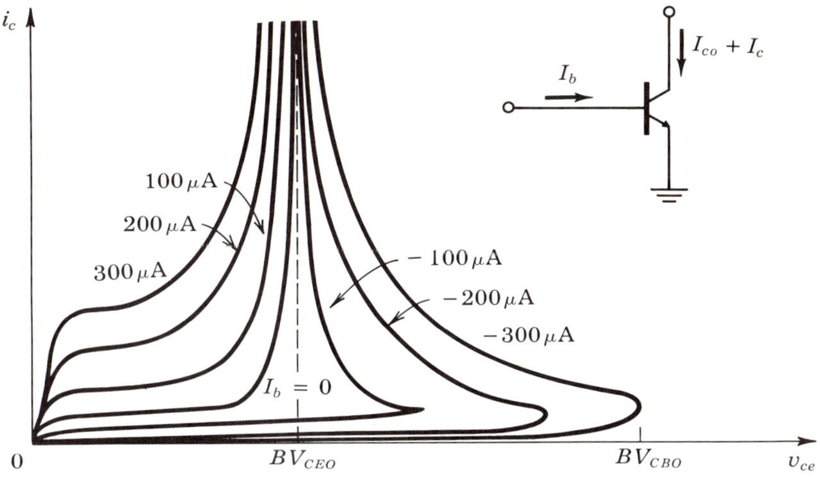

Fig. 11-30 Volt-ampere characteristic showing avalanche-region operation of the bipolar transistor.

Once the collector current starts avalanching, we no longer satisfy the condition $I_b = -I_c$. Emitter current flows, and Eq. (11-46) again defines the collector current. Collector avalanche occurs at a higher voltage than the common-emitter breakdown voltage. As a result, the increase in collector current is accompanied by a drop in v_{ce}. The volt-ampere characteristic exhibits a current-controlled negative-resistance region.

Avalanche-region operation is ideally suited for extremely high-speed switching applications. Since the transistor is never saturated, there is no excess base-charge storage and no storage-delay time.

Tunnel diode

The tunnel diode, which was discovered by Dr. L. Esaki in 1957, consists of an extremely thin *pn* junction formed between two heavily doped regions (large amounts of added impurities) of a semiconductor. The shape of the volt-ampere characteristic (Fig. 11-31*a*), which exhibits a voltage-controlled negative-resistance region for small forward biases, can be explained from a consideration of the electron-wave propagation through the junction boundary. Because of the heavy doping, there exist relatively large numbers of conduction electrons in the *n* material and a wide range of empty states in the *p* material. The electron wave, which propagates freely within each region, cannot tunnel through the potential barrier at the junction unless the energy level on one side is matched by an equivalent empty state on the other side. During the transition, the electron wave is attenuated while the energy is conserved. Thus the junction must be thin in order for the electron wave to have an appreciable probability of transmission.

When the diode is back-biased, the energy level of the electrons in the n material is lowered below that of the free electrons in the p material. The reverse current, which is completely due to tunneling, can increase without limit. In this region, the volt-ampere characteristic looks exactly like that of a conducting diode.

An applied positive bias increases the potential energy of the electrons in the n material. As a consequence, the forward current will continue to increase until the complete range of free states in the p material is matched by the tunneling electron waves. Any further increase in the forward bias raises the energy level of the free electrons above that of the empty states and is, therefore, accompanied by a decrease in the terminal current. This is the negative-resistance zone. Eventually the increasing bias causes the injection of the minority carriers, the diode conducts in the normal manner, and the current again increases with increasing voltage.

The major advantage of the tunnel diode over all other negative-resistance devices is the high speed of the current transmission. The velocity of propagation approaches that of light. There is, however, a large capacity (20 to 60 pF) associated with the depleted junction as well as the series resistance and lead inductance shown in the active-region model of Fig. 11-31*b*. These parasitic elements tend to slow the switching time. In spite of this, the large value of negative conductance permits the generation

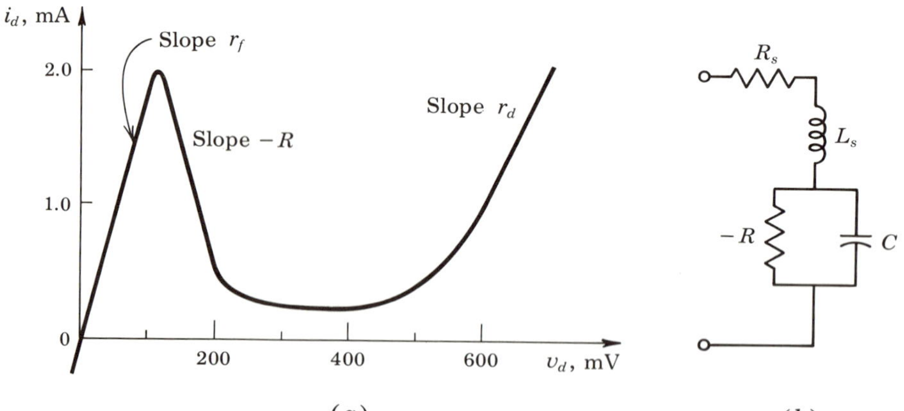

Fig. 11-31 (*a*) Tunnel-diode volt-ampere characteristic; (*b*) model holding in the negative-resistance region—L_s is due to lead inductance, R_s represents the lead resistance and ohmic losses in the semiconductor, and C is the junction capacity.

of pulses having rise times of less than 100 ps. Sinusoidal oscillation at frequencies in excess of 5 GHz is also possible (see Chap. 17).

It shound be noted that the negative-resistance region is quite narrow; the peak is normally located between 50 and 100 mV, and the valley point at 150 to 500 mV. Thus, if the junction is designed for large peak currents, the negative-resistance value will be extremely low. For example, approximately $-2\ \Omega$ is measured in the active region of a gallium arsenide tunnel diode having a peak current of 100 mA. When we examine a diode whose maximum current is less than 1 mA, we observe that the average negative resistance has increased to a few hundred ohms.

Since the source impedance of the power supply is of the same order of magnitude as the negative resistance, it is extremely difficult to bias properly and to stabilize the very-low-resistance devices in their negative-resistance region. Even a small amount of lead inductance may lead to undesired switching or spurious oscillations.

A two-terminal tunnel diode does not afford the same flexibility in obtaining an isolated output as do the three-terminal active devices. Consequently, care must be taken that the external load will not disturb the operation of the timing or switching circuits. In order to minimize possible interaction when multiple stages involving tunnel diodes are to be interconnected, some type of unilateral decoupling must be used. We may even be forced to associate a junction transistor with each tunnel-diode circuit. Because the actual rise is limited by the slowest stage, many of the advantages of the tunnel diode are negated. In some circumstances it becomes practical to employ fast-responding diodes to decouple the individual switching circuits. Diodes whose conduction is due to tunneling, but which do not exhibit negative resistance, are ideal for this purpose.

11-9 Frequency Dependence of NR Characteristics

Up to this point we have discussed only the static behavior of the various devices and circuits that exhibit negative-resistance regions. But to utilize these elements properly, their frequency limitations must also be known. Some of these were treated briefly in earlier chapters, e.g., the rise and fall times and the minority storage time of the solid-state devices. The major item which we must yet consider is the effect of frequency on the gain parameter.

In each case, the negative input resistance developed in the active region depended on the current or voltage amplification of the active elements employed. Equations (11-25), (11-30), (11-31), and (11-35), defining the negative-resistance regions, are all basically alike. Of course, the controlled-source parameter is A_v in two cases and A_c in the other two, but this is simply a detail dependent on the particular circuit considered. In a CNR, the term of interest is of the form

$$v_n = (1 - A)R_1 i_n$$

The VNR devices are defined by

$$i_n = (1 - A)G_1 v_n$$

where, in both cases, A is now used as a general amplification factor.

The parameter most affected by frequency is this amplification factor. If we assume that the system poles are widely separated, then the lowest-frequency pole will have the most significance in the variation of loop gain with frequency. This is a reasonable approximation in most of the amplifiers used for the NIC. In several examples, the gain was dependent on the product $\beta\alpha$, and certainly the α and β cutoff frequencies are widely separated. The frequency response of most of the devices treated in Sec. 11-8 may also be approximated by a single pole. Assuming a single significant pole, the low-frequency-amplification term A may be replaced by

$$A(j\omega) \approx \frac{A_0}{1 + j\omega/\omega_c} \tag{11-47}$$

$A(j\omega)$ decreases at the rate of 6 dB per octave as ω rises above the break at ω_c. The cutoff frequency f_c usually ranges from 100 kHz to 2 MHz in the unijunction and the *pnpn* transistors up to about 10 GHz in tunnel diodes and junction transistors that are specifically designed for high-frequency operation.

If we substitute Eq. (11-47) into the general negative-impedance term for a CNR, the resultant expression becomes

$$\begin{aligned} Z_{\text{in}} &\approx \left(1 - \frac{A_0}{1 + j\omega/\omega_c}\right) R_1 \\ &= R_1\left(1 - \frac{\omega_c{}^2 A_0}{\omega_c{}^2 + \omega^2}\right) + j\omega \frac{\omega_c R_1 A_0}{\omega_c{}^2 + \omega^2} \end{aligned} \tag{11-48}$$

The first part of Eq. (11-48) represents the resistive portion of the input impedance, which will remain negative up to

$$\omega_h = \omega_c \sqrt{A_0 - 1}$$

Above this frequency, the various transistors can no longer sustain the negative-resistance characteristics and are unable to supply any energy to the external load. It follows that switching operations would have to be restricted to the range of frequencies below ω_h.

The second term of Eq. (11-48) contributes an inductive component to the driving-point impedance. First of all, this prevents any instantaneous current jumps. Secondly, it may also resonate with the external capacity, generating an almost sinusoidal waveshape, provided that the circuit is biased within the active region.

Figure 11-32 shows an equivalent driving-point network which represents the incremental behavior of the device within its active region and which, of course, also satisfies Eq. (11-48). At low frequencies, the input inductance is effectively a short circuit, leaving only the negative-resistance term. At the higher frequencies, the presence of the additional energy-storage term complicates the network and forces us into a reevaluation of the conditions to be satisfied for absolute circuit stability.

Stability of nonideal CNR circuits

The problem of stability can be neither considered nor defined without treating the complete circuit before us. Instability simply means that sufficient surplus energy exists in the system to produce an increasing exponential response in any energy-storage element present. But if the external network also contains dissipative elements, then their effect on the system may even change one having an energy surplus into one that is completely dissipative. The particular conditions necessary to ensure this happening are those which produce absolute stability and prevent any regenerative switching.

Consider, for example, the network that represents the input characteristic of a current-controlled device (Fig. 11-32). The external load consists of a parallel combination of R and C together with a bias battery whose function is to set the intersection of the load line and the NR characteristic within the proper region. In the following discussion we shall assume intersection of the negative-resistance segment.

Since we are considering a CNR circuit, we are concerned with what will happen to the driving-point current. For stability, the poles of the

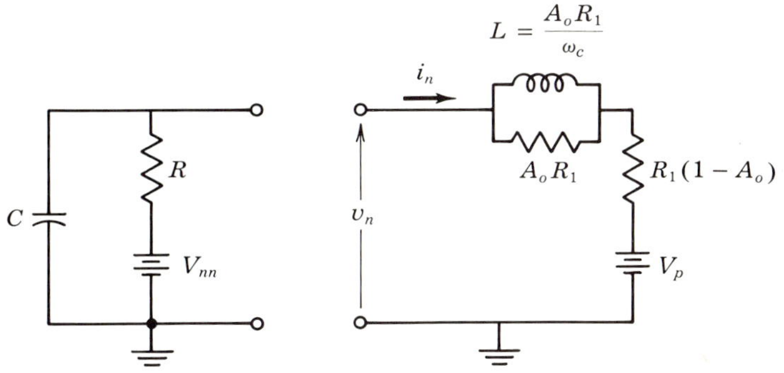

Fig. 11-32 Equivalent driving-point network of a CNR in the active region and the external load.

current function must not be allowed to lie in the right half of the complex plane, and this corresponds to restricting the zeros of the impedance function [Eq. (11-49)] to the left half plane:

$$Z(s) = (1 - A_0)R_1 + \frac{sLR_1A_0}{sL + A_0R_1} + \frac{R}{sRC + 1}$$

$$= \frac{R_1RCLs^2 + [(R_1 + R)L + CRR_1{}^2A_0(1 - A_0)]s + A_0R_1[R + R_1(1 - A_0)]}{(sL + A_0R_1)(sRC + 1)} \qquad (11\text{-}49)$$

In order to guarantee this restriction, all the coefficients of the polynomial of the numerator of Eq. (11-49) should have the same sign. This leads to the two inequalities, Eqs. (11-50) and (11-51), which we must satisfy to ensure absolute circuit stability:

$$R > -R_1(1 - A_0) = -R_n \qquad (11\text{-}50)$$

$$C < \frac{R_1 + R}{RR_1(A_0 - 1)\omega_c} \qquad (11\text{-}51)$$

The value of L given in the model of Fig. 11-32 was substituted while solving for the condition expressed in Eq. (11-51).

We can identify the right-hand side of the inequality of Eq. (11-50) as the magnitude of the negative resistance R_n. This inequality simply states the condition under which the load line will intersect the volt-ampere characteristic only in the negative-resistance region. A load less than $|R_n|$ would also intersect the two positive-resistance regions, and the circuit would operate as a bistable.

Previously we have contended that the reason that bistable operation can never stabilize within the negative-resistance region is because the operating point can never get there (see Sec. 11-4 and Fig. 11-11). Equation (11-50) is another way of stating this restriction.

Suppose that both inequalities are satisfied. The system is then completely stable and will not operate as a switching circuit. We conclude that by using a current sweep having a very low output capacity, it is possible to trace out the full volt-ampere characteristic for display on an oscilloscope.

Once C is larger than the minimum value given by Eq. (11-51), and if the inequality of Eq. (11-50) is still satisfied, the circuit becomes free-running. At the bounding value of C, the zeros of the impedance function lie on the imaginary axis; for C less than this value, the zeros are in the left half plane and the system is stable. As C increases, the roots move into the right half plane, toward the real axis (Fig. 11-33). If the zeros are complex conjugates lying close to the imaginary axis, the time response of the astable system is sinusoidal. This particular mode of operation will be discussed in Chap. 17.

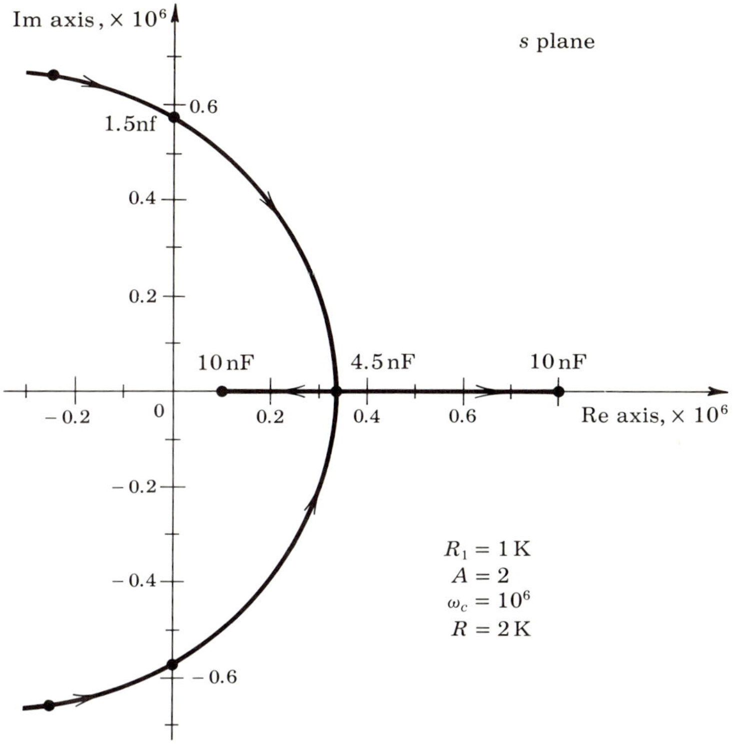

Fig. 11-33 Path of the zeros of $Z(s)$ [Eq. (11-49)] as a function of C. Specific circuit values are given above for astable operation.

The stability condition presented in Eq. (11-50) is exactly opposite to the condition found when we discussed the ideal NR in the earlier portion of this chapter [Eq. (11-6)]. This does not really contradict the earlier discussion, since a short circuit or a low resistance across the input terminals of the equivalent network of Fig. 11-32 results, essentially, in the inductivity-controlled behavior of the VNR (Fig. 11-14). The switching inductance is the internal inductance. Furthermore, when $\omega_c \to \infty$ or when $L = 0$, that is, when the circuit approaches the ideal, the stability conditions found from Eq. (11-49) will reduce to the single equation given in Eq. (11-6).

We conclude that a physical negative-resistance device cannot support a stable point within the negative-resistance region unless the two inequalities of Eqs. (11-50) and (11-51) are satisfied. Point H of Fig. 11-6 (the load-line intersection with the negative resistance) will, in general, be unstable. Away from the vicinity of this point, the system rapidly stabilizes at one of its two intersections in the completely dissipative regions.

Stability conditions are seen to depend on the particular circuit configuration. In order to avoid ambiguity, it is much better to refer always to the device used in terms of its single-valuedness, i.e., current or voltage, rather than its conditions for stability. Furthermore, when any changes are made in the external circuit, the reader must re-solve for the conditions necessary to ensure absolute stability.

VNR stability

The basic form of the negative conductance developed in the active region of a VNR is

$$Y_n = (1 - A)G_1$$

This equation is identical in form to the negative-impedance function derived for the CNR devices. It follows, from Eq. (11-48), that the substitution of $A(\omega)$ into the admittance term results in

$$Y_{\text{in}} = G_1\left(1 - \frac{\omega_c{}^2 A_0}{\omega_c{}^2 + \omega^2}\right) + j\omega\frac{\omega_c G_1 A_0}{\omega_c{}^2 + \omega^2} \tag{11-52}$$

Equation (11-52) yields the equivalent input network given in Fig. 11-34, which we note to be simply the dual of the network of Fig. 11-32.

To ensure that the poles of the controlling voltage function will not lie in the right half plane, the zeros of the admittance function must be restricted to the left half plane. The two required conditions for absolute system stability become

$$G > -G_1(1 - A_0) \tag{11-53}$$

$$L < \frac{G_1 + G}{GG_1(A_0 - 1)\omega_c} \tag{11-54}$$

Equations (11-53) and (11-54) are the duals of Eqs. (11-50) and (11-51), previously derived, and, as before, when neither equation is satisfied, the load line also intersects the positive-resistance regions, thus establishing bistable circuit operation. Satisfaction of only Eq. (11-53) allows astable behavior for L larger than the minimum value of Eq. (11-54). Just as with the CNR, this circuit will also exhibit almost sinusoidal oscillations for small values of L.

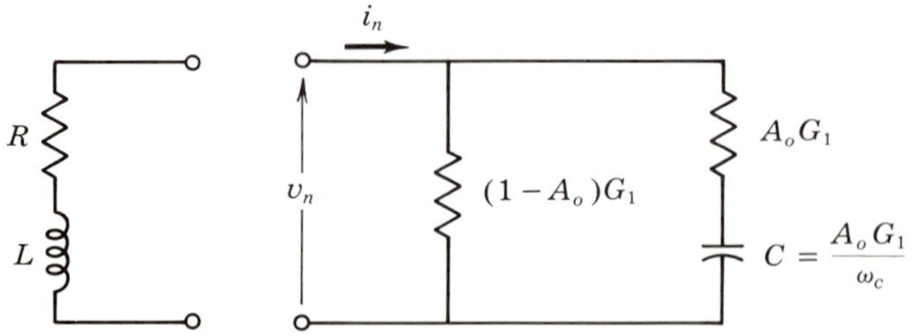

Fig. 11-34 Equivalent input network of the VNR for the negative-resistance region and the external load.

11-10 Improvement in Switching Time through the Use of a Nonlinear Load

Generally, if fast signals are to be generated and if narrow pulses are used for triggering, operation within the saturation region of semiconductor devices must be avoided. The long time delay introduced by the minority-carrier storage precludes all but the slowest switching intervals. In order to satisfy this restriction, we rely on nonlinear load lines, which are developed with the aid of appropriately biased diodes.

With one region forbidden us, both of the remaining circuit states must be used in a bistable; one stable point will be situated in the cutoff zone, and the other will lie in the negative-resistance region. A simple bistable circuit for high-speed switching, featuring this arrangement, is shown in Fig. 11-35. It makes use of a single external diode to insert the necessary break in the load line, ensuring that it intersects the volt-ampere characteristic only in these two regions.

The absolute stability of the point established in the completely dissipative cutoff region by the diode's conduction (point A) is not open to question. But for a true bistable, we must justify the stability of point F, which lies on the negative-resistance portion of the characteristic. At this point, the large load resistance R definitely satisfies one of the two stability conditions [Eq. (11-50)]. If the stray capacity is kept low enough or if the parameters of the CNR satisfy the second condition [Eq. (11-51)], then point F will also be absolutely stable. From Eq. (11-51) we see that the smaller the value of negative resistance, the larger the capacity that can be tolerated without changing the stable point to an unstable one.

Switching in the circuit of Fig. 11-35 will always be from A to F and from F back to A. We depend on the combination of the stray capacity and the inductive component present in the driving-point impedance of the

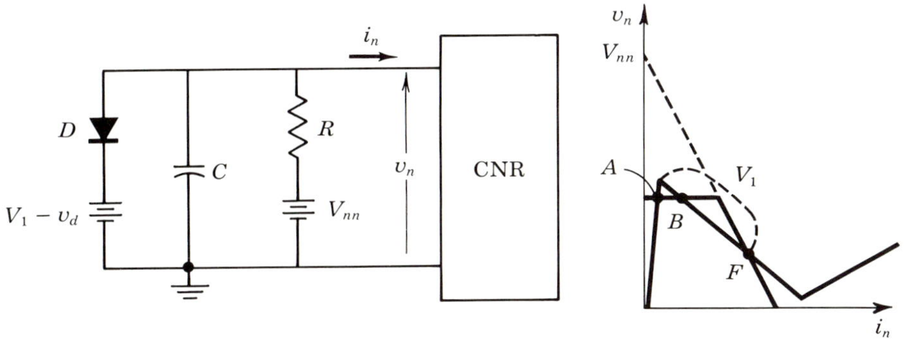

Fig. 11-35 Stabilization within the negative-resistance region—bistable operation.

CNR to slow the system response enough so that the path of operation will never enter the saturation region (see the dashed path of Fig. 11-35).

Astable-operation stabilization

An alternative technique for preventing the transistor from being driven into saturation, illustrated in Fig. 11-36, is most widely applied in astable systems. This circuit configuration requires the insertion of a resistor R_2 in series with C to stabilize the intersection occurring in the negative-resistance region. Suppose that the charge stored in C back biases the diode while the operating point is at A. Then the additional resistance increases the net circuit dissipation to a point where it will exceed the net energy supply.

In order to find the conditions which must be satisfied to stabilize point A when the diode is back-biased, we must solve the complete model of the system. Representing the CNR by the model of Fig. 11-32, the two conditions which will ensure absolute circuit stability in the negative-resistance region are

$$R > -R_1(1 - A_0)$$

$$R_2 > \frac{(A_0 - 1)RR_1}{R_1(1 - A_0) + R} - \frac{R_1 + R}{[R_1(1 - A_0) + R]C\omega_c}$$

where the negative resistance is $R_1(1 - A_0)$. The first condition is the same as that expressed in Eq. (11-50). From the second condition we see that the smallest value of R_2 that will maintain the circuit stable, regardless of the value of C, is a resistance equal in magnitude to the parallel combination of R and $(1 - A_0)R_1$. This makes the net resistance across the energy-storage element positive.

However, C eventually charges, and at the instant that the voltage at its lower terminal reaches 700 mV, the diode again conducts. The circuit becomes astable, and the circuit flips clockwise to point B. As discussed

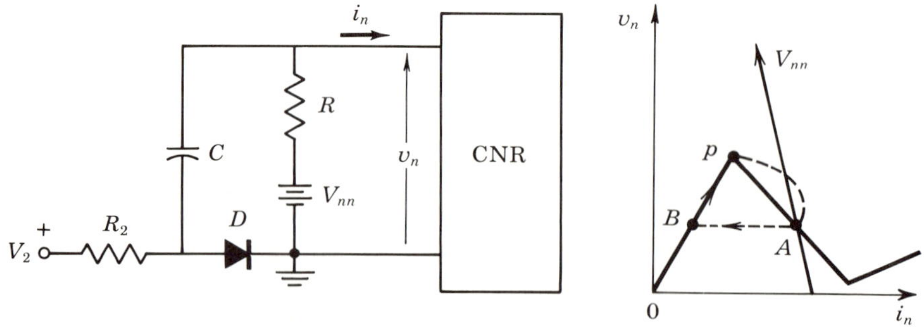

Fig. 11-36 Stabilization within the negative-resistance region—astable operation.

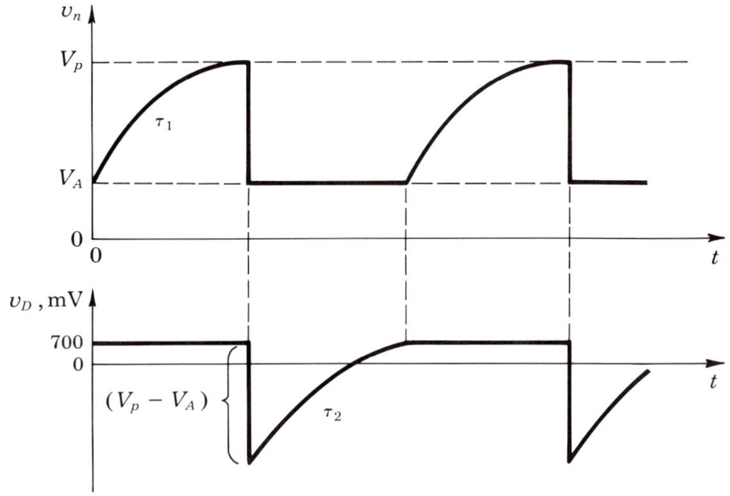

Fig. 11-37 Voltage waveshapes appearing at the input of the CNR and across the diode in the circuit of Fig. 11-36.

in Sec. 11-3, the capacitor charges toward the Thévenin equivalent voltage across its terminals, eventually reaching the peak, V_p.

In this circuit, the voltage across the terminals of the CNR can change instantaneously because this change will not appear across C but will be coupled by C to R_2. Any voltage drop back biases the diode. The operating point thus jumps from the peak to the now stable point A, and the diode is driven negative by this same voltage change. C again charges, with its bottom terminal rising from $V_D - (V_p - V_A)$ toward V_2 with the approximate time constant

$$\tau_2 \approx [R_2 + (R \parallel -R_n)]C$$

It eventually reaches 700 mV, the diode conducts, the circuit again becomes astable, and the cycle repeats. The approximate voltage waveshapes generated across the diode and at the input to the CNR are sketched in Fig. 11-37.

An arrangement similar to that shown in Fig. 11-35 might also be used in a monostable circuit. In this case, returning R_2 to a small negative voltage $-V_2$, instead of to a large positive one, will keep the diode back-biased and guarantee the stability of point F. After an external trigger forces the diode into conduction, the circuit momentarily becomes free-running and switches clockwise into the cutoff zone. If the circuit parameters are properly adjusted, the capacitor-charging current will maintain diode conduction during the complete interval that the operating point remains in this region.

Subsequent to reaching the peak value of the CNR, the drop in voltage, coupled by C, will turn off the diode and reestablish point A as the single stable point. The operating point drops from the peak to point A, and the circuit recovers. Since the steady-state voltage across the diode is $-V_2$, the diode remains back-biased, and the operating point finally stabilizes at A.

Problems

11-1 A CNR having the characteristic of Fig. 11-38 is used in the circuit of Fig. 11-3. The load resistance is 2 K and the external timing capacitor is 2 μF. Plot the current and voltage waveshapes after the circuit is triggered if V_{nn} is as given below. In each case, state the area of the current impulse which must be injected to cause switching.

(*a*) $V_{nn} = 30$ V
(*b*) $V_{nn} = -10$ V

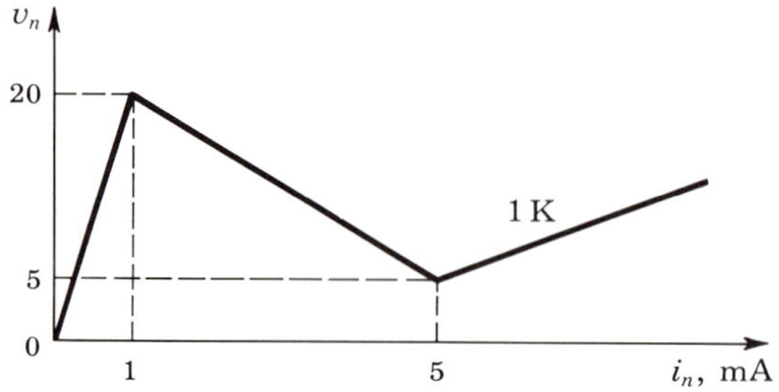

Fig. 11-38

11-2 (*a*) What range of values of load resistance will permit bistable operation of the CNR of Fig. 11-38?

(*b*) The load resistance used with the CNR of Fig. 11-38 is 10 K. What ranges of V_{nn} permit operation as a monostable?

(*c*) Repeat part *b* if the load resistance is now reduced to 1 K.

11-3 The CNR of Fig. 11-38 is loaded with a 5-μF capacitor shunted by a constant-current source of 3 mA. This establishes astable operation. Draw the voltage and current waveshapes, labeling them with respect to time constants, times, and voltage or current values. Over what range can the biasing-current source be varied without drastically changing the waveshape?

11-4 Draw the composite characteristic of the CNR of Fig. 11-38 and a biased load resistance of 5 K under the following two conditions. In both cases, C = 500 pF.

(*a*) $V_{nn} = 5$ V
(*b*) $V_{nn} = 40$ V

Does either condition appear to offer any particular advantages? Explain.

11-5 (*a*) Draw the composite characteristic of the four parallel CNR devices of Fig. 11-10*a*. Explain any anomalies.

(*b*) For $V_{nn} = 20$ V, give the range of R for which the circuit will have five stable points. Choose the average value of R and give the voltage and current coordinates of each stable point.

(*c*) R is increased to 500 ohms. V_{nn} is now a slowly increasing triangular wave, taking 10 s to reach the peak of 200 V and decreasing to zero in the next 10 s. Sketch the input and output voltages to scale, labeling all significant points.

11-6 At $t = 0$, the VNR of Fig. 11-39 is triggered by a 30-V voltage pulse applied in series with the 3-V supply. This pulse is twice the minimum duration needed for switching into region III. Show the path of operation. Sketch i_n and v_n, labeling all times, time constants, and voltage and current values.

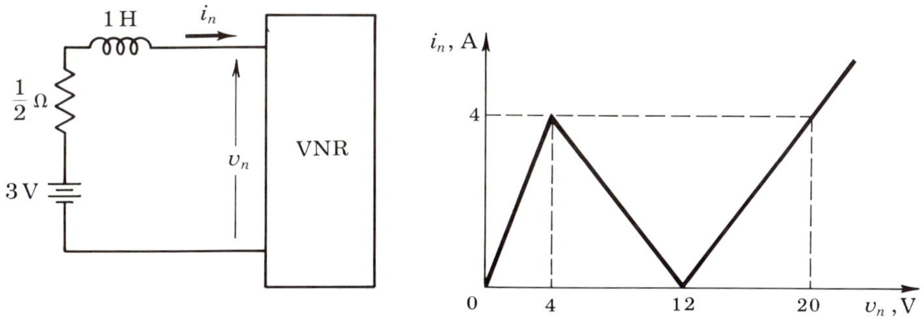

Fig. 11-39

11-7 (*a*) Give the range of external resistance that will make the circuit of Fig. 11-39 astable when the battery voltage is within the limits $5 < V_{nn} < 12$ V.

(*b*) Sketch and label completely the output waveshapes when $R = 1\ \Omega$ and $V_{nn} = 10$ V.

11-8 Calculate and plot the CNR characteristic seen when looking across C in the monostable circuit of Fig. 8-18. Make all reasonable approximations. Show how the characteristic is changed when we modify the circuit as shown in Fig. 8-20.

11-9 Plot the volt-ampere characteristic seen by R_1 in the Schmitt comparator of Fig. 10-22 (see Prob. 10-11). $V_z = v_1 = 4$ V.

11-10 Plot the CNR characteristic of the circuit of Fig. 11-15 when $R = 100$ K and $R_3 = 10$ K. All other elements remain as before. Superimpose this plot on a copy of Fig. 11-16.

11-11 (*a*) Plot the transfer characteristic, i.e., collector voltage of Q_2 vs. i_a, for the monostable multivibrator discussed in Example 11-2 (Fig. 11-15).

(*b*) Sketch one complete cycle of i_a after the circuit is triggered. Using the curve found in part *a*, draw to scale one complete cycle of v_{c2}.

11-12 Plot the volt-ampere characteristic of the monostable multivibrator of Fig. 11-15 when R_3 and R are both increased to 100 K. Compare your results with Fig. 11-16. Which regions are most affected by the change in R_3? By the change in R? What has happened to the mode of operation?

11-13 Calculate and plot the volt-ampere characteristic seen across the inductance in the circuit of Fig. 11-40. Using this plot, sketch and label the inductive current i_L after a trigger is applied.

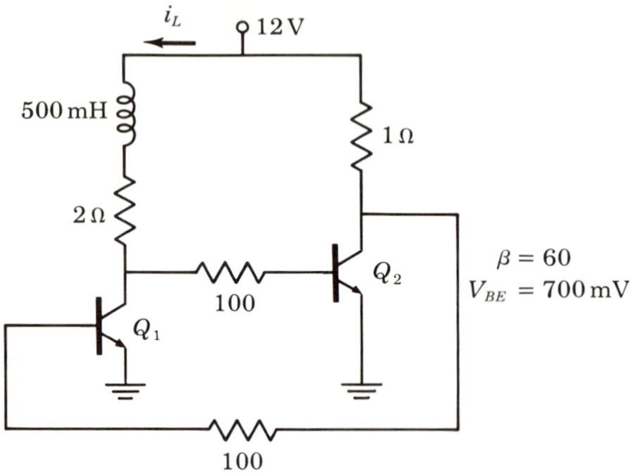

Fig. 11-40

11-14 Justify the volt-ampere characteristic of Fig. 11-19. Give the coordinates of both break points as well as the slopes of the line segments.

11-15 Figure 11-41 shows a circuit having a negative-impedance input characteristic. Draw the volt-ampere characteristic, giving the values of all break points and slopes. Use this circuit to form an astable multivibrator generating a 10-μs-period waveshape. Give all external component values. Sketch the waveshape produced.

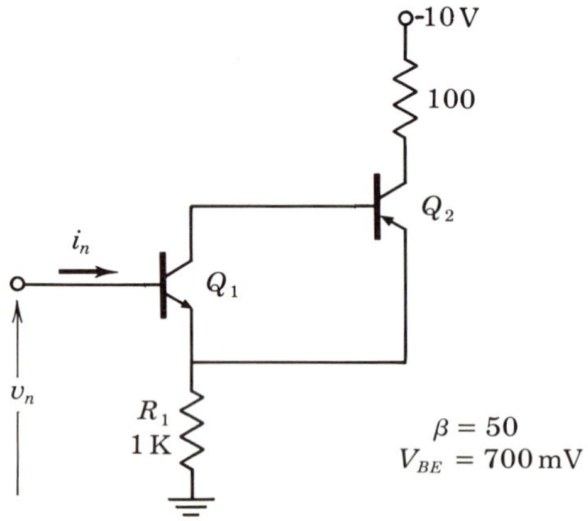

Fig. 11-41

11-16 The two transistors in the NIC of Fig. 11-21 have $\beta = 100$, $V_{BE} = 750$ mV, and $BV_{CBO} = 50$ V. Design an NIC circuit that will have the valley point at 5 mA, 50 V and the peak at 50 mA. Draw to scale the volt-ampere characteristic. Give all component values.

11-17 Draw the volt-ampere characteristic for the circuit of Fig. 11-24, giving the coordinates of all break points and the slopes of all segments. $R_1 = R_2 = R_3 = 10$ K, $R = 1$ K, $R_4 = 5$ K. $V_{CC} = 20$ V, $\beta = 100$, $V_{BE} = 700$ mV, $V_T = 3$ V, $I_D = 5$ mA at $V_{GS} = 4$ V.

11-18 Figure 11-42 shows an NIC which yields an input impedance of $-Z_L$. This negative impedance can exist at $v_n = 0$ and $i_n = 0$.

(*a*) Prove that this circuit converts Z_L into $-Z_L$. (Assume high-β transistors.)

(*b*) Plot the *vi* characteristic when $R = 10$ K, $V_{CC} = 10$ V, $V_{EE} = -10$ V, $V_{BE} = 700$ mV, $\beta = 100$, and $Z_L = 5$ K.

(*c*) What are the *vi* bounds of operation if $Z_L = 0.1\ \mu$F?

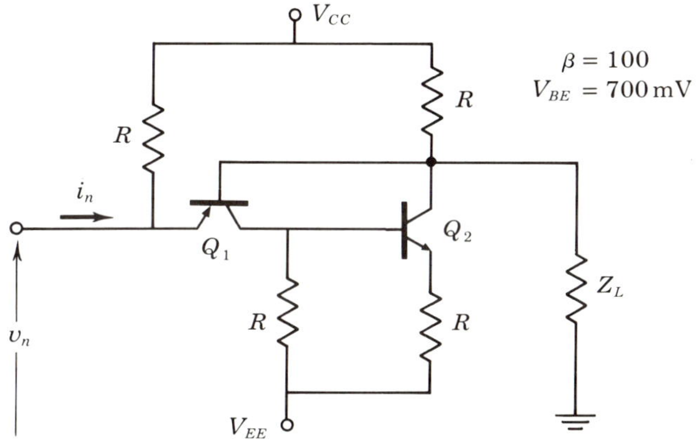

Fig. 11-42

11-19 Show how a single SCR can be used to create a characteristic similar to that of the two-transistor NIC of Fig. 11-18. Sketch the *vi* characteristic, giving all slopes and break points.

11-20 The unijunction transistor shown in Fig. 11-43 may be represented approximately by the linearized volt-ampere characteristic. (This includes the

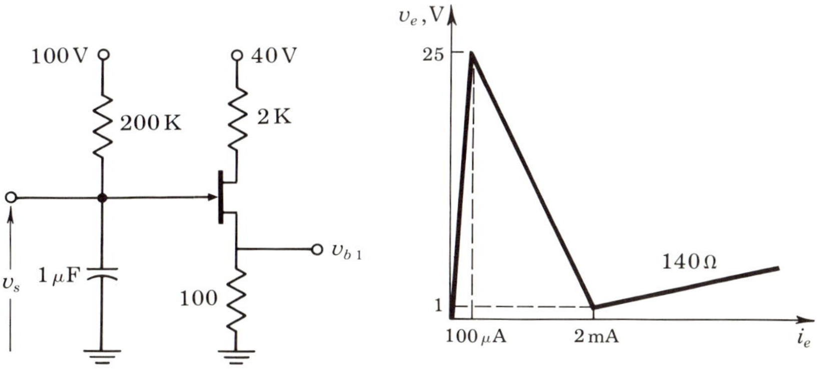

Fig. 11-43

effect of the 100-Ω resistor in series with base 1.) After the device switches into the high-current region, at least 2.5 μs is required for the recombination of excess carriers before the circuit switches back to the high-resistance region. Sketch the waveforms of v_s and v_{b1} as a function of time, labeling clearly all break points and time constants.

11-21 (*a*) Prove the equivalency of the driving-point network of Fig. 11-32.
(*b*) Derive Eq. (11-49) and verify Eqs. (11-50) and (11-51).
(*c*) Show that as $L \rightarrow 0$, the stability condition is satisfied by Eq. (11-6).

11-22 (*a*) The tunnel diode of Fig. 11-31*a* is connected to an external circuit consisting of a 10-mH inductance, a 10-Ω load resistance, and a 200-mV bias source. Approximate the characteristic by four line segments, and plot the terminal-voltage waveshape.
(*b*) If $R_s = 0.3\ \Omega$, $L_s = 2\ \mu\text{H}$, and $C = 50$ pF in the active-region model of Fig. 11-31*b*, what is the highest frequency of operation?
(*c*) Express the absolute stability conditions for the general tunnel-diode model of Fig. 11-31*b*.

11-23 Five tunnel diodes, each having the piecewise characteristic of Fig. 11-44*a*, are connected in series to form the voltage quantizer of Fig. 11-44*b*. Discrete levels of the continuous input voltage are identified by sequential jumps at the output. V_B must be set to produce jumps at approximately equal input voltage intervals.
(*a*) Specify the necessary value of V_B.
(*b*) Sketch the output voltage and its derivative when the input is a positive-going ramp increasing at the rate of 10 V/μs.

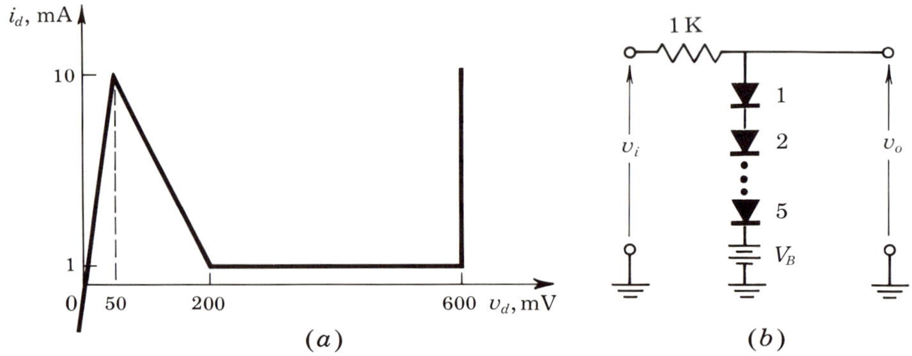

Fig. 11-44

11-24 A 4-μs periodic train of alternating 1-V 100-ns positive and negative pulses is applied to the circuit of Fig. 11-45. D_1 and D_2 are tunnel diodes having characteristics which can be linearized as shown in Fig. 11-44*a* except that these diodes have a peak current of 2 mA. D_3 is a fast-acting decoupling diode which will be considered ideal for the purposes of this problem.
(*a*) Sketch and label the output voltage over a 10-μs interval.
(*b*) Repeat part *a* when D_3 is removed and the coupling is directly through the 100-Ω resistor.

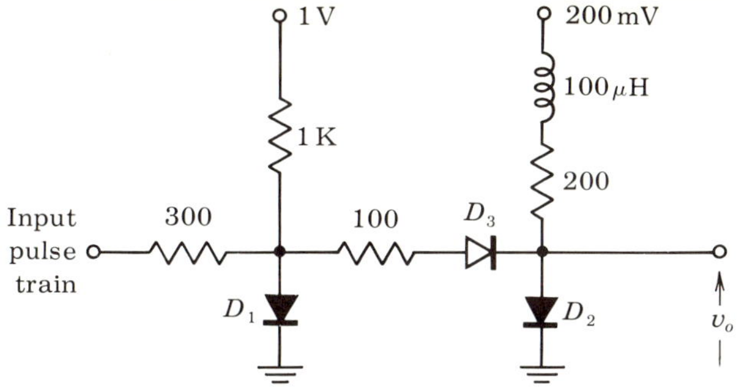

Fig. 11-45

11-25 The terminal characteristic of a certain CNR is shown in Fig. 11-46. Also shown is the equivalent circuit of the device in its active region.

(*a*) Determine i_n as a function of time after the switch is closed at $t = 0$. (*Hint:* Solve for the poles of the network in its active region and also for the steady-state component of i_n; then write $i_n = i_{\text{transient}} + i_{\text{steady-state}}$ and evaluate the necessary arbitrary constant.)

(*b*) How does the circuit function if C is reduced by a factor of 2? If it is increased by a factor of 10? Give a qualitative answer and roughly sketch the expected terminal waveshape.

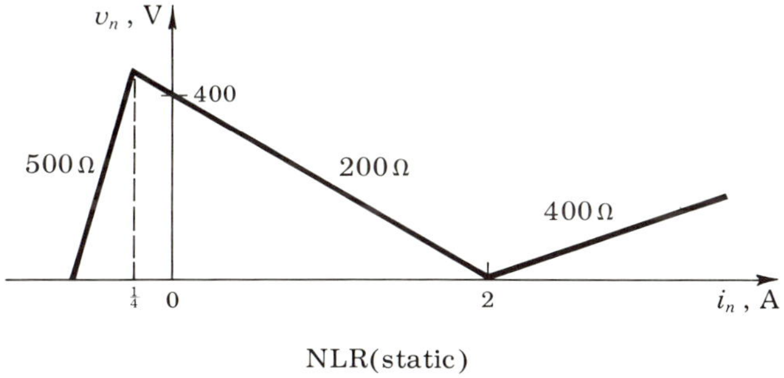

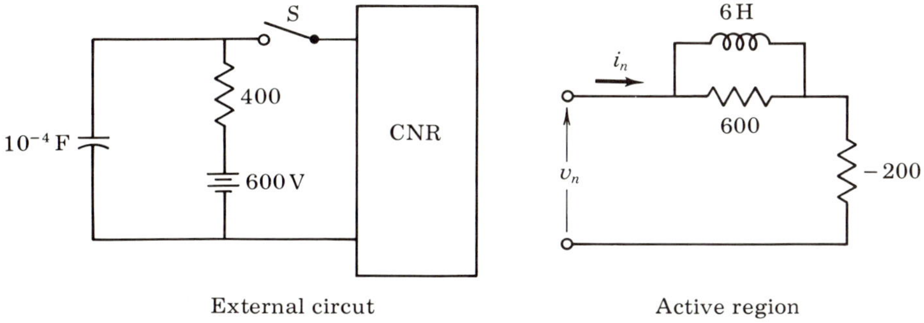

Fig. 11-46

11-26 Prove that the insertion of a resistor in series with C (as in Fig. 11-36) can make a normally unstable circuit stable. Start from the driving-point network of Fig. 11-32 and find the conditions necessary to restrict the zeros of $Z(s)$ to the left-hand plane when R_2 is in series with C. Compare these conditions with the stability requirements of Eqs. (11-50) and (11-51). Which condition is satisfied when R_2 is removed?

Bibliography

Beale, I. E. A., W. L. Stephenson, and E. Wolfendale: A Study of High Speed Avalanche Transistors, *Proc. IEE (London)*, pt. B, vol. 104, pp. 394–402, July, 1957.

Ebers, J. J.: Four-terminal *p-n-p-n* Transistors, *Proc. IRE*, vol. 40, no. 11, pp. 1361–1365, November, 1952.

Esaki, L.: Letter to the Editor, *Phys. Rev.*, vol. 109, pp. 603–604, Jan. 15, 1958.

Farley, B. G.: Dynamics of Transistor Negative Resistance Circuits, *Proc. IRE*, vol. 40, no. 11, pp. 1497–1508, November, 1952.

Hall, R. N.: Tunnel Diodes, *IRE Trans. Electron. Devices*, vol. ED-7, pp. 1–9, January, 1960.

Linvill, J. G.: Transistor Negative Impedance Converters, *Proc. IRE*, vol. 41, no. 6, pp. 725–729, June, 1953.

Lo, A. W.: Transistor Trigger Circuits, *Proc. IRE*, vol. 40, no. 11, pp. 1531–1541, November, 1952.

Merrill, J. L.: Theory of the Negative Impedance Converter, *Bell System Tech. J.*, vol. 30, no. 1, pp. 88–109, 1951.

Misawa, T.: Negative Resistance in *p-n* Junctions under Avalanche Breakdown Conditions, Parts I and II, *IEEE Trans. Electron. Devices*, vol. ED-13, no. 1, pp. 137–151, January, 1966.

Shockley, W., and J. F. Gibbons: Introduction to the Four-layer Diode, *Semicond. Prod.*, pp. 9–13, January–February, 1958.

Todd, C. D.: A Composite Circuit Exhibiting *S*-type Negative Resistance, *Semicond. Prod.*, pp. 24–28, October, 1962.

12. the blocking oscillator

Most of the multivibrators discussed in Chaps. 8, 9, and 11 depended for their timing on the energy-storage element maintaining the FET or transistor off for the required interval. Because an active element is essentially an open circuit within its cutoff region, the output pulse is available only at a relatively high impedance level and with a low power content. Furthermore, stray capacity slows the fast rising and falling edges, preventing the generation of extremely narrow pulses. One method of overcoming these disadvantages is to time the pulse within a high-current low-impedance region.

The blocking oscillator is one circuit which is so designed. Only a single active element is necessary, with a specially designed transformer used for timing as well as for phasing the regeneration. Because this deceptively simple circuit is so widely used, in computers, radar, television, etc., it will be treated in some detail. In doing so, we extend the approximation methods of analysis to multiple-energy-storage-element systems. We shall also be obliged to consider the problems posed when some of the passive components exhibit nonlinearities. No attempt will be made to obtain an exact solution, but the analysis will adequately explain the role of the various elements and the functioning of the circuit.

12-1 Some Introductory Remarks

Figure 12-1 shows one possible form of the blocking oscillator, a collector-emitter-coupled circuit. As possible alternative configurations, the transformer may couple either the collector or the emitter to the base. In brief, the sequence of monostable operation is the following. The transistor is initially back-biased, and in response to an externally applied trigger, it is forced into the active region. Provided that the transformer has the proper turns ratio and connections, the net loop gain will be positive and greater than unity. Regeneration drives the circuit into bottoming, indicated by the diode's conduction. The transistor acts as a switch, which operates with a minimum amount of input energy and which connects V_y across the primary of the transformer.

The full duration of the ON interval is controlled, as is all timing, by the energy-storage elements present (in this circuit by the magnetizing inductance L_m and the coupling capacitor C). When generating long pulses, an evaluation of the width may be further complicated if the large current buildup in the magnetizing inductance L_m drives the transformer into saturation. For short pulses, the timing will be influenced by the rate of flux penetration into the core (Secs. 12-3, 15-2, and 15-3). The solution of the complete problem is quite complicated, involving as it does multiple modes of energy storage, core properties, and hysteresis.

Eventually the circuit reenters the regenerative region and switches back off. The oscillator recovers and remains off until the next trigger is applied.

The blocking-oscillator transformer is a very important component, so critical, in fact, that the circuit is designed about its characteristics. By interleaving the coil windings and by using a high-permeability core material, the leakage inductance and stray capacity are minimized, and

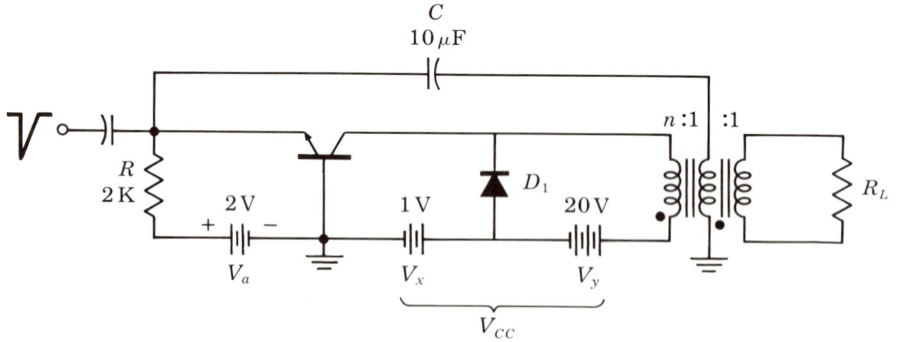

Fig. 12-1 Collector-emitter-coupled blocking oscillator incorporating a bottoming diode to avoid transistor saturation ($L_e = 30\ \mu\text{H}$, $L_m = 200\ \mu\text{H}$, $r_b = 200\ \Omega$, $\alpha = 0.98$, $n = 3$, $C_c = 25$ pF).

consequently so is the rise time. For example, a transformer designed to generate a 5-μs pulse may have a nominal magnetizing inductance of 200 μH, compared with a leakage inductance of only 10 to 30 μH. As it is quite small physically, the stray capacity would be only 10 pF. For nanosecond pulses, the transformers are even smaller; parameters of a 250-ns transformer are $L_m = 15\ \mu\text{H}$, $L_e = 6.5\ \mu\text{H}$, $n = 2$, $C_s = 5$ pF.

The transformer usually includes one or more tertiary windings, which are used to couple the generated pulse to the isolated external load.

12-2 Switching Times

The rise and fall times of a circuit establish the limiting conditions when attempting to generate an extremely narrow pulse. Since a blocking oscillator is primarily designed to generate a narrow high-power pulse having steep leading and trailing edges, we must examine the transition intervals in great detail. Figure 12-2*a* shows the simplified active-region model which we shall use for the initial analysis. In drawing this model, we have made a number of assumptions:

1. Because of the high-current operation, r_e will be very small and may be neglected.
2. The input impedance of the transistor in the active region is so very low that the stray transformer capacity C_s has only a second-order effect on the rise time.
3. R serves solely for biasing and is much greater than r_b.
4. The coupling capacitor C is very large, and we may therefore assume that its terminal voltage remains constant over the entire pulse interval.
5. R_L is very large compared with r_b.

The significant circuit capacitance is the total transistor collector-to-base feedback capacitance C_c. This is the capacitance which usually establishes the location of the high-frequency break point of a transistor amplifier.

During the short-duration switching interval, the small leakage inductance of the transformer controls the current buildup. Since $L_m \gg L_e$, we can solve for the current in L_e first, omitting the magnetizing inductance from the model.

The resulting active-region model of Fig. 12-2*a* can be further simplified as shown in Fig. 12-2*b* and *c*. We start by converting C_c and the controlled source αi_e to the Thévenin equivalent circuit. In Fig. 12-2*b* we see that we obtain a voltage source proportional to the current i_e/n flowing through it. Since the voltage drop is the same as that produced by the same current flowing through the appropriate impedance, we can replace the voltage source by the negative capacitance $-C_c/\alpha n$.

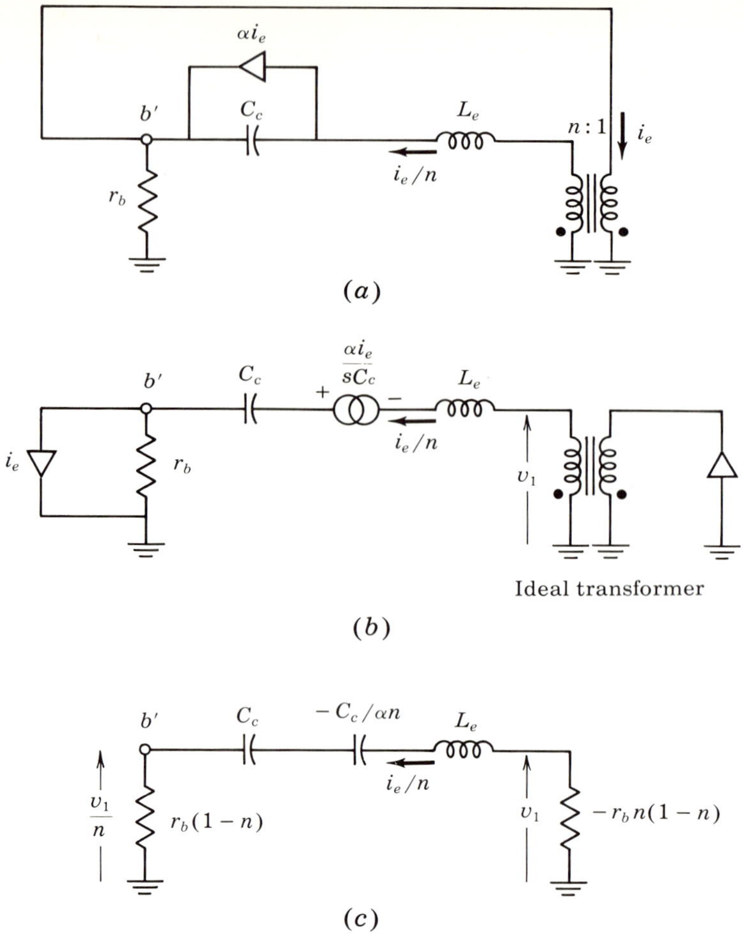

Fig. 12-2 (*a*) Active-region model of blocking oscillator used for calculation of the buildup transient; (*b*) first step in reduction of the model; (*c*) final reduced model.

When the controlling current i_e is replaced by two independent sources, as shown in Fig. 12-2*b*, we change nothing in the circuit. At the left side of Fig. 12-2*b*, we do, however, again obtain a parallel combination of an impedance and a current source, where the current source is proportional to the loop current. As before, the i_e generator can be combined with the impedance r_b to obtain the equivalent resistance $r_b(1-n)$. This substitution appears at the left-hand side of Fig. 12-2*c*.

The voltage drop across the equivalent resistor $r_b(1-n)$ is

$$\frac{v_1}{n} = \frac{i_e}{n} r_b(1-n) \qquad (12\text{-}1)$$

At the collector side of the ideal transformer, the voltage drop is n times as large, or v_1. The primary current is the loop current i_e/n. Thus from Eq. (12-1) we have

$$v_1 = i_e r_b(1 - n)$$

When we finally substitute the equivalent resistance for the drop v_1, taking into account the direction of current flow and the voltage polarities, we finally arrive at the single loop of Fig. 12-2c.

The poles of the blocking oscillator are found from the circuit loop equation

$$s^2 + s\,\frac{r_b(n-1)^2}{L_e} + \frac{1 - \alpha n}{C_c L_e} = 0 \tag{12-2}$$

When the net circuit capacity is positive ($\alpha n < 1$), Eq. (12-2) defines the response of a tuned circuit having a pair of complex poles.

In order for the circuit to be regenerative, i.e., to have a pole in the right half plane, the sign of at least one of the coefficients of Eq. (12-2) must be negative. Only when

$$\alpha n > 1 \tag{12-3}$$

can this condition be satisfied. The statement of Eq. (12-3) corresponds to a loop current gain greater than unity.

Once the circuit is adjusted to satisfy Eq. (12-3), the active transistor creates the net negative capacity $(1 - \alpha n)/C_c$. The presence of the transistor rotates the complex poles through 90°. The two poles of Eq. (12-2) will then lie on the real axis, symmetrical about $-r_b(n-1)^2/2L_e$. The approximate locations of the two poles are given by

$$s_{1,2} \approx -\frac{r_b(n-1)^2}{2L_e} \pm \sqrt{\frac{\alpha n - 1}{L_e C_c}} \tag{12-4}$$

The exact locations of the poles are readily found from Eq. (12-2) once numbers are substituted. Using the numbers of the circuit of Fig. 12-1, the approximation of Eq. (12-4) yields a positive-pole location at $s_p = 3.77 \times 10^7$, compared with the exact solution of $s_p = 3.906 \times 10^7$.

The positive pole controls the rate of the regenerative switching through the active region, and it determines the rise time. But to find the time required to switch through the active region, we must know the starting point of the positive exponential. Since this depends on the energy content of the excitation pulse, which, in general, is unknown, the problem is not completely defined. Instead of solving for the exponential response, we define the rise time of the pulse by a 10:1 change in the collector current. This corresponds to 2.2 time constants of the positive pole of Eq. (12-4):

$$T_R = 2.2\tau_R = \frac{2.2}{s_p} = 56.4 \text{ ns} \tag{12-5}$$

The rise time given by our approximation of Eq. (12-5) may be on the order of magnitude of the cutoff frequency of the transistor. If it is, then the solution will be too fast. We must now substitute

$$\alpha = \frac{\alpha_0}{1 + s/\omega_\alpha}$$

into Eq. (12-2), and we obtain the cubic

$$s^3 + \left[\omega_\alpha + \frac{r_b(n-1)^2}{L_e}\right]s^2 + \left[\frac{1}{L_eC_c} + \frac{\omega_\alpha r_b(n-1)^2}{L_e}\right]s + \omega_\alpha\frac{1-\alpha_0 n}{L_eC_c} = 0 \tag{12-6}$$

In Eq. (12-6), when $\alpha_0 n > 1$, the coefficients of the s^3, s^2, and s terms are positive while the constant is negative. Only one of the poles will lie in the right half plane, and this pole will control the pulse rise time.

In any specific problem, the location of the right-half-plane pole is easily found by one of the standard methods of solution of a cubic, for example, Newton's method. For a polynomial of any degree, such as

$$f(s) = s^3 + as^2 + bs + c = 0$$

we first assume a suitable trial solution s_1 and find $f(s_1)$ and $f'(s_1)$. The second trial is

$$s_2 = s_1 - \frac{f(s_1)}{f'(s_1)} \tag{12-7}$$

We substitute s_2 of Eq. (12-7) into $f(s)$ and continue the process. Convergence is rapid. The maximum bound of the right-half-plane pole of Eq. (12-6) is defined by the inequality

$$s_{p,\max} \leq \sqrt[3]{\omega_\alpha \frac{\alpha_0 n - 1}{L_eC_c}}$$

Either $s_{p,\max}$ or the positive root of the quadratic [Eq. (12-4)] is an appropriate choice as the first trial solution in the Newton method.

ω_α will have a significant effect on pole location only if it is on the order of magnitude of or smaller than the pole location given by Eq. (12-4). If it is much larger, the pole location will correspond exactly to our previously assumed constant α. For example, with $\omega_\alpha = 50 \times 10^7$, the new pole location is at $s_p = 3.66 \times 10^7$ (compared with the old location at 3.9×10^7). However, if $\omega_\alpha = 0.5 \times 10^7$, the positive pole shifts to $s_p = 0.737 \times 10^7$. The rise time increases from 56 to 300 ns.

Effects of R_L on the rise time

R_L is reflected through the ideal transformer to the emitter winding, where it appears in shunt with r_b. For other than a 1:1 turns ratio, R_L must be

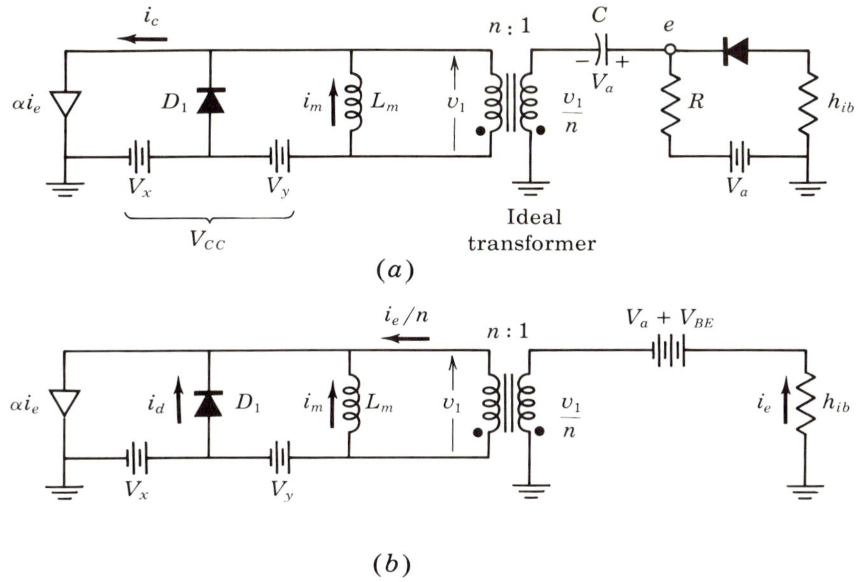

Fig. 12-3 (*a*) Model holding in the timing region; (*b*) simplified model.

multiplied by the square of the appropriate turns ratio. If we now replace r_b by $r_b \parallel R'_L$, all the preceding arguments still apply. Since R'_L reduces the net series resistance, the approximations made in finding the pole location become even more valid.

12-3 Inductively Timed Pulse Duration

Once the initial current buildup is complete, the conducting diode connects the supply V_y directly across the transformer's magnetizing inductance. (The diode D_1 prevents transistor saturation and the attendant long storage-delay time during turn-off. When generating wide pulses, D_1 may be omitted. Transistor saturation will thus perform the switching function.)

Figure 12-3*a* represents the circuit after the current through L_e has reached its final value. The emitter input circuit is linearized to simplify the analysis. Furthermore, since the coupling capacitor C is very large, we can assume that its voltage V_a remains constant over the entire pulse. Figure 12-3*b* shows the simplified model which will be used to define the timing of the pulse duration.

Initial conditions for the timing interval are found from Fig. 12-3*b*, with $i_m = 0$ (L_m out of the circuit). The constant voltage across the primary,

$V_1 = -(V_y + V_D)$, is reflected into the emitter winding, resulting in a constant emitter current of

$$I_{ep} = \frac{1}{h_{ib}}\left(\frac{V'_y}{n} - V'_a\right)$$

$$= \frac{1}{5}\left[\frac{20 + 0.7}{3} - (2 + 0.7)\right] = 840 \text{ mA} \qquad (12\text{-}8)$$

We use the primed notation $V'_y = V_y + V_D$ and $V'_a = V_a + V_{BE}$ to indicate the addition of a diode drop to the supply voltages V_y and V_a. Note that with this high current, $r_e \approx 0$, thus justifying our decision to neglect r_e. Because of the very small r_e, $h_{ib} \approx (1 - \alpha)r_b + R_E$ (R_E is the series lead resistance, which is approximately 1 Ω, making $h_{ib} = 5$ Ω).

The primary current is related to I_{ep} by the transformer turns ratio n:

$$I_1 = \frac{I_{ep}}{n} = \frac{840}{3} = 280 \text{ mA} \qquad (12\text{-}9)$$

But the collector current is αI_{ep}. Thus the initial diode current must be

$$i_d(0^+) = I_{dp} = I_{ep}\left(\alpha - \frac{1}{n}\right) = 546 \text{ mA} \qquad (12\text{-}10)$$

The pulse duration is timed by the current buildup in the magnetizing inductance. Because the conducting diode has connected a constant voltage across L_m, the magnetizing current will increase linearly with time:

$$i_m = \frac{V'_y}{L_m} t \qquad (12\text{-}11)$$

Since the collector current remains constant at αI_{ep}, as i_m increases the diode current must decrease linearly at the same rate; current is transferred from the diode to the magnetizing inductance:

$$i_d(t) = I_{dp} - \frac{V'_y}{L_m} t \qquad (12\text{-}12)$$

Once the magnetizing current builds up to I_{dp} [Eq. (12-10)] and the diode current drops to zero, the shunting of the transistor by the diode ceases. The loop gain again becomes greater than unity. Regenerative switching turns the blocking oscillator back off. The pulse duration is found from Eqs. (12-12), (12-10), and (12-8):

$$T_p = \frac{I_{ep}L_m(\alpha n - 1)}{nV'_y} = \frac{0.84 \times 200 \times 10^{-6}(2.94 - 1)}{3(20 + 0.7)} = 5.2\ \mu\text{s}$$

The above equation may also be written in the form

$$T_p = \frac{(V'_y - nV'_a)L_m(\alpha n - 1)}{n^2 h_{ib} V'_y} \qquad (12\text{-}13)$$

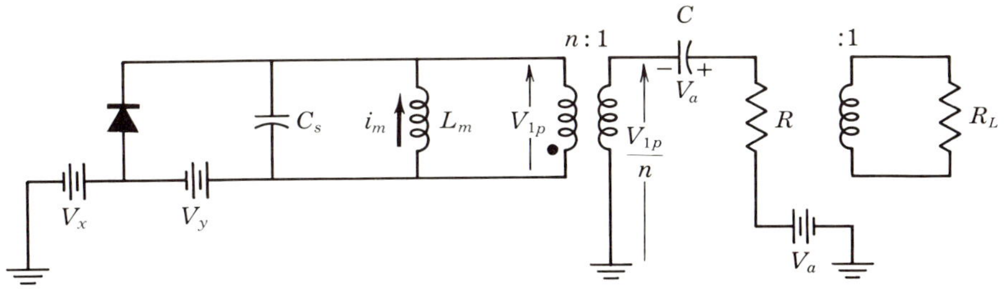

Fig. 12-4 Recovery-region model for the blocking oscillator of Fig. 12-1.

We see from Eq. (12-13) that when all voltages are derived from the same power supply, the pulse duration, to a very good first approximation, is independent of the power supply. Because T_p must be positive and real for any combination of supply voltages, Eq. (12-13) gives us a second design condition:

$$V'_y > nV'_a$$

In the special case where $V_a + V_{BE} = 0$ at the beginning of the pulse, we obtain the maximum duration of

$$T_{pm} = \frac{L_m(\alpha n - 1)}{n^2 h_{ib}} \tag{12-14}$$

From Eq. (12-14) we see the linear dependence of T_{pm} on L_m and the inverse dependence on h_{ib}. A resistance inserted in series with either the emitter or the base can be used to shorten the pulse period.

Figure 12-5 shows the current and voltage waveshapes illustrating the linearly increasing current and the constant primary voltage during the timing interval.

12-4 Recovery Region

After the transistor switches off, in much the same manner as it turned on, the blocking oscillator begins its recovery. As the current in L_m decays, the change in the sign of di_m/dt produces a large backswing of voltage, which, coupled by the transformer, aids in back biasing the emitter.

The model of Fig. 12-4 defines the recovery-region response. This process is identical with the recovery of the switched inductance described in Chap. 7. Recovery can be of the damped-oscillatory type, it can be critically damped, or it can even be overdamped. A small R_L reflected across L_m will overdamp the recovery, and this is the only case we shall consider.

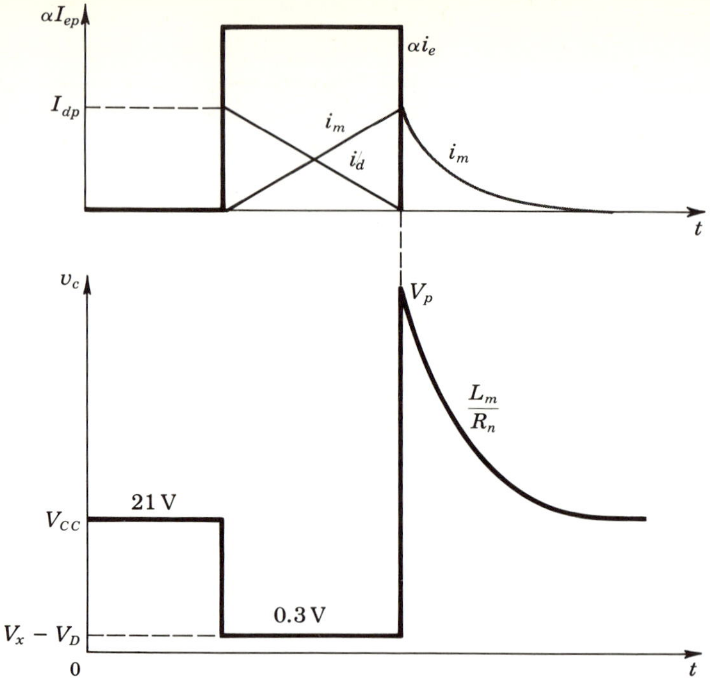

Fig. 12-5 Current and voltage waveshapes for the inductively timed blocking oscillator of Fig. 12-1.

From the model of Fig. 12-4, we see that the primary voltage would be given by the product of the peak value of the magnetizing current and the reflected resistance. From Eq. (12-10),

$$V_{1p} = I_{mp}n^2(R \parallel R_L) = I_{ep}\left(\alpha - \frac{1}{n}\right)n^2(R \parallel R_L)$$

The collector voltage jumps from $V_x - V_D$ to

$$V_{cp} = V_{CC} + V_{1p}$$

with the emitter voltage going to

$$V_{ep} = V_a + \frac{V_{1p}}{n}$$

Recovery toward the initial conditions is with the fast-time constant

$$\tau_o = \frac{L_m}{n^2(R \parallel R_L)} = 220 \text{ ns} \tag{12-15}$$

Even with $R_L = 100\ \Omega$, $V_{1p} = 491$ V, which will certainly cause collector-junction breakdown. To protect the transistor, we must shunt the transformer with a diode or a zener diode, as shown in Fig. 12-6. Diode D_2

conducts on the backswing once the voltage exceeds its junction drop. Since it shunts the coil with the low series resistance R_R, the discharge-time constant will be greatly increased. In the solution of the recovery interval, for a circuit containing a zener diode, two time constants must be considered—a long one while D_Z conducts and a faster one [Eq. (12-15)] after the voltage decays below the zener break.

With a large R_L and no diode, instead of the current and voltage decaying monotonically toward the steady-state values, the magnetizing inductance may resonate with the stray capacity of the circuit, producing ringing in the output. The initial amplitude depends on the amount of energy previously stored in the transformer. If the oscillations damp out slowly, the next change in polarity may force the transistor back into its active region, resulting in a second, undesired output pulse—the circuit would become astable.

12-5 Transformer Core Properties and Saturation

The inductance of the blocking-oscillator transformer is proportional to the ratio of flux linkages and the current establishing this flux:

$$L = N \frac{d\Phi}{di} 10^{-8} \text{ H} \qquad (12\text{-}16)$$

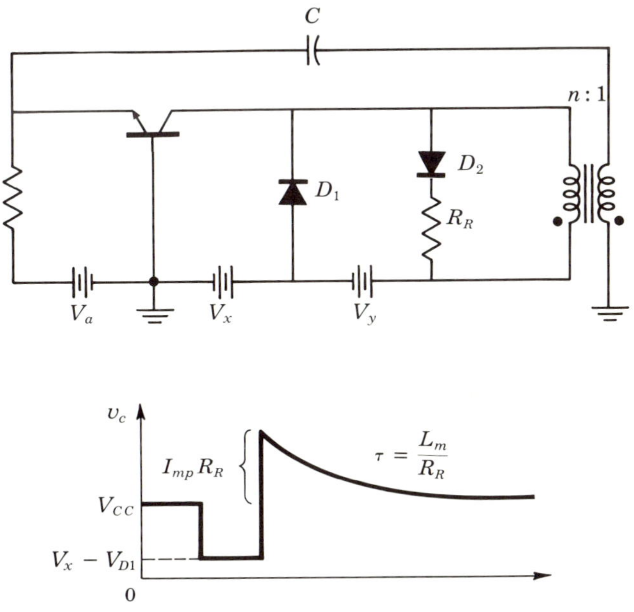

Fig. 12-6 Diode limiting of recovery-region voltage of blocking oscillator.

At low frequencies or even at dc, the relationship between the flux and the excitation is adequately expressed by the hysteresis curve (Fig. 12-7). Assuming ideal conditions, B is simply the flux per unit area and H is proportional to the ampere-turns per unit length. For large signals, $B = \mu H$, and hence the inductance is proportional to the permeability of the core. From the hysteresis loop of Fig. 12-7*a*, we can see that μ varies over rather wide limits as the core flux increases from zero to saturation.

If the hysteresis loop is approximated by the rectangular plot of Fig. 12-7*b*, we avoid the necessity of treating a circuit containing a continuously changing inductance. After lumping the core losses together with the external resistance, it can be assumed that

$$L = L_m \qquad |i| < |i_s|$$

and

$$L \approx L_s \qquad |i| > |i_s|$$

where i_s is the value of the coil current corresponding to the saturation magnetizing force H_s. The inductance in the saturation region approximates that of the coil in air; since $L_s \ll L_m$, it is often taken as zero.

Because it requires finite time for the applied excitation field H to propagate into the core and establish and align the randomly oriented magnetic domains, the above formulation is not completely correct. With a pulse excitation, the flux is initially only at the surface, and consequently the effective permeability is quite low. It increases as the field penetrates into the core. Even if the excitation is well above the saturation value, it may still take several microseconds to saturate a thin core. The penetration time may be long compared with the width of the pulse generated in the blocking oscillator, and consequently the assumption of a constant L_m is not valid.

In order to avoid this inconsistency, we would have to postulate a magnetizing inductance whose value increases with time from slightly above

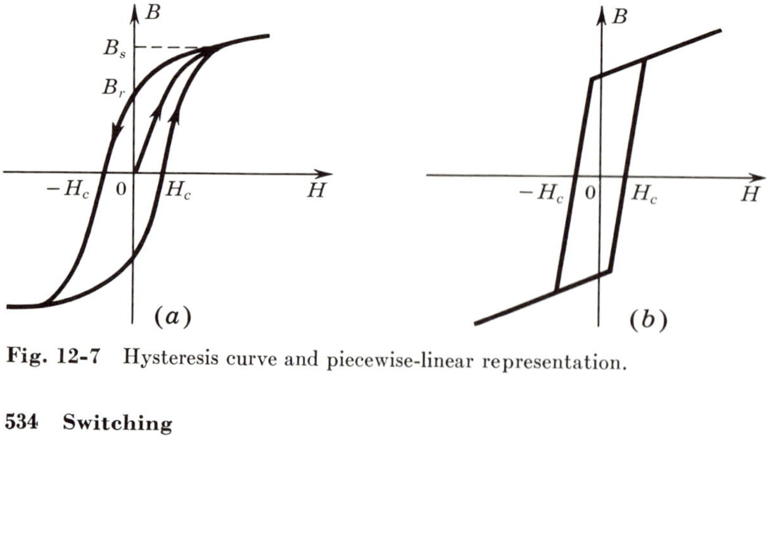

Fig. 12-7 Hysteresis curve and piecewise-linear representation.

that of an air-core coil to a peak given by the steady-state core characteristics. A circuit element which exhibits a response of this nature is impossible to use in any simple approach to circuit analysis. We shall therefore continue to explain the blocking-oscillator operation on the basis of a fixed L_m and accept the resulting errors. However, the accuracy of calculations can be improved if the inductance value used is not that at direct current but rather that measured under pulsed conditions.

To summarize:

1. For pulse durations greater than several microseconds, the assumption of a constant L_m gives reasonably correct times.
2. When generating pulses of less than 1 or 2 μs, the duration cannot be accurately calculated from any simple model. If a constant L_m is assumed, then the actual pulse will terminate sooner than calculated.
3. A finite time is required to saturate the core, and therefore this problem need be considered only when generating wide pulses.

Returning to the blocking oscillator, we shall now examine the operation when the core does saturate. Up to this point, the response will be identical with that discussed in Sec. 12-3. Once the magnetizing current reaches the saturation current i_s, L_m drops to a very small value. The primary current will increase extremely rapidly [Eq. (12-11)]. At the same time, saturation inhibits the transformer action, causing the secondary voltage to fall and the emitter current to drop. The circuit returns to the active state (diode D_1 off) almost immediately.

As the stored energy is dissipated, the left-hand branch of the hysteresis curve is traversed by the operating point. As a result, a large reverse voltage pulse appears at the collector. The time required for the decay depends on the nature of the transformer core. (This problem is treated in more detail in Chap. 15.) Because of the extremely nonlinear behavior, it is difficult to define the decay time in terms of the specific circuit parameters. In general, the manufacturer's specifications as to the rise and fall times and pulse width would be used in the circuit design.

12-6 Capacitively Timed Blocking Oscillator

The coupling capacitor C is now reduced in size until it alone will control the pulse duration. Timing no longer depends on the complete transfer of diode current to the magnetizing inductance bringing the circuit back into the regenerative region. Now, instead, the decreasing emitter current due to the charging of C will eventually turn off the transistor.

To simplify the following analysis, we shall assume, quite incorrectly, that the magnetizing current remains zero over the complete pulse interval. Moreover, since the Ch_{ib} time constant may be of the same order of magnitude as the time constant associated with the positive pole given by Eq. (12-4), the approximation that the initial buildup is complete before the timing element starts charging is no longer completely valid. We shall, however, assume that the peak values expressed in Eqs. (12-8) and (12-10) are still correct and shall begin our analysis from this point. By doing so, the nature of the reduction in the pulse duration may be seen most readily.

From the model of Fig. 12-8*a*, we can see that C charges from its initial value of V_a toward the Thévenin equivalent voltage

$$v_q(\infty) \approx \frac{V'_y}{n} + V'_a \frac{h_{ib}}{R} - V_{BE} \tag{12-17}$$

The time constant is

$$\tau_c = C(R \parallel h_{ib}) \approx Ch_{ib}$$

During this interval, the emitter current decays from its peak [Eq. (12-8)] toward the steady-state value of

$$i_e(\infty) = \frac{V'_a}{h_{ib} + R} \approx -\frac{V'_a}{R}$$

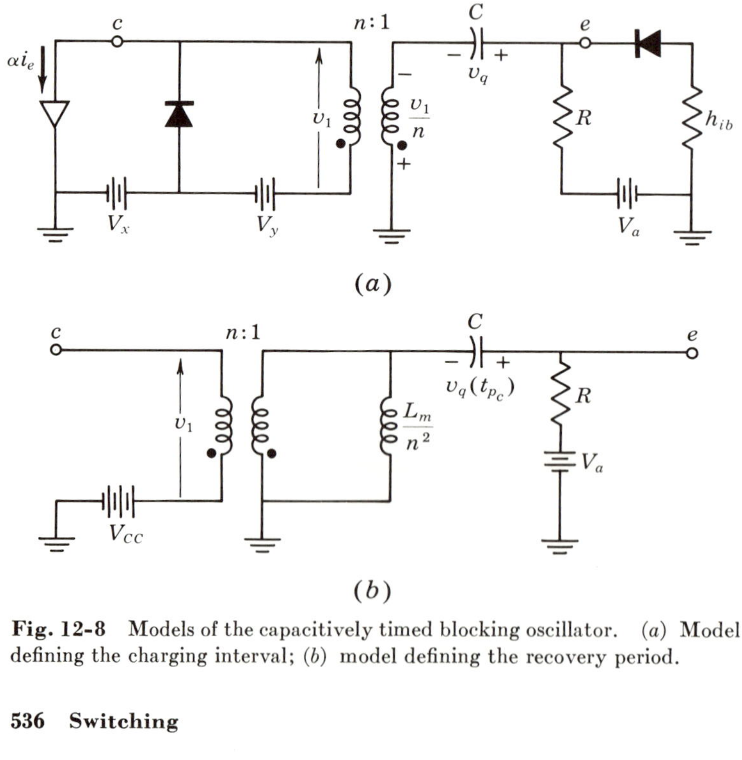

Fig. 12-8 Models of the capacitively timed blocking oscillator. (*a*) Model defining the charging interval; (*b*) model defining the recovery period.

The primary current will always be $1/n$ times as large. Consequently, because the diode is conducting at the beginning of the pulse, it will remain conducting for the full pulse interval. The end of the pulse will occur once $i_e = 0$ and the transistor turns off.

The duration of the pulse, as found from the exponential charging equation, is given by

$$t_{p_c} = \tau_c \ln \frac{i_e(\infty) - I_{ep}}{i_e(\infty)}$$

or by substituting the appropriate terms,

$$t_{p_c} = Ch_{ib} \ln \left[1 + \frac{R}{nh_{ib}} \left(\frac{V'_y}{V'_a} - n \right) \right] \tag{12-18}$$

Since $R \gg h_{ib}$, this portion of the pulse will consist of virtually the complete exponential. To avoid the ambiguity arising when the switching point is close to the steady-state value, V_a might be made highly negative and R might be reduced.

At the termination of the pulse, the transistor switches off and the recovery model reduces to the one shown in Fig. 12-8*b*. Because the approximations made in the course of the above analysis neglected the energy storage in the transformer, the exact nature of the response over this portion of the cycle is somewhat difficult to define. One type of response would be expected when the secondary magnetizing inductance limits the rate of capacitor-current flow, and yet another type where the transformer saturates rapidly. Regardless of the particular controlling element, the energy storage will almost always produce a large spike of voltage, and then either ringing (which must be suppressed) or an exponential decay back toward the original state. Waveshapes of the emitter current and collector voltage, illustrating the type of recovery to be expected, appear in Fig. 12-9.

We shall now consider the change in the response of the blocking oscillator previously discussed when C is reduced from 10 to 0.05 μF. From Eq. (12-18), the duration of the final portion of transistor saturation is

$$t_{p_c} = Ch_{ib} \ln \left[1 + \frac{2{,}000}{15} (7.7 - 3) \right]$$

Since this is almost the complete exponential,

$$t_{p_c} \approx 4Ch_{ib} = 4(0.05 \times 10^{-6}) \times 5 = 1 \ \mu\text{s}$$

The pulse width, under this condition, is approximately one-fifth the time calculated when L_m controlled the timing.

We should note that the maximum pulse duration is due to the current buildup in the magnetizing inductance. Either the charging of C or the saturation of the core can only reduce this time. Furthermore, the simul-

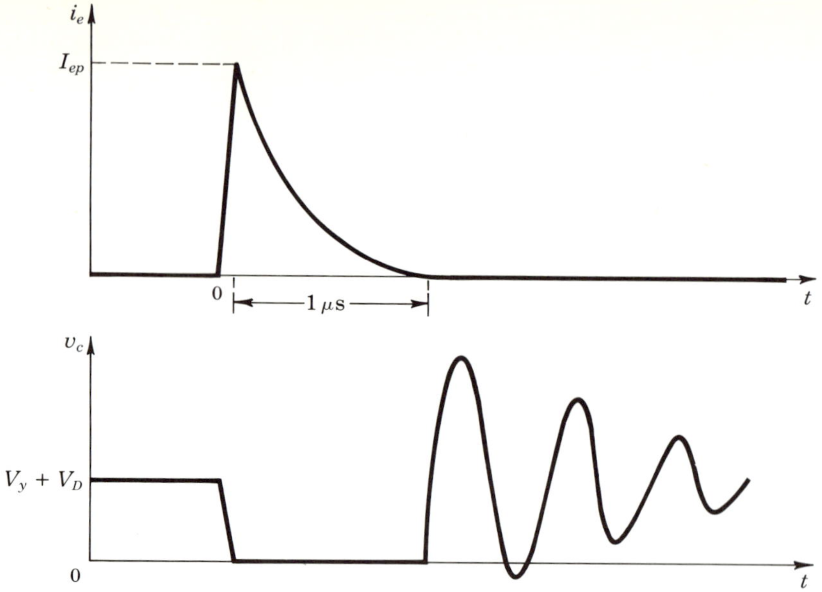

Fig. 12-9 Waveshapes appearing in the capacitively timed blocking oscillator. The recovery region illustrates ringing.

taneous magnetic-flux buildup means that the peak currents that are reached and the actual pulse duration will usually be less than calculated on the basis of the idealizations assumed above. C is almost never used to control the pulse width, but it is used to time the interval between pulses in a free-running blocking oscillator (Sec. 12-7).

12-7 Astable Operation

As might be anticipated from the discussion of monostable and astable operation in Chaps. 8, 9, and 11, shifting the quiescent operating point from the cutoff to the active region will make the blocking oscillator free-running. For example, the astable circuit of Fig. 12-10, except for the reversal of the emitter battery, is identical with the monostable circuit of Fig. 12-1.

During the pulse interval, the large emitter current charges C slightly positive. After the circuit switches off, the accumulated charge maintains the emitter junction back-biased. Eventually the voltage from emitter to ground decays to zero and the circuit generates another pulse.

At the instant of switching on, the net voltage on C must be $-V_{BE}$. The voltage at the transformer secondary has not as yet built up to V'_y/n, and therefore the emitter junction state depends solely on the voltage across C. If this voltage were more positive, the emitter would be back-biased, and if it were more negative, the circuit would have switched on earlier. Except for the change in the initial condition, the operation of the blocking oscillator during the pulse interval is identical with that previously discussed. The peak negative voltage at the emitter and the peak current flow are now given by

$$V_{em} = -\frac{V'_y}{n} - V_{BE} \qquad I_{ep} = \frac{V'_y}{nh_{ib}} \tag{12-19}$$

For the purposes of computation, we shall assume that this current remains constant over the complete pulse interval. Since I_{ep} flows through C, it will produce a net voltage change of

$$\Delta V = \frac{1}{C} I_{ep}\, \Delta t \tag{12-20}$$

where Δt is the width of the pulse [$\Delta t = T_p$ as defined in Eq. (12-13)]. The capacitor thus charges linearly from its initial value of $-V_{BE}$ to

$$V_{qp}(\Delta t) = -V_{BE} + \Delta V$$

as shown in Fig. 12-11. In order to sustain a constant emitter current over the pulse interval, the change in the capacitor voltage must be small in comparison with the voltage across the transformer secondary:

$$|V_{qp}| \ll \frac{V'_y}{n} \tag{12-21}$$

Combining Eqs. (12-19) and (12-20) and the inequality of Eq. (12-21), we can express one of the limiting conditions as

$$C \gg \frac{\Delta t}{h_{ib}} \tag{12-22}$$

At the end of the pulse interval, the transistor is driven hard off by the inductive backswing. After the initial recovery with the fast time con-

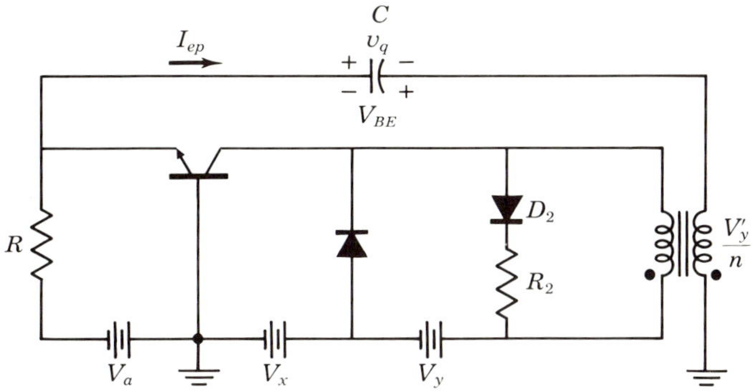

Fig. 12-10 Astable blocking oscillator—voltages shown at the beginning of the pulse timing region.

stant $\tau_o = L_m/n^2(R \parallel R_L)$, the charge in C will maintain the cutoff condition. As C discharges toward $-V_a$, with the relatively long time constant $\tau_c = RC$, the emitter voltage decays from V_{qp}. Once it reaches $-V_{BE}$, the transistor again becomes active and the cycle repeats. Thus the duration of the interval between adjacent pulses is given by

$$T = RC \ln \frac{-V_a - V_{qp}}{-V_a - V_{BE}} \tag{12-23}$$

and, for a given period, Eq. (12-23) determines the size of R.

The astable blocking oscillator is commonly used to generate trains of narrow pulses having relatively wide separation. The time defined in Eq. (12-23) very nearly represents the complete period of the generated signal.

As with all free-running circuits, the blocking oscillator may be synchronized to an external control signal. It must, of course, have a frequency and amplitude placing it within the regions of synchronization. Because the pulse is so very narrow when compared with the full period, the solution of the synchronization problem would be much closer to that found for the sawtooth sweep in Chap. 6 than to any regions which might be constructed for a multivibrator.

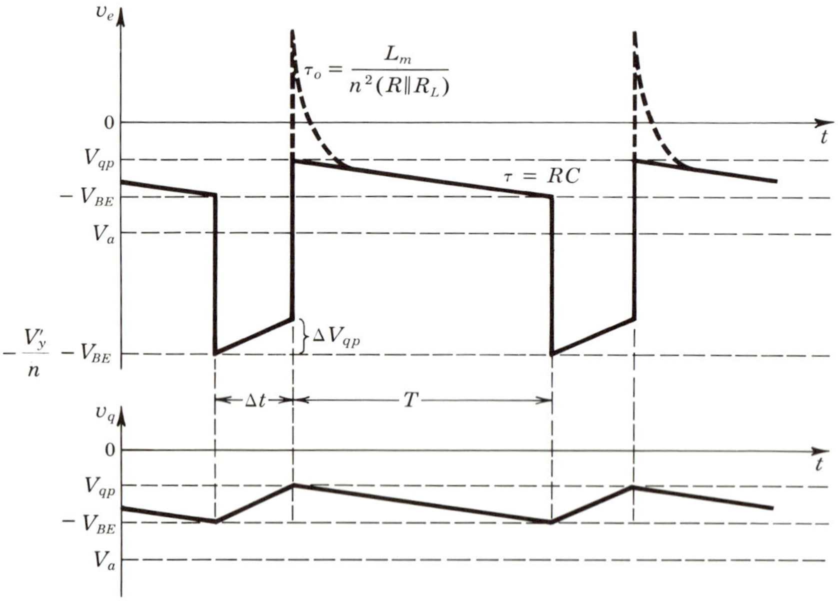

Fig. 12-11 Emitter and capacitor voltage waveshapes in the astable blocking oscillator.

Example 12-1

In this problem we wish to calculate the values of C and R necessary to establish a pulse spacing of 2.5 ms. The blocking oscillator generates a 5-μs pulse. The significant parameters for the recovery interval are $h_{ib} = 10\ \Omega$, $V'_y = 20$ V, $V_a = -1$ V, $n = 2$, and $L_m = 4$ mH. $R_L = 1$ K.

From these parameters, Eq. (12-22) yields

$$C \gg \frac{5 \times 10^{-6}}{10} = 0.5\ \mu\text{F}$$

In order to keep its size within bounds, we shall choose $C = 2\ \mu$F. The peak voltage developed across it is

$$V_{qp} = \frac{V'_y}{Cnh_{ib}}\,\Delta t - V_{BE} = 0.625 - 0.700 = -0.075\ \text{V}$$

The remaining circuit component R is now found from Eq. (12-23):

$$R = \frac{T}{C \ln \dfrac{-V_a - V_{qp}}{-V_a - V_{BE}}} = \frac{2.5 \times 10^{-3}}{2 \times 10^{-6} \ln 3.58} = 980\ \Omega$$

As a final step, we should verify that the voltage backswing due to the energy stored in the transformer will have damped out long before the transistor turns back on. This time constant is

$$\tau_o = \frac{L_m}{n^2(R \parallel R_L)} = \frac{4 \times 10^{-3}}{4 \times 495} = 2.02\ \mu\text{s}$$

and thus the backswing occupies an insignificant portion of the 2.5-ms recovery interval. If a zener diode were included, this might no longer be true.

Problems

12-1 (*a*) Find the exact location of the poles of the blocking oscillator. Define the rise time in terms of the circuit parameters. What conditions are necessary to justify the approximation of Eq. (12-4)?

(*b*) Find the right-half-plane pole of Eq. (12-6) for the circuit of Fig. 12-1 by Newton's method. Take $\omega_\alpha = 5 \times 10$. Start your solution from $s_{p,\max}$. Find the location of the other two poles.

12-2 Plot the path of operation of the blocking oscillator on the collector volt-ampere characteristic. Label each region, showing rise, pulse timing, and recovery. Show how the transistor junction breakdown may affect the recovery path. Give the new recovery-region time constant and draw the waveshape.

12-3 Repeat the calculations for the circuit of Fig. 12-1 for $n = 4$. What happens to the pulse width if $n = 5$?

12-4 Sketch and label completely the waveshapes appearing in the circuit of Fig. 12-1 when the transformer is replaced by one having $L_e = 15\ \mu H$, $L_m = 6.5\ \mu H$, $n = 2$, $C_s = 5$ pF. $R_L = 50\ \Omega$. All other terms remain the same.

12-5 Since the ideal-blocking-oscillator pulse should have a very fast rise time compared with the width of the pulse, the ratio of T_p/T_R can serve as a figure of merit for the circuit. This ratio can tell us what type of transformer is optimum. For simplicity, assume $\alpha = 1$ and $r_b = 0$.

(a) Evaluate $T_{p,\max}/T_{R,\min}$. State the assumptions made.

(b) What L_m, L_e, and n combination is the best choice for a blocking oscillator of the type of Fig. 12-1?

12-6 Figure 12-12 shows a collector-to-base-coupled blocking oscillator with the external load connected to a tertiary winding.

(a) What is the minimum n for proper operation? Indicate the polarity of the transformer connections by dots.

(b) Assume that the transformer turns ratio is three times the bounding value. In addition, the ratio $L_m/L_e = 15{:}1$. Specify the transformer parameters for a 5-μs pulse duration.

(c) Sketch and label the collector-voltage and load-current waveshapes.

(d) Compare the advantages and disadvantages of this configuration with those of the circuit of Fig. 12-1.

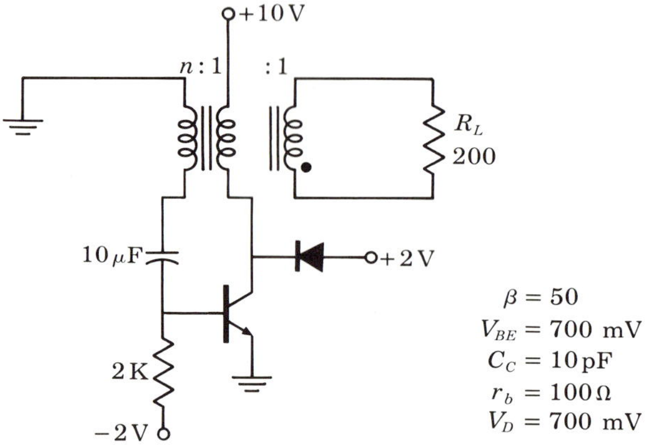

Fig. 12-12

12-7 Draw the circuit for an emitter-to-base-coupled blocking oscillator. Use the parameters of Fig. 12-12 with leakage and magnetizing inductance (referred to the emitter winding) of $L_m = 500\ \mu H$, $L_e = 50\ \mu H$, and $n = 3$ (base winding). Calculate the load-current and emitter-voltage waveshapes.

12-8 Show three methods of triggering the blocking oscillator of Fig. 12-1. Specify the pulse height, polarity, and order of magnitude of the source impedance. Does any one method chosen have overwhelming advantages? Explain.

12-9 The transformer of the circuit of Fig. 12-1 is changed to one which saturates at 200 mA (all parameters remain the same). After saturation, L_m drops from 200 to 50 μH. Plot the new waveshapes, paying particular attention to operation after saturation. Are all the assumptions made in the text still valid? What happens after the transistor turns off?

12-10 Figure 12-13 shows a blocking oscillator designed so that the pulse width can be controlled by a constant current injected into a fourth winding. This current establishes a bias flux and thus influences the saturation point. Saturation occurs at a field strength of 5 A-turns. V_B varies from 0 to 50 V. Specify the L_m, C, and R that will permit the adjustment of the pulse duration from 1 to 10 μs (ignore the rise time).

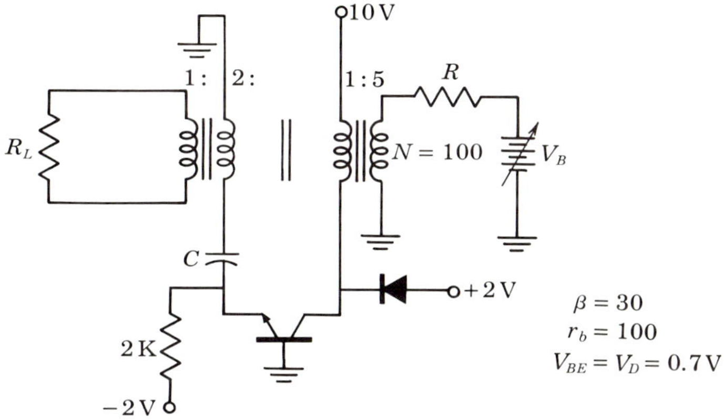

Fig. 12-13

12-11 (*a*) The circuit of Fig. 12-12 is made astable by reversing the polarity of the base bias supply. In addition, $L_m = 1$ mH and $n = 0.1$. Calculate the value of C that will set the period at 10 ms. Neglect I_e in your calculations. Show the recovery portion of the collector waveshape and the capacitor current.

(*b*) By how much does the charging current of C reduce the pulse duration? Make all reasonable approximations in your calculations.

Bibliography

Benjamin, R.: Blocking Oscillators, *J. IEE (London)*, pt. IIIA, vol. 93, pp. 1159–1175, 1946.

Linvill, J. G., and R. H. Mattson: Junction Transistor Blocking Oscillator, *Proc. IRE*, vol. 43, no. 11, pp. 1632–1639, December, 1955.

Narud, J. A., and M. R. Aaron: Analysis and Design of a Transistor Blocking Oscillator Including Inherent Nonlinearities, *Bell System Tech. J.*, vol. 38, no. 3, pp. 785–852, 1959.

PART 4

memory

13. the bistable as a logic element

A bistable multivibrator is simply two cross-coupled inverters; the output of the OFF inverter at the logical-1 level coupled to the input of the second inverter turns it ON, with the latter's logical-0 output keeping the first inverter off. The state of either inverter output thus registers the digital information stored. Our problem is to ensure that the correct inverter is in the ON position. This can be accomplished by externally driving the appropriate inverter from the OFF to the ON state or from the ON to the OFF state by making the inverter part of a logic array and then satisfying the logic by the appropriate combination of logical-0 and logical-1 inputs. Any particular switching requirement is thus determined by the inputs to the logic array.

In integrated circuits, the inverter is the output stage of the NAND or NOR gate that is the basic element of the particular class of logic chosen for use in the overall system. The inverter varies from the simple transistor stage in an RTL logic (Fig. 4-16), to the MOSFET stage (Fig. 5-18), to the totem-pole output of the DTL or TTL logic (Fig. 4-28). The input logic (NANDs or NORs) follows the logic of the class and will be driven from the outputs of similar gates or bistables.

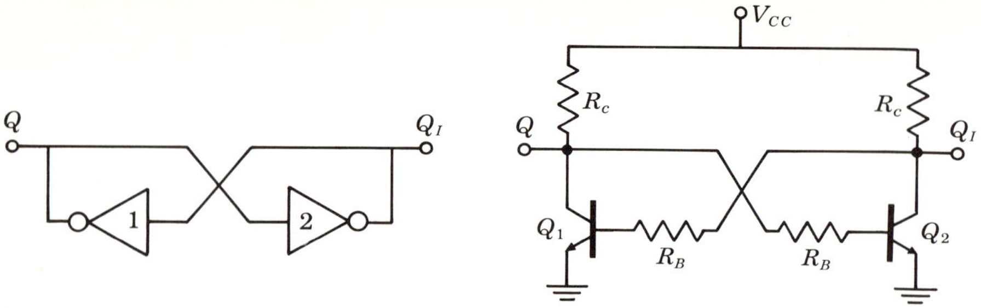

Fig. 13-1 Bistable multivibrator shown as two cross-coupled inverters.

13-1 *S-R* Multivibrator Switching: NOR Inputs

The simple two-inverter multivibrator of Fig. 13-1 satisfies the following set of equations, which, we note, are completely self-consistent:

$$Q = \bar{Q}_I \tag{13-1}$$

$$Q_I = \bar{Q} \tag{13-2}$$

The inverse output $Q_I = \bar{Q}$. $\bar{Q}$ and Q_I will be used interchangeably. Q_I will be used primarily where we apply the inverse signal to another inverter and wish to keep track of the origin of the signal. $\bar{Q}$ will be used where the logic statement is the significant factor and the particular source of the input signal is not important.

By itself, the bistable is useless. We must be able to set the state of the inverters by external signals, the application of which should not interfere with the normal bistable operation. A two-input NOR gate which replaces the simple inverter (Fig. 13-2) permits two separate signals to set the common-collector output to logical 0; one signal is coupled from the other NOR output, and the other signal is the external input used to set the stage to a logical 0. The logic statements at the common-collector nodes now become

$$Q = \overline{Q_I + R} = \bar{Q}_I\bar{R} \tag{13-3}$$

$$Q_I = \overline{Q + S} = \bar{Q}\bar{S} \tag{13-4}$$

Figure 13-2 shows the block diagram of the logic as well as typical bistables in both the bipolar and the MOSFET forms.

If Q is the logical-1 state (which means that Q_1 is off), then a logical-1 input at R will bring the common-collector node (drain node) Q to the logical-0 level. The reset input turns the OFF device on. From Eq.

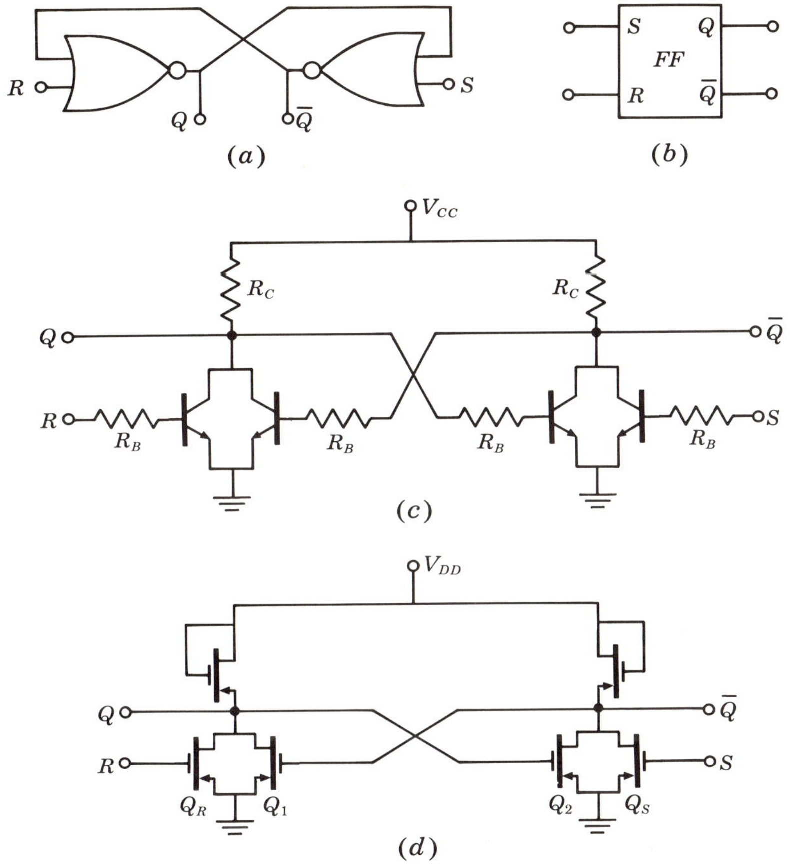

Fig. 13-2 *S-R* NOR-input flip-flop. (*a*) Logic; (*b*) circuit symbol; (*c*) bipolar-transistor RTL NOR-gate bistable; (*d*) MOSFET NOR-logic bistable.

(13-4) we see that now, with $Q = 0$ and with no set input, $Q_I = 1$. The multivibrator operation sustains the new state after the reset pulse drops back to the logical-0 level. A logical 1 at the set input S causes Q_I to become 0, and because of the cross-coupling, $Q = 1$. Thus in this circuit the Q output follows the set input and the inverse output Q_I follows the reset input.

In the case of $S = 0$ and $R = 0$, the multivibrator remains in the state previously existing. For $S = 1$ and $R = 0$, the Q output is a logical 1. For $S = 0$ and $R = 1$, $Q = 0$. When both R and S are simultaneously set to logical 1, both NOR-gate outputs are forced to the logical-0 level by the saturation of the gating transistor. The bistable's state upon removal of the excitation is ambiguous since it depends on the random unbalance of the circuit. It is, consequently, not defined (ND).

The truth table for the S-R bistable is

S	R	Q	$\bar{Q}$
0	0	Q	$\bar{Q}$
0	1	0	1
1	0	1	0
1	1	ND	

Not all manufacturers define the inputs in the same way. Some use the term *clear* rather than *reset*. Others even interchange the set and reset inputs, causing the Q output to follow the reset and $\bar{Q}$ output to follow the set input. An examination of the specific circuit and its truth table will indicate the correct sequence of responses.

More complex input logic can be constructed for special purposes, and to treat these cases, we simply consider the final logic output applied for switching as S or $\bar{S}$ (or R or $\bar{R}$), depending on whether or not the applied signal is inverted.

Clocked bistables

One special case is where we wish to limit switching to an interval defined by the presence of a clock pulse. Since we want switching only upon the coincidence of two pulses, we apply both pulses to the input of an AND gate, with the gate output going to the set or reset terminal of the bistable. The switching logic becomes

$$Q = \overline{Q_I + R_C C_p} \tag{13-5}$$

$$Q_I = \overline{Q + S_C C_p} \tag{13-6}$$

Clocked switching is also called *synchronous switching*, in contrast to the *asynchronous* direct set-reset switching. Figure 13-3 shows one possible circuit based on the NOR logic of Fig. 13-2. Only when both S_C and C_p are at the logical-1 level will the AND-gate output, $Y = S_C C_p$, be at the logical-1 level. This signal saturates the NOR stage of the bistable and thus brings Q_I to the 0 state and Q to a logical 1.

The AND gate is an inhibit circuit. When the clock pulse is at the logical-0 level, the conducting input diode effectively grounds the gate input and inhibits the transmission of all other inputs. Only after priming the gate with $C_p = 1$ will the set pulse be transmitted to change the state of the bistable. Once the clock returns to the logical-0 level, the gate is again inhibited and the multivibrator stays in the switched state, storing the set state.

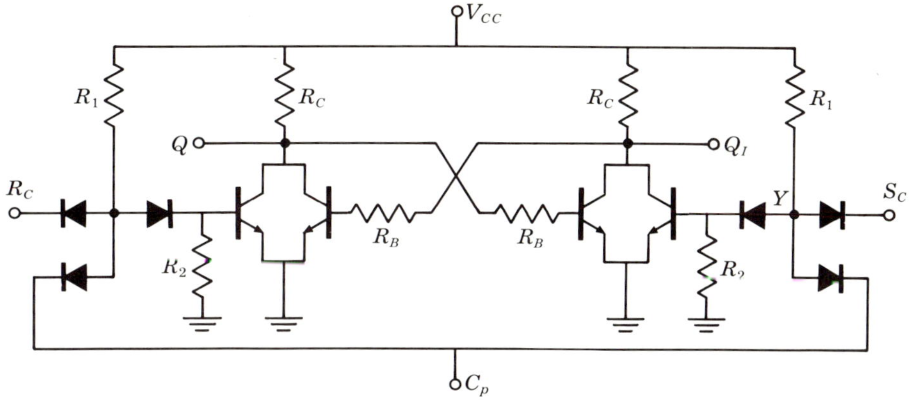

Fig. 13-3 Clocked S-R flip-flop.

13-2 S-R Bistable Switching: NAND Inputs

A bistable may also be made by cross-coupling two NAND gates (Fig. 13-4). As we have seen, the NOR input circuit combines the inverter states at the output node with any single logical-1 input switching the output node to ground. With NAND gates, the signals are combined at the input and any single logical 0 inhibits the inverter by grounding the input node; it sets the output to a logical 1. We must, therefore, apply the inverse signal $\bar{S}$ to the NAND inputs. At the input of the Q inverter, the combined AND-gate input signals are $\bar{S}Q_I$. The output Q becomes

$$Q = \overline{Q_I\bar{S}} = \bar{Q}_I + S = Q + S \tag{13-7}$$

and, by symmetry, the Q_I switching logic is

$$Q_I = \overline{Q\bar{R}} = \bar{Q} + R \tag{13-8}$$

A logical 0 at $\bar{S}$ sets Q to logical 1, and a logical 0 at $\bar{R}$ resets Q to logical 0. The $\bar{S}$ and $\bar{R}$ functions may be obtained directly from the output of other NAND gates in the overall logic system or may be derived by inverting the positive-logic S and R signals.

$\bar{S}$	$\bar{R}$	Q
0	0	*ND*
0	1	1
1	0	0
1	1	Q

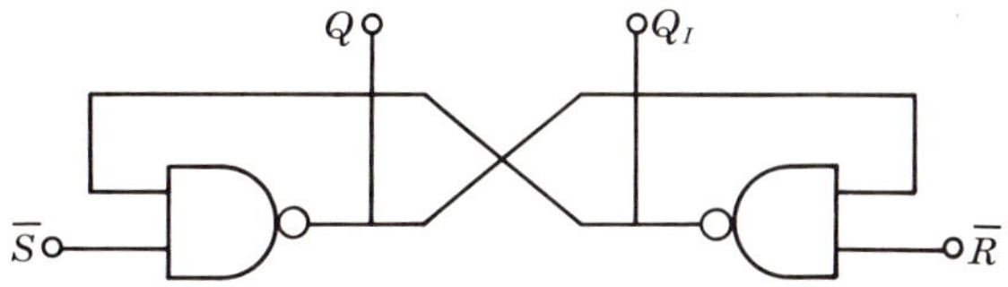

Fig. 13-4 NAND-gate S-R bistable. Truth table shown for positive logic.

We note that in the normal state all inputs except those used for the cross-coupling must be left in a primed state, that is, at the logical-1 level. Think what would happen if one input of each gate were left at the logical-0 level. The conducting input diodes would cause both inverters to be cut off, and both outputs would go to the logical-1 level. There would be no multivibrator operation. Thus external signals used for set or reset must be negative-logic signals, i.e., normally at a logical 1, dropping to a logical 0 for switching, and, for such signals, the input AND gate works as an OR gate. We conclude that the NAND bistable may be considered as a negative-logic NOR bistable.

Figure 13-5 shows a TTL S-R clocked flip-flop. The basic TTL logic is a NAND gate where the output of the input AND, Q_1, is coupled to the phase splitter and inverter Q_3, which, in turn, drives the totem-pole output

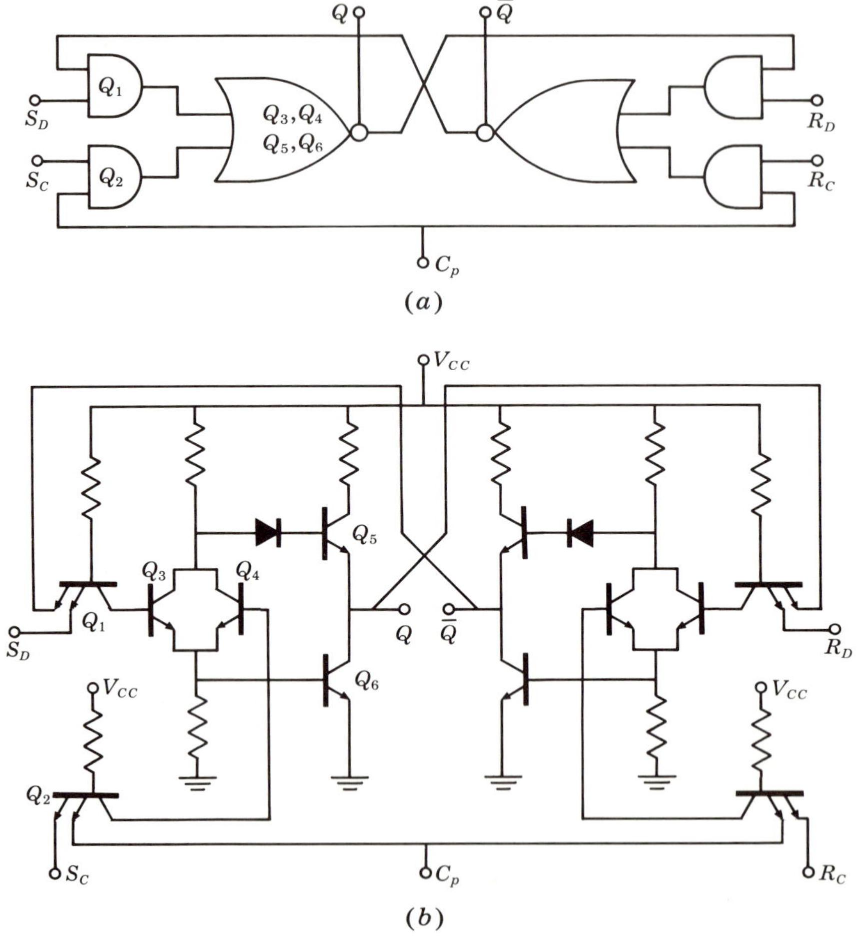

Fig. 13-5 TTL clocked S-R flip-flop. (*a*) Logic; (*b*) circuit.

stage of Q_5 and Q_6. Q_3, Q_5, and Q_6 may be treated as a single inverter stage which inverts the logic appearing at the base of Q_3.

The logic of the basic direct set-reset multivibrator is given by Eqs. (13-7) and (13-8), where $\bar{S}_D$ and $\bar{R}_D$ are the direct set and reset inputs.

Transistor Q_2 is the second AND gate. It ANDS the clock pulse and clocked set inputs. The output is coupled by Q_4 and combined with the basic multivibrator inverter through the wired NOR function at the common collector and emitter nodes of Q_3 and Q_4. Q_3 and Q_4 NOR the output of the AND gates Q_1 and Q_2. The total switching logic becomes

$$Q = \overline{Q_I\bar{S}_D + S_C C_p} \tag{13-9}$$

$$Q_I = \overline{Q\bar{R}_D + R_C C_p} \tag{13-10}$$

These equations lead to the logic array shown in Fig. 13-5*a*. Thus the state of the flip-flop may be set or reset either by the combination of the clock pulse and the set (S_C) or reset (R_C) inputs or by the direct set (S_D) and reset (R_D) inputs. From Eqs. (13-9) and (13-10) we see that the $\bar{R}_D$ and $\bar{S}_D$ inputs override all other inputs.

13-3 *S-R* Flip-Flop: Toggle-mode Operation

The simple set-reset flip-flops change state immediately upon satisfying the input logic. If the output is coupled to another bistable's input, then, after the switching delay, the next stage will change state. After a second switching delay, a third logic stage will switch state, etc. In a complex network, the state of some later bistable may affect the logic of the input to the first stage and cause it to reswitch. We see that the input pulse "races" through the system, creating the possibility of overall logic-network instability.

We can eliminate racing by storing, during the presence of the clock pulse, the input signal that will cause the change in the bistable's output. Switching or "toggling" the actual flip-flop output to the new state occurs only after the clock pulse returns to the logical-0 level.

Toggling, or time-sequential switching, is the second degree of switching complexity. The implication of circuitry where events are expected to occur sequentially is that the circuit contains a memory. The various bits of input information are introduced into the short-time memory through the input logic, which is turned on by the clock pulse. The clock pulse also activates a controlled gate or inhibit circuit, which disconnects the memory from the basic bistable during the storage interval. At the end of the clock pulse, the input logic is disabled. Slightly later, the memory is connected to the bistable and the stored information is used for switching. A block diagram illustrating the basics of toggle-mode switching appears in Fig. 13-6.

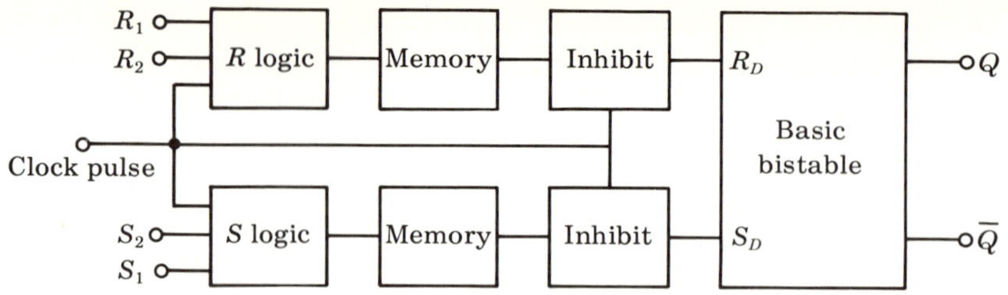

Fig. 13-6 Block diagram of toggle-mode bistable operation.

Many different inhibit circuits may be used. For example:

1. An AND gate, where a logical-0 input can be used to inhibit the gate operation.
2. A MOSFET used as a series gate—an open circuit when off and a low resistance when on.
3. A diode back-biased for inhibition and forward-biased for conduction.
4. A transistor where the emitter-to-base junction is back-biased for inhibit operation and forward-biased for transmission.

These forms of inhibit circuits are used in the various *S-R* toggle-mode flip-flops that will be discussed.

We can use the following for short-term information storage:

1. One bistable as a memory element to trigger a second bistable in a master-slave circuit.
2. Capacitor-charge storage. Under this category, we can include a regular capacitor, the gate capacitance of a MOSFET, and the depletion-region capacitance of a *pn* junction.
3. Base-charge storage in an ON transistor or charge storage in a conducting diode.

As our first example, we shall examine the master-slave circuit of Fig. 13-7, where each multivibrator is individually a clocked set-reset flip-

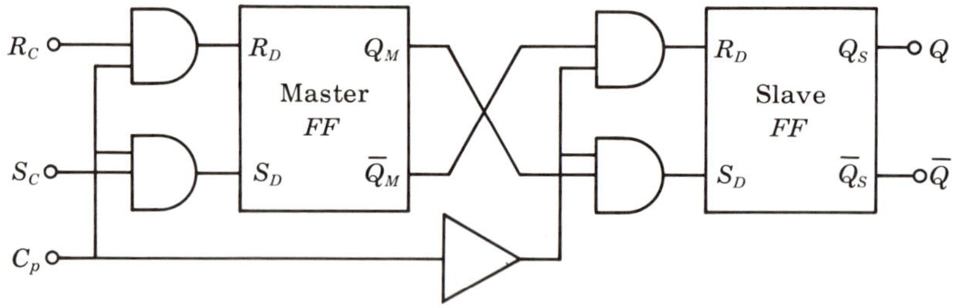

Fig. 13-7 Master-slave *S-R* flip-flop.

flop. The AND logic at the input to the master is turned on by the clock pulse, and the Q_M and $\bar{Q}_M$ outputs change to the state set by the set (or reset) input gate. The logical-1 clock state sets the inverter output to logical 0, which is used to inhibit the AND gates at the set and reset inputs of the slave multivibrator. After the clock pulse returns to the logical-0 level, the clock inverter output, $\bar{C}_p = 1$, turns on the slave input logic and the slave switches, or "toggles," to the same state as the master.

The proper sequence can be ensured by setting the threshold of the input logic gate (logical-1 level) above the inhibit level. Looking at the clock pulse (Fig. 13-8), we see that during the turn-on interval, the 0 level occurs before the logical-1 state. Thus the time sequence in the master-slave circuit is:

t_1—Slave flip-flop inhibited.
t_2—Master input logic primed and master flip-flop switched to the state determined by $S = 1$ or $R = 1$.
t_3—Master logic off.
t_4—Slave input logic primed and slave switched to the same state as master.

In order to indicate the sequential nature of toggle-mode switching, we express the logic truth table as

S_C	R_C	Q_{n+1}
0	0	Q_n
0	1	0
1	0	1
1	1	ND

Q_{n+1} indicates the output in the $n + 1$ time interval, and Q_n is the output in the preceding nth time interval. These intervals correspond to two successive clock pulses. The S_C and R_C inputs are the inputs during the coincidence of the nth clock pulse. Because of the toggling action, switching

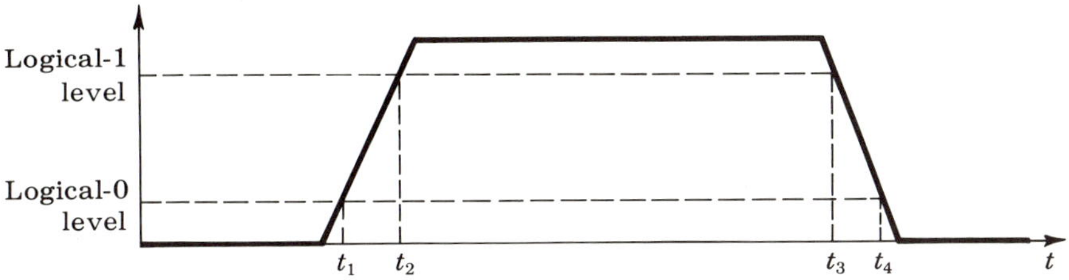

Fig. 13-8 Clock pulse used for toggling operation indicating switching times.

actually occurs on the clock-pulse 1 → 0 transition at the beginning of the $n + 1/2$ time interval.

13-4 Other Toggling Techniques

Figure 13-9 illustrates capacitor information storage. The transistor is normally saturated, with the capacitor charged to the transistor's base-to-emitter voltage drop. When a positive pulse V_1 is applied at the input, the capacitor rapidly charges to $V_1 - V_{BE}$. Since one end of the capacitor is connected to the pulse source and the other end to the base of a saturated transistor, the charging-time constant is extremely short. Upon termination of the pulse, the left-hand side of the capacitor drops to zero. Because the voltage across the capacitor cannot change instantaneously, it follows that the base side of the capacitor will also change by $-V_1$ to the new value $-(V_1 - V_{BE})$. This negative pulse turns off the transistor, and the output rises to V_{CC}.

The transistor's base recovers toward V_{CC} with the time constant R_BC. When the voltage reaches V_{BE}, the transistor turns on and the positive pulse at the inverter output drops back to the logical-0 level. Thus we see that the capacitor has stored the input pulse. At the end of the applied signal, the discharge of the charged capacitor provides a delayed negative pulse which can be used directly to turn off the ON transistor in a multivibrator. This is exactly the coupling network used in the monostable multivibrator of Chap. 8.

Since we define any signal that turns off an inverter as a logical 0, we can interpret capacitor storage as delayed inverter operation. A logical-1

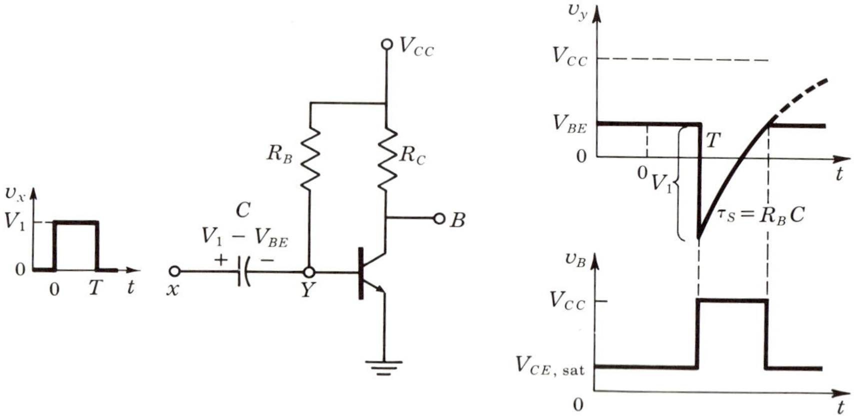

Fig. 13-9 Capacitor short-time storage, illustrating toggling action.

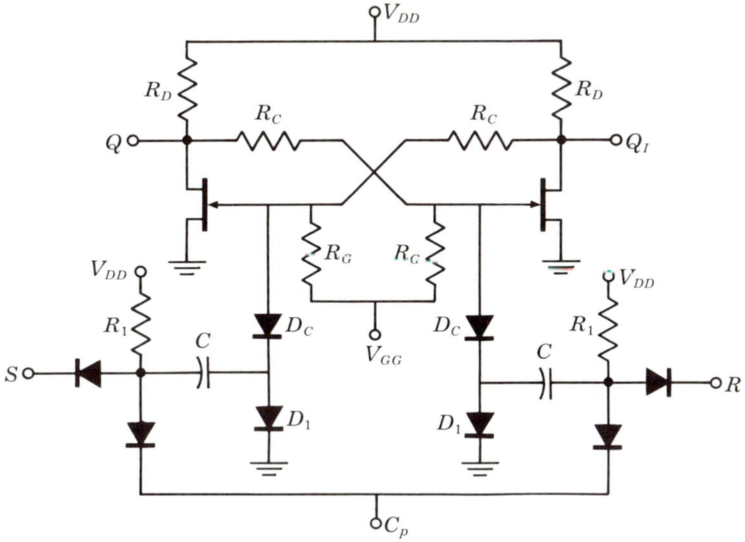

Fig. 13-10 A junction-FET clocked S-R bistable. Capacitive coupling of the output of the input AND gate causes toggle operation.

input generates a delayed logical-0 output. For switching-logic purposes, we can express the base voltage signal as

$$Y_{n+1} = \bar{X}_n \tag{13-11}$$

where X is the pulse applied to the input side of the capacitor at time $t = n$.

Figure 13-10 shows how the negative pulse produced by the capacitor (at the trailing edge of the input clock pulse) is used to toggle a junction-FET bistable. The two input diodes form an AND gate. During the interval when the clock is at the logical-1 level, a simultaneous positive pulse at the S input will cause the output of the AND gate to rise to a logical 1. The coupling capacitor charges through R_1 and diode D_1 to V_{DD}. During charging, diode D_C, connected from the capacitor to the gate of the FET, is back-biased. It prevents the capacitor-charging current from turning on a gate biased beyond pinchoff.

At the trailing edge of the clock pulse, one of the diodes in the input AND gate again conducts, connecting the input side of the capacitor to the logical-0 level. The voltage at the other end of the capacitor drops by the same amount, and the negative pulse generated is coupled through the now conducting diode D_C to the ON FET's gate. The charging diode D_1 becomes back-biased. The negative pulse brings the gate below pinchoff and turns off the ON FET, switching the state of the bistable. A negative pulse applied to the gate of the OFF FET will have no effect on the bistable's state.

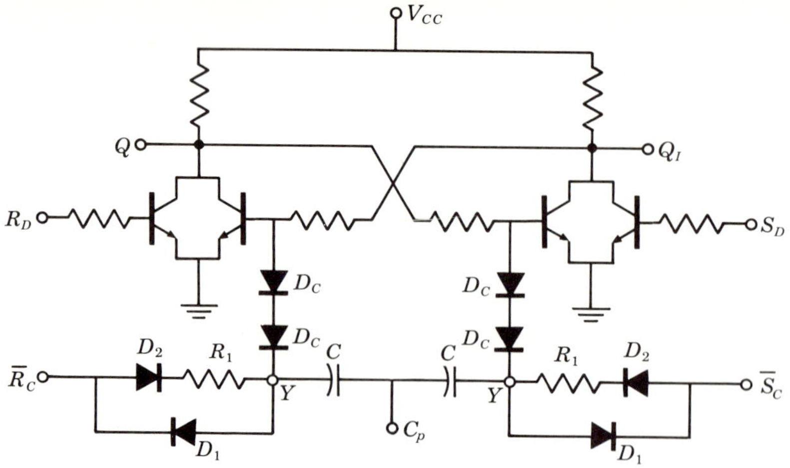

Fig. 13-11 *S-R* toggle-mode bistable multivibrator.

Since the capacitor storage may be considered to invert the gate output, we obtain the signal $\bar{S}_C$, which is applied to the FET gate, where it is effectively ANDed with the signal from the Q_1 output. Thus the clocked switching logic will be expressed by the NAND gate equations (13-7) and (13-8). Because of the toggling-mode operation, we rewrite Eqs. (13-7) and (13-8) as

$$Q_{n+1} = \overline{Q_{I,n}\bar{S}_C} = Q_n + S_C \tag{13-12}$$

$$Q_{I,n+1} = \overline{Q_n\bar{R}_C} = Q_{I,n} + R_C \tag{13-13}$$

where S_C and R_C represent, respectively, the coincidence of the set and reset inputs and the clock pulse during the nth clock interval.

The identical switching technique can be used in a bipolar-transistor multivibrator by coupling the negative pulse to the transistor's base.

Figure 13-11 shows a second method of using capacitor information storage for toggling-mode operation. Here the capacitor replaces one of the diodes in the AND gate. We want to generate a negative pulse at the node Y after the termination of the clock pulse. This means that the capacitor must be charged during the coincidence of the clock pulse and the correct set or reset input and must discharge when $C_p = 0$.

In order to obtain a negative pulse on discharge, the capacitor must be charged with a logical 1 on the clock-pulse side and with a logical 0 on the set or reset side. To achieve the correct coincidence of input states, the set and reset are driven by the inverse functions $\bar{S}_C$ and $\bar{R}_C$. When $C_p = 1$ and $\bar{R}_C = 0$, the capacitor charges through diode D_1 to the difference between the logical-1 clock-pulse level and the $\bar{R}$ logical-0 level.

When C_p drops back to 0, the capacitor couples the negative transition in the clock levels to node Y. The negative-node voltage back biases diode D_1.

Both decoupling diodes D_C, now forward-biased, couple the negative pulse to the transistor's base, where it turns an ON transistor off. The capacitor recovers by discharging through diode D_2 with the time constant CR_1.

When C_p and $\bar{R}_C$ are both equal to logical 1, we have the identical voltage on each side of the capacitor and the capacitor does not charge. The same situation exists when C_p and $\bar{R}_C$ are both at the logical-0 level.

In the final case, where $C_p = 0$ and $\bar{R}_C = 1$, the capacitor charges through the long time constant CR_1. Once $\bar{R}_C$ goes to 0, node Y is positive and the capacitor discharges through diode D_1. The maximum output will be one diode drop. But because there are two decoupling diodes and a transistor base-emitter junction in the coupling path, we need at least three diode drops to switch the multivibrator. Consequently, the bistable will not toggle.

Inhibit-mode toggling

In the previous methods of capacitor-charge storage, we used the input logic to connect one end of the capacitor at time t_n to V_{CC} for charging and then one-half of a clock cycle later, at time $t_{n+1/2}$, to ground for discharging. Thus, we generated an opposite-polarity pulse at the other end of the capacitor.

Suppose that, instead, we simply store the logical-1 output of the input switching logic in a capacitor while at the same time inhibiting the coupling to the bistable. When the inhibit signal is removed, we reestablish the path which connects the stored logical-1 signal to the flip-flop, and the stored signal switches the state of the bistable.

This technique of short-time storage is illustrated in the skeleton circuit of Fig. 13-12, where the charge is stored in the depletion region of a back-biased *pn* junction. In a practical circuit, the multivibrator could be com-

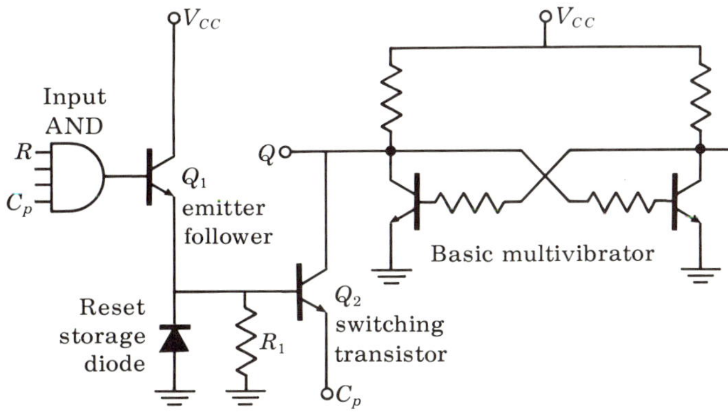

Fig. 13-12 Charge storage in a back-biased *pn* junction used for toggling.

posed of more complex cross-coupled NAND or NOR logic elements and there would be a separate identical pulse storage and switching circuit for each half of the bistable.

Upon satisfying the AND-gate conditions (and when $C_p = 1$), the logical-1 AND-gate output is coupled through the emitter follower (Q_1) and stored in the depletion region of the charge-storage diode. During the readin interval, transistor Q_2 is inhibited by the logical-1 clock pulse applied at its emitter. Because of the base-to-emitter drop in Q_1, the voltage across the storage diode and at the base of Q_2 will be at least 700 mV less than the AND-gate logical-1 input. The emitter of Q_2 is at the logical-1 level. Thus the base-emitter junction of Q_2 will be back-biased by the 700-mV junction drop of Q_1. Since Q_2 is off, this isolates both the AND gate and the storage diode from the basic bistable; the pulse being stored will not change the state of the bistable.

At the termination of the clock pulse, $C_p = 0$ and the emitter of Q_2 is returned to ground. Transistor Q_2 forms a NOR gate with one of the inverters of the basic bistable. The logical 1 stored in the diode now turns on Q_2 and pulls one of the collectors in the basic bistable down to the logical-0 level. The Q output goes to a logical 0.

Because of the low output impedance of the emitter follower, we have a very fast time constant during the storage interval. However, the discharge-time constant R_1C_D is relatively long. This does not bother us during the $C_p = 0$ time interval when the flip-flop changes state, but if the information to be stored during the next $C_p = 1$ interval is a logical 0, any residual charge will back bias the emitter follower and may result in an incorrect storage state. The long discharge time limits the maximum clock frequency.

Figure 13-13 illustrates a technique for removing charge rapidly from one storage diode during the time that charge is being stored in the other diode. The two transistors form a cross-coupled pair of inverters powered by the charge stored in the diodes. Suppose that a logical 1 is being stored

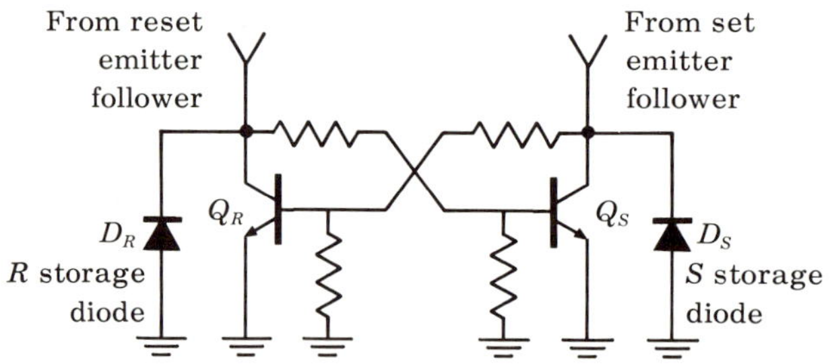

Fig. 13-13 Cross-coupled storage diodes for fast discharge of stored charge.

in diode D_R. The cross-coupling to the base of transistor Q_S will turn it on, and if a logical 1 has previously been stored in D_S, the collector current of Q_S will rapidly discharge the diode junction capacitance. In a similar manner, charge storage in D_S will discharge D_R.

A MOSFET storage circuit, which is equivalent to the junction diode storage of Fig. 13-12, appears in Fig. 13-14. Q_1, Q_2, and Q_3, together with the drain load Q_4, form an input NAND gate. The clock pulse switches Q_5 into the ohmic region, and it then connects the gate of Q_6 to the output of the input NAND logic. The output state of the input logic is stored in the gate-to-substrate capacitance of Q_6. The information-pulse storage logic statement is

$$P_{S,n} = (\overline{ABC})C_p \tag{13-14}$$

During the storage period, we have $\bar{C}_p = 0$ present at the gate of Q_7. Since this is below the threshold, it turns off Q_7. Q_7 forms an AND gate with the storage MOSFET. By inhibiting Q_7, we disconnect the storage-MOSFET device from the basic bistable and thus separate storage and switching functions.

After the end of the clock pulse, Q_5 turns off. The high-resistance path disconnects the storage MOSFET Q_6 from the input NAND logic. Q_7 turns on when $\bar{C}_p = 1$. If we simultaneously have a logical 1 stored in the gate-to-substrate capacitance at the gate of Q_6, the NAND combination

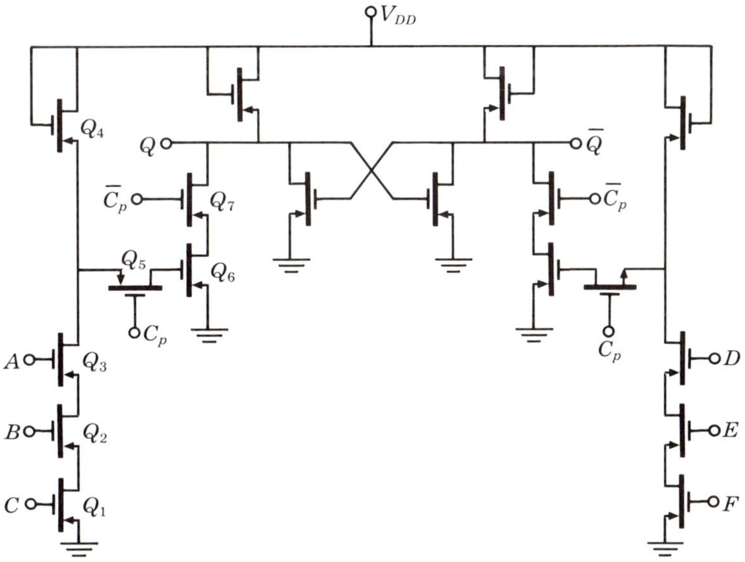

Fig. 13-14 Toggling-mode MOSFET flip-flop. Information is stored in the gate input capacity during the clock pulse.

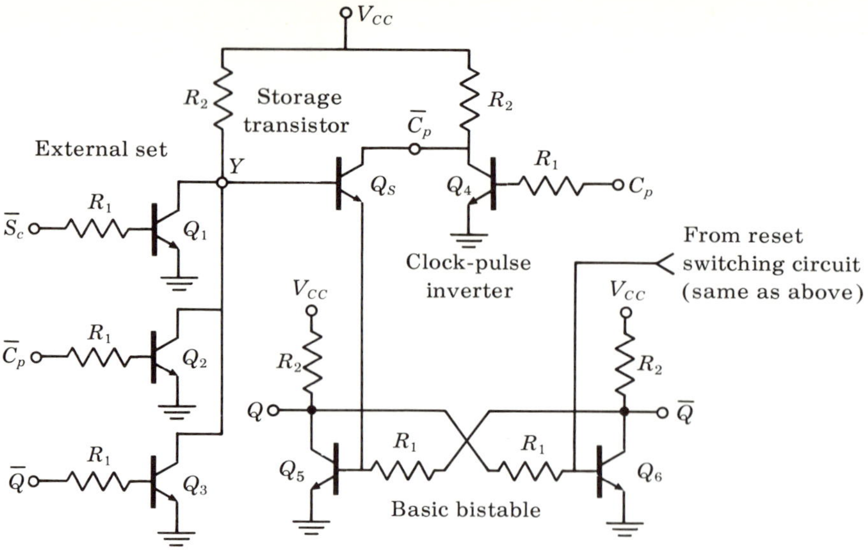

Fig. 13-15 RTL logic—base-charge-storage switching. Switching circuit shown for set side only.

of Q_6 and Q_7 is satisfied. These two ohmic MOSFETs pull the Q output of the bistable down to the logical-0 level. The switching logic at the Q output is given by

$$Q_{n+1} = \overline{\bar{Q}_n + P_{S,n}\bar{C}_{p,n+\frac{1}{2}}} \qquad (13\text{-}15)$$

A similar equation may be written for the $\bar{Q}$ output.

Figure 13-15 illustrates a toggle-switching mode where, during the clock pulse, charge is stored in the base-collector diode of the storage transistor Q_S. Once the clock pulse drops back to the logical-0 level, the stored charge is transferred through the base-emitter diode of Q_S to the base of the OFF transistor of the bistable, turning it on. Except for the mode of information storage, the techniques are similar to the capacitor switching previously discussed. We must, therefore, concentrate our attention on the switching logic needed on each side of the transistor to establish the correct charge and discharge paths.

To store charge in the base of Q_S, we forward bias the base-collector junction. This requires that all inputs to the NOR gates on the base side of Q_S be at a logical 0. The NOR output will then be at a logical 1. If, at the same time, the clock pulse is at a logical 1, then the transistor Q_4 on the collector side of Q_S will be saturated, returning the collector of Q_S to ground. The junction is now forward-biased; current flows from V_{CC} through R_2, through the base-collector junction, and finally through the saturated transistor Q_4. During the storage interval n, the base side of Q_S

(at point Y) satisfies the logic statement

$$Y_n = \overline{\bar{S} + \bar{C}_p + \bar{Q}} \tag{13-16}$$

To forward bias the base-collector junction, we must also have $\bar{C}_p = 0$ at the collector of Q_S. Thus, for storage,

$$P_{S,n} = Y_n C_p \tag{13-17}$$

We see that during the storage interval the base-emitter diode of Q_S is back-biased.

The negative NOR logic expressed in Eq. (13-16) is equivalent to

$$Y_n = SC_pQ_n \tag{13-18}$$

Alternatively, we could connect the base of Q_S to an AND gate driven by the positive-logic inputs of Eq. (13-18).

At the end of the clock pulse, the $\bar{C}_p = 1$ input to the NOR (Q_2) saturates the NOR gate and returns the base of the storage transistor to ground. Simultaneously, $C_p = 0$ turns off transistor Q_4, returning the collector of Q_S to V_{CC}. The base-collector junction is now back-biased by V_{CC}.

The charge previously stored in the base must be swept out. It will flow to the emitter and into the base of the OFF transistor of the basic bistable, turning it on.

Only if Q_5 is the OFF transistor in the bistable and if we intend to switch it on, do we have to store charge in Q_S. That is, only if $Q = 1$ will the bistable be in the state from which it can be switched. In Eq. (13-14) we note that one of the NOR inputs is $\bar{Q}$ and the condition for storage is that $\bar{Q} = 0$. If Q is already at a logical 0, $\bar{Q} = 1$ and the storage condition does not hold. The input NOR gate is now saturated by the $\bar{Q} = 1$ input, inhibiting storage in Q_S. No storage means that there will not be any toggling at $t = n + \frac{1}{2}$.

13-5 *J-K* Flip-Flops

We have seen that whenever the logic at both inverter inputs is simultaneously satisfied, the output of an *S-R* bistable is not uniquely defined. For example, if the set and reset inputs are applied through NOR logic and are simultaneously at the logical-1 level, then both of the inverter outputs drop to logical 0. In a NAND gate bistable, simultaneous logical-0 inputs inhibit both gates and drive the two outputs to the logical-1 state. Upon the simultaneous removal of the R and S inputs, the final output state will depend on any stray unbalance present in the circuit.

In order to eliminate possible ambiguous states, we construct a *J-K* flip-flop, where the J and K inputs correspond in most essentials to the S and

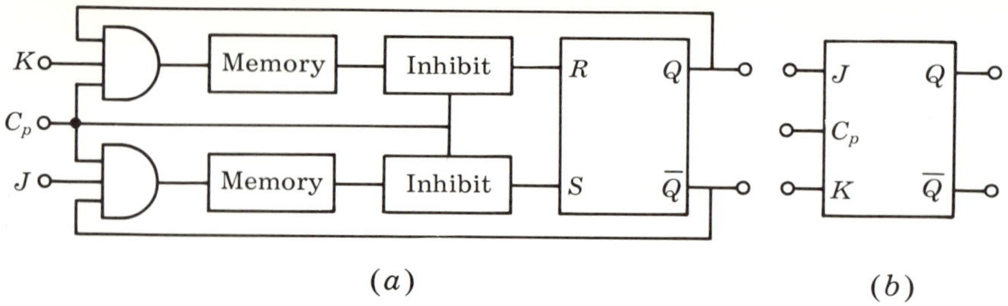

Fig. 13-16 (a) Basic J-K flip-flop operation (compare with Fig. 13-6); (b) circuit symbol for J-K bistable.

R inputs, respectively. The bistable toggles in the same manner as the S-R flip-flop except when the J and K inputs are simultaneously present. In the J-K configuration, then, the bistable always toggles, reversing the current state. The truth table becomes

J	K	Q_{n+1}
0	0	Q_n
0	1	0
1	0	1
1	1	$\bar{Q}_n$

We note from the truth table that J is equivalent to the S input and K is equivalent to the R input in the S-R flip-flop. The switching logic may be expressed as

$$Q_{n+1} = J\bar{Q}_n + \bar{K}Q_n \tag{13-19}$$

$$\bar{Q}_{n+1} = KQ_n + \bar{J}\bar{Q}_n \tag{13-20}$$

The simplest way of synthesizing this switching logic is by coupling the bistable outputs to the input logic through AND gates. The basic circuit is diagramed in Fig. 13-16. Most master-slave J-K bistables, such as the one shown in Fig. 13-17, operate in this manner. Any multiple-AND-input clocked S-R bistable can be converted to J-K operation by feeding each output back to the appropriate S or R input logic.

If we did not have the feedback to the input AND gate, a logical 1 at the K input would always set the Q output to logical 0.

Let us consider what happens with feedback when the Q_n output is in the logical-1 state. This output at the logical-1 level, together with the clock pulse and a logical 1 at the K input, satisfies the input AND gate. Information stored in the memory during this storage interval triggers the basic bistable R input during the next $C_p = 0$ interval, forcing the Q_{n+1} output down to the logical-0 level.

Suppose that the bistable is already in the $Q_n = 0$ state. The Q_n logical-0 output applied to the input AND inhibits the gate. Even when $K = 1$, the AND output will be 0 and a logical 0 will be stored in the memory. This will not cause switching. Since the bistable starts in the correct state, there is no need to switch the state. We see that the $Q_n = 0$ signal fed back to the input logic automatically inhibits the K memory.

Finally, when the J and K inputs are simultaneously applied, only one of the input AND gates will have a logical-1 signal fed back from the Q or $\bar{Q}$ output and only this AND gate will transmit the input pulse. Thus the one

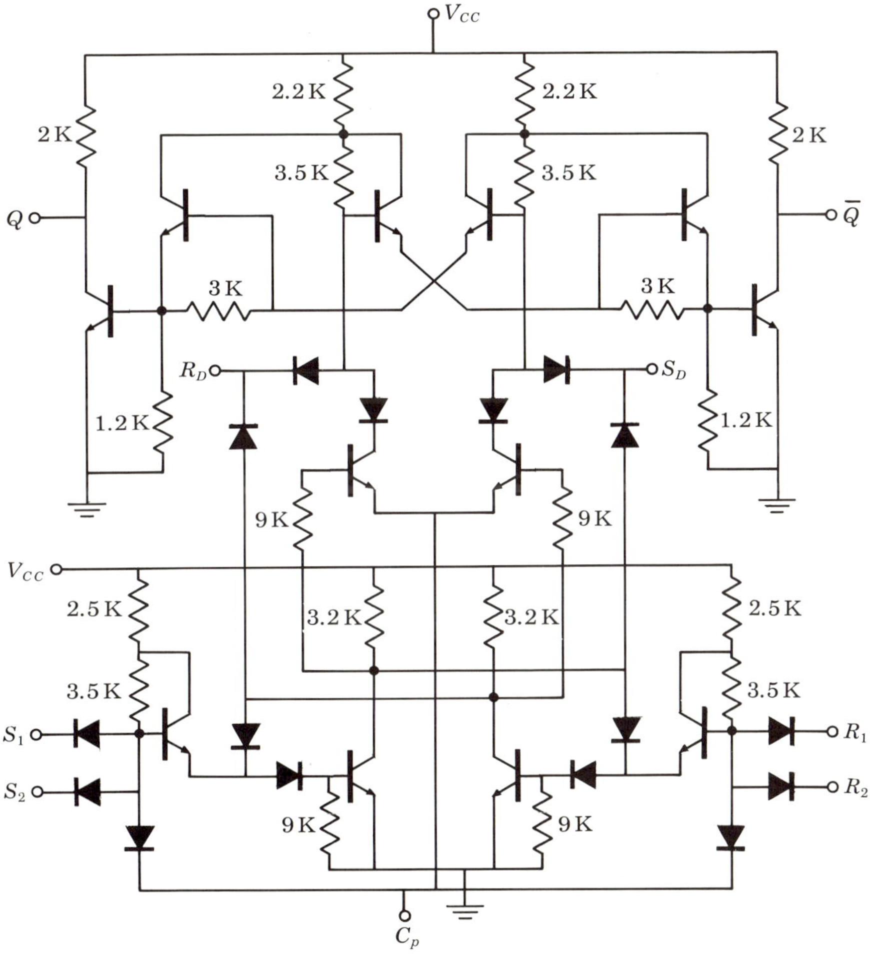

Fig. 13-17 DTL master-slave flip-flop. For J-K operation, connect $\bar{Q}$ to S_1 and Q to R_1. S_2 becomes the J input, and R_2 the K input.

memory which can later switch states will be charged with a logical 1. This memory will invert the bistable's state once the clock pulse drops back to 0. The other output at logical 0 has inhibited the other input gate. We see that we have used inhibit logic to establish a situation where we have access only to the memory cell which will later cause switching.

Figure 13-18 shows a TTL logic operating in the mode described above. The Q and $\bar{Q}$ outputs are coupled back to the K and J input AND gates. In this circuit, the memory is the capacitance of the back-biased diode junction which was discussed in Sec. 13-3 (see Figs. 13-12 and 13-13).

The bistable of Fig. 13-15, discussed in Sec. 13-4, is actually a J-K flip-flop using inverted logic applied to the NOR input gate. The storage is

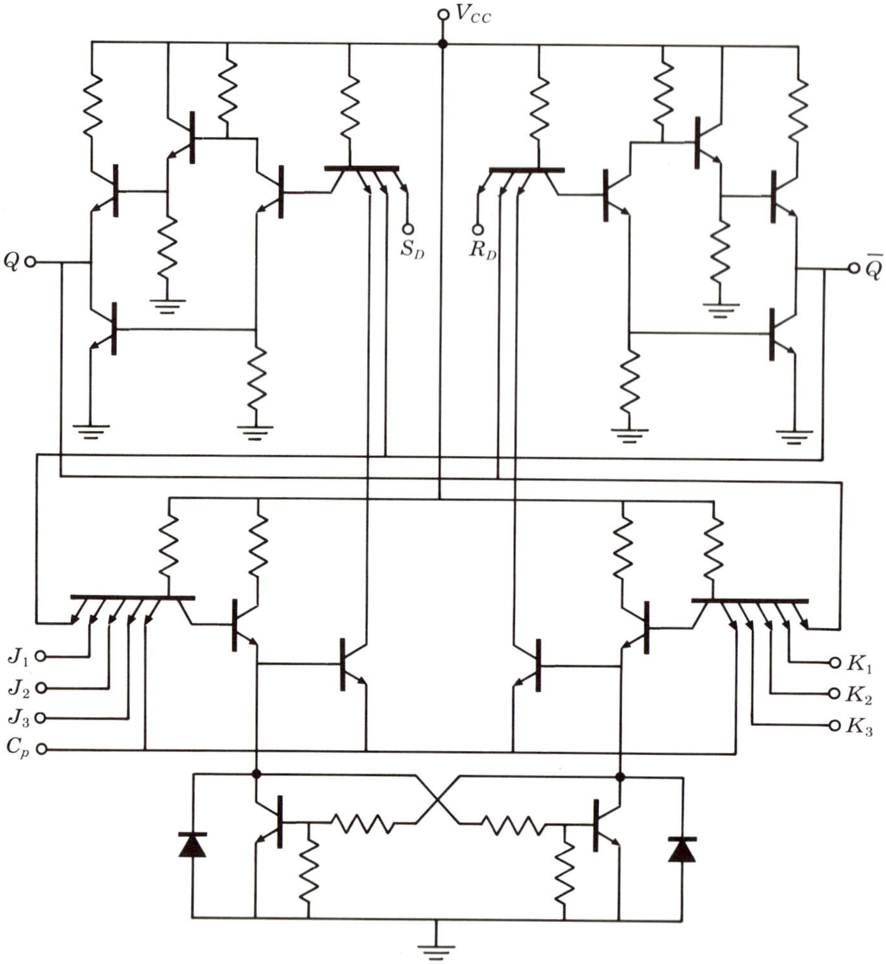

Fig. 13-18 TTL logic J-K flip-flop. This circuit uses the diode charge storage of Figs. 13-12 and 13-13.

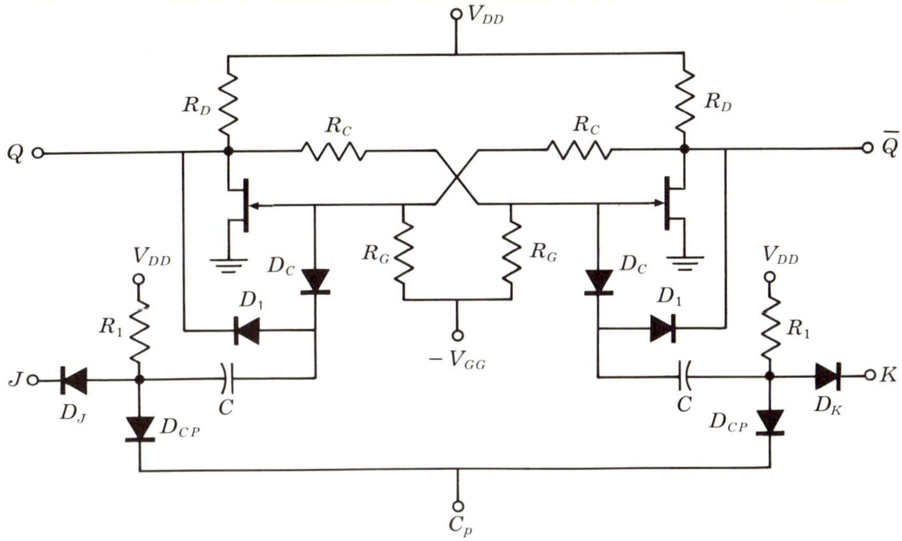

Fig. 13-19 Junction-FET *J-K* flip-flop. The output of the basic bistable is used to inhibit input information storage.

determined by the AND statement given in Eq. (13-18). The circuit is in all respects the NOR equivalent of the basic *J-K* flip-flop of Fig. 13-16.

Instead of using the output state to inhibit the input logic, we can accomplish the same purpose by using the bistable's output to inhibit the storage circuit. Figure 13-19 shows how the junction-FET *S-R* flip-flop of Fig. 13-10 can be modified to form a *J-K* bistable. The diode D_1, which was previously connected to ground and served as the charging path for the storage capacitor C, is now connected to the bistable output for feedback. When the particular output (say Q) is at the logical-0 level, the diode is connected to ground. If the J AND-gate conditions are satisfied, the storage capacitor charges as before, and once $C_p = 0$, the bistable switches. If, however, Q is already at the logical-1 level, the J-circuit storage capacitor will have both sides returned to V_{DD}. The capacitor C does not charge, and there will not be any toggle action.

When J and K are both equal to 1, the particular output at a logical 0 will permit the associated storage capacitor to charge and eventually to toggle the bistable.

ECL *J-K* flip-flop

A similar technique, where the level of storage is controlled by the bistable output state, is used in the current-mode bistable of Fig. 13-20. In this circuit, all switching logics are coupled to the basic bistable through NOR-

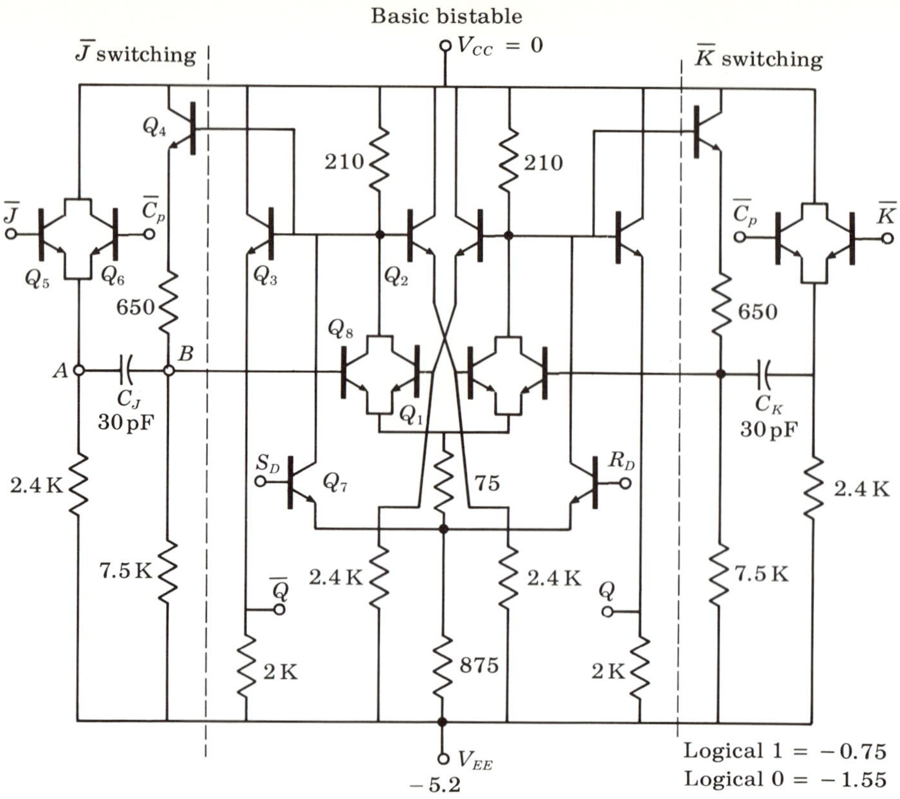

Fig. 13-20 ECL *J-K* flip-flop.

gate connections. To switch a logical-1 output to the logical-0 state, we require a transition from the 0 to the 1 state at the NOR input. This signal can be generated by coupling the 0-to-1 transition appearing at the input logic, at the end of the clock pulse, through a capacitor to the NOR-gate input.

On the input-logic side of the capacitor C_J, we have two inverted logic signals coupled to a common mode through emitter followers. The OR input logic at node A is

$$A = \bar{C}_p + \bar{J} \tag{13-21}$$

The state of the $\bar{Q}$ output is coupled through the emitter follower Q_4 and the resistive divider to the bistable side of C_J. The resistive divider shifts the level of the $\bar{Q}$ signal down by about half the difference between the 0- and 1-levels. When $\bar{Q} = 1$, the output signal at B is low enough to keep the NOR gate transistor Q_8 cut off. We can say that node B is roughly at the logical-0 level. When $\bar{Q} = 0$, the signal at B is approximately one pulse transition lower, or at a -1 level.

During the nth clock interval, a $\bar{J} = 0$ pulse at the AND input pulls node A down to the logical-0 level. At the end of the interval, the 0-to-1 transition which occurs is coupled by C_J to the NOR gate. If the Q_8 gate input was originally at a logical-0 level (because $\bar{Q}_n = 1$), then the positive input transition raises the input to the logical-1 level. The $Q_1 - Q_8$ NOR-gate action drives the output to a logical 0.

Suppose that $\bar{Q}_n = 0$. Then, because of the level shifting, the NOR-gate input is at a -1 level. The 0-to-1 input transition brings the gate only to the logical-0 level, which will not cause switching. Thus the $\bar{Q}$ output controls the gating at the bistable input and permits switching only when in the logical-1 state. After switching, the change in the $\bar{Q}$ state inhibits the NOR input. The switching logic for the J input is

$$\bar{Q}_{n+1} = \overline{(C_pJ)\bar{Q}_n} \tag{13-22}$$

and for the K input,

$$Q_{n+1} = \overline{(C_pK)Q_n} \tag{13-23}$$

The overall switching logic satisfies Eqs. (13-19) and (13-20).

Most of the available integrated-circuit J-K flip-flops operate as master-slave bistables or in one of the other toggling modes discussed above. There are, however, some J-K bistables which switch state as soon as the logic is satisfied on the leading edge of the clock pulse. The switching action is then used to inhibit the input logic. Recovery of the inhibit action takes place during the $C_p = 0$ interval.

Direct-storage and toggling techniques are generally faster than the master-slave circuits, where two bistables must switch in sequence. Propagation delay times of 3 to 10 ns and clock rates of better than 20 MHz can be expected for the faster bistables of the TTL and current-controlled logic classes. Commercial-grade RTL and DTL bistables are much slower, having propagation delays in the 25- to 100-ns range and maximum clock rates of 1 to 4 MHz. MOSFET devices are even slower, with typical clock rates of 500 kHz to 2 MHz.

13-6 Type-*D* Edge-triggered Flip-Flop

A third type of bistable is triggered by the leading edge of the clock to an output state set by the data present at the D input. After triggering, selective feedback of inhibit logic disengages the input circuitry from the output slave bistable. Because the interval devoted to storage of pulse data in a buffer memory can be eliminated, leading-edge triggering is

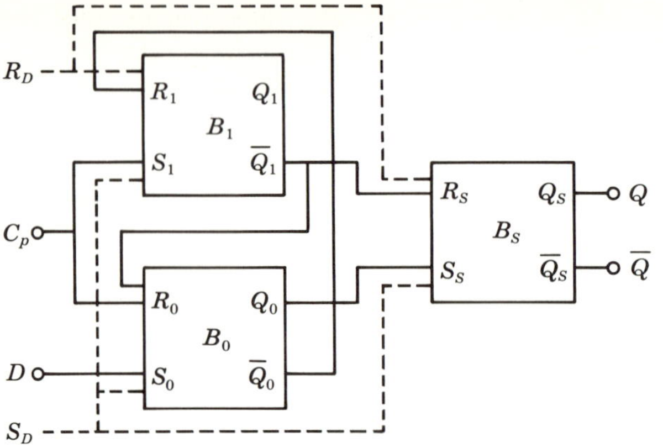

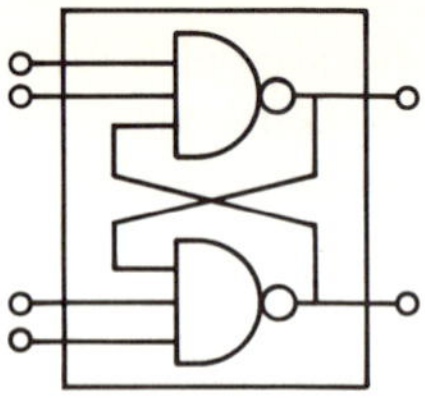

n	$n+1$	
D	Q	$\bar{Q}$
0	0	1
1	1	0

Fig. 13-21 Type-D edge-triggered bistable constructed out of NAND gates (also see Fig. 13-22). The dashed lines indicate the direct set and reset inputs.

faster than toggle-mode triggering. Switching speeds of 5 ns and propagation delays of less than 20 ns are typical. The logic does, however, become more complicated.

Figure 13-21 shows the circuit of a type-D flip-flop in block form. The actual circuit appears in Fig. 13-22. Each of the three bistables is a standard S-R type made by cross-connecting TTL NAND gates (Fig. 13-5). One bistable is used to register a logical 1 (B_1), another is used to register a logical-0 input (B_0), and the third is the slave output stage (B_S). This stage isolates the output loading from the input switching circuitry.

A single logical 0 at the input to any gate sets the corresponding output to a logical 1. Following our convention, a 0 at the S input sets $\bar{Q}$ to 1 and a 0 at the R input sets Q to 1. A single logical-1 input to any gate will not affect the state of any bistable. Internal regeneration will always sustain the previously set condition. Thus, in analyzing this circuit, we must look for the 0's present at the various inputs.

The two register bistables B_0 and B_1 are cross-coupled, $\bar{Q}_0$ to R_1 and $\bar{Q}_1$ to R_0, thus preventing ambiguous output states. Whenever $\bar{Q}_0 = 0$, it pulls down the R_1 input, setting $\bar{Q}_1 = 1$. Whenever $\bar{Q}_1 = 0$, it pulls down the R_0 input, setting $Q_0 = 0$ ($\bar{Q}_0 = 1$). It follows that the two inputs used to drive the slave bistable will always be complementary; we cannot have an undefined state.

A logical-0 clock inhibits switching by holding one AND input of each bistable at a logical 0. Only on the positive transition does the clock permit switching.

The first possibility for a data input is $D = 0$. This input pulls down the S_0 gate and sets $\bar{Q} = 1$. On a positive clock transition, the inhibits are removed. $\bar{Q} = 0$ coupled into R_1 sets $\bar{Q} = 1$. We have now satisfied the R_0 NAND-gate input conditions ($\bar{Q}_0 C_p \bar{Q}_1 = 1$), thus switching Q_0 to a logical 0. The slave output is $Q_S = 0$, following the data input.

Suppose that during the $C_p = 1$ interval, D changes from 0 to 1. What effect will this have on the B_0 state? None. The internal regeneration supplies the logical 0 at the S_0 input gate; this sustains the previously set output condition.

Next let us examine the triggering when $D = 1$. During the $C_p = 0$ interval, $Q_0 = 1$. Thus at S_0 we satisfy the AND-gate conditions ($Q_0 D = 1$), setting $\bar{Q}_0 = 0$. Bistable B_1 is inhibited by $C_p = 0$. During

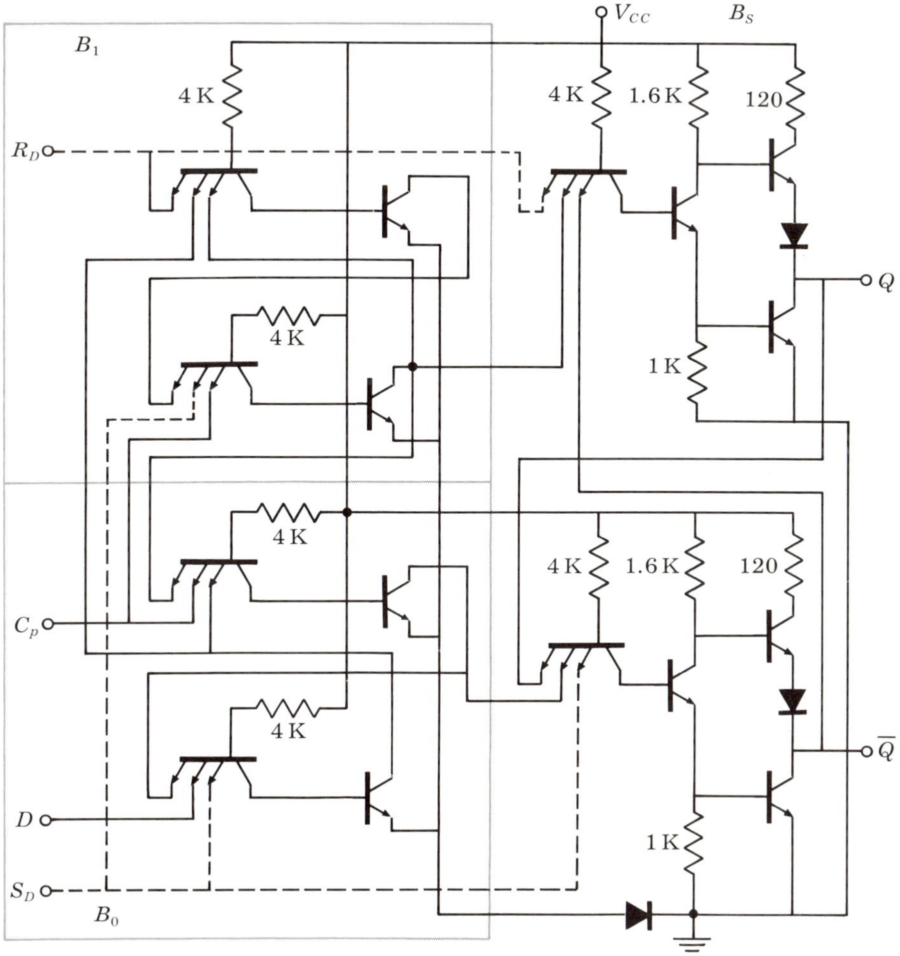

Fig. 13-22 TTL D edge-triggered bistable.

the 0 to 1 clock transition, the logical 0 at R_1 switches B_1's state to $\bar{Q}_1 = 0$. Because of the direct coupling to B_S, the output immediately goes to $Q_S = 1$. Again, the output follows the data input D.

In the actual circuit of Fig. 13-22, we see that only the slave bistable has totem-pole output stages. The input bistables are direct-coupled TTL AND gates feeding a single inverter output. Internal-logic levels and noise immunity are preserved by including an additional diode drop in series with the emitters of the input bistables.

The techniques described above can be combined with a different input logic to create a J-K edge-triggered flip-flop. Triggering occurs on the coincidence of the appropriate J-K logic and the leading edge of the clock pulse.

13-7 Bistable Applications

A bistable is a memory element which stores one bit of information (0 or 1), indicating the state of the stored information by the logic of the Q and $\bar{Q}$ outputs. These outputs serve as inputs to the logic-system NAND and NOR gates. To store more than one bit, a more complex memory requiring a more elaborate network of interconnected bistables and logic elements is necessary.

Simple S-R bistables cannot be directly interconnected for multiple bit storage. Since the output changes state immediately upon the application of the input set signal, the change at one bistable causes the next to switch, which switches the next, and so on down the chain. The input signal "races" through the system of bistables, setting them all to the same state.

Shift registers

If, instead, the bistables are switched on by the transition of states, racing can be prevented. For example, the $\bar{Q}$ output can be coupled to the next stage's S input through an RC network, as shown in Fig. 13-23a. Let us assume that all stages are initially in the logical-0 state ($Q = 0$) and that we apply, in time sequence, the binary word 1101 to the A bistable set input. The first bit (logical 1) sets the state of Q_A to 1. At $\bar{Q}_A$, we have the negative transition $1 \rightarrow 0$, which is coupled by the capacitor and causes a negative pulse to appear at the set input of B. The negative pulse will not affect the state of the B bistable. A reset pulse following the information storage sets Q_A to 0 and causes the positive transition at $\bar{Q}_A$ of 0-to-1. At the end of the reset pulse, the coupled positive pulse sets $Q_B = 1$. The reset input has shifted the first-word bit from one bistable to the next.

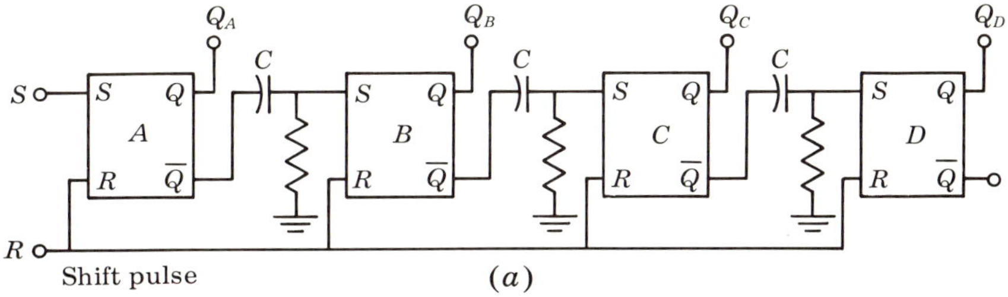

(a)

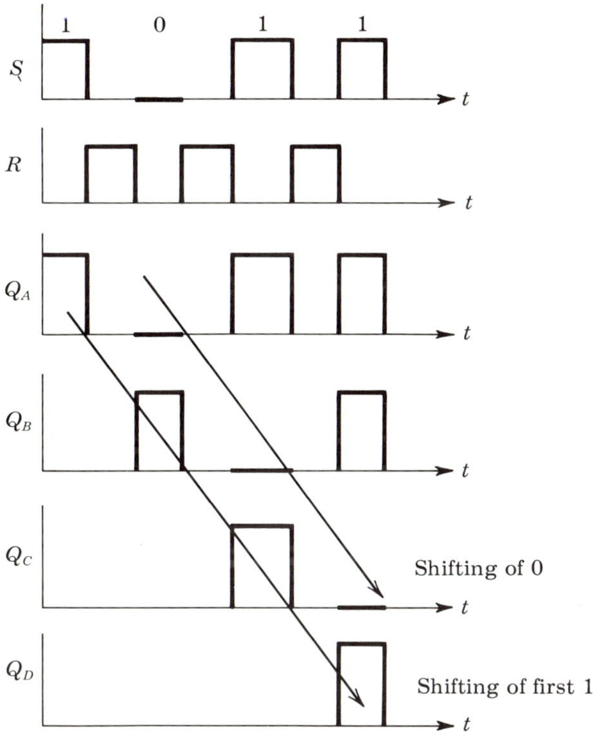

(b)

Fig. 13-23 (*a*) Four-stage shift register; (*b*) time sequence of operation, showing pulse position shifting.

The second input bit, a logical 0, does not change the state of A. Since there is no $0 \rightarrow 1$ transition on the second reset pulse, stage B will remain at the reset logical-0 state. But stage B previously had stored the logical-1 first bit, and on its reset, the $\bar{Q}_B$ transition transfers the 1 into stage C. In a similar manner, the other input bits are first read in at stage A and then shifted along the shift register by the reset pulses. Figure 13-23*b* shows both the input and the shift sequence along the shift register.

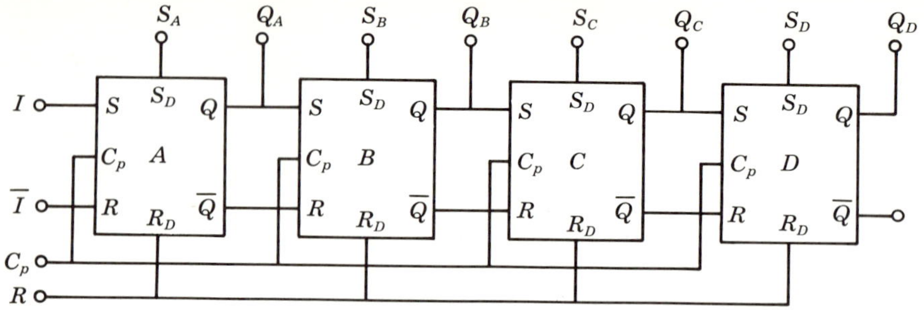

Fig. 13-24 *S-R* toggling-mode bistables used in a four-stage shift register. Inputs I and $\bar{I}$ required.

Information which has been read into the shift register serially can be transferred out of the shift register's final stage serially by applying more shift pulses. After entry, the total binary word is available at the Q outputs and may also be read out simultaneously (parallel readout).

The RC network introduces toggle-mode operation into the bistable chain. To ensure proper shifting, the $\bar{Q}$ transition must remain stored in the capacitor after the end of the reset pulse. Thus

$$\tau = RC > T_R \tag{13-24}$$

where T_R is the width of the reset pulse. We observe that the RC network limits the speed of operation.

Clocked S-R flip-flops which already operate in the toggle mode and which have provisions for fast storage and discharge are directly interconnected to form the shift register of Fig. 13-24. The input word $I = 1101$ and its complement $\bar{I} = 0010$ must be simultaneously available to set or, when necessary, to reset the first flip-flop. We can always form the complement by inverting I. After each bit is read in, the clock-pulse $1 \rightarrow 0$ transition causes toggling and shifts the input bit along the chain.

Each bit in the binary word may be applied simultaneously directly to the appropriate stage's direct-set inputs (and the complementary word to the direct-reset inputs) for parallel readin. Readout may be either in parallel or serially.

For nondestructive serial readout, the change of states of the D bistable is fed back as inputs to the A bistable. After shifting the information through four steps, the original stored word is back in the correct sequence. Usually the closed path is formed by AND gates, so that an additional steering input can be used to determine whether the readout will be destructive or nondestructive.

Now suppose we connect a closed chain of 10 bistables (Fig. 13-25) and set the first to logical 1 and the rest to logical 0 (with the external set and

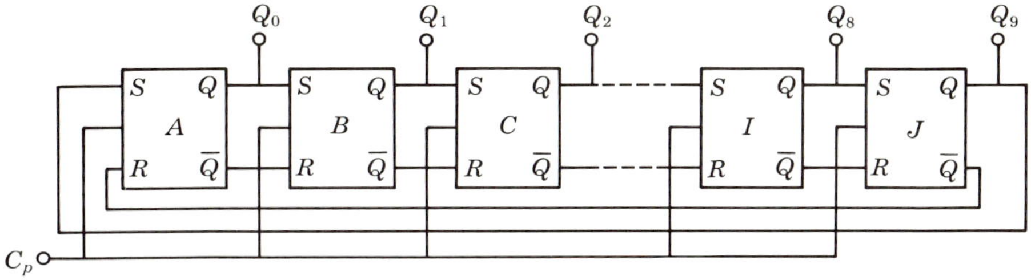

Count	Q_A	Q_B	Q_C	Q_D	Q_E	Q_F	Q_G	Q_H	Q_I	Q_J
0	1	0	0	0	0	0	0	0	0	0
1	0	1	0	0	0	0	0	0	0	0
2	0	0	1	0	0	0	0	0	0	0
3	0	0	0	1	0	0	0	0	0	0
4	0	0	0	0	1	0	0	0	0	0
5	0	0	0	0	0	1	0	0	0	0
6	0	0	0	0	0	0	1	0	0	0
7	0	0	0	0	0	0	0	1	0	0
8	0	0	0	0	0	0	0	0	1	0
9	0	0	0	0	0	0	0	0	0	1

Fig. 13-25 Decade ring counter.

reset). Each successive clock pulse shifts the stored logical 1 to the next position in the ring. The position of the logical-1 output indicates the count of clock pulses. This special kind of shift register is called a *ring counter.*

Ripple counters

A straight-through ripple counter registers, in binary form, the number of pulses applied. The input flip-flop switches state on every input pulse, and the output transitions are used to toggle the next counting stage (Fig. 13-26). *J-K* flip-flops toggle on every clock pulse whenever a logical-1 input is simultaneously present at each input. Thus, in our counter, we must either use a *J-K* directly or form a *J-K* by feedback from the Q and $\bar{Q}$ outputs to the S and R inputs (as in Fig. 13-26). If *J-K* bistables (Fig. 13-27) are used, all unused AND inputs are left either open-circuited or in the primed state by being connected to the logical-1 level.

The first clock pulse causes the A bistable output to switch from 0 to 1. The second clock pulse resets the A stage to 0, and on the transition the B bistable toggles to the logical-1 state. The third clock pulse again sets A to 1. Both A and B are now in the 1 state. The fourth clock pulse resets A and B. The B-stage output transition now causes bistable C to register a logical 1. Thus each stage counts down by a factor of 2. The outputs

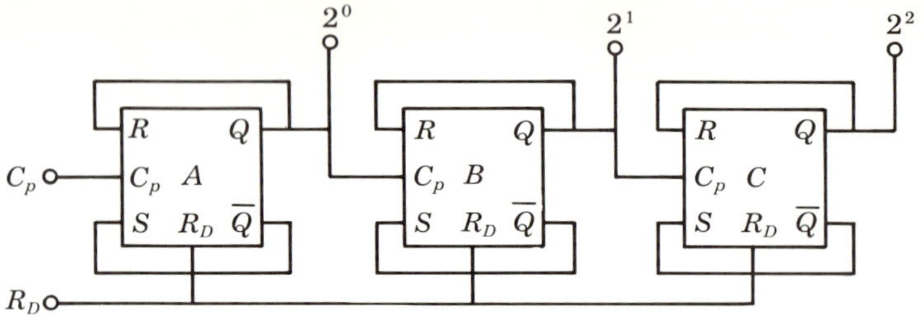

$t = n$	$t = n + 1$		
C_p	Q_C	Q_B	Q_A
0	0	0	0
1	0	0	1
2	0	1	0
3	0	1	1
4	1	0	0
5	1	0	1
6	1	1	0
7	1	1	1

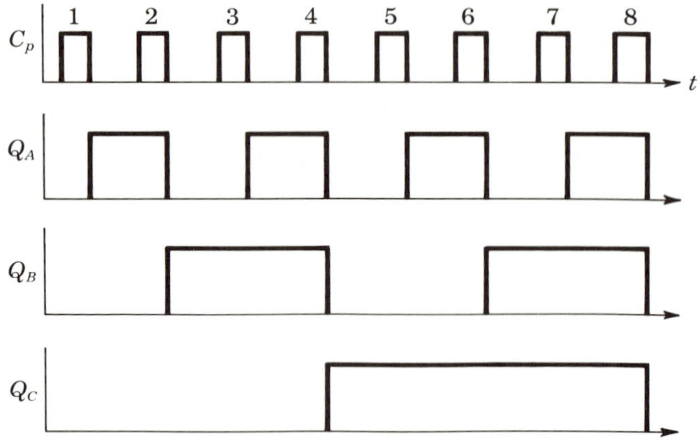

Fig. 13-26 Three-stage straight-through ripple counter (toggle-mode bistables). *J-K* flip-flop formed from *S-R* by feedback from output back to input.

register, in binary form, the total count N of input pulses:

$$N = Q_C \times 2^2 + Q_B \times 2^1 + Q_A \times 2^0 \tag{13-25}$$

With four input pulses, the bistables register 100 (binary four).

The three-stage counter of Fig. 13-26 can register a maximum count of 7. On the eighth pulse, it resets to 000, generating one output pulse. We say that the circuit counts down by 8.

The frequency of a periodic input signal used for clocking would be reduced to

$$f_o = \frac{f_{\text{in}}}{2^n} \tag{13-26}$$

where n is the number of stages in the binary counter. Note that each stage counts down by 2 and that the total countdown is the product of the individual stage counts.

Counters can also be made to count down by nonbinary factors through the use of feedback. For example, the output pulse of the scale-of-8 counter can be fed back to the direct-set inputs of stages A and B, resetting them to

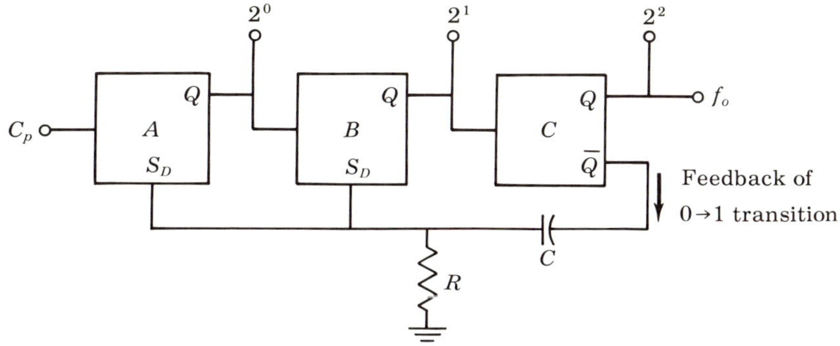

Fig. 13-27 Scale-of-5 counter using feedback to preset count.

register a count of 3 every time the output pulse is generated (Fig. 13-27). The RC network couples the $0 \rightarrow 1$ transition at the $\bar{Q}$ output to the direct-set inputs of stages A and B. Only five additional input pulses are necessary to generate another output pulse. The counter now counts down by

$$N = 2^n - S_D \tag{13-27}$$

where S_D is the total feedback reset.

Inhibit feedback counters

A second and more general class of feedback counter is illustrated by the circuit of Fig. 13-28. J-K flip-flops having multiple AND inputs are used for each stage. In the final stage, the clock inputs for the J and K portions of the bistable are separated and used to independently toggle the J or K switching logic (see Fig. 13-16). Counting from 0 to 3 follows the operating sequence described above. The fourth input sets Q_C to logical 1. The logical 0 at $\bar{Q}_C$ is fed back to the AND input of stage A, where it inhibits the gate and prevents the A bistable from switching to 1 on the

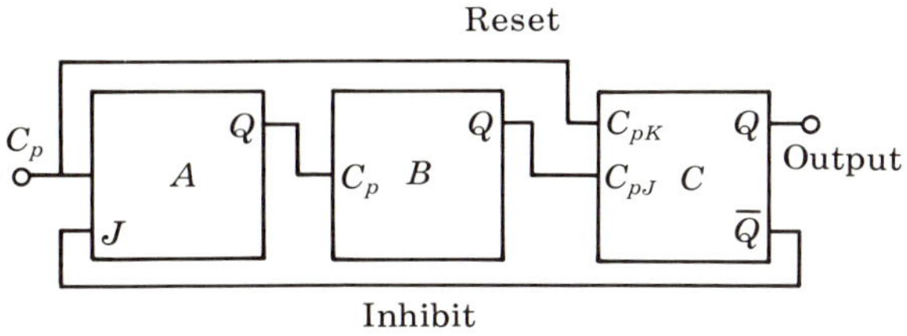

Count	Q_C	Q_B	Q_A
0	0	0	0
1	0	0	1
2	0	1	0
3	0	1	1
4	1	0	0
5	0	0	0

Fig. 13-28 Scale-of-5 counter using inhibit feedback.

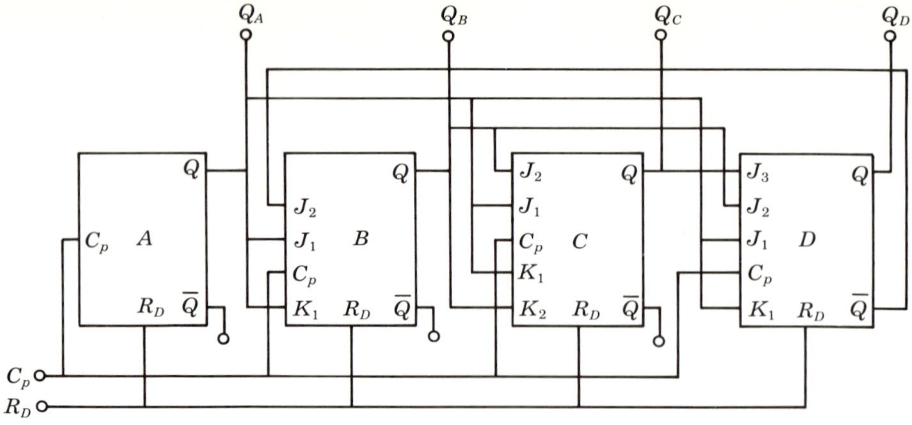

Input at $t = n$	Counter logic Output $t = n + 1$				Output decoding logic			
Bit weight	8	4	2	1				
0	0	0	0	0	$\bar{D}$	$\bar{C}$	$\bar{B}$	$\bar{A}$
1	0	0	0	1	$\bar{D}$	$\bar{C}$	$\bar{B}$	A
2	0	0	1	0	$\bar{D}$	$\bar{C}$	B	$\bar{A}$
3	0	0	1	1	$\bar{D}$	$\bar{C}$	B	A
4	0	1	0	0	$\bar{D}$	C	$\bar{B}$	$\bar{A}$
5	0	1	0	1	$\bar{D}$	C	$\bar{B}$	A
6	0	1	1	0	$\bar{D}$	C	B	$\bar{A}$
7	0	1	1	1	$\bar{D}$	C	B	A
8	1	0	0	0	D	$\bar{C}$	$\bar{B}$	$\bar{A}$
9	1	0	0	1	D	$\bar{C}$	$\bar{B}$	A

Fig. 13-29 BCD counter using *J-K* flip-flops having AND-gate inputs. Unused *J* and *K* inputs may be left open or may be connected to the logical-1 level.

next input pulse. The fifth clock pulse is applied to stage C at the K clock input. It resets Q_C to logical 0, generating a single output pulse for every five input pulses.

The output count is due to the binary count of the first $n - 1$ bistables plus one count from the nth stage:

$$N = 2^{n-1} + 1 \tag{13-28}$$

Thus the frequency division may be expressed as

$$f_o = \frac{f_{\text{in}}}{2^{n-1} + 1} \tag{13-29}$$

Inhibit feedback always yields an odd-number counting ratio.

Feedback counting can be combined with straight ripple counters to obtain the count

$$N = 2^m(2^{n-1} + 1) \tag{13-30}$$

m is the number of stages of regular binary counters, and n is the number of stages inside the feedback loop.

If two feedback loops are used, then the total count becomes

$$N = 2^{p-n-1}(2^{n-1} + 1) + 1 \tag{13-31}$$

where p is the total number of bistables in the major overall feedback loop and n is the number of stages in the minor internal loop.

Figure 13-29 shows a special case of feedback counter: a binary-coded-decimal (BCD) counter constructed from *J-K* flip-flops having AND inputs. This counter counts up to 9, resetting to 0 on the tenth count. The state of each bistable registers the decimal input number in binary form.

The count can be decoded into decimal form by connecting the inputs of AND gates to the appropriate bistable outputs. Each decimal count satisfies one of the gates of the decoding logic given in Fig. 13-29. Figure 13-30 shows the diode decoding network.

Other decimal coding commonly used includes the 1-2-4-2 (ABCD) weighted code and an excess three (XS3) code (binary-coded decimal preset to a count of 3 and counting up from 3 to 12). These two counters are self-complementary; that is, the 9 output is the complement of the 0 output, etc.

Large system integration (LSI)

All the systems described in this section can be synthesized on a single chip. Each of the separate circuits, *S-R* and *J-K* flip-flops, logic arrays, coding and decoding matrices, are formed on the appropriate sections of the chip. The final metallization completes the interconnections that form the system. For very complicated systems, two levels of metallization may be necessary. A separate insulation layer between the wirings prevents short circuits.

Whenever the system is designed as a whole, great simplification results. Since the internal system interconnections interface with each other only in a well-defined way, the circuitry of one stage can often be combined with the next. Only those circuits terminating on contact pads need high-power totem-pole output stages.

A second type of LSI is based on organizing the system on a generalized substrate containing a maximum number of gates and bistables. After designing the system in block form, one or two metallization masks are drawn; the first is superimposed on the circuit interconnections and the second is the top level wiring. By means of this technique, it becomes

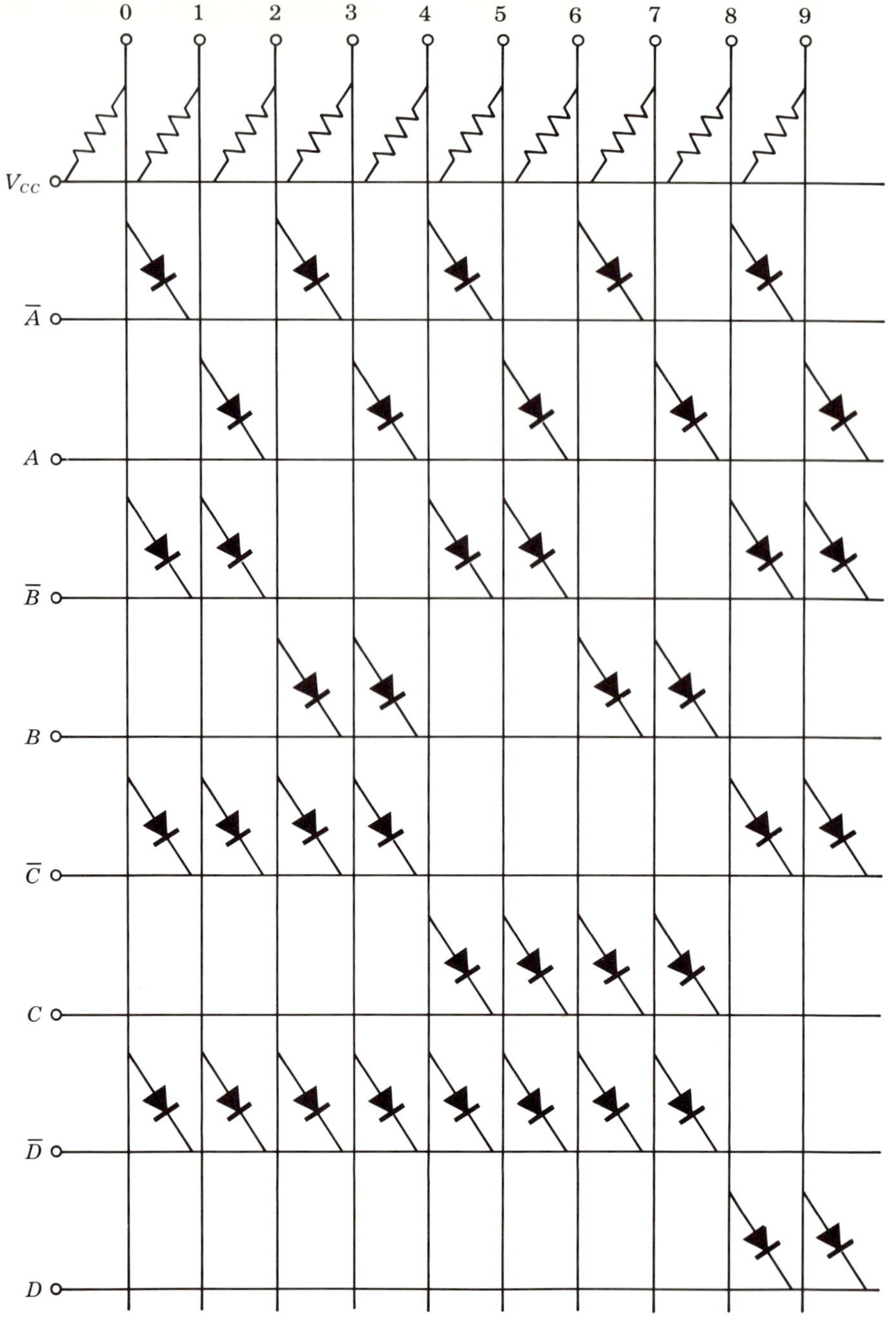

Fig. 13-30 An AND diode decoding network for the output of the BCD counter of Fig. 13-29 (8-4-2-1 code).

economically feasible to integrate logic systems even for relatively specialized applications.

Problems

13-1 The simple TTL bistable of Fig. 13-31 is employed as a memory cell in a large-scale integrated-circuit memory array. It operates on half-level S and R pulses.

(*a*) Give the limits of the input S and R pulse levels and the output logic levels for S-R operation ($\beta = 30$, $V_{BE} = 700$ mV, $V_{CE,\text{sat}} = 100$ mV).

(*b*) Show how this circuit can be modified into a clocked S-R bistable. Give the clock-pulse amplitude. Will switching occur on the leading or on the trailing edge of the clock?

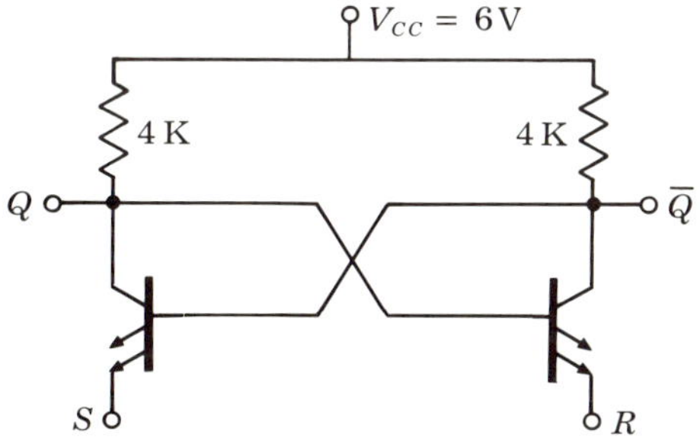

Fig. 13-31

13-2 Use the multiple-input ECL logic inverters of Fig. 4-29 to form a clocked S-R logic. Switching should occur on the trailing edge of the clock pulse. Specify all external parameters for operation with clock pulses from 100 kHz to 2 MHz. Explain the operation of the selected system.

13-3 Figure 13-32 shows how feedback through an RC toggle network converts an S-R flip-flop into a monostable. Explain the general operation in terms of the logical-0 and logical-1 states of the bistable. Is there any restriction as to the type of input logic, i.e., can it be either OR or AND? Give the approximate pulse duration in terms of RC and the logic levels. Select a specific logic for which

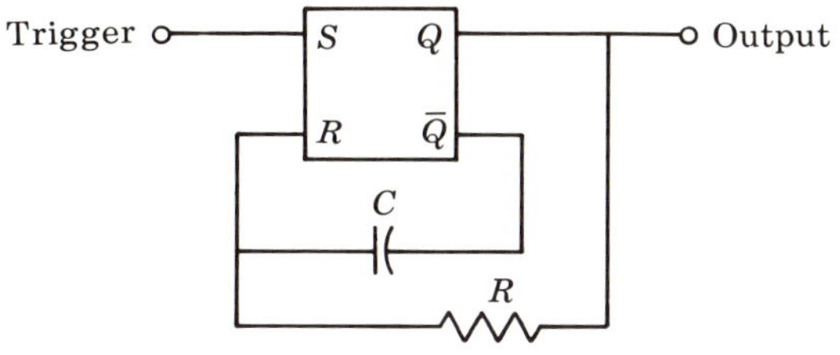

Fig. 13-32

this configuration will work and calculate the actual levels and the pulse duration. Specify any restrictions on R.

13-4 Start from the S-R flip-flop of the type shown in Fig. 13-2. Introduce the switching logic needed to convert the circuit into a clocked S-R. Switching should occur only on the leading edge of the clock pulse. Show the actual circuit used for triggering and the feedback-inhibit path which will disconnect the input logic after switching.

13-5 Tabulate the sequence of states of all diodes in the J-K flip-flop of Fig. 13-19. For each truth-table condition, specify the states at $t = n - \frac{1}{2}$, $t = n$, and $t = n + \frac{1}{2}$. Give the equivalent logic statement at each time interval.

13-6 Figure 13-33 is a MOSFET complementary-symmetry J-K flip-flop.

(*a*) Draw the basic S-R flip-flop.

(*b*) Separate the circuit used for toggling. Explain its operation. On which edge of the clock pulse does the circuit switch?

(*c*) Explain the J-K sequence of operation.

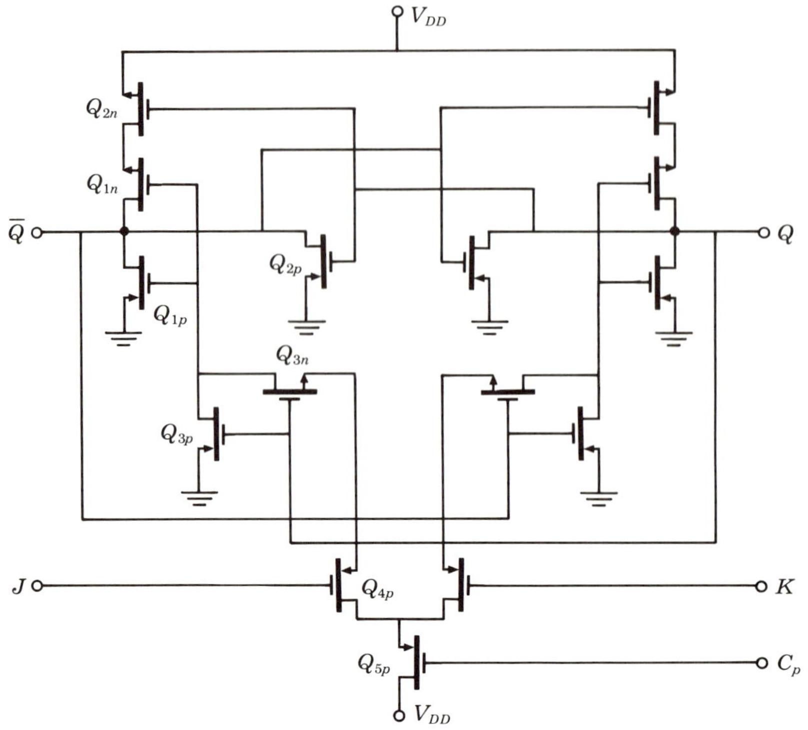

Fig. 13-33

13-7 Tabulate the advantages and disadvantages of the S-R, J-K, and D flip-flops when used in shift registers, counters, and as a storage element. Does any one type offer exceptional advantages for any application?

13-8 (*a*) Draw the logic diagram of a four-bit recirculating shift register. Under normal conditions, the stored information is continuously shifted around a closed loop. However, upon the application of a stored pulse, new data must be inserted serially into the register, replacing the previously stored data. Show the time sequence of applied control, clock, and information pulses.

(*b*) Show how to modify the shift register of part *a* to permit parallel data input.

13-9 Draw the logic diagram of a four-bit open-loop shift register where the data can be shifted in either direction. A shift-right steering pulse establishes the conditions for forward shifting upon clocking. For reverse shifting, we apply a shift-left steering pulse.

13-10 Verify Eqs. (13-30) and (13-31). Tabulate the permissible counts between 1 and 32, using a single feedback loop, a double feedback loop, and single and double feedback loops combined with binary stages. Are any counts forbidden? What could be done to obtain these counts?

13-11 Design a counter that counts down by 12. Make a table showing the state of each bistable for each count.

13-12 Design a counter that counts down by 13. Make a table showing the state of each bistable for each count.

13-13 Prove that placing the binary stage at the output of any even-number counter makes the output waveshape symmetrical. Consider as an example a scale-of-6 counter. Show the relative waveshapes at the input and output for 13 input pulses when the binary stage is at the input and when it is shifted to the output.

13-14 Design a 1-2-4-2 self-complementing-code decade counter. Show the decoding network that will yield the appropriate decimal output information.

13-15 Figure 13-34 shows a clocked pseudorandom-code generator. The half adder (Fig. 4-4) is used as an exclusive OR gate.

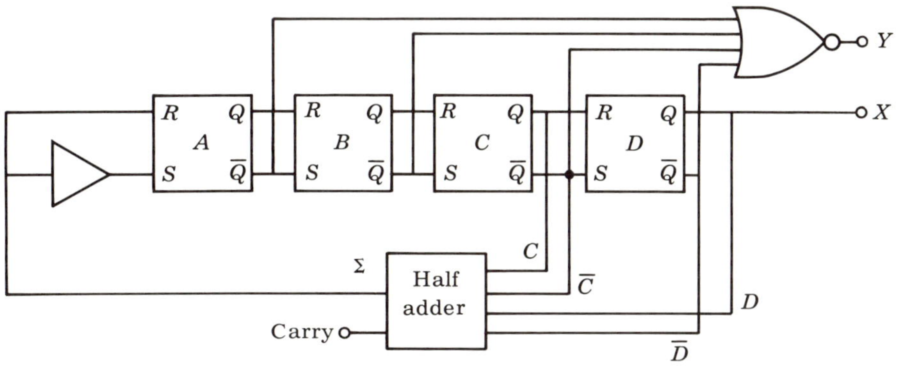

Fig. 13-34

(*a*) Show the output pulse train at both *X* and *Y*.

(*b*) How many bits long is the generated code? Why is it less than 2^n, where n is the number of binary stages?

(*c*) Give the code generated when the output of the *B* bistable replaces the output of the *C* bistable at the half adder.

Bibliography

Davies, D. C., R. B. Seeds, and S. H. Shou: An Integrated Charge-control *J-K* Flip-Flop, *IEEE Trans. Electron. Devices,* vol. ED-11, no. 12, pp. 556–562, December, 1964.

Manufacturers' Application Notes and Data Sheets, especially from Motorola, Sylvania, Texas Instrument, and Fairchild.

14. active-element memories

In this chapter we shall examine the nature of active-element arrays used to store information in digital systems. There are several different types of memories, each having its own special application. But in all, we must be able to store information rapidly for various lengths of time, we must be able to retrieve the stored information rapidly, and we must be able to separately select the various storage locations.

In digital computers, rapid storage and readout are essential for the solution of complex problems. Besides storing data and answers, the program controlling the sequence of the solution is fed into the memory banks. As each step is completed, the computer looks for the further instructions stored in the next memory location.

14-1 Characteristics of a Memory System

A generalized memory system may be represented by the schematic diagram of Fig. 14-1. Several such systems are interconnected to facilitate the storing of a relatively large amount of information. While the degree of complexity increases with the total required storage, the basic system remains the same.

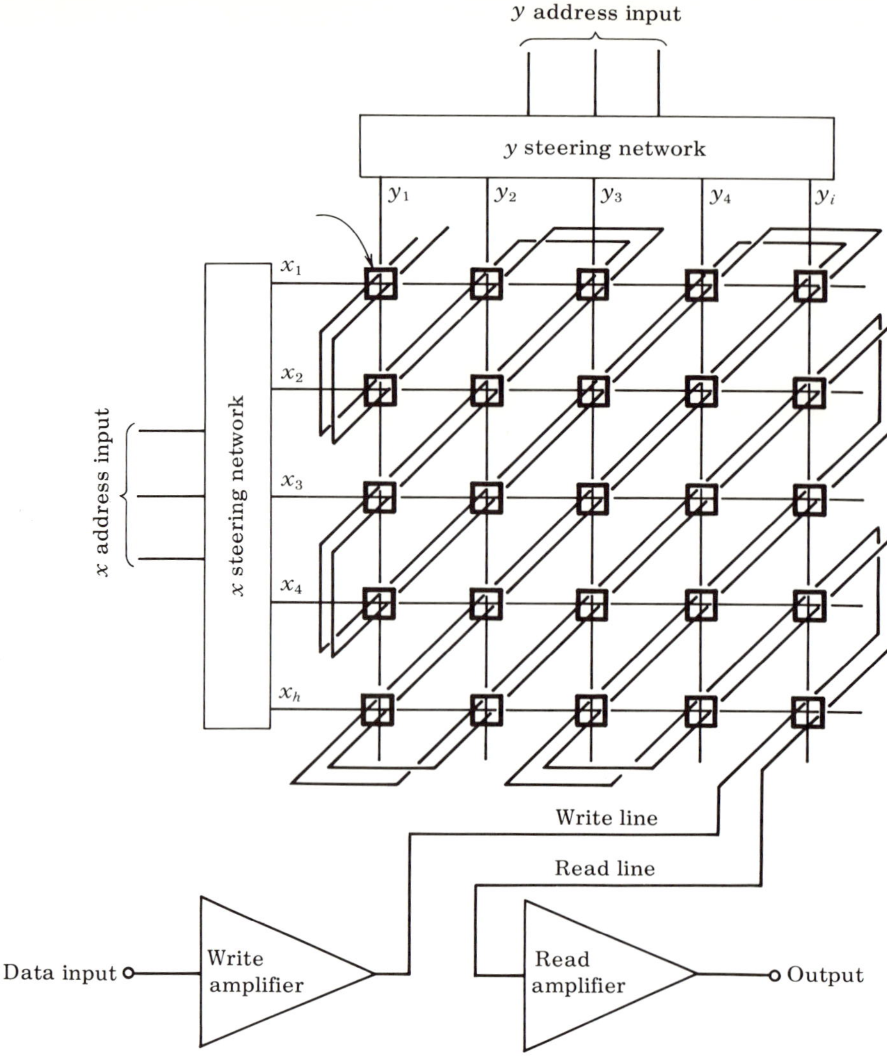

Fig. 14-1 Basic form of memory array.

The total memory may be broken into four parts:

1. An array of bistate memory cells, each storing a single bit of information by registering a logical-0 or a logical-1 level.
2. Steering circuitry which permits the selection of any particular bit or group of bits (*word*). The location of the bit at the intersection of a particular row and column is called its *address*.

3. A readout amplifier which acts as an interface between the particular memory cell and the external logic network. Readout may be either destructive, erasing previously stored information (as in shift registers), or nondestructive.
4. Write circuitry which sets the state of the addressed memory cell. The write amplifier translates the input logic levels into the signals needed to switch the addressed memory cell to the required logical-0 or logical-1 state.

Multiple systems of the type shown in Fig. 14-1 expand the memory to a three-dimensional array having individual cells located at the intersection of x, y, z address lines. Often all the z_1, z_2, z_3, z_j cells are driven in parallel with a single xy address, simultaneously selecting z_j bits. Thus a full word containing j bits can be stored in the z direction at each xy location. Parallel write and read is, of course, much faster than serial storage.

Memory cells

We shall now consider the particular characteristics required in the individual memory unit.

1. It should have at least two definitive, widely separated levels. Once excited from the lower to the higher, the memory element should maintain the new state after the initial storage pulse has disappeared. This precludes the use of all nonlinear devices having continuous characteristics. We cannot identify the state of the diode (curve a in Fig. 14-2) by looking at the zero-excitation point.

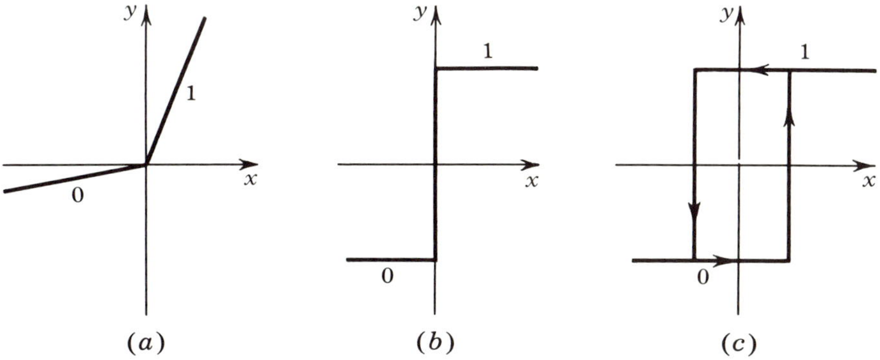

Fig. 14-2 Three typical characteristics of bistate elements.

2. Both the process of storage and readout should require a finite amount of energy. If the device switches levels with an absolute-minimum input, what is to prevent it from switching erratically in the presence of noise? The two states of curve *b* in Fig. 14-2 are interchangeable at zero excitation; a device having this characteristic would be unreliable for use as a memory element.
3. At any instant, the state of the device must represent the past history of the system as well as the present excitation.
4. In general, storage and readout should require different types of excitation. This is necessary because we do not want the memory to shift from 1 to 0 when a storage pulse is applied or to falsely store a readout pulse by switching from 0 to 1. The differentiation might be as simple as using positive and negative pulses for storage and readout, respectively.

It follows directly from the above statements that hysteresis is essential in any device used for long-term information storage (curve *c* in Fig. 14-2). It can be a natural property of the device, as in ferromagnetic cores or in various ferroelectric dielectrics (Chap. 15), or it can be simulated by the use of diodes and energy-storage elements, as discussed in Sec. 2-9. Bistable multivibrators, constructed either with bipolar or unipolar transistors or with the two-terminal negative-resistance devices of Chap. 11, also exhibit hysteresis, and as we know, these circuits are used as shift registers and memory cells.

For short-time computation steps, the digital word may be stored in a shift register composed of bistables (Chap. 13). If only short-time storage is required, the word may also be stored in a bank of charged capacitors. Charge will eventually leak off, and the memory will decay. Since this type of storage is volatile, it is obviously suitable only when the decay time is long compared with the required storage duration. Section 14-2 discusses some MOSFET applications of this mode of storage.

Among the disadvantages of active-device memories is their large size per stored bit. Even when using integrated circuits containing multiple bistables on a single chip, the sheer bulk of the units needed to store the 10^5 to 10^9 bits of information required in a large-scale computer would force us to turn to some other storage device. Furthermore, since each bistable always has one element conducting, a large memory bank consumes an excessive amount of power. Their extremely fast read and write times (20 to 50 ns) make bistable memories suitable for smaller, random-access, "scratch-pad" memories of 10^4- to 10^5-bit capacity.

Chapter 15 will discuss the slower core memories used for high-capacity storage as well as other applications of the static hysteresis exhibited by ferromagnetic and ferroelectric materials.

Steering or decoding

Steering circuitry translates the coded address information to the decimal form needed to select a particular path. By individually selecting a row (say x_h) and a column (say y_i), the coincidence of two logical-1 signals opens an AND gate at the intersection hi and connects the reading or writing amplifier to the particular cell location. Often the AND gate is part of the memory cell itself.

Steering networks are logic decoders similar to the one illustrated in Fig. 13-30. They decode the binary-word input and translate it to decimal row or column selection. Particular types of decoding networks will be discussed in more detail in Sec. 14-4.

Write and read amplifiers

Write and read circuitry is determined by the nature of the particular memory cell. The circuits may range from the simple logic which triggers the bistable memory cells (Chap. 13) and buffer amplifiers to read the output states to the much more complex sense amplifiers (Chap. 10) and current drivers needed with magnetic-core memories. Specialized amplifiers will be discussed in connection with particular memory arrays.

14-2 Short-time MOSFET Memory

Short-time memories are normally used for storage between computational steps or for storage while subroutines are carried out. The digital word is fed into a shift register, either in parallel or bit by bit serially, and is fed out, when required, either in parallel or serially. When constructed by directly interconnecting clocked bistables, as discussed in Sec. 13-7, a shift register can also serve as a long-time low-capacity memory.

Much-larger-capacity short-time dynamic shift registers can be accommodated on a single silicon chip if, instead of being stored in a separate bistable, each bit is stored in the gate-to-substrate capacity of a MOSFET. Dynamic shift registers of the type shown in Fig. 14-3 having a total storage capacity of several hundred bits can be fabricated on a single chip, together with all the necessary steering logic.

Information is serially gated into the gate-to-substrate capacitance of the first MOSFET inverter stage by a logical-1 clock pulse applied to the series-gate element Q_3. The following inverse clock pulse transfers the inverted bit to the C_{GS} of the second inverter. During the storage interval, the transfer path is disabled, and during the transfer interval, the storage path is open. To prevent false storage or transfer, the clock pulses and inverted clock pulses are usually separated by a small dead time.

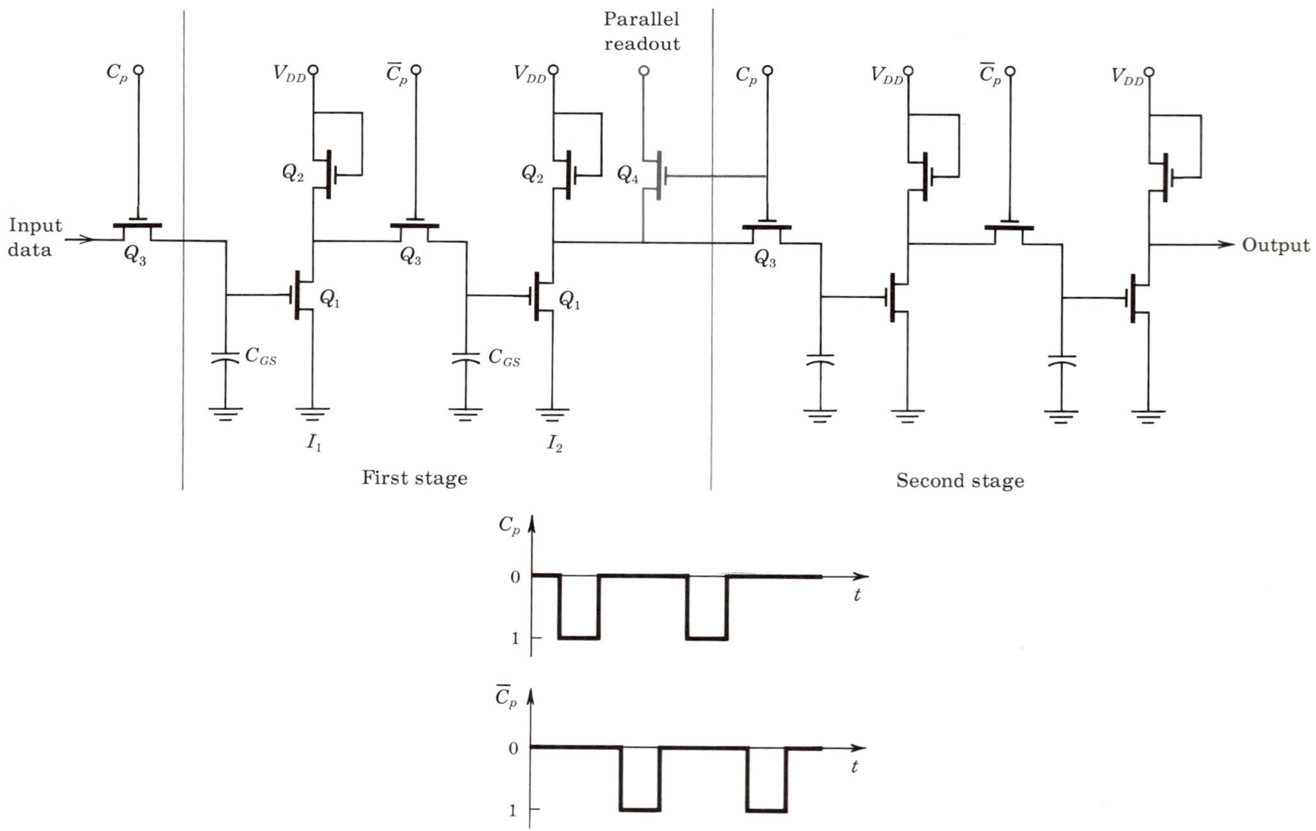

Fig. 14-3 Basic MOSFET, dynamic shift register (induced-p-channel devices). Short-time information storage is in the gate-to-substrate capacitance. MOSFET Q_4 would be included in each stage for parallel readout.

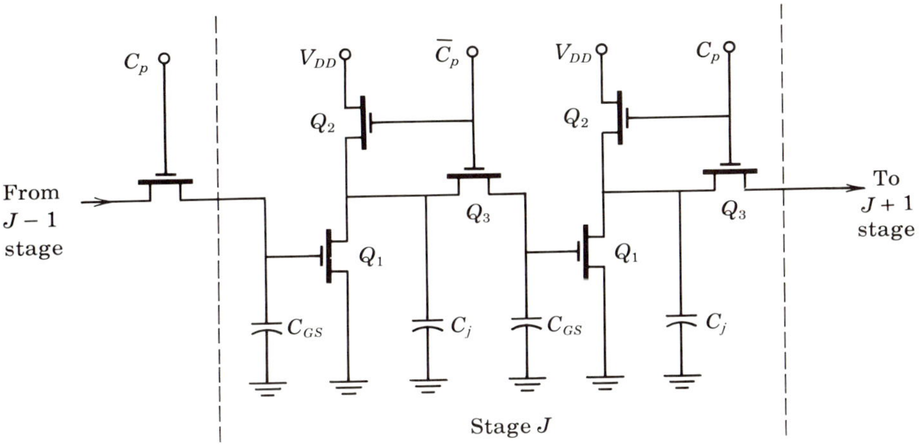

Fig. 14-4 Single state of modified dynamic shift register. ($V_{DD} = -20$ V, $V_T = -3$ V. Logic levels: logical 0 ≈ -1 V, logical 1 ≈ -17 V.)

Successive clock pulses write successive bits of the input word into the first stage of the register. Previously stored information is simultaneously transferred along the chain of stages. To ensure that each bit leaves a storage stage in the same phase as it entered, each stage is constructed from two cascaded inverters.

Additional transfer gates (Q_4 in Fig. 14-3) may be incorporated for parallel readout.

The inverter stages of the shift register of Fig. 14-3 need be operative only during transfer and readout clock intervals. By also gating the inverter load MOSFET (Q_2), an inverter will be on only when a stored bit is transferred to the next inverter. With a large number of stages on a single chip, the reduction in power dissipation will be significant. One stage of the modified circuit is shown in Fig. 14-4.

Typical induced-p-channel MOSFET elements, where $V_T \approx -3$ V, operate with negative-logic-voltage bounds of

$$\text{Logical } 0 > -2 \text{ V}$$

$$\text{Logical } 1 < -10 \text{ V}$$

The clock pulses are usually more negative than the supply voltage V_{DD}. By this means, we guarantee the largest possible swing at each inverter and improve the circuit's noise immunity.

The transient delay of each inverter pair establishes the upper limit of the clock frequency for the dynamic shift register of Fig. 14-3 or 14-4. To evaluate the rise and decay times, we apply the procedures of Sec. 5-7 to the models of Fig. 14-5.

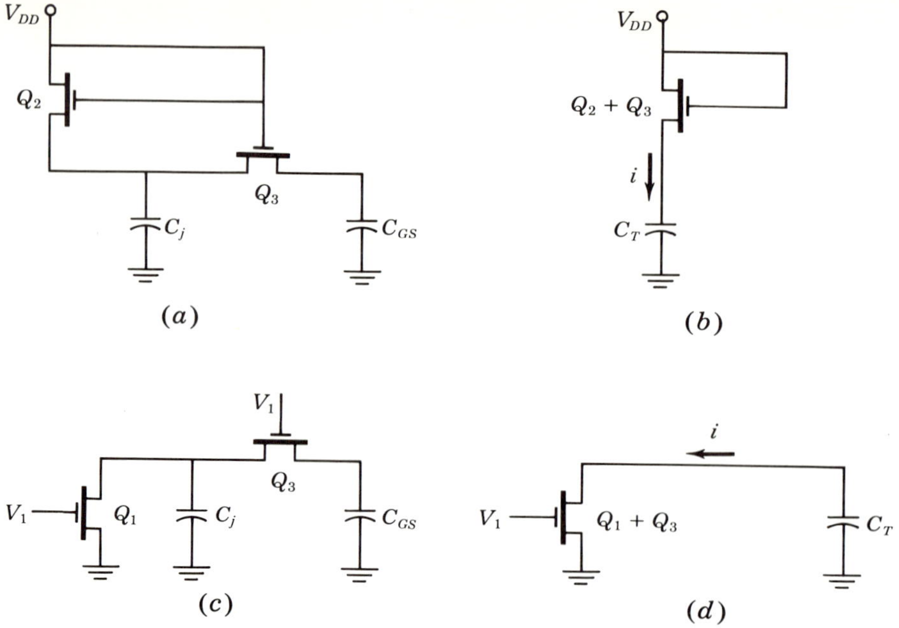

Fig. 14-5 Models for calculation of charge and discharge times for the shift-register stage of Fig. 14-4. (*a*) Charging of C_j and C_{GS} (input logical 0) to a logical 1; (*b*) reduced charging model; (*c*) discharge of C_{GS} and C_j to a logical 0 (input logical 1); (*d*) reduced discharge model.

Figure 14-5*a* shows the path for charging the junction and stray capacitance C_{j+s} to a logical 1 and then transferring the charge to the C_{GS} of the next MOSFET. If the drain-to-source capacity of Q_3 is neglected and if C_{j+s} ($\sim$0.6 pF) is lumped with C_{GS} ($\sim$0.3 pF), the circuit analysis is reduced to a calculation of the charging of a single capacitor through the series path of the two saturated MOSFETs, Q_2 and Q_3. It is obvious from the reduced model of Fig. 14-5*b* that we have placed the single capacitor at the point where we shall obtain the worst-case rise time.

The solution of the source follower charging the total capacitance C_T to a logical 1 is given by Eq. (5-51):

$$V_{GS}(t) = V_1 \frac{t/\tau_c}{1 + t/\tau_c}$$

Since, in this particular case, C_T is charged through two series MOSFETs, the proportionality constant τ_c [Eq. (5-49*b*)] becomes

$$\tau_c = 2(R_{\mathrm{ON}2,\min} + R_{\mathrm{ON}3,\min})C_T \tag{14-1}$$

To obtain a reasonable inverter gain, Q_2 must be physically small compared with Q_1. We shall see below that for optimum low-frequency response, the size of Q_3 must also be minimized. Thus both Q_2 and Q_3 will have relatively high ON resistances, about 4,000 Ω. With $C_T = 0.9$ pF,

$$\tau_c = 2(4{,}000 + 4{,}000) \times 0.9 \times 10^{-12} = 14.4 \text{ ns}$$

The slow rise time to a logical 1 is given by

$$t_R = 8.9\tau_c = 128 \text{ ns} \tag{14-2}$$

The time for decay from a stored logical 1 to a new input at a logical 0 is controlled by the current flow through the ON Q_1 and the series gate Q_3 (Fig. 14-5*c*). Again, by lumping all capacitors together, we place the energy storage at a single node. In Sec. 5-7, it was shown that the discharge starts with the short linear rundown of a constant-current discharge followed by [Eq. (5-55)]

$$v(t) = V_1'\left(1 - \tanh\frac{t}{\tau_D}\right)$$

This second portion predominates, and we shall let it define the full decay interval.

The proportionality constant τ_D can be expressed in the form

$$\tau_D = 2(R_{\text{ON}_1} + R_{\text{ON}_3})C_T \tag{14-3}$$

Q_1 is the physically large amplifier device (Fig. 5-15), and it will have a relatively small R_{ON}. Assuming $R_{\text{ON}_1} = 500$ Ω,

$$\tau_D = 2(500 + 4{,}000) \times 0.9 \times 10^{-12} = 8.1 \text{ ns}$$

and the total decay time is [Eq. (5-56)]

$$t_D = 1.47\tau_D = 12 \text{ ns}$$

Within each cycle, C_{GS} of one stage of the inverter pair discharges and C_{GS} of the other stage charges. Each half of the clock pulse must last long enough for the completion of the slower charging process. Thus the maximum clock frequency is given by

$$f_{\max} = \frac{1}{2t_R} = 3.9 \text{ MHz} \tag{14-4}$$

If we also set a dead time t_0 between the clock and the inverted clock pulse, then the maximum frequency will be reduced to

$$f'_{\max} = \frac{1}{2(t_R + t_0)}$$

In order to accommodate the increase in t_R at elevated temperatures, the $f_{\max}$ of an actual shift register is limited to about 2 MHz.

The low-frequency limit of the dynamic shift register is determined by the rate of charge leakage from the gate storage capacitance. Since the series logic element is off, the reverse current flow through the drain-substrate junction of Q_3 is the mechanism that drains the stored charge. Assuming a reverse saturation current flow I_{rs} of 250 pA, the time for the stored -17-V charge to decay linearly by 7 V to the -10-V logical-1 lower limit is found to be

$$T_D = \frac{C_{GS}\,\Delta V_s}{I_{rs}} = \frac{0.3 \times 10^{-12} \times 7}{250 \times 10^{-12}} = 8.4 \text{ ms} \tag{14-5}$$

The lower operating frequency is set by Eq. (14-5). The 8.4-ms decay time corresponds to a frequency of 120 Hz. Practical MOSFET dynamic shift registers must allow for worst-case operation. Usually the lower limit is set at several kHz instead of at the calculated 120 Hz.

Readin and readout stages

The input stage of the MOSFET shift register usually is zener-diode-protected to prevent large electrostatic fields from damaging the gate-to-substrate insulation (Fig. 5-10). The total input capacity, including the capacities of the diode, leads, and header, reaches about 5 pF shunted by about 10^{10} Ω leakage resistance.

The final stage of the shift register must supply the fan-out current to charge the input capacity of many other logic elements. For this reason, the output stage uses physically larger devices having lower saturation and ohmic-region resistance. Because of the lead and header contributions, the output capacity will also be about 5 pF.

If this MOSFET dynamic shift register is to be used with bipolar transistor logic, external interface amplifiers which translate logic levels are required at both the input and the output.

Regenerative MOSFET memory—dc to 2-MHz shift register

Figure 14-6 shows how the basic dynamic memory cell of Fig. 14-3 can be converted from a short-term to a long-term memory. A clocked feedback path is periodically closed, restoring the charge that is drained out of the gate-to-substrate capacitor during the storage interval. Since the output and input logic states are the same, we create a sequentially switched equivalent of a regenerative loop.

Three separate clock pulses are required for proper sequencing. The normal clock pulse C_{p1} is applied only during the interval when each bit is read into the input stage. The following inverse pulse C_{p2} transfers the inverted input bit into the gate capacitance of the second inverter. During the whole readin and transfer interval, the regenerative-path switch must

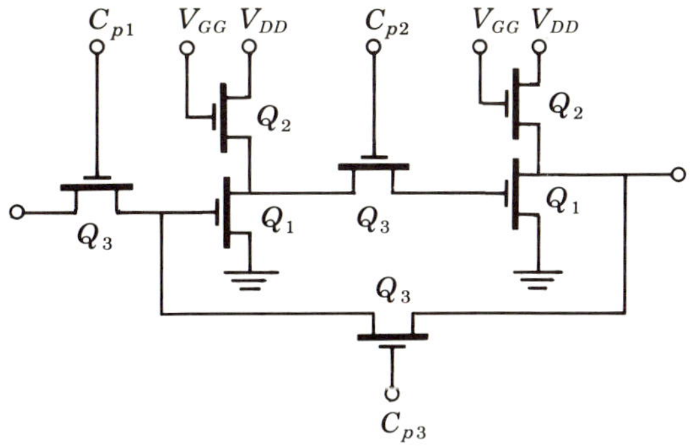

Fig. 14-6 One clocked regenerative stage of a MOSFET long-term memory-storage register.

be open ($C_{p3} = 0$). If it is closed, feedback of previously stored information will interfere with the storage of the new bit.

To satisfy the logic conditions described above, the input clock pulse C_{p1} would be ANDed with an external READ signal before being applied to the register. The regenerative-path clock pulse C_{p3} would be ANDed with the inverse READ signal.

After the bit is stored, the periodic application of C_{p2} and C_{p3} reconnects the gates of the MOSFET inverter stages to the appropriate logic levels found at the inverter output. Thus the two gate-to-substrate capacitances are periodically recharged. C_{p2} and C_{p3} can be the same pulse, or they can be different. There is no particular need for these pulses to be synchronized. It is only essential that each individually have a periodicity less than the decay-time interval of a stored logical 1; the frequencies of both C_{p2} and C_{p3} must be above the minimum operating frequency of the dynamic shift register.

Four-phase clocked shift registers

Multiphase clocking of MOSFET dynamic shift registers decreases the net power consumption and improves the switching speeds. Figure 14-7 shows the basic register stage (a single inverter pair), together with the sequence of clock inputs. We note that this pair has no ground return. It follows that current will flow through the MOSFET only into the uncharged node capacitor. There is no steady-state dissipation; only the power necessary to charge the node capacitance is required, and this power is supplied by the various clock inputs.

During the first interval t_1, defined by $C_{p1} = 1$, Q_2 turns on and the intermediate-node capacitance C_A is primed by being charged to a logical 1 (Fig. 14-8a). If the data input is also at a logical 1, then Q_1 will also be

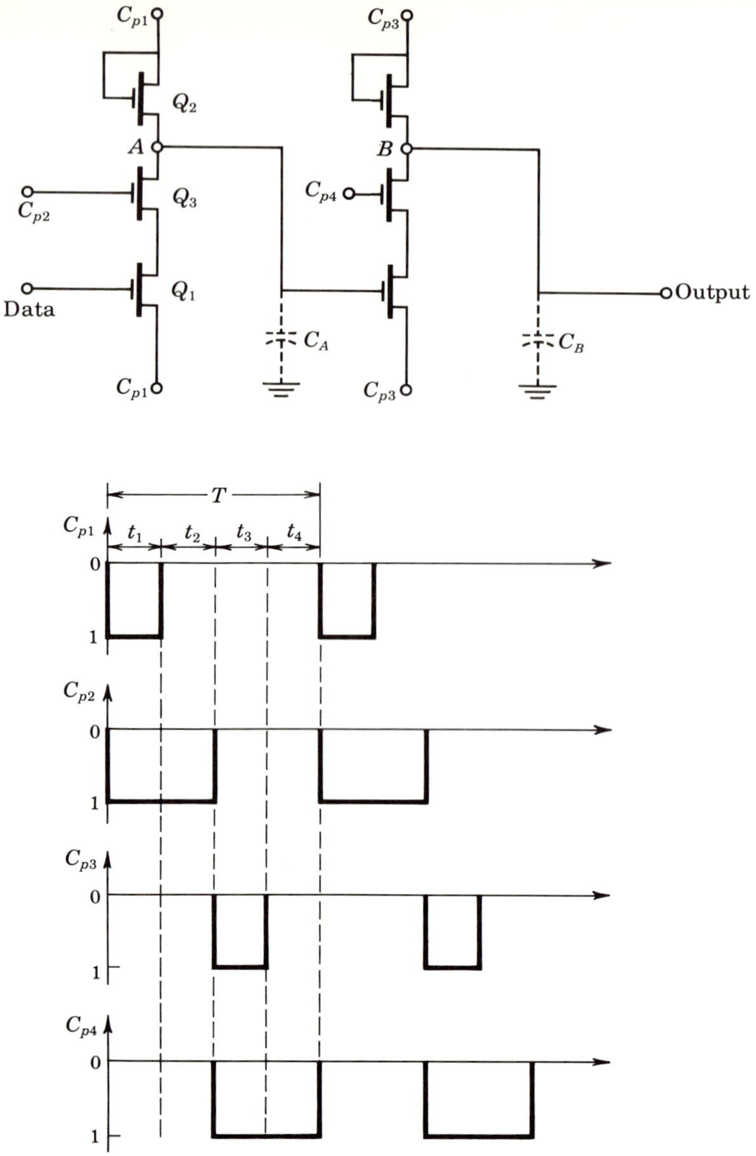

Fig. 14-7 Single stage of a shift register using four-phase logic and clocking sequence.

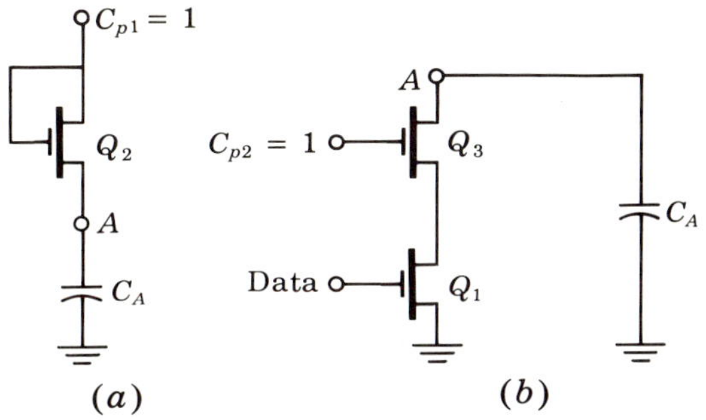

Fig. 14-8 (a) Path for priming the node capacitance C_A; (b) path for the discharge of C_A.

clocked on. Q_1 and Q_3 contribute a second parallel charging path, which reduces the time required for storage. Since C_A is charged through a maximum of one, instead of two, series MOSFETs, we gain a factor of 2 in the charging time [Eq. (14-1)].

During the second interval t_2, C_{p1} drops back to 0, grounding the source of Q_1. C_{p2}, still at the logical-1 level, keeps the gating transistor (Q_3) on. Figure 14-8b shows the circuit state. If the input is at a logical 0, then Q_1 will be back-biased and C_A cannot discharge. The node-capacitance voltage remains at the primed logical 1. If, however, the input data is at a logical 1, then with $C_{p2} = 1$, we have satisfied the NAND conditions of the Q_1-Q_3 series gate. Both devices conduct, establishing a fast-discharge path for C_A. Node A rapidly decays to the logical-0 level.

During the next two intervals, t_3 and t_4, the priming and transfer process is repeated in the second half of the shift-register stage. At the end of the clock sequence (T_{n+1}), the inverter-pair output finally registers the input data present at T_n. While transferring the information stored in C_A, C_{p2} is set to 0. This opens Q_3 and disconnects the input data from the primed intermediate node. Changes in the input data cannot cause changes in the output data. We say that the output data are *firm*.

A two-dimensional shift register

Because of the high packing densities possible in regular arrays of MOSFETs (see Fig. 14-14), multiple-shift registers can be interconnected, through the appropriate logic, on a single chip, thus forming extremely complex logic networks. The possible systems are limited solely by the imagination of the engineer. For example, Fig. 14-9 illustrates a 2×3 bit section of a 16×32 bit two-dimensional shift register. Each rectangle includes one register stage of the type shown in Figs. 14-3, 14-4, or 14-6.

When a clock logical 1 is ANDed with a horizontal control signal and applied to the gates of the horizontal shift MOSFETs (Q_3), information

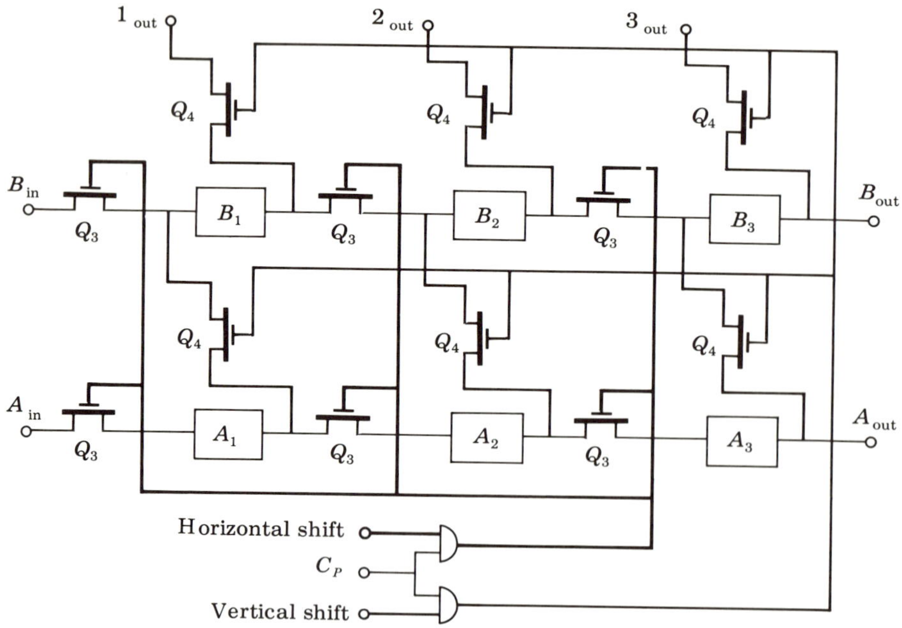

Fig. 14-9 A 2 × 3 bit section of a 16 × 32 bit two-dimensional shift register.

stored in each register is shifted in the normal manner to the input inverter of the next register on the right. The following $\bar{C}_p = 1$ interval transfers this information into the register's output inverter.

If, instead, the logical-1 clock pulse is ANDed with a vertical control pulse and applied to the gates of the vertical shift MOSFETs (Q_4), then the stored information in each register is transferred, in parallel, vertically to the next higher register element.

Additional steering circuitry can be incorporated into the system for down shifting as well as for back shifting.

Appropriately programmed transfers of the shift paths in the network of Fig. 14-9 may be used to code input information before transmission. The scrambled output ensures privacy of the communication channel. An inverse process at the receiver end, performed by a similar two-dimensional array, reverses the process and decodes the scrambled message.

14-3 A Bipolar-bistable Matrix

Bipolar-transistor bistables are the logical choice for fast-responding memory cells which are able to keep up with the computational speeds of on-line systems. Read and write and access times of 10 to 20 ns are compatible with the switching speeds of ECL and TTL gates. We trade off the bits-per-chip capacity of similar MOSFET memories for the faster speed (100-ns switching time is typical in a MOSFET bistable). Storage capacities of 16 or 32 bits per chip can be extended by interconnecting several chips within a single package. A per-package storage capacity of 64, 128, or 256 bits, together with all steering, read, and write circuitry, is realized through multilevel wiring.

Figure 14-10 shows the general array of a 4 $\times$ 4 direct-address memory. A bipolar TTL bistable having AND inputs is used as the memory cell at each bit location. Any particular bistable may be addressed by exciting the appropriate x and y lines with logical-1 pulses. Information is then read out of, or written into, the addressed bistable through the 0 and 1 read-write lines common to all cells.

Cross talk and interaction between any of the address lines and the read-write lines can be avoided by separating the address and set functions within the AND gate of each bistable. To do so, we take advantage of the fact that in a TTL AND gate only the emitter connected to the lowest potential conducts—all other emitters are back-biased. We shall show below that by using different voltage levels for the address and for the read-write instructions, we can individually query each bistable.

Under quiescent conditions, all bistables must be disconnected from the read-write lines. Since the address lines are at the low logical-0 level, at least one pair of emitters of each transistor of each bistable can conduct. The bistable remains in the previously set state. For isolation, the read-write 0 and read-write 1 lines are held at a higher dc potential, back biasing the associated emitters. Memory cells are thereby isolated from each other and from the output amplifier by the back-biased read-write emitter junctions.

Address logical-1 levels are used to select an individual cell within the matrix. Logical-1 pulses on an x and a y line back bias the address emitters. The bistable at the intersection has two pairs of emitters back-biased, leaving the single remaining emitter pair conducting through the read-write 0 and read-write 1 lines. Because all the unaddressed bistables have at least one pair of emitters still at the logical-0 address level, they remain isolated from the read-write circuitry.

Condition of 0-line readout amplifier

Figure 14-11 shows the read-write amplifier used in both the 0 and 1 lines. This amplifier sets the dc conditions for the read-write lines, as well as indicating the stored state of the addressed bistable.

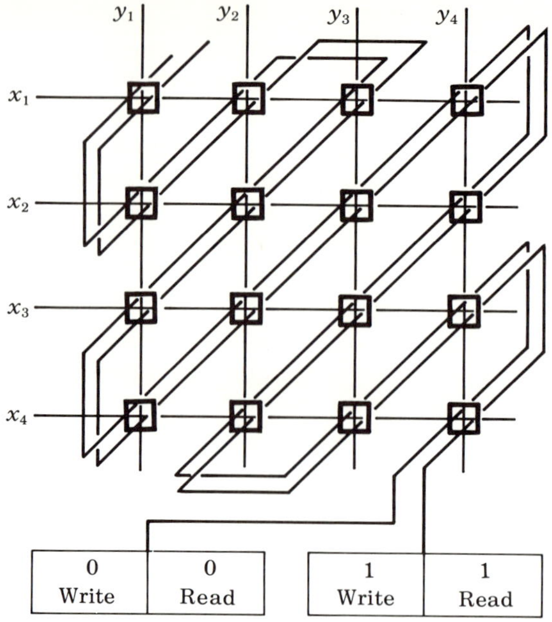

(*a*)

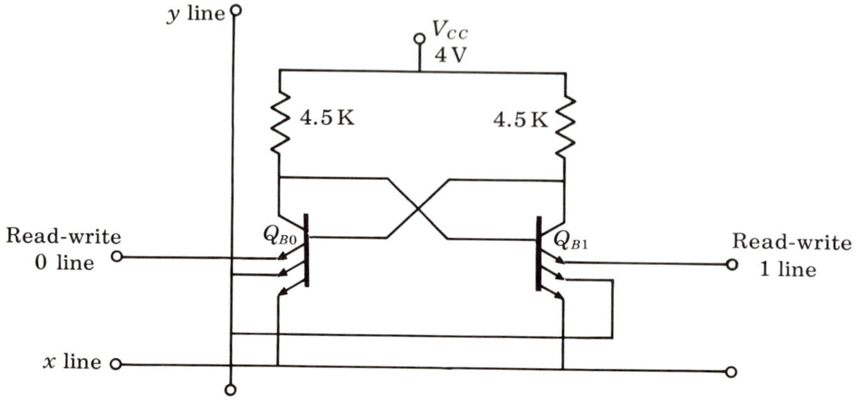

(*b*)

Fig. 14-10 (*a*) 16-bit bipolar-bistable memory matrix; (*b*) single bipolar bistable, showing address and read-write inputs.

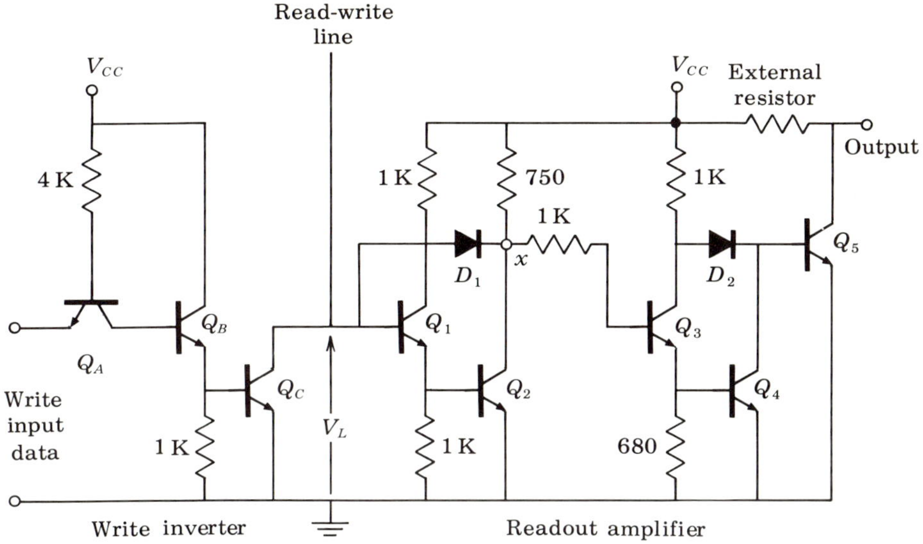

Fig. 14-11 Read-write amplifier for memory matrix of Fig. 14-10. Separate amplifiers are required for the 1 line and for the 0 line. (Q_4 speeds switching time by drawing out excess charge stored in the base of Q_5. $V_{CC} = 4$ V.)

Suppose that this particular bistable has a stored 1. Q_{B1} of Fig. 14-10*b* is the ON transistor, and Q_{B0} is the OFF transistor of the bistable. Current flows only in the read-write 1 line from the conducting emitter of Q_{B1}.

Since no current flows in the read-write 0 line, both the input emitter follower Q_1 and the following transistor Q_2 are cut off. The line-voltage level is established by the diode D_1 returned to the collector of the OFF Q_2 (point x). The 0-line voltage is at the high level

$$V_H = V_x + V_{D_1} = \left(V_{CC} - \frac{V_{CC} - 2V_{BE}}{1{,}750} 750\right) + V_{D_1}$$
$$= 3.68 \text{ V} \tag{14-6}$$

which we can conceive of as corresponding to a logical 1. The high line voltage of Eq. (14-6) keeps the bistable's OFF transistor Q_{B0} turned off.

We see from the amplifier circuit of Fig. 14-11 that as a result of Q_1 and Q_2 being off, Q_3 and Q_4 conduct and Q_5 is cut off. The readout amplifier converts the 0-current input into the high-level logical-1 output.

Condition of 1-line readout amplifier

The read-write 1 line conducts the emitter current of the ON transistor (Q_{B1}) in the addressed bistable. This current turns on Q_1 and Q_2. Hence the line-1 voltage is at the low level established by the two conducting base-emitter diodes:

$$V_L = 2V_{BE} \tag{14-7}$$

The saturation of Q_2 back biases diode D_1, which sets the high line level for the nonconducting bistable. Note that since V_L of Eq. (14-7) is above the address line-0 level, cell-to-cell isolation of the individual bistables is ensured.

Now, with Q_2 saturated, Q_3 is off and Q_5 saturates. The stored logical 1 reads out inverted as a logical 0.

It is interesting to observe that the two different dc levels of Eqs. (14-6) and (14-7) guarantee that the readout is nondestructive. The low voltage level (V_L) at the emitter of the conducting bistable transistor and the high voltage level (V_H) at the emitter of the nonconducting transistor are exactly the conditions necessary for switching into the state corresponding to the data already stored.

Bit storage

A bit is stored in the addressed bistable by turning on the appropriate emitter. Since the bistable is composed of cross-coupled TTL elements, a logical-0 trigger is used for switching. Each of the write amplifiers (Fig. 14-11) is a simple one-input TTL inverter consisting of the three transistors Q_A, Q_B, and Q_C. A WRITE-instruction logical-1 pulse turns off Q_A and saturates Q_C. The saturated output transistor pulls the read-write line down to the logical-0 level.

If the instruction is to write a 0, the read-write 0 line is brought to the low logical-0 potential and the memory-cell transistor Q_{B0} turns on. The higher voltage at the uninstructed read-write 1 line keeps Q_{B1} off.

14-4 Coding and Decoding

Increasing the memory storage from tens to hundreds of bits capacity also increases the number of individual address lines beyond the number that can be accommodated on the integrated-circuit header. A direct-address $J \times K$ matrix requires $J + K$ lines. For example, a 512-bit memory arranged in a 16×32 matrix requires $16 + 32 = 48$ separate input lines for direct address to each bit. Besides the large number of external connections, each lead internal to the package must be bonded to the silicon chip. Each bond is a potential source of trouble. The bonds occupy a large area, reducing the net surface area available for the memory array.

If, for example, the decimal input information is first externally coded into a weighted binary code, then the memory unit can be addressed by parallel binary digits. Every n-digit binary word can be used to carry address information for 2^n separate lines. The binary word must be decoded on the chip into the decimal information required to select a particular memory cell. We can compute the saving in leads by observing that a $J \times K$ matrix may be expressed in binary terms as

$$J \times K = 2^n \times 2^m = 2^{n+m} \tag{14-8}$$

We restrict J and K to powers of 2. The total number of parallel address lines required is

$$L = n + m \tag{14-9}$$

A 16×32 matrix is expressed by

$$16 \times 32 = 2^4 \times 2^5 = 2^9$$

Only four parallel inputs (a four-bit word) are needed to address a particular row, and a five-bit word to find a particular column. The use of external coding in combination with an internal decoding (steering) network requires 39 fewer input connections than a direct-address memory.

A weighted binary code is expressed in the form

$$N = d_4w_4 + d_3w_3 + d_2w_2 + d_1w_1 \tag{14-10}$$

In Eq. (14-10), N is the decimal number to be coded, w is the binary weighting given to the particular position in the coded word, and d is the appropriate binary digit (0 or 1). Many codes are in common use; two include the BCD (binary-coded decimal), where the weighting is 8-4-2-1, and a 2-4-2-1 weighted code which can be made self-complementing.

Decimal numbers are encoded into a binary code through OR-gate outputs. Figure 14-12 shows an encoding procedure using a diode array to obtain a self-complementing 2-4-2-1 code. Each column of the network corresponds one-to-one to the OR-gate synthesis of the appropriate column of the truth table. Inverters are incorporated to supply the complementary functions $\bar{A}$, $\bar{B}$, $\bar{C}$, and $\bar{D}$.

Any other binary code will also be synthesized by an OR-gate expansion of the new truth-table columns.

Decoding

Decoding the binary input code into a decimal line selection is accomplished by expanding the truth table by rows. The resulting matrix is composed of AND gates excited by the binary code and its complement. The complementary function may be available externally or may be created internally by inverters contained in the decoding matrix.

	2 D	4 C	2 B	1 A
0	0	0	0	0
1	0	0	0	1
2	0	0	1	0
3	0	0	1	1
4	1	0	1	0
5	0	1	0	1
6	1	1	0	0
7	1	1	0	1
8	1	1	1	0
9	1	1	1	1

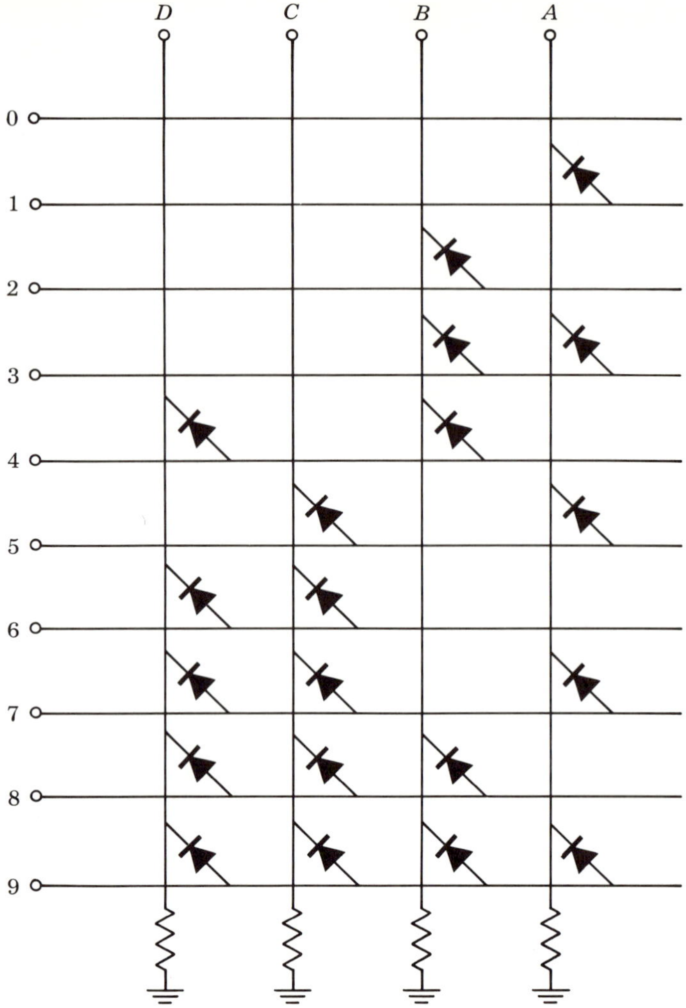

Fig. 14-12 Diode encoder matrix for decimal to self-complementing 2-4-2-1 code. Synthesis is based on the OR statement of each column of the truth table.

Figure 14-13 shows a MOSFET AND array which decodes a BCD binary-word input into the decimal output lines. The input A-B-C-D and $\bar{A}$-$\bar{B}$-$\bar{C}$-$\bar{D}$ lines are connected only to the gates of the series AND array. This array can be extended to a 16-line decoder without requiring additional input address lines. In order to reduce the number of bonded contacts, the inverters in Fig. 14-13 are usually included on the chip. Figure 13-30 showed a diode AND matrix used to decode the BCD output of a bistable counter.

	A	B	C	D
0	0	0	0	0
1	0	0	0	1
2	0	0	1	0
3	0	0	1	1
4	0	1	0	0
5	0	1	0	1
6	0	1	1	0
7	0	1	1	1
8	1	0	0	0
9	1	0	0	1

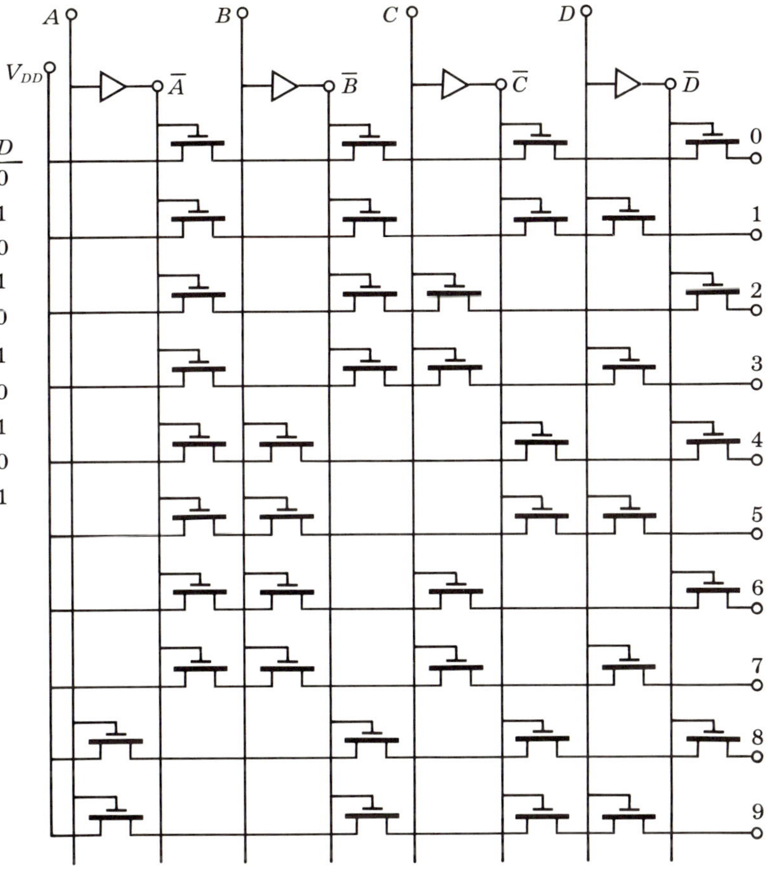

Fig. 14-13 A MOSFET BCD decoding network. Four address lines are needed to select 1 out of 10 lines.

Internal machine subassemblies may not all operate with the same binary code. Interface stages are then required for machine-to-machine translation of one binary-coded output to the different code needed at the next input. For example, an 8-4-2-1 code might be converted to a self-complementing 2-4-2-1 code for certain arithmetic operations and then reconverted back to the 8-4-2-1 code. Each translator is synthesized by an array of AND and OR gates and inverters.

Final-readout coding again requires translation from machine language to some form of decimal display. The particular readout device employed determines the nature of the decoding. Alphanumeric or decimal displays often accept BCD information directly and perform the code translation within the display unit. Some readout display devices require a quibinary-

Table 14-1 Two seven-bit weighted binary codes

	QUIBINARY 02468-01	BIQUINARY 01234-05
0	10000-10	10000-10
1	10000-01	01000-10
2	01000-10	00100-10
3	01000-01	00010-10
4	00100-10	00001-10
5	00100-01	10000-01
6	00010-10	01000-01
7	00010-01	00100-01
8	00001-10	00010-01
9	00001-01	00001-01

coded input (02468-01), a biquinary code (01234-05) (Table 14-1), or some other arbitrary seven-bit code. Various integrated-circuit code translators are available for use as the required interface networks.

14-5 MOSFET Read-only Memory

The extremely small size of an integrated MOSFET device makes possible the fabrication of circuits containing 1,500 to 5,000 active devices on a single silicon chip. These high densities are realizable only when the circuit is synthesized as an iterative array with minimum area allotted to interconnection metallization. Large-scale integration arrays are particularly suitable for read-only memories (ROM).

ROMs contain permanently stored information such as sine or tangent tables, correction factors, constants, and other information needed during computation. These memories may also be used to store particular subroutines and sequencing programs. The stored information pattern is impressed on the chip during the final metallization, when the interconnection matrix is formed. Steering logic for decoding the address information and buffer-amplifier output stages are also included on the chip.

Figure 14-14 is a microphotograph of a General Instrument Corp. ROM matrix having a total storage capacity of 2,048 bits arranged in a 64×32 array. Address and readout time is less than 1 μs. With minor modification during the final metallization, this particular array can be changed to a word-organized memory storing 256 eight-bit words.

Fig. 14-14 A 2,048-bit random-access read-only memory. (*Microphotograph courtesy of General Instrument Corp.*)

The storage devices occupy the major portion of the chip area in a regular array of 64 columns and 32 rows. Storage of a logical 1 at any bit position is accomplished by connecting the gate to the address row during metallization. This particular location contains a MOSFET which can conduct during readout. A logical 0 is stored by not metallizing the gate region. Since, without a gate, the intersection region is simply two back-to-back *pn* junctions, the drain-to-source path of the logical-0 device is effectively an open circuit.

Clearly evident on the right side of the photograph is the row-decoding matrix. A five-bit row address word is inverted on the chip. The address word and its complement are fed into the AND decoding logic, where 1 out of 32 rows is selected.

At the bottom of the chip is the column-decoding network. Here a six-bit address word and its complement (produced on the chip) select 1 out of the 64 columns.

At the top of the photograph we see the large-area MOSFETs of the eight buffer amplifiers (and loads) used for readout. In this particular bit-oriented memory, all of the output amplifiers are wired NORed at the contact pads. For a word-oriented geometry, the wired connections would be removed, permitting the parallel readout of each of the 256 stored eight-bit words.

Just below the output buffers we have the additional steering logic needed to connect one out of eight columns to each amplifier. Also visible is some of the clocking logic, which turns the queried circuit paths on and off. Two- or four-phase clocking (Fig. 14-8) may be employed to improve switching time and reduce power consumption.

The ROM of Fig. 14-14 requires complex interconnections. Two-level wiring is used. The top level is the metallization, which is clearly visible in the photograph. The lower-level wiring is within the *n* substrate and can be faintly seen through the top oxide layer. Extensions of the high-conductivity diffused-*p*-type source and drain regions are used to interconnect various drains and sources. The *p*-region diffusions are also used for wiring to the contact pads (lower right corner). We can make use of the *p* lands for wiring because, when no gate region exists, the several *p* regions are separated from each other by at least one back-biased junction.

Even if it were desirable, it is obviously impractical to analyze in detail a circuit containing over 4,000 MOSFET devices. However, the quantitative operation of an ROM can be understood from the greatly simplified array of Fig. 14-15. Here we have a 4×4 memory matrix of two-bit words.

Each address input is inverted on the chip. The address and its inverse are applied to the gates of various MOSFET devices arranged in a series AND decoding network.

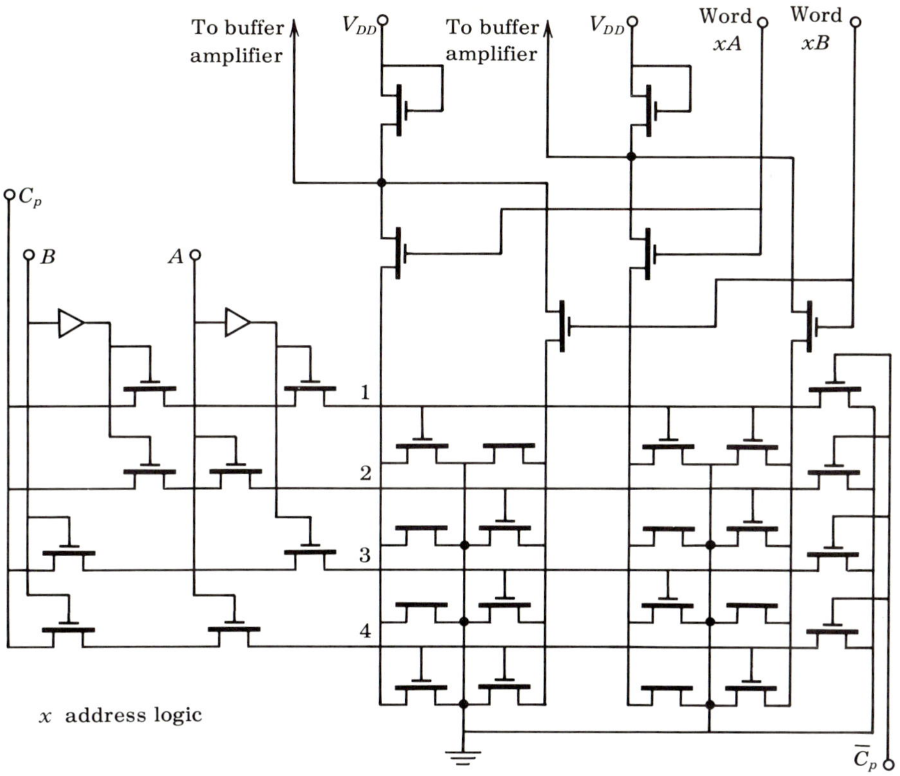

Fig. 14-15 Simplified random-access MOSFET ROM with a word-oriented geometry. The 4 × 4 array stores eight separate two-bit words. This memory is similar in structure to that shown in the photograph of Fig. 14-14.

Let us assume that we address row 3 by applying the address word 10 (BA). Logical-1 signals on lines B and $\bar{A}$ satisfy the series AND conditions at row 3. Along all other rows, at least one series MOSFET has an applied logical 0 which keeps the path open. Figure 14-16*a* shows the complete circuit existing along this row. We see that the conducting row 3 in the decoding matrix completes the path from the clock pulse to the gates of all storage elements along this row. During the $C_p = 1$ interval, the logical 1 queries the row elements. Q_D removes any stored charge during the following $\bar{C}_p = 1$ interval. (Four-phase clocking can also be employed to shorten the readout time.)

At every row and column intersection in the memory matrix, there is a MOSFET region where the drain is connected to the column and the source to a common line. The gate, if any, is connected to the address row. All elements in a column are OR-connected. We have shown above that

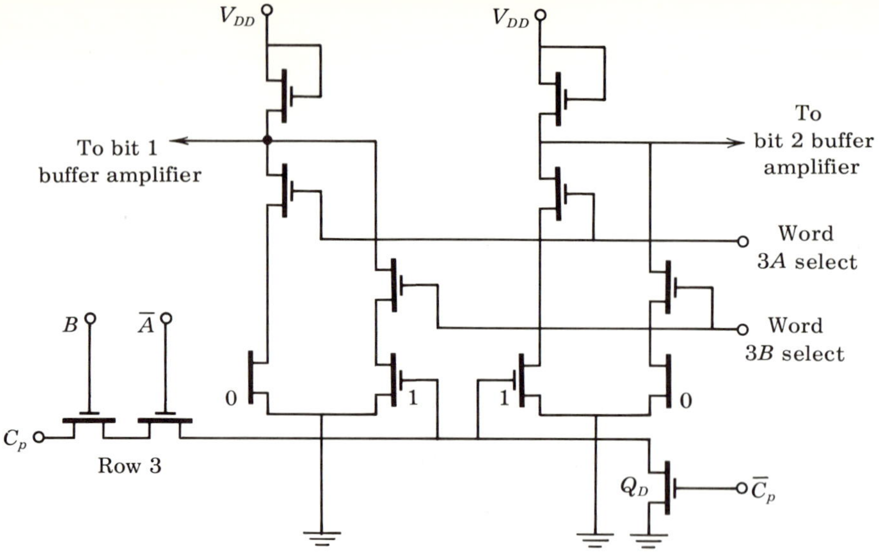

(a)

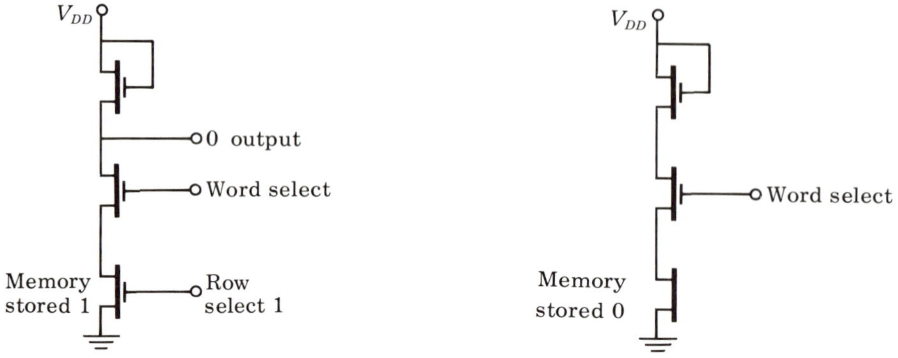

(b) (c)

Fig. 14-16 (a) Circuit along row 3 of memory of Fig. 14-15, showing intersecting-column logic; (b) column structure when a stored logical 1 is selected; (c) column structure when a stored logical 0 is selected.

selection of the particular column OR element is a function of the AND row address.

Any stored 1 (gate metallized, as in bit 2 of word 3*A* and bit 1 of word 3*B*) completes the conduction path in the column once the logical 1 is connected to the gate (Fig. 14-16*b*). The logical-0 NOR output is the input to the output buffer amplifier.

Any stored 0 (gate not metallized, as in bit 1 of word 3*A* and bit 2 of word 3*B*) results in the open-circuit region of two back-to-back junctions (Fig. 14-16*c*). The output of the column NOR is a logical 1.

Geometric considerations show that chip area will be conserved when several words are stored in parallel columns. Equivalent bits feed the same output buffer. The selection of a particular word is by addressing series word-select gates in each column (Fig. 14-16*a*). The word columns are wired NORed at the common load.

Problems

14-1 Figure 14-17 illustrates a very simple memory which may be used for intermediate-length storage. The silicon diode has a drop $V_D = 750$ mV at $I_D = 1$ mA.

(*a*) Plot the transfer characteristic of the circuit. Repeat the plot when the physical diode is replaced by an ideal diode.

(*b*) The input is a logical-1 20-V pulse applied through a 500-Ω source impedance. The minimum acceptable logical-1 level is 15 V. Specify C for a 5-ms storage interval; for a 30-s interval.

(*c*) Under the two conditions of part *b*, specify the minimum-duration logical-1 pulse for proper storage.

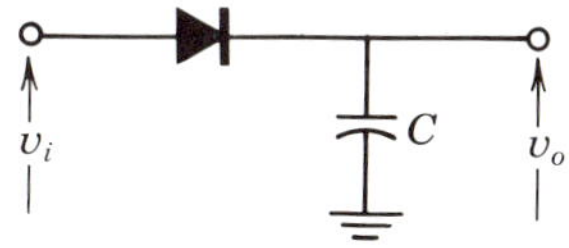

Fig. 14-17

14-2 Figure 14-18 shows one stage of a shift register using pull-up capacitors as the inverter loads. Use the parameters for Q_1 and Q_3 given in the text in Sec. 14-2. $C_1 = 10$ pF, and $C_2 = 1$ pF. C_1 is returned to $V_{DD} = -20$ V.

(*a*) Plot V_{GS} at the second inverter stage if the input is a logical-1 pulse followed by a logical-0 pulse.

(*b*) Specify the maximum and minimum clocking rates.

(*c*) What advantages (if any) does this circuit have over the shift registers of Figs. 14-3 and 14-4?

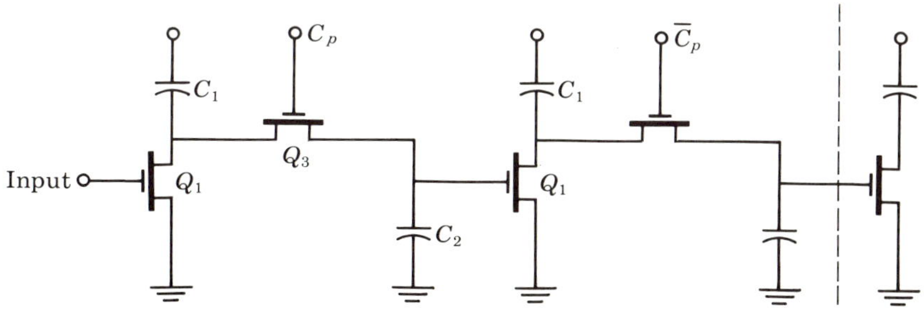

Fig. 14-18

14-3 The shift-register stage of Fig. 14-18 is converted into a clocked regenerative stage similar to Fig. 14-7. Will this circuit function properly? What are the minimum and maximum clocking rates?

14-4 Solve the circuit of Fig. 14-7 for the minimum time required to shift a logical-1 pulse through one stage. Use the MOSFET and circuit parameters given in the text with respect to the circuit of Fig. 14-4. Specify the minimum clock-pulse width and the total clocking period.

14-5 (*a*) Plot the transfer characteristic for both the read and the write amplifiers of Fig. 14-11. ($\beta = 50$, $V_{BE} = 750$ mV, $R_{\text{external}} = 240\ \Omega$, and $V_{CC} = 5$ V.)

(*b*) What is the write noise immunity?

(*c*) What is the readout fan-out N for 500-mV noise immunity when feeding a TTL logic?

14-6 Show the arrangement of a 16×8 bit, word-organized memory (8-bit words) based on the 4×4 TTL bistable matrix of Fig. 14-10*a*. Use a minimum number of read-write amplifiers. Show a simple address matrix. If all circuitry is fabricated on a single substrate, what is the total number of external connections?

14-7 Design a coding array which codes a decimal input into a self-complementing 1-2-4-2 binary code.

14-8 Design a MOSFET translation network from a BCD (binary-coded-decimal) input to quibinary output lines.

14-9 A simple MOSFET read-only memory is used to store a sine table (every 10°). Each sine is stored as an eight-bit word. Express the sines in binary terms. Show the geometry of the ROM, indicating a stored logical 1 by a □ and a stored logical 0 by a ○. Show the sine-selection matrix and the readout logic.

Bibliography

Manufacturers' Application Notes and Data Sheets, especially from General Instrument Corp., Philco, and Fairchild.

Watkins, B. G.: A Low-power Multiphase Circuit Technique, *IEEE J. Solid-state Circuits*, vol. SC-2, pp. 213–220, December, 1967.

15. magnetic and dielectric devices as passive memories and switching elements

In this chapter we shall be making use of just those properties of magnetic and dielectric materials that are so annoying in linear circuits. One of the more important is hysteresis, i.e., the dependency of the present state on the past excitation, the material's memory. The other characteristic, saturation, which allows several regions of operation, permits the use of these solid-state devices in controlled gates and other circuits previously restricted to diodes and transistors. Not only are special materials used, but these are also treated to exaggerate the hysteresis and saturation regions before being fabricated into cores which are used for high-capacity memories.

The behavior of magnetic materials has been studied extensively, and consequently there is a much wider range of properties available than can be found in the dielectric mediums. For this reason, we shall concentrate mainly on those circuits making use of ferromagnetic devices. The last sections of this chapter will consider briefly certain ferroelectric devices and circuits.

15-1 Ferromagnetic Properties

In order to explain, even cursorily, the salient features of the *B-H* curve of Fig. 15-1, we shall start with a microscopic region and increase our range

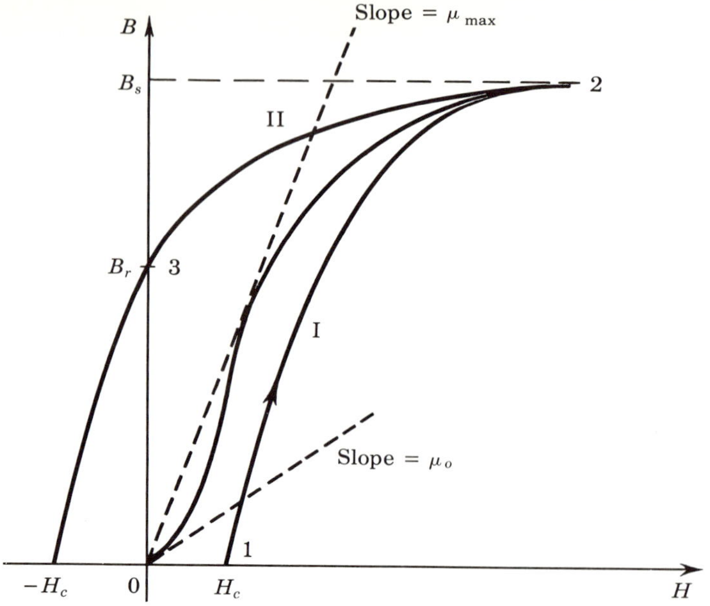

Fig. 15-1 Portion of the magnetization curve, showing some of the important constants of the material.

until it encompasses the whole core. There exist, within the material, small domains where all the molecular magnetic moments are aligned in some preferred direction. At room temperature, iron has six and nickel has eight easy directions of magnetization, which are related to the crystal structure. Separating the domains are walls composed of atoms whose moments are skew to the principal direction. Since, from a macroscopic viewpoint, the distribution of magnetic moments is random, the material is normally unmagnetized.

An external magnetic field increases the area of those domains magnetized in its direction and causes the contraction of those which have other magnetic moments. As the excitation is increased, portions of the walls become unstable, jumping to the direction of forced orientation. The finite field required to produce this change results in the hysteresis shown in Fig. 15-1. When all the atomic moments are essentially in alignment, we say that the material is saturated; it then supports a magnetic flux density B_s

The presence of the external field is necessary to maintain the alignment of the magnetic domains. After the excitation is removed, they relax to the nearest easy direction of magnetization. Instead of the original random orientation, there now exists a component of flux in the direction of the applied field, and this is called the remanence B_r. To completely demagnetize the material, we must apply a reverse field that will overcome the internal coercive force $-H_c$.

One important factor used to relate the field and flux is the magnetic permeability,

$$\mu = \frac{\Delta B}{\Delta H} \tag{15-1}$$

which is not a constant but increases from a low value near the origin to a maximum near the peak of the magnetization curve, decreasing as the core finally saturates. Even though B_s, B_r, H_c, and σ (the volume conductivity) are also needed for a complete description, μ_{max} serves as a convenient measure of the magnetic quality. Those materials having a high permeability almost always have a narrow hysteresis loop. For example, iron, which has a maximum permeability of 5,000, requires a magnetizing force of 3.2 Oe to develop the saturation flux density of 16,000 G. A high-permeability alloy, such as permalloy (79 percent nickel, 17 percent iron, and 4 percent molybdenum), may have a maximum μ of 100,000 or higher. This material would saturate at a flux density of close to 8,000 G, when the field strength is approximately 0.1 Oe. It has a coercive force of 0.05 Oe and a retentivity of 6,000 G. Note that the hysteresis loop of this second sample (Fig. 15-2) begins to approach a rectangle, so much so that it becomes valid to approximate the curve in the piecewise-linear manner shown.

The above discussion was based on the response either to direct current or to relatively low frequencies where the flux in the core has time to build up to the maximum value permissible with the applied excitation. If the dynamic hysteresis loop were to be plotted by using a very-high-frequency excitation signal, then the area of loop, which represents the losses, would be greatly increased. This effect is noticeable even at 400 Hz (Fig. 15-2).

In pulse applications we are interested, not in the steady state, but in the much more complicated transient B-H response. With any suddenly applied magnetization force H_a, we should expect an immediate flux density B_a. It would not be present. The H field must produce the alignment of the magnetic domains before the flux given by the B-H curve can be sustained. In addition, the H field propagates from the surface into the core with a finite velocity. Initially, only the surface layer contributes, and therefore the effective μ is an extremely small percentage of the final value. It will, of course, increase as the flux builds up toward the steady-state value.

Both the coercive force and the retentive field are structural properties, depending on such factors as grain orientation and microscopic imperfections, i.e., voids and inclusions. If, in working the magnetic alloy, the

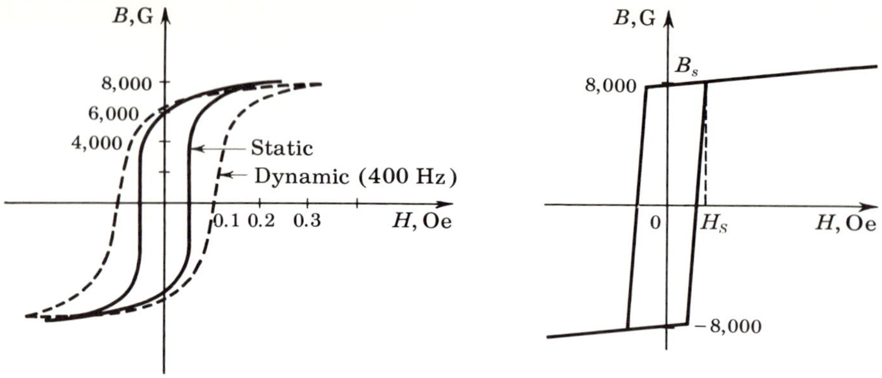

Fig. 15-2 Typical hysteresis curves and equivalent static piecewise-linear representation (4-79 Mo permalloy).

grain size is controlled and aligned to favor one easy direction of magnetization, μ_{max} can be greatly increased. One method used to produce almost square loop materials (Fig. 15-2) is to cool the hot-rolled sheet slowly in a hydrogen atmosphere with the magnetic field applied in the direction of rolling. The high permeability is measured only in the direction of grain orientation. To obtain the full advantage, the thin tapes produced are wound into toroidal cores. Deltamax (50 percent nickel and 50 percent iron), which has a maximum permeability of 70,000, has a squareness factor (B_r/B_s) of 0.98.

Energy must be supplied to the excitation coil to overcome the losses involved in establishing and aligning the magnetic domains. The input per unit volume required to switch between two points on the hysteresis loop is given by

$$w = \oint_{H_1}^{H_2} H\, dB \tag{15-2}$$

The terminal characteristics of the coil depend on the amount of stored, and hence recoverable, energy. In switching from point 1 to 2 in Fig. 15-1, the net energy supplied is proportional to the area between branch I and the B axis. Once the excitation is removed, the operating point traverses segment II from point 2 to point 3. Only the energy represented by the small area between branch II and the B axis is recoverable. The area of the hysteresis loop lying in the first quadrant represents the dissipated power and, consequently, the resistive component of the terminal characteristic.

In square-loop materials (Fig. 15-3), all the supplied energy is dissipated and none can be recovered. We thus interpret the driving-point characteristic of the core as that of a nonlinear resistance. Since this case is of great interest, it will be considered in some detail below.

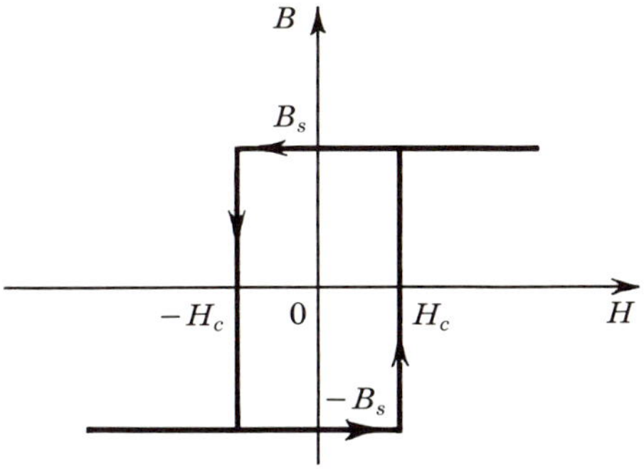

Fig. 15-3 Square hysteresis loop such as is found in a tape-wound toroidal core of Deltamax.

Ferrites

Some of the disadvantages of the thin-strip cores have been overcome by the use of various molded ceramic semiconductors. These materials, such as $NiO \cdot Fe_2O_3$ (nickel ferrite) or $MgO \cdot Fe_2O_3$ (magnesium ferrite), have a much lower electric conductivity and consequently lower losses. The nature of the material is such that the domain switching time would be much less than that of the tape cores. Even though the maximum permeability is small ($\mu_{max} = 1{,}000$), it is almost independent of frequency. The easily fabricated square-loop ferrite cores, of almost any desired configuration, are widely used in the high-speed memory arrays discussed below (in Sec. 15-6) and in various switching applications.

15-2 Terminal Response of Cores

Before we can make use of cores in specific circuits, we must know their terminal responses to various excitations. We have seen that this is a function of both the core geometry and the shape of the hysteresis loop. Because the general problem is quite complex, our discussion will concentrate on a square-loop core of relatively simple geometry. The core of Fig. 15-4 is constructed by winding a thin tape of grain-oriented magnetic material on a bobbin. By making the following assumptions, the subsequent analysis will be greatly simplified:

1. At every point the radius of curvature is large compared with the thickness d. In any small region, the core may be treated as an infinitely long plane sheet and the geometry reduces to two dimensions.
2. The width h is much greater than the thickness d. Fringe effects may be ignored. Consequently, the excitation can be assumed to be a current sheet flowing on the outer and inner surfaces of the core.

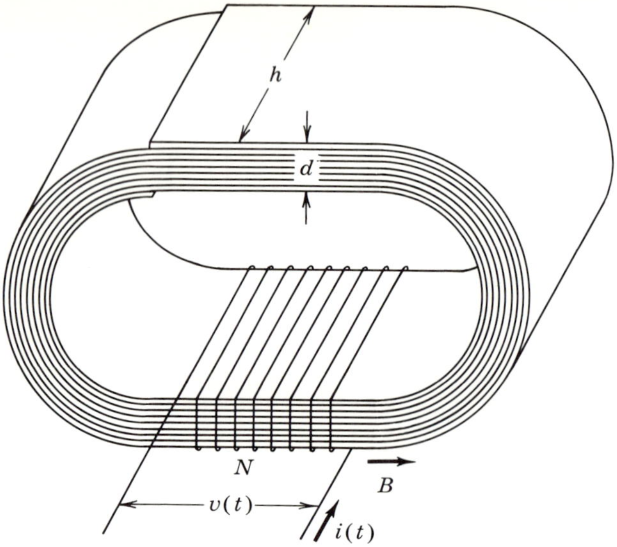

Fig. 15-4 Tape-wound core having a square hysteresis loop used for the investigation of switching response.

3. The hysteresis loop is perfectly rectangular. It follows that the core, which is initially magnetized to $-B_s$, switches to $+B_s$ as the H field propagates from the surface toward the center.
4. Because of the symmetry, the field will penetrate at equal rates from both the inside and outside of the core.

Once the magnetic field is established on the surface of the core, it creates the domain wall, which propagates toward the center, dividing the region having a flux density of $+B_s$ from that at $-B_s$. Throughout the complete material, the flux will have one of these two values. Only across the domain wall will there be any change of flux with respect to either time or space. Figure 15-5 illustrates the conditions in the core before the field penetrates to the center and completes the switch in the saturation direction.

Response to a known voltage excitation

The terminal voltage response of the excitation coil is given by

$$v(t) = N\frac{d\Phi}{dt} = 2Nh\frac{d}{dt}\int_0^{d/2} B\,dx \tag{15-3}$$

But from Fig. 15-5, we see that

$$\int_0^{d/2} B\,dx = B_s x_s - B_s\left(\frac{d}{2} - x_s\right) \tag{15-4}$$

and consequently Eq. (15-3) becomes

$$v(t) = 4NhB_s \frac{dx_s}{dt} \tag{15-5}$$

Solving Eq. (15-5) yields the spatial variation of the domain wall in terms of the known voltage excitation:

$$x_s(t) = \frac{\int_0^t v(\tau)\, d\tau}{4NhB_s} \tag{15-6}$$

Saturation occurs when the domain wall reaches the center of the core. Setting $x_s = d/2$, we obtain

$$\int_0^{t_s} v(\tau)\, d\tau = 2NB_shd = 2NB_sA \tag{15-7}$$

where A = cross-section area, m^2
B_s = saturation flux density, Wb/m^2

Equation (15-7) states that the time required for the complete reversal of the core state depends on the volt-time area of the applied voltage signal. Regardless of waveshape, equal area signals will result in equal

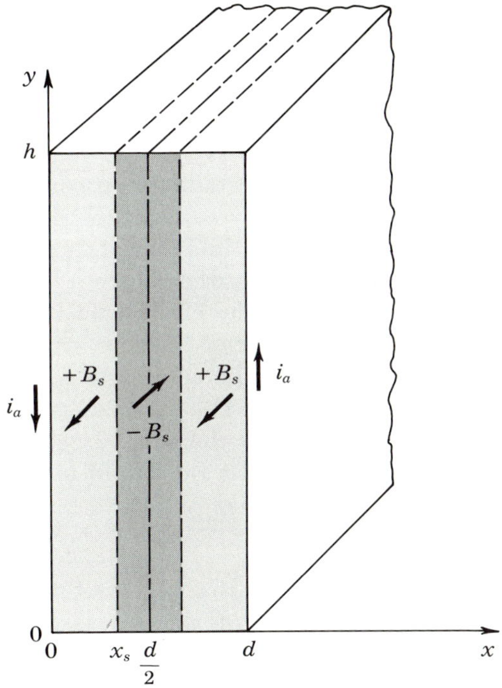

Fig. 15-5 Cross section of the core, illustrating the flux penetration and the direction of the current flow.

switching times. Or expressing this result in another way, under ideal conditions a volt-time area greater than that given by Eq. (15-7) cannot be sustained across the excitation coil.

In the special case where $v(t)$ is a constant V, Eq. (15-6) reduces to

$$x_s(t) = \frac{Vt}{4NhB_s}$$

and we can see that the domain wall propagates at a constant velocity from the surface to the center. The time required for complete saturation is

$$t_s = \frac{2NAB_s}{V} \tag{15-8}$$

In order to calculate the current flow, we turn to Maxwell's equations. Because of the simple geometry, they reduce to

$$\frac{\partial H}{\partial x} = -\sigma E \tag{15-9}$$

$$\frac{\partial E}{\partial x} = -\frac{\partial B}{\partial t} \tag{15-10}$$

where E = electric field intensity

σ = volume conductivity of the core material

From Eqs. (15-10), (15-4), and (15-5),

$$E(t) = -\frac{d}{dt}\int_0^{d/2} B\,dx = -2B_s\frac{dx_s}{dt} = -\frac{v(t)}{2Nh} \tag{15-11}$$

The presence of $E(t)$ results in an eddy-current flow in the excited portion of the core. This current may be considered as establishing a retarding field which limits the time rate of flux reversal.

B is constant everywhere except across the domain wall. Hence the spatial variation of the electric field must be zero everywhere except at $x = x_s$. In the center of the core, where the magnetic field has not as yet penetrated, $E(t)$ will be zero ($x > x_s$). For $x < x_s$, the electric field will be given by Eq. (15-11). Since E is constant with respect to x, H must decrease linearly from the surface value $H_a(t)$ to the coercive force H_c, at $x = x_s$. It follows that Eq. (15-9) may be rewritten as

$$\frac{H_a(t) - H_c}{x_s} = 2\sigma B_s\frac{dx_s}{dt} = \frac{\sigma v(t)}{2Nh} \tag{15-12}$$

where $H = Ni/l$, A-turns/m. l is the effective magnetic path length.

Substituting $B_s\,dx_s/dt$ and x_s, as given by Eqs. (15-5) and (15-6), into Eq. (15-12), we obtain the solution for the terminal current due to the known voltage excitation:

$$i_a(t) = \frac{\sigma l v(t)}{8N^3h^2B_s}\int_0^t v(\tau)\,d\tau + I_c \tag{15-13}$$

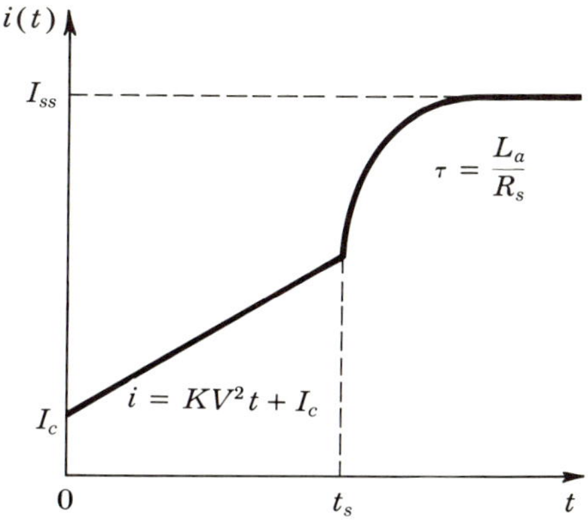

Fig. 15-6 Terminal-current response of a square-loop core to a constant-voltage excitation.

The first term in Eq. (15-13) is the total eddy-current flow in the core; the second term corresponds to the current needed to overcome the internal coercive force. We might note that the higher the conductivity of the ferromagnetic material, the greater the core loss. If the applied voltage is a constant, the current will increase linearly with time, as shown in Fig. 15-6:

$$i_a(t) = \frac{\sigma l}{8N^3h^2B_s} V^2t + I_c$$

After the core saturates, the current will continue to increase, but at a rate controlled by the inductance of the driving coil in air, its winding resistance, and the internal resistance of the voltage source.

We might further note that the constant-current term in Eq. (15-13), due to the coercive force, may be treated as a separate bias generator in parallel with a postulated ideal core (zero-width hysteresis loop). This leads to the conclusion that the response of the core may be represented schematically by the models of Fig. 15-8.

Response to a known current excitation

As our starting point, we shall solve Eq. (15-12) for the location of the domain wall in terms of the applied magnetic field:

$$x_s(t) = \sqrt{\frac{1}{\sigma B_s} \int_0^t [H_a(\tau) - H_c]\, d\tau} \tag{15-14}$$

Substituting $x_s(t)$ back into Eq. (15-12) yields

$$v(t) = \frac{2Nh[H_a(t) - H_c]\sqrt{B_s}}{\sqrt{\sigma \int_0^t [H_a(\tau) - H_c]\, d\tau}} \tag{15-15}$$

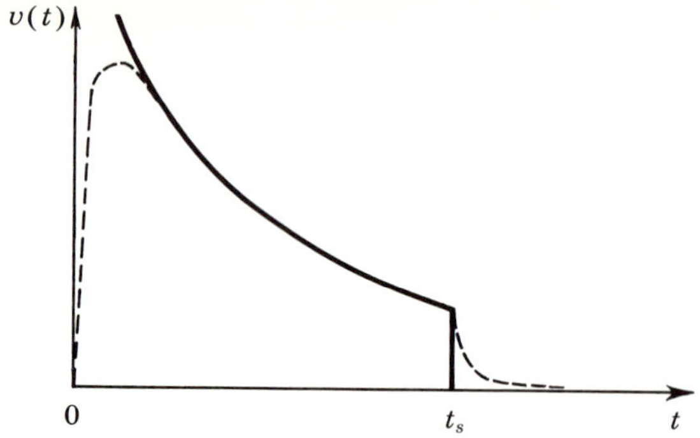

Fig. 15-7 Voltage response of a square-loop core to a constant-current excitation.

When the input is a constant current I_a, the terminal voltage and the saturation time are given by

$$v(t) = \sqrt{\frac{4h^2N^3B_s(I_a - I_c)}{\sigma l t}} \tag{15-16}$$

$$t_s = \frac{\sigma l B_s d^2}{4N(I_a - I_c)} \tag{15-17}$$

According to Eq. (15-16), the terminal voltage must be infinite at $t = 0$. This impossible solution is a result of the assumed ideal core characteristics. In practice, a finite time must elapse before the domain wall is established. Furthermore, the drive-line inductance and the stray capacity present also limit the time rate of voltage buildup. The actual response will look like the dashed rather than the solid line of Fig. 15-7. After the core saturates at $t = t_s$, the two terminals should appear to be a short circuit and the voltage should drop to zero. However, the large series impedance of the current source, together with the small drive-line and coil inductance, actually produces the fast exponential decay shown.

From Eq. (15-17), we can see that an external excitation of I_c would not allow the core to saturate until an infinite time had elapsed. The larger the excess of current over the internal bias value, the faster the core will switch between the two saturation values. Here also the coercive force may be treated as an external bias generator, as shown in Fig. 15-8.

Response of an externally loaded core

During the interval that the external excitation acts to switch the core state, the flux changes will link any other windings on the same core. Because of the high permeability, the coefficient of coupling is almost unity. The linked coils will act as a transformer only during the transition from one to the other saturated state. The voltage appearing across a load R connected to a secondary winding of N_2 turns is

$$v_R(t) = nv(t)$$

where $n = N_2/N_1$ is the turns ratio and $v(t)$ is the terminal voltage developed across the excitation winding. An additional retarding field of

$$H_2(t) = \frac{n^2v(t)}{Rl} \tag{15-18}$$

is produced by the current flow in R. $H_2(t)$ acts in the same direction as the internal coercive field, and it is simply added to H_c in all the integral and differential equations previously derived.

We can also interpret the effect of the load by dividing the total driving-point current into two parts: One component is transformer-coupled to the load. The second current component is the current flow in the core as the domains switch their orientation.

In the case where the applied voltage is a known function of time, the input current given by Eq. (15-13) would include, on the right-hand side, the reflected load-current term

$$i_R = \frac{nv(t)}{R}$$

On the other hand, when a known current waveshape is used for excitation, the time-varying H_2 must be included inside the integral sign in Eq. (15-14). The voltage response would have to be recalculated from Maxwell's equations. We would find that the smaller the effective load resistance, the larger the additional retarding field and the longer the time required to switch the core's state. The load current subtracts from the total applied current and results in a smaller net excitation of the core.

From an examination of the final response to both constant-voltage and constant-current excitation [Eqs. (15-13) and (15-16)], we see that the nonsaturated core might be approximated by a nonlinear resistance of the form

$$R_c = \frac{K}{tv(t)} \tag{15-19}$$

in parallel with the current generator I_c (Fig. 15-8). K has one value for constant-current excitation and yet another value when a step of voltage is applied.

The total resistance seen at the driving terminal is the parallel combination of the time-and-voltage-dependent core resistance R_c and the reflected load resistance:

$$R_{\text{in}} = R\left(\frac{N_1}{N_2}\right)^2 \parallel R_c \qquad \text{at } t = 0,\ R_c = \infty,\ R_{\text{in}} = \left(\frac{N_1}{N_2}\right)^2 R$$

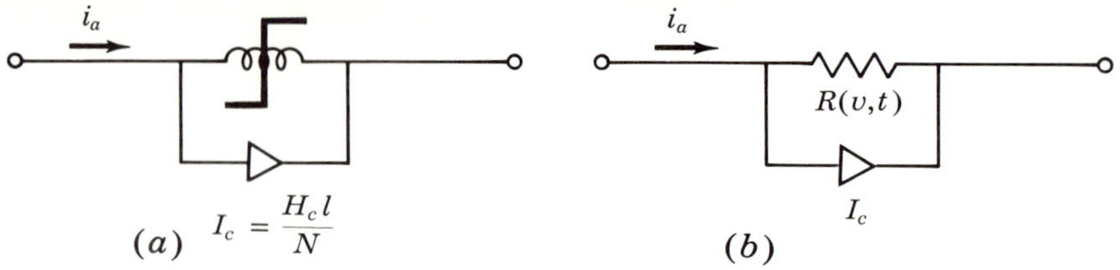

Fig. 15-8 Two possible equivalent-model representations for the square-loop core.

The now finite resistance limits the initial value of the voltage. The parallel combination is always smaller than the input resistance of the unloaded core, and consequently the voltage produced by a constant-current drive will be less than that found in Eq. (15-16). In fact, if $R(N_1/N_2)^2 \ll R_c$ over the complete time and voltage range of interest, then the input resistance would remain essentially constant and a rectangular voltage pulse would be the result of a constant-current excitation. Since the switching time is controlled by the volt-time area [Eq. (15-7)], the external loading will have the adverse effect of increasing t_s.

To a good approximation, the ferromagnetic core can be considered to be a reasonably large resistance when it is unsaturated and a very small resistance (practically a short circuit) when saturated. It thus acts as a bivalued circuit element with, however, its switching action dependent on its own past history. If multiple windings are available, then the current flow in one can be used to control the terminal response at the others. For example, by alternately reflecting a high and a low resistance into the transmission path, the magnetic-circuit element can replace the diode bridge of Sec. 4-10 for use as a controlled gate (Fig. 15-14).

A note on units

The parameters of the ferromagnetic sample are usually expressed in the electromagnetic units (emu) of oersteds (H) and gauss (B). However, all equations used and derived are given in mks units. The important conversion factors are:

	SYMBOL	Emu	Mks
Magnetizing force	H	Oersteds $\times 10^3/4\pi$	A-turns/m
Flux	Φ	Maxwells $\times 10^{-8}$	Wb
Flux density	B	Gauss $\times 10^{-4}$	Wb/m^2
Permeability	μ	Gauss/oersted $\times 4\pi \times 10^{-7}$	H/m

Example 15-1

In this example we shall compare the current responses of two cores to an excitation of 20 V. The first core is toroidally wound on a ½-in.-diameter form of 20 turns of 1-mil-thick 0.25-in.-wide Permalloy tape. The second core is a molded ceramic ferrite of the same size. Both cores have a 200-turn excitation winding.

	B_s, gauss	B_r, gauss	H_c, oersteds	ρ,* $\mu\Omega$-cm
Permalloy	8,000	7,200	0.07	55
Ferrite	4,500	3,400	0.18	10^5

* ρ is the volume resistivity, the reciprocal of the volume conductivity σ.

Solution (*a*) The first step in the solution is the conversion of the mixed units given in the statement of the problem into the mks system, using the conversion factors given above. The parameters of permalloy are $H_c = 5.56$ A-turns/m, $B_s = 0.8$ Wb, and $\rho = 55 \times 10^{-8}$ Ω-m. Expressing the dimensions in meters and substituting into the appropriate form of Eq. (15-13) result in

$$i_a(t) = 1.45 \times 10^4 t + 1.16 \times 10^{-3} \text{ A}$$

From Eq. (15-8), the saturation time of the permalloy core is

$$t_s = 51.8 \ \mu\text{s}$$

At the end of the switching interval, the coil current has increased to

$$i_a(t_s) = 760 + 1.16 \text{ mA}$$

where the first term in the above expression is the eddy-current flow and the second is due to the coercive force. We might note that the low conductivity of the permalloy sample results in relatively large eddy currents. This means that additional power is dissipated during the switching interval above and beyond that necessary to overcome the coercive force.

(*b*) We shall assume that the results derived for a tape-wound core apply equally well to the sintered ferrite core. They actually do not, but the answers obtained are of the correct order of magnitude and may be used for comparison. The solution of Eq. (15-13), for this sample, is

$$i_a(t) = 14.1t + 2.9 \times 10^{-3} \text{ A}$$

The ferrite core switches state in only 29 μs. Hence the peak current flow becomes

$$i_a(t_s) = 0.41 + 2.9 \text{ mA}$$

Because of the much higher resistivity of the ferrite sample, the eddy current of 410 μA is almost completely negligible.

15-3 Magnetic Counters

The basic magnetic-core counter of Fig. 15-9 consists of two components: the core itself, where the count is registered by the change in the magnetic state, and an energy-storage element which automatically resets the core after the required count has been entered. One of the more common applications of this circuit is as a binary, where one output pulse is produced for every two inputs. In this respect, the magnetic counter behaves similarly to the transistor bistable multivibrator circuits of Chap. 8. The circuit of Fig. 15-9 does offer one major advantage in that no power is required to sustain the excited state.

So that we may make use of the results of Sec. 15-2, we shall assume that the core is constructed of perfect square-loop material. In its unexcited condition, it is at $-B_s$; we shall call this state 0. A positive input pulse drives it toward $+B_s$ (state 1). The differential equation describing the response to the applied excitation is

$$v_{\text{in}} = Ri + \frac{1}{C}\int i\,dt + v_s \tag{15-20}$$

where v_s is the voltage drop across the coil.

In Sec. 15-2 we have seen that the core appears to be a very high resistance during the interval when it is switching from 0 to 1 or from 1 to 0. Since R is the small series combination of the input-source impedance and the winding resistance of the coil, almost the full excitation voltage will be developed across the driving coil. Equation (15-20) reduces to $v_{\text{in}} = v_s$. Hence the initial response is due to the known voltage excitation of the

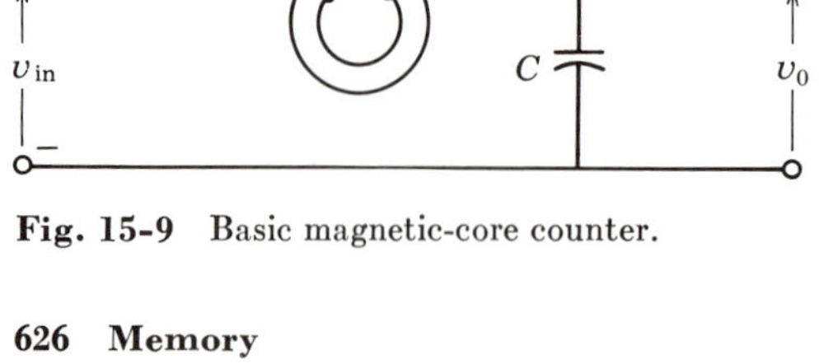

Fig. 15-9 Basic magnetic-core counter.

core and the current is given by Eq. (15-13). Until such time as the core saturates, the very small output may be approximated by

$$v_o(t) = \frac{1}{C} \int i\,dt$$

In the special case of a constant-amplitude pulse, this reduces to

$$v_o(t) = \frac{1}{C}(KV^2t^2 + I_ct) \tag{15-21}$$

where K is evaluated with the aid of Eq. (15-13).

The volt-time area of each input pulse $V\,\Delta t$ is adjusted so that the core state will not reverse completely. The first one primes the core, leaving it in the partially magnetized condition corresponding to point a in Fig. 15-10a. After the pulse terminates, the capacitor discharges through the high coil resistance. But since only a small amount of charge was stored when the pulse was present, the demagnetization path follows the dashed line in Fig. 15-10a from a to b.

The primed core also looks like a high resistance to the first portion of the second pulse. Consequently, in the interval from 0 to t_1 (Fig. 15-10b), the output will be given by Eq. (15-21). Once the core saturates at $+B_s$, its terminal response is that of a short circuit. The capacitor continues charging toward V with the small time constant RC. It is only here that a large amount of energy may be stored in C in a short time interval. It follows that R must be kept as small as possible. Other-

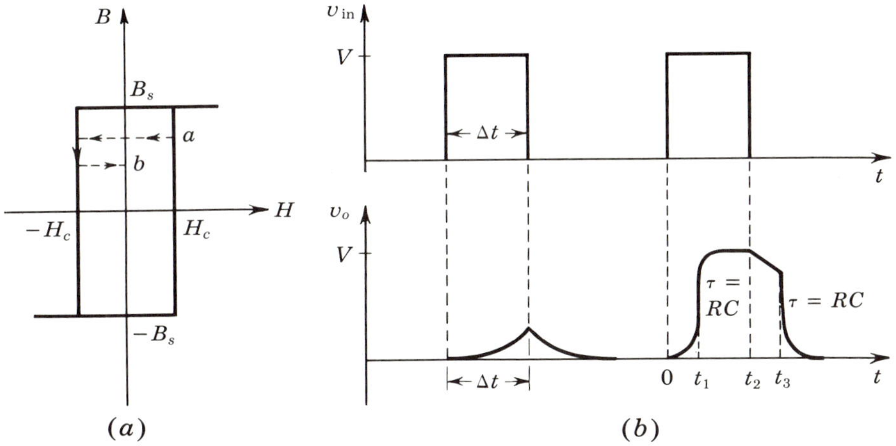

Fig. 15-10 Response of the counter of Fig. 15-9. (a) Hysteresis curve, showing the partial demagnetization during the discharge of the first pulse; (b) excitation pulses and the resultant output.

wise the single output pulse, which represents the count of 2 (Fig. 15-10*b*), may never even reach V.

At the end of the second pulse, the core is left in state 1. But as the capacitor starts discharging from its peak voltage V, the reverse current flowing in the driving coil switches the core back to state 0. Any energy remaining after the core resaturates at $-B_s$ is dissipated in the series resistance R. This portion of the cycle is identified by the double-segmented decay waveshape shown in Fig. 15-10*b*. The interval from t_2 to t_3 represents the flux reversal, and the remaining portion shows the final discharge of C. To ensure the automatic reset after every second input, the energy stored in C must be greater than that dissipated in the core and the series resistance. Neglecting the small power loss in the series resistance, we obtain the necessary inequality

$$\tfrac{1}{2}CV^2 \geq 2B_sH_cAl$$

or

$$C \geq \frac{4B_sH_cAl}{V^2} \qquad (15\text{-}22)$$

If Eq. (15-22) is not satisfied, then the discharge of C will not completely reset the core to state 0. The next applied pulse starts the switching process from a partially excited condition. It may even drive the core directly to state 1, leading to a false count at the output.

We conclude that the core in the magnetic counter acts as a controlled-series gate, with its bivalued state dependent on its own past history. When unsaturated, it acts as a high series resistance which prevents the charging of the storage capacitor. After the core finally saturates, its resistance drops to zero, thus closing the charging path.

Counting to an arbitrary base

If the volt-time area of the input pulse is carefully controlled, then this same circuit may be used to count down by an arbitrary factor. Assuming that the minor hysteresis loops are also reasonably square, each input pulse shifts the core state in discrete jumps along the hysteresis curve, as shown in Fig. 15-11*b*. The final excitation produces saturation and resetting. For example, a scale-of-10 counter will result when each input pulse has a volt-time area that is slightly more than one-tenth of the total needed to reverse the saturation direction. The extra area of each pulse causes the core to saturate near the beginning of the tenth input and consequently permits the storage of the reset energy in C during the final count. Reset is provided by the discharge of C through a separate winding.

Provided that the input signal is larger than the minimum necessary to switch the core state, the volt-time area of the signal developed across a secondary winding will depend solely on the turns ratio and the core prop-

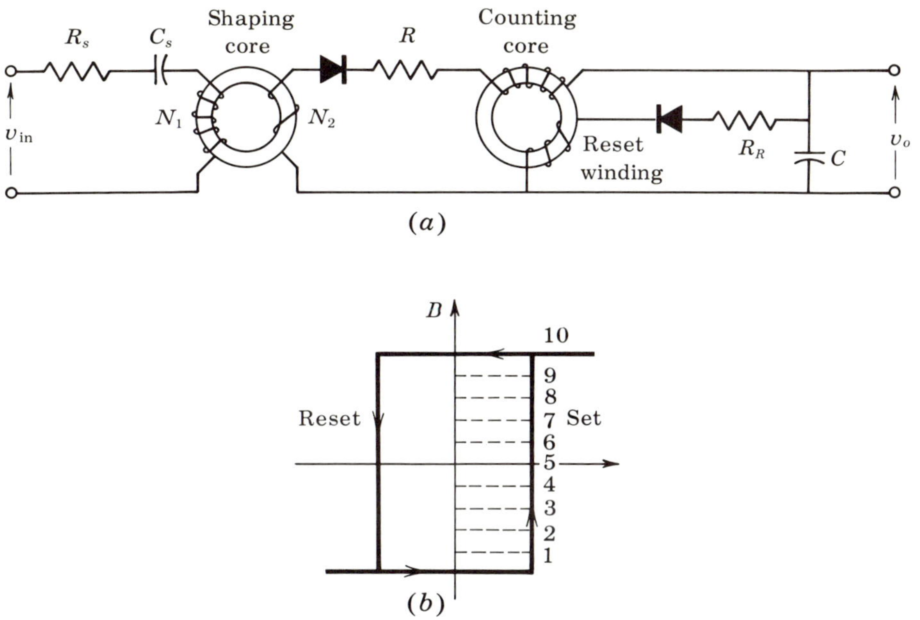

Fig. 15-11 (a) An arbitrary base counter using a pulse shaper; (b) path of operation followed in a scale-of-10 counter.

erties [Eq. (15-17)]. Only in the unsaturated condition does the ferromagnetic device act as a transformer; once it saturates, the driving voltage will be developed across the series resistance instead of across the low input impedance. This means that a second core may be used to shape the variable input pulses to a constant volt-time area. The charge stored in C_s in Fig. 15-11*a* serves to reset the shaping core after each pulse.

The diode determines the polarity of those pulses applied to the counting core; it prevents the reset pulse of the input core from registering as a negative count in the storage core. If we wish, a separate winding could be used for negative pulses, thus creating an add-subtract counter.

Cascading counters

In order to achieve higher counts than are possible with a single core, the output of one stage may be coupled to a second, and so on. Each core represents the appropriate binary or decimal place. Because of the difficulty in reading the counts corresponding to the unsaturated states of the core (numbers between 1 and 9), most cascaded counters use a binary rather than a decimal base.

When several counters are directly coupled, as shown in Fig. 15-12, the single input must be able to supply the energy needed to reset all cores

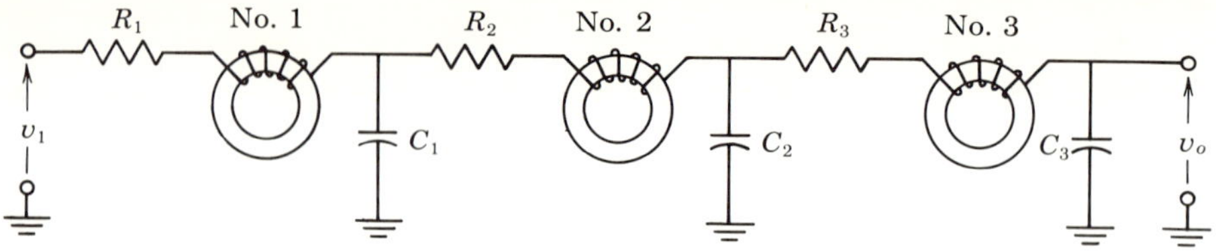

Fig. 15-12 Three-stage cascaded counter.

simultaneously. Consider the operation when the three cores are in state 1 (just below saturation) and the count is $1 \times 2^0 + 1 \times 2^1 + 1 \times 2^2$, reading from left to right. The eighth pulse saturates core 1, and as C_1 charges, core 2 saturates, and so on, until finally C_3 is also fully charged. Thus we see that the volt-time area of the input pulse must be able to change the state of all cores and that it must also supply the energy that will reset all three cores.

If more than two or three stages are cascaded, they may impose a severe load on the signal source. Consequently, an active regenerative circuit is usually included within the transmission path to decouple and to reshape the transmitted pulse. These circuits are discussed in Sec. 15-4.

Shift registers

The binary-state cores can be used for storage purposes in exactly the same manner as the transistor bistables discussed in Chap. 13. For example, we can interconnect cores and *RC* delay networks to form a five-bit shift register (Fig. 15-13).

When the five digits are simultaneously available, each can be steered directly to a separate readin winding on the appropriate core. This form of data processing, where only one pulse width is needed to register the information, is called *parallel readin.*

On the other hand, the five-bit word 10011 may be presented serially (the serial input is pulse, pulse, blank, blank, pulse). Separating each digit, there is sufficient space to insert the pulses that control the switching of the shift register of Fig. 15-13*a*. The first digit, i.e., the one farthest to the right in the word to be stored, is inserted in the normal way into core *a*, changing its state from 0 to 1. The decoupling diode prevents the resulting negative output from switching the second core in the chain.

Before the next digit appears, we apply a separate shift pulse to all cores (Fig. 15-13*c*). Those which are in the 1 state will be switched back to 0, while those already in the 0 state will be unaffected. We see that the counter is set by the input and reset by the shift pulse. The positive voltage, which will now be developed across the output winding of the switched

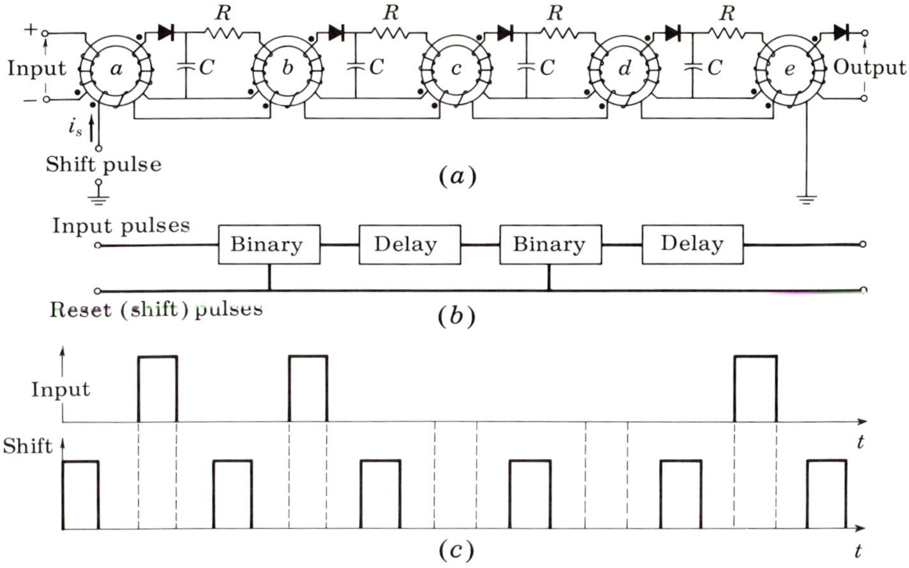

Fig. 15-13 (*a*) Magnetic-core shift register; (*b*) block diagram of any shift register; (*c*) time relationship between the input pulses (10011) and the shift pulses.

core, charges the storage element C. Its discharge, through R and the primary of core b, switches the second core to state 1. In order to prevent the shift pulse from interfering with the transferred digit, there must be a time delay in the forward-transmission path. In the shift register of Fig. 15-13*a*, this delay is obtained by inserting a resistance in series with the driving coil of the core.

The next digit of the word, input digit 1, again changes the state of core a without influencing any other core. A shift pulse will now set the excited cores a and b back to zero. As both capacitors discharge, the two digits are transferred to the right, changing the states of cores b and c from 0 to 1.

The third digit of the input word is 0. After its application the state of the cores is 011−−, where the dashes represent the unknown states due to any previous excitation. A shift pulse transfers the core states to 0011−. Continuing the process of first switching the state of core a and then transferring the stored information to the right, the full word is finally registered.

For readout, five shift pulses will sequentially feed the stored digits to the output terminal. This process is identical with the normal shifting which transfers the input forward from core to core. In fact, as a new word is entered (starting with a shift rather than a digit pulse), the old word is automatically shifted out on the right.

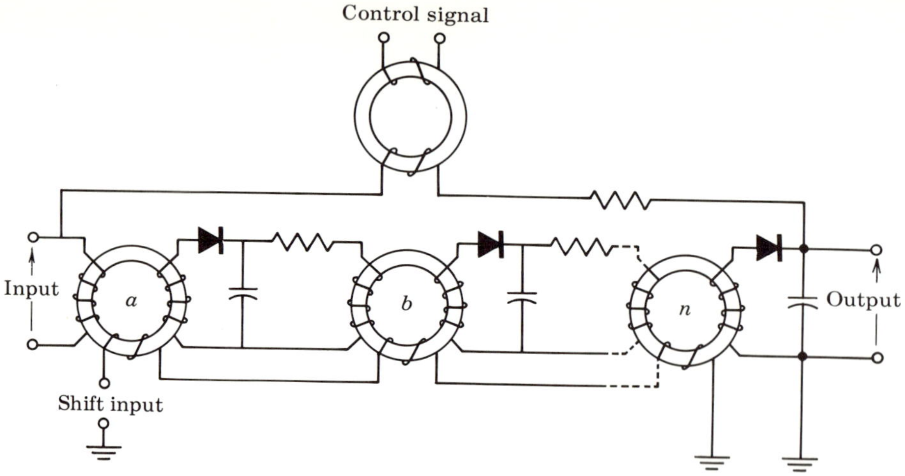

Fig. 15-14 Closed-loop shift register; a switched core is used as a controlled gate.

One obvious disadvantage is that the stored word is erased as it is fed out for examination or use. However, if the shift register is formed into a closed loop, then the information fed out on the right can be automatically re-stored in the left-hand core. After the appropriate number of shift pulses are applied, the register again contains the original word.

A convenient means of closing the self-storage path is by way of a controlled gate such as that shown in Fig. 15-14. This circuit makes use of the bivalued resistance properties of the core to interrupt or close the feedback path. During storage the gate is kept open (the gating core is in its active region) and the register functions normally. When reading out information temporarily stored, the gate is also left open. But if we desire more permanence, then the gate must be closed (core-saturated) during the intervals that the output pulses are produced. It is implicitly assumed that the amplitude of the transferred pulse will not change the state of the magnetic gate and that the reflected control signal will not register as a storage pulse.

15-4 Core-Transistor Counters and Registers

The number of stages which can be directly coupled in a counter or shift register is limited by the power capabilities of the pulse source. In the counter, the input must be able to switch all cores simultaneously. In the register, the shift pulse not only resets all cores but must also supply the energy that transfers the stored information to the right. And since the switching time depends on the excess of magnetizing current over the coercive force [Eq. (15-17)], the problem is further complicated when the counting speed becomes critical. For example, in order to switch a typical core in a 0.5-MHz register, the 0.5-μs shift pulse applied must have a peak power of almost 1 W. The 330-mA pulse current will develop an average of 3 V—and a peak which may be many times as large—across the shift coil. Under these conditions we can expect a single transistor amplifier to switch, at the most, three or four stages.

One obvious means of improving the response is to use an amplifier to decouple the individual cores. An even better way is by inserting a regenerative circuit between two cores; the output of the first core can be made to trigger the generation of a fast-rising high-power pulse for propagation in the forward direction.

Figure 15-15 shows a counter which is a combination of a core and a transistor. This circuit employs a single core both as a memory (winding N_1) and as a transformer for the blocking oscillator. Besides the economy in using one rather than two cores, the sharing improves the response because the oscillator pulse also resets the counter. In the normal state, the core is at 0 and the transistor is cut off by the external bias. The first pulse brings the core to the verge of saturation, and the second one saturates it and charges C_1. With the transformer connections shown, these input pulses act in a direction that keeps the transistor off.

After the termination of the second pulse, the voltage across N_1 reverses as C_1 starts discharging. This pulse, coupled to N_2, brings the transistor into its active region. And as the regenerative circuit goes through one

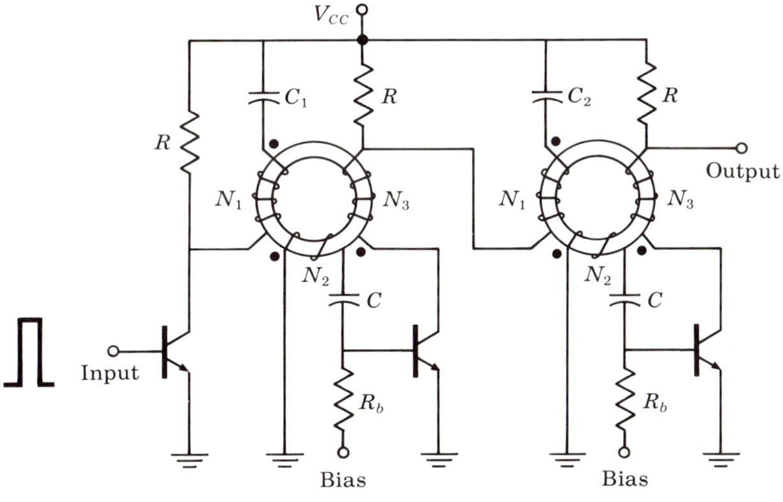

Fig. 15-15 Cascaded counter using a transistor blocking oscillator for regeneration and a core for storage. The input interface amplifier stage is also shown.

complete cycle, the extra energy supplied by the collector current rapidly returns the core to a logical 0.

In the counter of Fig. 15-15, the functions of regeneration and storage are separately handled by the transistor and the core. The active element is operative only during the reset interval. At all other times, the passive core controls the circuit behavior. The capacitor only triggers the regenerative circuit; it does not have to supply the energy for reswitching. Hence the restrictions on C are much less severe, and this circuit will recover much faster than those depending only on the stored charge for reset.

From which point in the circuit shall we take the output? If the collector is chosen, all the transformed input pulses will also be coupled to the next stage. However, as these are of the polarity opposite to that of the pulse produced by the blocking oscillator, a series diode will prevent them from registering falsely at the next core. This diode may be eliminated if the output is taken at the junction of the collector resistor and the transformer winding, as shown in Fig. 15-15. Here the cutoff transistor effectively replaces the decoupling diode.

As an alternative, the output may be taken from a fourth winding on the core. Since the switching during recovery is between the two saturation levels, this pulse would have the stabilized volt-time area which is especially necessary when counting to a base other than 2. By properly choosing the turns ratio, the output volt-time area can be adjusted to any desired percentage of the reset pulse. This avoids the use of a separate core for shaping the counting pulses that are applied to the next stage.

If, instead, the fourth winding is used as an input for the shift pulses, then the train of cores and transistors will function as a shift register. When the core is in the zero state, the low impedance offered to the shift input will not permit the development of a pulse large enough to bring the transistor out of cutoff. After the information pulse primes the core, the opposite-acting shift pulse can trigger the regenerative circuit. During reset, the previously stored information is transferred one position to the right.

15-5 Magnetic-core Memory Arrays

Small ferrite cores are ideally suited for the 10^5- to 10^7-bit storage capacity required in the main-frame memories of modern computers. Smaller-capacity matrices are also employed as scratch-pad memories and for storage purposes in industrial on-line control systems. Cores as small as 20 mils outside diameter can be switched from one saturated state to the other in less than 200 ns. Larger cores switch more slowly; a 50-mil core takes 3 μs to switch state. After a core is set into the $+B_s$ (logical-1) or

$-B_s$ (logical-0) state, the information remains stored without need for any external power supply.

2D memory plane

Figure 15-16 illustrates one possible geometry for a large-capacity memory unit. A typical memory contains tens of thousands of cores in each one of a stack of identical planes. This particular 4×4 memory plane, arranged in a word-oriented geometry, is called a 2D array. A single four-bit word is stored along each row, one bit in each core. Each core is threaded by three wires. The single x wire is the word-select line. One of the two y wires is used to write in the particular bit. The other y wire is connected to the input of a sense amplifier and serves to read out the stored bit. In some 2D arrays, a single y lead alternatively performs both the read and write functions.

Suppose that we start with all cores at $-B_s$ (logical 0) and we wish to store the word 1001 in a particular word location. A positive-current

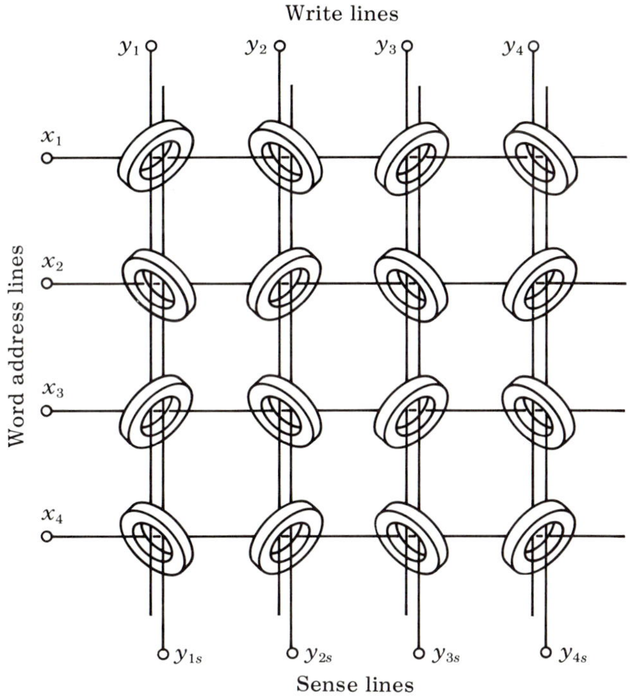

Fig. 15-16 A four-word four-bit word-oriented magnetic-core memory (2D three-wire geometry).

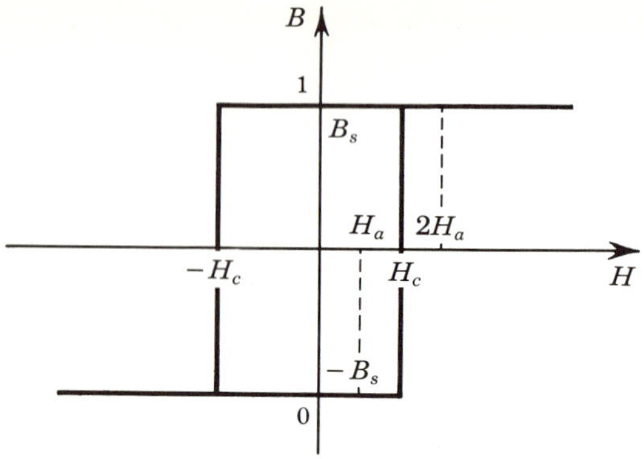

Fig. 15-17 Hysteresis loop, showing excitation of switched and nonswitched cores of the memory array of Fig. 15-16.

pulse of amplitude I_A is applied to the selected word row (say x_2). Simultaneously, equal positive pulses are applied to each column in which we wish to store a logical 1. The total field in the two cores y_1 and y_4 is due to the addition of the two half write pulses. The effective excitation of $2I_A$ must cause these cores to switch to $+B_s$ saturation. However, the other two cores in the word row (22 and 23), which are excited by only a single pulse, must remain in the original $-B_s$ state. This is possible only so long as the pulse amplitude satisfies the conditions

$$\tfrac{1}{2}I_c < I_A < I_c \tag{15-23}$$

In small cores, I_A will be on the order of 450 mA.

Only the cores 21 and 24 are excited past H_c, to point $2H_a$ in Fig. 15-17, and only these switch state to $+B_s$. All the other cores in the addressed word and in the logical-1 excited columns have an applied field strength of only $H_a < H_c$. After the termination of the write pulses, they relax back to the original logical-0 state. We depend on the effective amplitude ANDing of the two write pulses within the core for the required write-select logic. Figure 15-18 shows the coincidence of the applied write pulses.

In order to read out the stored word, the memory matrix must be interrogated. A current pulse more negative than $-I_c$ (for example, a pulse equal to $-2I_A$) injected into a word line will cause all cores to switch into the 0 state. Cores in which a 1 was stored will experience a flux reversal from $+B_s$ to $-B_s$. The accompanying voltage pulse appears in the column sense line. We can expect a peak signal strength of approximately 50 mV when switching a 20-mil core with a 900-mA read current. Each sense line is connected to the input of a sense amplifier of the type described in Chap. 10. The 50-mV sense signal registers as a logical 1 at the amplifier output.

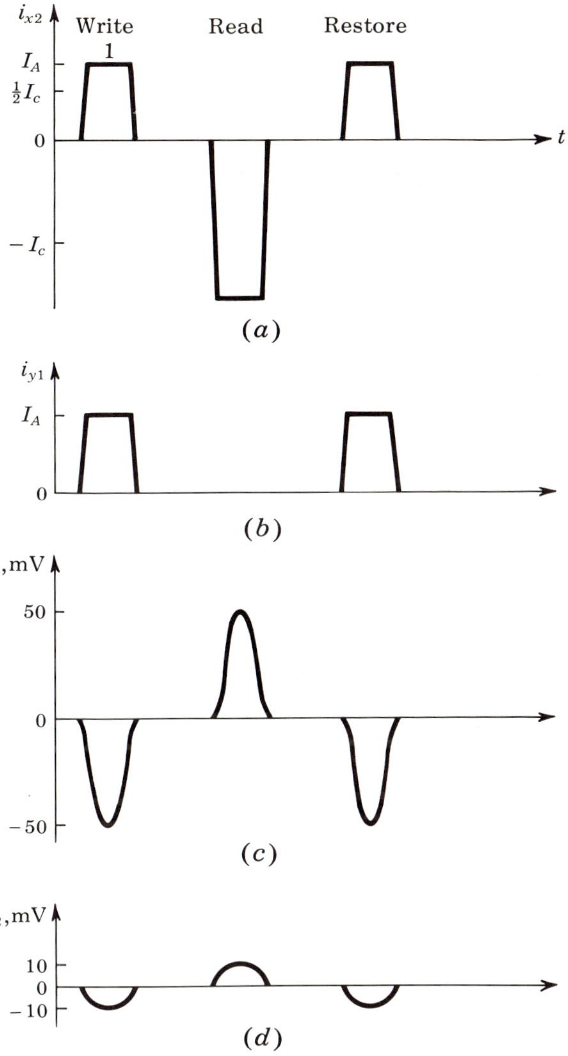

Fig. 15-18 Drive and response waveshapes for the core memory of Fig. 15-16. (*a*) Word-line address current for write, read, and restoration; (*b*) write-line current to write and restore a logical 1; (*c*) output voltage in sense line when writing or reading a logical 1; (*d*) sense output in logical-0 cores due to word-line drive.

Suppose that a particular word bit is a 0. Since the core in which this bit is stored is already in the 0 state, there is no change of stored flux during interrogation. This core presents a very low impedance to the read-current pulse. The very small voltage developed in the sense line (about 10 mV) could not possibly be mistaken for the larger logical-1 output. The sense-amplifier threshold would be adjusted to yield a logical-0 output.

Only the bits in the single word addressed are queried by the read pulse. Because no read current flows in any other row, large-amplitude read pulses may be used to decrease the switching time [Eq. (15-17)] and to increase the sense-voltage output [Eq. (15-16)].

We must note that the readout process is destructive; the read-current pulse sets all word cores back to logical 0, and we detect the flux transitions. The presence of a logical-1 sense signal in any column may be used to trigger the generation of a delayed half write pulse in the same column. If the word-row negative read pulse is followed by a delayed positive half write pulse, then it will sum with the returned column half write pulse at the core, reswitching it. Information will be restored in the logical-1 cores. Figure 15-18 shows the pulse sequence used for writing, reading, and restoration. The sense voltages observed during the positive flux transition of writing a logical 1 are of polarity opposite to that of the sense voltage generated during the negative readout flux transition.

3D memory plane

Figure 15-19 illustrates a core plane based on a bit-oriented geometry. Every core can be addressed separately by the coincidence of currents in the x and y address lines. To write a 1, half write-current pulses are simultaneously applied to the x and y lines. They sum at the intersection, and the inherent AND function causes the addressed core to switch into the 1 state. All the other cores in the addressed row and column are only excited by a single half-current pulse (I_A) satisfying Eq. (15-23), and their states will be unaffected.

To write a zero into a previously erased core, half-current pulses are supplied to the x and y address lines. Simultaneously, a negative half-current pulse is injected into the inhibit line which threads all cores. The addressed core has a net half-current pulse, while all other cores in the addressed row and column have a net zero current. No core changes state. The use of an inhibit line permits the separation of address and write circuitry. Address pulses select the particular core, and the negative-current logic on the inhibit line determines the final core state.

Reading is accomplished by addressing the selected row and column with negative half-current pulses. These sum at the intersecting core, resetting it to the logical-0 state. The voltage pulse generated appears in the single sense line which threads all cores. Since only a single-bit readout is

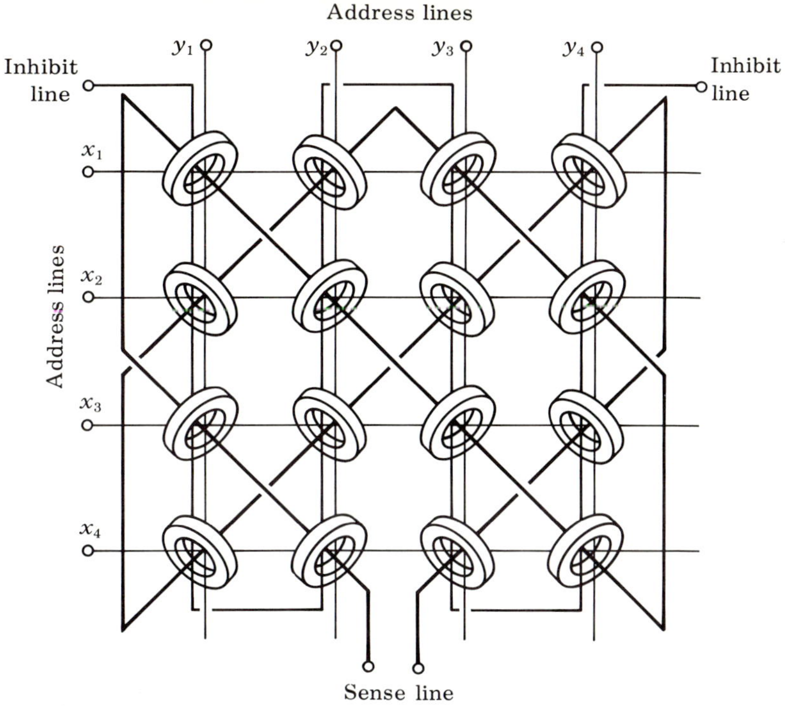

Fig. 15-19 A 3D four-wire core-memory matrix.

possible during each read condition, any word stored in the plane must be read out serially.

Normally all of the bits of the stored word are located in parallel memory planes. All the planes in the z direction may be simultaneously addressed with the individual-plane sense lines reading out the word in parallel.

In a three-wire 3D array, a single line alternately serves for both the sense and the inhibit functions.

3D memories are slower than 2D memories. The cores must be larger to accommodate more wires and more complex wiring. They will take longer to switch states. Cycle times of 1 to 3 μs are typical. However, since the electronic circuitry needed for the 3D serial readout is much simpler than that required for the parallel-operating 2D array, 3D memories are more economical.

2½D memory

A 2½D memory is a compromise between the speed of a word-oriented geometry and the economy of a bit-oriented geometry. Only three wires thread each core; one is used for the x address, the second is used for the y address, and the third is a sense wire which passes through every core in the plane. In this respect, the memory is very similar to a three-wire 3D array. Information is entered in parallel; each row stores one word. But because there is only a single sense line (as in the 3D array), the stored word must be read out serially. Per-bit cycle times of 0.3 to 1.0 μs are realizable.

Film memories

Very-high-speed magnetic memories cannot employ cores for storage. They just cannot be made small enough. Instead, the storage depends on switching the direction of stored flux in small regions of ferrite or magnetic film. Several techniques have been proposed. One stores information in 5,000- to 10,000-Å-thick layers of permalloy plated on the surface of a conducting sense line. Circumferential flux is switched from clockwise to counterclockwise directions by current flow in the surrounding address lines. These are either woven or wound around the storage wires. Switching times of 20 to 30 ns can be expected.

Planar-sandwich memories composed of a printed-wire matrix containing address and sense lines with insulated magnetic regions deposited at the intersections are another possibility. Current flow in the address lines rotates the flux within the magnetic regions. Special geometries can be incorporated to obtain a nondestructive readout. The planar-sandwich memory is most suitable in large arrays. Memories of over 10^6 bits have been produced.

Both of these memories lend themselves to economical automatic-production techniques.

15-6 Core-Transistor Multivibrator

In Sec. 15-2 we saw that the saturation time of a square-loop core, excited by a constant voltage, is very clearly defined. The abrupt change in the driving-point impedance, which marks the end of the flux buildup, permits the core to be used as the timing element in a monostable or an astable circuit.

Figure 15-20 shows a single-core multivibrator employing resistive coupling to complete the regenerative loop. In this circuit, the winding direction is such that when Q_1 is saturated, the current through N_1 drives the core from $-B_s$ to $+B_s$. Current through the second winding N_2, due to saturation of Q_2, resets the core state to $-B_s$. As in all astable multivibrators, the transistors furnish the greater-than-unity loop gain

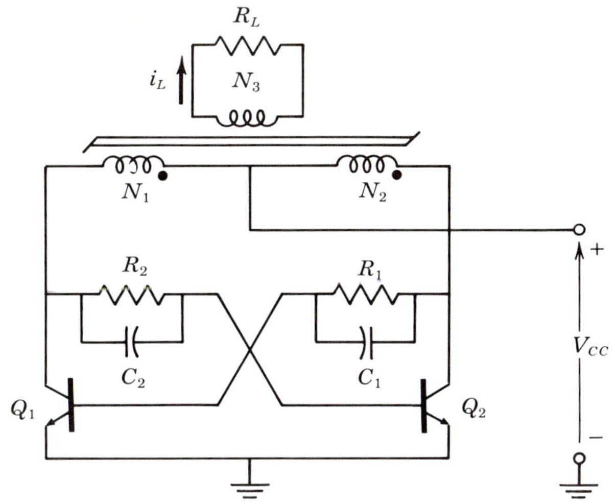

Fig. 15-20 Core-transistor multivibrator used as a dc-to-ac or a dc-to-dc converter.

needed to switch between the two quasi-stable states. They become active only during the small switching interval corresponding to the saturation of the core. At all other times the individual transistor is either saturated or cut off.

We shall now detail the operation over one complete cycle, assuming, as the starting point, that Q_1 is off, just turning on, and Q_2 is on, just turning off. This is equivalent to saying that the core is on the verge of saturating at $-B_s$. So long as the flux is still changing, the full supply voltage will appear across the high driving-point impedance of N_2 and the collector of Q_2 will be approximately zero. Hence no current can flow into the base of Q_1; it is cut off.

Once the core saturates, there is a marked decrease in the coil's driving-point impedance. Under these circumstances, an appreciable voltage is developed across the collector-to-emitter saturation resistance of the series transistor Q_2. Current can now flow into the base of the OFF transistor, and as Q_1 turns on, the current through the winding N_1 starts the flux reversing. V_{CC} is connected across this winding, as shown in the model of Fig. 15-21*a*. The transformer action reverses the voltage across N_2. The resultant large increase in the base current of Q_1 rapidly drives this transistor into saturation. At the same time, the drop in the collector voltage of Q_1, coupled through the biasing and speedup capacitor C_2, will turn Q_2 off (Fig. 15-21*b*).

The time that Q_1 remains saturated is controlled by the flux buildup in the core. If the small demagnetizing base current is neglected, then

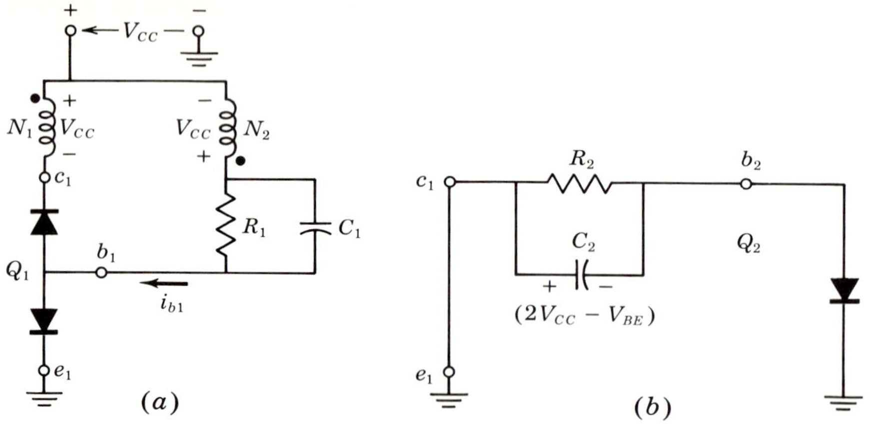

Fig. 15-21 Models holding during the change of flux density from $-B_s$ to B_s in the core of the converter of Fig. 15-20. (a) Saturated transistor Q_1; (b) back biasing of cutoff transistor Q_2.

the core switching time, as given by Eq. (15-8), is

$$t_1 = \frac{2N_1AB_s}{V_{CC}}$$

At time t_1, the core saturates at $+B_s$ and the circuit switches to its other quasi-stable state.

In Fig. 15-21a we see that so long as Q_1 is saturated, C_1 is in series with the transformer winding N_2 and V_{CC}. When $N_1 = N_2$, C_1 will charge toward $2V_{CC} - V_{BE}$. At the end of the half cycle, the saturation of the core causes the circuit to start switching. Because of the charge stored in the speedup capacitor (Fig. 15-21b), the OFF transistor's base is brought to well below 0 (Fig. 15-22). This speeds the switching process. C_2 now discharges to 0 through R_2. C_2 should be fully discharged within the half cycle. Then the full positive transition at the collector of Q_1 at the end of the half cycle will be coupled into the base of Q_2, turning it on hard; again C_2 speeds the switching process.

If the two timing coils are identical, this circuit will generate the square wave shown in Fig. 15-22. The frequency of oscillation will be given by

$$f = \frac{1}{2t_1} = \frac{V_{CC}}{4NAB_s} \qquad (15\text{-}24)$$

when the transistor switching time is neglected. It is interesting to note that the square-wave frequency is linearly dependent on the supply voltage.

The output is usually obtained from a third winding; hence the transformer action not only affords isolation but also permits the amplitude to be scaled up or down.

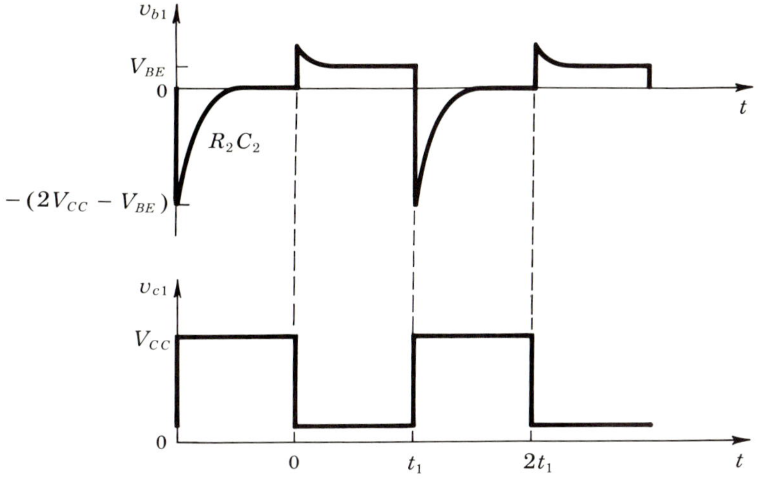

Fig. 15-22 Transistor collector and base waveshapes—dc-to-ac converter of Fig. 15-20.

Figure 15-23 shows another core-timed multivibrator, one which utilizes transformer rather than resistive coupling between the active elements. The resistor network R_1 and R_2 supplies turn-on bias for the transistors. When the switch is thrown, the transistors start out active. The loop regeneration rapidly drives one to cutoff and the other into saturation. The turn-on bias makes the circuit self-starting.

In this circuit, the voltages developed across the auxiliary windings N_2 and N_2' sustain the one transistor in cutoff and the other in saturation during the flux buildup interval. Consider the conditions when the saturated transistor Q_1 connects V_{CC} across winding N_2. With the directions shown, winding N_2 supplies the saturation base current of Q_1. The voltage appearing across N_2' back biases Q_2. These voltages are easily calculated once the turns ratios are known.

The half cycle terminates when the saturation of the core permits the transistors to reenter their active regions. Any decrease in voltage across the driving coil is coupled by the transformer into the base of Q_2, turning it on. Since the circuit is regenerative, switching continues until Q_2 saturates and Q_1 cuts off. The flux then starts reversing, and the cycle repeats.

These same circuits are widely used as dc-to-ac or as dc-to-dc converters. The primary peak-to-peak voltage of $2V_{CC}$ (measured from collector to collector) is first multiplied by the primary-to-secondary winding ratio n. If direct current is desired, the square wave, which is present at the output, must be rectified. Thus a 12-V battery may be used to develop a 120-V (or even higher) ac or dc output. Of course, the primary source impedance

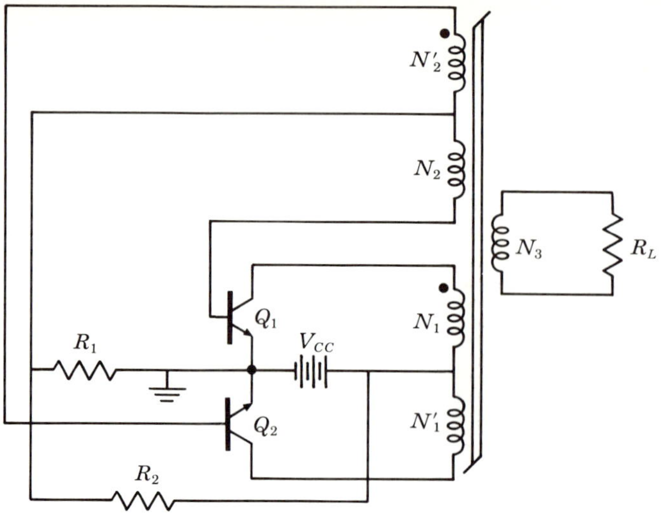

Fig. 15-23 A transformer-coupled core-transistor multivibrator. Resistor network R_1 and R_2 supplies the necessary turn-on bias for the transistors and makes the circuit self-starting.

is multiplied by n^2 as it is reflected into the output circuit while the voltage is increased only by the factor n. Thus the I^2R losses increase faster than the voltage. This limits the permissible step-up ratio, especially when large amounts of power must be supplied to the load. In any case, power transistors having a very low saturation resistance should be chosen.

When this circuit is used as a dc-to-ac converter, the frequency of operation is usually the critical factor. It will even determine the transformer design, i.e., the turns ratio, the cross-section area, and the type of magnetic material employed. Since the frequency is inversely proportional to the product of the transformer parameters, the greater the turns ratio, the smaller the volume of iron needed. However, increasing the number of turns also increases the winding resistance and the losses; therefore some compromise must be selected.

On the other hand, when the circuit is operating as a dc-to-dc converter, the higher the frequency of oscillation, the easier it is to filter out any ac components appearing in the output. As a further advantage, both the amount of transformer iron and the number of turns needed decrease with increasing frequency. But if the frequency of operation becomes excessively high, the stray circuit capacity will adversely affect the waveshape. Since the transistor will remain active over a larger portion of the cycle, the effective source impedance also increases, resulting in poorer regulation with respect to load changes.

For example, a typical converter used in an automobile will operate from the 12-V battery and will supply a maximum of 1 A to the load at a nominal 120 V and 60 Hz. In this application the total primary resistance may be estimated at 0.1 Ω, 0.05 Ω due to the saturation of the power transistor and the other 0.05 Ω contributed by the winding and battery resistance. Since the transformer's turns ratio is 10:1, 10 Ω will be reflected into the output. At full current flow, the output drops to 110 V. The primary voltage falls by the same percentage and causes a frequency shift down to 55 Hz [Eq. (15-24)]. While this converter might be used to operate many small appliances, the frequency change under load would prevent its use for phonographs or other small constant-speed motors.

15-7 Properties of Ferroelectric Materials

Dielectrics are those substances where all charged particles are relatively tightly bound to the atomic nucleus or the molecular region. In the presence of an external field, these charges cannot move freely through the material to the surface as do the relatively loosely bound electrons in conductors. Instead, there will be a slight separation in the individual microscopic region; the positive charge moves in the direction of the **E** field, and the negative charge moves in the opposite direction. Hence the local effect on the field pattern is that of an elementary dipole, i.e., an associated positive and negative charge (q) separated by some small distance d (Fig. 15-24). In certain dielectrics, these dipoles exist even before the external field is applied. But since they are randomly oriented, the material does not exhibit a net field of its own. It follows that the driving-point characteristics of the dielectric may be determined from the motion and alignment of the elementary dipoles under the influence of the applied field.

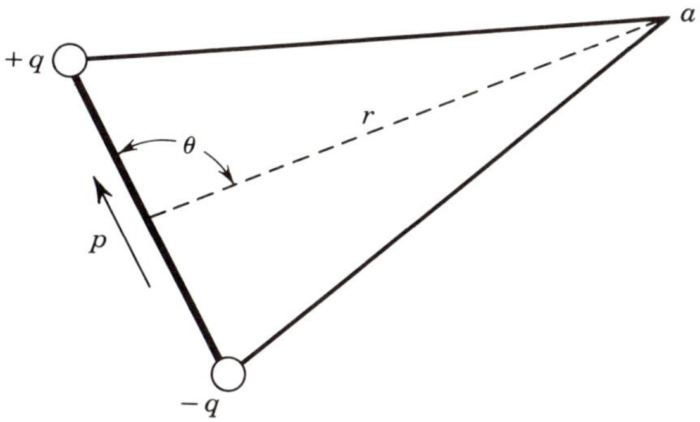

Fig. 15-24 Defining an individual dipole.

Figure 15-24 shows an individual dipole from which the potential field at any point can be calculated. If the point is relatively far from the charge, the potential is given by

$$V_a = \frac{qd \cos \theta}{4\pi\epsilon_0 r^2} \qquad \text{volts} \tag{15-25}$$

where ϵ_0, the permittivity of free space, is equal to 8.85×10^{-12} F/m in the mks system. The dipole is defined as

$$\mathbf{p} = q\mathbf{d}$$

and $\mathbf{p}$ is directed from the negative to the positive charge. This vector is called the polarization of the dipole. In any small volume, where there are n dipoles, the total polarization is the vector sum of the individual dipole moments. When they are all aligned, $\mathbf{P} = n\mathbf{p}$, and when they are in random array, $\mathbf{P} = 0$.

In order to express the terminal characteristics in simple terms, we define the displacement vector

$$\mathbf{D} = \epsilon_0\mathbf{E} + \mathbf{P} \tag{15-26}$$

The relationship between **P** and **E** is, in general, quite complex; for example, the polarization in a single direction in a crystal may depend on all three components of the applied field. However, for the ideal dielectric, we can assume the linear relationship

$$\mathbf{P} = \epsilon_0 k\mathbf{E} \tag{15-27}$$

which leads to

$$\mathbf{D} = \epsilon_0\epsilon_r\mathbf{E}$$

where $\epsilon_r = 1 + k$ is called the relative permittivity (or dielectric constant) of the medium. In the ferroelectric materials with which we shall be concerned, $k \gg 1$, and the electric displacement is almost exactly equal to the polarization.

One class of dielectrics can be defined as those materials where the polarization is produced solely by the external field. For these to be useful in capacitors, the relationship between **P** and **E** should be the linear one expressed in Eq. (15-27). If it is not, then the dielectric constant, and hence the capacity, will vary with the applied voltage.

Certain organic waxes exhibit a quasi-permanent residual polarization when they are solidified in the presence of an electric field. These are called electrets, and they are the closest electrostatic analogs to the permanent magnet currently existing.

In the third and, for our purposes, the most important class of dielectrics, the polarization is a nonlinear function of E. Here the material is characterized by spontaneous polarization, saturation, and hysteresis

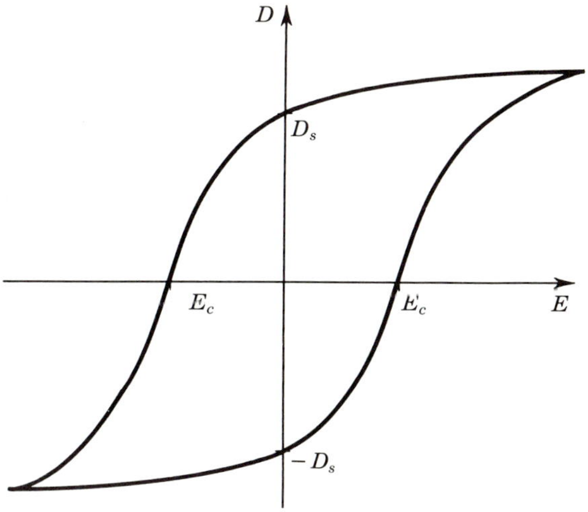

Fig. 15-25 Typical hysteresis loop of ferroelectric materials.

(Fig. 15-25). So as to recognize the duality with the magnetic materials, these dielectrics are inappropriately called ferroelectric. They contain no iron, nor is the duality complete.

Ferroelectricity was first observed in rochelle salts, $NaK(C_4H_4O_6)\cdot 4H_2O$, by J. Valasek, in 1921. The usable temperature range of this substance is only 40°C, and because of its complicated crystal structure, it develops ferroelectric properties in one direction only. Barium titanate, $BaTiO_3$, is much more useful since it is ferroelectric below its Curie point of 120°C. The crystal has a simple perovskite structure. Many other substances, such as triglycine sulfate and guanidine aluminum sulfate hexahydrate (GASH), also exhibit ferroelectric properties. Some of these other materials have squarer hysteresis loops, higher Curie temperatures, or better long-term storage properties than barium titanate.

The switching behavior of the ferroelectric crystal has been extensively investigated with the aid of polarized light and optical examination. In the unsaturated sample, small domains are found where the net dipole moment is in one of the permissible directions. Studies with a thin plate cut from a single crystal of barium titanate indicate that adjacent domains may have polarization directions either 180 or 90° apart. In the 180° domains, the application of a field $E > E_c$ causes the formation of spikes of new domains in the preferred direction (Fig. 15-26*a*). These then extend across the whole crystal. The 90° domains are reversed by the formation of wedges skewed to the cathode wall (Fig. 15-26*b*), which grow and spread sideways, eventually covering the entire region.

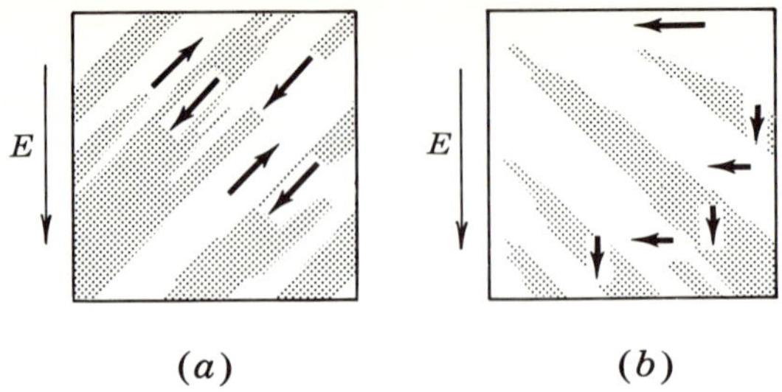

Fig. 15-26 Domain switching. (*a*) 180° domains, showing the spike formation; (*b*) 90° domains, showing the wedge formation. The shaded regions have been switched by the external field.

The number of both the 90 and 180° wedges formed, and hence the switching time, is a function of the excess of field strength over the internal coercive force.

15-8 Ferroelectric Terminal Characteristics

The practical form of the ferroelectric device is a thin wafer of barium titanate, or some other ferroelectric material, cut from a single crystal. The two electrodes must be carefully deposited on the flat surfaces because any air gap that exists will decrease the effective dielectric constant and seriously degrade the squareness of the observed hysteresis loop.

In a parallel-plate capacitor of thickness d and area A,

$$D = \frac{Q}{A} \qquad \text{and} \qquad E_a = \frac{V_a}{d} \tag{15-28}$$

where V_a is the terminal voltage. Furthermore, the charge flow is simply

$$Q = \int_0^t i\,dt$$

For a known current excitation, the time required to reverse the saturation direction of a perfect square-loop dielectric (Fig. 15-27) is given by

$$Q_{sw} = 2Q_s = \int_0^{t_s} i\,dt = 2D_sA \tag{15-29}$$

Equation (15-29) states that a fixed current-time area (charge) is required to switch the dielectric. Or, expressing the relationship of Eq. (15-29) in a more useful manner, the ferroelectric device acts as a series charge regulator. It permits only the fixed charge packet Q_{sw} to flow during

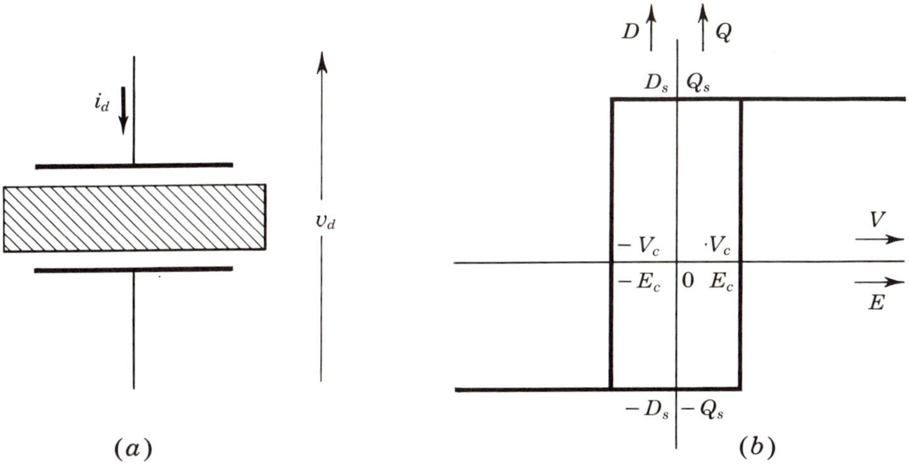

Fig. 15-27 (a) Schematic representation of a ferroelectric device; (b) ideal square-loop characteristics. Both the microscopic (E, D) and macroscopic (Q, V) coordinates are shown.

the switching intervals. This equation should be compared with the equivalent relationship previously derived for the ferromagnetic core [Eq. (15-7)]. Under the special condition where the applied current is constant ($i = I_a$), the time required to switch from $-D_s$ to $+D_s$, or vice versa, is

$$t_s = \frac{2D_sA}{I_a}$$

During this interval the terminal voltage increases linearly with time, starting from the coercive value V_c.

After switching, the ideal ferroelectric device can no longer transfer any charge. Consequently, it seems reasonable to approximate the saturated response by an open circuit.

When a voltage excitation is applied, the switching time is found to be inversely proportional to the excess over the coercive force. For a constant voltage V_a, the relationship is

$$t_s = \frac{K}{V_a - V_c} \tag{15-30}$$

where K is a constant that depends on the geometry and properties of the ferroelectric medium. Equation (15-30) is exactly analogous to Eq. (15-17).

To a very good approximation, the voltage and current responses of the ferroelectric device during switching will look like the current and voltage responses, respectively, of the ferromagnetic core (Fig. 15-28).

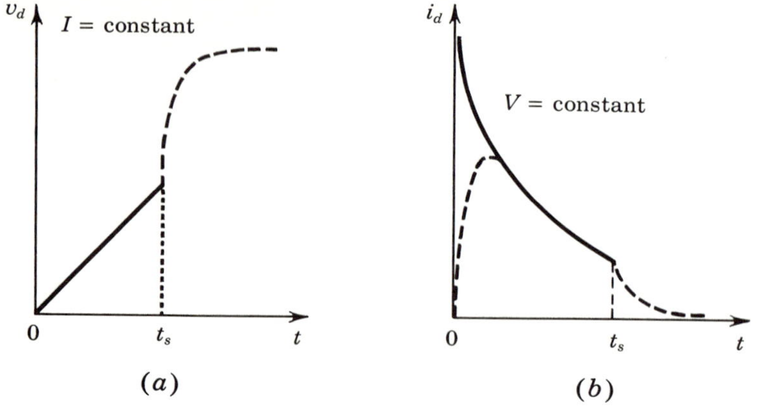

Fig. 15-28 Terminal response of an ideal ferroelectric device. (a) Constant-current excitation; (b) constant-voltage excitation.

Where our concept of the core was a time-varying resistor decreasing from infinity as the core approached saturation, our model of the square-loop ferroelectric device is a resistance that increases from zero with time and current. It appears in series with a bucking voltage V_c equal to the internal coercive force. Sometimes the model of the ferroelectric element is further simplified, by neglecting the energy loss during switching, and it becomes a bivalued capacitor. In fact, when the termination of the excitation leaves the ferroelectric element unsaturated, its incremental response is most like that of a large capacitor. Since the relative dielectric constant drops from several hundred to unity as the material saturates, the capacity is reduced to that of the two deposited electrodes in free space.

The widely used ferroelectric material barium titanate has a coercive field strength of 1,500 V/cm. Only by using extremely thin slabs (0.033 to 0.167 mm) can the coercive voltage V_c be kept between 5 and 25 V. Over the transition interval, the average resistance of the ferroelectric element ranges between 100 and 1,000 Ω. Because of the large remnant polarization, 22×10^{-6} C/cm^2, a large charge packet is transferred during switching, even from quite small elements.

15-9 The Ferroelectric Counter

The principle of charge monitoring is employed in the counter of Fig. 15-29. As the input square-loop ferroelectric element (FE_1) is switched from negative to positive saturation by the leading portion of the input pulse,

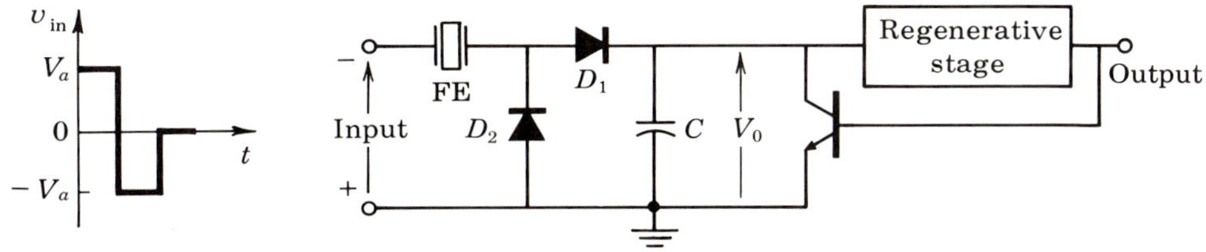

Fig. 15-29 Ferroelectric counter.

a charge packet of $2Q_s$ flows through D_1 into the storage element C. This increases the output voltage by the definitive increment

$$\Delta V_o = \frac{2Q_s}{C} \tag{15-31}$$

The negative portion of the input pulse resets FE_1 to negative saturation, with D_2 establishing the current path. During this part of the cycle the back-biased diode D_1 prevents the discharge of the previously stored count.

Note that the input pulses must be properly shaped. They must have positive peaks greater than $V_c + V_o$, as well as a duration greater than the switching interval. The negative peak can be smaller, but its magnitude must still exceed V_c. Otherwise the ferroelectric device will not switch and reswitch between the two saturation limits and will not meter the full charge packet to the storage capacitor.

During the final count, the change in the output voltage brings the regenerative circuit into conduction. For example, the final change of ΔV_o may be used to trigger either a monostable multivibrator or a blocking oscillator which will propagate a single count in the forward direction into the next stage. This same pulse will turn on the shunting transistor, discharge C, and reset the counter.

The switching action of the counter can be greatly improved by the simple modification shown in Fig. 15-30, i.e., the replacement of C by a second ferroelectric element. Each input pulse will be shaped into a

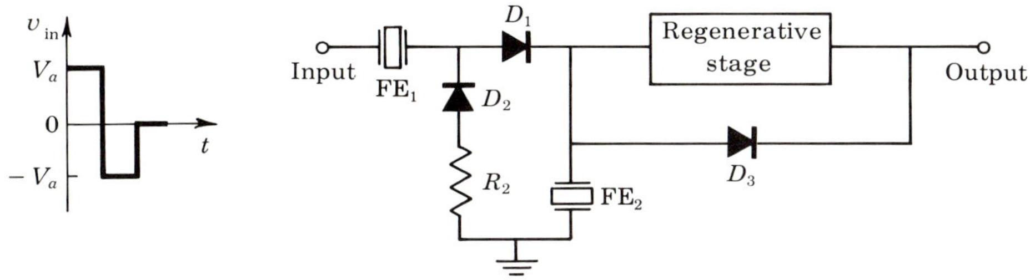

Fig. 15-30 Improved ferroelectric counter.

charge packet of $2Q_{s1}$ by FE_1. This charge also flows into FE_2 and starts it switching from $-Q_{s2}$ toward $+Q_{s2}$. If the second element is larger, n charge packets (or input pulses) will be needed to saturate FE_2. Until such time as it does saturate, the ferroelectric element appears to be a low impedance and only a very limited voltage can be developed across it.

If FE_2 becomes saturated during the nth input, then it may be replaced by a very high impedance for the remaining portion of the pulse. The full excitation will appear across the output storage element, and instead of depending on identifying the small change of ΔV_o, the large voltage jump presents a definitive trigger to the regenerative stage. With the modification of Fig. 15-30, this counter can accurately register many more counts than possible with the simpler circuit. Furthermore, since the count is registered by a change in the state of the element and not by the storage of charge, it will be less affected by leakage resistance. Counters having rates as low as one per day and counting ratios as high as 30:1 or 40:1 have been successfully operated.

In order to reset the storage element, a negative pulse can be applied through a diode from some point in the regenerative stage. However, unless resistance is inserted in series with D_2, the two input diodes would act as a short circuit, preventing the resetting of FE_2. Current will also flow into FE_1 during the reset interval, but this only aids its recovery toward negative saturation.

Example 15-2

Another interesting application of the pulse-area shaping by the ferroelectric and ferromagnetic devices is the tachometer shown in Fig. 15-31. The sharp input pulses obtained from the distributor in a six-cylinder two-cycle engine are assumed shaped to switch and reset the charge-metering ferroelectric element. Their amplitude and duration will, of course, be erratic. When switching from negative to positive polarization, the input element delivers a charge packet of 20×10^{-6} C. We wish the highly damped dc voltmeter reading of 5 V to correspond to a speed of 5,000 rpm.

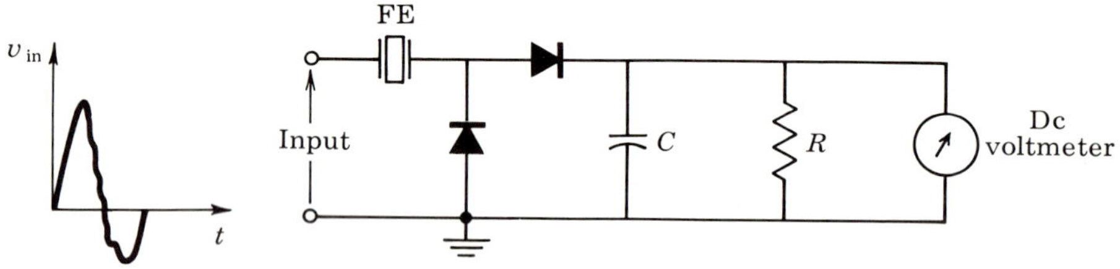

Fig. 15-31 A ferroelectric tachometer (Example 15-2).

Three input pulses are generated on each revolution. Hence the minimum spacing between the individual inputs, which occurs at the maximum velocity, is

$$T_m = \frac{60 \text{ s}}{3 \times 5{,}000} = 4 \text{ ms}$$

So that the storage capacitor can discharge completely between the adjacent pulses, the output time constant must be much less than T_m. Assuming that the charging time is insignificant and that $T > 4\tau$, the average voltage read on the dc meter is given by

$$V_{\text{dc}} = \frac{1}{T} \int_0^T \Delta V\, e^{-t/\tau}\, dt = \frac{\Delta V\, \tau}{T}$$

where ΔV is the peak output voltage due to the metered charge packet. Note that the output RC circuit provides a dc voltage inversely proportional to the spacing between the pulses and directly proportional to the motor rpm.

At 5,000 rpm,

$$\Delta V\, \tau = \frac{Q_{sw}}{C} RC = 20 \text{ V-ms}$$

Thus $R = 1{,}000\ \Omega$. Setting $\tau = 500\ \mu\text{s}$ yields $C = 0.5\ \mu\text{F}$. Since $\Delta V = 40$ V and $V_c = 10$ V, for proper switching the peak positive input must be greater than 50 V. The diode allows satisfactory resetting with negative peaks greater than 10 V.

15-10 Coincident Memory Arrays

The ferroelectric element is somewhat more difficult to use in a memory matrix than the ferromagnetic core. Since only two electrodes are available, these must be used not only for storage and interrogation but also for readout. It is not possible to place a third electrode at the memory position, as it was to thread the complete core array with a third winding. Consequently, various ingenious circuits have been devised to obtain the output digit.

Consider the memory plate of Fig. 15-32, where the x and y direction leads are printed at right angles on the opposite faces of the ferroelectric plate. The intersections are the memory positions, and they may be treated as individual ferroelectric elements. In order to obtain the coincident switching, voltage inputs must be applied to the two leads. To store a digit in a particular location, that memory element is switched from negative to positive polarization by applying a positive pulse to the

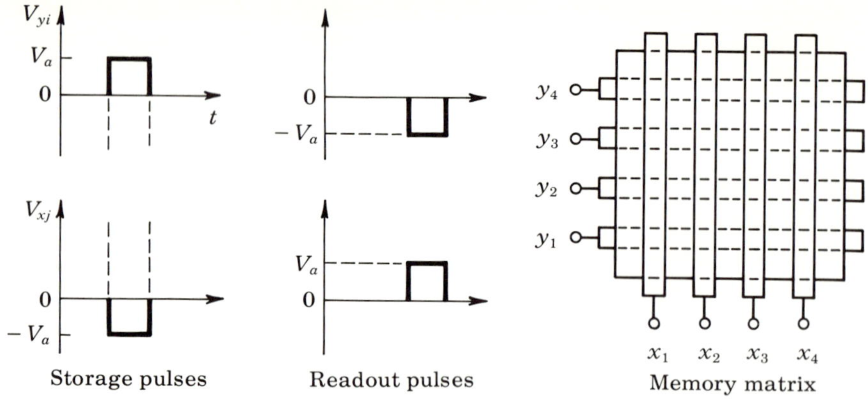

Fig. 15-32 A ferroelectric memory matrix.

y_i row and a negative pulse to the x_j column. Each pulse must have an amplitude falling within the limits

$$\tfrac{1}{2}V_c < |V_a| < V_c$$

and a duration sufficient to cause switching.

It is quite simple to reset the memory element from 1 to 0. If it contains a stored bit, pulses of the polarity opposite to those used for storage will sum at the intersection and reverse the polarization. All other positions remain unaffected. But how may the bit be registered after interrogation? The switching process results in the transfer of a charge packet which must somehow be monitored. In one method, shown in Fig. 15-33*a*, a resistor is connected in the ground return of either the x or y pulse generator. The current flow during storage will cause a negative voltage to appear across

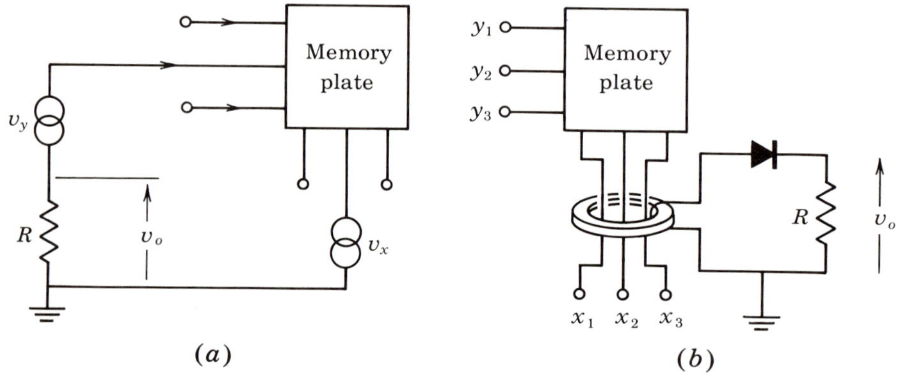

Fig. 15-33 Two methods of monitoring the readout pulse.

the resistor. During reset, the opposite polarity signal is developed. The two output pulses are easily distinguished, and if desired, a shunting diode may be incorporated to eliminate the unwanted negative pulse.

Figure 15-33*b* illustrates a somewhat superior method of producing an output after interrogation. All the x leads (or y leads) are threaded through a single readout core. The current flow during the transfer of the metered charge packet induces a pulse into the output winding. One polarity output is produced on storage, and the opposite on reset; the diode shown selects the correct signal.

One major disadvantage of the ferroelectric memory matrix is that the state of the particular element is affected by single-voltage pulses of less than the coercive force. In barium titanate it was discovered that multiple queries in any one row caused the partial depolarization of all memory positions subjected to the half-voltage pulse. After several hundred interrogations, the polarization may even be reduced to half of the saturation value. It has been found, however, that crystals of other materials, such as triglycine sulfate, appear to have more nearly ideal characteristics. As these sustain full polarization under repeated questioning, they will no doubt be used in practical memory banks.

Problems

15-1 A rectangular-cross-section permalloy core measuring 1 cm wide by 5 mm thick has an effective magnetic-path length of 5 cm. It is excited by a 10-V peak-to-peak square wave applied across a 20-turn winding.

(*a*) What is the minimum-period square wave that just drives the core between the two saturation limits?

(*b*) Plot the spatial penetration of the field into the core and the terminal-current buildup as a function of time when the square-wave period is twice the length found in part *a*.

In the course of the calculations assume a source impedance of 10 Ω and a winding resistance of 1 Ω where necessary to keep the solution within bounds. The saturation-region inductance is only 10 μH.

15-2 Plot the spatial variation of the field and the terminal-current buildup for the core of Prob. 15-1 for the following excitations. After the core saturates, the current is limited by the 11-Ω series resistance. This resistance may be neglected during the interval that the core remains unsaturated.

(*a*) A 25-V peak-to-peak 50-μs-period sine wave.

(*b*) A 25-V peak-to-peak 30-μs-period triangular wave.

15-3 The core of Prob. 15-1 is excited by a square wave of current having a peak-to-peak excitation ten times as large as the coercive current. For this case, plot the location of the field edge and the terminal-voltage response as a function of time.

15-4 Calculate and sketch the voltage response of the core of Prob. 15-1 to the following current excitations. The core is initially at negative saturation, and each signal reverses the flux direction.

(*a*) A current ramp which increases at the rate of 20 A/ms.

(*b*) A single cycle of a sine wave having a peak-to-peak amplitude of 50 mA and a duration of 25 μs.

15-5 Plot the incremental driving-point resistance as a function of time for the two cores described in Example 15-1. Consider two different excitations.

(*a*) Constant voltage (20 V)

(*b*) Constant current ($I = 10I_c$)

15-6 (*a*) Solve for the general expression of the voltage response of a core winding to a current step ($I \gg I_c$) when the core is loaded across a secondary winding N_2 with a resistor R. The applied step drives the core from negative to positive saturation.

(*b*) Plot the response of the permalloy core of Example 15-1 to a 50-mA current step. The external load is a 1-K resistor. $N_2 = 10$.

(*c*) What happens to the response when the load is reduced to 100 Ω?

15-7 Consider the coil of Prob. 15-1 excited by a 10-V step. We wish to take into account the effects of the inductance of the leads and winding on the core switching response. The inductance of driving coil and leads is 5 μH when the core is saturated, increasing to 100 μH during the interval that the core state is changing. Set up the basic equations defining the switching time and terminal-current response. Without solving these equations, indicate the effect of L on the switching time and on the magnitude of the current response. Does L introduce any switching-time delay? Would a current drive improve the response? Explain.

15-8 In the counter of Fig. 15-9, the ferrite core is wound with 10 turns. It has an effective radius of 1 cm and a 1-cm by 5-mm cross-section area. The magnetic parameters are given in Example 15-1. $C = 1$ nF, and $R = 500\ \Omega$.

(*a*) Specify the range of volt-time areas of the input pulses for which this counter will count down by the ratio 5:1. What is the minimum spacing between pulses?

(*b*) Sketch and label the output waveshape for the condition of part *a*.

15-9 The ferrite core of Prob. 15-8 is used as the memory element in the simple counter of Fig. 15-9. One output pulse should be produced for every three of the 25-V 10-μs input pulses. During the priming inputs, the output must remain below 5 V.

(*a*) Specify the minimum value of C that will properly reset the core on the termination of the count.

(*b*) Using twice the minimum value of C, specify the range of core turns for which this counter will work. (The RC time constant is 1 μs.)

(*c*) Sketch and label the output voltage when C is twice the minimum value and when N lies in the center of the acceptable range.

15-10 Two identical ferrite cores, each having the dimensions given in Prob. 15-8, are used in the scale-of-10 counter of Fig. 15-11. The input core has $N_2 = 10$, and the counting core has 10 turns. $R_s = R = 500\ \Omega$, and $C = 2$ nF. Complete the design by specifying N_1 and the C_s needed to reset the input core. Calculate the minimum input volt-time area and the pulse spacing necessary for proper counting. Discuss what circuit modification is necessary to ensure the automatic reset of the counting core. Can the counting core be preset to a count of 3? How?

15-11 Figure 15-34 shows a magnetic-core AND gate incorporating automatic reset. Each coil consists of five turns wound on a single ferrite core (effective diameter is 1 cm and cross section is 1 × 0.2 cm).

(a) Sketch the output voltage when 1-ms current pulses, each having an amplitude equal to $0.4I_c$, are simultaneously applied to two inputs; to three inputs.

(b) If $C = 1$ nF, what is the minimum pulse width that will ensure the automatic reset of the core? The loss in R may be neglected in the calculation.

(c) How might this circuit be converted into an OR gate?

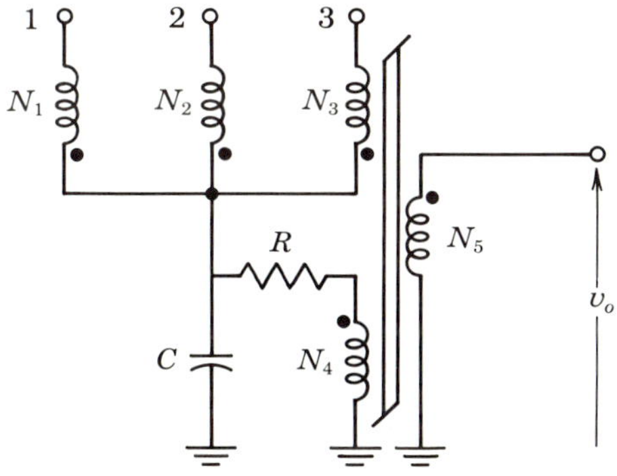

Fig. 15-34

15-12 Ferrite cores 2 mm ID, 3 mm OD, and 0.5 mm thick are strung in the 3D four-wire memory array of Fig. 15-19. Each line through a core may be considered a single-turn winding. A single core is to be switched from the logical-0 to the logical-1 state by the simultaneous excitation of an x, a y, and an inhibit line. The drive amplifiers are transistor current sources.

(a) Give the appropriate drive currents for each line.

(b) Find the approximate switching time of the core. Sketch the approximate waveshape appearing in the sense line. Explain the effects on the waveshape of stray capacity and lead inductance.

(c) Explain what signal might be observed if the selected core is already in the logical-1 state.

15-13 The transistor-core multivibrator of Fig. 15-20 is used as a dc-to-dc converter producing 150 V from a 24-V source. It operates at approximately 1,000 Hz, which simplifies the filtering at the output of a full-wave bridge rectifier. Each of the 50-turn primary windings has 0.2-Ω resistance. The saturation resistance of the transistor is 0.3 Ω, and at the boundary between the active and saturation regions, β may be taken as 20. ($V_{BE} = 750$ mV.) Grain-oriented silicon steel, having the following parameters, is used for the square cross-section core: $B_s = 2$ Wb, $H_c = 12$ A-turns/m, $\rho = 40 \times 10^{-8}$ Ω-m, and $l = 30$ cm.

(a) What is the core cross section for operation at 1,000 Hz?
(b) Calculate the total power dissipated in the converter when the external load current is 200 mA. How much power is dissipated under no-load conditions?
(c) Sketch and label the current and voltage waveshapes at the base and collector under loaded conditions. Specify the maximum $R_1 = R_2$.

15-14 Design a core-transistor dc-to-ac converter which operates from a 12-V car battery and which supplies 1 A at 120 V (60 Hz). The transformer is constructed out of grain-oriented silicon steel having the parameters given in Prob. 15-13. The magnetic-path length is 50 cm, and the core cross section is rectangular, with the sides in a ratio of 2:1. Both collector primary windings are 20 turns, having a resistance of 0.05 Ω. The power transistors have $\beta = 25$, $V_{BE} = 750$ mV, and $r_{sc} = 0.1$ Ω. Complete the design. Draw the output and the transistor waveshapes. Specify the output frequency change and voltage change from no load to full load.

15-15 Figure 15-35 shows a single-transistor dc-to-dc converter. Reset of the core is provided by the discharge of C_1. R_1 and R_2 supply the turn-on bias for the transistor. The core is grain-oriented silicon steel having a cross section of 2×4 cm and a magnetic-path length of 30 cm. $N_1 = 20$ turns and has a resistance of 0.3 Ω. $N_3 = 100$ turns, and its resistance is 5 Ω. C_2 is a filter capacitor and is very large. $\beta = 30$, $V_{BE} = 700$ mV, and $r_{sc} = 0.2$ Ω.
(a) Specify the minimum number of turns for N_2.
(b) What is the minimum value of C_1 for proper reset?
(c) Set N_2 at twice the minimum value and C_1 to five times the minimum. Sketch the voltage waveshapes at the collector and base of the transistor. Sketch the collector-current waveshapes. Give all significant values and times.

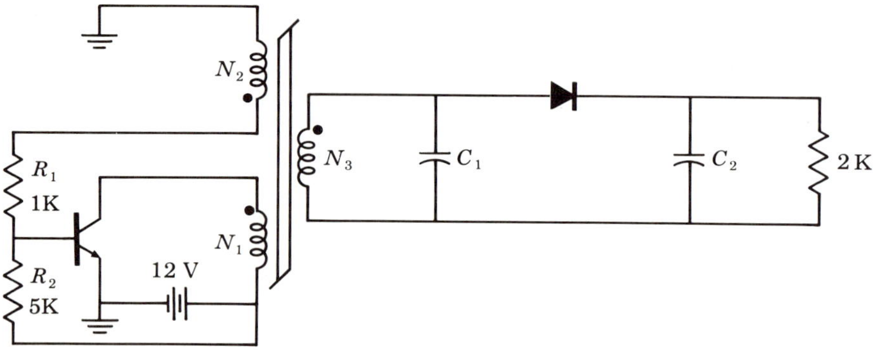

Fig. 15-35

15-16 The incremental capacity of the wedge-shaped ferroelectric element shown in Fig. 15-36 is a function of the applied dc voltage. Calculate the capacity variation, neglecting any fringe effects. The dielectric used has $\epsilon_r = 2{,}000$ and $E_c = 1.5 \times 10^5$ V/m, and its square cross section is 0.25 cm². (*Hint:* First show that the

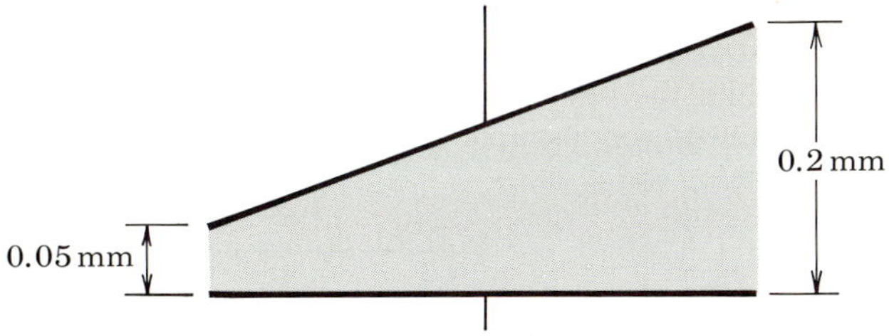

Fig. 15.36

capacity of a large parallel-plate capacitor is $C = \epsilon_o \epsilon_r A/d$.) How will the resonant frequency of a tank circuit using this element vary with the dc bias voltage?

15-17 (*a*) Prove that the charge packet flowing during the switching interval in a ferroelectric element is equal to the charge transferred in a parallel-plate capacitor of the same dimensions as it charges from $-V_c$ to V_c.

(*b*) The ferroelectric element FE_1 of Prob. 15-18 is charged from a 20-V source through a 5-K resistor. After saturating, ϵ_r drops to unity. Plot the current and voltage response. Give the values at all significant coordinates.

15-18 The counter of Fig. 15-37 is excited by two separate inputs, each connected to the single storage capacitor through a separate charge-metering element. FE_1 consists of a 0.03-mm-thick slab having an effective surface area of 0.02 cm^2, while FE_2 is twice as thick and has only half the electrode area. The ferroelectric material has a coercive field strength of 2×10^5 V/m and a remnant polarization of 8×10^{-6} C/cm^2. We may further assume that the constant in the switching equation (15-30) is $K = 10$ V-μs.

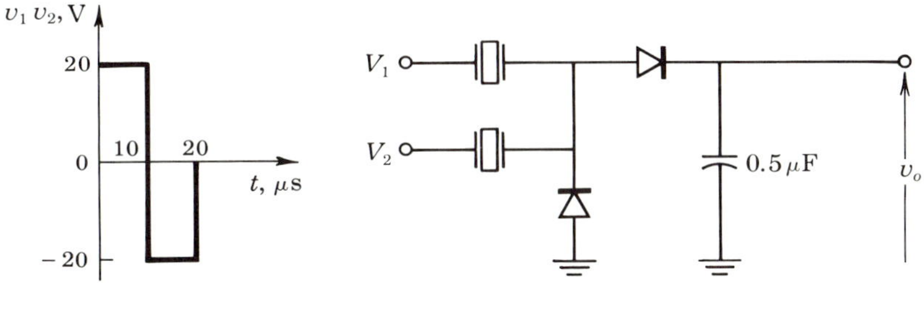

Fig. 15-37

Sketch the output, giving all significant values if the shaped pulses are applied at the two inputs at

$$v_1 \text{ at } t = 0, 10, 20, \ldots \text{ ms}$$

$$v_2 \text{ at } t = 5, 15, 25, \ldots \text{ ms}$$

15-19 Figure 15-38 shows a monostable multivibrator where the duration of the quasi-stable state is controlled by the switching time of a ferroelectric element. In order to simplify the calculations, it will be represented by a 5-nF capacitor which decreases by a factor of 1,000 after saturating at ± 5 V. Sketch and label the voltage at the collector of Q_1 and at the collector and base of Q_2 after a pulse is injected into the base of Q_1. Does this type of timing offer any advantage over that obtained from a simple capacitor-timed multivibrator?

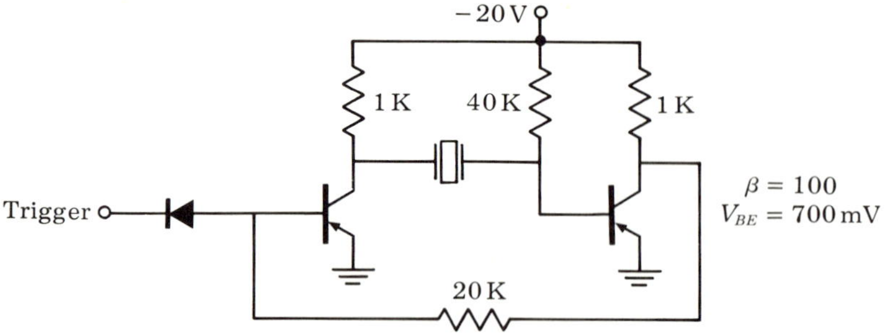

Fig. 15-38

15-20 Show that the circuit of Fig. 15-38 can be modified so that it can make use of a ferromagnetic core to control the timing. Sketch and label the important waveshapes when the core of Prob. 15-8 is employed.

Bibliography

Chen, Kan, and A. J. Schiere: A Single Transistor Magnetic Coupled Oscillator, *Trans. AIEE Commun. Electron.*, pt. I, vol. 75, pp. 396–399, September, 1956.

Chen, T. C., and A. Papoulis: Domain Theory in Core Switching, *Proc. Symp. Role Solid State Phenomena Elec. Circuits,* Polytechnic Institute of Brooklyn, New York, April, 1957; also in *Proc. IRE,* vol. 46, no. 5, pp. 839–849, 1958.

Katz, H. W.: "Solid State Magnetic and Dielectric Devices," John Wiley & Sons, Inc., New York, 1959.

Little, C. A.: Dynamic Behavior of Domain Walls $BaTiO_3$, *Phys. Rev.,* vol. 98, no. 4, pp. 978–984, 1955.

Rajchman, J. A.: A Myriabit Core Matrix Memory, *Proc. IRE,* vol. 41, no. 10, pp. 1407–1421, 1953.

———: A Survey of Magnetic and Other Solid-state Devices for the Manipulation of Information, *IRE Trans. Circuit Theory,* vol. CT-4, no. 3, pp. 210–225, September, 1957.

Royer, G. H.: A Switching Transistor D-C to A-C Converter Having an Output Frequency Proportional to the D-C Input Voltage, *Trans. AIEE Commun. Electron.*, pt. I, vol. 74, pp. 322–326, July, 1955.

Sands, E. A.: An Analysis of Magnetic Shift Register Operation, *Proc. IRE,* vol. 41, no. 8, pp. 993–999, 1953.

Wolfe, R. M.: Counting Circuits Employing Ferroelectric Devices, *IRE Trans. Circuit Theory,* vol. CT-4, no. 3, pp. 226–228, 1957.

Yovits, M. C. (ed.): "Symposium on Large Capacity Memory Techniques for Computing Systems," The Macmillan Company, New York, 1962.

PART 5
oscillations

16. almost sinusoidal oscillations—the linear approximation

The earlier chapters of the text showed how the generation of many periodic waveshapes (rectangular pulses and linear sweeps) was predicated on driving the active elements of the circuit far into their saturation and/or cutoff regions. The timing function depended upon the nonactive regions of the system, with the active zone simply supplying the energy necessary for switching between the two exponential timing regions.

As we start discussing the generation of almost sinusoidal signals, it should be pointed out that the nonlinearity of the system will now play a subsidiary role. The basic timing is due to a frequency-determining network, with the amplifier simply setting the necessary conditions for oscillation. To avoid distorting the output signal, the degree of nonlinearity must be small, and consequently most of the subsequent discussion can concern itself with the linear approximation of the sinusoidal oscillator.

16-1 Basic Feedback Oscillators

Figure 16-1 presents the basic configuration of a feedback oscillator as a block diagram. Even though each of the three essentials shown is not completely isolated, it is convenient to think of them as separate entities.

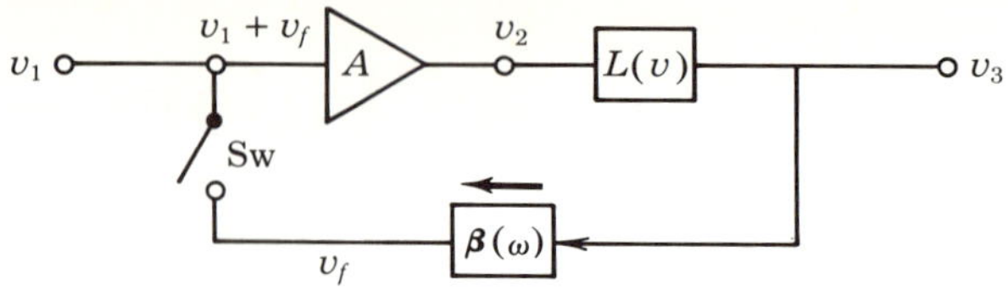

Fig. 16-1 Basic-feedback-oscillator configuration showing the ideal amplifier A, the amplitude limiter $L(v)$, and the frequency-determining network $\beta(\omega)$.

It is especially helpful to segregate the complete nonlinearity so that it can be given individual attention. The remaining elements in the circuit may then be treated by the standard methods of linear analysis.

The behavior of the circuit of Fig. 16-1 is defined by two equations, one for the forward transmission and the other giving the feedback voltage:

$$v_3 = (v_1 + v_f)AL \qquad v_f = \beta v_3$$

From these equations, we find that the overall closed-loop gain is

$$G = \frac{v_3}{v_1} = \frac{AL}{1 - \beta AL} \tag{16-1}$$

For self-sustained oscillations, there must be an output without any external excitation. This becomes possible only when the denominator of Eq. (16-1) vanishes. It follows that the frequency and amplitude of oscillation must satisfy

$$1 - \beta(\omega)AL(v) = 0 \tag{16-2}$$

where βAL is simply the open-loop transmission (i.e., the switch in Fig. 16-1 is open). Equation (16-2) implies that the feedback must be regenerative.

Limiting is often accomplished by driving one or more of the amplifier's active elements into their saturation or cutoff regions for a portion of the cycle. Alternatively, an amplitude-controlled resistor or other passive nonlinear element may be included as part of the amplifier or in the frequency-determining network. If gross nonlinear operation is permitted, the limiter will distort the signal and the output will be far from sinusoidal.

For small signals, no limiting will occur and $L(v)$ will take on its maximum value of unity. It follows that if $A\beta > 1$ in the small-signal region, the amplitude will build up until the limiter stabilizes the system at an output level that satisfies Eq. (16-2). Thus, in the active region, the threshold loop transmission that permits self-sustained oscillation is

$$\beta A_t = 1$$

This equation is called the Barkhausen criterion for oscillation.

Unity loop gain at a single frequency is a necessary but not a sufficient condition for self-sustained oscillations. If the network characteristics are such that the net phase shift is zero at several frequencies, then the criterion for oscillation can be arrived at only from an examination of the complete amplitude and phase portrait of the system. Nyquist stated that the polar plot of $A\beta$ for $-\infty < \omega < \infty$ must encircle the point $1 + j0$ for oscillation. However, for many of the simpler configurations, the Barkhausen condition will be both necessary and sufficient. Below we shall consider circuits which are known to oscillate and which can be analyzed on the basis of their having unity loop gain.

To ensure our objective, the generation of a sinusoidal signal, the frequency should be primarily determined by a network whose characteristics can be rigidly controlled. This network must contain at least two energy-storage elements so that the system response [Eq. (16-2)] will have the necessary pair of complex conjugate roots which give the natural frequency of oscillation. For simplicity, A may be assumed to be a constant, with its frequency dependence lumped together with that of β.

Since A in Eq. (16-2) now represents the gain of an ideal amplifier, it will be a pure number having a sign corresponding to the stages of amplification (minus for an odd number and plus for an even number). β represents the transmission characteristics of the frequency-selective network and will therefore be a function of ω. To satisfy Eq. (16-2), the imaginary part of βA must vanish. Since the complete frequency variation is included in the β network, the imaginary part depends only on β. Consequently the frequency at which

$$\text{Im}\ \beta = 0 \tag{16-3}$$

will be the frequency of oscillation.

At the oscillation frequency, the remaining portion of Eq. (16-2), that is, Re (βA), must be identically equal to unity. The value of gain that will just ensure the sinusoidal oscillations will be given by

$$A_t\beta(\omega_0) = 1$$

or

$$A_t = \frac{1}{\beta(\omega_0)} \tag{16-4}$$

This value of threshold gain is measured with the amplifier output loaded by the input impedance of the β network. The open-circuit gain must, of course, be somewhat larger.

Equations (16-3) and (16-4), taken together, state that the overall loop gain must be unity and the overall phase shift must be zero at the frequency of oscillation.

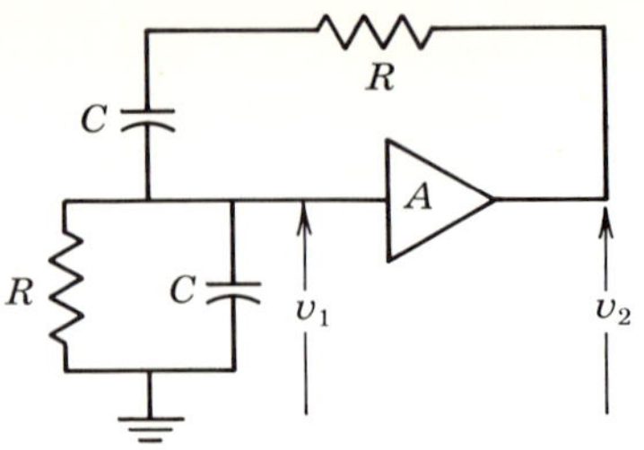

Fig. 16-2 Example of a feedback oscillator.

Consider the circuit of Fig. 16-2, where we shall assume that the ideal voltage amplifier is adjusted so that it will just sustain the oscillations. The frequency-determining network is characterized by

$$\beta(s) = \frac{v_1}{v_2} = \frac{\dfrac{R}{sCR + 1}}{R + \dfrac{1}{sC} + \dfrac{R}{sCR + 1}} = \frac{1}{3 + sCR + \dfrac{1}{sCR}} \tag{16-5}$$

Substituting $j\omega$ for s results in

$$\beta = \frac{1}{3 + j(\omega/\omega_0 - \omega_0/\omega)} \tag{16-6}$$

where $\omega_0 = 1/CR$. The imaginary portion of the denominator of Eq. (16-6) vanishes at

$$\omega = \omega_0 = \frac{1}{CR}$$

At this frequency, $\beta = \frac{1}{3}$, and from Eq. (16-4) we find that the gain necessary to sustain oscillation is $A_t = 3$. To achieve this positive gain, the amplifier must contain an even number of stages.

Because they determine the nature of the transient response, it is interesting to examine the location of the roots of $1 - \beta A$, and hence the poles of the complete feedback system as a function of A. The roots of the circuit of Fig. 16-2 are found by substituting Eq. (16-5) into $1 - \beta A = 0$. This leads to

$$s^2 + \frac{3 - A}{CR} s + \left(\frac{1}{CR}\right)^2 = 0 \tag{16-7}$$

The oscillator of Fig. 16-2 has two poles that migrate as shown in Fig. 16-3. As A increases from 0 to 1, they first coalesce at $-1/CR$ on the negative real axis and then separate along the semicircular paths shown, finally reaching $\pm j/CR$ on the imaginary axis for $A = 3$. This condition corresponds to the threshold of oscillation, with the location of the complex conjugate roots giving the frequency.

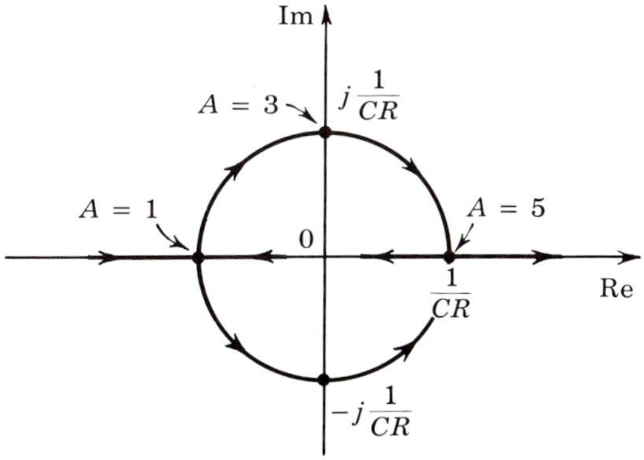

Fig. 16-3 Migration of the roots of Eq. (16-7) as a function of A. The arrows indicate increasing gain.

When the system has two distinct complex conjugate poles, i.e., for A between the limits $1 < A < 5$, the transient response will be given by

$$v(t) = Ke^{\alpha t} \cos (\omega_1 t + \phi)$$

and the two poles are located at

$$s_{1,2} = \alpha \pm j\omega_1$$

where $\omega_0 = \sqrt{\alpha^2 + \omega_1{}^2}$. For small values of α, the natural frequency is almost ω_0.

If A is less than 3, then the real part of the root will be negative and any oscillatory response will damp out. The degree of damping depends on the distance from the poles to the imaginary axis and hence on the value of A.

Once the poles move into the right half plane, the real part forces an increasing exponential buildup. But the voltage is bounded by the limiter; the output eventually stabilizes at a peak amplitude that makes the average gain over the cycle equal to 3. At the peaks of the sine wave, the limiter reduces the loop gain, driving the system poles from the right into the left half plane. The amplitude decays, the circuit reenters the active region, and the roots move back into the right half plane.

If the poles are initially close to the imaginary axis, then the buildup will be quite slow and small changes in the root location will have little influence over any single cycle of the sine wave. When they are located far into the right half plane, the exponential buildup makes itself felt during each cycle; the waveform will include a greater degree of distortion. Even with a small gain margin over the threshold value, the location of the roots will be such that the oscillator will generate a signal of slightly lower frequency than that calculated on the basis of the borderline behavior.

The oscillator will be self-starting only if initially the poles lie in the right half plane. It follows that the necessary small-signal gain must be somewhat in excess at A_t. Any infinitesimal disturbance will cause an amplitude buildup until the limiter stabilizes the output. The threshold gain sustains existing oscillations but does not provide a margin for buildup. Excess gain will also stabilize the circuit in that any slight reduction in A will not stop the oscillations when $A > A_t$ but will cause them to damp out when the circuit operates marginally.

If the amplifier gain ever becomes large enough to drive the roots onto the positive real axis, that is, $A > 5$, the system response will no longer be oscillatory. The exponential buildup will lead to a relaxation phenomenon similar to that found for the astable multivibrator.

16-2 Characteristics of Some *RC* and *LR* Frequency-determining Networks

The self-starting oscillator drives itself into limiting and in the process distorts the sinusoidal output signal. Harmonics, introduced by the limiter, may be treated as additional signals injected within the feedback loop. Each harmonic term will also be affected by the feedback present; it will be acted upon by the factor

$$H_n = \frac{1}{1 - A\beta_n} \tag{16-8}$$

where β_n is the feedback factor evaluated at the particular harmonic under investigation.

Since we wish to maintain an almost sinusoidal output, we should ensure that the amplitude of the harmonics produced in the limiter will be small compared with the fundamental. Each distortion term is further modified by H_n. With a properly designed network, the feedback will change from positive to negative as the loop excitation frequency increases from ω_0 to $2\omega_0$. The magnitude of H_n will become less than unity at the second-harmonic frequency and even smaller at the higher-harmonic frequencies. Thus the nature of the existing feedback is changed and is used to reduce the amount of harmonic distortion.

The more selective the network, the greater the degree of limiting which can be tolerated without excessively distorting the final sinusoidal output. Consequently, $|H_2|$ serves as a measure of the network's quality for oscillator application. For consistency, it will always be evaluated under threshold conditions, instead of using the slightly larger value of the actual small-signal gain. The best network, all other things being equal, is the one having the lowest value of H_2.

The output signal, which is of the form

$$v_3 = V_1 \cos (\omega_0 t + \phi_1) + \sum_{n=2}^{\infty} H_n V_n \cos (n\omega_0 t + \phi_n) \tag{16-9}$$

is transmitted back to the input through the β network. Each term is also multiplied by β_n, resulting in an effective driving signal of

$$v_1 = \beta_1 V_1 \cos (\omega_0 t + \phi_1) + \sum_{n=2}^{\infty} \beta_n H_n V_n \cos (n\omega_0 t + \phi_n)$$

In order to have a true basis of comparison as to the relative amplitudes of the distortion components at the input and at the output, it is advisable to use the normalized input harmonic factor

$$K_n = \frac{\beta_n H_n}{\beta_1} \tag{16-10}$$

Equation (16-10) accounts for the relative attenuation of the fundamental component, which would be ignored if $\beta_n H_n$ were chosen instead as the figure of merit. When $|K_n| = |H_n|$, the relative harmonic contents of the input and output are equal. In any specific circuit, the best signal, i.e., the one most nearly sinusoidal, will appear at the output for $|K_n| > |H_n|$ and at the input for $|K_n| < |H_n|$.

Phase-shift network

One network commonly used in the so-called "phase-shift oscillator," which we must therefore name "the phase-shift network," appears in Fig. 16-4*a*. Either the three series branches (Z_1) will be energy-storage elements L or C and the shunt branches (Z_2) resistors, or vice versa. In general, all series branches and all parallel branches are composed of equal elements; however, this is not an essential condition. Satisfactory oscillators have been constructed with widely unbalanced sections and with networks of more than three sections.

From Fig. 16-4*a*, the network transfer characteristics can be found to be

$$\beta = \frac{v_f}{v_3} = \frac{1}{(Z_1/Z_2)^3 + 5(Z_1/Z_2)^2 + 6Z_1/Z_2 + 1} \tag{16-11}$$

Since either Z_1 or Z_2, but not both, is an energy-storage element, the odd-power terms in the denominator of Eq. (16-11) contribute the imaginary part of $\beta(\omega)$. For it to vanish,

$$\left(\frac{Z_1}{Z_2}\right)^3 + 6\frac{Z_1}{Z_2} = 0 \tag{16-12}$$

In the specific case of the RC network of Fig. 16-4b, $Z_1 = 1/j\omega C$ and $Z_2 = R$; substituting these values into Eq. (16-12) yields

$$\omega_0 = \frac{1}{\sqrt{6}\,CR} \tag{16-13}$$

As the elements in the network change, so will the form of the equation defining ω_0.

By either substituting the value of ω_0 from Eq. (16-13) into Eq. (16-11) or making the simpler substitution of

$$\left(\frac{Z_1}{Z_2}\right)^2 = -6$$

from Eq. (16-12) into Eq. (16-11) and by noting that the odd powers of Z_1/Z_2 have vanished, the threshold value of β becomes

$$\beta(\omega_0) = \beta_1 = -\tfrac{1}{29}$$

Thus the amplifier must have a minimum gain of -29 for sustained oscillations. With this particular type of network, the required gain is independent of the elements comprising Z_1 and Z_2. Any other RC or LR combination will give the same value of gain but a different operating frequency and network input impedance. Since the necessary gain is

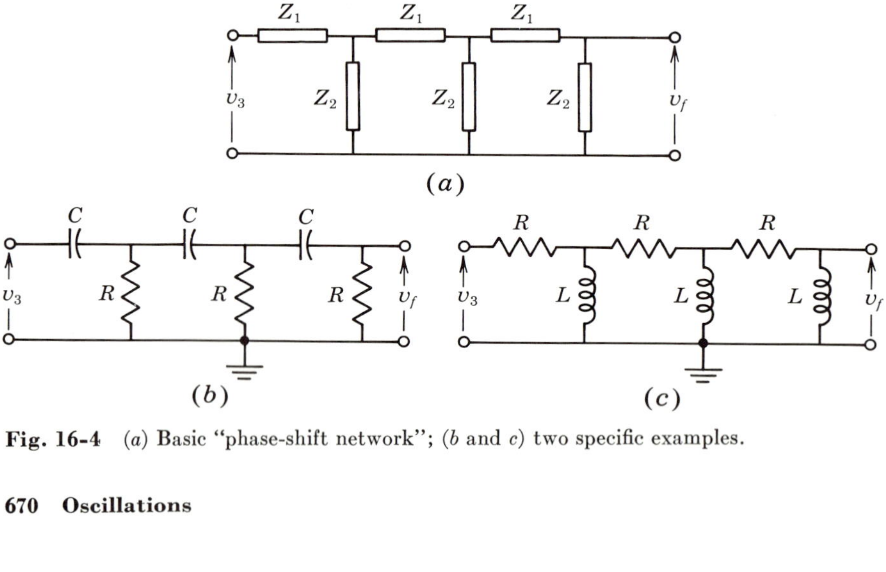

Fig. 16-4 (a) Basic "phase-shift network"; (b and c) two specific examples.

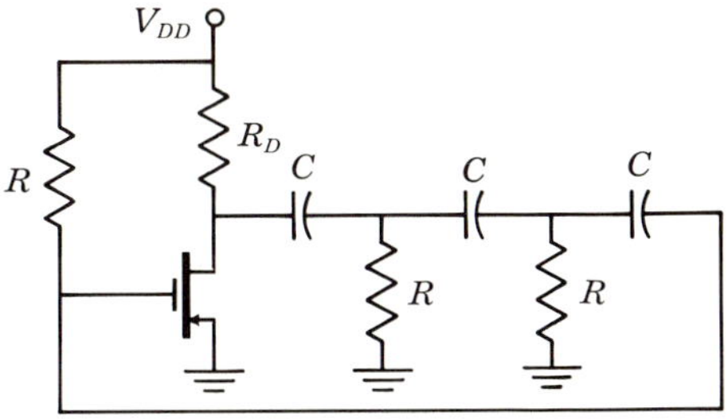

Fig. 16-5 A single-MOSFET phase-shift oscillator.

negative, a single stage or an odd number of stages is required. Figure 16-5 illustrates one possible circuit.

In order to evaluate $|H_n|$ and $|K_n|$ for the *RC* network of Fig. 16-4*b*, β_n may be expressed as

$$\beta_n = \frac{1}{1 - 30/n^2 - j(6\sqrt{6}/n)(1 - 1/n^2)}$$

where n represents the order of the harmonic. Taking $A_t = -29$ and substituting into Eqs. (16-8) and (16-10) at the second harmonic, we obtain

$$|H_2| = 0.368$$
$$|K_2| = 1.25$$

For this network, the waveshape containing the smallest amount of harmonics appears at the amplifier output. If the positions of R and C are interchanged, this will no longer be true; the best signal would now appear at the input to the base amplifier.

Variation of the oscillator frequency over the widest frequency range requires the simultaneous adjustment of three similar elements. In the network of Fig. 16-4*b*, any changes in R will also change the driving-point impedance. This may load the amplifier to a point where oscillations can no longer be sustained. If C is varied, then we face the problem of tracking three independent variable capacitors. Because of these difficulties, the phase-shift network is generally used for a fixed-frequency oscillator. Here only minor calibration adjustments are needed, and they may be made by means of a single small trimmer capacitor or padding resistor.

The Wien bridge

The simple *RC* network used as an example in Fig. 16-2 exhibits extremely poor selectivity. Any second-harmonic components introduced not only are not reduced but are actually increased in amplitude. In this network, $|H_2| = 2.24$ and $|K_2| = 2$.

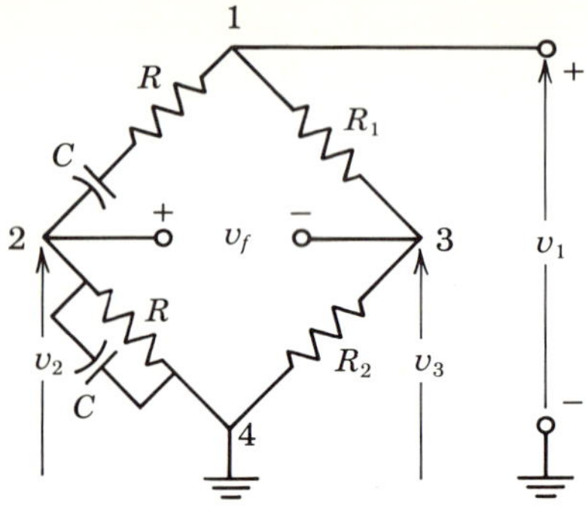

Fig. 16-6 A Wien-bridge frequency-selective network.

But all is not yet lost. We can convert the simple network into a null balance bridge (Fig. 16-6) and exchange necessary loop gain for selectivity. In the bridge

$$v_f = v_2 - v_3$$

but $v_2/v_1 = \beta$ of the simple network of Fig. 16-2 [Eq. (16-6)]. The resistor combination R_1 and R_2 is adjusted to satisfy the condition

$$\frac{v_3}{v_1} = \frac{R_2}{R_1 + R_2} = \frac{1}{3} - \frac{1}{\delta}$$

Thus the overall β' of the bridge of Fig. 16-6 may be expressed as

$$\beta' = \frac{v_f}{v_1} = \beta - \left(\frac{1}{3} - \frac{1}{\delta}\right) \tag{16-14}$$

where $$\beta = \frac{1}{3 + j(\omega/\omega_0 - \omega_0/\omega)}$$

The amplitude and phase responses of β' are plotted in Fig. 16-20. At the resonant frequency of the system ($\omega_0 = 1/CR$), β' reduces to

$$\beta' = \frac{1}{\delta}$$

and the threshold gain A_t is equal to δ. The term $1/\delta$ may be considered as the degree of bridge unbalance; when $\delta \to \infty$, the bridge is perfectly balanced at ω_0, and when $\delta = 3$, the circuit reduces to the simple network of Fig. 16-2. If follows that the greater the degree of balance, the larger the necessary loop gain to sustain oscillations. At perfect balance, an infinite gain is required; with any physical amplifier, the circuit cannot oscillate.

The improvement in selectivity may best be seen by substituting β' [given by Eq. (16-14)] and $A_t = \delta$ into the expression for H_n [Eq. (16-8)]. The result is

$$H'_n = \frac{3}{\delta(1 - 3\beta_n)} \tag{16-15}$$

where β_n is the transmission factor of the RC network at the harmonic frequency. Taking the ratio of H'_n for the bridge to H_n found for the completely unbalanced case ($\delta = 3$) shows the dependence of the selectivity on δ:

$$\frac{H'_n}{H_n} = \frac{3/[\delta(1 - 3\beta_n)]}{1/(1 - 3\beta_n)} = \frac{3}{\delta} \tag{16-16}$$

From Eq. (16-16) we conclude that the percentage of second harmonics present at the output is reduced in proportion to the increase in required amplifier gain. With a two-stage amplifier having a gain of 300, the value of $|H'_2|$ is only 0.0224, a hundredfold reduction from the unbalanced case.

At the output of the bridge, i.e., at v_f in Fig. 16-6, the improvement in selectivity is not as impressive. K'_n may be expressed by

$$K'_n = \frac{\beta'_n}{\beta'_1} H'_n = \frac{3 - \delta(1 - 3\beta_n)}{\delta(1 - 3\beta_n)} \tag{16-17}$$

As δ gets large, the magnitude of Eq. (16-17) approaches unity as a limit. For other than very large gains, $|K'_n|$ will always be slightly above unity.

To incorporate the Wien bridge into an oscillator, the base amplifier must have two separate input terminals: a noninverting input for the positive feedback through the frequency-determining network and an inverting input for the negative feedback which increases the circuit selectivity. An integrated differential amplifier of the type discussed in Chap. 10 is ideally suited for this purpose (Fig. 16-7). At ω_0, the positive

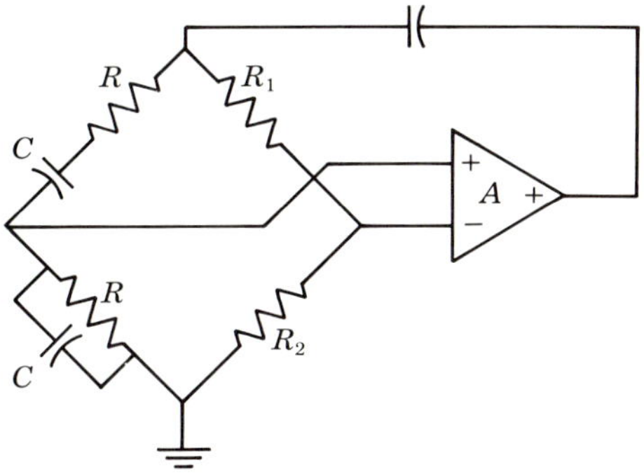

Fig. 16-7 Wien-bridge oscillator employing an integrated-circuit amplifier.

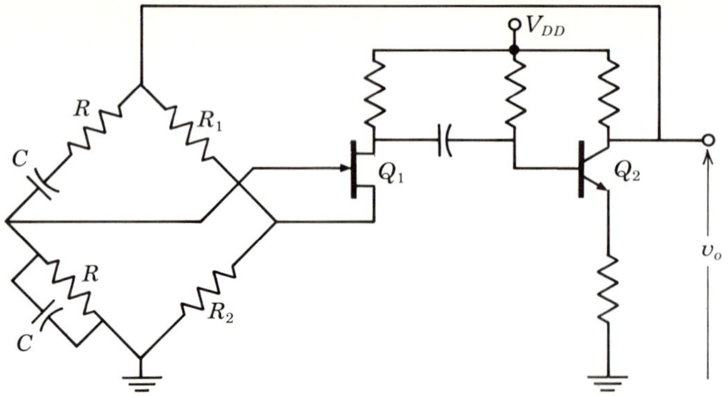

Fig. 16-8 A composite FET-bipolar Wien-bridge oscillator.

feedback predominates, and at the harmonics, the net negative feedback reduces the distortion components.

Figure 16-8 shows a lumped-element composite FET-bipolar transistor oscillator. In this circuit, the positive feedback is applied at the gate of the input FET while the negative feedback is injected into the source.

The Wien bridge is employed in many commercial wide-range oscillators operating from 10 Hz to 2 MHz or even higher. To adjust the frequency, usually both resistors are changed by steps and C is changed smoothly as a fine control. We can also do the reverse—switch C in steps and vary R continuously. Both arrangements perform satisfactorily.

Other null networks

Almost any three-terminal null network may be used as the frequency-determining pair of branches in a bridge. The other two arms are composed of resistors adjusted for the proper degree of unbalance.

All arguments employed with respect to the Wien bridge will also apply to these other networks, provided that care is taken as to the direction of the unbalance. For some networks, the greatest reduction in harmonic content comes when δ is positive, as in the Wien bridge, and for others, δ must be negative. Depending on the network and the sign of unbalance at ω_0, in some cases the resistive branches must be used for the positive feedback, and in others the reactive branches so serve.

Figure 16-9 shows two networks which give zero transmission at their null frequency. The feedback factors and further defining relationships are given in Table 16-1. A plot of the amplitude and phase response appears in Fig. 16-20. At the oscillation frequency, there cannot be any feedback through the reactive elements and the necessary positive feed-

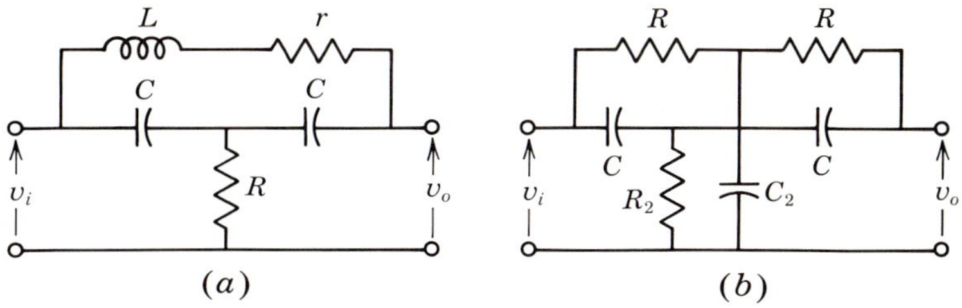

Fig. 16-9 Two null networks. (*a*) Bridged-T network; (*b*) twin-T network.

back must be supplied through the resistive path. The basic circuit configuration needed to satisfy this condition is shown in Fig. 16-10. Here the value of β' may be expressed as

$$\beta' = \frac{R_2}{R_1 + R_2} - \beta$$

and at the null point, the required threshold gain is

$$A_t = 1 + \frac{R_1}{R_2}$$

As in the Wien bridge, when A_t is large, the harmonic content of the output will be small.

To adjust the frequency of these two networks, we must vary three elements simultaneously. The difficulty that this entails makes these networks best suited for fixed-frequency operation. For the same degree of unbalance, the bridged-T circuit will usually give the cleanest waveshape.

Table 16-1

NETWORK	TRANSFER FUNCTION	DEFINING TERMS
Bridged-T	$\beta = \dfrac{1}{1 - j\dfrac{2}{Q_0}\dfrac{\omega\omega_0}{\omega^2 - \omega_0^2}}$	$\omega_0 = \dfrac{2}{LC}$ $Q_0 = \dfrac{\omega_0 L}{r}$ $R = \dfrac{L}{2rC}$
Twin-T	$\beta = \dfrac{1}{1 - j2\dfrac{k+1}{\sqrt{k}}\dfrac{\omega\omega_0}{\omega^2 - \omega_0^2}}$	$\omega_0^2 = \dfrac{1}{2RR_2C^2}$ $k = \dfrac{R}{2R_2} = 2\dfrac{C}{C_2}$

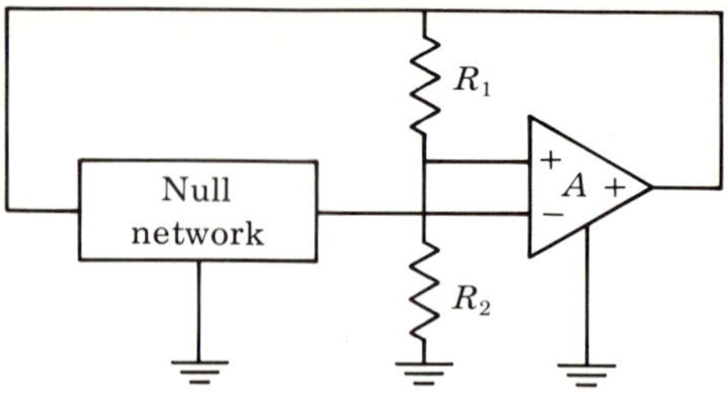

Fig. 16-10 Oscillator connection when using a null network for frequency determination. An operational amplifier is used as the active device —note the feedback-connection polarities.

However, since it is more difficult and expensive to obtain a high-quality inductance than it is to find extremely good resistors and capacitors, the twin-T network is easier to synthesize.

16-3 Current-controlled Oscillations

If the transistor is used in its natural mode as a current amplifier, with the appropriate form of feedback, then the oscillations produced may be said to be current-controlled. The basic circuit configuration is shown in Fig. 16-11. Except that i replaces v, it is identical with the voltage-controlled block diagram of Fig. 16-1. To avoid possible confusion between the voltage and current feedback factors and the forward-current-amplification factor of the transistor, we shall designate the current feedback factor by $\gamma(\omega)$. This term is defined by

$$i_f = \gamma(\omega) i_2$$

and as before, the complete frequency dependence of the system is assigned to $\gamma(\omega)$.

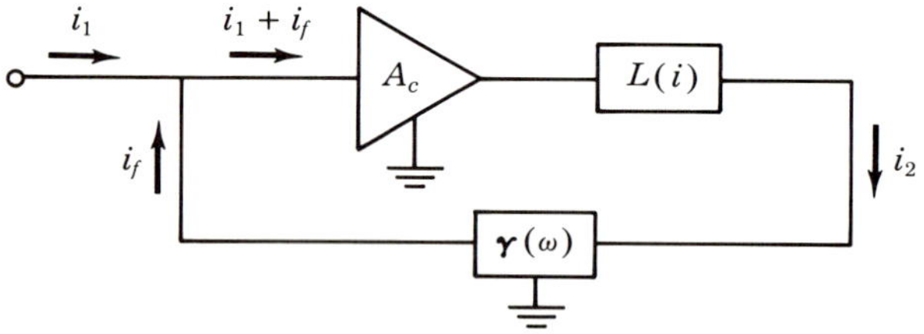

Fig. 16-11 Basic current-controlled oscillator: current amplifier A_c, limiter $L(i)$, and frequency-determining network $\gamma(\omega)$.

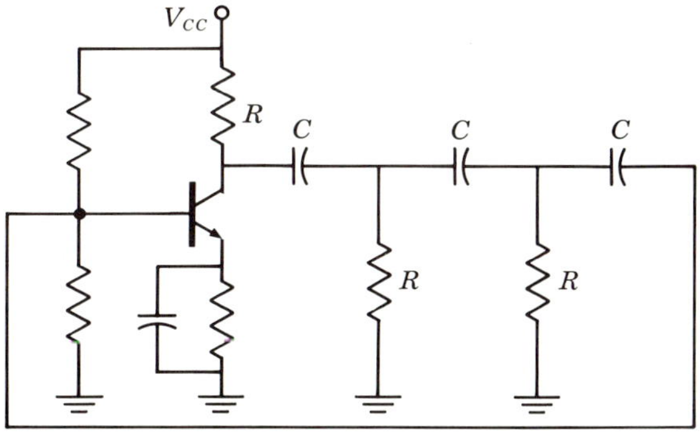

Fig. 16-12 Current-mode *RC* phase-shift oscillator.

The conditions for self-sustained oscillations ($i_1 = 0$) may be written by analogy with Eq. (16-2):

$$1 - \gamma(\omega)A_cL(i) = 0 \tag{16-18}$$

Equation (16-18) states that the net loop-current gain must be unity at the frequency and amplitude of oscillations.

Figure 16-12 illustrates a circuit that is the current-controlled equivalent of the simple phase-shift oscillator of Fig. 16-5. By assuming that the biasing resistors are large compared with h_{ie}, the full current transmitted by the feedback network will flow into the base of the transistor. For the ideal case where h_{ie} can be assumed to be zero,

$$\gamma(s) = \frac{1}{(Y_2/Y_1)^3 + 5(Y_2/Y_1)^2 + 6(Y_2/Y_1) + 1}$$

where $Y_1 = sC$ and $Y_2 = G$. [Compare this equation with Eq. (16-11).] The frequency of oscillation is $\omega_0 = 1/(\sqrt{6}\, RC)$ and the threshold current gain must be -29.

The only condition which we must impose on the various *RC* and *RL* networks, which may be used for current feedback, is that $\gamma(\omega)$ must yield a clearly defined oscillation frequency and good harmonic rejection. Also the oscillator's current amplifier should approach the ideal, i.e., a zero input impedance, a high output impedance, and a constant gain over the frequency range of interest.

16-4 Tuned-circuit Oscillators

At frequencies above 50 kHz, it becomes practical to use a high-Q tuned circuit for frequency selection. Because such a network is resistive only at its resonant frequency and is reactive everywhere else, the oscillation

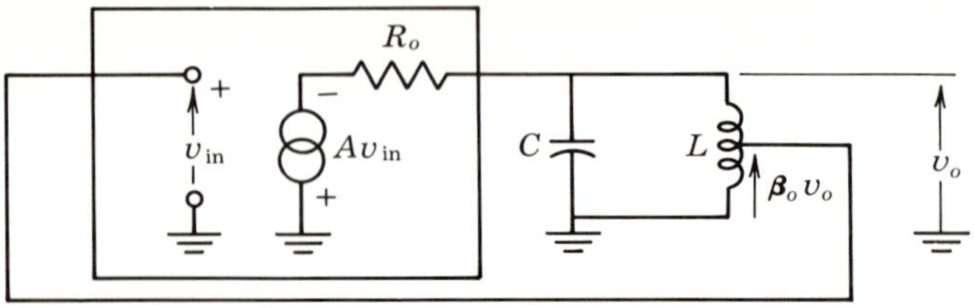

Fig. 16-13 An oscillator using a tuned circuit as a frequency-selective network.

condition of zero net phase shift can be satisfied only at this particular frequency. Figure 16-13 illustrates how a portion of the developed voltage may be fed back to the input of the amplifier by tapping the tuned circuit. Other methods of feedback are also in use, and some of these will be discussed below.

Even if the excitation current is rich in harmonics, the low impedance of the tank, everywhere but at its resonant frequency, will not permit the development of other than an almost pure sinusoidal voltage. The feedback will further improve the selectivity, and consequently we shall not be particularly concerned with the harmonic response. We are, however, quite interested in generating a signal whose frequency will not be affected by external factors.

All voltage-controlled tuned-circuit oscillators may be represented by the basic circuits of Fig. 16-14*a* and *b*, where, in order to achieve the proper phase relationship with a single-stage amplifier, the tap of the tuned circuit is returned to the emitter (or source). Any convenient node in the circuit may be grounded. For simplicity, the biasing and power supplies have been omitted. In any specific circuit, these connections would depend on the nature of the three impedance elements.

We might note that the network of Fig. 16-14 includes two modes of feedback, a direct transmission path from the drain (collector) to the gate (base) through Z_3 and the mutual coupling linking the input and output loops. Some oscillators use purely capacitive elements for Z_1 and Z_2, and in others two isolated coils are used. When we treat these cases, M would not appear. However, many of the more commonly used circuits do contain coupled coils, and therefore M must be included in any general treatment. Figure 16-14 also shows the various models with which we shall be concerned in the analysis of this type of oscillator. In the interests of simplicity, the small reverse-voltage-transmission term has been omitted from the transistor model. If this term ever becomes significant, it can always be included by inserting a voltage generator $h_{re}v_{ce}$ in series with h_{ie}.

Since the single FET or bipolar transistor introduces an effective phase shift of 180°, there must be an additional 180° contributed by the network,

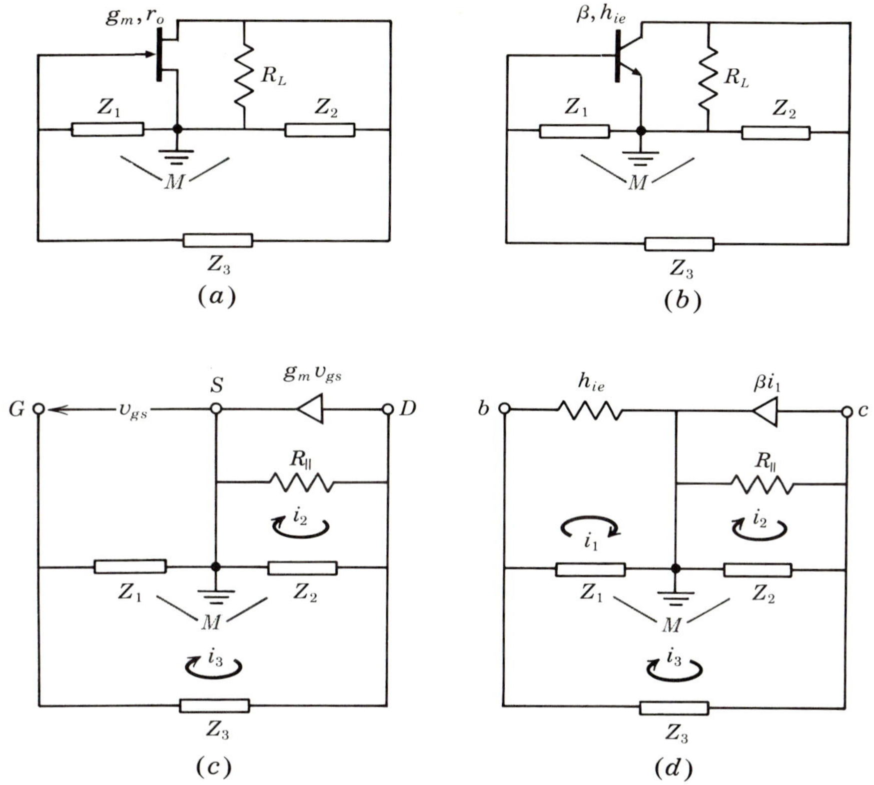

Fig. 16-14 Basic tuned-circuit oscillators. (*a*) FET used as an active element; (*b*) bipolar transistor used as an active element; (*c*) FET small-signal model; (*d*) bipolar-transistor small-signal model.

composed of Z_1, Z_2, and Z_3. These models may, of course, be treated as before by evaluating $A\beta$, where A is the gain when the output is loaded by the input impedance of the feedback network and β is the transmission from the drain (collector) to the gate (base).

By observing that the criteria for self-sustained oscillation are inherent in the vanishing of $1 - \beta(\omega)A$ or $1 - \gamma(\omega)A_c$ in Eqs. (16-2) and (16-18), we can consider an alternative method of obtaining these criteria. The nature of the overall response of any system may be found by solving a set of mesh or node equations. This response is undefined when the system determinant vanishes. Thus, by writing the loop equations (or, where appropriate, the node equations), finding the system determinant, and setting it equal to zero, we satisfy

$$1 - \beta(\omega)A = 0 \qquad \text{or} \qquad 1 - \gamma(\omega)A_c = 0$$

The vanishing of the imaginary part of the determinant gives the frequency of oscillation. From the real part, we obtain the circuit conditions which must be satisfied if the oscillator is to be self-starting.

FET tuned oscillators

Since the circuit of Fig. 16-14*a* is simpler than the circuit of Fig. 16-14*b*, we shall consider it first. After taking the Thévenin equivalent in the drain, where $R_\parallel$, the shunt resistance at the drain, is the parallel combination of r_o and the external load R_L, the two loop equations become

$$\begin{aligned} (R_\parallel + Z_2)i_2 - (Z_2 + Z_m)i_3 &= -g_m R_\parallel v_{gs} \\ -(Z_2 + Z_m)i_2 + (Z_1 + Z_2 + Z_3 + 2Z_m)i_3 &= 0 \end{aligned} \tag{16-19}$$

One additional defining equation is needed before the set of equations in (16-19) can be solved. The control voltage must be expressed in terms of the loop currents:

$$v_{gs} = -Z_m i_2 + (Z_m + Z_1)i_3 \tag{16-20}$$

Substituting Eq. (16-20) into Eq. (16-19) and collecting terms yield the system determinant

$$\Delta = \begin{vmatrix} R_\parallel + Z_2 - g_m R_\parallel Z_m & -(Z_2 + Z_m - g_m R_\parallel Z_m - g_m R_\parallel Z_1) \\ -(Z_2 + Z_m) & Z_1 + Z_2 + Z_3 + 2Z_m \end{vmatrix} \tag{16-21}$$

If we initially assume that the impedance elements are purely reactive ($Z_i = jX_i$), then the imaginary portion of the determinant will contain only odd powers of X_i. The sign of the reactance is assigned to the element (plus for inductance and minus for capacity). It follows that the frequency of oscillation will be given by

$$\text{Im}\,\Delta = R_\parallel(X_1 + X_2 + X_3 + 2X_m) = 0 \tag{16-22}$$

Equation (16-22) can be satisfied only when the sign associated with one reactance is opposite to that associated with the other two. We shall see below [Eq. (16-23)] that Z_1 and Z_2 must be the same type of element to satisfy the gain condition. Thus if Z_1 and Z_2 are inductive at the oscillation frequency, Z_3 must be capacitive, and vice versa.

Since the sum of the loop reactances vanishes in Eq. (16-22), the real part of the determinant of Eq. (16-21) may be expressed as

$$\text{Re}\,\Delta = g_m R_\parallel(X_1 + X_m)(X_2 + X_m) - (X_2 + X_m)^2$$

By setting this equal to zero, we find that the threshold condition for self-starting is

$$g_m R_\parallel \geq \frac{X_2 + X_m}{X_1 + X_m} = \frac{-1}{\beta_0} \tag{16-23}$$

We identify $g_m R_{\|}$ as the magnitude of the stage gain. Equation (16-23) can be satisfied only when Z_1 and Z_2 are the same type of component: either both are capacitive or both are inductive. If they are not the same, we would have negative rather than positive feedback.

Unless the ratio of impedances [Eq. (16-23)] is somewhat less than $g_m R_{\|}$, the overall loop gain will not be greater than unity and the oscillator will not be self-starting. Furthermore, with threshold operation, the system is very sensitive to changes in the parameters of the active element and on the bias point. On the other hand, if this ratio becomes very much less than $g_m R_{\|}$, the limiting may introduce an excessive amount of harmonics, with a corresponding deterioration of the waveform.

The following possibilities arise where the major mode of feedback is through X_3 and the mutual coupling either does not exist at all or is purely incidental.

1. Colpitts oscillator (Fig. 16-15*a*):
   X_1 and X_2 are capacitors.
   X_3 is an inductance.
   $X_m = 0$.
2. Hartley oscillator (Fig. 16-15*b*):
   $X_1 + X_2 + 2X_m$ is a single tapped coil.
   X_3 is a capacitor.
3. Tuned-drain tuned-gate oscillator (Fig. 16-15*c*):
   X_1 and X_2 are tuned circuits adjusted to be somewhat inductive (tuned slightly below ω_0).
   X_3 is the stray gate-to-drain capacity.
   $X_m = 0$.

When the major mode of coupling is through the mutual inductance and X_3 represents the insignificant parasitic coupling, the following two circuits are also possible:

4. Tuned-drain oscillator (Fig. 16-15*d*):
   X_2 is a tuned circuit in the drain.
   X_1 is the gate coil.
   $2X_m$ is the mutual coupling between the two coils.
5. Tuned-gate oscillator:
   Virtually identical with the tuned-drain oscillator except that the single tuned circuit is moved into the gate.

Various hybrid oscillators combining significant features of the five listed above are also in use. The above tabulation only indicates the range of possibilities; it makes no attempt to exhaust them.

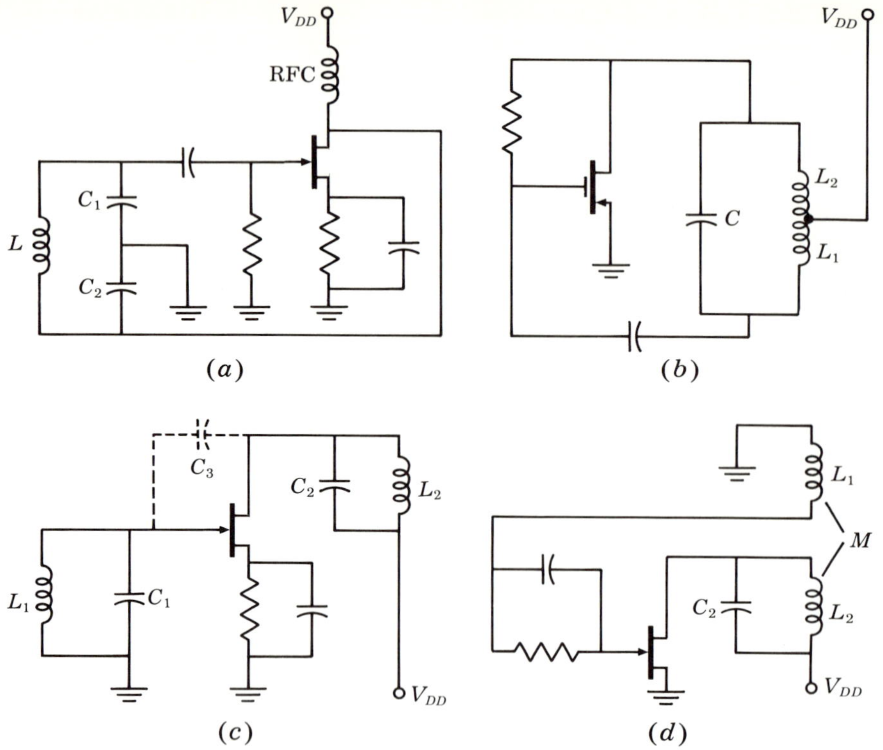

Fig. 16-15 Some practical FET tuned-circuit oscillators. (*a*) Colpitts oscillator; (*b*) Hartley oscillator; (*c*) tuned-drain tuned-gate oscillator; (*d*) tuned-drain oscillator. The various components not labeled are used for bias and decoupling and would be chosen to have a negligible influence on the frequency of oscillation.

In the Colpitts oscillator of Fig. 16-15*a*, the frequency of oscillation as found from Eq. (16-22) is

$$\omega_{0C} = \sqrt{\frac{1}{LC_T}}$$

where $C_T = C_1C_2/(C_1 + C_2)$. For self-starting, $g_mR_\| \geq C_1/C_2$ [from Eq. (16-23)]. The Hartley circuit (Fig. 16-16*b*) oscillates at

$$\omega_{0H} = \sqrt{\frac{1}{C_3(L_1 + L_2 + 2M)}}$$

and here
$$g_mR_\| \geq \frac{L_2 + M}{L_1 + M}$$

In both circuits the frequency of oscillation is simply the resonant frequency of the *LC* combination. The capacitive term in the Colpitts equation is the series combination of C_1 and C_2. If the FET has a reasonably large gain, $C_2 \ll C_1$ and $C_T \approx C_2$. In the Hartley circuit, the inductive term is the total tank inductance and, with a high-gain amplifier, $L_2 \gg L_1$.

The Hartley and the single tuned-circuit oscillators (circuits 2, 4, and 5 above) are best suited for variable-frequency operation. In these, the frequency may be changed by varying a single capacitor. Moreover, this adjustment will not affect the conditions for self-starting operation, which depends on the mutual coupling between two coils and on the location of the coil tap. In the other oscillators, either two LC circuits must be retuned or two capacitors must be simultaneously tracked.

If the tank elements are dissipative, or if they are loaded by the external circuit when coupling to the next stage, then the associated resistance will change both the frequency of oscillation and the conditions for self-starting. Provisions for retuning and for adjusting the amount of feedback allow us to compensate for these changes. There is no point in calculating the new condition for threshold operation. It is, however, of some interest to recalculate the frequency of oscillation. By doing this we shall see that the circuit dissipation also makes the frequency dependent on the parameters of the active element.

Example 16-1

Let us consider, for example, the basic Colpitts oscillator of Fig. 16-15*a*, where the nominal unloaded frequency is $\omega_0 = 10^7$ rad/s. The FET used is biased at a point where $g_m = 5$ mA/V and $r_o = 25$ K. The tuned circuit is so designed that $Q\sqrt{L/C_T} = 25$ K and, in addition, its loaded $Q = 10$.

From the above conditions, the tank components are found to be $C_T = 100$ pF and $L = 100$ μH. To ensure self-starting and to allow for changes in the FET parameters, we shall satisfy

$$g_m R_{\|} \approx 5 \frac{C_1}{C_2}$$

instead of the threshold condition. Solving yields $C_2 = 108$ pF and $C_1 = 1{,}350$ pF. The final parameter—the equivalent series coil resistance—is

$$r = \frac{10^7 L}{Q} = \frac{10^7 \times 10^{-4}}{10} = 100\ \Omega$$

By substituting $Z_3 = r + j\omega L$, $Z_1 = -j/\omega C_1$, and $Z_2 = -j/\omega C_2$ into Eq. (16-21) and collecting the imaginary terms, the frequency of oscillation is found from

$$R_{\|}\left(\omega L - \frac{C_1 + C_2}{\omega C_1 C_2}\right) - \frac{r}{\omega C_2} = 0$$

The frequency becomes

$$\omega_0' = \sqrt{\frac{C_1 + C_2}{LC_1C_2}\left(1 + \frac{C_1}{C_1 + C_2}\frac{r}{R_{\|}}\right)}$$

$$= \sqrt{10^{14}\left(1 + \frac{1{,}350}{1{,}458}\frac{100}{12.5\text{ K}}\right)} = 10^7\sqrt{1.0074}$$

or the frequency increases by approximately 0.36 percent over its nominal value as a direct result of the resistive loading of the tank. The conditions for self-starting will also change; this calculation is left as an exercise for the reader.

Unless $R_{\|} \gg r$, the circuit will be quite sensitive to the external loading; any changes will change the frequency of oscillation. External-loading effects may be minimized by using an amplifier to decouple the final load from the oscillator stage and by loosely coupling this stage to the tank.

Bipolar-transistor tuned oscillators

From the defining equations for the transistor model of Fig. 16-14*d*, the overall system determinant is found to be

$$\Delta = \begin{vmatrix} h_{ie} + Z_1 & Z_m & -(Z_1 + Z_m) \\ \beta R_{\|} + Z_m & R_{\|} + Z_2 & -(Z_2 + Z_m) \\ -(Z_1 + Z_m) & -(Z_2 + Z_m) & Z_1 + Z_2 + Z_3 + 2Z_m \end{vmatrix} \tag{16-24}$$

When the impedances are all purely reactive, the imaginary part of the determinant will reduce to

$$R_{\|}h_{ie}(X_1 + X_2 + X_3 + 2X_m) - X_3(X_1X_2 - X_m^2) = 0 \tag{16-25}$$

where each reactance term includes the appropriate sign. We might note that the first part of Eq. (16-25) contains the total series impedance of the self-resonant circuit composed of the three reactive elements. The last portion is a correction term due to the nature of the external loading of the tank. In general, we shall have to minimize the influence of these external factors if the oscillator is to have good frequency stability.

The significant portion of the frequency-determining equation is identical with that found for the FET circuit. As discussed above, the sign of Z_3 must be opposite to Z_1 and Z_2. Consequently, the possible circuit variations are those previously tabulated.

By assuming that the actual frequency of oscillation is very close to the natural frequency of the tuned circuit, the term

$$X_1 + X_2 + X_3 + 2X_m \approx 0$$

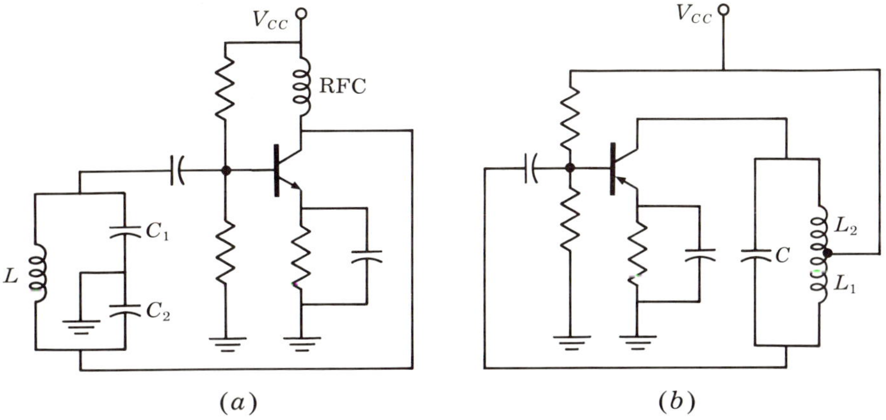

Fig. 16-16 (a) Transistor Colpitts oscillator; (b) Hartley oscillator.

and the determinant may be simplified before solving for the condition for self-starting. It will be found by setting the real part of the remaining terms equal to zero:

$$h_{ie}(X_2 + X_m)^2 + R_{\|}(X_1 + X_m)^2 - \beta R_{\|}(X_1 + X_m)(X_2 + X_m) = 0 \quad (16\text{-}26)$$

Because the second term on the left-hand side is relatively insignificant, Eq. (16-26) may be further simplified. The resulting criterion for self-starting oscillations becomes

$$\frac{X_2 + X_m}{X_1 + X_m} \leq \frac{\beta R_{\|}}{h_{ie}} = g_m R_{\|} \quad (16\text{-}27)$$

Equation (16-27) is identical with the conditions previously found for the FET circuit. To justify the use of this equation, we shall solve Eq. (16-27) for $X_1 + X_m$ and substitute back into the original expression of Eq. (16-26):

$$h_{ie}(X_2 + X_m)^2 + \frac{h_{ie}}{\beta^2 R_{\|}}(X_2 + X_m)^2 - h_{ie}(X_2 + X_m)^2 = 0$$

The neglected second term is only $h_{ie}/\beta^2 R_{\|}$ times as large as the first term. Since h_{ie} is the small input impedance, $\beta^2 R_{\|} \gg h_{ie}$ and the self-starting condition expressed in Eq. (16-27) is essentially correct.

From Eqs. (16-25) and (16-27) the frequency and gain requirements of the Colpitts oscillator of Fig. 16-16a are

$$\omega_{0C} = \sqrt{\frac{C_1 + C_2}{LC_1C_2} + \frac{1}{C_1C_2R_{\|}h_{ie}}}$$

where $\sqrt{(C_1 + C_2)/LC_1C_2}$ is the resonant frequency of the unloaded tank. For self-starting, the Colpitts oscillator must be adjusted so that

$$g_m R_\| \geq \frac{C_1}{C_2}$$

The Hartley circuit of Fig. 16-16*b* oscillates at

$$\omega_{0H} = \frac{1}{\sqrt{C(L_1 + L_2 + 2M) - \dfrac{L_1L_2 - M^2}{R_\| h_{ie}}}}$$

and for self-starting,

$$g_m R_\| \geq \frac{L_2 + M}{L_1 + M}$$

In both cases the resistive loading of the tuned circuit causes a small change in frequency from the nominal resonant point. This can be minimized by maximizing the $h_{ie}R_\|$ product and by properly choosing the ratio L to C. Furthermore, by using tightly coupled coils in the Hartley circuit, the numerator of the correction term will be reduced; with unity coupling, it will reach zero and the circuit will oscillate exactly at ω_0.

Crystal oscillators

A suitably clamped quartz crystal may be made to resonate by exciting it with an electrical signal of the proper frequency. The mechanical oscillations are fixed by the relative dimensions of the crystal, and the damping depends on the characteristics of the mounting. As the damping factor can be made quite small, the electromechanical Q will be proportionately large; expected values are between 1,000 and 50,000. To use the crystal in a fixed-frequency oscillator, we simply substitute it for one of the reactances in the basic tuned-oscillator configurations of Fig. 16-14.

Sinusoidal oscillation is possible at one of the two modes of electromechanical resonance corresponding to the electrical resonances of the equivalent circuit of Fig. 16-17. The parallel-resonant mode depends on the mass factor L_i and the capacity of the mounting plates using the crystal as a dielectric, C. At this frequency, the terminal impedance is purely resistive and is extremely large. In its other mode, the mass and spring constant of the crystal, L_i and C_i, are series-resonant. Here the crystal is essentially a short circuit. Since the dissipation term r_i is small, the network appears reactive everywhere but at these two frequencies. Many crystals are cut to have several resonant points, and these might be incorporated into the electrical model by adding, in parallel, additional $L_ir_iC_i$ series arms.

Even though the frequency of oscillation is essentially that of the crystal, minor adjustments are possible, without reducing the Q. By adding series

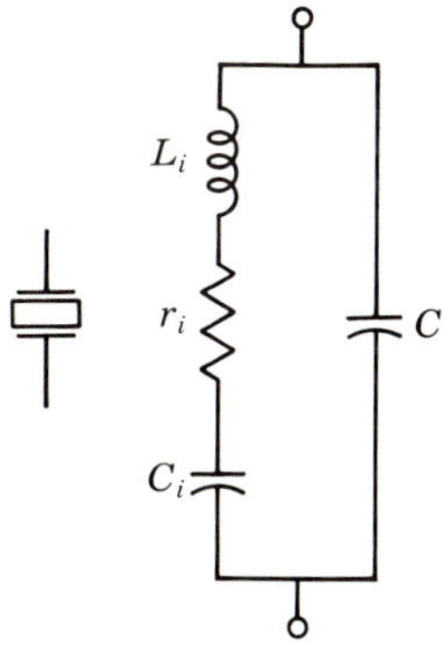

Fig. 16-17 Equivalent electrical circuit of a quartz crystal.

capacity or inductance, the frequency can be changed by a few parts in a million. For wider ranges, the variable-frequency circuits would be used instead.

Two crystal-controlled oscillators are shown in Fig. 16-18. In the Pierce circuit, the crystal controls the amount of feedback applied. At its series-resonant frequency, the low impedance presented increases the amount of feedback to a point where oscillations can be sustained. Since the gate input is capacitive, the crystal must be slightly inductive for the proper phase of the feedback signal. To satisfy the conditions found above, the plate tank would also be tuned somewhat capacitively. To all other frequencies the crystal presents an extremely high reactance. It thus reduces the loop gain well below the threshold value. Furthermore, the net phase shift of the feedback voltage will no longer be the 180° needed for oscillation.

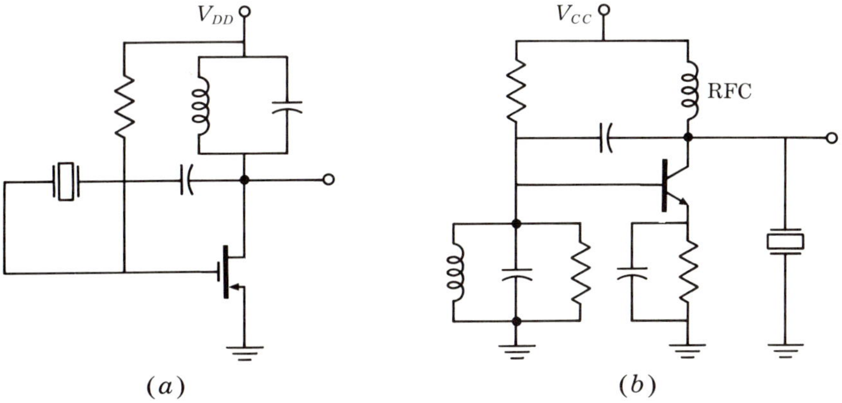

Fig. 16-18 (*a*) A Pierce oscillator using a quartz crystal in a series-resonant mode; (*b*) a transistor oscillator where the crystal operates as a parallel-resonant circuit.

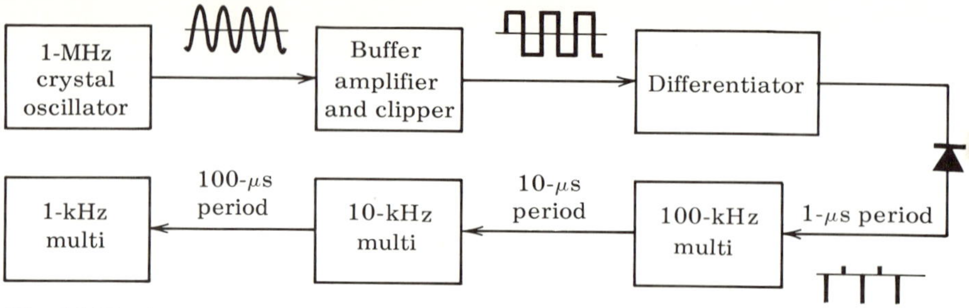

Fig. 16-19 A crystal-controlled oscillator system designed for precise timing measurements.

The second circuit of Fig. 16-18, the transistor oscillator, is a tuned-input tuned-output circuit with capacitive feedback. Here the output will be developed when the crystal is a high impedance, i.e., at its parallel-resonant frequency. At all other frequencies the small reactance presented will load the amplifier and will also cause the feedback voltage to be in quadrature to that needed for oscillation.

Crystal-controlled oscillators are widely used as frequency standards. By amplifying, clipping, and finally differentiating a 1-MHz sinusoid, we derive a train of negative pulses spaced 1 μs apart. As shown in Fig. 16-19, these are next applied as a trigger source to synchronize an astable multivibrator normally operating at 100 kHz. The 10-μs-period pulse train may be further divided down by the other multivibrators shown. From this chain we drive pulse outputs spaced by 10 μs, 100 μs, 1 ms, etc. Since the oscillator frequency can be maintained to better than 1 part in 10^9, this system has been used for highly precise timing.

16-5 Frequency Stabilization

All oscillators are to some degree susceptible to frequency changes caused by the resistive loading of the selective network. If the loading were to remain constant, then the initial calibration of the instrument would correct for the deviation from the nominal tank resonance frequency. Unfortunately some of the dissipative elements may be expected to change with time, sometimes in one direction and sometimes erratically. All the circuit components contribute to this drift; even slight changes in the power-supply voltage will shift the operating point and thus affect the parameters of the active element. Therefore any oscillator which is designed as a precise frequency standard should be compensated with respect to all such variations.

We implicitly assume that the nature of the elements comprising the selective network is such that they themselves are completely stable with respect to aging and ambient-temperature variations.

Since it would be identical with detuning the tank, we are also neglecting the effects of any reactive loading of the tuned circuit. Some reactance would always appear across a portion, or across the complete tuned circuit, because of the junction and stray capacity associated with the FET and the bipolar transistor. Choosing a large tuning capacity would, of course, make the frequency much less dependent on any changes in the parasitic-circuit elements.

Our first objective, then, is to eliminate, as far as possible, any other factors producing a frequency shift away from ω_0. Even though special provisions may be made in each individual circuit, such as using tightly coupled coils in the Hartley oscillator, in Sec. 16-4 we saw that the performance of all oscillators will be materially improved by minimizing the external loading of the network. First, a buffer amplifier may be used to decouple the oscillator from the varying output load, and second, since the harmonics produced in the limiter are reflected as additional loading, the circuit can be adjusted close to the threshold of oscillation. Finally, we should investigate the mechanism of frequency instability with a view toward selecting the circuit which best suits our particular needs.

Frequency stabilization factor

The frequency of oscillation is identically that frequency at which the net phase shift around the amplifier and feedback loop is zero. Since the feedback network contains reactive elements, the insertion of any resistance will change the phase response and also the null frequency. For example, assume that a change in one parameter, external to the tuned circuit, produces a net phase shift of 5° at the nominal frequency of oscillation f_0. For sustained oscillations, the remainder of the circuit must contribute the complementary shift of $-5°$. If the normal rate of phase variation in the vicinity of f_0 is $-10°$ per 1,000 Hz, then a frequency increase of only 500 Hz will reestablish the condition of zero net phase shift. The new frequency of oscillation would be $f_0 + 500$. From this simple example, we conclude that the highest degree of stability would correspond to the most rapid change of phase with frequency.

If, in an oscillator, we can find one set of elements whose phase varies most rapidly with frequency, then these elements would have the most pronounced effect on the overall frequency stability. This will, in general, be the frequency-selective network. As a measure of frequency stability, we define the normalized sensitivity factor

$$S_f \equiv \frac{\Delta\phi}{\Delta\omega/\omega_0} \tag{16-28}$$

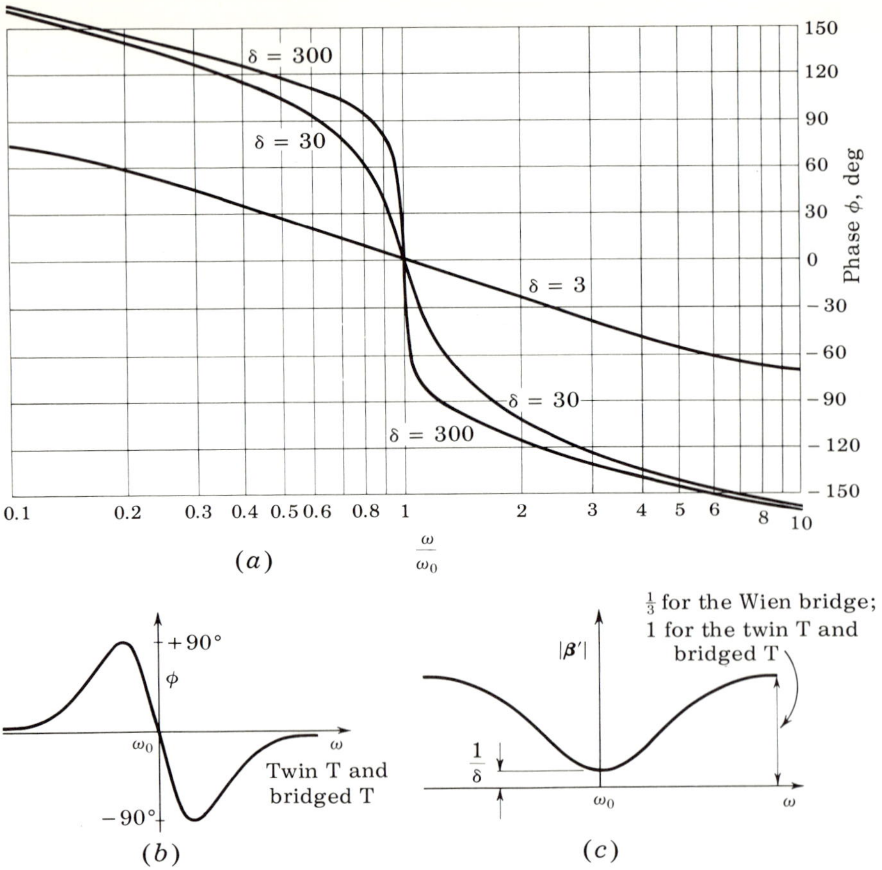

Fig. 16-20 Amplitude and phase characteristics of null networks. (*a*) Phase response of the Wien bridge; (*b*) phase response of an unbalanced bridge using a twin-T or a bridged-T network for one pair of arms; (*c*) the general amplitude response of null networks.

The larger the value of S_f, the more stable the oscillator with respect to all changes in the circuit which may produce an undesirable frequency shift. When, in the limit, S_f becomes infinite, the frequency of oscillation is completely independent of all other sections of the system.

Figure 16-20*a* is a plot of the phase behavior of a Wien bridge. In the vicinity of its null frequency, the response is similar to that seen in all other unbalanced null networks (Fig. 16-20*b*). For this particular example, the phase of β', as found from Eq. (16-14), is

$$\phi = \tan^{-1}\frac{3-\delta}{9}\left(\frac{\omega}{\omega_0}-\frac{\omega_0}{\omega}\right) - \tan^{-1}\frac{1}{3}\left(\frac{\omega}{\omega_0}-\frac{\omega_0}{\omega}\right) \qquad (16\text{-}29)$$

This is plotted as a function of ω/ω_0 for three different degrees of unbalance. One of the three is the completely unbalanced bridge ($\delta = 3$). It is readily apparent that S_f reaches a maximum at ω_0 and that the closer the bridge is to balance, the larger its value. Differentiating Eq. (16-29) with respect to ω and multiplying by ω_0 yield

$$S_f = \omega_0 \frac{d\phi}{d\omega} = \frac{\frac{3-\delta}{9}\left[1 + \left(\frac{\omega_0}{\omega}\right)^2\right]}{1 + \left(\frac{3-\delta}{9}\right)^2\left(\frac{\omega}{\omega_0} - \frac{\omega_0}{\omega}\right)^2} - \frac{\frac{1}{3}\left[1 + \left(\frac{\omega_0}{\omega}\right)^2\right]}{1 + \frac{1}{9}\left(\frac{\omega}{\omega_0} - \frac{\omega_0}{\omega}\right)^2}$$

Since $\delta \geq 3$, both terms will have the same sign and each will reach its maximum at $\omega = \omega_0$. Evaluating S_f at this frequency,

$$S_{f_0} = -\tfrac{2}{9}\delta \tag{16-30}$$

The minus sign indicates the direction of the phase change with frequency. Equation (16-30) shows that the bridge improves the circuit stability in proportion to its degree of balance. We might note that previously we had shown that this same condition was necessary for the optimum harmonic rejection [Eq. (16-16)].

Unless a sharp null is exhibited by the network, the phase changes relatively slowly in the vicinity of ω_0. The curve for $\delta = 3$ in Fig. 16-20*a* illustrates a variation of this nature. Since this curve represents the behavior of a completely unbalanced bridge, it also typifies the response of such *RC* and *RL* circuits as the phase-shift oscillator.

The phase response of a tuned-circuit oscillator is related to the circulating current through the three reactive components Z_1, Z_2, and Z_3. Near ω_0 the relative phase of the feedback voltage, which is developed at the active-elements input, may be approximated as

$$\phi = \tan^{-1} Q_0\left(\frac{\omega}{\omega_0} - \frac{\omega_0}{\omega}\right) \tag{16-31}$$

If this were to be plotted, we would have a family of curves similar to those drawn for the Wien bridge. After all, a tuned circuit is also a null network. The phase shift starts at 90°, reaches the 180° necessary for the positive feedback at ω_0, and then continues increasing toward 270° at the higher frequencies. The frequency-sensitivity factor of this phase response, evaluated at ω_0, is

$$S_{f_0} = 2Q$$

For optimum stability we should use a high-Q circuit and should set the oscillation frequency as close to ω_0 as possible. An extremely good choice for one of the elements is a crystal which will give a frequency-

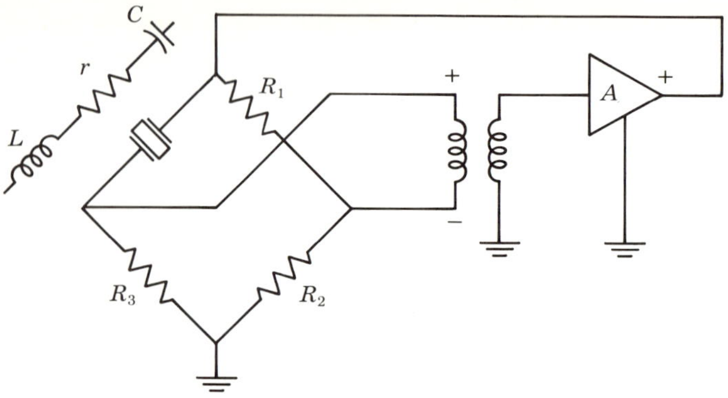

Fig. 16-21 Meacham-bridge oscillator.

stability factor of 10^3 to 10^5. With a normal tuned circuit, the best that can be expected is about 10^2.

For an extremely precise frequency standard, we can combine the improvement found in a high-Q crystal-controlled oscillator with the selectivity multiplication of an almost balanced bridge. The Meacham oscillator of Fig. 16-21 has the highest stability of any circuit yet devised. The bridge nulls at exactly the series-resonant frequency of the crystal. At this frequency the crystal appears completely resistive, which makes the bridge easy to balance. By using a small degree of unbalance $(1/\delta)$ and approximately equal resistance arms, the stability factor will be

$$S_{f_0} = 2\delta Q$$

and for $Q_0 = 10^5$, a stabilization factor of 1 part in 10^8 is possible. This means that if the amplifier introduces a phase shift of 0.1 rad (a relatively large value), the frequency would change by only 1 part in 10^9.

Reactive stabilization

Llewellyn showed that it is possible to stabilize an oscillator by inserting reactances in series with the terminals of the active element or in series with the applied load. The theory behind such stabilization is that the tank circuit appears resistive at its own resonant frequency and that only the need to compensate for the additional loop phase angles causes the frequency to shift. Now if the external resistance sees a completely resistive internal impedance, it would not affect the phase of the system; it would only increase the loading and the gain required for threshold operation. At many points in the circuit the two-terminal impedance has a reactive part. When this is loaded, the complete phase response of the system changes. By inserting in series with the external load a reactance of the type opposite to that originally seen, it is possible to tune the subsidiary loop and thus reduce the net phase shift to zero. The load now sees a pure resistance, and the frequency becomes independent of the load and of load variations.

The nature of this stabilizing reactance depends on the terminal at which it is introduced and on the configuration of the overall network. It is also possible, by inserting a single reactance at an appropriate point, to compensate for the loading at several different parts of the network. But, in general, compensation elements would have to be inserted in series with each of the resistive components of the external circuit.

As an example, we shall insert such a reactance in series with the base in the Hartley transistor oscillator described in Sec. 16-4. For simplicity, the mutual coupling will be taken as zero. Furthermore, since the stabilization will return the frequency to ω_0, the series impedance of the tank will also be zero. The simplified system determinant will now be [from Eq. (16-24)]

$$\Delta = \begin{vmatrix} h_{ie} + Z_s + Z_1 & 0 & -Z_1 \\ \beta R_{\|} & Z_2 + R_{\|} & -Z_2 \\ -Z_1 & -Z_1 & 0 \end{vmatrix}$$

where Z_s represents the stabilizing reactance. Setting the imaginary part equal to zero (when all impedances are purely reactive) yields

$$\text{Im}\,\Delta = X_s X_2^2 + X_1 X_2^2 + X_1^2 X_2 \doteq 0$$

and this equation enables us to find the value of X_s:

$$X_s = -X_1\left(1 + \frac{X_1}{X_2}\right)$$

But if the active element employed in the oscillator has a high gain, then, from Eq. (16-27), we see that $1 \gg X_1/X_2$ and the stabilizing reactance

$$X_s \approx -X_1 \qquad (16\text{-}32)$$

The element to be inserted in series with the base is a capacitor whose reactance at ω_0 is given in Eq. (16-32).

In order to obtain an order of magnitude for C_s, consider a Hartley oscillator where $C = 250$ pF, $L_1 = 10\ \mu$H, $L_2 = 990\ \mu$H, and $\omega_0 = 2 \times 10^6$. Substituting into the expression for X_s yields

$$C_s = \frac{1}{\omega_0^2 L_1} = \frac{1}{4 \times 10^{+12} \times 10 \times 10^{-6}} = 0.025\ \mu\text{F}$$

This value is quite reasonable. C_s would appear in the circuit as an input coupling capacitor. Note that C_s depends both on the frequency and on the value of L_1. For variable-frequency oscillators, stabilization of this nature is obviously impractical.

16-6 Amplitude of Oscillation—Part I, Active-element Limiting

The exponential buildup of the self-starting oscillator will continue so long as the system's poles remain in the right half plane. Eventually the amplitude of the output sinusoid becomes bound as the peaks force the instantaneous operating point into the limiting regions of the active device, where the loop gain drops to below the threshold value. The oscillatory circuit will stabilize with an *average* gain over the cycle exactly equal to the threshold value and with an *average* location of the pair of poles on the imaginary axis.

As a crude zeroth-order approximation, we can say that the amplitude of an oscillator adjusted close to the threshold will be equal to the linear capabilities of the system. For a more exact solution we must examine, in detail, the nature of the device nonlinearities. Fortunately, the exact signal level is much less significant than the exact frequency, and therefore we can even justify relatively gross approximations in treating this problem.

Our technique of solution makes use of a describing function. In any good oscillator, the selectivity of the frequency-determining network is such that there exists an almost pure sinusoid at one point, at least, in the circuit. It follows that we can break the loop at this point, inject a sinusoidal driving signal, plot the distorted response, calculate the magnitude of the fundamental component, and reject the harmonics. The plot of the amplitude of the open-loop sinusoidal response as a function of the sinusoidal driving signal is called the system's describing function. Superimposing the threshold gain line and finding the point of intersection finally yield the amplitude of oscillation satisfying the closed-loop response.

Linear-amplifier oscillators

For our first example, we shall consider the class of almost linear oscillators, where the amplitude is bounded by abrupt circuit nonlinearities. As the signal amplitude increases, the active device will switch from active operation to the zero-gain regions of cutoff and saturation. To illustrate the limiting present in linear amplifiers, we shall linearize the square-law *vi* transfer characteristic of the fixed-bias junction-FET Colpitts oscillator shown in open-loop form in Fig. 16-22*a*. (This gross piecewise-linear representation may, in fact, be realized through the use of a FET load for the amplifier device.) To avoid changing the overall response, the tuned network must be loaded by the impedance it normally sees looking into the drive terminal of the active device and the sinusoidal signal must be injected through a source impedance of the appropriate magnitude.

For the pseudolinear case treated, the drain current i_D follows the input voltage up to the point of clipping, as illustrated in Fig. 16-22*d*. One clipping boundary corresponds to cutoff, and the other one to the limiting of the gate voltage by the gate-substrate diode. It follows that the current waveshape is a truncated sinusoid having its zero crossings at the fixed bias point (I_Q, V_{GG}).

In our calculations, the full sinusoid is the term produced by the gate driving signal as if there were no clipping:

$$I_m \cos \omega_0 t = g_{m0} V_{G1m} \cos \omega_0 t$$

g_{m0} is the slope of the linearized transfer function of Fig. 16-22c. The magnitudes of the two limiting half angles shown in Fig. 16-22d are given by

$$\cos \theta_1 = \left| \frac{I_{DSS} - I_Q}{I_m} \right| = \left| \frac{V_{GG}}{V_{G1m}} \right| \qquad (16\text{-}33a)$$

$$\cos \theta_2 = \left| \frac{I_Q}{I_m} \right| = \left| \frac{V_{GG} - V_p}{V_{G1m}} \right| \qquad (16\text{-}33b)$$

The amplitudes of the dc, fundamental, and higher-harmonic components are found from the drain-current waveshape of Fig. 16-22d. Since the

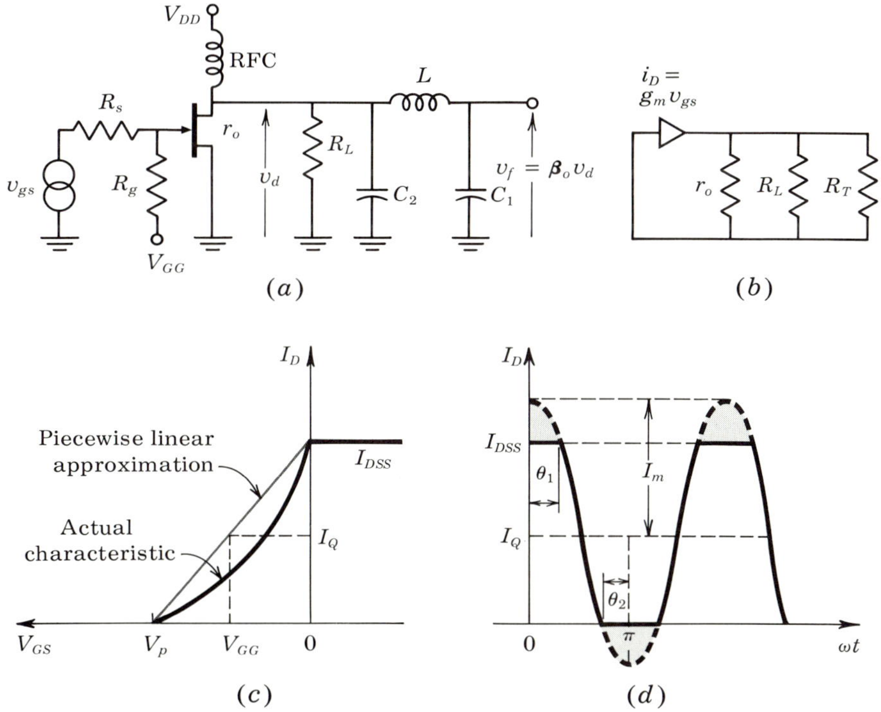

Fig. 16-22 (a) Open-loop FET Colpitts oscillator, fixed bias; (b) equivalent small-signal model at resonance; (c) linearized transfer function for approximate calculation of amplitude; (d) clipped-current waveshape, half clipping angles θ_1 and θ_2.

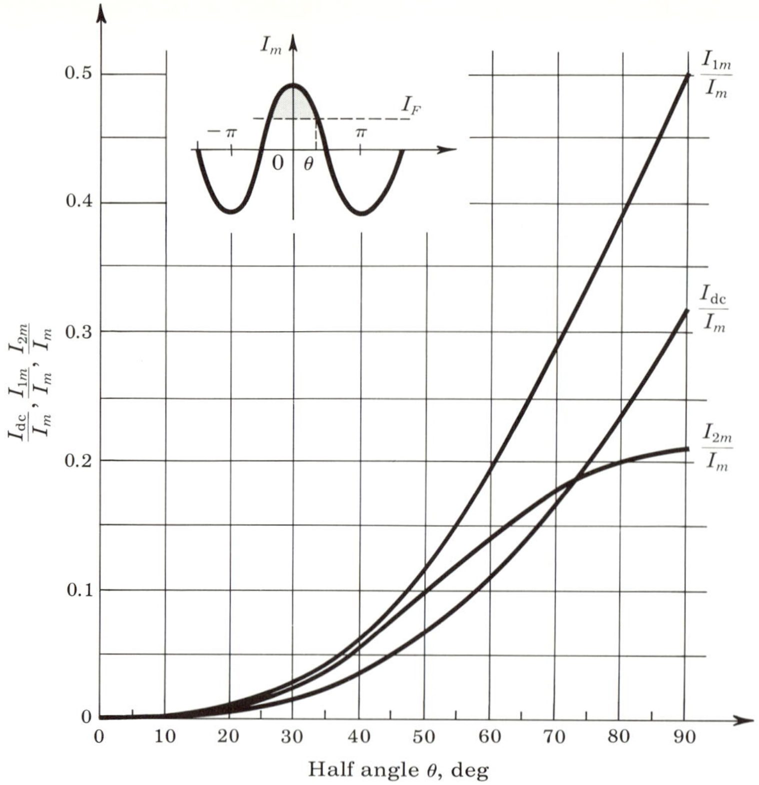

Fig. 16-23 Harmonic content of a sine-wave tip as a function of the conduction half angle θ.

truncated sinusoid can be resolved into the subtraction of two sinusoid tips from a full sinusoid centered at I_Q, instead of solving one specific waveshape, centered at one specific bias point, we shall solve the more general problem of finding the harmonic components of a sinusoidal tip. Subtracting the contributions of the two tips from the full sinusoid will then yield the Fourier-series coefficients of the actual waveshape.

A sinusoid tip, starting at I_m and ending at the truncation value I_F (Fig. 16-23), is defined by

$$I_{st} = I_m \cos \omega_0 t - I_F \Big|_{-\theta}^{\theta}$$

$$= I_m(\cos \omega_0 t - \cos \theta) \Big|_{-\theta}^{\theta} \tag{16-34}$$

where the bounds are

$$\cos \theta = \frac{I_F}{I_m}$$

Making use of the symmetry present, we obtain the normalized Fourier coefficients of the sine-wave tip:

$$\left.\frac{I_{dc}}{I_m}\right|_{st} = \frac{1}{\pi}\int_0^\theta (\cos\omega_0 t - \cos\theta)\,d\omega_0 t$$

$$= \frac{1}{\pi}(\sin\theta - \theta\cos\theta) \tag{16-35a}$$

$$\left.\frac{I_{1m}}{I_m}\right|_{st} = \frac{2}{\pi}\int_0^\theta (\cos\omega_0 t - \cos\theta)\cos\omega_0 t\,d\omega_0 t$$

$$= \frac{2}{\pi}\left(\frac{\theta}{2} + \frac{1}{4}\sin\theta - \cos\theta\sin\theta\right) \tag{16-35b}$$

$$\left.\frac{I_{2m}}{I_m}\right|_{st} = \frac{2}{\pi}\int_0^\theta (\cos\omega_0 t - \cos\theta)\cos 2\omega_0 t\,d\omega_0 t$$

$$= \frac{2}{\pi}\left(\frac{1}{2}\sin\theta + \frac{1}{6}\sin 3\theta - \frac{1}{2}\cos\theta\sin 2\theta\right) \tag{16-35c}$$

These functions are plotted in Fig. 16-23.

When both sinusoidal tips are truncated by circuit limiting, the contributions of each must be included in the total Fourier coefficients. It can readily be seen that the even-order terms (dc, second-harmonic, etc.) from the two opposite-polarity tips subtract and that the odd-order contributions from each tip sum. Thus at the fundamental frequency the normalized drain current is given by

$$\frac{I_{D1m}}{I_m} = 1 - \left.\frac{I_{1m}}{I_m}\right|_{\text{pos tip}} - \left.\frac{I_{1m}}{I_m}\right|_{\text{neg tip}} \tag{16-36}$$

Once we define θ_1 and θ_2 from Eqs. (16-33*a*) and (16-33*b*), the contribution from each tip is found from the plot of Fig. 16-23. The combined function of Eq. (16-36) is plotted in Fig. 16-24 for two different values of gate bias. Curve 1 is for a fixed bias of $V_{GG} = -1$ V, compared with $V_p = -3$ V. The limiting due to the active-device constraints is clearly evident. For curve 2, the FET is biased beyond pinchoff at $V_p = -3.5$ V.

The ordinate of the describing functions of Fig. 16-24 can be identified as

$$\frac{I_{D1m}}{I_m} = \frac{I_{D1m}}{g_{m0}V_{G1m}} = \frac{g_{mA}}{g_{m0}} \tag{16-37}$$

Equation (16-37) is the normalized transconductance of the active device. At any given amplitude, the average transconductance over a cycle of the fundamental is

$$g_{mA} = \frac{I_{D1m}}{V_{G1m}}$$

Thus the describing function of Fig. 16-24 is a plot of the average transconductance as a function of the gate driving signal. In Secs. 17-2 and 17-3,

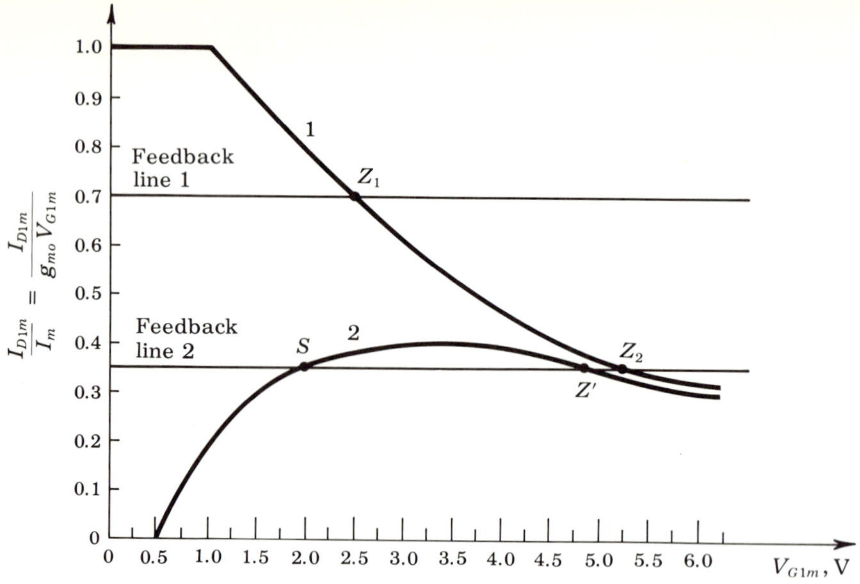

Fig. 16-24 Conductance-describing functions for the circuit of Fig. 16-22*a*.

we shall present a second interpretation of the conductance-describing function.

Once we combine the contributions from the total current in the form of a Fourier series, the voltage across the tank is found by multiplying each current term by the impedance at the drain at that particular frequency:

$$v_D = V_{DD} - \sum_{n=1}^{\infty} [r_o \parallel R_L \parallel Z_T(n\omega_0)] I_{Dnm} \cos n\omega_0 t$$

A high-Q tuned circuit is a high pure resistance at resonance and a low capacitive reactance at the harmonics. It follows that even a truncated sinusoidal current rich in harmonics will develop only a fundamental voltage. Only the fundamental-current term is significant.

At resonance, the purely resistive input impedance of the tuned circuit of Fig. 16-22*a* is expressed by

$$R_T = \left(\frac{C_1}{C_1 + C_2}\right)^2 Q \sqrt{\frac{L}{C_T}}$$

where C_T is the effective tank capacity $C_1C_2/(C_1 + C_2)$. From the self-starting condition for the Colpitts oscillator, $C_1 < g_m R_\parallel C_2$, we see that C_2 is the smaller capacity. When $g_m R_\parallel > 10$, C_2 will completely predominate in C_T. Thus the nominal resistance of the tank at resonance becomes

$$R_T \approx Q \sqrt{\frac{L}{C_2}}$$

Our equivalent model of the open-loop circuit at resonance reduces to a resistive load ($R_{\|} = R_T \| R_L \| r_o$) driven by a current source (Fig. 16-22*b*).

Stable oscillations in the closed-loop oscillator correspond to the feedback statement

$$V_{G1m} = \beta_0 V_{D1m} = \beta_0 I_{D1m} R_{\|}$$

where β_0 is given by Eq. (16-23). Normalizing the feedback condition into the same ordinate plotted in Fig. 16-24, we obtain the balance condition

$$\frac{I_{D1m}}{g_{m0} V_{G1m}} = \frac{1}{\beta_0 R_{\|} g_{m0}} = K_B \tag{16-38}$$

By rearranging terms we can recognize that the constant given by Eq. (16-38) is an alternative form of the threshold-gain condition of Eq. (16-23):

$$g_{mA} R_{\|} = \frac{I_{D1m}}{V_{G1m}} R_{\|} = \frac{1}{\beta_0}$$

The amplitude of oscillation is the intersection of Eq. (16-38) and the conductance-describing function.

In the circuit of Fig. 16-22*a*, with $g_{m0} R_{\|} = 33$, $I_{DSS} = 10$ mA, and $R_{\|} = 10$ K, when we adjust $\beta_0 = 0.0435$ so that the feedback constant of Eq. (16-37), K_B, is 0.7 (feedback line 1), the intersection with the describing function is at $V_{G1m} = 2.5$ V. The amplitude at the drain is $1/\beta_0$ times as large, or $V_{D1m} = 57.5$ V. If we increase the feedback by a factor of 2 to $\beta_0 = 0.087$ (feedback line 2), the peak amplitude of the gate signal increases to 5.25 V while the drain amplitude rises only to 60.5 V. Because of the increased limiting necessary under the conditions of the higher feedback, the waveshape distortion will be much more pronounced.

When the small-signal gain is only slightly greater than the threshold value, the stable point Z_1 will fall close to the upper limit of the linear range of the circuit. Thus the area between the feedback line and the describing function gives a qualitative measure of the harmonic content of the output—if the area is small, the signal will be an acceptable sinusoid; if it is large, the output will be somewhat distorted.

That Z of Fig. 16-24 (curve 1) is the sole stable point may be verified by considering the response to a small change in the input excitation. If the input is momentarily reduced, then the new value of loop gain becomes greater than unity, indicating that the signal must now increase with time. Beyond point Z, the average loop gain is less than the minimum value needed to sustain the oscillations and the amplitude begins decreasing toward the single stable value.

Curve 2 in Fig. 16-24 illustrates the shape of a describing function of the non-self-starting oscillator which exists when the FET is biased beyond pinchoff (at -3.5 V) instead of in its linear region. At small-signal levels, the device is cut off. But as the driving signal increases, operation within the active region for a portion of a cycle results in an effective g_{mA} greater than zero. At a large enough excitation, g_{mA} becomes greater than $1/\beta_0 R_{\|}$ and the circuit can sustain self-oscillations under proper feedback (line 2 but not line 1). To start this oscillator, an external excitation must force the operating point past S. Here the oscillatory criterion is satisfied. The amplitude will continue to increase toward the stable point Z'.

We could just as readily have plotted the current-describing function of the unnormalized I_{D1m} vs. V_{G1m}. By multiplying through by $R_{\|}$, we obtain the voltage-describing function $V_{D1m} = I_{D1m}R_{\|}$ vs. V_{G1m}. In each case, the equation of the feedback line must be written in terms of the coordinates of the plot. For the voltage-voltage describing function, the equation becomes

$$V_{D1m} = \frac{1}{\beta_0} V_{G1m}$$

which is a straight line through the origin having the slope $1/\beta_0$ (see Fig. 16-36).

If some form of self-biasing, instead of fixed biasing, is employed, then the dc component of the distorted drain current will aid in developing the final bias voltage. Since the final "quiescent" point will be a function of the amplitude of oscillations, the dc and fundamental amplitude problems must be simultaneously solved. In all cases, however, the direction of the dc current terms will always cause a shift in bias toward more symmetrical operation.

Bias limiting and sine-squared operation

Clamping by the gate-substrate diode is used to establish the self-bias conditions in the junction-FET oscillator of Fig. 16-25. Conduction by the diode on the peaks of the positive half cycle charges the clamp capacitor

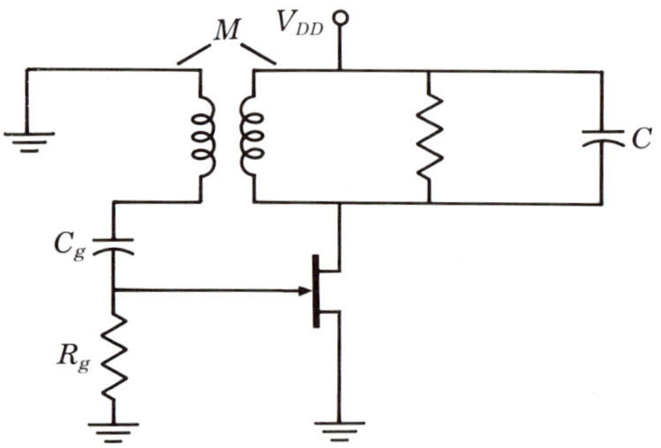

Fig. 16-25 FET tuned-drain oscillator—clamped-gate biasing.

C_g to the sine-wave voltage peak. (The large R_g is used to ensure the presence of a discharge path.) We can therefore say that the circuit automatically shifts the gate operating point to maintain the positive peak at approximately zero.

The clamped sinusoidal gate signal is written as

$$v_{GS} = V_{G1m}(-1 + \cos \omega_0 t)$$

Substituting into the square-law current equation

$$I_D = I_{DSS}\left(\frac{V_{GS}}{V_p} - 1\right)^2$$

and normalizing, we obtain

$$\frac{I_D}{I_{DSS}} = [(x - 1) - x \cos \omega_0 t]^2 \tag{16-39}$$

where the normalization factor is

$$x = -\frac{V_{G1m}}{V_p}$$

Drain current follows Eq. (16-39) up to the clipping point, which occurs once the negative peaks of the gate signal drop below V_p. This limit corresponds to $x = 0.5$. Larger drive signals cut off the FET for the half angle θ shown in Fig. 16-26.

$$\cos \theta = \frac{V_p + V_{G1m}}{V_{G1m}} = 1 - \frac{1}{x} \tag{16-40}$$

By making use of symmetry, the fundamental component of drain current is expressed by the integral

$$\begin{aligned}\frac{I_{D1m}}{I_{DSS}} &= \frac{2}{\pi}\int_0^\theta [(x - 1) - x \cos \omega_0 t]^2 \cos \omega_0 t \, d\omega_0 t \\ &= \frac{2}{\pi}\left[(x - 1)^2 \sin \theta - x(x - 1)\theta - \frac{x(x - 1)}{2} \sin 2\theta \right. \\ &\qquad \left. + \frac{3}{4} x^2 \sin \theta + \frac{x^2}{12} \sin 3\theta\right]\end{aligned} \tag{16-41}$$

By substituting the value of x from Eq. (16-40), it can be shown that the first and third terms of Eq. (16-41) equal zero:

$$(x - 1)^2 \sin \theta - \frac{x(x - 1)}{2} \sin 2\theta = 0$$

Equation (16-41) simplifies to the current-describing function

$$\frac{I_{D1m}}{I_{DSS}} = \frac{2}{\pi}[-x(x - 1)\theta + \tfrac{3}{4}x^2 \sin \theta + \tfrac{1}{12}x^2 \sin 3\theta] \tag{16-42}$$

Equation (16-42) is plotted in Fig. 16-26.

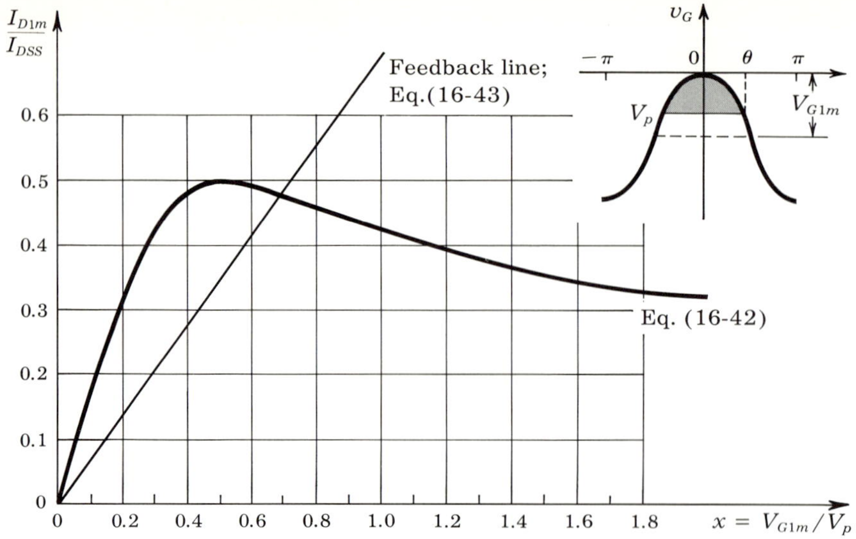

Fig. 16-26 General describing function for clamped-sine squared operation [Eq. (16-42)].

Figure 16-26 is a universal plot for any FET clamped-bias oscillator. In order to find the amplitude of oscillation in any particular case, we need the equation of a normalized feedback line:

$$\frac{I_{D1m}}{I_{DSS}} = \frac{V_p}{\beta_0 R_{\|} I_{DSS}} x \tag{16-43}$$

The intersection of the line defined by Eq. (16-43) with the describing function of Eq. (16-42) is the stable oscillation amplitude.

For example, using the parameters previously given in the linear case—$I_{DSS} = 10$ mA, $V_p = -3$ V, $R_{\|} = 10$ K, and $\beta_0 = 0.0435$—the point of intersection is located at $x = 0.625$. This point corresponds to a peak oscillator amplitude of 1.87 V at the gate, while at the drain, $V_{D1m} = 43.1$ V.

Squegging

When the output amplitude is stabilized on the decreasing segment of the describing function, the circuit is prone to intermittent oscillations, or "squegging." On the positive peaks, the large pulses of grid current charge the input capacitor. If the net charge accumulated in C_g on the positive peaks is greater than the energy dissipated in R_g over the remaining portion of the sinusoid, the bias voltage will build up to a point where it pinches off the FET. The larger the net charge accumulating, the fewer the number of cycles contained in the oscillation interval. As the oscillations damp out

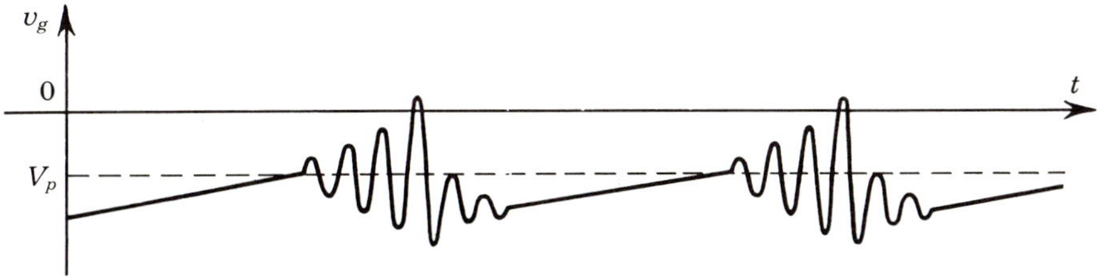

Fig. 16-27 Intermittent oscillation in a gate-bias-limited oscillator.

(Fig. 16-27), C_g recovers toward zero, with the long time constant R_gC_g. The FET eventually turns on, and the cycle repeats. Reducing the time constant will often allow the circuit to oscillate properly.

Exponential limiting—bipolar transistor

The significant nonlinearity in a bipolar-transistor oscillator is the exponential volt-ampere response of the base-emitter junction. It therefore seems logical to start the analysis of this case by calculating the emitter current. Superimposing the sinusoidal driving signal on the dc bias point (V_{BE}) and substituting into the exponential current equation result in

$$\begin{aligned} i_e &= I_{es}e^{(V_{BE}+V_{b1m}\cos\omega_0 t)/V_T} \\ &= I_{es}e^{V_{BE}/V_T}e^{x\cos\omega_0 t} \end{aligned} \qquad (16\text{-}44)$$

where x is the amplitude-normalization factor

$$x = \frac{V_{b1m}}{V_T}$$

The second exponential term in Eq. (16-44) may be expanded into the series

$$e^{x\cos\omega_0 t} = \mathbf{I}_0(x) + 2\sum_{n=1}^{\infty}\mathbf{I}_n(x)\cos n\omega_0 t \qquad (16\text{-}45)$$

$\mathbf{I}_n(x)$ is a modified Bessel function of the first kind of order n and argument x. For the case with which we are concerned ($x \geq 0$ and $n \geq 0$), all functions $\mathbf{I}_n(x)$ are positive and increase monotonically. The lowest-order term $\mathbf{I}_0(x)$ starts at 1. All higher terms start from 0 (Table 16-2).

Selecting the dc and fundamental terms from the expansion of Eq. (16-45), because these are the only ones of significance in determining the amplitude, we obtain

$$\begin{aligned} I_c &= \alpha I_{es}e^{V_{BE}/V_T}\mathbf{I}_0(x) + \alpha I_{es}e^{V_{BE}/V_T}2\mathbf{I}_1(x)\cos\omega_0 t \\ &= I_{dc}\left[1 + 2\frac{\mathbf{I}_1(x)}{\mathbf{I}_0(x)}\cos\omega_0 t\right] \end{aligned} \qquad (16\text{-}46)$$

Table 16-2 Modified Bessel functions

x	$\mathbf{I}_0(x)$	$\mathbf{I}_1(x)$	$\mathbf{I}_2(x)$	$\mathbf{I}_3(x)$
0	1.0000	0	0	0
0.5	1.0635	0.2579	0.0319	0.0026
1.0	1.2661	0.5651	0.1357	0.0222
1.5	1.6467	0.9817	0.3378	0.0808
2.0	2.2796	1.5906	0.6889	0.2127
2.5	3.2898	2.5167	1.2765	0.4744
3.0	4.8808	3.9534	2.2452	0.9597
3.5	7.3782	6.2058	3.8320	1.8264
4.0	11.3019	9.7595	6.4222	3.3373
4.5	17.4812	15.3892	10.6415	5.9301
5.0	27.2399	24.3356	17.5056	10.3312
5.5	42.6946	38.5882	28.6626	17.7426
6.0	67.2344	61.3419	46.7871	30.1505
6.5	106.2929	97.7350	76.2205	50.8301
7.0	168.5939	156.0391	124.0113	85.1755
7.5	268.1613	249.5844	201.6055	142.0614
8.0	427.5641	399.8731	327.5958	236.0752
8.5	683.1619	641.6199	532.1925	391.1763
9.0	1,093.5884	1,030 9147	864.4962	646.6941
9.5	1,753.4810	1,658.4531	1,404.3330	1,067.1550
10.0	2,815.7166	2,670.9983	2,281.5190	1,758.3807

Equation (16-46) points up the interdependence of the fundamental and dc current terms: to find the fundamental current, we must know the dc current. But the dc current itself depends on the amplitude parameter x.

In the simplest case, where the transistor is biased by a constant-current drain (Fig. 16-28), I_{dc} remains independent of the amplitude of oscillation. The fundamental component of collector current is given by the ratio

$$\frac{I_{c1m}}{I_{dc}} = 2\frac{\mathbf{I}_1(x)}{\mathbf{I}_0(x)} \tag{16-47}$$

Equation (16-47) is the normalized describing function for the bipolar oscillator (Fig. 16-29). As before, the fundamental amplitude is given by the intersection with the describing function of the reflected feedback line, which is defined by

$$\frac{I_{c1m}}{I_{dc}} = \frac{V_T}{\beta_0 R_{\|} I_{dc}}\,x = \frac{1}{\beta_0 R_{\|} g_{mQ}}\,x \tag{16-48}$$

$R_{\|}$ must include any resistance appearing across the base-emitter circuit, multiplied by the impedance-transformation factor of the tuned circuit, as it is reflected back into the collector circuit.

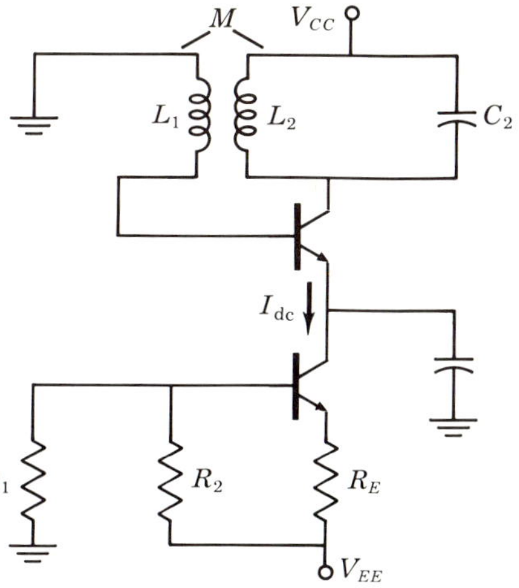

Fig. 16-28 Fixed constant-current biasing of a bipolar oscillator.

As might be expected, the bias point shifts with the increase of signal amplitude. From the dc-current component of Eq. (16-46),

$$I_{C_{dc}} = \alpha I_{es} e^{V_{BE}/V_T} \mathbf{I}_0(x)$$

we find that

$$V_{BE} = V_{BE_Q} - V_T \ln \mathbf{I}_0(x) \tag{16-49}$$

where V_{BE_Q} is the junction voltage under quiescent conditions [$x = 0$, $\mathbf{I}_0(x) = 1$]. As $\mathbf{I}_0(x)$ increases, a larger portion of the total dc current will be supplied from the distortion of the current waveshape. To maintain

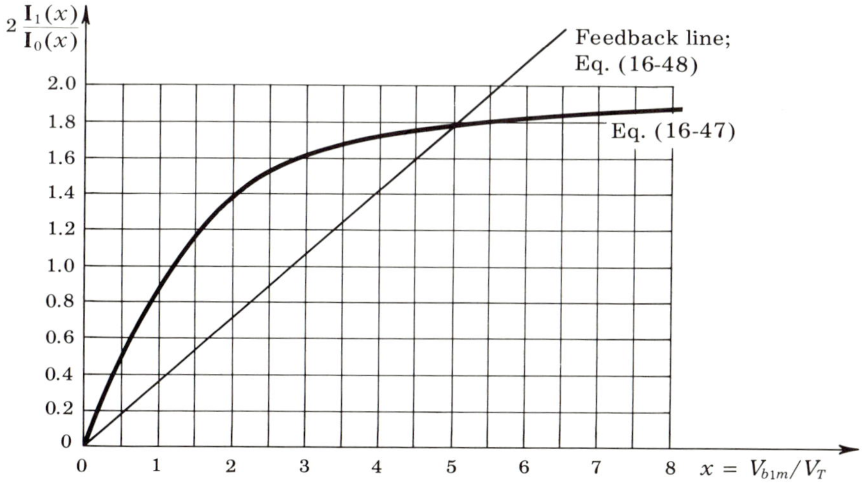

Fig. 16-29 General bipolar-transistor describing function.

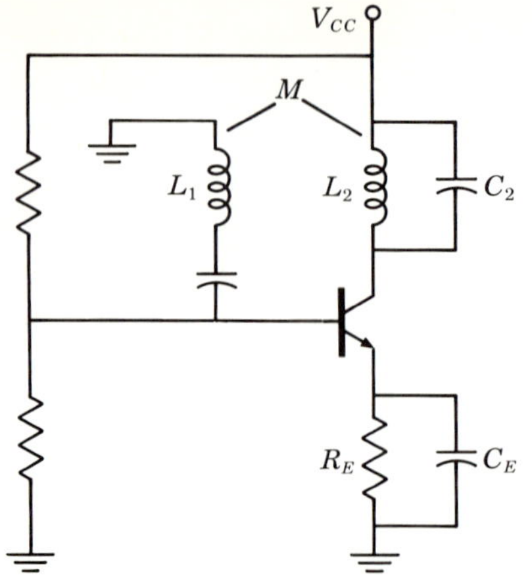

Fig. 16-30 Self-biasing version of the oscillator of Fig. 16-28.

the same total I_{dc}, V_{BE} must shift toward zero. Since this down shift is in the direction toward a lower g_m, it aids the self-limiting action of the circuit.

Whenever the bipolar transistor is biased through a resistive network, our previous calculations must be modified to account for the bias shift. The emitter current in the circuit of Fig. 16-30 can be written as

$$I_{edc} = \frac{V_{EE} - V_{BE}}{R_E} = \frac{V_{EE} - V_{BE_Q}}{R_E} + \frac{\Delta V_{BE}}{R_E}$$

Substituting ΔV_{BE} from Eq. (16-48), we obtain

$$I_{edc} = I_{dcQ}\left[1 + \frac{V_T}{I_{dcQ}R_E} \ln \mathbf{I}_0(x)\right]$$

In any specific oscillator, this equation must be used to modify the describing function previously calculated.

16-7 Amplitude of Oscillation—Part II, Automatic Gain Control

Feedback limiting

In this case, a passive nonlinear element is included in the feedback network at a point where it can change the magnitude of β but where it will not affect the null frequency. For example, in the Wien-bridge oscillator of Fig. 16-31, the degree of bridge unbalance is controlled by the thermal characteristics of the tungsten lamp, which is used as one of the resistive arms. Its resistance increases with temperature, and hence with the voltage applied across the branch and across the bridge (Fig. 16-32). Normally the bridge is slightly unbalanced ($\beta' = 1/\delta$), adjusted so that the amplifier can just sustain the given amplitude of oscillation. Any increase in the output voltage will increase the resistance of the lamp and bring the bridge into closer balance, thus increasing δ. The loop gain will drop below unity, and the oscillations will start damping out. Any decrease in the output would establish the conditions necessary for an amplitude buildup. We conclude that this form of voltage-controlled resistive limiting can maintain a nearly constant output amplitude with a high degree of stability.

The normal operating condition is now given by

$$A\beta'(v,\omega) = 1 \tag{16-50}$$

where A is simply the fixed gain of the voltage amplifier. For the Wien bridge at ω_0, Eq. (16-50) reduces to

$$\frac{A}{\delta(v)} = 1$$

and the particular voltage v satisfying this equation is the oscillation amplitude. Since the limiting process readjusts the β' of the network, this oscillator will always operate under the threshold condition expressed in Eq. (16-50). Moreover, as the amplifier is never driven into its nonlinear region, the output will be almost completely free of harmonic distortion.

The lamp has thermal inertia, and consequently it will respond only to the mean value of the voltage applied over some small time interval. A standard small bulb, operating in the range of 600 to 1000°K, far below the temperature used for illumination, has a thermal time constant between 20 and 75 ms. The mass of the filament and the construction of the bulb are the determining factors; the smaller the nominal power rating of the lamp, the shorter the time constant. With respect to any single cycle, the tungsten lamp may be treated as a fixed resistance. At the lower audio frequencies, where the period is comparable to the thermal time

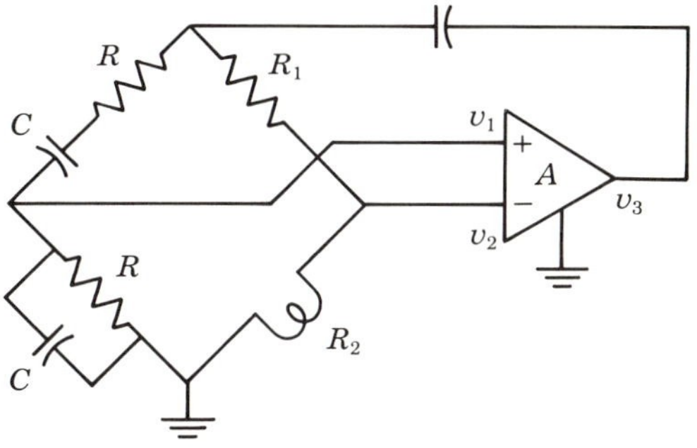

Fig. 16-31 Amplitude-controlled Wien-bridge oscillator.

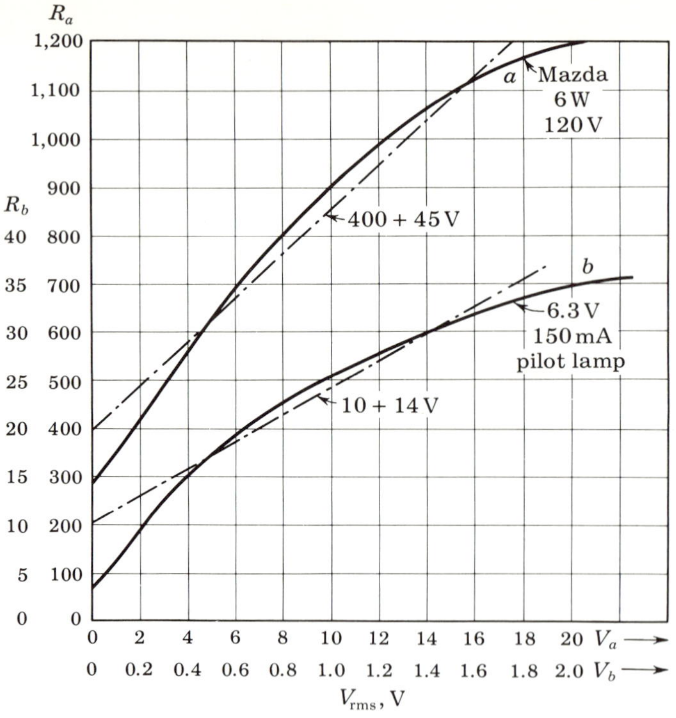

Fig. 16-32 Resistance characteristics of two typical tungsten lamps and their linear approximations.

constant (below 20 Hz), the change in resistance over the individual cycle distorts the output sinusoid and sets a lower frequency limit for this method of amplitude limiting.

Example 16-2

Let us consider the design of the amplitude-limited bridge of Fig. 16-31. The circuit contains an amplifier which has a nominal gain of 300 and a dynamic range of 60 V peak to peak. The 6-W Mazda lamp of Fig. 16-32 will be used for amplitude limiting.

To allow some latitude for variations in the amplifier response, we shall limit the output sinusoid to 15 V rms. With the large gain available, the bridge will be almost completely balanced. For threshold operation, $\delta_t = A_t = 300$, leading to

$$\frac{R_2}{R_1 + R_2} = \frac{1}{3} - \frac{1}{\delta} = \frac{1}{3} - \frac{1}{300} = 0.330$$

The voltage across the lamp is

$$V_{R_2} = 0.33V_3 = 5 \text{ V rms}$$

Since a high degree of quality control cannot be expected in a device made for illumination and not originally designed as a control element, the actual lamp characteristics will vary over very wide limits. Any reasonable approximation will serve for design purposes; for simplicity we shall use a straight line. Within the range of interest ($2 < V < 18$), the Mazda lamp will have a resistance variation given by

$$R_2 \approx 400 + 45V_{R_2}$$

At 5 V across the lamp, the effective resistance is 625 Ω. Substituting into the bridge-arm equation,

$$R_1 = \frac{1.0 - 0.33}{0.33} R_2 = 1{,}250 \ \Omega$$

R_1 would be made somewhat adjustable to allow for variation in the lamp resistance.

We shall now assume that the bridge is balanced as calculated above and that the amplifier gain decreases by 33 percent because of a change of active devices. To sustain the oscillations, R_2 must also change until it satisfies

$$\frac{R_2'}{R_2' + R_1} = \frac{1}{3} - \frac{1}{200} = 0.3283$$

Thus R_2 becomes

$$R_2' = (1 - 0.0076)R_2 = 621.2 \ \Omega$$

The decrease of 0.76 percent in the lamp resistance corresponds to a new lamp voltage of

$$V_{R_2}' = 4.916 \text{ V}$$

The new stable output is 14.75 V rms. A gain reduction of 33 percent is accompanied by an amplitude change of only 1.7 percent. Consequently, this method of limiting also ensures amplitude stability. At the new output, the oscillator again operates under threshold conditions.

The technique of limiting by automatically controlling the feedback has also been applied to the Meacham bridge (Fig. 16-21), where the lamp replaces R_2, and to several tuned-circuit oscillators, where a resistance bridge containing lamps might be used for feedback. Figure 16-33 illustrates an amplitude-controlled bridge used to separate the feedback control from the frequency-selective network. Limiting may be assigned solely to the bridge. Other tuned circuits have used small lamps in series with the coil to control the effective Q of the tank.

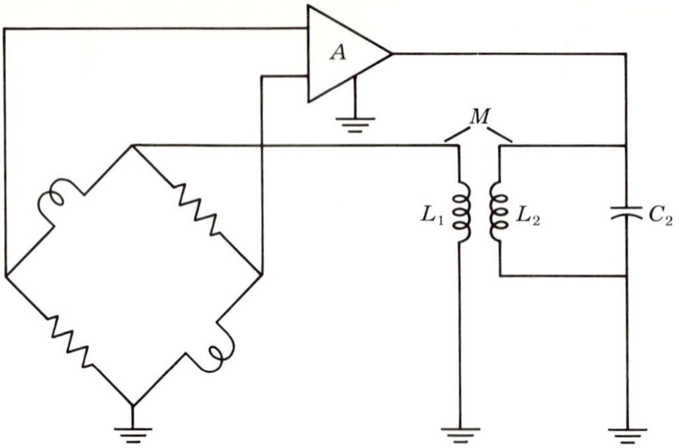

Fig. 16-33 Bridge-controlled amplitude—feedback oscillator.

Besides tungsten lamps, most semiconductor materials (e.g., carbon, silicon, and germanium) and many compounds, such as silicon carbide, exhibit temperature-sensitive resistance properties. Since semiconductor thermistors have a negative temperature coefficient of resistance, in contrast to the positive one of the pure metals, they replace the complementary arm of the bridge. In the Wien-bridge circuit of Fig. 16-31, the thermistor would substitute for R_1 instead of R_2. Thermistors are fabricated in a much wider range of resistance, and rate of resistance variation with temperature, than can be expected from a tungsten filament. Operating as they do at a lower temperature (300 to 375°K), they are extremely sensitive to changes in the ambient temperature. For acceptable amplitude stability, some form of temperature compensation would be necessary.

A FET as an amplitude stabilizer

In Fig. 16-34, amplitude control is provided by the gate voltage-controlled resistance of a FET (Sec. 5-10). Within the ohmic region, we have

$$R_2 = R_{ON} = \frac{R_{ON_0}}{1 - |V_{GS}/V_p| - |V_{DS}/V_p|}$$

The output of the differential amplifier is rectified, scaled, and applied to the gate of the FET used as one arm of the Wien bridge. The FET stabilizes the amplitude of oscillation in exactly the same manner as did the tungsten lamp in the circuit of Fig. 16-31. An increase in the output voltage increases the device resistance, thus automatically bringing the bridge into closer balance. This decreases the net loop gain, returning the amplitude to the stable value. Independent control over the time constant in the gate

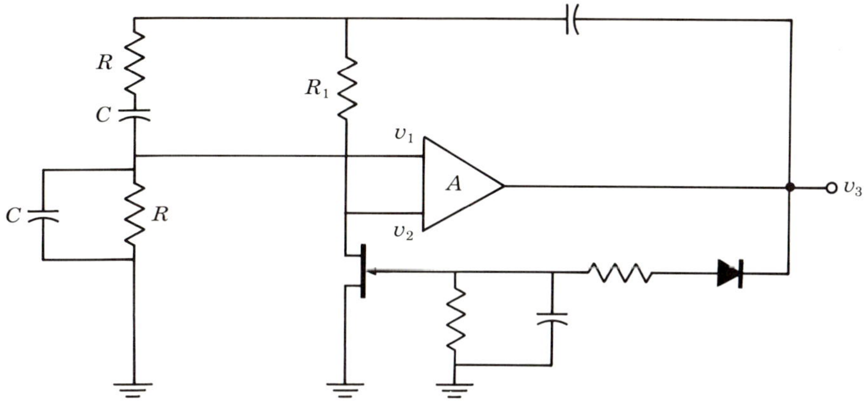

Fig. 16-34 Wien-bridge oscillator showing a FET as a voltage-controlled resistor for amplitude stabilization.

circuit permits amplitude-stabilized oscillator operation at extremely low frequencies.

Large drain-to-source signals cause the resistance of the FET to increase and may even bring it from ohmic to pinchoff operation. Since the bridge is brought into closer balance on the sinusoidal peaks, the device nonlinearity reduces the average loop gain over the cycle. This also helps to stabilize the amplitude. The nonlinearity does, however, increase the distortion present. Because of the nonlinearity, the amplitude would have to be found from a describing-function analysis.

Automatic gain control

Threshold operation of the oscillator may also be ensured by using the output amplitude to control the gain of the amplifier. With many active elements it is possible to find some dc voltage or current which will determine the magnitude of the forward transmission. For example, in a MOSFET, the gate-to-drain transconductance is a function of the gate bias voltage.

In a bipolar transistor, g_m increases with the quiescent emitter current. β also depends on the current; it drops off at very low and very high currents.

A voltage-controlled resistance such as a MOSFET may also be included within the amplifier as a gain-control element.

By deriving the gain-control voltage from the oscillator output, any increase in the amplitude will create the conditions that will automatically reduce the gain and the signal. The response to a decrease in the output would be just the opposite. This action is similar to the thermally adjusted bridge previously discussed. To avoid distorting the individual cycle,

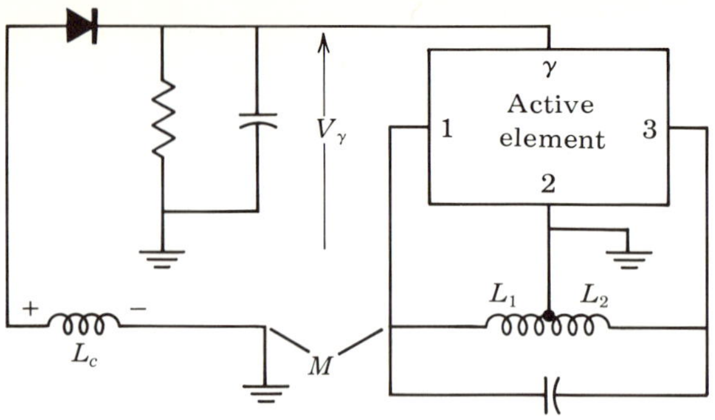

Fig. 16-35 Basic automatic-gain-control oscillator—Hartley circuit.

the sinusoidal output is rectified and applied through a relatively long time constant to the gain-control element. Therefore the amplitude-defining equation may be written

$$A(V)\beta_0 = 1$$

where V is the dc control signal.

Consider the general problem posed in Fig. 16-35, where the control terminal is indicated by γ. The output of the oscillator is coupled through the extra coil winding, rectified, and applied to this terminal. As shown, the gain is assumed to decrease with any increase in V_γ (or with an increase in the oscillation amplitude). For example, in a MOSFET operating in the pinchoff region, the transconductance is given by

$$g_m = K(V_{GS} - V_T)$$

By combining fixed biasing with feedback biasing of the appropriate polarity,

$$g_m = g_{m0} - K'V_\gamma \tag{16-51}$$

where K' is a positive constant. If g_m were to increase with V_γ, then the control loop would be regenerative instead of degenerative and the output amplitude would increase until the active-element limits.

At any particular amplitude, the threshold of oscillation, as given by Eq. (16-23), is

$$g_m R_\| = \frac{L_2 + M}{L_1 + M}$$

Taking the various coupling factors and amplitude transformations into account, this condition may be expressed as

$$g_{m0}R_\| - K''V_o = \frac{L_2 + M}{L_1 + M}$$

V_o is the peak value of the sinusoidal output voltage and will be found from the solution of this equation.

In all the amplitude-controlled circuits, the variation of the controlled source with respect to the variable bias is difficult to express in general terms and will have to be evaluated in each individual case. Because the range of operation is severely restricted, a few measurements are usually sufficient. The best region in which to operate is the one where the variation of the controlling element with voltage is most pronounced. Even though satisfactory limiting has been obtained by this method, the selectivity multiplication of the bridge makes it superior when the output amplitude must be maintained constant. For an extremely high degree of stability, both methods of limiting may be incorporated in a single circuit.

16-8 Amplitude Stability

As a second figure of merit for the oscillator, we might define a normalized amplitude-stability factor

$$S_V^\mu = \frac{d\mu/\mu}{dV/V} \tag{16-52}$$

where μ is the variable having the most pronounced effect on the amplitude. We interpret S_V^μ as meaning that a 10 percent change in μ produces only a $10/S_V^\mu$ percent change in the amplitude of oscillation.

In those circuits where the limiting depends on the nonlinearity of the active elements, S_V^μ may be found graphically by plotting a family of describing functions for various values of μ. This process is illustrated in Fig. 16-36, where the variable of interest is the g_m of the active device. At the intersection with the $1/\beta_0$ line, the stability factor would be given by

$$S_V^{g_m} \approx \frac{\Delta g_m/g_{m,\text{av}}}{\Delta V/V_{\text{av}}} = \frac{(g_{m1} - g_{m2})/(g_{m1} + g_{m2})}{(V_1 - V_2)/(V_1 + V_2)}$$

For the values shown in Fig. 16-36,

$$S_V^{g_m} = 3.8$$

If the amount of feedback were to be increased, then the stable point of operation would fall on the more nearly horizontal portion of the describing function and the stability would improve. However, under these conditions, the harmonic content of the output would also increase, and as this represents a loading on the frequency-selective network, the frequency stability would be adversely affected.

In the automatically stabilized circuits, such as the Wien-bridge oscillator of Fig. 16-31, any change in the gain is compensated for by controlling

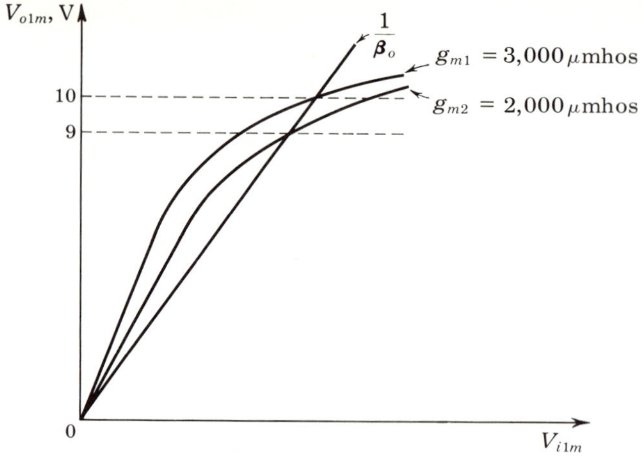

Fig. 16-36 Amplitude stability illustrated by describing functions.

the degree of bridge unbalance. Since A is always equal to δ, the stability factor of interest is S_V^δ. For the Wien bridge,

$$\frac{R_0 + KV}{R_0 + KV + R_1} = \frac{1}{3} - \frac{1}{\delta} \tag{16-53}$$

By taking the differential, there results

$$\frac{R_1 K}{(R_1 + R_0 + KV)^2}\,dV = \frac{1}{\delta^2}\,d\delta \tag{16-54}$$

But from Eq. (16-53),

$$\frac{R_1}{R_1 + R_0 + KV} = \frac{2}{3} + \frac{1}{\delta} \tag{16-55}$$

Since $\delta \gg 1$, the result of substituting Eq. (16-55) into Eq. (16-54) and of solving for the stability factor is

$$S_V^\delta = \tfrac{2}{3}\delta\,\frac{KV}{R_1 + R_0 + KV} \tag{16-56}$$

As K, the rate of the resistance variation with voltage, increases, S_V^δ will approach $2\delta/3$ as an upper limit. Thus the amplitude stability is directly proportional to the amplifier gain and to the degree of bridge unbalance. Maximizing Eq. (16-56) by using a high-gain amplifier also optimizes the frequency stability [Eq. (16-30)] and the harmonic-rejection ratio [Eq. (16-16)].

For the particular values used in Example 16-1 ($\delta = 300$), the amplitude stabilization becomes

$$S_V^\delta = 19.8$$

This is many times the stability that could be obtained in the pure resistance bridge, unless the signal is grossly distorted by the limiting of the active device.

The automatic-gain-control circuit (Fig. 16-35) has a stability factor directly proportional to the rate of change of g_m with the control voltage. Since

$$g_m = g_{m0} - K'V_\gamma$$

$$S_{V_\gamma}^{g_m} = -K' \frac{V_\gamma}{g_m}$$

For optimum stability, a device having the most pronounced voltage control over the transconductance should be chosen. Moreover, if the signal were to be amplified before being applied for control, the stability factor could be increased to almost any desired value.

Problems

16-1 (*a*) Find the frequency and necessary gain for sustained oscillation if the phase-shift network of Fig. 16-5 consists of four equal RC sections instead of three.

(*b*) Evaluate $|H_2|$ and $|K_2|$ for this network. Compare the answers with the harmonic coefficients found for the three-section network.

(*c*) Can the conditions for oscillation be satisfied with a two-section network? Explain your answer.

16-2 (*a*) The network of Fig. 16-4*c* is used in the phase-shift circuit. Calculate the necessary amplifier gain, the frequency of oscillation, and the two harmonic factors.

(*b*) Repeat part *a* if R and L are interchanged.

16-3 The frequency of the phase-shift oscillator of Fig. 16-5 is to be controlled by varying only one of the three series capacitors over the range $0.1C \leq C_a \leq 10C$. In which section should this capacity be placed if the range of frequency variation is to be maximized? What are the two bounding values of ω, expressed in terms of $\omega_0 = 1/(\sqrt{6}\,RC)$?

16-4 Calculate the input impedance of the three-RC-section phase-shift network at ω_0. Under what conditions will it be invariant as the null frequency is changed? What is the oscillation frequency when the output impedance of the amplifier is equal to R? What is the new value of the threshold gain?

16-5 Verify Eqs. (16-15) and (16-17). Find H_n and K_n for the Wien bridge for $2 \leq n \leq 5$. (Let $\delta = 300$.)

16-6 The integrated-circuit amplifier of Fig. 16-7 has an unloaded gain of 1,000 and an output impedance of 500 Ω. The differential input stage has an input impedance of 20 K. If $\omega_0 = 10^5$ and $R_2 = 3$ K, specify the remaining parameters of the bridge. What is the actual value of δ? Make all reasonable approximations in the course of your solution, but be sure to take account of the loading of the bridge by the amplifier and the loading of the amplifier by the bridge.

16-7 Design a composite Wien-bridge oscillator of the type of Fig. 16-8. The circuit is to oscillate at $\omega_0 = 10^4$. Use a FET where $I_{DSS} = 10$ mA, $V_p = -2$ V and a bipolar having $\beta = 100$ and $V_{BE} = 700$ mV. $V_{DD} = 40$ V. Specify all circuit components.

16-8 Figure 16-37 shows another version of a phase-shift current-controlled oscillator. Calculate the frequency of oscillation, the minimum transistor β for proper operation, and the harmonic-rejection factors. Show the locus of the oscillator poles as a function of the transistor β. Make all reasonable approximations.

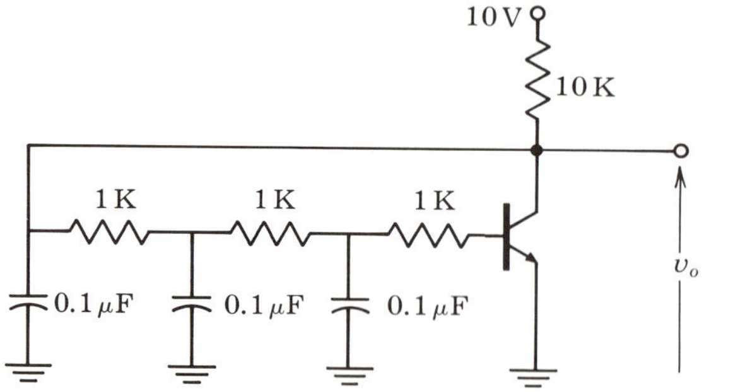

Fig. 16-37

16-9 A FET tuned-drain oscillator (Fig. 16-15*d*) has a drain-tank circuit which consists of a 10-mH coil and a 100-pF capacitor. $Q_o = 50$ (this includes the effect of r_o). The gate coil has an inductance of 5 mH, while the mutual inductance is only 1 mH.

(*a*) Specify the frequency of oscillation and the minimum g_m needed for oscillation. Take into account the tank resistance.

(*b*) The tank is loaded by $R_L = 10$ K. What is the new minimum value of g_m required? Does the frequency change?

16-10 A FET having $I_{DSS} = 10$ mA, $V_p = -2$ V, and $r_o = 20$ K at I_{DSS} is used in a Hartley oscillator. For simplicity we shall assume that the two coils are completely isolated. This circuit is to be tuned over the range of 500 to 1,500 kHz by a capacitor that is adjustable from 300 down to 30 pF. If the Q of the complete coil is 20 at 1 MHz, specify the two inductances that will permit operation over the complete range of frequencies. Allow a feedback safety factor of 5. Be sure to check the end points and to account for the loading effect of the tank resistance.

16-11 Verify Eqs. (16-25) and (16-27) and the frequency and self-starting conditions given in the text for the bipolar-transistor Hartley and Colpitts oscillators.

16-12 Figure 16-38 shows a grounded-base Colpitts oscillator. Complete the design by specifying a C_2 that makes the feedback factor five times that needed for threshold operation. Give the frequency of operation.

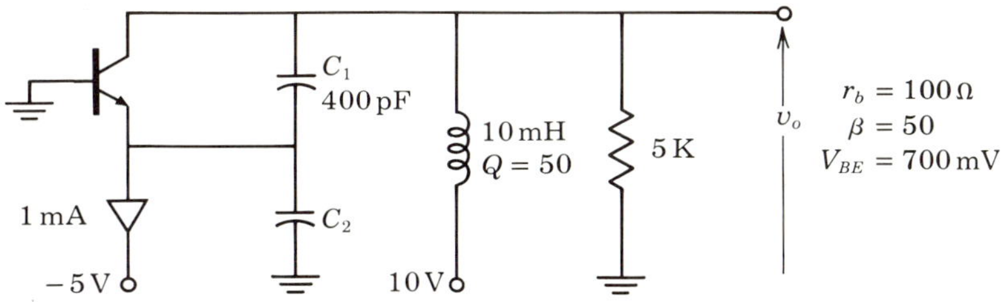

Fig. 16-38

16-13 A bipolar Hartley oscillator employs a silicon transistor having $\beta = 100$ and $r_b = 100\ \Omega$. The nominal frequency of oscillation is $\omega_0 = 10^7$ ($C = 40$ pF), and at this frequency the unloaded tank $Q_0 = 50$.

(*a*) Plot the required feedback factor for oscillation as a function of the quiescent emitter current.

(*b*) For a dc emitter current of 1 mA, specify the coil ratios. Allow a feedback safety factor of 5 to ensure proper oscillations. Neglect the mutual inductance in your calculations.

(*c*) Plot the deviation of frequency from ω_0 as a function of the external load from a value of R_L that just stops oscillation to a maximum $R_L = 100$ K.

16-14 An integrated-circuit differential amplifier, having a voltage gain of 1,000, an output impedance of 100 Ω, and an input impedance of 20 K, is to be used as the active device in a tuned-circuit oscillator. Investigate the two different configurations shown in Fig. 16-39. Solve for the conditions for oscillation and the frequency of oscillation. Is either configuration especially suitable? Explain.

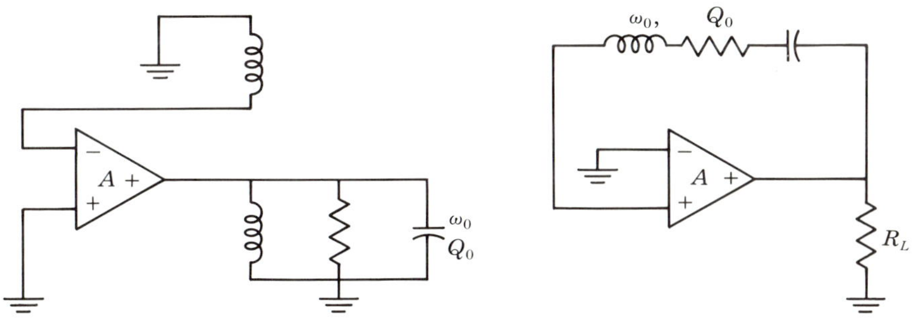

Fig. 16-39

16-15 Figure 16-40 shows a three-phase transistor oscillator where the three secondaries are symmetrically loaded.

(*a*) Calculate the frequency of oscillation, assuming threshold operation.

(*b*) What is the maximum load that can be applied at the secondary without causing the cessation of oscillation?

(*c*) Calculate the frequency-stability factor and the harmonic-rejection factor (H_2).

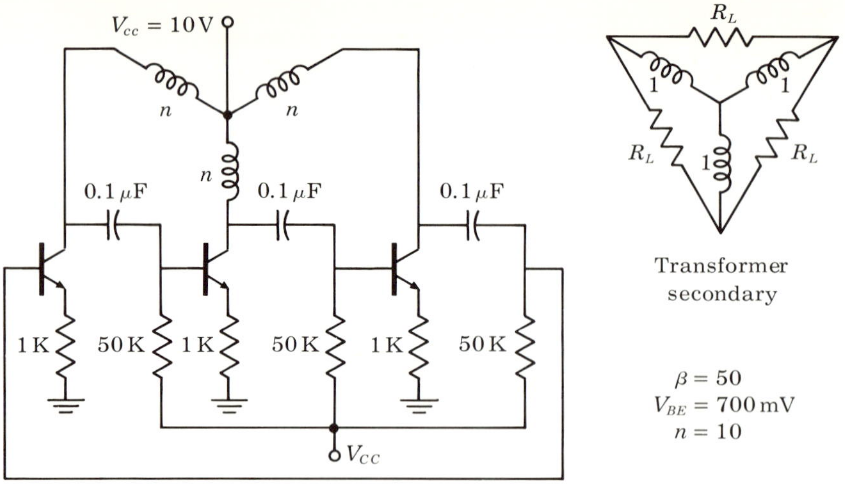

Fig. 16-40

16-16 Compare the frequency-stability factor of the phase-shift oscillator of Fig. 16-5, when the frequency-determining network consists of three sections, with one where four *RC* sections are used.

16-17 (*a*) Verify the frequency-stability factor for the Meacham bridge.

(*b*) Two matched crystals are used in this bridge instead of one. The second replaces R_2 in Fig. 16-21. If each has $Q = 10^5$ and the degree of bridge unbalance is $\delta = 100$, what is the new frequency-stability factor? The nominal resonant frequency of the crystal is 1 MHz, and the change in the amplifier phase shift may reach a maximum of 10°; what is the effective change in frequency?

16-18 A transistor used in a Colpitts oscillator has the following parameters at the operating point: $h_{ie} = 1{,}500\ \Omega$, $\beta = 50$, and $V_{BE} = 700$ mV. The oscillator uses a 5-mH coil ($Q = 100$) to set the nominal frequency at $\omega_0 = 5 \times 10^6$. We can expect stray capacity of 5 pF between the base and the collector.

(*a*) If the feedback is initially adjusted to four times the threshold, by how much will the frequency shift as a result of a 2-pF change in the stray capacity?

(*b*) If the tank is loaded to $Q = 20$ and if stray elements cause a total 10° shift in the amplifier, by how much will the frequency shift?

16-19 The tuned circuit in the Colpitts oscillator of Prob. 16-18 has a loaded $Q = 10$. The feedback is adjusted to five times above threshold. Without accounting for the effects of stray capacity, calculate the reactance that would have to be inserted into the base lead in order to return the frequency of oscillation to ω_0. Repeat the calculation for a stabilizing reactance placed in series with the emitter.

16-20 Calculate the magnitude of the second- and third-harmonic components of the tank voltage of the linearized oscillator of Fig. 16-22*a*. Use the two feedback factors defined in the text. At the resonant frequency of $\omega_0 = 10^6$, $Q_0 = 25$.

16-21 Draw the generalized describing function for the fixed-bias FET oscillator, taking account of the square-law characteristic. For your example, take the same two bias points and the same feedback factors used in the text for the linearized case. Is the linearized approximation ever reasonably valid?

16-22 Given the Colpitts oscillator of Fig. 16-22 adjusted to three times the feedback threshold, the circuit parameters are $V_{DD} = 40$ V, $V_p = -3$ V, $I_{DSS} = 5$ mA, $r_o = 20$ K at I_{DSS}, $V_{GG} = -1.5$ V. The tank circuit has a loaded $Q = 20$ and an input impedance of 20 K at its resonant frequency of 10 MHz.

Plot the voltage-describing function by finding first the maximum value of the fundamental component of the drain voltage in the nonlimiting region, next the output when the cosinusoidal signal is clipped by $\pm 15°$, then when it is clipped by $\pm 30°$, etc. Represent the describing function by a sequence of straight lines connecting these points. Find the approximate amplitude of oscillation. Calculate $S_V^{R_L}$ at the operating point.

16-23 Repeat Prob. 16-22 when the circuit is changed so that the amplitude limiting is controlled by assumed perfect gate-circuit clamping.

16-24 Calculate the amplitude of oscillation of the clamped-FET phase-shift oscillator of Fig. 16-41. $V_p = -3$ V, $I_{DSS} = 5$ mA. Make all reasonable approximations.

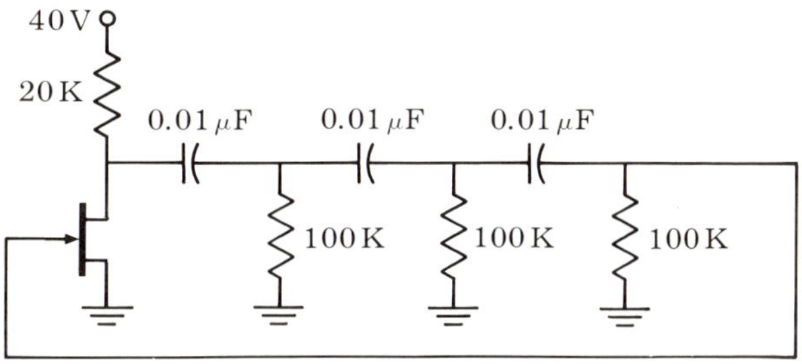

Fig. 16-41

16-25 Find the amplitude of oscillation of the circuit of Fig. 16-38:

(*a*) When $C_2 = 4$ nF

(*b*) When $C_2 = 40$ nF

(*c*) When $C_2 = 400$ nF

For each case, give the amplitude of the second- and third-harmonic voltages appearing at the collector. Make all reasonable approximations.

16-26 (*a*) Verify Eq. (16-49) and solve for I_{edc}.

(*b*) Plot the describing function on the same graph as that of Fig. 16-29 for

$$R_E I_{dc_Q} = 0.5 \text{ V}$$

$$R_E I_{dc_Q} = 1.0 \text{ V}$$

(*c*) Solve for the amplitude of oscillation in the circuit of Fig. 16-38 when $C_2 = 40$ nF and when the emitter biasing is through a 1-K resistor returned to -1.7 V. Since $V_{BE} = 700$ mV at $I_e = 1$ mA, this sets $I_{dc_Q} = 1$ mA.

16-27 Design an amplitude-stabilized Meacham-bridge bipolar-transistor oscillator. Use the 6.3-V pilot lamp of Fig. 16-32 as the control element. The peak amplitude across the bridge should be 4.5 V. At its series-resonant frequency, the crystal may be replaced by a 20-Ω resistor. Specify all elements of the bridge. Draw the circuit of the voltage amplifier giving values of all components. Pay particular attention to the impedance levels. Specify the amplitude-stability factor of this circuit.

16-28 Find the output amplitude of a Wien-bridge oscillator where the bridge is initially unbalanced by a factor $\delta = 200$ and where the amplitude-control element is a thermistor having the characteristic $R \approx 10{,}000 - 250V_R$ Ω. The associated pure-resistance bridge arm is 3 K. Evaluate S_V^δ.

16-29 Repeat Prob. 16-28 if the remaining resistance arm is replaced by a tungsten lamp having the approximate characteristic $R' \approx 1{,}000 + 100V_{R'}$ Ω. Thus two temperature-dependent elements are used in the bridge. Evaluate S_V^δ.

16-30 The integrated-circuit amplifier in the circuit of Fig. 16-34 has a gain of 1,000, an input impedance of 25 K, and an output impedance of 100 Ω. The FET used for amplitude stabilization has $V_p = -5$ V and $I_{DSS} = 5$ mA. $R_1 = 5$ K. Design the circuit for a 5-V peak output at $\omega_0 = 10$ Hz. Specify all circuit parameters. Give reasons for your choice. Make all reasonable approximations in your analysis.

Bibliography

Anderson, F. B.: Seven-league Oscillator, *Proc. IRE*, vol. 39, no. 8, pp. 881–890, 1951.

Bode, H. W.: "Network Analysis and Feedback Amplifier Design," D. Van Nostrand Company, Inc., Princeton, N.J., 1945.

Bollman, J. H., and J. G. Kreer, Jr.: Application of Thermistors to Control Networks, *Proc. IRE*, vol. 38, no. 1, pp. 20–26, 1950.

Bothwell, F. E.: Nyquist Diagrams and the Routh-Hurwitz Stability Criterion, *Proc. IRE*, vol. 38, no. 11, pp. 1345–1348, 1950.

Chance, B., et al.: "Waveforms," M.I.T. Radiation Laboratory Series, vol. 19, McGraw-Hill Book Company, New York, 1949.

Clarke, K. K.: Design of Self-limiting Transistor Sine-wave Oscillators, *IEEE Trans. Circuit Theory*, vol. CT-13, no. 1, pp. 58–66, March, 1966.

Edson, W. A.: "Vacuum-tube Oscillators," John Wiley & Sons, Inc., New York, 1953.

Ginzton, E. L., and L. M. Hollingsworth: Phase-shift Oscillators, *Proc. IRE*, vol. 29, no. 1, pp. 43–49, 1941; also corrections, vol. 32, no. 10, p. 461, 1944.

Hooper, D. E., and A. E. Jackets: Current Derived Resistance-Capacitance Oscillators Using Junction Transistors, *Electron. Eng.*, vol. 28, pp. 333–337, August, 1956.

Keonjian, E.: Variable Frequency Transistor Oscillators, *Elec. Eng.*, vol. 74, no. 8, pp. 672–675, 1955.

Llewellyn, F. B.: Constant-frequency Oscillators, *Proc. IRE*, vol. 19, no. 12, pp. 2063–2094, 1931.

Meacham, L. A.: The Bridge Stabilized Oscillator, *Bell System Tech. J.*, vol. 17, pp. 574–590, 1938; also *Proc. IRE*, vol. 26, no. 10, pp. 1278–1294, 1938.

Nyquist, H.: Regeneration Theory, *Bell System Tech. J.*, vol. 11, pp. 126–147, 1932.

17. negative-resistance oscillators

The energy needed to sustain oscillations can be supplied by shunting the dissipative tuned network with the negative driving-point resistance of an active element. The solution of this problem is well defined, and from it we can draw some general conclusions as to the behavior of even those circuits where it is difficult to isolate the two terminals across which the negative impedance is developed. This chapter serves not only as an extension of the previous techniques but also as an introduction to the difficult problems posed by nonlinear differential equations.

17-1 Basic Circuit Considerations

A two-terminal self-starting oscillator must be biased within the negative-resistance region of the device. For a VNR, this implies constant-voltage biasing. Since the bias network should not affect the tuned circuit, in Fig. 17-1*a* we employ series biasing through the low resistance of the tank inductance.

The response of the voltage-controlled circuit of Fig. 17-1*a*, with which we are initially concerned, is defined by the single-node equation

$$C\frac{dV}{dt} + GV + \frac{1}{L}\int V\,dt + I_n = 0 \tag{17-1}$$

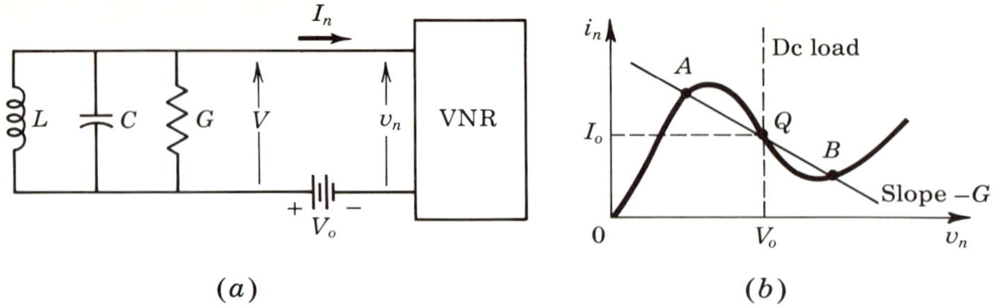

Fig. 17-1 (a) Basic negative-resistance oscillator; (b) typical volt-ampere characteristic with superimposed dc and conductive load lines.

where the bottom of the GLC circuit is taken as the reference node. Since

$$V = v_n - V_o \qquad \text{and} \qquad I_n = I_o + i_n(t)$$

by differentiating Eq. (17-1) and collecting terms, the time-varying response will be given by the solution of

$$\frac{d^2v_n}{dt^2} + \frac{1}{C}\left(G\frac{dv_n}{dt} + \frac{di_n}{dt}\right) + \frac{v_n}{LC} = 0 \tag{17-2}$$

For small-signal operation,

$$\frac{di_n}{dt} = \frac{di_n}{dv_n}\frac{dv_n}{dt} = g\frac{dv_n}{dt}$$

where g is the incremental conductance of the VNR measured at the Q point of the volt-ampere characteristic (point Q in Fig. 17-1b). With the above substitution, the basic incremental defining equation may finally be written as a quasi-linear equation:

$$\ddot{v}_n + \frac{1}{C}(G + g)\dot{v}_n + \frac{1}{LC}v_n = 0 \tag{17-3}$$

The linearization of the differential equation, which results from the implied assumption of a constant value of g, will not be valid once the signal amplitude carries the operating path into the dissipative regions of the volt-ampere characteristics. To find the amplitude of oscillation, the nonlinear differential equation of the system must be solved. Before treating this more difficult problem, we shall find the necessary and sufficient conditions for the circuit of Fig. 17-1 to sustain almost sinusoidal oscillations from the quasi-linear equation.

Equation (17-3) has two roots, which are located at

$$\begin{aligned} s_{1,2} &= -\frac{1}{2C}(G + g) \pm j\sqrt{\frac{1}{LC} - \frac{1}{4C^2}(G + g)^2} \\ &= \alpha \pm j\omega \end{aligned} \tag{17-4}$$

These roots determine the nature of the time-varying response of the negative-resistance oscillator, and therefore we shall investigate their location as a function of g. In order to have a growing transient, $\alpha \geq 0$, which corresponds to

$$G < -g \tag{17-5}$$

Equation (17-5) can be satisfied only when the conductive load line intersects both the dissipative and the negative conductance portions of the VNR characteristic. If there is only a single intersection of the negative portion of the curve, the response will decay instead of increasing with time.

The second restriction which must be imposed on the roots, if the response is to be oscillatory, is that ω must be real. From Eq. (17-4),

$$\frac{4C}{L} > (G + g)^2 \tag{17-6}$$

The boundaries delineating the inequality of Eq. (17-6) are two straight lines which satisfy

$$-G + \sqrt{\frac{4C}{L}} \geq g \geq -G - \sqrt{\frac{4C}{L}}$$

These regions are shown in Fig. 17-2. We are interested in only that portion of the plane below the $g = -G$ line, because it is here that a self-

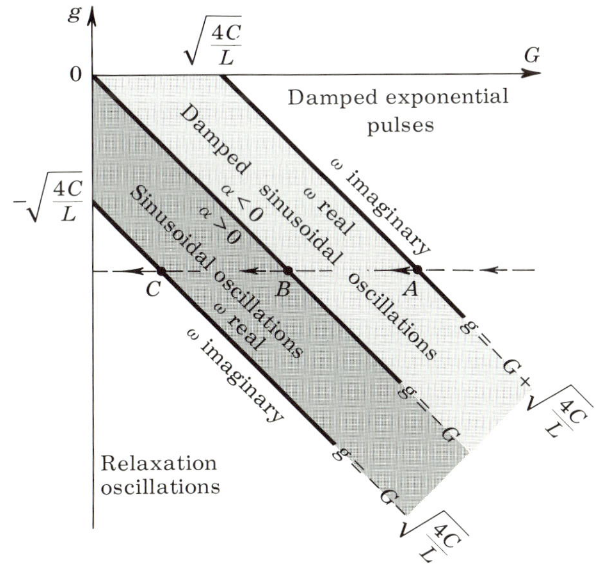

Fig. 17-2 Modes of operation of the circuit of Fig. 17-1 as a function of the external conductance G and the internal negative conductance g.

excited signal is generated. We are primarily concerned with the narrow segment defining the sinusoidal oscillatory response.

At the threshold of sustained oscillations, i.e., when $G = -g$, the external conductance is tangent, at the Q point, to the volt-ampere characteristic and the negative conductance of the active element exactly cancels the positive conductance of the tuned circuit. The purely imaginary poles are located at

$$s_{1,2} = \pm j \frac{1}{\sqrt{LC}}$$

For any other value of the external conductance lying in the oscillatory zone, the load line also intersects the positive-resistance regions, and as the oscillations build up, these segments will limit the amplitude to a finite value.

When G is small compared with the magnitude of the negative conductance, ω becomes imaginary and the circuit operates as an astable multivibrator. In the relaxation region, the response is similar to that discussed in Chap. 11, where the second energy-storage element appeared as part of the driving-point impedance of the active element instead of as part of the external load. To return the operating point to the sinusoidal oscillation region, we can load down the tuned circuit and hence reduce its effective Q.

We might further note that as $\sqrt{L/C}$ is increased, the width of the oscillatory region shrinks. A circuit operating properly under load may switch into the relaxation zone once the load is removed. Since the design of the tank controls the width of the oscillatory region, by minimizing $\sqrt{L/C}$, the performance of the oscillator becomes much less critical.

The possible modes of system operation may also be explained by following the migration of the system's poles as G is reduced toward zero (as along the dashed line of Fig. 17-2). Initially, the circuit is dissipative and the two poles lie on the real axis (Fig. 17-3). As G decreases, the poles first coalesce and then separate along the arcs of a circle. This occurs at point A in Figs. 17-2 and 17-3. When $G = -g$, they lie on the imaginary axis. A further reduction in G permits an exponentially increasing output, with the poles moving into the right half plane; the negative conductance now predominates. Eventually they again coalesce (point C in Fig. 17-2) and separate along the positive real axis. Here the circuit operates as a relaxation oscillator. Consider also the plot of the roots of the negative-resistance switching circuit presented in Fig. 11-33.

CNR oscillator

If a CNR is employed as the energy source in a negative-resistance oscillator, then the normal mode of operation is satisfied by an external series-resonant

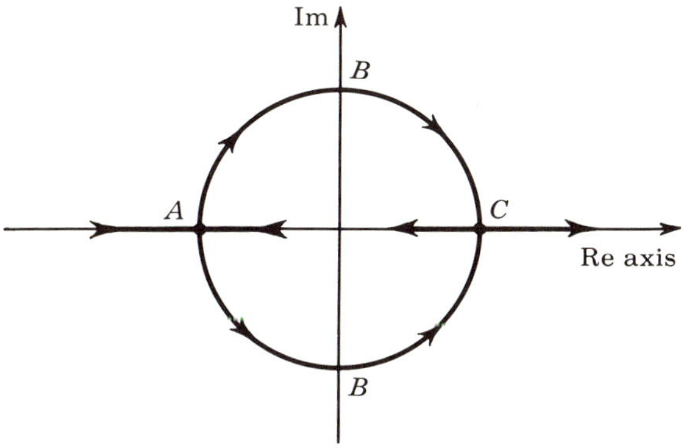

Fig. 17-3 Pole migration along the dashed path of Fig. 17-2, with decreasing values of G. Points of interest correspond to those labeled in Fig. 17-2.

circuit (Fig. 17-4). In the circuit of Fig. 17-4, we have assumed constant-current biasing in the negative-resistance region. Since this is the dual of Fig. 17-1, the single defining equation is written on a loop basis:

$$L\frac{dI}{dt} + RI + \frac{1}{C}\int I\,dt + v_n = 0$$

where $I = I_0 + i_n$. After differentiating and collecting terms, we obtain

$$\frac{di_n^2}{dt^2} + \frac{1}{L}\left(R + \frac{dv_n}{di_n}\right)\frac{di_n}{dt} + \frac{i_n}{LC} = 0 \tag{17-7}$$

which is the dual of Eq. (17-2). All arguments employed with respect to the voltage solution of Eq. (17-2) are immediately applicable to the current solution of Eq. (17-7).

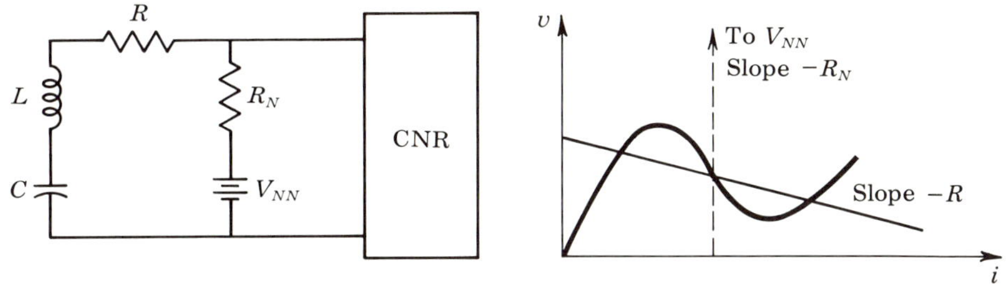

Fig. 17-4 A CNR oscillator using a series-resonant tuned circuit. Current biasing sets the Q point in the negative-resistance region.

For example, the roots of the quasi-linear differential equation, where we define

$$\frac{dv_n}{di_n} \equiv r$$

are now located at

$$s_{1,2} = -\frac{1}{2L}(R + r) \pm j\sqrt{\frac{1}{LC} - \frac{1}{4L^2}(R + r)^2}$$

and the new boundary conditions are defined by the inequalities

For $\alpha > 0$: $$R < -r$$

For ω real: $$-R + \sqrt{\frac{4L}{C}} > r > -R - \sqrt{\frac{4L}{C}}$$

These should be compared with the conditions previously derived for the VNR circuit and with the plot of Fig. 17-2.

17-2 A FET Tuned Oscillator as a Negative-resistance Circuit

In order to place some of the thoughts to be presented in this chapter in their proper perspective, we shall now present as a negative-resistance circuit the FET tuned-circuit oscillator which was treated from a feedback viewpoint in Sec. 16-4. Consider the general configuration of Fig. 17-5*a*, where all loss is lumped in $R_{\|}$. Making use of the fact that at resonance [Eq. (16-22)] the purely reactive tank elements are

$$X_1 + X_2 + X_3 + 2X_m = 0$$

the feedback factor becomes

$$\beta_0 = \frac{v_1}{v_2} = \frac{X_1 + X_m}{X_1 + X_m + X_3} = -\frac{X_1 + X_m}{X_2 + X_m}$$

Since X_1 and X_2 must be the same kind of element, β_0 is a negative number.

If we now substitute $\beta_0 v_2$ for v_1 in the controlled source, we find that the magnitude of the current-controlled source is proportional to the voltage across its terminals (Fig. 17-5*b*). Thus the source can be replaced by an equivalent negative conductance, as shown in Fig. 17-5*c*. The FET shunts the external $G_{\|} = 1/R_{\|}$ by the internal negative conductance

$$g_{mA} = -\frac{X_1 + X_m}{X_2 + X_m} g_m$$

For sinusoidal oscillations, this negative conductance must lie within the appropriate region of Fig. 17-2.

The amplitude of oscillation is found by means of a describing function. We can base these calculations on a piecewise-linear approximation to the

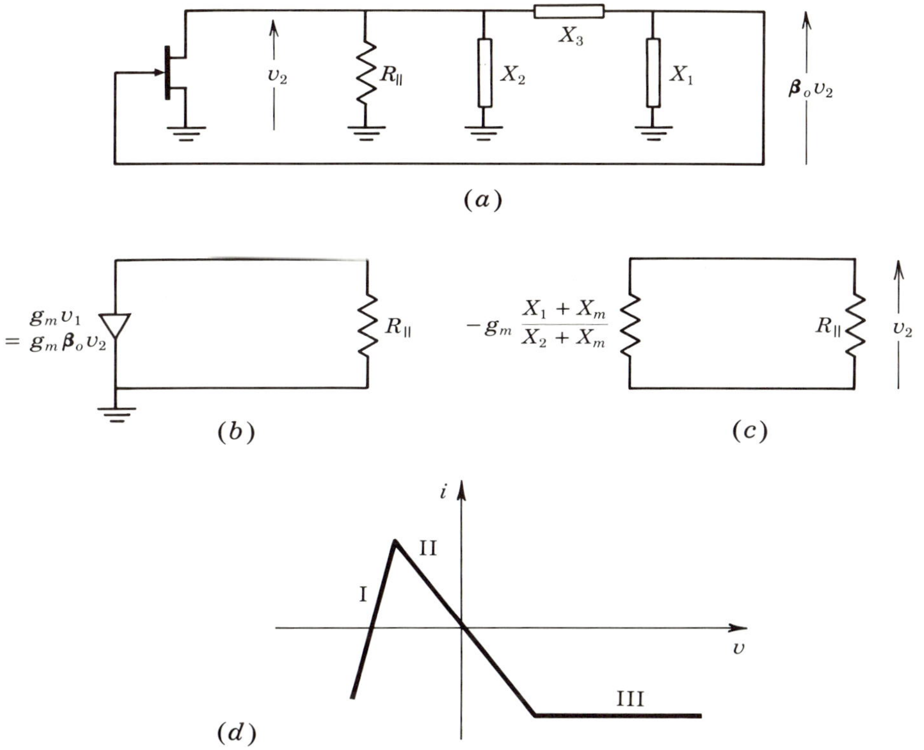

Fig. 17-5 (a) General FET tuned oscillator, all loss assumed lumped into $R_{\|}$; (b) reduced small-signal model at resonance; (c) further model reduction with the controlled source appearing as a negative conductance; (d) piecewise-linear approximation of the volt-ampere characteristic showing the two dissipative bounding regions.

actual characteristics plotted about the Q point in Fig. 17-5d. This sketch illustrates how the negative-conductance region II is bounded by the dissipative segments. Region I is where the gate-substrate diode conduction limits the amplification, and region III is where the FET is driven into cutoff.

17-3 First-order Solution for Frequency and Amplitude

In order to keep the nonlinearity in the forefront, Eq. (17-2), defining the complete response of the basic voltage-controlled negative-resistance oscillator of Fig. 17-1, may be rewritten as

$$\ddot{v}_n + \frac{G}{C}\left(1 + \frac{1}{G}\frac{di_n}{dv_n}\right)\dot{v}_n + \frac{1}{LC}v_n = 0 \tag{17-8}$$

This is in the form of the Van der Pol nonlinear differential equation

$$\ddot{v} + \epsilon f(v)\dot{v} + \omega_0^2 v = 0$$

For almost sinusoidal oscillations, $\epsilon f(v)$ must remain small over the range of interest. When it is zero, the Van der Pol equation reduces to the equation of simple harmonic motion,

$$\ddot{v} + \omega_0^2 v = 0$$

which has the solution

$$v = A \cos (\omega_0 t + \phi)$$

If the nonlinear term is large, then Eq. (17-8) can be easily solved only for some specific analytic functions of $\epsilon f(v)$.

To find the frequency and amplitude to a first approximation, we shall apply the method of equivalent linearization. Over any single cycle, we shall assume that the gross behavior of the system may be represented by an equivalent linear circuit. Any variations taking place during the individual cycle may, therefore, be neglected.

Choosing as a suitable trial solution for Eq. (17-8)

$$v_n = V_1(t) \cos \omega t$$

and substituting into Eq. (17-8) and collecting terms yield

$$\left[\frac{d^2V_1}{dt^2} - \omega^2 V_1 + \frac{V_1}{LC} + \frac{G}{C}\left(1 + \frac{1}{G}\frac{di_n}{dv_n}\right)\frac{dV_1}{dt}\right] \cos \omega t - \left[2\omega \frac{dV_1}{dt} + \omega V_1 \frac{G}{C}\left(1 + \frac{1}{G}\frac{di_n}{dv_n}\right)\right] \sin \omega t = 0 \qquad (17\text{-}9)$$

For almost sinusoidal oscillations, we may further assume that $V_1(t)$ remains essentially constant over any single cycle of the steady-state response. Since

$$\frac{dV_1}{dt} \approx 0 \qquad \text{and} \qquad \frac{d^2V_1}{dt^2} \approx 0$$

Eq. (17-9) reduces to

$$\left(-\omega^2 + \frac{1}{LC}\right) \cos \omega t - \frac{\omega G}{C}\left(1 + \frac{1}{G}\frac{di_n}{dv_n}\right) \sin \omega t = 0 \qquad (17\text{-}10)$$

To evaluate the average response, Eq. (17-10) is now multiplied by $\cos \omega t$ and integrated over a complete cycle:

$$\int_0^{2\pi} \left(-\omega^2 + \frac{1}{LC}\right) \cos^2 \omega t \, d(\omega t) - \int_0^{2\pi} \frac{\omega G}{C}\left(1 + \frac{1}{G}\frac{di_n}{dv_n}\right) \sin \omega t \cos \omega t \, d(\omega t) = 0$$

If the circuit is adjusted close to the threshold of oscillation (to minimize the distortion of the sinusoid), then the degree of limiting, and hence the nonlinearity, is small. The coefficient of $\sin \omega t \cos \omega t$ in the second integral is this very nonlinearity. It plays a major role in determining the amplitude, but since it is small, it will have a relatively minor effect on the frequency. By assuming that this coefficient is essentially constant over the period, the second definite integral would be identically equal to zero. Consequently, the approximate frequency of oscillation, given by the vanishing of the coefficient in the first integral, is the natural frequency of the tuned circuit,

$$\omega_0 = \frac{1}{\sqrt{LC}}$$

Oscillation amplitude

We shall now investigate $V_1(t)$ by rewriting the fundamental circuit equation. With the assumed cosinusoidal solution, Eq. (17-2) becomes

$$-\frac{di_n}{dt} = \left[C\frac{d^2V_1}{dt^2} + G\frac{dV_1}{dt} + \left(\frac{1}{L} - \omega^2 C\right)V_1\right]\cos \omega t - \left(2\omega C\frac{dV_1}{dt} + \omega G V_1\right)\sin \omega t \qquad (17\text{-}11)$$

By again assuming that all coefficients are approximately constant over the cycle, the first expression on the right side of Eq. (17-11) may be eliminated by multiplying through by $\sin \omega t$ and integrating over a complete cycle. The equation of interest reduces to

$$-\omega \int_0^{2\pi} \frac{di_n}{d(\omega t)} \sin \omega t \, d(\omega t) = -\left(2\omega C\frac{dV_1}{dt} + \omega G V_1\right)\int_0^{2\pi} \sin^2 \omega t \, d(\omega t) \qquad (17\text{-}12)$$

The left side of Eq. (17-12) may now be integrated by parts:

$$-\omega \int_0^{2\pi} \frac{di_n}{d(\omega t)} \sin \omega t \, d(\omega t) = -\omega(i_n \sin \omega t)\Big|_0^{2\pi} + \omega \int_0^{2\pi} i_n \cos \omega t \, d(\omega t)$$

$$= \omega \int_0^{2\pi} i_n \cos \omega t \, d(\omega t) \qquad (17\text{-}13)$$

But the right side of Eq. (17-13) is proportional to the peak value of the fundamental component of the current flow into the VNR. Thus the solution of Eq. (17-12) is

$$\omega\pi I_{1m} = -\left(2\omega C\frac{dV_1}{dt} + \omega G V_1\right)\pi$$

or

$$-\frac{dV_1}{dt} = \frac{1}{2C}(I_{1m} + GV_1) \qquad (17\text{-}14)$$

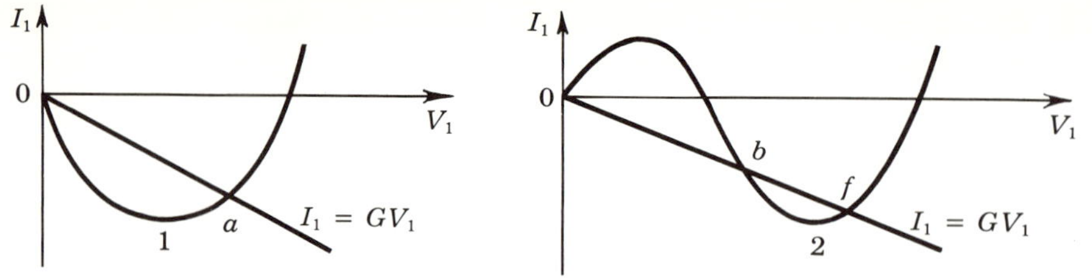

Fig. 17-6 Two current-describing functions for the basic nonlinear oscillator of Fig. 17-1: curve 1 results from a circuit biased in the center of the negative-conductance region, and curve 2 results from a circuit biased in the positive-resistance region near the peak point.

At equilibrium, the oscillations are constant, $dV_1/dt = 0$, and

$$I_{1m} = -GV_1 \tag{17-15}$$

The amplitude satisfying Eq. (17-15) is found by plotting a describing function, i.e., the fundamental component of the current into the VNR, vs. the applied-voltage excitation and solving for the intersection with the conductance line. Two typical curves appear in Fig. 17-6. Curve 1 appears when the circuit is normally biased within the negative-conductance region. For small signals, the current and voltage are out of phase. As the amplitude of oscillation increases, the operating point is driven into the dissipative regions for a portion of the cycle and the fundamental current begins decreasing. The second curve corresponds to a device normally biased within the dissipative region, such as at point A or B in Fig. 17-1. For small signals, the driving-point impedance is positive and the current is in phase with the applied excitation. A larger amplitude, where the average conductance over a cycle is negative, permits sustained oscillations.

In order to see which of the multiple intersections represent stable points of operation, we return to Eq. (17-14) and consider the effects of a small variation in V_1 about the amplitude satisfying Eq. (17-15).

$$\frac{dV_1}{dt} = -\frac{1}{2C}[I_e + \Delta I + G(V_{1e} + \Delta V_1)]$$

and I_e and V_{1e} are the values at the equilibrium point. From Eq. (17-14) we know that $I_e + GV_{1e} = 0$. Consequently,

$$\frac{dV_1}{dt} = -\frac{1}{2C}\left(\frac{\Delta I}{\Delta V_1} + G\right)\Delta V_1 \tag{17-16}$$

For stability, any change in V_1 must establish the conditions that will force a return to the equilibrium point. It follows that if the initial perturbation ΔV_1 is positive and momentarily increases the output, dV_1/dt must be negative in order for the voltage to decrease with time back to the stable amplitude. This is possible only when the term in the parentheses of Eq. (17-16) is positive. The condition for stability becomes

$$\frac{dI}{dV_1} + G > 0 \tag{17-17}$$

Equation (17-17) says that the total incremental conductance at the stable point must be positive, or in other words, the circuit must appear dissipative with respect to small variations about the equilibrium point.

In order for any oscillator to be self-starting, the origin must be a point of unstable equilibrium. The oscillator of curve 1 will build up to point a, the single stable point. However, if the quiescent point is along the positive-resistance segment, then the origin will be stable and the circuit cannot start by itself (curve 2). Once the circuit is externally excited to, or past, the unstable point b, the buildup will continue to the second point of stable equilibrium, point f.

To evaluate the describing function analytically, the characteristic of the VNR may be approximated by a power series which is written with respect to the quiescent point:

$$i_n = a_1 v_n + a_2 v_n^2 + a_3 v_n^3 + \cdots \tag{17-18}$$

where the various coefficients are found from the curve. Because of the nature of the external circuit (which is a relatively high-Q tank), only the fundamental component of voltage may be developed across the terminals of the nonlinear element. Furthermore, when the nonlinearity is small, the fundamental is the only current component of interest.

Substituting $v_n = V_1 \cos \omega t$ into Eq. (17-18) and collecting the coefficients of the $\cos \omega t$ terms result in

$$I_1 = a_1 V_1 + \tfrac{3}{4} a_3 V_1^3 + \tfrac{5}{8} a_5 V_1^5 + \cdots \tag{17-19}$$

For each value of the assumed voltage excitation, the current response term is now known. Plotting I_1 as a function of V_1 gives the describing functions shown in Fig. 17-6.

The stable amplitude may also be found by realizing that at the threshold of oscillation, the negative conductance of the active element exactly cancels the damping of the tuned circuit. Again, only the fundamental components are of interest; at the harmonics the current and voltage are in quadrature and do not represent power supplied to the tank. From Eq. (17-19), the average conductance of the VNR over a cycle is

$$G_n = \frac{I_1}{V_1} = a_1 + \frac{3}{4} a_3 V_1^2 + \frac{5}{8} a_5 V_1^4 + \cdots \tag{17-20}$$

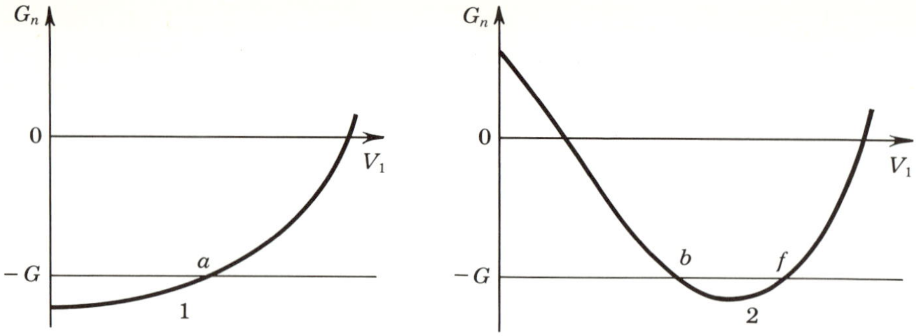

Fig. 17-7 Two conductance describing functions. Curve 1 holds when the quiescent point is at the center of the negative-conductance region, and curve 2, when the quiescent point is in the positive-conductance region near the knee of the curve. These curves correspond to the two current-describing functions of Fig. 17-6.

Equation (17-20) may be plotted as shown in Fig. 17-7 (the two curves drawn correspond to the two current-describing functions of Fig. 17-6). By superimposing the G line on the negative-conductance-describing function, the amplitude of oscillation satisfying $G_n + G = 0$ is finally found.

Example 17-1

A device having the almost symmetrical piecewise-linear characteristics of Fig. 17-8 is used in the basic oscillator circuit of Fig. 17-1. It is initially biased in the center of the negative-resistance region, at Q_1, where $I_o = 40$ mA and $V_o = 15$ V.

By taking three coordinates on the curve, measured with respect to the Q point, we can construct a cubic that will adequately approximate the characteristic. Choosing the following points: (-10 V, 0 mA), (-5 V, 20 mA), and ($+5$ V, -20 mA), the approximating polynomial becomes

$$i_1 = -5.33v + 0.0533v^3 \qquad \text{mA}$$

Thus the conductance-describing function, as given by Eq. (17-20), is

$$G_{n1} = -5.33 + 0.04V_1^2 \qquad \text{mmhos}$$

and it is plotted in Fig. 17-8b. When, for example, the conductance of the tank is 3.5 mmhos, the circuit will oscillate with a peak amplitude of 6.75 V.

Suppose that the quiescent point is shifted to Q_2, where $V_o = 12$ V; by how much would the amplitude change? Rather than solve for a new describing polynomial, for the purposes of this discussion we can simply

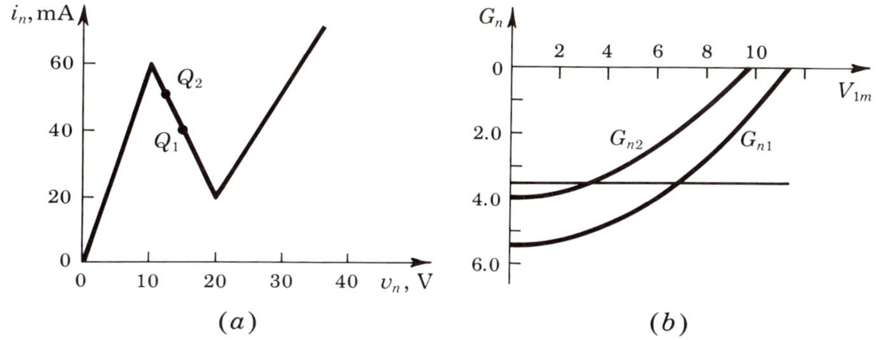

Fig. 17-8 (a) Volt-ampere characteristic; (b) the conductance-describing functions for Example 17-1.

shift the previously derived curve to correspond to the new origin. The shifted curve is defined by

$$i_2 \approx -5.33(v - 3) + 0.0533(v - 3)^3 + A$$

where the constant A is included to make $i = 0$ when $v = 0$. This equation reduces to

$$i_2 \approx -3.89v - 0.48v^2 + 0.0533v^3 \qquad \text{mA}$$

Consequently, the new conductance-describing function is characterized by

$$G_{n2} \approx -3.89 + 0.04V_1^2 \qquad \text{mmhos}$$

and the new peak amplitude stabilizes at 3.5 V.

Whenever the device characteristics can be approximated in a piecewise-linear manner, the current response to a sinusoidal voltage drive can be resolved into a truncated sinusoid over the negative-resistance region and two sinusoid tips within the bounding dissipative regions. The harmonic analysis of this waveshape will follow the techniques discussed in Sec. 16-6. After the current tips are redefined in terms of the device constraints, the plots of Fig. 16-23 can be used to find the harmonic content.

17-4 Frequency of Oscillation to a Second Approximation

First method

In order to obtain a more accurate expression as to the frequency of oscillation, we shall now solve the Van der Pol equation with the simple assumption that the voltage waveshape is periodic. This equation is

$$\ddot{v} + \epsilon f(v)\dot{v} + \omega_0^2 v = 0$$

where $\omega_0^2 = 1/LC$ and where

$$\epsilon f(v) = \frac{1}{C}\left(G + \frac{di_n}{dv_n}\right)$$

for the general circuit of Fig. 17-1. As the first step toward the evaluation of the oscillation frequency, the Van der Pol equation is multiplied by the unknown solution and integrated over the unknown period:

$$\int_0^{2\pi} \ddot{v}v\, d\omega t + \epsilon \int_0^{2\pi} f(v)\dot{v}v\, d\omega t + \omega_0^2 \int_0^{2\pi} v^2\, d\omega t = 0 \qquad (17\text{-}21)$$

Let us now consider each term of Eq. (17-21) individually, starting with the one that contains the system's nonlinearity,

$$F(v) = \epsilon \int_0^{2\pi} f(v)\dot{v}v\, d\omega t \qquad (17\text{-}22)$$

By expanding the volt-ampere characteristic into a power series,

$$i = a_1 v + a_2 v^2 + a_3 v^3 + a_4 v^4 + \cdots$$

the nonlinearity becomes

$$f(v) = G + \frac{di}{dv} = G + a_1 + 2a_2 v + 3a_3 v^2 + \cdots \qquad (17\text{-}23)$$

The necessary condition of a periodic solution leads to the expression of v and $\dot{v}$ as Fourier series:

$$v = \sum_{k=1}^{\infty} V_k \cos(k\omega t + \phi_k) \qquad (17\text{-}24a)$$

and

$$\dot{v} = -\sum_{k=1}^{\infty} k\omega V_k \sin(k\omega t + \phi_k) \qquad (17\text{-}24b)$$

Consequently $v\dot{v}$ consists solely of terms which are the product of two time-varying signals in quadrature, a sinusoid and a cosinusoid; it does not contain any dc terms. The substitution of v into Eq. (17-23) and the collection of terms will result in a dc component and a series of cosinusoidal terms, i.e., the fundamental and all harmonics. The integral of interest [Eq. (17-22)] reduces to the evaluation of various factors of the form

$$\int_0^{2\pi} (\text{dc} + \cos n\omega t)(\cos k\omega t)(\sin m\omega t)\, d\omega t$$

over a complete cycle of the fundamental, where k, m, and n are nonzero positive integers. But such an integral is always equal to zero. Thus we have proved that

$$F(v) = \epsilon \int_0^{2\pi} f(v)\dot{v}v\, d\omega t = 0$$

Returning to Eq. (17-21), the first integral may be evaluated by parts:

$$\int_0^{2\pi} \ddot{v}v \, d\omega t = \omega \int_0^{2\pi} \frac{d\dot{v}}{d\omega t} v \, d\omega t = \omega \int_{\omega t=0}^{\omega t=2\pi} v d(\dot{v})$$

$$= \omega v\dot{v} \Big|_{\omega t=0}^{\omega t=2\pi} - \omega \int_{\omega t=0}^{\omega t=2\pi} \dot{v}^2 \, dt \tag{17-25}$$

It follows directly from the argument employed with respect to the integral of Eq. (17-22) that

$$v\dot{v} \Big|_{\omega t=0}^{\omega t=2\pi} = 0$$

and as a result

$$\int_0^{2\pi} \ddot{v}v \, d\omega t = - \int_0^{2\pi} \dot{v}^2 \, d\omega t \tag{17-26}$$

The right-hand integral may be identified as the mean-square value of the derivative. Equation (17-21) finally reduces to

$$\int_0^{2\pi} \dot{v}^2 \, d\omega t = \omega_0{}^2 \int_0^{2\pi} v^2 \, d\omega t \tag{17-27}$$

Equation (17-27) is equivalent to equating the system's kinetic energy to its potential energy, normalizing, and averaging over a cycle. This equation, subject to appropriate boundary conditions, could well have been the starting point for the analysis of the nonlinear oscillator.

When the Fourier series of Eqs. (17-24*a*) and (17-24*b*) are substituted into Eq. (17-27), only the squared terms will contribute to the final answer; all cross products integrated over a cycle equal zero. The evaluation of Eq. (17-27), consistent with the assumed periodicity of the solution, results in

$$\tfrac{1}{2} \sum_{k=1}^{\infty} k^2\omega^2 V_k{}^2 = \tfrac{1}{2}\omega_0{}^2 \sum_{k=1}^{\infty} V_k{}^2$$

or

$$\left(\frac{\omega}{\omega_0}\right)^2 = \frac{\sum_{k=1}^{\infty} V_k{}^2}{\sum_{k=1}^{\infty} k^2 V_k{}^2} \tag{17-28}$$

Equation (17-28) proves that when limiting introduces distortion terms, the frequency of oscillation is always depressed below the nominal resonant frequency of the tuned circuit, and that the greater the distortion, the greater the frequency depression. This equation is somewhat difficult to use in the form derived, since for small amounts of distortion, $\omega \approx \omega_0$. To simplify the expression, let us consider

$$1 - \frac{\omega^2}{\omega_0{}^2} = 1 - \frac{\sum_{k=1}^{\infty} V_k{}^2}{\sum_{k=1}^{\infty} k^2 V_k{}^2} = \frac{\sum_{k=1}^{\infty} (k^2 - 1) V_k{}^2}{\sum_{k=1}^{\infty} k^2 V_k{}^2} \tag{17-29}$$

However,

$$1 - \frac{\omega^2}{\omega_0{}^2} = \frac{(\omega_0 + \omega)(\omega_0 - \omega)}{\omega_0{}^2} \approx \frac{-2\Delta\omega}{\omega_0}$$

where the approximation holds when ω is close to ω_0 and where the frequency deviation is $\Delta\omega \equiv \omega - \omega_0$. Furthermore, by restricting the simplification of Eq. (17-29) to signals containing only slight distortion, the fundamental term greatly predominates. Because

$$V_1{}^2 \gg \sum_{k=2}^{\infty} k^2 V_k{}^2$$

the denominator of Eq. (17-29) may be approximated by $V_1{}^2$.

The frequency deviation may finally be expressed in the form

$$\frac{\Delta\omega}{\omega_0} \approx -\frac{1}{2} \sum_{k=2}^{\infty} (k^2 - 1) \frac{V_k{}^2}{V_1{}^2} \tag{17-30}$$

For example, when the VNR characteristics are symmetrical about the Q point, only odd-harmonic terms will be present in the output. If the principal one is the third harmonic and if it is 5 percent of the fundamental,

$$\frac{\Delta\omega}{\omega_0} = -\frac{1}{2} \times 8 \times \left(\frac{0.05V_1}{V_1}\right)^2 = -1.0\%$$

Contributions of the higher-harmonic components will depress the frequency even further.

Second method

Our second method of finding the frequency depression caused by the nonlinearity of the negative-conductance characteristic will be based on a power balance of the circuit. The ideal negative-resistance device, by its very nature, is unable to store any energy and is also unable to supply any reactive power to the external circuit. Like any characteristic that is a single-valued function of the independent variable, it does not exhibit hysteresis; consequently the line integral of $i\,dv$ over a cycle is

$$\oint i\,dv = 0 \tag{17-31}$$

We shall therefore sum up the reactive energy terms at each harmonic and set the result equal to zero. The particular frequency at which this condition is satisfied will be the frequency of oscillation.

As before, the unknown periodic voltage may be expressed as a Fourier series. When it excites the tuned circuit of Fig. 17-1a, the reactive component of the current at each harmonic will be given by

$$I'_k = I_k \sin \phi_k = B_k V_k \tag{17-32}$$

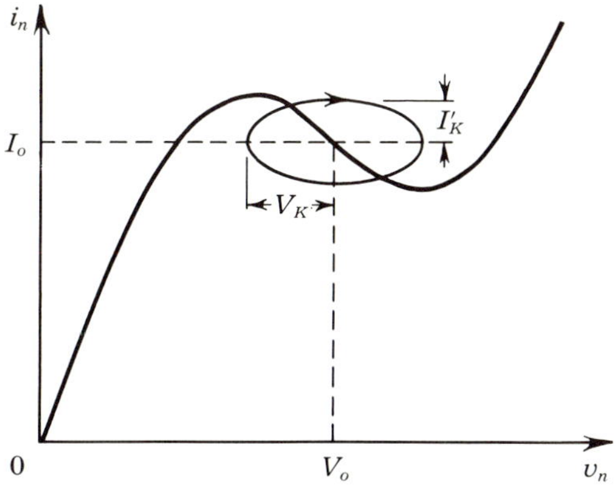

Fig. 17-9 Negative-resistance characteristic showing the path traversed by the harmonic components of voltage and current.

where B_k is the susceptance at the kth harmonic. I'_k is in quadrature with the voltage component of the same frequency. The trajectory traversed on the volt-ampere characteristic, due to this current, is an ellipse having the area

$$A'_k = \pi V_k I'_k = \pi V_k I_k \sin \phi_k$$

This path is shown in Fig. 17-9.

During one period of the fundamental, the harmonic circumscribes this path k times, tracing out the area

$$A_k = \pi k V_k I_k \sin \phi_k = \pi k V_k^2 B_k \tag{17-33}$$

But this is simply $2\pi k$ times the reactive energy. Since no energy is stored by the VNR, Eq. (17-31) must be satisfied. That condition may now be expressed as

$$\sum_{k=1}^{\infty} A_k = 0 = \pi V_1^2 B_1 + \pi \sum_{k=2}^{\infty} k V_k^2 B_k \tag{17-34}$$

or, in terms of the reactive energy,

$$P_{1R} + \sum_{k=2}^{\infty} k P_{kR} = 0 \tag{17-35}$$

At some sufficiently high harmonic, usually at and above the second, the tuned circuit appears capacitive. In order for Eq. (17-34) to hold, B_1 must have a sign different from that associated with B_2, B_3, etc. This is possible only if the tank appears inductive at the fundamental frequency. We have therefore proved once again that the frequency of oscillation must be below the resonant frequency of the tuned circuit.

The susceptance of the tuned circuit may be expressed as

$$B = C\omega_0 \left(\frac{\omega}{\omega_0} - \frac{\omega_0}{\omega}\right) \tag{17-36}$$

In the vicinity of resonance, Eq. (17-36) may be approximated by

$$B_1 \approx 2C\,\Delta\omega$$

and at the harmonics, Eq. (17-36) reduces to

$$B_k \approx \omega_0 C \frac{k^2 - 1}{k}$$

Substituting these two terms into Eq. (17-34) and solving result in

$$\frac{\Delta\omega}{\omega_0} = -\frac{1}{2} \sum_{k=1}^{\infty} (k^2 - 1) \frac{V_k^2}{V_1^2}$$

which is identical with Eq. (17-30). Furthermore, since from Eq. (17-34)

$$I_1 V_1 \sin \phi_1 = -\sum_{k=2}^{\infty} k I_k V_k \sin \phi_k$$

the depression in frequency is also given by

$$\frac{\Delta\omega}{\omega_0} = -\frac{I_1 \sin \phi_1}{2V_1\omega_0 C} \tag{17-37}$$

where ϕ_1 is the phase angle of the fundamental current with respect to the fundamental voltage.

Synchronization

The automatic balancing of the circuit, which shifts the frequency to a point where the net reactive energy is equal to zero, may be utilized for frequency entrainment or synchronization. Suppose that an external signal that is some rational ratio of the desired fundamental is applied to the basic negative-resistance oscillator. Because the tank is primarily reactive at this frequency, the additional energy injected (P_{sk}) will be reactive and Eq. (17-35) will become

$$P_{1R} + \sum_{k=2}^{\infty} kP_{kR} + kP_{sk} = 0 \tag{17-38}$$

Depending on the phase of the injected signal, the effective reactive term P_{sk} can be either positive or negative; the frequency at which Eq. (17-38) is satisfied may be shifted in either direction.

For example, if we wish to return the frequency of the oscillator signal to ω_0, the injected synchronizing signal at $k\omega_0$ must be inductive. P_{sk} must be large enough to cancel all of the harmonic capacitive reactance terms introduced by the oscillator's nonlinearity. On the other hand, if we wish to depress the frequency slightly more, to some $\omega_1 < \omega_0$, then the injected signal at $k\omega_1$ must introduce additional capacitive reactive energy. Essentially, the injection of the external signal changes the effective susceptance of the tuned circuit.

Equation (17-38) establishes the minimum conditions for synchronization at a subharmonic of the injected signal P_{sk}. Since any increase in the energy of the synchronizing signal P_{sk} must be balanced by additional reactive energy at the synchronized frequency of oscillation, the phase of the fundamental must shift relative to the injected harmonic signal. Even a small increase in P_{sk} above the minimum needed for frequency entrainment will cause large shifts in the phase of oscillation without, however, further affecting the frequency.

17-5 Introduction to Topological Methods

Nonlinear systems which are described by a differential equation of the second order may also be studied with the aid of various topological constructions. This form of representation is important because it enables us to gain perspective as to the totality of the possible behavior of the system without having to solve the actual differential equation for the time response. The periodic solution will appear as a closed contour on some graph. And as the graphically presented nonlinearity will be used directly—without conversion to an approximating polynomial—grossly nonlinear as well as quasi-linear circuits may be treated. This is not to say that the topological solutions do not have their own limitations; they do. In the course of the following arguments some of the drawbacks will become apparent.

As a convenient starting point, consider the second-order linear differential equation

$$\ddot{x} + b\dot{x} + \omega_0^2 x = 0 \qquad (17\text{-}39)$$

Two initial conditions which give information as to the initial energy in the system are necessary for the complete solution; usually the position and velocity at $t = 0$ are specified.

For the simplest case of Eq. (17-39), we set $b = 0$, which reduces the problem to one of simple harmonic motion. The well-known solution for x in terms of time is

$$x = K \cos (\omega_0 t + \phi) \qquad (17\text{-}40)$$

But in order to evaluate the two constants, we usually also find the derivative:

$$y = \dot{x} = -\omega_0 K \sin (\omega_0 t + \phi) \qquad (17\text{-}41)$$

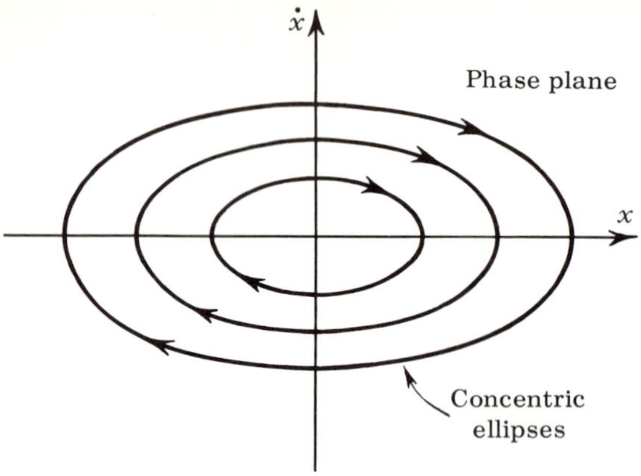

Fig. 17-10 Phase-plane portrait of simple harmonic motion [Eq. (17-42) with $b = 0$].

Since time appears in Eqs. (17-40) and (17-41) merely as a parameter, as an alternative form of presenting the answer, $\dot{x}$ may be plotted as a function of x.

Equations (17-40) and (17-41) are parametric equations of an ellipse. Eliminating t yields, as the solution of the original differential equation,

$$\frac{x^2}{K^2} + \frac{y^2}{\omega_0{}^2K^2} = 1 \tag{17-42}$$

The starting point of the plot on the x, $\dot{x}$ plane (hereafter called the *phase plane*) is simply the coordinates given by the initial conditions. Since two pieces of information uniquely determine any second-degree curve, one and only one ellipse, satisfying Eq. (17-42), may be drawn through each point on the plane. The totality of all such curves (Fig. 17-10) is called the phase portrait of the system.

Rotation of the trajectory about the origin is always in a clockwise direction. Counterclockwise rotation would indicate a decreasing displacement when the velocity is still positive and an increasing displacement for negative velocities—an obviously impossible situation. Furthermore, these ellipses must cross the x axis at right angles. The x intercepts are the points at which the velocity changes direction, and consequently they must also be the points of the displacement maximums. The above arguments lead directly to two additional conclusions:

1. All closed paths must encircle the origin.
2. All closed curves correspond to periodic motions.

The equilibrium point of the system is where $\dot{x}$ and $\ddot{x}$ vanish. In this example only the origin is stable, and it is therefore called a *singular* point.

If friction (or damping) is included in the system, then $b > 0$ and Eq. (17-39) has two possible solutions:

1. The damped oscillatory wave:

$$x = Ke^{-\alpha t} \cos (\omega_1 t + \phi) \tag{17-43}$$

which exists when the roots are complex conjugates,

$$s_{1,2} = -\alpha \pm j\omega_1$$

The phase portrait corresponding to Eq. (17-43) and its derivative appears in Fig. 17-11*a*. As the oscillatory wave damps out, the trajectory describes a logarithmic spiral about the origin, circling it once for each cycle of the cosinusoidal portion of the solution. Since the spiral eventually terminates at the origin, this singularity is now called a stable *focal* point.

2. The damped aperiodic wave which corresponds to the overdamped case:

$$x = K_1 e^{-\alpha_1 t} + K_2 e^{-\alpha_2 t} \tag{17-44}$$

The damped aperiodic wave of Eq. (17-44) cannot have more than one point of zero velocity (excluding the trivial point at $t = \infty$). Therefore the phase trajectory will intercept the x axis once—it cannot encircle the

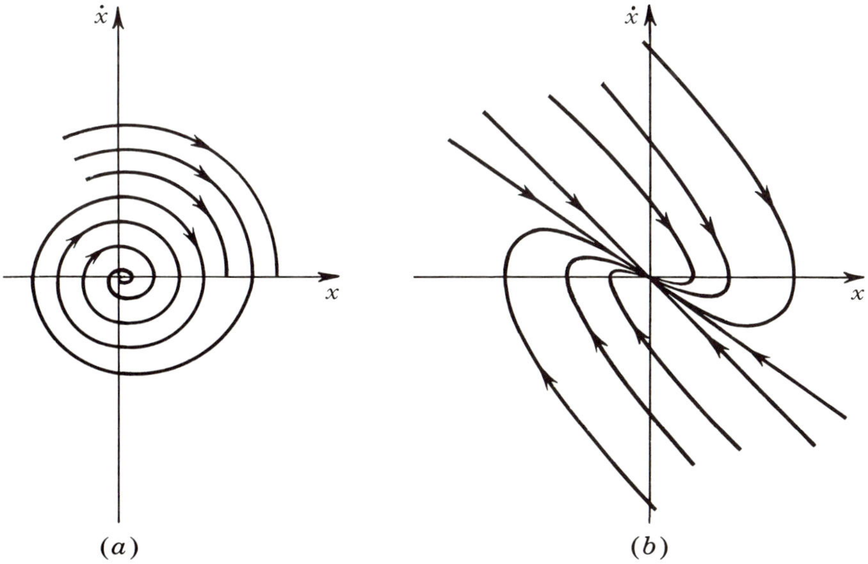

Fig. 17-11 Phase-plane trajectories of the linear second-degree equation. (*a*) Damped oscillatory response; (*b*) the aperiodic solution.

origin. In Fig. 17-11*b*, the origin becomes the stable *nodal* point of the system.

Isocline construction

The trajectories that represent the solution of the differential equation on the phase plane may be plotted directly from the equation by the method of isoclines. These are the locus of all points in the phase plane where the curve has the same slope. By making the substitution

$$y = \frac{dx}{dt}$$

Eq. (17-39) may be rewritten

$$\frac{dy}{dt} = -by - \omega_0^2 x \tag{17-45}$$

But in order to find the slope, Eq. (17-45) will be divided through by y, resulting in

$$S = \frac{dy}{dx} = \frac{dy}{dt}\frac{dt}{dx} = -b - \omega_0^2 \frac{x}{y} \tag{17-46}$$

Thus all points having the same slope S lie on a straight line through the origin, satisfying the equation

$$y = -\frac{\omega_0^2}{S + b} x \tag{17-47}$$

Once the isoclines are drawn, short-line segments of the appropriate slope may be marked off along each one, as shown in Fig. 17-12. These will serve as guides when drawing the phase portrait. Starting at any initial point, a short section of the curve can be drawn having the slope defined by the isocline passing through the starting point. From the end of the first section, the next section is drawn, and so on, each intersecting the next clockwise isocline with the proper slope until the curve is completed. The totality of the individual segments is the actual phase trajectory satisfying the initial conditions.

Such a construction is carried out in Fig. 17-12 for the damped oscillatory case. For the purpose of illustration, we set $\omega_0 = 1$ and $b = \frac{1}{4}$, reducing Eq. (17-47) to

$$y = -\frac{4}{4S + 1} x$$

The initial conditions, which give the starting point of the single trajectory drawn, are $x_i = -3$ and $\dot{x}_i = 5$.

Returning to Eq. (17-47), we might observe that the condition of an infinite slope is satisfied along the x axis. This proves the contention that the trajectories cross the x axis perpendicularly.

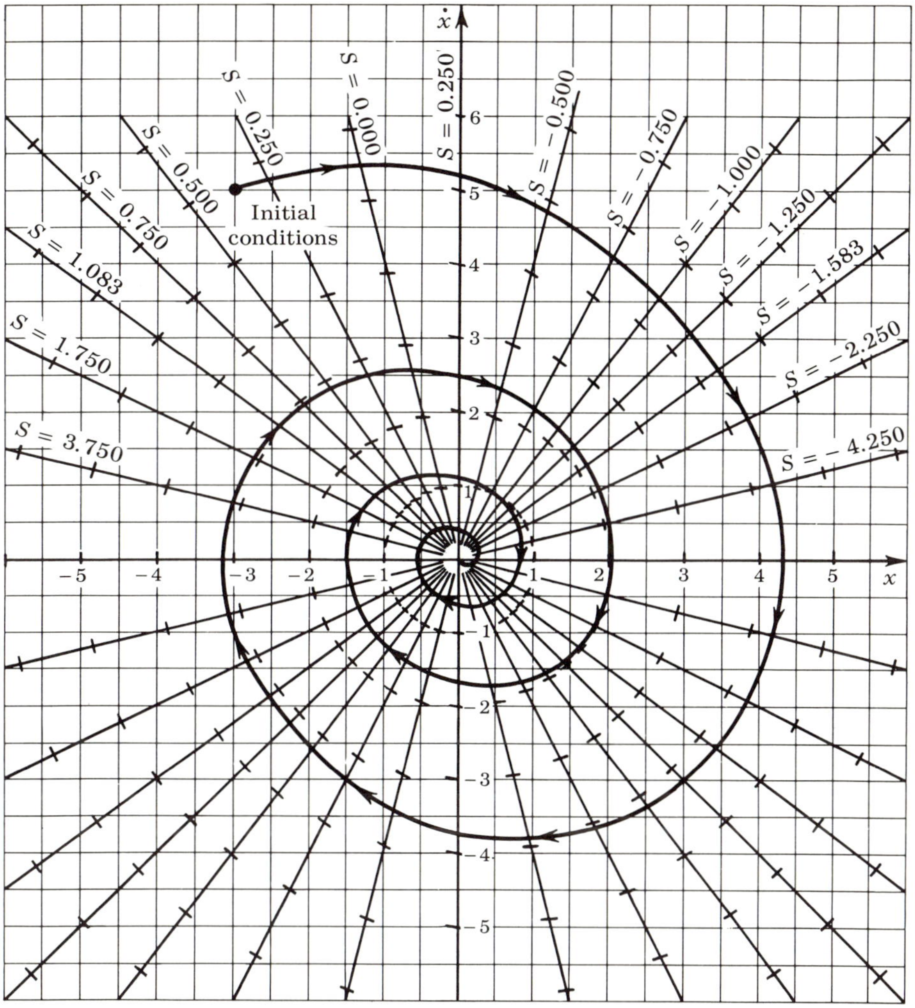

Fig. 17-12 Isocline method of constructing a phase-plane trajectory.

The most direct path to the stable point would be along an isocline. Since the slope of the trajectory is identical with that of the isocline, from Eq. (17-47) we see that the necessary condition for such a path is

$$S = -\frac{\omega_0^2}{S + b}$$

But by multiplying out and collecting terms, there results the quadratic

$$S^2 + bS + \omega_0^2 = 0$$

Except that it is a function of S, this quadratic is identical with the characteristic equation from which the roots of the original differential equation (17-39) are found.

Straight lines having a slope equal to the roots are feasible only when these roots are real, i.e., when the response is aperiodic. Two such lines exist on each phase plane, as shown in Fig. 17-11*b*. As they head toward the origin, all trajectories approach the line corresponding to the smallest root. It predominates, since it represents the long time decay of the system. The larger root determines the initial rise, and its isocline separates the trajectories that finally approach the origin in the second quadrant from those which terminate in the fourth quadrant.

Unstable equilibrium

If energy is supplied to the system at the proper rate, for example, by connecting the negative driving-point resistance across a *GLC* parallel circuit, the amplitude would increase rather than decrease with time. In the region where the circuit may be approximated by a quasi-linear equation, i.e., before limiting, the trajectories would diverge from the singularity at the origin, rotating clockwise as they did so. The two cases of most interest are shown in Fig. 17-13. The increasing spiral represents the buildup of an almost sinusoidal oscillator, and the aperiodic trajectory illustrates the switching path of a multivibrator. If the system has some initial energy, the single trajectory of interest would be the one passing through the point on the phase plane corresponding to the initial

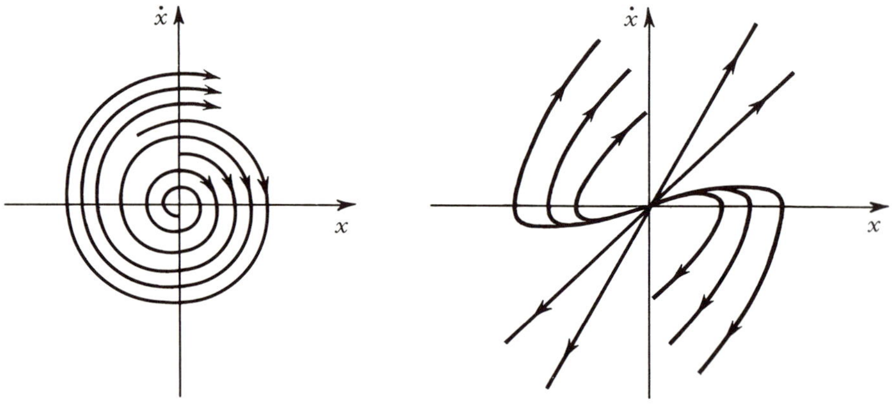

(*a*) (*b*)

Fig. 17-13 Trajectories for negative damping. (*a*) Increasing oscillatory response; (*b*) increasing aperiodic solution.

conditions. In the case of an increasing exponential, the trajectories will approach the isocline corresponding to the larger root of the second-order equation, regardless of how they diverge from the origin. If we are interested in the response very far away from the point corresponding to the initial conditions, then this particular isocline would closely approximate the path. Of course, nonlinearities in the system may distort the trajectories long before they ever reach this asymptote.

Period and waveshape

Topological methods may also be employed to convert the trajectory on the phase plane into the equivalent time-varying signal. We can construct a small line segment in the time domain having the slope and position given by the corresponding point on the phase plane. To the end of the first one, a second segment is added, then a third, and so on. As we work our way along the trajectory, the totality of all segments will give a rough approximation to the waveshapes generated. This process is, of course, quite tedious, but it is necessary where simpler methods cannot yield a satisfactory solution.

Figure 17-14 shows just such a construction for slightly more than one complete encirclement of the origin by the trajectory of Fig. 17-12. The initial point of the trajectory was chosen as the start of the construction. Wherever the slope is large, large increments of position are permissible over the associated line segment. Since this section of the curve does not contribute much toward the period, relatively gross approximations are justified. But when the operating point is moving very slowly, the positional increments must be quite close together. To further reduce the timing error, the average slope over each segment of the trajectory was used in the construction of Fig. 17-14.

The timing accuracy of this construction is very poor, and therefore it should be used mainly to identify the waveshape. In the example shown, the time for one complete cycle is almost 1 s too long.

The time required for the operating point to pass between any two points on the trajectory is found from the line integral:

$$\oint_a^b \frac{1}{y}\,dx = \oint_a^b \frac{dt}{dx}\,dx = \oint_a^b dt = T_b - T_a \tag{17-48}$$

In general, Eq. (17-48) is evaluated by means of a graphical or a numerical integration. For the complete period of any closed contour on the phase plane, the integral becomes

$$T = \oint \frac{1}{y}\,dx \tag{17-49}$$

and it is taken over the complete trajectory.

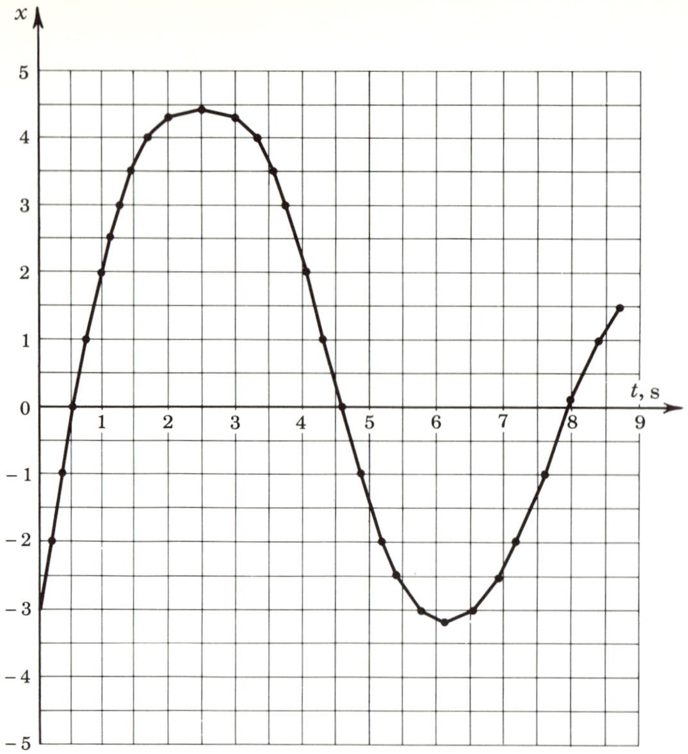

Fig. 17-14 Waveshape constructed from the trajectory of Fig. 17-12.

17-6 The Liénard Diagram

As the first step in finding the topological solution of the Van der Pol nonlinear differential equation, we shall normalize the basic equation (17-8) by dividing through by ω_0^2; this leads to the result

$$\frac{d^2v}{d(\omega_0 t)^2} + \frac{\epsilon}{\omega_0} f(v) \frac{dv}{d(\omega_0 t)} + v = 0 \tag{17-50}$$

Since the variables of interest are the voltage across the terminals of the VNR and its derivative, there should not be any ambiguity or confusion of terms if we redefine $\dot{v} = dv/d(\omega_0 t)$ and rewrite Eq. (17-50):

$$\ddot{v} + \epsilon' f(v)\dot{v} + v = 0 \tag{17-51}$$

And following the procedure outlined in Sec. 17-5, the slope of the trajectory, at each point in the phase plane, will be given by

$$S = \frac{d\dot{v}}{dv} = -\frac{\epsilon' f(v)\dot{v} + v}{\dot{v}} \tag{17-52}$$

The isoclines defined by Eq. (17-52) are no longer straight lines but are curves which depend on the nature of the nonlinearity. With any given volt-ampere characteristic, $f(v)$ may be evaluated for each value of v. The isoclines are finally constructed, if necessary, by plotting individual points.

As this process is extremely tedious, we shall instead turn to the simpler construction developed by A. Liénard. By means of a linear transformation from the phase plane to the Liénard plane, the problem becomes one of finding the normal to the trajectory and then simply striking an arc to find a segment of the topographical portrait. The transformation chosen is

$$z = \dot{v} + \epsilon' F(v) \tag{17-53}$$

where
$$f(v) = \frac{dF(v)}{dv} \tag{17-54}$$

By differentiating Eq. (17-53) with respect to v and rearranging terms, an alternative expression for the isoclines results:

$$S = \frac{d\dot{v}}{dv} = \frac{dz}{dv} - \epsilon' f(v)$$

Substituting S into Eq. (17-52) leads to

$$\frac{dz}{dv} = -\frac{v}{\dot{v}}$$

From the linear transformation of Eq. (17-53), the slope of the trajectory on the Liénard plane finally becomes

$$\frac{dz}{dv} = -\frac{v}{z - \epsilon' F(v)} \tag{17-55}$$

and the slope of the normal to the trajectory is given by

$$N = -\frac{dv}{dz} = \frac{z - \epsilon' F(v)}{v} \tag{17-56}$$

Before we can construct the trajectories, we must evaluate $\epsilon' F(v)$. For the basic VNR circuit of Fig. 17-1,

$$\frac{dF(v)}{dv} = f(v) = G + \frac{di}{dv}$$

Integrating both sides with respect to v yields

$$F(v) = Gv + i = i - (-Gv) \tag{17-57}$$

where v and i are measured from the Q point on the volt-ampere characteristics and where $-Gv$ is the conductive load line passing through the

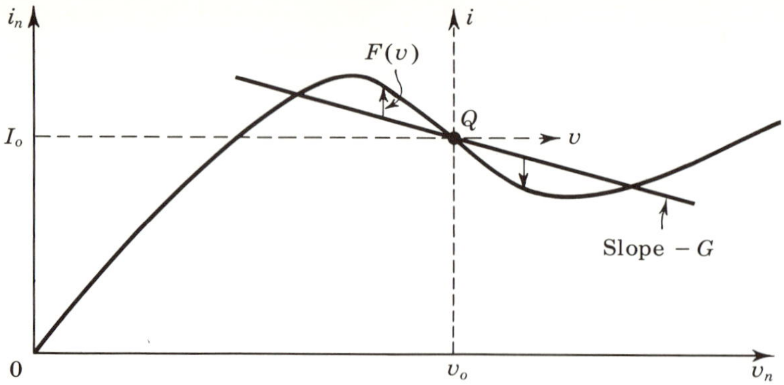

Fig. 17-15 Illustrating the means of evaluating $F(v)$ from the volt-ampere characteristic.

Q point. Equation (17-57) says that $F(v)$ is the difference in current between the load line and the characteristic (as shown by the vectors in Fig. 17-15).

In replotting $F(v)$, the ordinate would also be multiplied by ϵ' to scale the curve. This factor is dependent on the design of the tank and may, in fact, be considered a defining parameter:

$$\epsilon' = \frac{1}{\omega_0 C} = \sqrt{\frac{L}{C}} \tag{17-58}$$

The degree of the circuit nonlinearity is proportional to ϵ'; the system becomes quasi-linear for very small values of ϵ' and grossly nonlinear for very large values. We conclude that for an almost sinusoidal oscillator, L/C should be made as small as possible consistent with the other design requirements. This supports the arguments made earlier with respect to Figs. 17-2 and 17-3.

The final plot of $\epsilon' F(v)$, which serves as the basis for the construction of the Liénard diagram, appears in Fig. 17-16. The procedure followed in constructing first the normal and then the trajectory is based on Eq. (17-56). The various steps illustrated in Fig. 17-16 are:

1. Plot the nonlinearity $\epsilon' F(v)$, following the geometric construction outlined above, on the Liénard plane. This plot cannot be scaled in either coordinate but must be drawn 1:1.
2. At any point on the plane p_1, drop a perpendicular to $\epsilon' F(v)$ at q_1.
3. Draw a line from q_1 to the z axis at r_1.
4. The vertical line segment is

$$p_1 q_1 = z_1 - \epsilon' F(v_1)$$

at that particular value of z_1 and v_1. The horizontal line segment is

$$q_1r_1 = v_1$$

Hence, from Eq. (17-56), the hypotenuse of the triangle $r_1p_1q_1$ is the normal to the trajectory passing through point p_1.

5. Strike small arcs intersecting line segment p_1q_1 using point r_1 as the center. These arcs are segments of the various trajectories on the Liénard plane.
6. Starting at any initial point on the plane, strike an arc; from the end of this arc strike another one; and then continue the construction from one point to the next in a clockwise direction until the trajectory converges into a closed contour.

A construction for four points appears in Fig. 17-16, and complete trajectories, cycling to the closed curve representing a periodic solution, appear in Figs. 17-17 and 17-18. In Fig. 17-17 three different starting points, two close to the origin and one at $z = 30$, $v = 5$, were chosen. By following the procedure outlined above, three trajectories are constructed, and these all eventually form the same closed contour encircling the origin. This curve represents the only periodic solution of the system, and it is called a *limit cycle.* The path leading to the limit cycle, from any starting point, is the transient portion of the solution. Two trajectories in Fig. 17-17 show the exponential buildup. One, which

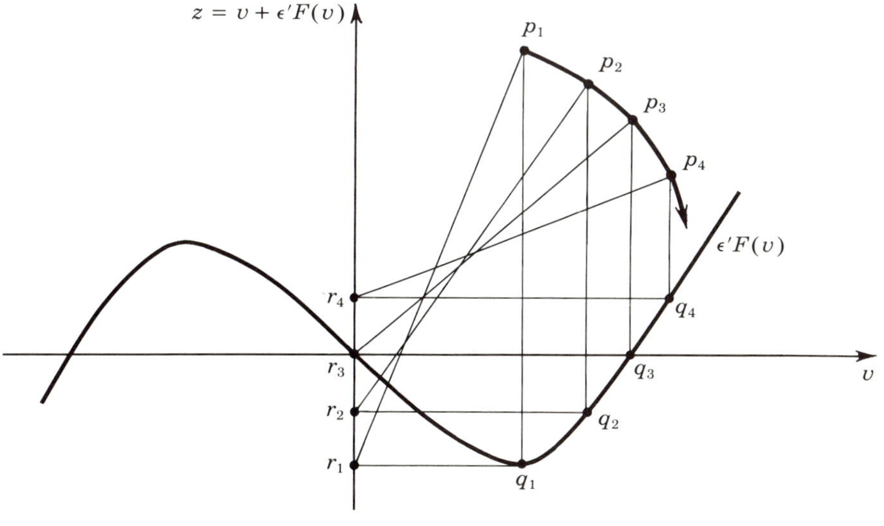

Fig. 17-16 Steps in the construction of the Liénard diagram.

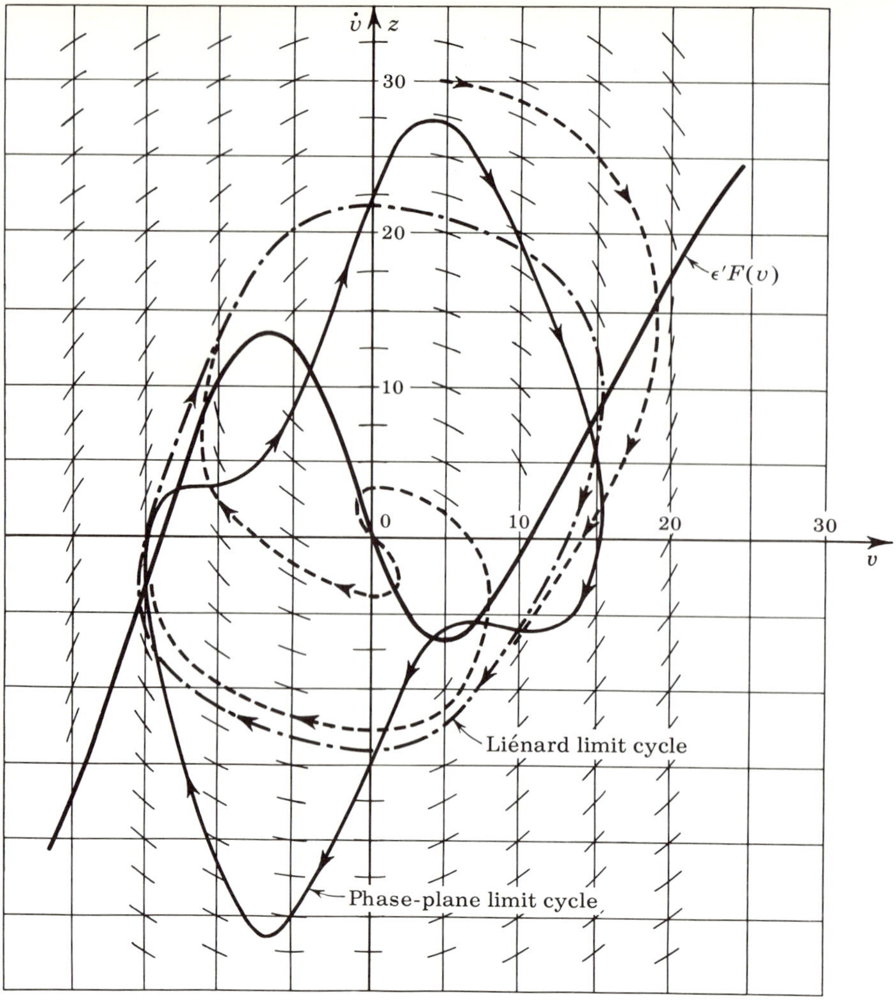

Fig. 17-17 Liénard-plane construction of the trajectory of the limit cycle and of the phase-plane limit cycle.

starts with excessive amplitude, damps down to the final steady-state solution.

The plot on the Liénard plane can be converted into a phase portrait by performing the inverse transformation

$$\dot{v} = z - \epsilon' F(v)$$

which involves subtracting from each point on the Liénard trajectory the value of $\epsilon' F(v)$. This process is also carried out in Fig. 17-17. Because of the large degree of nonlinearity, it leads to the grossly distorted contour shown.

The phase portrait, as well as the Liénard plot of an oscillatory system, will exhibit one or more limit cycles. If all adjacent trajectories converge on one contour (as in Fig. 17-17), then it represents a stable oscillation.

Every unstable equilibrium condition is associated with some closed contour from which the trajectories representing the transient response diverge. It follows that stable and unstable limit cycles separate each other. The origin may also be regarded as a solution which may be either stable or unstable, depending on the nature of the trajectories terminating on it. If the origin is unstable, the circuit is self-starting. If the origin is stable, the circuit must be triggered past the adjacent unstable limit cycle.

Figure 17-18 illustrates the nature of the unstable limit cycle, which arises when the negative-resistance device is normally biased in its dis-

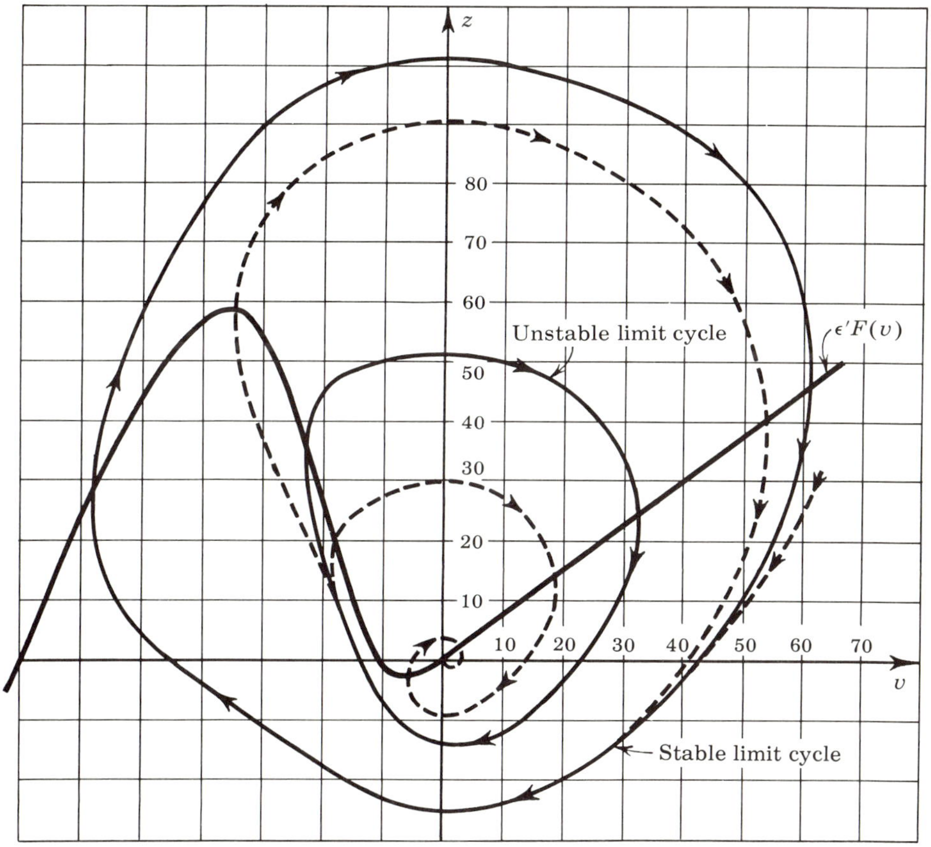

Fig. 17-18 Liénard-plane construction, illustrating the conditions yielding both stable and unstable limit cycles.

sipative region near the knee of the volt-ampere characteristics. For small signals, the circuit is stable and all trajectories cycle toward the origin. However, after the signal becomes large enough so that the circuit exhibits an average negative resistance, the buildup will proceed toward the next stable limit cycle. These limit cycles correspond to the fundamental amplitudes V_b and V_f in the describing-function solutions of Figs. 17-6 and 17-7.

The shape of the limit cycle, which depends on the size of ϵ' for any given volt-ampere characteristic and frequency, indicates the nature of the time response. When ϵ' is small, the limit cycle approaches a circle or ellipse and the signal generated is almost sinusoidal (Fig. 17-19*a*). When ϵ' is large, the closed trajectory becomes almost rectangular and the time response is that of a relaxation oscillator (Fig. 17-19*b*).

Since the current through a capacitor is proportional to the derivative of the applied voltage, the phase portrait is readily obtained experimentally.

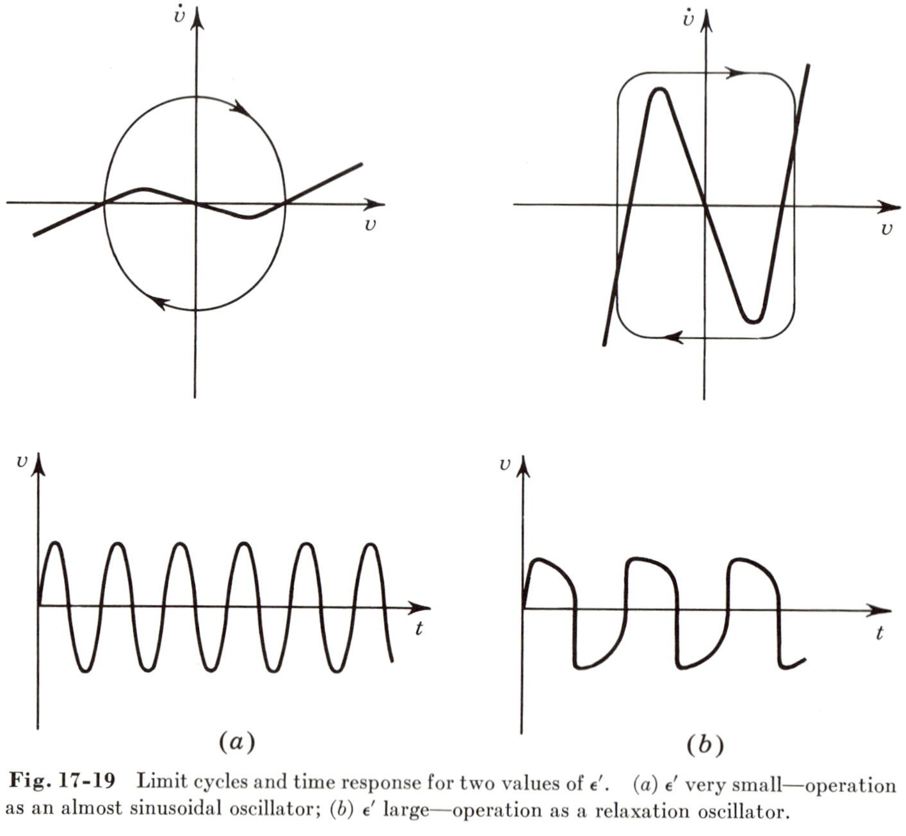

Fig. 17-19 Limit cycles and time response for two values of ϵ'. (*a*) ϵ' very small—operation as an almost sinusoidal oscillator; (*b*) ϵ' large—operation as a relaxation oscillator.

The VNR voltage is applied to the x input of an oscilloscope and a voltage proportional to the capacitor current of the tuned circuit is applied to the y deflection circuit.

At this point we might observe that the various topological constructions discussed in this chapter may also be applied to the solution of the negative-resistance switching circuits of Chap. 11. At the low frequencies of operation, the external inductance in the VNR switching circuit predominates. This corresponds to a very large value of ϵ'. A trajectory similar to that shown in Fig. 17-19*b* would be obtained. Analogous response appears in the CNR switching circuit. At the higher frequencies, the effective internal energy-storage element of the device plays a more pronounced role, reducing ϵ' and introducing curvature into the trajectory. Thus the phase portrait explains the changes in the behavior of the multivibrator with frequency and even its almost sinusoidal time response when operating at the elevated frequencies.

Furthermore, the phase- or Liénard-plane construction can be used to illustrate the transient switching path between the two states of a bistable circuit. Since the stable points are the nodal points of the system, two constructions would be necessary, one with each singular point as the origin.

17-7 Summary

The complete solution of the nonlinear oscillator, i.e., the waveshape, amplitude, and period, cannot be arrived at by either the topological or the analytic method alone. These two techniques should not be regarded as independent, but rather as complementary: where one is weak, the other is strong. By combining the results, we are able to obtain almost any required information as to the behavior of the system, information which cannot readily be obtained from either method individually.

To find the frequency of oscillation, including the depression due to the harmonic content of the output, we must turn to the analytic solution where an explicit expression was derived [Eq. (17-30)]. However, to evaluate this equation, we must have some knowledge as to the waveshape produced. If this is to be obtained analytically, then the nonlinearity must be expressed as a power series. After the fundamental amplitude is found from the describing function, the harmonic content is evaluated by substituting back into this series. In simple cases, for example, where the volt-ampere characteristic is almost symmetrical, the accuracy requirements are satisfied by approximating the nonlinearity by a cubic polynomial. The coefficients are relatively easy to evaluate.

However, in most cases a fifth- or higher-degree equation is necessary for an adequate description, and it becomes extremely tedious to solve for the coefficients. Finally, we should note that the analytic method does not give the waveshape of the output directly, but only as a Fourier series.

On the other hand, the topological solution yields the peak-to-peak amplitude of the stable oscillation quite rapidly but supplies almost no accurate information as to the frequency of operation. A good approximation to the waveshape generated is easily constructed. From it, one could find the relative harmonic content of the output either by a graphical integration or by some schedule method.

Suppose that we assume that the period of the waveshape constructed can be scaled to satisfy the frequency found from the analytic expression. Then, from the topological construction, we obtain the waveshape and the harmonic content and use this to find the corrected frequency. Thus each mode of solution is employed where it best serves, and without actually arriving at an analytic expression for the output voltage, we are able to understand the operation of the sinusoidal oscillator.

Problems

17-1 Prove that the describing function of Fig. 16-26 can be drawn to represent the shunting of the tank circuit by a negative conductance. Draw the conductance-describing function and show that the feedback line becomes the tank conductance at resonance.

17-2 Repeat Prob. 17-1 for the bipolar-transistor describing function of Fig. 16-29.

17-3 (*a*) Specify the bounds and linearized slopes of the regions of operation of the circuit of Fig. 17-5*d*. The FET has $I_{DSS} = 5$ mA, $V_p = -3$ V, $r_o = 20$ K at I_{DSS}. The conductance of the tank is 10 mmhos, and $V_{DD} = 30$ V. State and justify all approximations made in the course of the analysis.

(*b*) Will possible positive gate voltage have any effect other than reduction in gain? Explain.

(*c*) If this is a tuned-drain oscillator where $M = 0.2L_1$, what restrictions must be placed on the other tank parameters for self-starting operation?

17-4 Two negative-resistance characteristics have power-series expansions about the Q point given by

1. $i_n = 3.0v - 0.2v^3 + 0.001v^5$ mA.
2. $i_n = -3.0v + 0.2v^3 + 0.001v^5$ mA.

(*a*) Plot the current-describing functions of these two curves on the same graph and find the amplitude of oscillations if $G = 2$ mmhos; if $G = 1$ mmho.

(*b*) Repeat part *a* with respect to the conductance-describing function.

17-5 This problem is designed to investigate the response of various negative-resistance oscillators from a consideration of the power-series expansion. Assume that the terms of interest are

$$i_n = a_1 v + a_2 v^2 + a_3 v^3 + a_4 v^4 + a_5 v^5$$

(*a*) Prove that in order for any oscillator to be self-starting,

$$a_1 < -G$$

(*b*) Prove that in this case,

$$a_5 > 0$$

for any values of the other coefficients.

(*c*) Under what conditions will the describing function have a minimum at some nonzero value of V_1? State the answer in terms of the relative size of the coefficients.

(*d*) Under what conditions will the describing function have both a maximum and a minimum value of i for some nonzero value of V_1?

17-6 Figure 17-20 shows the piecewise-linear volt-ampere characteristic of a semiconductor device. Find the approximating cubic polynomial holding when the quiescent point is in the center of the negative-resistance region. Solve for the amplitude of oscillation when $R = 400$ and $\sqrt{L/C} = 1{,}000$.

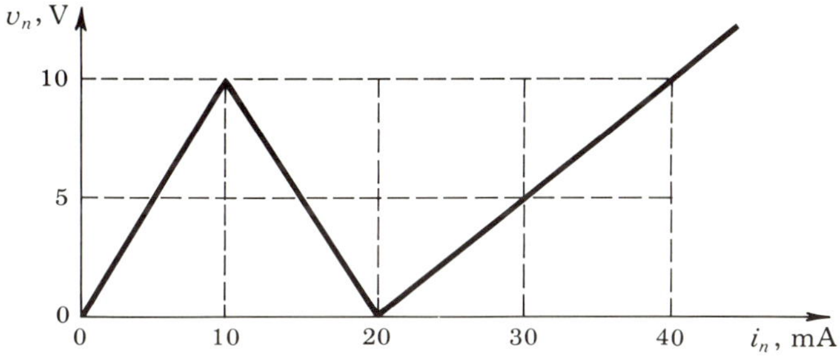

Fig. 17-20

17-7 Express the frequency deviation in terms of the harmonic current components of the VNR. Give the answer in a form analogous to Eq. (17-30). State the reason for all approximations made in the course of the solution.

17-8 Consider a negative-resistance oscillator using a device described by the polynomial

$$i = -2.0v + 0.5v^3 \qquad \text{mA}$$

(*a*) Find the amplitude of the fundamental term when $G = 1.0$ mmho.

(*b*) Calculate the amplitude of the third-harmonic current component with the assumed sinusoidal excitation found in part *a*.

(*c*) When $C = 100$ pF and when $\omega_0 = 10^6$, evaluate the depression in frequency due to the harmonic voltage developed.

(*d*) Repeat the above calculations for $G = 1.8$ mmhos.

17-9 Use Eq. (17-28) to calculate the approximate depression in frequency if the distortion is great enough so that the terminal voltage is almost a square wave. Even though the answer will be far from exact, it will still give a rough order of magnitude.

17-10 An oscillator with a nominal resonant frequency of $\omega_0 = 10^6$ and a signal amplitude of 20 V has a 5 percent third-harmonic and a 1 percent fifth-harmonic component. The tuned circuit has a Q of 50 at ω_0.

(*a*) By what percentage will the frequency be depressed?

(*b*) If $C = 100$ pF, what is the phase angle of the fundamental?

(*c*) What is the amplitude of the injected seventh-harmonic component that will just return the frequency to ω_0?

(*d*) The frequency is to be increased to $1.03\omega_0$ by injecting a second-harmonic signal. What is the necessary amplitude of the synchronizing signal? Calculate the new phase angle at the fundamental.

17-11 Plot the phase portrait of the following equation on a normalized phase plane; i.e., the coordinates are $\dot{x}/\omega_0$ and x.

$$\ddot{x} + 2A\omega_0\dot{x} + \omega_0^2 x = 0$$

(*a*) Set $A = 1.1$.

(*b*) Set $A = 3$.

Pay particular attention to at least one path that falls between the straight-line trajectories.

17-12 Repeat Prob. 17-11 for

(*a*) $A = -1.1$.

(*b*) $A = -3$.

17-13 The behavior of an undamped pendulum in the vicinity of its unstable point is defined by

$$\ddot{x} - bx = 0$$

where $b > 0$. In this case the origin is called a saddle point.

(*a*) Prove that the trajectories are hyperbolas.

(*b*) Plot two pairs of trajectories when $b = 4$. Sketch the asymptotes and indicate the direction of rotation of all curves.

17-14 Assume that the plot drawn in Fig. 17-17 corresponds to $\epsilon' = 1$. We now wish to consider the response when $\epsilon' = 0.1$ and when $\epsilon' = 10$.

(*a*) Construct the limit cycle on the Liénard plane for these two cases.

(*b*) Sketch the approximate waveshapes produced.

(*c*) By how much does the peak-to-peak amplitude change as ϵ' increases from 0.1 to 1 to 10?

17-15 Accurately construct, out of small line segments, the time-varying signal represented by the limit cycle of Fig. 17-17.

17-16 Plot a Liénard-plane trajectory and the equivalent phase-plane portrait for the oscillator employing a device having the piecewise-linear volt-ampere characteristic of Fig. 17-21. The quiescent point is located at $V_o = 50$ V and $I_o = 5$ mA, and the tank parameters are $\sqrt{L/C} = 3{,}000$ and $R = 5{,}000\ \Omega$. Sketch the waveshape.

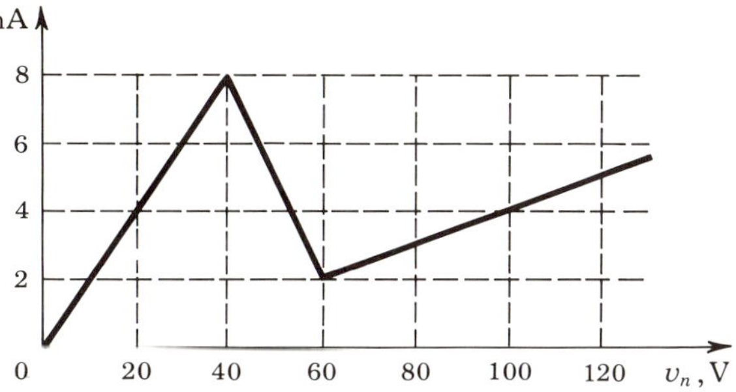

Fig. 17-21

17-17 The load resistance of Prob. 17-16 is changed to 10 K, and the Q point to $V_o = 80$ V and $I_o = 3$ mA.

(*a*) Plot at least two Liénard-plane trajectories showing the transient decay toward the stable point.

(*b*) Is it possible to make this circuit oscillate? Give the expected peak-to-peak output if it is possible. If it is impossible, justify your answer.

17-18 Solve Prob. 17-6 by the method of the Liénard plane.

Bibliography

Andronov, A. A., A. A. Vitt, and S. E. Khaikin: "Theory of Oscillators," Addison-Wesley Publishing Company, Inc., Reading, Mass., 1966.

Belevitch, V.: "Théorie des circuits non-linéaires en régime alternatif," Gauthier-Villars, Paris, 1959.

Edson, W. A.: "Vacuum-tube Oscillators," John Wiley & Sons, Inc., New York, 1953.

Kryloff, N., and N. Bogoliuboff: "Introduction to Non-linear Mechanics," Princeton University Press, Princeton, N.J., 1943.

le Corbeiller, P.: The Nonlinear Theory of the Maintenance of Oscillations, *J. IEE (London)*, vol. 79, pp. 361–378, 1936.

Liénard, A.: Étude des oscillations entretenus, *Rev. gén. élec.*, vol. 23, pp. 901–946, 1928.

Minorsky, N.: "Nonlinear Oscillations," D. Van Nostrand Company, Inc., Princeton, N.J., 1962.

Oser, E. A., R. O. Enders, and R. P. Moore, Jr.: Transistor Oscillators, *RCA Rev.*, vol. 13, pp. 369–385, September, 1952.

van der Pol, B.: The Nonlinear Theory of Electric Oscillations, *Proc. IRE*, vol. 22, no. 9, pp. 1051–1086, 1934.

index

Aaron, M. R., 543
Abraham, H., 407
Activation energy of dopants, 41
Adder, logic of, 153–154
Address in memory, 586, 603–605, 609
  (*See also* Coding; Decoding)
Adler, R. B., 98
α, variation of, with Q point, 498
Amplifier:
  for current bootstrap, 344
  differential (*see* Differential amplifier)
  inverter (*see* Inverter)
  junction FET, 221–223
  MOSFET, 223–224
    with MOSFET load, 224–230
    multistage, 229–230
  operational (*see* Operational amplifier)
  rise time of, 23
  transient response, RC-coupled, 20–23
  transistor, 108–110, 119–122
Amplitude-controlled bridge, 707, 710
Amplitude gating (*see* Chopper)
Amplitude limiting (*see* Clipper; Describing functions; Oscillations, amplitude of; Topological solutions)
Amplitude-stability factor, 713
Analog-to-digital converter, 459–460
AND gate:
  diode, 156–159
    integrated-circuit geometry, 172, 173
  input stage as: DTL, 159–160
    TTL, 175
  logic, 151–152
Anderson, F. B., 720
Andronov, A. A., 757
Angelo, E. J., Jr., 147
Astable multivibrator:
  bipolar-transistor, 385–387
    nonsaturating, 388–389
    resistive bounds in, 388
  blocking oscillator as, 538–541
  core-transistor, 640–645
  differential amplifier as, 447–449
  emitter-coupled, 426–431
  as frequency-modulated source, 406
  incomplete recovery in, 394–396
  integrated logic connected as, 393
  junction FET, 390–391
  MOSFET, 391–392
  negative-resistance circuit, 471–472
  non-self-starting of, 388

Astable multivibrator:
single timing element, 397–400, 426–431
synchronization of, 400
Asynchronous switching, 550
Attenuator:
compensated (*see* Compensated attenuators)
voltage-controlled, 256–257
Automatic gain control, 706–713
Avalanche breakdown, 49
in inductive switching, 322–334
for negative resistance, 504–505
(*See also* Miller multiplication)

Barium titanate, 647, 650
depolarization of, 655
Barkhausen criterion, 664, 665
Base current, direction of, 100
as recombination current, 102
Base-padding resistance, 114, 136
Base-width modulation, 115
BCD counter, 578
BCD decoding matrix, 605
Beale, I. E. A., 522
Belevitch, V., 757
Benjamin, R., 543
Bessel function, $\mathbf{I}_n(x)$, 703–704
B-H curve, 614–615, 617
Bias:
amplitude limiting by, 700–702
by clamping, 700–701
constant current, 434–435, 705
feedback, 440
of junction FET, 220–221
of MOSFET, 223–224
shift in oscillator, 705–706
of transistor, 108–110
Bias stability factor, 110
Bilateral transmission bridge, 194
Binary (*see* Bistable multivibrator)
Binary coded decimal (*see under* BCD)
Binary counter (*see* Counter)
Bipolar transistor (*see* Transistor)
Bipolar-unipolar pair, 253–254
Biquinary code, 606
Bistable multivibrator:
balance conditions for, 355–356
bipolar memory matrix, 598–602
complementary symmetry MOSFET, 402
Bistable multivibrator:
as cross-coupled inverters, 354–355, 547–548
D edge-triggered, 569–571
direct-coupled, 355, 359
J-K, 563–569
basic circuit, 564
DTL master-slave, 565
ECL, 567–569
junction-FET, 567
logic of, 564
MOSFET complementary symmetry, 582
RTL, 562
TTL, 566
junction-FET, 359–362
level-shifting network in, 360–361, 363–365
master-slave, 554, 565
MOSFET, 358–359
negative-resistance circuit as, 470
composite characteristic of, 479
RTL inverters as, 354–358
speedup capacitor in, 362–364
S-R, 548–563
clocked, 550–553
NAND inputs, 551–553
NOR inputs, 548–550
S-R toggle mode, 553–563
time sequence in, 555
stable points in, 355–356, 470
triggering of, 365–367
(*See also* Counter; Shift registers)
Bivalued devices (*see* Diode; Ferroelectric devices; Magnetic core)
Bloch, E., 407
Blocking oscillator:
astable operation, 538–541
synchronization of, 540
core-transistor, 633–634
monostable operation, 529–538
capacitively timed, 535–538
inductively timed, 529–531
transformer saturation timed, 533–535
pole location in, 527–528
recovery region of, 531–533
ringing in, 533
switching time of, 525–528
effect of ω_α on, 528

Bode, H. W., 720
Bogoliuboff, N., 757
Bollman, J. H., 720
Boothroyd, A. R., 324
Bootstrap sweep:
  capacitive voltage supply in, 300–301
  current sweep, 341–345
  double bootstrap, 301–302
  gain dependence of, 298
  voltage sweep, 296–302
    emitter follower as, 299–300
Bothwell, F. E., 720
Bottoming diode, 529
Bottoming voltage, MOSFET inverter, 227
Break point, 34, 36
  in multiple diode network, 71–79
Breakdown:
  avalanche, 49, 504–505
  during current sweep recovery, 332–334
  in multivibrator, 374–375
  in transistors, 136, 140
  zener, 50
  (*See also* Miller multiplication)
Bridge modulator (*see* Diode bridge)
Bridged-T network, 675, 690
Briggs, B. H., 324
Burns, J. R., 263

Camenzind, H. R., 147
Capacitance:
  base storage, 119, 144
  depletion region, 83–85, 118–119
    of graded junction, 85, 172
    isolation region, 172, 177
  diode forward, 97
  in field-effect transistor, 218–219
  parallel-plate, 659
  in transistor, 118–119
    typical values, 136
Capacitor:
  energy storage in, 335
  initial conditions due to, 6
  in integrated circuits, 138
  as power supply, 61, 300–301
  as short-time memory, 556–557
    toggle storage, 556–558
  voltage variable, 85
Carrier:
  buildup: in diode, 85–87
    in transistor, 125
  densities in integrated-circuit transistor, 134
  densities in semiconductor, 40–42
  diffusion, 42–44, 47
    profiles, 135
    in transistor, 100–102
  distribution of, 45–47
    in base, 106
Carrier lifetime, 43
Catching diode, 277–278
CCL (*see* ECL)
Chance, B., 324, 349, 433, 720
Charge distribution:
  in depletion region, 84
  in transistor base, 106, 125
Charge metering, 648
Charge storage delay, 87, 130
Chen, K., 660
Chen, T. C., 660
Chopper:
  diode bridge, 189–193
    transient response of, 194–198
  field-effect transistor, 249–253
    transient response of, 230–234
  transistor, 194–195
Clamp:
  capacitor as battery, 61
  charge transfer in, 64
  conduction angle of, 66–67
  current, 69–71
  shunt resistance, need for, 63
  voltage, 61–68
    distortion in, 65–66
    transfer characteristic, 67
    voltage area ratio, 64
Clamping for self-bias, 700–701
Clarke, K. K., 148, 720
Clipper:
  current, 68–69
  voltage, 54–58
    optimum resistance in, 56
Clocked logic, 245–246
  four-phase, 595–597
  (*See also* Bistable multivibrator; Shift registers)
Code:
  BCD, 578
  biquinary, 606

Code:
  quibinary, 606
  weighted binary, 579, 603
Code generator, pseudorandom, 583
Coding:
  BCD counter, 578–580
  diode matrix, 604
  for information transmission, 598
  in memory address, 586, 602–606
  weighted binary, 603
Coercive force, 615, 620
  load current as, 623
Collector saturation resistance, 136
Collector time constant, 126
Colpitts oscillator:
  field-effect transistor, 681–684
  transistor, 685–686
Compensated attenuators, 23–26
  in multivibrator, 362–363
  in RCTL, 170–171
  in regenerative comparator, 445
  to speed inverter switching, 130–131
Complementary MOSFET pair:
  bistable, 402
  as chopper, 252
  inverter gain of, 248–249
  J-K bistable, 582
  logic, 246–249
Complementary transistor pair:
  as chopper, 194
  in integrated circuits, 140
    logic, 165–166
  monostable multivibrator, 405
  in negative-impedance converter, 489–490, 493–494
  as *pnpn* structure, 498
  for sweep recovery, 278–279
Composite devices:
  junction FET-bipolar, 253–254
    as negative-impedance converter, 495
  junction FET-MOSFET, 262–263
  multiple negative-resistance devices, 478–481
  multiple transistors, 122–124
Conductance describing function (*see* Describing functions)
Conduction angle:
  in clamp, 66–67
  of sine-wave tip, 696
Controlled source, circuit reduction:
    with blocking oscillator, 526
  with feedback amplifier, 441
  with Miller sweep, 280–281
  with source follower, 222
Conwell, E. M., 98
Core (*see* Magnetic core)
Core-transistor multivibrator, 640–645
Counter:
  bistable: BCD, 578
    ring, 575
    ripple, 575–579
  feedback: inhibit, 577–579
    reset, 576–577
  ferroelectric, 650–652
  magnetic core, 626–630
    to arbitrary base, 628–629
    core-transistor, 632–634
  (*See also* Shift registers)
Crawford, R. H., 263
Critical damping, 329, 331
Crystal-controlled oscillator, 686–688
Current clamp, 69–71
Current clipper, 68–69
Current-controlled negative resistance (*see* Negative resistance)
Current-controlled oscillations, 676–677
Current-coupled logic (*see* ECL)
Current feedback:
  for current sweep, 341–343
  for NIC, current-controlled, 488–492
  for voltage sweep, 280–282
Current-limiting diode, 254–255
Current source:
  as current sweep, 345
  MOSFET as, 238–242
  transistor as, 275–278, 434–435
  in voltage sweep, 268–269
Current sweep:
  basic, 325–327
  bootstrap, 341–345
  current source as, 345
  emitter follower in, 340–341
  initial delay in, 326–327, 334
    minimization of, 338–339
  linearization of, 338–340
  Miller, 344
  recovery region of, 328–331, 333–334
    damping diode role, 334
    figure of merit, 330–331
    oscillatory, 337

Current sweep:
switched periodic, 334–337
energy recovery in, 335–336
transistor, 331–337

D edge-triggered flip-flop, 569–571
Damping:
in blocking oscillator, 532–533
in current sweep, 328–330
Darlington composite transistor, 122
Data coder, 598
Davies, D. C., 584
Dc-ac converter, 640–645
Dc-dc converter, 640–645
DCTL logic, 166–168
Dead zone, 72–82, 97
Decimal coding (*see* Coding)
Decoding:
AND diode network, 580
in memory, 586, 589, 603–607
MOSFET BCD matrix, 605
Deflection coil, 326, 337
Deltamax, 617
Depletion region:
broadening: in base, 115
in field-effect transistor, 208
capacitance of, 83–84, 118–119, 136, 172
for storage, 559
width of, 83
Describing functions:
clamped sine-squared limiting, 702
defined, 694
exponential limiting, 705
linear clipping, 698
negative-resistance oscillator, 729–733
stable intersection of, 730–731
Differential amplifier:
as astable multivibrator, 447–449
constant current biasing of, 434–435
current division in, 436–437
ECL input stage, 185–186
field-effect transistor, 461–462
integrated circuit, 132–133, 139, 444
transconductance of, 438
(*See also* Operational amplifier; Voltage comparator)
Differentiation:
of current waveshapes, 30
of voltage waveshapes, 13–15, 18
Diffusion of carriers (*see* Carrier)
Diffusion profiles, 135
Diode:
AND gate, 156–159
biasing of, 35–36
breakdown in, 48–50
capacitance: depletion region, 83–85
forward biased, 97
current limiting, 254–255
dynamic response, 85–87
equation of, 48
ideal, 34–35, 52
in integrated circuits, 138, 140
geometry, 172–174
models for, 50–54
resistance in, 53
OR gate, 154–156
physics of, 40–50
to prevent saturation, 529
for recovery improvement, 278, 383–384
storage delay, 87
tunnel, 505, 507
for voltage catching, 278
voltage limiting by, 334
voltmeter probes, 93
zener (*see* Zener diode)
Diode bridge, 189–193
transient response of, 194–198
(*See also* Chopper)
Diode clamp (*see* Clamp)
Diode clipper (*see* Clipper)
Diode networks:
continuity in, 38, 39
dead zone, 80–81
hysteresis simulation, 82–83
for multiplier, 95
piecewise-linear model of, 35–39, 74–75
RC, transient response of, 58–61
sine-wave shaping, 95
transfer function, arbitrary, 71–79
for triggering, 366
volt-ampere characteristic, arbitrary, 71–79
Diode-transistor logic (*see* DTL)
Dipole, 645–646
Domain behavior:
in ferroelectric materials, 647–648
in magnetic materials, 613–615
Dopants, 40–41
Drift current, 42, 44, 206

DTL:
master-slave bistable, 565
modified logic, 165–166
NAND gate, 159–166
complete circuit, 184
fan-out, 161–162
noise immunity, 162–164
propagation delay, 164
transfer characteristic, 160
Dynamic response:
of diode, 85–87
of transistor, 124–131
(*See also* Transient response)

Ebers, J. J., 103, 148, 522
Ebers-Moll model, 103–108
saturation conditions from, 112–114
Eccles, W. H., 407
Eccles-Jordan circuit (*see* Bistable multivibrator)
ECL, 183–187
J-K bistable, 567–569
noise immunity, 187
propagation delay in, 183
as regenerative trigger circuit, 452–453
Edson, W. A., 720, 757
Einstein equation, 46
Electrets, 646
Electromagnetic units, 624
Emitter-coupled logic (*see* ECL)
Emitter-coupled multivibrator:
astable operation, 426–431
bistable (*see* Schmitt trigger)
monostable, 408–415
modes of operation, 415–421
potentiometer impedance, effects on, 422–423
pulse duration of, 421–422
pulse variation in, 424–426
transistor states in, 415, 421
Emitter current:
biasing in terms of, 108–109
from carrier distribution, 101
variation of α with, 498
Emitter efficiency, 102
Emitter follower:
in bootstrap sweep, 299–300
to charge capacitor, 180–181
as comparator output stage, 439, 442
in current sweep, 340–341
for sweep recovery, 279–280
Encoder (*see* Coding)
Enders, R. O., 757
Energy:
balance: reactive, 736–739
real, 735
recovery in current sweep, 335–336
storage: in capacitor, 335
in inductance, 336
switching, in magnetic material, 616
Enhancement, 206, 208–210
Epitaxy, 131
Esaki, L., 505, 522
Exponential:
expansion: Bessel functions, 703
power series, 20, 274
linear approximation, 269
nature of, 51

Fairchild, 139, 584, 612
Fan-out:
DTL NAND gate, 161–162
noise immunity limitations, 164
ECL, 186
RTL, 170
Farley, B. G., 522
Feedback:
amplitude control by, 706–713
biasing, 440
in counter (*see* Counter)
current: for current-controlled NIC, 488–492
for Miller sweep, 280–282
oscillator (*see* Oscillator)
regenerative comparator, 445–447
voltage: in bootstrap sweep, 297
in Miller sweep, 286–293
in multivibrator, 353–354
for voltage-controlled NIC, 492–496
Ferrites, 617, 625
in memory planes, 634–640
Ferroelectric devices:
for charge metering, 648, 653
as counter, 650–652
hysteresis in, 649
as memory, 653–655
model of, 650
in multivibrator, 660
response to excitation, 648–649
as tachometer, 652–653

Ferroelectric materials:
depolarization of, 655
domain formation in, 647–648
properties of, 645–648
Ferromagnetic properties, 613–616, 625
(*See also* Magnetic core)
Field-effect transistor:
amplifier, 221–224
with MOSFET load, 224–230
as amplitude stabilizer, 710–711
biasing of, 220–221, 223–224
capacitances in, 218–220
as memory, 559–561, 589–594
as current diode, 254–255
junction, 205–209
characteristics of, 209
pinchoff in, 206
models of, 219–220
parameter variation with I_D, 217–218
for transient response, 236–242
MOSFET, 210–214
complementary symmetry, 246–249
in depletion mode, 214–215
geometry of, 209, 210, 216, 225
inverter pair, 224–230, 246–249
saturation boundary in, 213, 238
substrate, role of, 215
as switch, 293–294, 561, 589–590
MOSFET logic, 242–246
clocked, 245–246, 595–597
complementary, 249
noise immunity, 245
transient response: approximate, 230–234
exact, 236–242
as voltage-controlled resistor, 255–257
Figure of merit, inductive recovery, 330–331
Flip-flop (*see* Bistable multivibrator)
Flyback voltage, 328–331, 531
Focal point, 741
Four-layer diode, 497–500
Four-phase clocked logic, 595–597
Frequency determining networks (*see* Oscillator)
Frequency division, 576, 578
Frequency modulation generator, 406
Frequency stabilization, 686–693
Frequency stabilization factor, 689
Gas tubes, as negative resistance, 500
Gate, 149–150
as logic (*see* Logic)
as switch (*see* Switch)
Gate-substrate capacitance, 218–219
as memory, 559–561, 589–594
General Instrument Corporation, 607, 612
Gibbons, J. F., 522
Giles, J. N., 463
Ginzton, E. L., 720
Glegg, K., 433
Goodrich, H. C., 349
Grain-oriented materials, 615–616
Graphical summation:
composite negative-resistance characteristics, 478–481
for diode networks, 35–36
in Liénard construction, 748
Gray, P. E., 98, 148

h-parameters model:
common base, 115–117
common emitter, 117–118
Half adder, 153
Harmonic components:
exponential limiting, 703
negative-resistance oscillator, 734, 754
sine-squared clamped limiting, 701
sine-wave tips, 696
Harmonic rejection factor, 668–669
Harris, J. N., 204
Hartley oscillator, 681–683, 686
automatic gain control in, 712
frequency stabilization of, 693
Heiman, F. P., 263
Herzog, G. B., 349
Hofstein, S. R., 263
Hollingsworth, L. M., 720
Hooper, D. E., 720
Hybrid π model, 119, 144, 369
Hysteresis:
in bistable multivibrator, 480
diode network representation, 82–83
in ferroelectric materials, 646–647
in magnetic materials, 614, 616–617
frequency dependence, 616
in memory, 587–588
in regenerative comparator, 445–447
transformer, 533–534

Ideal diode, 34–35, 52–53
Impulse, 338
Incomplete recovery in multivibrator, 394–396
Inductance:
  energy stored in, 335
  initial conditions due to, 9
  of square-loop cores, 621
  sweep recovery, 328–331
  as timing element: in blocking oscillator, 529–531
    in multivibrator, 379–380
    in negative-resistance circuit, 481–483
Inductive load, switched, 331–334
Information storage (*see* Memory)
Inhibit:
  in *D*-type bistable, 569–570
  in *J-K* bistable, 563–569
  in *S-R* bistable, 553–555
  toggling for, 559–563
Initial conditions:
  across switching, 7, 11
  due to capacitor, 6
  due to inductance, 9
Integrated circuit:
  capacitors, 138
  complementary transistors, 140, 165
  diffusion profile of, 135
  diodes, 138, 140, 172–173
  geometry, 132–134, 137, 172–174, 225
  isolation in, 134, 173–174
  large system, 578, 581, 607–608
  metallization, 137–138, 579, 608
  microphotograph: operational amplifier, 139
    read-only memory, 607
  resistors in, 136, 137
  two-level wiring, 579, 608
  (*See also* Bistable multivibrator; Logic; Memory; Operational amplifier; Shift registers; Voltage comparator)
Integrator:
  current, 30
  voltage, 15–16, 18
    periodic response of, 16
Interface:
  of logic classes, 202
  MOSFET to bipolar, 261, 594
Inverter:
  junction-FET, transient response of, 230–234
  as logic element, 152
  MOSFET complementary, 246–249
  MOSFET pair, 224–230
    logic levels in, 226, 227
    transient response of, 236–242
  RTL: high-frequency break, 369
    dynamic response, 124–130
    as logic output, transient response, 178–180
  toggle capacitor as, 556–557
  transistor, 110–112
    dynamic response, 124–130
    as logic output, transient response, 178–180
    transfer characteristic, 110
Iron, magnetic constants of, 615
Isocline, 742–744

Jackets, A. E., 720
J-K bistable (*see* Bistable multivibrator)
Johnson, H., 263
Johnson, M. H., 433
Jonscher, A. K., 148
Jordan, F. W., 407
Joyce, M. V., 148
Junction, *pn* (*see pn* junction)
Junction breakdown (*see* Breakdown)
Junction field-effect transistor (*see* Field-effect transistor)
Junction voltage:
  α variation with, 498
  built-in, 44
  shift due to I_e, 109, 705

Katz, H. W., 660
Keep-off timing, 370–372, 381–382, 399
  in emitter-coupled multivibrator, 427–428
  with TTL gates, 392
Keep-on timing, 375–377, 398
  in emitter-coupled multivibrator, 428–429
Keonjian, E., 720
Khaikin, S. E., 757
Kirchhoff's laws in diode network, 34–36

Kreer, J. G., Jr., 720
Kryloff, N., 757

Lamp, characteristics of, 708
Large system integration, 578, 581, 607–608
le Corbeiller, P., 757
Level-shifting network, 360–361
  in differential amplifier, 442
Liénard, A., 757
Liénard diagrams, 750–752
  construction of, 746–749
  (*See also* Phase plane)
Limit cycles, 749–752
Limiting in oscillator (*see* Oscillations)
Linear sweep (*see* Bootstrap sweep; Current sweep; Miller sweep; Voltage sweep)
Linearity of sweep, 275
Linvill, J. G., 407, 522, 543
Little, C. A., 660
Llewellyn, F. B., 692, 720
Lo, A. W., 522
Logic:
  adder, 153–154
  AND, 151–152, 156–159
  as astable multivibrator, 392–393
  as bistable multivibrator, 354–355
    (*See also* Bistable multivibrator)
  DCTL, 166–168
  DTL (*see* DTL)
  ECT (*see* ECL)
  MDTL, 165–166
  MOSFET (*see* Field effect transistor)
  NAND, 152
  nonsaturated (*see* ECL)
  NOR, 152
  OR, 151–152, 154–156
  output stages: emitter-follower, 180–181
    inverter, 178–180, 224–230
    totem-pole, 181–183
  propagation delay, 164
  RCTL, 170–171
  RTL (*see* RTL)
  statements: identities, 152
    J-K bistable, 564
    S-R bistable, 548, 549, 551, 553
    for storage, 561–563
    subtractor, 198
Logic:
  TTL (*see* TTL)
  wired, 187–189, 244, 608
  (*See also* Counter; Inverter; Memory; Shift registers)
Lohman, R. O., 349

MacLean, W. R., 324
Magnetic core:
  as bivalued element, 624, 627
  as gate, 632, 636, 657
  as memory, 634–640
  model of, 624
  reset of, 628, 633–634
  response of: current excited, 621–622
    externally loaded, 622–623
    voltage excited, 618–621
  saturation time of, 620, 625, 626
  switching energy, 616, 628
  tape wound, 616–618
  as timing element, 642
  as transformer, 622–623
  with transistor: blocking oscillator, 633–634
    multivibrator, 640–645
  for volt-time area shaping, 619–620, 628–629
Magnetic counters, 626–630
Magnetic domains, 613–615
Magnetic properties, 613–617, 624–625
Magnetizing current, 530
Master-slave bistable, 554, 565
Mattson, R. H., 543
MDTL, 165–166
Meacham, L. A., 720
Meacham-bridge oscillator, 692
  amplitude control in, 709
Memory:
  address of, 586, 603–605, 609
  bistable multivibrator, random access, 598–602
  capacitor short-time, 556–567, 588
    logic for storage, 561–563
  characteristics of, 587–588
  ferroelectric element as, 653–655
  magnetic core arrays, 634–640
    2D plane, 635–638
    2½D plane, 639–640
    3D plane, 639–640
  magnetic film, 640

Memory:
read-only, 606–611
system characteristics, 585–587
for toggling, 554, 556–563
(*See also* Bistable multivibrator; Shift registers)
Merrill, J. L., 522
Metal oxide semiconductor (*see* Field-effect transistor)
Metallization, 137–138, 579, 608
Meyer, J. A., 204
Microphotographs:
bipolar operational amplifier, 139
MOSFET read-only memory, 607
Middlebrook, R. D., 463
Miller, S. L., 49, 98, 103
Miller multiplication, 103
in *pnpn* diode, 499
Miller sweep:
current feedback, 280–286
for current sweep, 344
operational amplifier for, 286–288
source follower for fast recovery, 293–296
voltage control over sweep time, 292–293
voltage feedback, 286–293
Millman, J., 407
Minority carriers:
in diode, 45–48
due to doping, 42
in transistor, 100–102, 106, 107
Minorsky, N., 757
Misawa, T., 522
Mobility, 40
in MOSFET, 213
temperature dependence of, 42
Models:
for diode, 50–54
FET small-signal, 216–220
negative-resistance device, high-frequency, 506, 509, 512
transistor: Ebers-Moll, 103–108, 112–114
small signal: *h*-parameter, 115–118
hybrid π, 119, 144, 369
T, 114–116, 118
Moll, J. L., 103, 148
Monostable multivibrator:
blocking oscillator as, 529–538
complementary symmetry bipolar, 405
in luctively timed, 379–380, 481–483
junction breakdown in, 374–375
junction-FET, 384–385
keep-off timing, 370–373
keep-on timing, 375–378
MOSFET, 380–383
NAND gates as, 392–393
negative-resistance circuit, 468–469, 475–478, 482
collector-base coupled as, 483–487
toggle network in, 370
Moore, R. P., Jr., 757
MOSFET (*see* Field-effect transistor)
Motorola, 148, 204, 584
Multiemitter transistor, 174
(*See also* TTL)
Multiplexing, MOSFET commutation, 253
Multiplier, diode network for, 95
Multivibrator:
astable (*see* Astable multivibrator)
balance conditions in, 353–358
bistable (*see* Bistable multivibrator)
core-transistor, 640–645
emitter-coupled (*see* Emitter-coupled multivibrator)
gain in, 353–354, 357, 361
junction breakdown in, 374–375
monostable (*see* Monostable multivibrator)
negative-resistance, 466–471
state transition time of, 367–369
triggering of, 365–367
tristable, 481

Nambiar, K. P. P., 324
NAND gates:
as bistable multivibrator, 551
D edge-triggered, 570
DCTL, 166–168
DTL (*see* DTL)
logic, 152
MDTL, 165–166
MOSFET, 242–244, 249
TTL (*see* TTL)
wired, 187–188
Narud, J. A., 204, 543

Negative-impedance converters, 488–496
CNR from current feedback, 488–492
composite FET-bipolar as, 495
VNR from voltage feedback, 492–496
Negative logic, 150, 245, 552
Negative resistance:
characteristics, 464–466, 481
equivalent network of, 506, 509, 512
frequency dependence of, 508–509
stability of: CNR, 472, 509–512
VNR, 482, 512
Negative-resistance devices:
avalanche transistor, 502–504
pnpn diode, 497–500
silicon controlled rectifier (SCR), 500–502
thyristor (*see* SCR)
TRIAC, 502
tunnel diode, 505–507
unijunction transistor, 502–504
Negative-resistance oscillator:
amplitude: by describing function, 729–733
by topological solutions, 746–753
conditions of operation: CNR, 724–726
VNR, 721–724
energy balance in: reactive, 736–739
real, 735
frequency: depression of, 733–738
to first approximation, 727–728
phase shift in, 738, 739
pole migration in, 725
synchronization of, 738–739
trajectories, 750–752
tuned oscillator as, 726–727
(*See also* Oscillations; Oscillator)
Negative-resistance switching circuits:
astable operation, 471–472, 514–516
composite characteristic, 480
basic circuit conditions, 464–472
bistable operation, 470
composite characteristic, 478–479
hysteresis in, 480
collector-base monostable as, 483–487
current-controlled, 465–478
monostable operation, 468–469, 475–478
nonlinear load, 513–515
Negative-resistance switching circuits:
stable points in, 467, 470, 472, 513
trajectories of, 469–471, 476
tristable operation, 481
voltage-controlled, 466, 481–482
waveshapes in, 473–475
Network:
bridged-T, 675, 690
negative-resistance equivalent, 506, 509, 512
phase shift, 669–671
twin-T, 675, 690
voltage-tuned, 257
Wien bridge, 671–674, 690
Newton's method, 528
Nillesen, A. H., 349
Nodal point, 742
Noise immunity:
DTL NAND gate, 162–164
ECL logic, 186
MOSFET inverter, 245
RTL NOR gate, 168–169
Nonlinear load:
diode arbitrary, 71–79
MOSFET as, 224–230
in negative-resistance circuit, 513–515
Nonlinearity:
of exponential sweep, 275
ideal, 34
Nonsaturated logic (*see* ECL)
NOR gates:
as bistable multivibrator, 549
ECL (*see* ECL)
logic, 152
MOSFET, 242–243, 249
RCTL, 170–171
RTL (*see* RTL)
wired, 187–189, 244, 608
Null networks, 674–676, 690
Nyquist, H., 720
Nyquist plot, 665

ON resistance, 218, 237
Operational amplifier:
as differential amplifier (*see* Differential amplifier)
Fairchild μA702, 139
in Miller sweep, 286–288
in oscillator, 673, 711
(*See also* Voltage comparator)

Optimum resistance:
in chopper, 251
in clipper, 56
in diode bridge, 193
OR gates:
diode, 154–156
ECL (*see* ECL)
logic, 151, 152
Oscillations:
amplitude of: by automatic gain control, 706–713
feedback limiting, 706–710
FET amplitude stabilizer, 710–711
by describing function (*see* Describing functions)
by trajectory, 746–752
amplitude stabilization factor, 713–715
Barkhausen criterion for, 664–665
current-controlled, 676–677
frequency of, 665, 680, 684, 727–728
depression in, 733–738
effect of tank loading on, 685–686
frequency stabilization, 686–693
harmonic components, 698, 703, 734–736
harmonic rejection, 668–669, 694
intermittent, 702–703
non-self-starting, 699, 751
Nyquist condition for, 665
self-starting of, 668, 680, 685, 750
trajectories of, 750, 751, 752
Oscillator:
blocking (*see* Blocking oscillator)
crystal, 686–688
Meacham-bridge, 692
Pierce, 687
current-controlled, 676–677
feedback, 663–677
networks for, 669–676
phase shift, 669–671, 677
three-phase, 718
Wien-bridge, 666–667, 671–674
squegging in, 702–703
tuned-circuit, 677–686
Colpitts, 681–686
frequency stabilization factor, 691
Hartley, 681–683, 686, 693
Oscillator:
tuned-circuit: as negative-resistance oscillator, 726–727
reactive stabilization for, 692–693
tuned-drain, 681, 682
tuned-drain tuned-gate, 681, 682
tuned-gate, 681
(*See also* Negative-resistance oscillator; Oscillations)
Oser, E. A., 657
Output stages:
complementary symmetry, 202
emitter-follower, 180–181
in comparator, 439, 442
inverter, 178–180
for read-out, 594, 599–602
totem-pole, 181–184

Papoulis, A, 660
Parabola, nature of, 425–426
Parallel readout, 574, 590–591
Parasitic transistor:
diode-to-diode, 173
emitter-to-emitter, 175, 176
Periodic response:
of clamp, 64–65
as condition for solution, 11–13
negative-resistance oscillator, 734
current sweep, 334–337
of integrator, 18–19
of *RL* network, 11–13
voltage sweep, 270–273
Permalloy, 615, 625
Permeability, 615, 624
Permittivity, 646
Phase plane, 740–746
isocline, 742–744
limit cycle, 750
oscilloscope display of, 752–753
trajectory, 741, 743, 744, 750
waveshape from, 745–746
(*See also* Liénard diagrams)
Phase-shift oscillator, 669–671
current-controlled, 677
harmonic rejection factor of, 671
Philco, 612
Phillip, R. H., 433
Piecewise-linear model:
diode, 35, 50

Piecewise-linear model:
FET inverter, transient response by, 230–234
oscillator amplitude from, 694–700
transistor: Ebers-Moll, 103–108
inverter, 110
Piecewise-linear networks, 36–38, 71–79
break points of, 36, 72, 76–77
for multiplier, 95–96
as negative-resistance load, 513–515
to shape sine wave, 95
Pierce oscillator, 687
Pinchoff, 206
pn junction, 44–48
breakdown of, 49–50, 136, 140
carrier distributions, 45, 47
depletion region, 83
capacitance of, 83–85, 118–119, 136, 172
in FET, 206–207, 209
for isolation, 134
potential barrier, 44
pnpn diode, 497–500
Polarization, 646
of barium titanate, 650
Poles, 4
location of: in blocking oscillator, 527–528
in current sweep, 328–330
in multivibrator, 368
migration of: in negative-resistance oscillator, 725
in Wien-bridge oscillator, 667
Positive logic, 150–151
in ECL, 185–186
Propagation delay, 164, 170, 178
Puckle, O. S., 324
Pulse generator (*see* Astable multivibrator; Blocking oscillator; Monostable multivibrator)
Pulse-height analyzer, 459
Pulse transformer, 524–525
saturation in, 533–535
Pulse variation:
in emitter-coupled multivibrator, 424–426
in Miller sweep, 292–293

Quartz crystal, 687
Quasi-stable point, 358, 369–370, 385
Quibinary code, 606

Racing in bistables, 553, 572
Rajchman, J. A., 660
Ramp generator (*see* Current sweep; Voltage sweep)
Random access memory (*see* Memory)
Random-code generator, 583
Raster, 325, 337
RC-coupled amplifier, transient response of, 16–18
RCTL, 170–171
Read amplifier, 586–587, 589
for TTL memory, 599–602
Read-only memory, 606–611
microphotograph, 607
(*See also* Memory)
Recombination centers, 43
Recombination current, 44
in base, 101–102
time constant of, 42–43, 85, 102, 124–131
Recovery-region improvement:
catching diode for, 278
clocked logic, 245–246
of current sweep, 328–334, 337
emitter follower, 279–280
source follower for, 293–296
Regenerative comparator, 445–447
hysteresis in, 445, 446
Schmitt trigger, 449–458
Regions of synchronization (*see* Synchronization)
Remanence, 614
Research Triangle Institute, 148, 263
Resistor:
in integrated circuit, 136–137
voltage-controlled, 255–257
Resistor-transistor logic (*see* RTL)
Reverse transistor, 103, 106, 176
Ring counter, 575, 632
Ringing in inductance, 328–329, 337
Ripple counter, 575–579
Rise time:
of blocking oscillator, 527–528
exponential, 23, 235
linear function, 238, 239
MOSFET inverter, 236–241
multivibrator, 367–369
$t/(1 + t)$, 237, 239
tanh function, 239, 240
Rochelle salts, 647
Royer, G. H., 660

RTL, 168–170
as bistable, 354–355
J-K, 562
S-R, 549
fan-out, 170
high-frequency response, 369
noise immunity, 168, 169
propagation delay, 170
rise time of, 369
transfer characteristic, 110, 355

Sah, C. T., 263
Sands, E. A., 660
Saturation:
of ferroelectric element, 646, 649
of magnetic core, 619, 622, 625
in transformer, 533–535, 622–623
of transistor, 112–114
delay time, 130
prevention of, by diode, 529
Saturation region of MOSFET, 213
Schiere, A. J., 660
Schmitt, O. H., 433, 463
Schmitt trigger, 449–458
direct-coupled, 453
ECL as, 452–453
FET-bipolar, 454–455
modes of operation, 456–458
SCR:
double-ended (TRIAC), 502
as negative-resistance device, 500–502
volt-ampere characteristic, 271, 501
in voltage sweep, 270–273
Searle, C. L., 148
Seeds, R. B., 584
Semiconductor:
carrier densities in, 40, 42, 89
constants of, 40
doping of, 40–41, 135
(*See also* Diode; Field-effect transistor; *pn* junction; Transistor)
Sense detector, 458
Sevin, L. J., 263
Sheet resistivity, 134–137
Shift registers:
bistable, 572–575
magnetic core, 630–632
MOSFET: dynamic, 589–594
regenerative, 594–597
two-dimensional, 597–598
Shockley, W., 263, 522
Shou, S. H., 584
Silicon-controlled rectifier (*see* SCR)
Sine-squared limiting, 700–702
Sine-wave tip, harmonic components of, 696
Singular point, 740
Sinusoidal oscillations (*see* Oscillations; Oscillator)
Soller, T. M., 349
Source-coupled multivibrator, 433
Source follower:
capacitor charging from, 236–238
for recovery-time improvement, 293–296
Speed-up capacitor (*see* Compensated attenuators)
Square loop (*see* Hysteresis)
Square-wave amplifier response, 23
Square-wave generator, 387
Squegging, 702–703
S-R multivibrator (*see* Bistable multivibrator)
Stability:
of describing function solution, 730–731
in negative-resistance circuit, 472, 478–479, 482, 509–512
Stability factor, bias, 110
Starr, A., 349
Steering logic (*see* Decoding)
Stephenson, W. L., 522
Storage (*see* Memory)
Storage delay, 87, 130
Storage diodes, 559, 560
Substrate:
effect on MOSFET, 215
for integrated circuit, 131
Subtractor logic, 198
Summation, 17, 18
Sweep:
current, basic, 325–327
linearity of, 273–275
synchronization of (*see* Synchronization)
voltage, basic, 268–270
(*See also* Bootstrap sweep; Current sweep, Miller sweep; Negative-resistance switching circuit)
Switch:
FET as, 293–294, 303

Switch:
magnetic core as, 632
MOSFET as, 293–294, 561, 589–590, 594–598
transistor as, 302–303
(*See also* Chopper; Diode bridge)
Sylvania, 584
Synchronization, 303–318
of astable multivibrator, 400
in blocking oscillator, 540
regions of, 308–312
pulse, 312, 323
sine-wave, 311, 316
square-wave, 309–311
triangular-wave, 313–315
in sinusoidal oscillator, 738–739
sweep stability improvement, 305–307
Synchronous switching, 550
Sziklai, G. C., 349

T model of transistor, 114–116, 118
Tachometer, 652–653
Taub, H., 407
Television, sweep for, 325, 337–338
Temperature dependence:
of bias point, 109, 110
of mobility, 42
Texas Instruments, Inc., 204, 349, 584
Thermistors, 710
Threshold voltage, 212
effect of substrate on, 215
Thyristor (*see* SCR)
Time constant, 6, 12
Tilt:
in clamp, 63–64
in *RC* amplifier, 23
Time response, normalized, 239
Time-sequential solution:
linear, 4–9
nonlinear, 58–60, 124–130
Todd, C. D., 522
Toggle network, 370
as inverter, 556–557
Topological solutions:
Liénard diagrams, 746–753
phase plane, 739–744, 750
waveshape from, 745–746
Totem-pole output stage, 181–184
Transfer characteristic:
of clamp, 67
of clipper, 55, 57
diode arbitrary piecewise-linear, 71–79
diode dead-zone, 81
diode hystersis, 82
DTL, 160
ECL, 185
inverter: junction-FET, 209, 221
MOSFET complementary pair, 248
MOSFET pair, 229
transistor, 110
regenerative comparator, 445
Transformer, blocking oscillator, 524–525
saturation in, 533–535
Transient response:
amplifier, *R-C*-coupled, 20–23
coefficients of equation, 4–5
compensated attenuator, 23–26
of diode, 85–87
diode bridge, 194–198
emitter-follower, 180, 181
inverter: FET, approximate, 230–234
MOSFET pair, exact, 236–242
transistor, 124–131, 178–180
of nonexponential type, 236–242
RC network, 5–11
with diode, 58–61
RL network, 11–13
Transistor:
α variation with quiescent conditions, 498
amplifier stage, 108–110
small-signal calculations, 119–122
avalanche region operation, 502–504
base-width modulation, 115
biasing of, 108–109
constant current, 434–435
feedback, 440
temperature stability in, 109–110
breakdown voltages of, 136, 140
chopper, 203
composite, 122–124
as current source, 275–278, 434–435
dynamic response of, 124–131
integrated circuit, 131–134, 136
complementary, 140
inverter, 110–112, 124–131
logic (*see* DTL; ECL; RTL; TTL)
models, 103–108, 114–119

Transistor:
  models: Ebers-Moll, 103–108
    h-parameter, 115–118
    hybrid π, 119, 144, 369
    T, 114–115, 116, 118
  multiemitter, 174
  parameters of, 136
  parasitic, 173–176
  physics of, 100–103
  reverse operation, 106–107, 175–176
  saturation of, 112–114, 388
    prevention of, 529
  storage-delay time, 130
  volt-ampere characteristic, 104
Transistor-core counters, 632–634
Transistor-transistor logic (*see* TTL)
Transmission gates (*see* Chopper; Diode bridge)
Transport factor, 102
Trapazoidal drive, 339, 343
TRIAC, 502
Triggering:
  of multivibrators, 365–367
    (*See also* Bistable multivibrator)
  of negative-resistance switching circuit, 468–469, 475–477
Tristable operation, 481
Truth table, 151–152
  (*See also* Logic)
TTL, 171–178, 184
  advanced type, 201
  bistable multivibrator: *D*-type, 571
    J-K, 566
    in memory matrix, 598–600
    S-R, 552
  geometry of input stage, 174
  as monostable multivibrator, 392–393
  parasitic transistors in, 175, 176
  propagation delay, 178
Tuned circuit:
  impedance at resonance, 698
  susceptance of, 738
Tuned-circuit oscillator, 677–686
  as negative-resistance circuit, 726–727
Tungsten lamp, as amplitude limiter, 706–710
Tunnel diode, 505–507
Twin-T network, 675, 690
  voltage-tuned, 256–257

Unijunction transistor, 502–504
Unipolar transistor (*see* Field-effect transistor)

Vadasez, L., 263
Valasek, J., 647
Valley, G. E., Jr., 349
Van der Pol, B., 757
Van der Pol equation, 728
Varicaps, 85
Vitt, A. A., 757
Volt-ampere characteristic:
  arbitrary piecewise-linear, 74–78
  diode, 49
  field-effect transistor: expanded, 256
    junction, 209
    MOSFET, 213, 214, 228–229
    MOSFET complementary, 247
  negative-resistance: current-controlled, 456, 476
    voltage-controlled, 466
  pnpn diode, 498
  SCR, 271, 501, 502
  transistor, 104
    avalanche region, 505
  tunnel diode, 506
Volt-time area, 619
Voltage comparator, 434–460
  as analog-to-digital converter, 459–460
  difference amplifier stage, 434–439
  integrated circuit: Fairchild μA702, 139
    type 710, 439–444
  as pulse-height analyzer, 459
  regenerative, 445–447, 449–458
  as sense detector, 458–459
Voltage control over sweep time, 277–278, 292–293
Voltage-controlled resistor, 255–256
Voltage converter, 640–645
Voltage feedback:
  for NIC, voltage-controlled, 492–496
  for voltage sweeps, 286–293, 296–302
Voltage sweep:
  basic, 268–273
  recovery-time improvement, 278–280, 293–296

Voltage sweep:
voltage control over time, 278
(*See also* Bootstrap sweep; Miller sweep; Negative-resistance switching circuits)
Voltmeter probe, 93

Wallmark, J. T., 263
Watkins, B. G., 612
Widlar, R. J., 463
Wien-bridge oscillator, 666–667, 671, 674
amplitude-controlled, 706–709
amplitude stability factor for, 714
characteristics of bridge, 690
frequency stabilization factor, 691
harmonic rejection factor, 673
Wired logic, 187–189, 244, 608
Wolfe, R. M., 660
Wolfendale, E., 522
Write amplifier, 589, 601–602

XS3 code, 579

Yovits, M. C., 660

Zener breakdown, 50
Zener diode:
in diode network, 81–83
for level shifting, 360–361, 442
for MOSFET gate protection, 215–216
voltage limiting by, 58, 533
Zero-crossing detector, 458

This book was set in Scotch Roman by The Maple Press Company, and printed on permanent paper and bound by The Maple Press Company. The designer was Paula E. Tuerk; the drawings were done by B. Handelman Associates, Inc. The editors were Michael Elia and Albert Shapiro. Les Kaplan supervised the production.